Nonlinear Optical Waves

Fundamental Theories of Physics

An International Book Series on The Fundamental Theories of Physics: Their Clarification, Development and Application

Editor:
ALWYN VAN DER MERWE, *University of Denver, U.S.A.*

Editorial Advisory Board:
LAWRENCE P. HORWITZ, *Tel-Aviv University, Israel*
BRIAN D. JOSEPHSON, *University of Cambridge, U.K.*
CLIVE KILMISTER, *University of London, U.K.*
PEKKA J. LAHTI, *University of Turku, Finland*
GÜNTER LUDWIG, *Philipps-Universität, Marburg, Germany*
NATHAN ROSEN, *Israel Institute of Technology, Israel*
ASHER PERES, *Israel Institute of Technology, Israel*
EDUARD PRUGOVECKI, *University of Toronto, Canada*
MENDEL SACHS, State *University of New York at Buffalo, U.S.A.*
ABDUS SALAM, *International Centre for Theoretical Physics, Trieste, Italy*
HANS-JÜRGEN TREDER, *Zentralinstitut für Astrophysik der Akademie der Wissenschaften, Germany*

Volume 104

Nonlinear Optical Waves

by

A. I. Maimistov

and

A. M. Basharov

Moscow State Engineering Physics Institute (MEPhE),
Moscow, Russia

KLUWER ACADEMIC PUBLISHERS
DORDRECHT / BOSTON / LONDON

A C.I.P. Catalogue record for this book is available from the Library of Congress.

ISBN 0-7923-5752-3

Published by Kluwer Academic Publishers,
P.O. Box 17, 3300 AA Dordrecht, The Netherlands.

Sold and distributed in North, Central and South America
by Kluwer Academic Publishers,
101 Philip Drive, Norwell, MA 02061, U.S.A.

In all other countries, sold and distributed
by Kluwer Academic Publishers,
P.O. Box 322, 3300 AH Dordrecht, The Netherlands.

Printed on acid-free paper

Printed in the Netherlands.

TABLE OF CONTENTS

PREFACE

A non-linear wave is one of the fundamental objects of nature. They are inherent to aerodynamics and hydrodynamics, solid state physics and plasma physics, optics and field theory, chemistry reaction kinetics and population dynamics, nuclear physics and gravity. All non-linear waves can be divided into two parts: dispersive waves and dissipative ones. The history of investigation of these waves has been lasting about two centuries. In 1834 J.S. Russell discovered the extraordinary type of waves without the dispersive broadening. In 1965 N.J. Zabusky and M. D. Kruskal found that the Korteweg-de Vries equation has solutions of the solitary wave form. This solitary wave demonstrates the particle-like properties, i.e., stability under propagation and the elastic interaction under collision of the solitary waves. These waves were named solitons. In succeeding years there has been a great deal of progress in understanding of soliton nature. Now solitons have become the primary components in many important problems of nonlinear wave dynamics. It should be noted that non-linear optics is the field, where all soliton features are exhibited to a great extent.

This book had been designed as the tutorial to the theory of non-linear waves in optics. The first version was projected as the book covering all the problems in this field, both analytical and numerical methods, and results as well. However, it became evident in the process of work that this was not a real task. The project had be reduced at least at the first stage. So we will dwell on the basic aspects of the theory, on the analytical methods and on the non-dissipative phenomena. Nevertheless, the work appeared to have been more bulky than it was expected to be. Besides, there exist remarkable books embracing some of the problems listed above. So, we have restricted ourselves to the fundamental and up-to-date problems which hold interest nowadays. We shall also consider only one-dimensional waves. The book has grown out of a series of lectures delivered to graduate students and post-graduates in the frame of the courses "*Photonics*" "*Quantum Electronics*", "*Nonlinear and Integrated Optics*" and "*Quantum optics*" in the Moscow Engineering Physics Institute.

This book has the following structure. Chapter 1 is devoted to the basic mathematical aspects of non-linear optics theory. In general case all optical processes can be regarded either as resonant or non-resonant ones. The non-resonant processes are described in terms of non-linear susceptibilities contained in the Maxwell equations or reduced Maxwell equations. Moreover, the responsibility of the resonant medium should be considered in the frame of the density matrix formalism leading to the Maxwell-Bloch equations or to generalisation of these equation. We derive in a unified manner the basic equations describing both non-resonant and resonant processes by means of unitary transformation of initial atomic Hamiltonian. The broad class of resonant phenomena arises from the coherent transient processes. They are considered in Chapter 2. There is optical nutation, free induction decay, photon echoes, and superfluorescence. General characteristic of all these phenomena is the short time duration in relation to the relaxation times of the resonant medium. The coherent transient phenomena are often discussed in the case of an optical thin medium, where the propagation effects are of minor importance. If one considers the optical wave propagation in long-haul medium, for instance, optical fibres, then the space-time variations of the envelope wave become an essential feature of the processes under consideration. The greatest advance in the non-linear wave theory was made after the discovery of the powerful method of non-linear evolution equations analysis by C.S.Gardner, J.M.Greene, M.D.Kruskal, and R.M.Miura in 1967. Chapter 3 has been written to sketch out the method named inverse scattering transform method or the method of inverse scattering problem. In the following chapters we shall exploit this effective technique to obtain useful results.

The self-induced transparency phenomenon sets the bright example of an important role of solitons in non-linear optics. The self-induced transparency theory by McCall-Hahn and several generalisations of this theory are represented in Chapter 4. The advance in the investigation of the ultra-short pulse propagation (when the self-induced transparency takes the place) has been achieved under consideration of the pulses of the polarised radiation. A self-induced transparency phenomenon at the two-photon resonance condition is the non-trivial example of the coherent pulse propagation too. More complicated examples of the ultra-short propagation are considered in Chapter 5. The steady-state solitary waves in these examples are not solitons in the strict sense. However, such pulses are often referred to as optical solitons.

The non-linear wave propagation in non-resonant medium also occupies an important place in non-linear optics. Nowadays the optical solitons propagation in fibres has attracted considerable attention. The number of important results in this field are represented in Chapter 6. The picosecond and femtosecond pulses, two- and multi-component solitons have been treated. The bright illustration of the effective application of the inverse scattering transform method yields the problem of the soliton formation from the initially chirped pulses. Here we have restricted ourselves to several exactly analytical approaches to investigate solitary wave propagation. As an example of the approximate method, we consider the variational approach to the studies of the pulse envelope evolution. The role of higher nonlinearity and birefringence of the fibres can be studied in the frame of this method.

The classical problem of non-linear optics is the parametric interaction of waves. The harmonic generation, stimulated by both Raman and Brillouin scattering, parametric amplifications, sum-frequency mixing, and four-wave mixing have been the subject of many investigations in this field. Three- and four-wave parametric interactions theory is represented in Chapter 7. The three-wave interaction process is of particular interest because it gives us a new instance of the application of the inverse scattering transform method to non-linear system without any dispersion.

A new field of the non-linear optics is non-linear integrated optics. The parametric interactions and coherent transient phenomena there can be considered similarly to the case of the bulk medium. The specific feature of non-linear integrated optics is the existence of the surface and guided waves. We shall discuss these waves in Chapter 8. The non-linear wave in the directed couplers and the wave in the distributed feedback structures are also worthy of our attention. The coherent transient phenomena, optical bistability and parametric wave interaction into the thin film of resonant atoms at the interference of two linear dielectrics have been the subject of the recent investigation in integrated optics. These phenomena are considered in Chapter 9. We present here an example of the application of the inverse scattering transform method to a system of ordinary differential equations.

At the beginning of each chapter there is a summary of the results obtained in the theory of the considered phenomena. Here we also try to represent the references review as completely as possible. Several important theoretical approaches to problems of non-linear optics are represented in Appendixes.

We are particularly grateful to Dr. S.O. Elyutin, who has read the manuscript of this book. The text has been greatly improved by his criticism. We would like to express sincere respect to Pr. E.A. Manykin and Pr. A.I. Alekseev, who encouraged our interest in the problems of non-linear optics and non-linear wave phenomena. We are also thankful to our colleagues for helpful remarks.

November, 1998

A.I. Maimistov,
A.M. Basharov

CHAPTER 1

BASIC EQUATIONS

The theoretical analysis of various coherent optical effects in gas media, liquids, dielectrics, semiconductors in exciton spectrum bands can be carried out semiclassically by use of classical Maxwell equations for electromagnetic fields and quantum-mechanical equation for atomic density matrix [1-3]. For non-magnetic media without carriers of current, which we will consider, the Maxwell equations reduce to a wave equation for an electric field strength $\vec{E}$:

$$\Delta\vec{E} - \frac{1}{c^2}\frac{\partial^2}{\partial t^2}\vec{E} = \frac{4\pi}{c^2}\frac{\partial^2\vec{P}}{\partial t^2} + \text{graddiv}\,\vec{E}\,, \tag{1.0.1}$$

$$\text{div}(\vec{E} + 4\pi\vec{P}) = 0\,,$$

where the vector of polarisation of a medium (the dipole momentum of a unit volume)

$$\vec{P} = \text{Sp}(\rho\,\vec{d}) \tag{1.0.2}$$

is determined by a density matrix ρ and atomic dipole moment operator $\vec{d}$. We normalise the density matrix to make $\text{Sp}\,\rho$ be a density of atoms in the unit volume. The equation for density matrix (or master equation) in the electric-dipole approximation is given by

$$i\hbar\left(\frac{\partial}{\partial t} + \hat{\Gamma}\right)\rho = \left(H_0 - \vec{E}\vec{d}\right)\rho - \rho\left(H_0 - \vec{E}\vec{d}\right), \tag{1.0.3}$$

where we neglect the difference between macroscopic field $\vec{E}$ and microscopic field acting on atom. This difference can be taken into account by the Lorentz field. In a non-resonant case it reduces to a simple factor in optical susceptibility (see sec. 1.2.2). In resonant interactions Lorentz field sometimes provides an essentially new effect (see section 9.2.5). Phenomenological account of the Lorentz field consists in the replacement of $\vec{E}$ in eq.(1.0.3) by $\vec{E} + \frac{4\pi}{3}\xi\vec{P}$, where factor ξ (the Lorentz parameter) is of order of unity.

The relaxation operator $\hat{\Gamma}$ allows for the damping processes. The simple and effective way for obtaining relaxation operator is the use of the quantum stochastic Ito equation [4]. In Appendix 1 we derive the master equation and relaxation operator for the resonant interaction of an atom with coherent and broadband squeezed fields by the quantum stochastic Ito equation. Below, the usual assumptions are made for relaxation operator $\hat{\Gamma}$:

$$\langle\alpha|\hat{\Gamma}\rho|\alpha\rangle = \gamma_\alpha\langle\alpha|\rho|\alpha\rangle - \gamma_\alpha\rho^0_\alpha,$$
$$\langle\beta|\hat{\Gamma}\rho|\alpha\rangle = \gamma_{\beta\alpha}\langle\beta|\rho|\alpha\rangle.$$

Labels $\alpha,\beta,...$, enumerate the eigenstates of Hamiltonian H_0 of an isolated atom in the absence of external fields

$$H_0|\alpha\rangle = E_\alpha|\alpha\rangle, \quad \langle\alpha|\beta\rangle = \delta_{\alpha\beta}, \quad \sum_\alpha\langle\alpha|\alpha\rangle = 1.$$

These states are distinguished from one another by their energy E_α and other quantum numbers, ρ^0_α represents the matrix element of a diagonal density matrix ρ of an atom in thermodynamic equilibrium. In other words, ρ^0_α is an equilibrium density of atoms at level $|\alpha\rangle$.

Usually the polarisation of a medium (1.0.2) can be expanded into a series in terms of degrees of an electrical field strength [5]

$$\vec{P}(\vec{r},t) = \sum_{n=1}^{\infty}\vec{P}^{(n)}(\vec{r},t), \tag{1.0.4}$$

$$P_i^{(n)}(\vec{r},t) = \sum_{i_1...i_n}\int d\vec{r}_1...d\vec{r}_n\int_0^{\infty} d\tau_1...d\tau_n\,\chi^{(n)}_{i,i_1...i_n}(\vec{r}_1...\vec{r}_n,\tau_1...\tau_n)\times$$
$$\times E_{i_1}(\vec{r}+\vec{r}_1,t-\tau_1)...E_{i_n}(\vec{r}+\vec{r}_n,t-\tau_n), \tag{1.0.5}$$

in which we take into account that the polarisation $\vec{P} = \vec{P}(\vec{r},t)$ in a point $\vec{r}$ and at time t depends on the magnitude of a field in some neighbourhood of $\vec{r}$ (space dispersion) in the preceding t instants (temporal dispersion). Factor $\chi^{(n)}_{i,i_1...i_n}$ is an optical susceptibility of the n-th order. Indices $i,i_1,...$ number the Cartesian projection of a vector on axes $x(i=1)$, $y(i=2)$ and $z\,(i=3)$. Expression (1.0.4) sometimes is written as $\vec{P} = \vec{P}^L + \vec{P}^{NL}$, where $\vec{P}^L \equiv \vec{P}^{(1)}$ and $\vec{P}^{NL}$ are linear and nonlinear (on a field) components of medium polarisation, respectively.

Optical susceptibilities are conveniently used to describe the non-resonant interaction of light with matter, when only few first terms in expansion (1.0.4) are essential. Thus, the solution of the equation for a density matrix can be obtained by means of perturbation theory. It will only establish the connection of $\chi^{(n)}$ with microscopic characteristics of a medium.

In resonance situation it is necessary to allow in (1.0.4) a large number of terms, so the standard using of optical susceptibilities appear ineffective. In this case, however, it is possible to simplify considerably the equations for density matrix by reducing them to the equations describing the interaction of light with a two-level quantum system. After such renormalisation of the atomic Hamiltonian, the non-resonant energy levels can be additionally taken into account by means of optical susceptibilities.

The reduction of the total Hamiltonian to the two-level one with regard to non-resonant levels by means of optical susceptibilities and parameters of renormalised Hamiltonian is the method of unitary transformation of Hamiltonian. The unitary transformation of Hamiltonian to simplify a problem is used from the first steps of quantum mechanics. The typical example from quantum mechanics, most closely related to the problem under consideration, is the canonical transformation of Foldy and Wouthuysen. They have reduced the Dirac equations to the pair of equations. One of this pair coincides with Pauli equation in nonrelativistic limit [6]. Thus, the unitary transformation method in question is a specific formulation of perturbation method. As to an analogy with classical mechanics, the unitary transformation method may be compared with the application of the Lie transformation in the Hamiltonian perturbation theory [7]. There are many branches of nonlinear optics where unitary transformations are applied. But for our purposes the method in the simplified form was firstly introduced in the resonance optics by Takatsuji [8] and was correctly applied by Grischkowsky *et al* [9] and by Ivanova and Melikyan [10]. The further development of the method of unitary transformation was done in papers [11-15]. In refs.[11,12] atomic level degeneracy and wave polarisation were taken into account. In ref.[13] the description of various relaxation processes was given on the base of the methods of unitary transformation and quantum stochastic Ito equation. Ref.[14] is devoted to establishing the relationships between the exactly integrable models of nonlinear optics, which are not gauge equivalent. We combine the unitary transformation method with the optical susceptibility [15] to describe the resonant and non-resonant atomic levels in a unified manner.

The equations derived in this chapter provide a basis for a semiclassical description of the various type of nonlinear optical phenomena. In the first section we develop the method of unitary transformation to reduce the resonance problems to two-level models. In section 1.1.1 we will discuss the various resonant situations such as one photon and two photon resonance with one or two light waves and we will obtain the renormalised Hamiltonians for their description. In sec. 1.1.2 the Bloch equations will be derived with the use of renormalised Hamiltonians. Section 1.1.3 is devoted to a proper presentation of a medium polarisation through transformed atomic density matrix and effective dipole moment operator. In paragraph 1.2 the optical susceptibilities are considered. After the review of some simple properties of optical susceptibilities (sec.1.2.1) we will give the common form of nonlinear optical susceptibilities and its representation by diagrams (sec.1.2.2). In paragraph 1.3 we consider the simplifications of wave equations. Firstly we will derive equations to describe non-resonant wave propagation, and then the various types of parametric processes will be considered without wave polarisation (sec. 1.3.1-1.3.6). Then, in 1.3.7 the Maxwell-Bloch equations for various types of resonant interactions are observed. Generalisation of the developed methods in order to include polarisation effects and level degeneracy is given in paragraph 1.4.

This chapter is accompanied by four Appendices. Appendix 1 is devoted to the derivation of the density matrix equation and the relaxation operator using the unitary transformation method and quantum stochastic Ito equation. In section A1.6.5 the quasiclassical description of the total angular momentum is developed and its use in deriving the density matrix equation, describing the resonant interactions with arbitrary polarised waves, is demonstrated. The density matrix equation of atoms moving in gas medium is developed in Appendix 2. It is shown how the recoil in radiation processes can be taken into account in gases. In Appendix 3 we consider the Bloch equations without allowance for level degeneracy and we describe their solution by the method of adiabatic following approximation. In Appendix 4 we establish, with the help of the method of unitary transformation, the relationships between the exactly integrable models of nonlinear optics, which are not gauge equivalent.

1.1 The method of unitary transformations

Let us now consider main processes of the resonance interaction of light with matter.

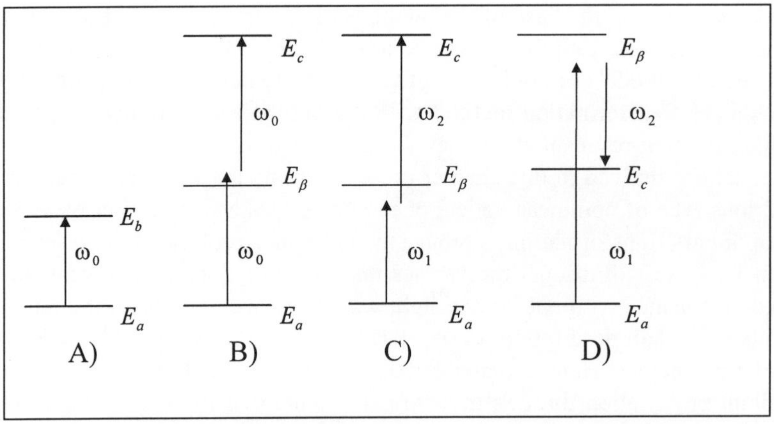

Fig. 1.1.1. Some typical cases of resonance interaction. The energy levels E_a and E_b relate to optically allowed transition, and E_a and E_c relate to two-quantum (optically forbidden) transition. E_β designates the non-resonant levels.

A. <u>One-photon resonance</u> (Fig. 1.1.1, A). In this case a carrier frequency ω_0 of the electromagnetic wave

$$\vec{E} = \vec{\mathcal{E}}(\vec{r},t)\exp[i(\vec{k}_0\vec{r} - \omega_0 t)] + c.c. \tag{1.1.1}$$

is close to the frequency $\omega_{ba} = (E_b - E_a)\hbar^{-1}$ of an optically allowed atomic transition $|b\rangle \to |a\rangle$, $\omega_0 \approx \omega_{ba}$. Characters *c.c.* designate a term, complex conjugated to previous.

B. <u>Two-photon degenerate resonance</u> (Fig. 1.1.1, B). In this case the double frequency of a carrying wave is close to the frequency $\omega_{ca} = (E_c - E_a)/\hbar^{-1}$ of an optically forbidden atomic transition $|c\rangle \to |a\rangle$, $2\omega_0 \approx \omega_{ca}$. Also, we will name this case as two-photon one-wave resonance.

C. <u>Two-photon nondegenerate resonance</u> (Fig. 1.1.1, C). The sum of carrier frequencies ω_1 and ω_2 of two waves

$$\vec{E}_1 = \vec{\mathcal{E}}_1(\vec{r},t)\exp[i(\vec{k}_1\vec{r} - \omega_1 t)] + c.c. \tag{1.1.2}$$

$$\vec{E}_2 = \vec{\mathcal{E}}_2(\vec{r},t)\exp[i(\vec{k}_2\vec{r} - \omega_2 t)] + c.c. \tag{1.1.3}$$

is close to the frequency ω_{ca} of an optically forbidden transition $\omega_1 + \omega_2 \approx \omega_{ca}$. Another name of this case is two-photon two-wave resonance.

D. <u>Raman resonance</u> (Fig. 1.1.1, D). In this case the difference of frequencies of two carrying waves (1.1.2) and (1.1.3) is close to the frequency ω_{ca} of an optically forbidden transition, $\omega_1 - \omega_2 \approx \omega_{ca}$.

The matrix element of atomic dipole moment operator for optically allowed transitions is different from zero $\langle b|d_i|a\rangle \neq 0$, whereas for optically forbidden transitions it is equal to zero. In the case of two-quantum optically forbidden transition $|c\rangle \to |a\rangle$ there are energy levels E_β, for which transitions $|c\rangle \to |\beta\rangle$ and $|\beta\rangle \to |a\rangle$ are optically allowed $\langle c|d_i|\beta\rangle\langle\beta|d_j|\alpha\rangle \neq 0$.

Along with the notation $|\alpha\rangle \to |\beta\rangle$ for the atomic transition, we will use the following $E_\alpha \to E_\beta$ or $j_\alpha \to j_\beta$, where j_α and j_β are the total angular momenta of upper and lower energy levels, respectively.

In this paragraph we use the notation ω_0 for carrier frequency of the wave propagating under the condition of one-photon (A) or two-photon (B) resonance to distinguish this frequency from the frequency variable in Fourier transformation. From section 1.3.4 we will designate this carrier frequency through ω keeping the notation ω_0 for the central frequency of inhomogeneously broadened spectral line of resonant transition. We hope misunderstanding does not take place.

In the cases mentioned above the interaction of light with levels E_a and E_b (or E_a and E_c) is considerably more intensive than with other quantum levels, which manifests in the unlimited growth of appropriate terms in optical susceptibilities $\chi^{(n)}_{i,i_1...i_n}$. This dominating role of the resonance levels enables to consider a medium approximately as an ensemble of two-level atoms, and the presence of the non-resonant levels is taken into account in a renormalised Hamiltonian of two-level atoms. So far created a two-level model of a medium is extremely useful in the various problems of resonance optics. Usually for a substantiation of a two-level model one uses rather bulky method of averaging over fast oscillations of field [16]. The method of unitary transformation is considerably more elegant and powerful.

We transform the density matrix using the unitary operator $\exp(iS)$ with some Hermitian operator $S = S^+$

$$\widetilde{\rho} = e^{-iS} \rho e^{iS}. \tag{1.1.4}$$

The equation for the transformed density matrix $\widetilde{\rho}$

$$i\hbar \left(\frac{\partial}{\partial t} + \widehat{\Gamma} \right) \widetilde{\rho} = \widetilde{H} \widetilde{\rho} - \widetilde{\rho} \widetilde{H}. \tag{1.1.5}$$

is determined by Hamiltonian

$$\widetilde{H} = e^{-iS} H_0 e^{iS} - e^{-iS} \vec{E} \vec{d} e^{iS} - i\hbar e^{-iS} \frac{\partial}{\partial t} e^{iS}. \tag{1.1.6}$$

The unitary transformation (1.1.4) affects also the relaxation operator $\widehat{\Gamma}$. The correct approach for deriving the transformed relaxation operator consists in the consequent development of the relaxation theory from the first principles. As an example of such development, the detailed analysis of spontaneous relaxation in external non-resonant monochromatic field was done in Appendix 1 on the basis of the methods of unitary transformation and quantum stochastic Ito equation. Further in the book we will neglect this influence and we consider the relaxation operator as a phenomenological one. In section (2.7.2) we describe the form of the relaxation operator corresponding to the model of depolarising atomic collisions. In section 9.2.4 the relaxation operator takes into account the atomic relaxation in broadband squeezed light.

Below, we will neglect for simplicity the level degeneracy due to the various orientation of the total angular momentum. Also, we will neglect the waves polarisation and omit the sign of the vector in electric field strength and dipole moment operator assuming that all of them are parallel to one axes and the values without the vector sign are their projection on this axis. The generalisation of the main equations to allow for the level degeneracy and wave polarisation will be done in paragraph 1.4.

1.1.1. THE EFFECTIVE HAMILTONIANS

According to representations of a two-level model we will require, that the matrix elements of a transformed Hamiltonian $\widetilde{H}$ describing the transitions between non-resonant levels and also between resonant and non-resonant levels, should be equal to zero:

For the one photon resonance

$$\widetilde{H}_{a\beta} = \widetilde{H}_{b\beta} = 0,\ \widetilde{H}_{\beta\beta'} = 0\,; \tag{1.1.7a}$$

For the two-photon resonance

$$\widetilde{H}_{a\beta} = \widetilde{H}_{c\beta} = 0,\ \widetilde{H}_{\beta\beta'} = 0, \tag{1.1.7b}$$

where the indices $\beta \neq \beta'$ number the non-resonant levels.

Identity [6]

$$e^{-B} A e^{B} = A + \sum_{n=1}^{\infty} \frac{(-1)^n}{n!} [B,[B,...,[B,A]..]],$$

where A and B are arbitrary operators, and $[B,A] = BA - AB$, provides the following

$$\widetilde{H} = H_0 - i[S,H_0] - \frac{1}{2}[S,[S,H_0]] - ... - Ed + i[S,Ed] +$$
$$+ \frac{1}{2}[S,[S,\vec{E}\vec{d}]] + ... - i\hbar \exp(-iS)\frac{\partial}{\partial t}\exp(iS).$$

We represent S and $\widetilde{H}$ by series in power of an electric field strength

$$S = S^{(1)} + S^{(2)} + ...,\qquad \widetilde{H} = \widetilde{H}^{(0)} + \widetilde{H}^{(1)} + \widetilde{H}^{(2)} + ...$$

($S^{(n)}$ and $\widetilde{H}^{(n)}$ are the terms of the n -th order over a field). Then we obtain the equations forming the basis for the subsequent construction of effective Hamiltonians:

$$\widetilde{H}^{(0)} = H_0,$$
$$\widetilde{H}^{(1)} = -Ed - i[S^{(1)},H_0] + \hbar\frac{\partial}{\partial t}S^{(1)}, \tag{1.1.8}$$
$$\widetilde{H}^{(2)} = \frac{i}{2}[S^{(1)},Ed] - \frac{i}{2}[S^{(1)},\widetilde{H}^{(1)}] - i[S^{(2)},H_0] + \hbar\frac{\partial}{\partial t}S^{(2)},$$
$$\cdots$$

In accord with (1.1.7) and (1.1.8) matrix $S^{(1)}$ satisfies the equations:

In a case of one photon resonance (A)

$$\frac{\partial S^{(1)}_{\alpha\alpha'}}{\partial t} + i\omega_{\alpha\alpha'}S^{(1)}_{\alpha\alpha'} = d_{\alpha\alpha'}\hbar^{-1}\left\{\mathcal{E}\exp[i(\vec{k}_0\vec{r} - \omega t)] + \mathcal{E}^*\exp[-i(\vec{k}_0\vec{r} - \omega t)]\right\},$$

$$\frac{\partial S^{(1)}_{ba}}{\partial t} + i\omega_{ba}S^{(1)}_{ba} = d_{ba}\hbar^{-1}\mathcal{E}^*\exp[-i(\vec{k}_0\vec{r} - \omega_0 t)], \tag{1.1.9}$$

$$S^{(1)}_{aa} = S^{(1)}_{bb} = 0;$$

For two-photon resonance (cases C and D)

$$\frac{\partial S^{(1)}_{\alpha\alpha'}}{\partial t} + i\omega_{\alpha\alpha'}S^{(1)}_{\alpha\alpha'} = d_{\alpha\alpha'}\hbar^{-1}\left\{\mathcal{E}_1\exp[i(\vec{k}_1\vec{r} - \omega_1 t)] + \mathcal{E}_2\exp[i(\vec{k}_2\vec{r} - \omega_2 t)] + \right.$$

$$\left. + \mathcal{E}_1^*\exp[-i(\vec{k}_1\vec{r} - \omega_1 t)] + \mathcal{E}_2^*\exp[-i(\vec{k}_2\vec{r} - \omega_2 t)]\right\}, \tag{1.1.10}$$

$$S^{(1)}_{aa} = S^{(1)}_{cc} = S^{(1)}_{ca} = 0.$$

In a case of two-photon degenerate resonance (case B) equation for $S^{(1)}_{\alpha\alpha'}$ and its solution are similar to those in case A, and $S^{(1)}_{aa} = S^{(1)}_{cc} = S^{(1)}_{ac} = 0$. Everywhere one of the indices α or α' can designate a resonance level.

We will solve equations (1.1.9) and (1.1.10) under condition of adiabatic switch of the field. Let us assume that the time of increase of the front of a field is much longer than a period of oscillation of an electromagnetic wave, and the spectral components of a wave do not affect non-resonant levels. We can write the solution of (1.1.9) as

$$S^{(1)}_{\alpha\alpha'} = d_{\alpha\alpha'}\hbar^{-1}\int_{-\infty}^{t} dt'\, e^{i\omega_{\alpha\alpha'}(t'-t)}\left\{\mathcal{E}(t')e^{i(\vec{k}_0\vec{r} - \omega_0 t')} + \mathcal{E}^*(t')e^{-i(\vec{k}_0\vec{r} - \omega_0 t')}\right\}, \quad \ldots .$$

Integration by parts gives a series

$$S^{(1)}_{\alpha\alpha'} = d_{\alpha\alpha'}\hbar^{-1}\left\{\frac{\mathcal{E}\exp[i(\vec{k}_0\vec{r} - \omega_0 t)]}{i(\omega_{\alpha\alpha'} - \omega_0)} + \frac{\mathcal{E}^*\exp[-i(\vec{k}_0\vec{r} - \omega_0 t)]}{i(\omega_{\alpha\alpha'} + \omega_0)}\right\} +$$

$$+ d_{\alpha\alpha'}\hbar^{-1}\left\{\frac{\frac{\partial\mathcal{E}}{\partial t}\exp[i(\vec{k}_0\vec{r} - \omega_0 t)]}{(\omega_{\alpha\alpha'} - \omega_0)^2} + \frac{\frac{\partial\mathcal{E}^*}{\partial t}\exp[-i(\vec{k}_0\vec{r} - \omega_0 t)]}{(\omega_{\alpha\alpha'} + \omega_0)^2}\right\} + \ldots,$$

...

We suppose that the amplitudes of the fields (1.1.1)-(1.1.3) vary slowly with respect to fast oscillating exponents. Then the solutions of (1.1.9) and (1.1.10) give:

for one photon resonance (A)

$$S^{(1)}_{\alpha\alpha'} = -id_{\alpha\alpha'}\hbar^{-1}\left\{\frac{\mathcal{E}\exp[i(\vec{k}_0\vec{r}-\omega_0 t)]}{\omega_{\alpha\alpha'}-\omega_0}+\frac{\mathcal{E}^*\exp[-i(\vec{k}_0\vec{r}-\omega_0 t)]}{\omega_{\alpha\alpha'}+\omega_0}\right\},$$

$$S^{(1)}_{ba} = -id_{ba}\hbar^{-1}\mathcal{E}^*\frac{\exp[-i(\vec{k}_0\vec{r}-\omega_0 t)]}{\omega_{ba}+\omega_0};$$

for two-photon resonance (cases C and D)

$$S^{(1)}_{\alpha\alpha'} = -id_{\alpha\alpha'}\hbar^{-1}\left\{\frac{\mathcal{E}_1\exp[i(\vec{k}_1\vec{r}-\omega_1 t)]}{\omega_{\alpha\alpha'}-\omega_1}+\frac{\mathcal{E}_1^*\exp[-i(\vec{k}_1\vec{r}-\omega_1 t)]}{\omega_{\alpha\alpha'}+\omega_1}+\right.$$

$$\left.+\frac{\mathcal{E}_2\exp[i(\vec{k}_2\vec{r}-\omega_2 t)]}{\omega_{\alpha\alpha'}-\omega_2}+\frac{\mathcal{E}_2^*\exp[-i(\vec{k}_2\vec{r}-\omega_2 t)]}{\omega_{\alpha\alpha'}+\omega_2}\right\}.$$

The sum of the terms in $\widetilde{H}^{(2)}_{ca}$, containing (after the substitution of explicit expressions for $S^{(1)}$) no factors

$$\exp[2i(\vec{k}_0\vec{r}-\omega_0 t)],\ \exp\{i[(\vec{k}_1+\vec{k}_2)\vec{r}-(\omega_1+\omega_2)t]\},\ \ \exp\{i[(\vec{k}_1-\vec{k}_2)\vec{r}-(\omega_1-\omega_2)t]\},$$

is zeroed for B, C and D cases accordingly.

We do not present here the expressions for $S^{(2)}$, which will not be needed. They can be easily obtained in the same manner as the $S^{(1)}$ terms. It is important to note that the values $S^{(1)}$ and $S^{(2)}$ contain no resonant denominator. This absence of resonant denominators confirms that the assumptions made concerning $\widetilde{H}$ are not contradictory.

Finally, we obtain the effective Hamiltonian in the form

for one-photon resonance $(\omega_0 \approx \omega_{ba})$

$$\widetilde{H}_{ba} = -\mathcal{E}d_{ba}\exp[i(\vec{k}_0\vec{r}-\omega_0 t)] = \widetilde{H}^*_{ab}, \qquad \widetilde{H}_{\alpha\alpha} = E_\alpha + E^{St}_\alpha,$$

$$E^{St}_a = |\mathcal{E}|^2\left\{\Pi_a(\omega_0) - \frac{|d_{ab}|^2}{2\hbar\omega_0}\right\}, \qquad E^{St}_b = |\mathcal{E}|^2\left\{\Pi_b(\omega_0) + \frac{|d_{ab}|^2}{2\hbar\omega_0}\right\},$$

$$E^{St}_\beta = |\mathcal{E}|^2\Pi_\beta(\omega_0);$$

for two-photon resonance (B) $\left(2\omega_0 \approx \omega_{ca}\right)$

$$\widetilde{H}_{ca} = -\frac{1}{2}\mathcal{E}^2\Pi_{ca}(\omega_0)\exp[2i(\vec{k}_0\vec{r} - \omega_0 t)] = \widetilde{H}^*_{ac}, \quad \widetilde{H}_{\alpha\alpha} = E_\alpha + E^{St}_\alpha,$$
$$E^{St}_\alpha = |\mathcal{E}|^2\Pi_\alpha(\omega_0);$$

for two-photon resonance (C) $\left(\omega_1 + \omega_2 \approx \omega_{ca}\right)$

$$\widetilde{H}_{ca} = -\tfrac{1}{2}\mathcal{E}_1\mathcal{E}_2[\Pi_{ca}(\omega_1) + \Pi_{ca}(\omega_2)]\exp\left\{i[(\vec{k}_1 + \vec{k}_2)\vec{r} - (\omega_1 + \omega_2)t]\right\} = \widetilde{H}^*_{ac},$$
$$\widetilde{H}_{\alpha\alpha} = E_\alpha + E^{St}_\alpha; \quad E^{St}_\alpha = |\mathcal{E}_1|^2\Pi_\alpha(\omega_1) + |\mathcal{E}_2|^2\Pi_\alpha(\omega_2);$$

for Raman resonance (D) $\left(\omega_1 - \omega_2 \approx \omega_{ca}\right)$

$$\widetilde{H}_{ca} = -\tfrac{1}{2}\mathcal{E}_1\mathcal{E}^*_2[\Pi_{ca}(\omega_1) + \Pi_{ca}(-\omega_2)]\exp\left\{i[(\vec{k}_1 - \vec{k}_2)\vec{r} - (\omega_1 - \omega_2)t]\right\} = \widetilde{H}^*_{ac},$$
$$\widetilde{H}_{\alpha\alpha} = E_\alpha + E^{St}_\alpha; \quad E^{St}_\alpha = |\mathcal{E}_1|^2\Pi_\alpha(\omega_1) + |\mathcal{E}_2|^2\Pi_\alpha(\omega_2).$$

Values E^{St}_α represent the Stark shifts of energy level $|\alpha\rangle$ due to the dynamical Stark effect. We see that the effective Hamiltonians are determined by the following parameters

$$\Pi_{ca}(\omega) = \sum_\beta \frac{d_{c\beta}d_{\beta a}}{\hbar}\left(\frac{1}{\omega_{\beta c} + \omega} + \frac{1}{\omega_{\beta a} - \omega}\right),$$
$$\Pi_\alpha(\omega) = \sum_\beta \frac{|d_{\alpha\beta}|^2}{\hbar}\left(\frac{1}{\omega_{\alpha\beta} + \omega} + \frac{1}{\omega_{\alpha\beta} - \omega}\right). \tag{1.1.11}$$

One should use the following symmetry properties of the parameter $\Pi_{ca}(\omega)$. In the case C of nondegenerate two-photon resonance $\omega_1 + \omega_2 \approx \omega_{ca}$ the equality $\Pi_{ca}(\omega_1) = \Pi_{ca}(\omega_2)$ is valid. In the case D of Raman resonance $\omega_1 - \omega_2 \approx \omega_{ca}$ we can obtain another equality $\Pi_{ca}(\omega_1) = \Pi_{ca}(-\omega_2)$ from the above definition of $\Pi_{ca}(\omega)$. We have not applied these properties above to emphasise a peculiar kind of symmetry to permutation of resonant waves.

In the case A (one photon resonance) the terms in Stark shifts E^{St}_a and E^{St}_b without factors Π_a or Π_b are named the Bloch-Siegert shifts [1]. When, as usual, there are many non-resonant levels the Bloch-Siegert shifts are negligibly small.

1.1.2. THE BLOCH EQUATIONS

The matrix elements of a density matrix $\widetilde{\rho}$, describing the resonance levels and the tran-

transitions between them, obey the equations

$$\begin{aligned}
&\left(\frac{\partial}{\partial t} - i\Delta + \gamma_{21}\right)R_{21} = i\Lambda\left(R_1 - R_2\right),\\
&\left(\frac{\partial}{\partial t} + \gamma_1\right)R_1 = i\left(\Lambda^* R_{21} - \Lambda R_{21}^*\right) + \gamma_1 N_1,\\
&\left(\frac{\partial}{\partial t} + \gamma_2\right)R_2 = -i\left(\Lambda^* R_{21} - \Lambda R_{21}^*\right) + \gamma_2 N_2.
\end{aligned} \tag{1.1.12}$$

Here, Δ is a detuning from resonance including the Stark shifts of resonant levels, constants γ_1 and γ_2 describe the relaxation of atomic densities at lower R_1 and upper R_2 levels whereas γ_{21} is the relaxation constant of an optical coherence R_{21}; N_1 and N_2 are stationary atomic densities at lower and upper resonant levels in the absence of all electromagnetic fields. The value $2|\Lambda|$ is a Rabi frequency; it represents the energy (in terms of $\hbar$) of resonance interaction of a radiation with atom.

The equations (1.1.12) are known in resonance optics as the Bloch equations. They play an important role in each theory describing any resonance interaction. For a concrete case they can be modified, for example to include level degeneracy (see section 1.4.1) or to describe an additional resonant interaction with a broadband squeezed light (see Appendix 1 and section 9.2.4).

For cases considered above we have the following notation :

A $(\omega_0 \approx \omega_{ba})$.

$$\begin{aligned}
&R_{21} = \tilde{\rho}_{ba}\exp[-i(\vec{k}_0\vec{r} - \omega_0 t)],\\
&R_1 = \tilde{\rho}_{aa}, \quad R_2 = \tilde{\rho}_{bb}, \quad \Lambda = \mathcal{E} d_{ba}/h,\\
&\Delta = \omega_0 - \omega_{ba} - (E_b^{St} - E_a^{St})/\hbar, \quad N_1 = \rho_a^0, \quad N_2 = \rho_b^0;
\end{aligned}$$

B $(2\omega_0 \approx \omega_{ca})$.

$$\begin{aligned}
&R_{21} = \tilde{\rho}_{ca}\exp[-2i(\vec{k}_0\vec{r} - \omega_0 t)], \quad R_1 = \tilde{\rho}_{aa}, \quad R_2 = \tilde{\rho}_{cc},\\
&\Lambda = \mathcal{E}^2\Pi_{ca}(\omega_0)/2\hbar, \quad \Delta = 2\omega_0 - \omega_{ca} - (E_c^{St} - E_a^{St})/\hbar,\\
&N_1 = \rho_a^0, \qquad N_2 = \rho_c^0;
\end{aligned}$$

C $(\omega_1 + \omega_2 \approx \omega_{ca})$.

$$\begin{aligned}
&R_{21} = \tilde{\rho}_{ca}\exp\left\{-i[(\vec{k}_1 + \vec{k}_2)\vec{r} - (\omega_1 + \omega_2)t]\right\},\\
&R_1 = \tilde{\rho}_{aa}, \quad R_2 = \tilde{\rho}_{cc}, \quad \Lambda = \mathcal{E}_1\mathcal{E}_2\Pi_{ca}(\omega_1)/\hbar,\\
&\Delta = \omega_1 + \omega_2 - \omega_{ca} - (E_c^{St} - E_a^{St})/\hbar,\\
&N_1 = \rho_a^0, \quad N_2 = \rho_c^0;
\end{aligned}$$

D $\left(\omega_1 - \omega_2 \approx \omega_{ca}\right)$.

$$R_{21} = \widetilde{\rho}_{ca} \exp\left\{-i[(\vec{k}_1 - \vec{k}_2)\vec{r} - (\omega_1 - \omega_2)t]\right\},$$
$$R_1 = \widetilde{\rho}_{aa}, \quad R_2 = \widetilde{\rho}_{cc}, \quad \Lambda = \mathcal{E}_1 \mathcal{E}_2^* \Pi_{ca}(\omega_1)/\hbar,$$
$$\Delta = \omega_1 - \omega_2 - \omega_{ca} - \left(E_c^{St} - E_a^{St}\right)/\hbar,$$
$$N_1 = \rho_a^0, \quad N_2 = \rho_c^0.$$

In writing the expressions for Λ in cases C and D we have taken into account the symmetry properties of parameters (1.1.11).

The Bloch equations (1.1.12) will be completed by the equations for field amplitudes in section 1.3.8.

Usually, the solution of the Bloch equations is represented as a sum of stationary solution of (1.1.12) and deviation from this stationary solution. The simple methods for obtaining the deviations from the stationary regime will be considered in Appendix 3 (the adiabatic following approximation), and in sections 2.2 and 2.6.1 (the given field approximation).

In a stationary case when all values in (1.1.12) do not depend on time, the derivatives in (1.1.12) can be omitted. Then a solution of obtained system of algebraic equations yields

$$R_{21} = \frac{(N_1 - N_2)\Lambda(-\Delta + i\gamma_{21})}{\Delta^2 + \gamma_{21}^2 + 2|\Lambda|^2 \gamma_{21}(\gamma_1 + \gamma_2)/\gamma_1\gamma_2}, \tag{1.1.13}$$

$$R_1 = N_1 - \frac{2|\Lambda|^2 (N_1 - N_2)\gamma_{21}/\gamma_1}{\Delta^2 + \gamma_{21}^2 + 2|\Lambda|^2 \gamma_{21}(\gamma_1 + \gamma_2)/\gamma_1\gamma_2}, \tag{1.1.14}$$

$$R_2 = N_2 + \frac{2|\Lambda|^2 (N_1 - N_2)\gamma_{21}/\gamma_2}{\Delta^2 + \gamma_{21}^2 + 2|\Lambda|^2 \gamma_{21}(\gamma_1 + \gamma_2)/\gamma_1\gamma_2}. \tag{1.1.15}$$

As the Stark shifts of levels are small, the density matrix $\widetilde{\rho}_{\beta\beta_1}$ for the non resonant levels coincides with an equilibrium density matrix $\delta_{\beta\beta_1}\rho_\beta^0$.

1.1.3. POLARISATION OF THE MEDIUM

By substituting expression $\rho = \exp(iS)\widetilde{\rho}\exp(-iS)$ in (1.0.2), we receive

$$\vec{P} = \mathrm{Sp}\left(\widetilde{\rho}e^{-iS}\vec{d}e^{iS}\right) = \mathrm{Sp}\left\{\widetilde{\rho}\left(\vec{d} - i\left[S,\vec{d}\right] - \frac{1}{2}\left[S,\left[S,\vec{d}\right]\right] - \ldots\right)\right\} = \tag{1.1.16}$$
$$= \vec{P}_{res} + \vec{P}_{nonres}.$$

Expression (1.1.16) is a sum of two terms. The first one $\vec{P}_{res}$ is due to resonance interaction of electromagnetic waves with two-level atoms. Under conditions of one- and two-photon resonances it can be written in the unified form

$$\vec{P}_{res} = \tilde{\rho}_{ea}\vec{D}_{ae} + c.c. + \tilde{\rho}_{aa}\vec{D}_{aa} + \tilde{\rho}_{ee}\vec{D}_{ee}, \tag{1.1.17}$$

where index $e = b$ corresponds to one-photon resonance with the atomic transition $E_b \to E_a$, whereas index $e = c$ describes all cases B-D of two-photon resonance with atomic transition $E_c \to E_a$ considered above. In (1.1.17) the effective operator of dipole momentum is introduced in the form:

$$\vec{D} = \exp(-iS)\vec{d}\exp(iS) = \vec{d} - i[S,\vec{d}] - \frac{1}{2}[S,[S,\vec{d}]] - \dots. \tag{1.1.18}$$

For our purposes it is sufficient to take the effective dipole moment operator with the accuracy up to the first order of electric field:

$$\vec{D} \approx d + i\left(\vec{d}S - S\vec{d}\right) \approx d + i\left(\vec{d}S^{(1)} - S^{(1)}\vec{d}\right). \tag{1.1.19}$$

For one photon resonance, $\vec{P}_{res}$ represents a wave packet with the same carrier frequency equal to the frequency of resonant field

$$\begin{aligned} P_{res} &= \tilde{\rho}_{ba}D_{ab} + c.c. + \tilde{\rho}_{aa}D_{aa} + \tilde{\rho}_{bb}D_{bb} = \\ &= \left(R_{21}d_{ab} - R_1\mathcal{E}\Pi_a(\omega_0) - R_2\mathcal{E}\Pi_b(\omega_0)\right)\exp[i(\vec{k}_0\vec{r} - \omega_0 t)] + c.c., \end{aligned} \tag{1.1.20}$$

Note that

$$\begin{aligned} D_{ab} = d_{ab}, \quad D_{aa} &= -\mathcal{E}\Pi_a(\omega_0)\exp[i(\vec{k}_0\vec{r} - \omega_0 t)] + c.c, \\ D_{bb} &= -\mathcal{E}\Pi_b(\omega_0)\exp[i(\vec{k}_0\vec{r} - \omega_0 t)] + c.c.. \end{aligned} \tag{1.1.21}$$

Resonance polarisation of a medium at two-photon interaction, is characterised by several carrier frequencies. Before we discuss them, let us write the matrix elements of the operator (1.1.19):

In the case B $\left(2\omega_0 \approx \omega_{ca}\right)$

$$D_{ac} = \mathcal{E}\Pi^*_{ca}(-\omega_0)\exp[i(\vec{k}_0\vec{r} - \omega_0 t)] + \mathcal{E}^*\Pi^*_{ca}(\omega_0)\exp[-i(\vec{k}_0\vec{r} - \omega_0 t)], \tag{1.1.22a}$$

$$D_{aa} = -\mathcal{E}\Pi_a(\omega_0)\exp[i(\vec{k}_0\vec{r} - \omega_0 t)] + c.c., \tag{1.1.22b}$$

$$D_{cc} = -\mathcal{E}\Pi_c(\omega_0)\exp[i(\vec{k}_0\vec{r} - \omega_0 t)] + c.c., \tag{1.1.22c}$$

In the cases C $(\omega_1 + \omega_2 \approx \omega_{ca})$ and D $(\omega_1 - \omega_2 \approx \omega_{ca})$

$$D_{ac} = \sum_{l=1}^{2} \left\{ \mathcal{E}_l \Pi^*_{ca}(-\omega_l) \exp[i(\vec{k}_l \vec{r} - \omega_l t)] + \mathcal{E}^*_l \Pi^*_{ca}(\omega_l) \exp[-i(\vec{k}_l \vec{r} - \omega_l t)] \right\}, \quad (1.1.23a)$$

$$D_{aa} = -\sum_{l=1}^{2} \mathcal{E}_l \Pi_a(\omega_l) \exp[i(\vec{k}_l \vec{r} - \omega_l t)] + c.c., \quad (1.1.23b)$$

$$D_{cc} = -\sum_{l=1}^{2} \mathcal{E}_l \Pi_c(\omega_l) \exp[i(\vec{k}_l \vec{r} - \omega_l t)] + c.c.. \quad (1.1.23c)$$

It can be seen from here that two-photon interaction induces polarisation $\vec{P}_{res}$ at the frequencies: ω_0 and $3\omega_0$ (case B), ω_1, ω_2, $2\omega_1 + \omega_2$ and $\omega_1 + 2\omega_2$ (case C) and ω_1, ω_2, $2\omega_1 - \omega_2$ and $|\omega_1 - 2\omega_2|$ (case D).

There are no reasons to write the polarisation of the medium $\vec{P}_{res}$ for each considered case. It is more useful to discuss the Raman scattering on an excited two-quantum (optically forbidden) atomic transition of any non-resonant wave

$$\vec{E} = \vec{\mathcal{E}}'(\vec{r},t) \exp[i(\vec{k}'\vec{r} - \omega' t)] + c.c. \quad (1.1.24)$$

If $\tilde{\rho}_{ca} \neq 0$, the additional polarisation of the medium P' arises not only at the carrier frequency ω' of non-resonant wave (1.1.24), but also at the combination frequencies $|\omega' \pm \omega_{ca}|$:

$$P' = \tilde{\rho}_{ca} D'_{ac} + c.c. + \tilde{\rho}_{aa} D'_{aa} + \tilde{\rho}_{cc} D'_{cc}, \quad (1.1.25)$$

where

$$D'_{ac} = \mathcal{E}' \Pi^*_{ca}(-\omega') \exp[i(\vec{k}'\vec{r} - \omega' t)] + \mathcal{E}'^* \Pi^*_{ca}(\omega') \exp[-i(\vec{k}'\vec{r} - \omega' t)], \quad (1.1.26a)$$

$$D'_{aa} = -\mathcal{E}' \Pi_a(\omega') \exp[i(\vec{k}'\vec{r} - \omega' t)] + c.c., \quad (1.1.26b)$$

$$D'_{cc} = -\mathcal{E}' \Pi_c(\omega') \exp[i(\vec{k}'\vec{r} - \omega' t)] + c.c.. \quad (1.1.26c)$$

As an example we consider the excitation of two-quantum transition $E_c \to E_a$ by the wave (1.1.1) under condition of two-photon degenerate resonance $2\omega_0 \approx \omega_{ca}$. Then the following representation of the density matrix elements by slowly varying functions R_{21}, R_1 and R_2 holds:

$$\tilde{\rho}_{ca} = R_{21} \exp[2i(\vec{k}_0 \vec{r} - \omega_0 t)], \quad \tilde{\rho}_{aa} = R_1, \quad \tilde{\rho}_{cc} = R_2.$$

The resonant polarisation $\vec{P}_{res}$, under the additional action of non-resonant wave (1.1.24), consists of two terms. The first one is due to two-photon interaction of the wave (1.1.1) with two-quantum (optically forbidden) transition $E_c \to E_a$. The second term is an additional term which appears to be due to the auxiliary propagation of non-resonant wave in the two-photon pre-excited medium. The additional polarisation has the form

$$\begin{aligned} P' = &\left\{R_{21}\mathcal{E}'\Pi^*_{ca}(-\omega')\exp\{i[(2\vec{k}_0 + \vec{k}')\vec{r} - (2\omega_0 + \omega')t]\} + c.c.\right\} + \\ &+ \left\{R_{21}\mathcal{E}'^*\Pi^*_{ca}(\omega')\exp\{i[(2\vec{k}_0 - \vec{k}')\vec{r} - (2\omega_0 - \omega')t]\} + c.c.\right\} + \\ &+ \left\{- R_{aa}\mathcal{E}'\Pi_a(\omega')\exp[i(\vec{k}\vec{r} - \omega't)] + c.c.\right\} + \\ &+ \left\{- R_{cc}\mathcal{E}'\Pi_c(\omega')\exp[i(\vec{k}\vec{r} - \omega't)] + c.c.\right\}. \end{aligned} \qquad (1.1.27)$$

Under conditions of spatially synchronism it determines the emission of wave packets with the carrier frequencies approximately equal to $|\omega' \pm 2\omega_0|$ or $|\omega' \pm \omega_{ca}|$ since $2\omega_0 \approx \omega_{ca}$. Further, the last two terms determine the reaction of the resonantly excited medium to the propagating of the low intensive non-resonant waves.

One should distinguish the additional polarisation of the resonantly excited by a weak non-resonant wave from the non-resonant polarisation $\vec{P}_{nonres}$ in (1.1.16)

$$\vec{P}_{nonres} = \sum_\beta \mathrm{Sp}\left\{\tilde{\rho}^0_\beta\left[-i\sum_\alpha\left(S_{\beta\alpha}\vec{d}_{\alpha\beta} - \vec{d}_{\beta\alpha}S_{\alpha\beta}\right) - \ldots\right]\right\}, \qquad (1.1.28)$$

which describes the interaction of the field with non-resonant energy levels. This term will be considered in section 1.2.2. Here, $\tilde{\rho}^0_\beta$ is the equilibrium density matrix of atoms on account of the Stark shifts of energy levels.

We emphasise that non-resonant propagation of wave (1.1.24) determines not only polarisation $\vec{P}'$ but makes contribution to the polarisation $\vec{P}_{nonres}$.

It is necessary to note that a nonlinear character of resonance polarisation $\vec{P}_{res}$ begins revealing at such electric field strengths, when $|\Lambda| \propto \gamma_{21}$ (see (1.1.13)), that is much less than electric field strength, causing nonlinear effects at non-resonant interaction. Therefore in a rather wide range of intensities the resonance interaction of light with substance is possible to regard as the interaction with two-level atoms, and the presence of non-resonant atoms and energy levels to take into account in a Hamiltonian of two-level atoms and in a linear dielectric constant of a medium.

1.2. The optical susceptibilities

We present an electric field strength and polarisation of a medium through Fourier integrals

$$\vec{E}(\vec{r},t) = \int \vec{E}(\vec{k},\omega)\exp\{i(\vec{k}\vec{r} - \omega t)\}\frac{d\vec{k}d\omega}{(2\pi)^4},$$

$$\vec{P}(\vec{r},t) = \int \vec{P}(\vec{k},\omega)\exp\{i(\vec{k}\vec{r} - \omega t\}\frac{d\vec{k}d\omega}{(2\pi)^4}.$$

The Fourier transformation of the expressions (1.0.4) and (1.0.5) gives

$$\vec{P}(\vec{k},\omega) = \sum_{n=1}^{\infty} \vec{P}^{(n)}(\vec{k},\omega), \tag{1.2.1}$$

$$\begin{gathered} P_i^{(n)}(\vec{k},\omega) = (2\pi)^4 \sum_{i_1 \ldots i_n} \int \frac{d\vec{k}_1 \ldots d\vec{k}_n d\omega_1 \ldots d\omega_n}{(2\pi)^{4n}} \chi^{(n)}_{i,i_1 \ldots i_n}(\vec{k}_1 \ldots \vec{k}_n, \omega_1 \ldots \omega_n) \times \\ \times \delta(\vec{k} - \vec{k}_1 - \ldots - \vec{k}_n)\delta(\omega - \omega_1 - \ldots - \omega_n)E_{i_1}(\vec{k}_1\omega_1)\ldots E_{i_n}(\vec{k}_n\omega_n) \end{gathered} \tag{1.2.2}$$

where the value

$$\begin{gathered} \chi^{(n)}_{i,i_1 \ldots i_n}(\vec{k}_1 \ldots \vec{k}_n, \omega_1 \ldots \omega_n) = \\ = \int d\vec{r}_1 \ldots d\vec{r}_n \int_0^{\infty} d\tau_1 \ldots d\tau_n \exp\{i(\vec{k}_1\vec{r}_1 + \omega_1\tau_1 + \ldots + \vec{k}_n\vec{r}_n + \omega_n\tau_n)\}\chi^{(n)}_{i,i_1 \ldots i_n}(\vec{r}_1 \ldots \tau_n) \end{gathered} \tag{1.2.3}$$

we shall also name as an optical susceptibility of $n-th$ order.

1.2.1. PROPERTY OF OPTICAL SUSCEPTIBILITIES

If the specific distance r_0 of variance of values $\chi^{(n)}_{i,i_1 \ldots i_n}(\vec{r}_1 \ldots \vec{r}_n, \tau_1 \ldots \tau_n)$ is much less than length of a wave $\lambda \approx 1/k$

$$r_0 << \lambda, \tag{1.2.4}$$

the dependence $\chi^{(n)}_{i,i_1 \ldots i_n}(\vec{k}_1 \ldots \vec{k}_n, \omega_1 \ldots \omega_n)$ on $\vec{k}_1, \ldots, \vec{k}_n$ (i.e. space dispersion) can be neglected. Note, that r_0 is typical size of the particle moving area in a medium, in which the particle keeps the memory about the action of a field in preceding instants. In dielectrics and semiconductors without carriers of a current r_0 is of the order of interatomic distance, and in gases it is defined by the length of free run. Therefore in optical and longer wavelength domain the inequality (1.2.4) is fulfilled. We shall neglect the space dispersion, and taking into account the presence of δ-function $\delta(\omega - \omega_1 - \ldots - \omega_n)$

in (1.2.2), we shall note the optical susceptibility of $n-th$ order as $\chi^{(n)}_{i,i_1...i_n}(-\omega,\omega_1...\omega_n)$, where $-\omega+\omega_1+...+\omega_n=0$.

From the form of (1.2.2) the relation follows

$$\chi^{(n)}_{i,i_1...i_k...i_s...i_n}\left(-\omega,\omega_1...\omega_k...\omega_s...\omega_n\right)=\chi^{(n)}_{i,i_1...i_s...i_k...i_n}\left(-\omega,\omega_1...\omega_s...\omega_k...\omega_n\right),$$

reflecting a permutation symmetry of $\chi^{(n)}$.

The optical susceptibilities $\chi^{(n)}_{i,i_1...i_n}$ are tensor values, their tensor property, including the number of nonzero components, is determined by point group of a symmetry of crystal (atom or molecule in case of gas). For crystals and molecules with a centre of inversion the obvious equality $\chi^{2n}_{i,i_1...i_{2n}}=0$, $n=1,2,...$ holds. The analysis of the tensor properties of the optical susceptibilities is given in [5].

Apart from restrictions, superimposed by group of a symmetry, for crystals, transparent in the whole range of frequencies of the nonlinear process, there are the additional restrictions, expressed by the equations

$$\chi^{(n)}_{i,i_1...i_k...i_n}\left(-\omega;\omega_1...\omega_k...\omega_n\right)=\chi^{(n)}_{i_k,i_1...i...i_n}\left(\omega_k;\omega_1...-\omega...\omega_n\right).$$

We shall underline, that the knowledge of a tensor structure of optical susceptibilities is necessary for studying polarisation properties of nonlinear phenomena.

Physical mechanisms, underlying to polarisation of medium and, therefore, defining optical susceptibilities, are various in each domain of frequencies of an electromagnetic field. They determine density matrix equations.

1.2.2. NONRESONANT OPTICAL SUSCEPTIBILITIES

We consider electromagnetic fields as a set of waves $l=1,2,...$ in the form

$$E_l=\mathcal{E}_l(\vec{r},t)\exp[i(\vec{\bar{k}}_l\vec{r}-\bar{\omega}_l t)]+c.c., \tag{1.2.5}$$

where $\mathcal{E}_l$ is slowly varying amplitude of an electrical field (in comparison with exponential factor); $\vec{\bar{k}}_l$ is a wave vector and $\bar{\omega}_l$ is the carrier frequency. The line over the characters ω_l and k_l distinguishes the indicated values from the variables in the Fourier components. We will omit them when it does not cause the mistakes.

Fourier transform of (1.2.5) distinguishes from zero in the vicinity of the points $\vec{k}=\vec{\bar{k}}_l$, $\omega=\bar{\omega}_l$ and $\vec{k}=-\vec{\bar{k}}_l$, $\omega=-\bar{\omega}_l$. The dimensions Δk_l and $\Delta\omega_l$ of these vicinities satisfy the inequalities $\Delta k_l<<\bar{k}_l$, $\Delta\omega_l<<\bar{\omega}_l$. Therefore, expression (1.2.5) represents the wave packet with a carrier frequency $\overline{\omega}_l$.

Assume that both the spectrum of electromagnetic waves $\overline{\omega}_l - \Delta\omega_l \le \omega \le \overline{\omega}_l + \Delta\omega_l$ is not superimposed with a absorption spectrum of atoms of a medium and characteristic frequency of interaction (interaction energy divided by $\hbar$) is much less than the minimum detuning Δ from the frequencies of atomic transitions

$$\gamma << \Delta \, , \; \mathcal{E}_l |d| / \hbar << \Delta \tag{1.2.6}$$

(d is the dipole matrix element of atomic transition corresponding to Δ, and γ is the effective width of this transition). Then the effective Hamiltonian (1.1.6) is diagonal with matrix elements distinguishing from that of H_0 on the values of Stark shifts.

We represent the unitary operator exp(iS) as

$$\exp(iS) = \exp(-iH_0 t/\hbar) U(\vec{r},t) \exp(i\widetilde{H}t/\hbar).$$

We assume the interaction with a field to be switched on adiabatically at $t \to -\infty$, which means we represent the electric field strength as $\vec{E}(\vec{r},t)\exp(\beta t)$, $\beta \to +0$. The equation for evolution operator U follows from (1.1.6) in the form

$$i\hbar \frac{\partial U(\vec{r},t)}{\partial t} = V(\vec{r},t) U(\vec{r},t), \quad U(\vec{r},t)\big|_{t=-\infty} = 1, \tag{1.2.7}$$

where

$$V(\vec{r},t) = -\exp(iH_0 t)\vec{E}\vec{d} \exp(-iH_0 t).$$

Expression (1.1.28) for nonresonant polarisation of a medium takes the form:

$$\vec{P}(\vec{r},t) = \mathrm{Sp}\left\{\widetilde{\rho}^0 U^+(\vec{r},t) e^{iH_0 t/\hbar} \vec{d} e^{-iH_0 t/\hbar} U(\vec{r},t)\right\}$$

or

$$\vec{P}(\vec{r},t) = \sum_{\substack{\alpha, \\ \alpha', \alpha''}} \widetilde{\rho}^0_\alpha U^*_{\alpha'\alpha}(\vec{r},t) \vec{d}_{\alpha'\alpha''} U_{\alpha''\alpha}(\vec{r},t) \exp\left(-i\omega_{\alpha''\alpha'} t\right), \tag{1.2.8}$$

where the diagonality of the matrix $\widetilde{\rho}^0$ and of the transformed Hamiltonian $\widetilde{H}$ was taken into account. The index α enumerates the eigenvalues and the eigenstates (quantum levels) of atom Hamiltonian

$$\omega_{\alpha''\alpha'} = \left(E_{\alpha''} - E_{\alpha'}\right)/\hbar, \;\; \vec{d}_{\alpha'\alpha''} = \langle\alpha'|\vec{d}|\alpha''\rangle, \;\; \widetilde{\rho}^0_\alpha = \langle\alpha|\widetilde{\rho}^0|\alpha\rangle.$$

For the sake of simplicity we suppose that all eigenvalues of H_0 are nondegenerate: each E_α corresponds to only one quantum level $|\alpha\rangle$, which means we may neglect the polarisation effects and change vectors $\vec{E}$, $\vec{P}$ and $\vec{d}$ parallel to one axis for their projections E, P and d on this axis.

After the Fourier transformation of (1.2.8) we have

$$P(\vec{k},\omega) = \sum_{\alpha\alpha'\alpha''} \int \frac{d\vec{k}'d\vec{k}''d\omega' d\omega''}{(2\pi)^4} \delta(\vec{k} + \vec{k}' - \vec{k}'')\delta(\omega + \omega' - \omega'' - \omega_{\alpha''\alpha'}) \times \\ \times \tilde{\rho}^0_\alpha U^*_{\alpha'\alpha}(\vec{k}',\omega') d_{\alpha'\alpha''} U_{\alpha''\alpha}(\vec{k}'',\omega''). \tag{1.2.9}$$

Now we will calculate the Fourier component of the matrix element of the evolution operator

$$U_{\alpha''\alpha}(\vec{k},\omega) = \int \exp\{-i(\vec{k}\vec{r} - \omega t)\} U_{\alpha''\alpha}(\vec{r},t) d\vec{r} dt.$$

Using condition (1.2.6), we present a solution of the equation (1.2.7) as a series

$$U(\vec{r},t) = 1 + \sum_{k=1}^{\infty} \left(\frac{1}{i\hbar}\right)^k \int_{-\infty}^{t} d\tau_1 \int_{-\infty}^{\tau_1} d\tau_2 \ldots \int_{-\infty}^{\tau_{k-1}} d\tau_k V(\vec{r},\tau_1) V(\vec{r},\tau_2) \ldots V(\vec{r},\tau_k).$$

Since

$$V_{\alpha_1\alpha_2}(\vec{r},t) = -d_{\alpha_1\alpha_2} \int E(\vec{k},\omega + \omega_{\alpha_1\alpha_2}) \exp[i(\vec{k}\vec{r} - \omega t)] \frac{d\vec{k}d\omega}{(2\pi)^4},$$

$$U_{\alpha''\alpha}(\vec{r},t) = 1 + \sum_{m=1}^{\infty} \int \frac{d\vec{k}_1 \ldots d\vec{k}_m d\omega_1 \ldots d\omega_m}{(2\pi)^{4m}} \exp\{i[(\vec{k}_1 + \ldots + \vec{k}_m)\vec{r} - (\omega_1 + \ldots + \omega_m)t]\} \times \\ \times \sum_{\alpha_1 \ldots \alpha_{m-1}} \left(\frac{-1}{\hbar}\right)^m \frac{d_{\alpha''\alpha_1} d_{\alpha_1\alpha_2} \ldots d_{\alpha_{m-1}\alpha}}{\omega_m(\omega_m + \omega_{m-1}) \ldots (\omega_m + \ldots + \omega_1)} E(\vec{k}_1, \omega_1 + \omega_{\alpha''\alpha_1}) \times \\ \times E(\vec{k}_2, \omega_2 + \omega_{\alpha_1\alpha_2}) \ldots E(\vec{k}_m, \omega_m + \omega_{\alpha_{m-1}\alpha})$$

it is easy to find $U^*_{\alpha'\alpha}(\vec{k}',\omega')$ and $U_{\alpha''\alpha}(\vec{k}'',\omega'')$. Below we will neglect the Stark shifts of the levels and we omit the sign tilde over the equilibrium density matrix.

By substituting the result for $U^*_{\alpha'\alpha}(\vec{k}',\omega')$ and $U_{\alpha''\alpha}(\vec{k}'',\omega'')$ in (1.2.9) and by executing an integration on $d\vec{k}'d\vec{k}''d\omega'd\omega''$ we receive, that $P(\vec{k},\omega)$ is identical to (1.2.1) and (1.2.2) with the following expressions for optical susceptibilities

$$\chi^{(n)}(-\omega;\omega_1 \ldots \omega_n) = \sum_{m+s=n} \tilde{\chi}^{(m+s)}(-\omega;\omega_1 \ldots \omega_m, \omega'_1 \ldots \omega'_s) \tag{1.2.10}$$

where $\omega'_1 = \omega_{m+1},..., \omega'_s = \omega_n$, and

$$\tilde{\chi}^{(m+s)}(-\omega;\omega_1...\omega_m,\omega'_1...\omega'_s) = \left(-\frac{1}{\hbar}\right)^{m+s}\sum_{\alpha,\alpha_1...}\rho^0_\alpha \times$$

$$\times\frac{d_{\alpha\alpha'_s}...d_{\alpha'_2\alpha'_1}d_{\alpha'_1\alpha_1}d_{\alpha_1\alpha_2}...d_{\alpha_m\alpha}}{(\omega_{\alpha\alpha'_s}-\omega'_s)...(\omega_{\alpha\alpha'_1}-\omega'_1-...-\omega'_s)(\omega_m+..+\omega_1-\omega_{\alpha_1\alpha})...(\omega_m-\omega_{\alpha_m\alpha})} \quad (1.2.11)$$

Each term in (1.2.11) can be represented graphically following [17,18]. Let us divide a straight line by $m+s$ points into $m+s+1$ segments. Let an output vector exit from the $m+1$ segment from the right end of the line and let the others enter the remaining segments (Fig. 1.2.1, a). The output vector is spoken about as a radiation of quantum with ω frequency while input vectors - as an absorption of the quanta of the frequencies $\omega_1,...,\omega_m$ (to the right from "radiation") and $\omega'_1,...,\omega'_s$ (to the left from "radiation").

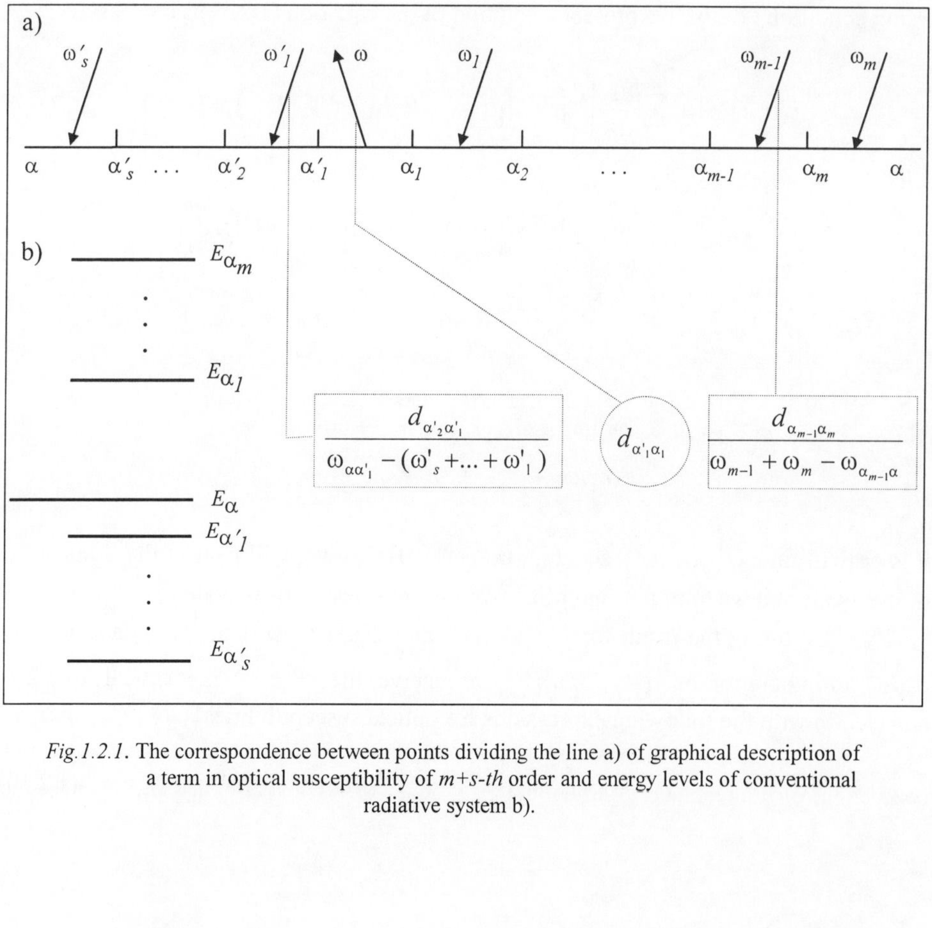

Fig.1.2.1. The correspondence between points dividing the line a) of graphical description of a term in optical susceptibility of *m+s-th* order and energy levels of conventional radiative system b).

Each grid point on the line is associated with the quantum level with energies $E_{\alpha_1},...,E_{\alpha_m}$ located above level E_α, and accordingly $E_{\alpha'_1},...,E_{\alpha'_s}$ located below level E_α (Fig. 1.2.1, b). A fraction is put in correspondence with each absorbed quantum from the right hand side from the radiated quantum. The numerator of this fraction is equal to the matrix element of dipole moment operator of transition between levels, where the absorption takes place. The denominator is the difference between a sum of frequencies of absorbed quanta from the right (including the given one) and the transition frequency from the lowest level to level E_α. Absorbed quantum from the left hand side from radiated one is represented by a fraction with a similar numerator but with a denominator equal to the difference between the transition frequency from level E_α to the highest level (excluding E_α) and a sum of frequencies of absorbed quanta from the left (including the given one). Matrix element of dipole moment operator of a transition between the two levels surrounding level E_α is assigned to the radiated quantum. Thus, eq.(1.2.11) represents a product of factor $(-1/\hbar)^{m+s}\rho_\alpha^0$ and all factors associated with the vectors from Fig. 1.2.1 a) with the summation over all atomic quantum states.

It is necessary to multiply the obtained expression (1.2.10) for optical susceptibility of the $n-th$ order by the Lorentz correction L-factor to take into account the interior field. This represents the fact that the field which enters the interaction operator $-E(\vec{r},t)\,d$ in the equation for a density matrix is macroscopic (i.e. averaged on physically infinitesimal volume) electrical field, whereas a microscopic field affects an atom. For liquids, gases and crystals with a cubic symmetry this factor has the form

$$L = \frac{\varepsilon(\omega)+2}{3}\prod_{i=1}^{n}\frac{\varepsilon(\omega_i)+2}{3},$$

where $\varepsilon(\omega_i)$ is linear dielectric susceptibility at frequency ω_i.

Equation (1.2.10) is the most effective to calculate optical susceptibilities of atomic and molecular gases, for which a sufficient amount of the matrix elements of an dipole moment operator is known. This applies to rare gases, alkaline metals vapour and some other media. In other cases it is convenient to determine optical susceptibilities experimentally.

It should be noted that when one of the carrier frequencies or the algebraic sum of carrier frequencies becomes equal to the frequency of some atomic transition, the denominator in (1.2.11) is zero. Expression (1.2.11) then tends to infinity. In this case the resonance interaction of a radiation with matter takes place. The resonant levels must be excluded from eq. (1.2.11) in accordance with the condition (1.2.6). The method of unitary transformation described above allows to separate the account of the resonant levels from one of nonresonant levels in polarisation. One can show that the Fourier component of nonresonant polarisation $\vec{P}_{nonres}$ has the form (1.2.1) and (1.2.2) with the optical susceptibilities in the form (1.2.10) and (1.2.11) without the resonant terms.

1.3. The wave equation

We discuss now the coupling of nonlinear polarisation (1.1.16) to Maxwell equations for fields. In each specific problem equations (1.0.1) and (1.1.16) can be noticeably simplified in different ways. The most straightforward way is to substitute expressions for the wave packets in (1.0.1) and to omit the higher derivatives of slow amplitudes. Apparently, the most preferable way is to derive the simplified equations for amplitudes of electrical field proceeding from the dispersing equations. It makes possible to receive the equations in need from rather common suggestions about the character of the dispersion relations [19]. Here we follow the elementary version of such approach.

1.3.1. LINEAR PARABOLIC EQUATION

First we shall explicitly discuss a derivation of the simplified equations for distribution of wave packet in linear isotropic medium. In the course of detailed treatment it will make clear how the nonlinear effects can be taken into account in the obtained equations. Then the nonlinear parabolic equation will be formulated for self-focusing and self-modulation. The equation for parametric interaction of waves and generation of optical harmonics and generalised Maxwell-Bloch equations will be discussed as well.

We begin from the equations for Fourier-components of an electric field strength $\vec{E}(\vec{k},\omega)$ and polarisation of a medium $\vec{P}_{nonres}$, which imply from (1.0.1):

$$k^2\vec{E}(\vec{k},\omega)-\vec{k}\left(\vec{k}\vec{E}(\vec{k},\omega)\right)-\frac{\omega^2}{c^2}\vec{E}(\vec{k},\omega)-\frac{4\pi\omega^2}{c^2}\vec{P}(\vec{k},\omega)=0, \qquad (1.3.1)$$

$$\vec{k}\left(\vec{E}(\vec{k},\omega)+4\pi\vec{P}(\vec{k},\omega)\right)=0. \qquad (1.3.2)$$

Assume that the Fourier-component $\vec{E}(\vec{k},\omega)$ distinguishes from zero in a narrow region of the points $\vec{k}=\vec{\bar{k}}$, $\omega=\overline{\omega}$ and $\vec{k}=-\vec{\bar{k}}$, $\omega=-\overline{\omega}$, that relates to the case of a wave packet, when

$$\int\vec{E}(\vec{k},\omega)\exp[i(\vec{k}\vec{r}-\omega t)]\frac{d\vec{k}d\omega}{(2\pi)^4}=\vec{E}(\vec{r},t)= \qquad (1.3.3)$$

$$=\vec{\mathcal{E}}(\vec{r},t)\exp[i(\vec{\bar{k}}\vec{r}-\overline{\omega}t)]+c.c.$$

Fourier-component of slowly varying amplitude

$$\vec{\mathcal{E}}(\vec{\kappa},\nu)=\int d\vec{r}dt\vec{\mathcal{E}}(\vec{r},t)\exp[-i(\vec{\kappa}\cdot\vec{r}-\nu t)]$$

is coupled to a Fourier-component of an electric field strength by the relation $\vec{E}(\vec{\bar{k}}+\vec{\kappa},\bar{\omega}+\nu)=\vec{\mathcal{E}}(\vec{\kappa},\nu)+\vec{\mathcal{E}}^*(-2\vec{\bar{k}}-\vec{\kappa},-2\bar{\omega}-\nu)$. It is possible to assign with the exponential accuracy that

$$\vec{E}(\vec{\bar{k}}+\vec{\kappa},\bar{\omega}+\nu)=\vec{\mathcal{E}}(\vec{\kappa},\nu). \tag{1.3.4}$$

It is worth noting, that the range of variation of $\vec{\mathcal{E}}(\vec{\kappa},\nu)$, i.e. the characteristic values of κ and ν is much less than $\bar{k}$ and $\bar{\omega}$, respectively.

After the replacement $\vec{k}=\vec{\bar{k}}+\vec{\kappa}$ and $\omega=\bar{\omega}+\nu$ in (1.3.1) and (1.3.2) we receive for (1.3.3) with the allowance made in (1.3.4)

$$\sum_j\left\{\left[(\vec{\bar{k}}+\vec{\kappa})^2-\frac{(\bar{\omega}+\nu)^2}{c^2}\right]\delta_{ij}-(\bar{k}_i+\kappa_i)(\bar{k}_j+\kappa_j)\right\}\mathcal{E}_j(\vec{\kappa},\nu)-$$
$$-\frac{4\pi(\bar{\omega}+\nu)^2}{c^2}P_i(\vec{\bar{k}}+\vec{\kappa},\bar{\omega}+\nu)=0, \tag{1.3.5a}$$

$$\sum_i(\bar{k}_i+\vec{\kappa}_i)\left[\mathcal{E}_i(\vec{\kappa},\nu)+4\pi P_i(\vec{\bar{k}}+\vec{\kappa},\bar{\omega}+\nu)\right]=0. \tag{1.3.5b}$$

We rewrite eq (1.3.5a) as

$$\sum_j\left\{\left[(\vec{\bar{k}}+\vec{\kappa})^2-\frac{(\bar{\omega}+\nu)^2}{c^2}\right]\delta_{ij}-(\bar{k}_i+\kappa_i)(\bar{k}_j+\kappa_j)\right\}\mathcal{E}_j(\vec{\kappa},\nu)-$$
$$-\frac{4\pi(\bar{\omega}+\nu)^2}{c^2}P_i^L(\vec{\bar{k}}+\vec{\kappa},\bar{\omega}+\nu)=\frac{4\pi(\bar{\omega}+\nu)^2}{c^2}P_i^{NL}(\vec{\bar{k}}+\vec{\kappa},\bar{\omega}+\nu). \tag{1.3.5'a}$$

In a linear $\vec{P}=\vec{P}^L$, $\vec{P}^{NL}=0$ isotropic medium

$$\vec{P}^L(\vec{\bar{k}}+\vec{\kappa},\bar{\omega}+\nu)=\chi^{(1)}(-\bar{\omega}-\nu;\bar{\omega}+\nu)\vec{\mathcal{E}}(\vec{\kappa},\nu),$$

so, introducing a linear dielectric susceptibility of a medium $\varepsilon(\omega)=1+4\pi\chi^{(1)}(-\omega;\omega)$, we can obtain (if $\varepsilon(\bar{\omega}+\nu)\neq 0$):

$$\left[(\vec{\bar{k}}+\vec{\kappa})^2-\frac{(\bar{\omega}+\nu)^2}{c^2}\varepsilon(\bar{\omega}+\nu)\right]\vec{\mathcal{E}}(\vec{\kappa},\nu)=0, \tag{1.3.6}$$

$$(\vec{\bar{k}}+\vec{\kappa})\vec{\mathcal{E}}(\vec{\kappa},\nu)=0. \tag{1.3.7}$$

The case $\vec{\kappa} = 0$ and $\nu = 0$ corresponds to the propagation of a plane monochromatic wave with wave vector $\vec{\bar{k}}$ and frequency $\bar{\omega}$ coupled by the dispersing relation

$$\vec{\bar{k}}^2 = \frac{\bar{\omega}^2}{c^2}\varepsilon(\bar{\omega}).$$

Equation (1.3.6) can be simplified. Let us involve $\vec{\mathcal{E}}$ and $\vec{\kappa}$ as a sum $\vec{\mathcal{E}} = \vec{\mathcal{E}}_\perp + \vec{\mathcal{E}}_{||}$ and $\vec{\kappa} = \vec{\kappa}_\perp + \vec{\kappa}_{||}$ of two terms - the transversal $\vec{\mathcal{E}}_\perp, \vec{\kappa}_\perp$ and the longitudinal $\vec{\mathcal{E}}_{||}, \vec{\kappa}_{||}$ components with respect to vector $\vec{\bar{k}}$. We assume that the dimension of the variation domain for $\vec{\mathcal{E}}$ in transversal direction is less than the one in longitudinal direction. That means that the inequality $\kappa_\perp >> \kappa_{||}$ holds. Thus (1.3.6) is reduced to the expression

$$\left[2\bar{k}\left(\kappa_{||} - \frac{\nu}{v_g}\right) + \kappa_{||}^2 + \frac{\bar{k}v_g' - 1}{v_g^2}\nu^2\right]\vec{\mathcal{E}}(\vec{\kappa},\nu) = 0\,, \tag{1.3.8}$$

where the terms, containing a group a velocity $v_g = d\bar{\omega}/d\bar{k}$ and its derivative $v_g' = \left(dv_g/d\omega\right)_{\omega=\bar{\omega}}$, came from the following expansion of $k^2(\omega) = \frac{\omega^2}{c^2}\varepsilon(\omega)$ in a series over degrees of a small parameter ν:

$$k^2(\bar{\omega}+\nu) = k^2(\bar{\omega}) + 2k(\bar{\omega})\left.\frac{dk}{d\omega}\right|_{\omega=\bar{\omega}}\nu + \left[\left(\left.\frac{dk}{d\omega}\right|_{\omega=\bar{\omega}}\right)^2 + k(\bar{\omega})\left.\frac{d^2k}{d\omega^2}\right|_{\omega=\bar{\omega}}\right]\nu^2 + \ldots$$

Here $k(\bar{\omega}) = \bar{k}$.

In a considered approximation a longitudinal component $\vec{\mathcal{E}}_{||}$ of the amplitude of an electrical field may be not taken into account, as it is much less than the transversal component since $\mathcal{E}_{||} = \kappa_\perp \mathcal{E}_\perp / \bar{k}$ as it follows from (1.3.7). Therefore it is necessary to put $\vec{\mathcal{E}} = \vec{\mathcal{E}}_\perp$ in (1.3.8).

The reverse Fourier transformation of (1.3.8) provides the linear parabolic equation for slowly varying amplitude $\vec{\mathcal{E}}(\vec{r},t)$.

$$\left[-2i\bar{k}\left(\frac{\partial}{\partial z} + \frac{1}{v_g}\frac{\partial}{\partial t}\right) - \Delta_\perp + \frac{1 - \bar{k}v_g'}{v_g^2}\frac{\partial^2}{\partial t^2}\right]\vec{\mathcal{E}}(\vec{r},t) = 0. \tag{1.3.9}$$

Here axis z coincides with the propagation direction of a wave packet $\vec{\bar{k}}/\bar{k}$, and $\Delta_\perp = \partial^2/\partial x^2 + \partial^2/\partial y^2$ is a transversal part of Laplacian.

Attention should be drawn to the correspondence of all coefficients in the formulae (1.3.5'a), (1.3.6), (1.3.8) and (1.3.9) that allows to generalise eq.(1.3.9) to account for any terms in polarisation of the medium.

Now the nonlinear part $\vec{P}^{NL}$ of a medium polarisation is involved into consideration. Several examples of nonlinear equations, describing the propagation of optical waves in the nonlinear media, will be discussed below.

1.3.2. NONLINEAR PARABOLIC EQUATION

In isotropic non-resonant medium the $\vec{P}^{NL}$ is reduced in the first approximation to a term of the third order of field. The optical susceptibility $\chi^{(3)}(-\overline{\omega};\overline{\omega},-\overline{\omega},\overline{\omega})$, characterising the dynamic self-action of a light wave, describes a single wave packet. Sufficiently intensive wave packet can generate a new wave packet with a carrier frequency $3\overline{\omega}$. Such process is called third harmonics generation. It can be described by $\chi^{(3)}(-3\overline{\omega};\overline{\omega},\overline{\omega},\overline{\omega})$. Two wave packets with the carrier frequencies $\overline{\omega}_1$ and $\overline{\omega}_2$ can interact. It is seen from the form of nonlinear susceptibilities $\chi^{(3)}(-\overline{\omega}_1;\overline{\omega}_2,-\overline{\omega}_2,\overline{\omega}_1)$ and $\chi^{(3)}(-\overline{\omega}_2;\overline{\omega}_1,-\overline{\omega}_1,\overline{\omega}_2)$. Such interaction is named the combinative one. Even more complicated parametric processes, involving three and four wave packets are possible. If the processes of frequency transformation are negligible, then the expression for nonlinear polarisation for the dynamic self-action of a wave packet writes

$$P_i^{NL}(\bar{\vec{k}}+\vec{\kappa},\overline{\omega}+\nu)=3\sum_{i_1i_2i_3}\int\frac{d\vec{\kappa}_1d\vec{\kappa}_2d\vec{\kappa}_3d\nu_1d\nu_2d\nu_3}{(2\pi)^8}\chi^{(3)}_{ii_1i_2i_3}(\overline{\omega}-\overline{\omega}-\nu;+\nu_1,-\overline{\omega}-\nu_2,\overline{\omega}+\nu_3)\cdot$$
$$\cdot\delta(\vec{\kappa}-\vec{\kappa}_1-\vec{\kappa}_2-\vec{\kappa}_3)\delta(\nu-\nu_1-\nu_2-\nu_3)\mathcal{E}_{i_1}(\vec{\kappa}_1\nu_1)\mathcal{E}^*_{i_2}(\vec{\kappa}_2\nu_2)\mathcal{E}_{i_3}(\vec{\kappa}_3\nu_3).$$

Factor 3 arises from the permutation symmetry of optical susceptibility. Since $\vec{P}^{NL}$ is the small addition to $\overline{P}^{L}$, it is possible to neglect both the dependence of

$$\frac{4\pi(\overline{\omega}+\nu)^2}{c^2}P_i^{NL}(\bar{\vec{k}}+\vec{\kappa},\overline{\omega}+\nu)$$

in (1.3.5'a) from ν and dependence of $\chi^{(3)}_{ii_1i_2i_3}(-\overline{\omega}-\nu;\overline{\omega}+\nu_1-\overline{\omega}-\nu_2,\overline{\omega}+\nu_3)$ from ν_1,ν_2,ν_3 and ν. Thus, for the self-action of a light wave we obtain the following equation for slowly varying amplitude:

$$\left[-2i\bar{k}\left(\frac{\partial}{\partial z}+\frac{1}{v_g}\frac{\partial}{\partial t}\right)-\Delta_\perp+\frac{1-\bar{k}v_g'}{v_g^2}\frac{\partial^2}{\partial t^2}\right]\mathcal{E}(\vec{r},t)=$$
$$=\frac{12\pi\bar{k}^2}{\varepsilon(\overline{\omega})}\chi^{(3)}(-\overline{\omega};\overline{\omega},-\overline{\omega},\overline{\omega})|\mathcal{E}(\vec{r},t)|^2\mathcal{E}(\vec{r},t), \tag{1.3.10}$$

where we restrict the consideration by linearly polarisation. The corresponded indices at $\chi^{(3)}$ are omitted.

We underline, that in the left hand side of (1.3.10) (as well as in (1.3.9)) certainly the term $-2i\bar{k}\left(\frac{\partial}{\partial z}+\frac{1}{v_g}\frac{\partial}{\partial t}\right)$ plays the main role. The remaining terms make small contributions. In a linear case they define a diffraction divergence ($\Delta_{\perp}$) and dispersing broadening ($\frac{1-\bar{k}v'_g}{v_g^2}\frac{\partial^2}{\partial t^2}$). In a nonlinear case both the non-resonant and resonant, they cause different sorts of instabilities. These terms should be thoroughly examined in the description of such basic effects as self-focusing (or self-defocusing) and self-modulation of a light wave. Nevertheless in the analysis of other nonlinear effects these terms can be often neglected.

It is worth noting that the nonlinear term in (1.3.10) is essentially inhomogeneous. It depends on coordinate and on time. It points out to a possible generalisation of the obtained equations in a case of inhomogeneous medium, where the linear dielectric constant can depend on coordinate and time as a result of, for example, sound waves propagation. It is only important that the scales, describing the mentioned dependencies, would be much less than the length of an electromagnetic wave and the period of its oscillations.

1.3.3. PARAMETRIC INTERACTIONS

Now we can write immediately the equations of parametric interaction of two wave packets

$$E_1 = \mathcal{E}_1(\vec{r},t)\exp[i(\bar{k}_1 z - \bar{\omega}_1 t)] + c.c.,$$
$$E_2 = \mathcal{E}_2(\vec{r},t)\exp[i(\bar{k}_2 z - \bar{\omega}_2 t)] + c.c.. \tag{1.3.11}$$

In isotropic media we have

$$-2i\bar{k}_1\left(\frac{\partial}{\partial z}+\frac{1}{v_1}\frac{\partial}{\partial t}\right)\mathcal{E}_1(\vec{r},t)=\frac{24\pi\bar{k}_1^2}{\varepsilon(\bar{\omega}_1)}\chi^{(3)}(-\bar{\omega}_1;\bar{\omega}_2,-\bar{\omega}_2,\bar{\omega}_1)|\mathcal{E}_2(\vec{r},t)|^2\mathcal{E}_1(\vec{r},t),$$
$$-2i\bar{k}_2\left(\frac{\partial}{\partial z}+\frac{1}{v_2}\frac{\partial}{\partial t}\right)\mathcal{E}_2(\vec{r},t)=\frac{24\pi\bar{k}_2^2}{\varepsilon(\bar{\omega}_2)}\chi^{(3)}(-\bar{\omega}_2;\bar{\omega}_1,-\bar{\omega}_1,\bar{\omega}_2)|\mathcal{E}_1(\vec{r},t)|^2\mathcal{E}_2(\vec{r},t). \tag{1.3.12}$$

Waves are linearly polarised along axis x; $\varepsilon(\bar{\omega}_l)$ and v_l are linear dielectric susceptibility and group velocity at frequency $\bar{\omega}_l, l=1,2$. Numerical factor before the nonlinear terms is twice as much as in (1.3.10) because of the greater number of variants of a permutation symmetry of tensor component of a nonlinear susceptibility

$\chi^{(3)}(-\overline{\omega}_1;\overline{\omega}_2,-\overline{\omega}_2,\overline{\omega}_1)$.

The terms describing diffraction divergence, dispersing broadening and self-action, are omitted. If necessary they can easily be restored by comparing with (1.3.10). This note also concerns all consequent equations for slowly varying amplitudes, which will be considered below.

Generation of the third harmonics

The case $\overline{\omega}_2 = 3\overline{\omega}_1$ corresponds to the process of the generation and the interaction with the third harmonics. The basic equations are slightly different:

$$-2i\overline{k}_1\left(\frac{\partial}{\partial z}+\frac{1}{v_1}\frac{\partial}{\partial t}\right)\mathcal{E}_1(\vec{r},t)=\frac{12\pi\overline{k}_1^2}{\varepsilon(\overline{\omega}_1)}\left\{\chi^{(3)}(-\overline{\omega}_1;3\overline{\omega}_1,-\overline{\omega}_1,-\overline{\omega}_1)\mathcal{E}_1^{*2}(\vec{r},t)\mathcal{E}_3(\vec{r},t)e^{i\varphi_3}+\right.$$
$$\left.+2\chi^{(3)}(-\overline{\omega}_1;3\overline{\omega}_1,-3\overline{\omega}_1,\overline{\omega}_1)\mathcal{E}_1(\vec{r},t)|\mathcal{E}_3(\vec{r},t)|^2\right\},$$
$$(1.3.13)$$
$$-2i\overline{k}_3\left(\frac{\partial}{\partial z}+\frac{1}{v_3}\frac{\partial}{\partial t}\right)\mathcal{E}_3(\vec{r},t)=\frac{4\pi\overline{k}_3^2}{\varepsilon(3\overline{\omega}_1)}\left\{\chi^{(3)}(-3\overline{\omega}_1;\overline{\omega}_1,\overline{\omega}_1,\overline{\omega}_1)\mathcal{E}_1^3(\vec{r},t)e^{-i\varphi_3}+\right.$$
$$\left.+6\chi^{(3)}(-3\overline{\omega}_1;\overline{\omega}_1,-\overline{\omega}_1,3\overline{\omega}_1)|\mathcal{E}_1(\vec{r},t)|^2\mathcal{E}_3(\vec{r},t)\right\}.$$

The value $\varphi_3 = (\overline{k}_3 - 3\overline{k}_1)z$ characterises the phase detuning of the waves (1.3.11). In order to emphasise, that it is third harmonics being considered now (it is the second wave packet in (1.2.11)), we have substituted index 2 by index 3 at the wave vector and group velocity.

Generation of the second harmonics

In uniaxial and other non centre symmetric media the expression for nonlinear part P^{NL} of polarisation of a medium starts with a term of the second order of a field, which is defined by optical nonlinear susceptibilities of the second order $\chi^{(2)}(-2\overline{\omega},\overline{\omega},\overline{\omega})$ and $\chi^{(2)}(-\overline{\omega},\overline{\omega}_1,\pm\overline{\omega}_2)$. This makes possible both the generation of second harmonics and parametric interaction of three waves with the carrier frequencies, satisfying the relation $\overline{\omega} = \overline{\omega}_1 \pm \overline{\omega}_2$. Thus, the basic equations should be derived from (1.3.5) accounting for the tensor nature of dielectric susceptibility $\chi^{(2)}$. However, if in the case of uniaxial crystal all waves spread along optical z axis of a crystal, then the generation of second harmonics $\overline{\omega}_2 = 2\overline{\omega}_1$ obeys the obvious equations:

$$-2i\overline{k}_2\left(\frac{\partial}{\partial z}+\frac{1}{v_2}\frac{\partial}{\partial t}\right)\mathcal{E}_2=\frac{4\pi\overline{k}_2^2}{\varepsilon(\overline{\omega}_2)}\chi^{(2)}(-2\overline{\omega}_1;\overline{\omega}_1,\overline{\omega}_1)\mathcal{E}_1^2e^{-i\varphi_2}, \qquad (1.3.14a)$$

$$-2i\bar{k}_1\left(\frac{\partial}{\partial z}+\frac{1}{v_1}\frac{\partial}{\partial t}\right)\mathcal{E}_1=\frac{8\pi\bar{k}_1^2}{\varepsilon(\bar{\omega}_1)}\chi^{(2)}(-\bar{\omega}_1;2\bar{\omega}_1,-\bar{\omega}_1)\mathcal{E}_1^*\mathcal{E}_2e^{i\varphi_2}, \tag{1.3.14b}$$
$$\varphi_2=(\bar{k}_2-2\bar{k}_1)z.$$

Parametric transformation of frequencies

For parametric interaction of three light waves

$$\begin{aligned}
E_1&=\mathcal{E}_1(\vec{r},t)\exp[i(\bar{k}_1z-\bar{\omega}_1t)]+c.c.,\\
E_2&=\mathcal{E}_2(\vec{r},t)\exp[i(\bar{k}_2z-\bar{\omega}_2t)]+c.c.,\\
E&=\mathcal{E}(\vec{r},t)\exp[i(\bar{k}z-\bar{\omega}t)]+c.c.
\end{aligned} \tag{1.3.15}$$

we have

$$\begin{aligned}
-2i\bar{k}_1\left(\frac{\partial}{\partial z}+\frac{1}{v_1}\frac{\partial}{\partial t}\right)\mathcal{E}_1&=\frac{8\pi\bar{k}_1^2}{\varepsilon(\bar{\omega}_1)}\chi^{(2)}(-\bar{\omega}_1;\bar{\omega}_1+\bar{\omega}_2,-\bar{\omega}_2)\mathcal{E}\mathcal{E}_2^*e^{-i\varphi},\\
-2i\bar{k}_2\left(\frac{\partial}{\partial z}+\frac{1}{v_2}\frac{\partial}{\partial t}\right)\mathcal{E}_2&=\frac{8\pi\bar{k}_2^2}{\varepsilon(\bar{\omega}_2)}\chi^{(2)}(-\bar{\omega}_2;\bar{\omega}_1+\bar{\omega}_2,-\bar{\omega}_1)\mathcal{E}\mathcal{E}_1^*e^{-i\varphi},\\
-2i\bar{k}\left(\frac{\partial}{\partial z}+\frac{1}{v}\frac{\partial}{\partial t}\right)\mathcal{E}&=\frac{8\pi\bar{k}^2}{\varepsilon(\bar{\omega})}\chi^{(2)}(-\bar{\omega}_1-\bar{\omega}_2;\bar{\omega}_1,\bar{\omega}_2)\mathcal{E}_1\mathcal{E}_2e^{i\varphi},
\end{aligned} \tag{1.3.16}$$
$$\varphi_2=(\bar{k}_1+\bar{k}_2-\bar{k})z.$$

We have considered here only the case $\bar{\omega}=\bar{\omega}_1+\bar{\omega}_2$. It would be useful to the reader to consider the case $\bar{\omega}=\bar{\omega}_1-\bar{\omega}_2$ himself. More details on harmonics generation processes and parametric interactions and also the solution of equations (1.3.13) - (1.3.15) can be found in the books [3,5,17,18].

1.3.4. THE MAXWELL-BLOCH EQUATIONS

Now we will discuss the simplified equations describing the resonant propagation of wave packets. In resonant conditions it is natural to represent the polarisation of a medium as a sum of two terms (1.1.16): the resonant and the non resonant one. In the cases of practical interest it is enough to restrict by the linear approximation for nonresonant terms and to neglect its frequency dispersion. Though the resonance interaction with two-level atoms also gives the substantial contribution to linear polarisation of a medium and determines a frequency dispersion, for convenience they assign $\vec{P}^L=\vec{P}_{nonres}$. When deriving the equations this supposition is equivalent to the identification of the quantity v_g appearing in the equations for amplitudes with the phase speed of light c' in a medium in the absence of resonance atoms.

Bearing in mind the results of sec.1.1.3 and 1.3.2 one can write the basic equations. Firstly, we consider the propagation of the wave

$$\vec{E} = \vec{\mathcal{E}}(\vec{r},t)\exp[i(\vec{k}r - \omega t)] + c.c.$$

under conditions of one-photon resonance. For the purpose of the further generalisation of the main equations to the case of inhomogeneously broadened spectral line we designate the carrier frequency as ω to keep the old notation ω_0 for the central frequency of inhomogeneously broadened spectral line. Also, in the notation of wave vector of the resonant wave we omit the index.

The polarisation of medium

$$P_{res} = \widetilde{\rho}_{ba} D_{ab} + c.c. + \widetilde{\rho}_{bb} D_{bb} + \widetilde{\rho}_{aa} D_{aa}$$

is presented in the following form

$$P_{res} = \mathcal{P}\exp[i(\vec{k}\vec{r} - \omega t)] + c.c..$$

Thus, the wave equations for the case of one-photon resonance can be written as

$$-2ik\left(\frac{\partial}{\partial z} + \frac{1}{c'}\frac{\partial}{\partial t}\right)\mathcal{E} = \frac{4\pi k^2}{\varepsilon_{nonres}}\mathcal{P},$$

where z-axis is chosen along $\vec{k}$, $\omega = kc'$, $c' = c/\sqrt{\varepsilon_{nonres}(\omega)}$. We shall write this equation in another way, omitting a prime sign at the phase velocity of a wave in a medium:

$$\left(\frac{\partial}{\partial z} + \frac{1}{c}\frac{\partial}{\partial t}\right)\mathcal{E} = 2\pi i\frac{\omega}{c}\mathcal{P}.$$

This is the equation for slowly varying wave amplitude and it is also named the reduced Maxwell equations. For each case of resonant interactions there are their own reduced Maxwell equations.

Equations (1.3.17) together with (1.1.12) are called the Maxwell-Bloch equations. Below we write them in a unified manner in order to include all the considered cases of resonant interactions. For this purpose we regard the resonant waves as propagating along the same direction. The Bloch equations remain the same

$$\left(\frac{\partial}{\partial t} - i\Delta + \gamma_{21}\right)R_{21} = i\Lambda(R_1 - R_2),$$
$$\left(\frac{\partial}{\partial t} + \gamma_1\right)R_1 = i(\Lambda^* R_{21} - \Lambda R_{21}^*) + \gamma_1 N_1, \qquad (1.3.17)$$
$$\left(\frac{\partial}{\partial t} + \gamma_2\right)R_2 = -i(\Lambda^* R_{21} - \Lambda R_{21}^*) + \gamma_2 N_2,$$

where the variables Δ, Λ, N_1, N_2, R_1, R_2 and R_{21} like the reduced Maxwell equations themselves should be specified for each case of resonant interactions. We have the following.

In the case of one photon resonance $\omega \approx \omega_{ba}$ (A)

$$\left(\frac{\partial}{\partial z} + \frac{1}{c}\frac{\partial}{\partial t}\right)\mathcal{E} = i2\pi k\{R_{21}d_{ab} - \mathcal{E}(\Pi_a(\omega)R_1 + \Pi_b(\omega)R_2)\}, \qquad (1.3.18)$$
$$R_{21} = \tilde{\rho}_{ba}\exp[-i(\vec{k}\vec{r} - \omega t)],\ R_1 = \tilde{\rho}_{aa},\quad R_2 = \tilde{\rho}_{bb},\quad \Lambda = \mathcal{E}d_{ba}/\hbar,$$
$$\Delta = \omega - \omega_{ba} - (E_b^{St} - E_a^{St})/\hbar,\quad N_1 = \rho_a^0,\quad N_2 = \rho_b^0;$$

In the case of two-photon degenerate resonance $2\omega \approx \omega_{ca}$ (B)

$$\left(\frac{\partial}{\partial z} + \frac{1}{c}\frac{\partial}{\partial t}\right)\mathcal{E} = i2\pi k\{R_{21}\mathcal{E}^*\Pi_{ca}^*(\omega) - \mathcal{E}(\Pi_a(\omega)R_1 + \Pi_c(\omega)R_2)\}, \qquad (1.3.19)$$
$$R_{21} = \tilde{\rho}_{ca}\exp[-2i(\vec{k}\vec{r} - \omega t)],\quad R_1 = \tilde{\rho}_{aa},\quad R_2 = \tilde{\rho}_{cc},\ \Lambda = \mathcal{E}^2\Pi_{ca}(\omega_0)/2\hbar,$$
$$\Delta = 2\omega - \omega_{ca} - (E_c^{St} - E_a^{St})/\hbar,\quad N_1 = \rho_a^0,\quad N_2 = \rho_c^0;$$

In the case of two-photon two-wave resonance $\omega_1 + \omega_2 \approx \omega_{ca}$ (C)

$$\left(\frac{\partial}{\partial z} + \frac{1}{c}\frac{\partial}{\partial t}\right)\mathcal{E}_1 = i2\pi k_1\{R_{21}\mathcal{E}_2^*\Pi_{ca}^*(\omega_2) - \mathcal{E}_1(\Pi_a(\omega_1)R_1 + \Pi_c(\omega_1)R_2)\}, \qquad (1.3.20a)$$
$$\left(\frac{\partial}{\partial z} + \frac{1}{c}\frac{\partial}{\partial t}\right)\mathcal{E}_2 = i2\pi k_2\{R_{21}\mathcal{E}_1^*\Pi_{ca}^*(\omega_1) - \mathcal{E}_2(\Pi_a(\omega_2)R_1 + \Pi_c(\omega_2)R_2)\}; \qquad (1.3.20b)$$

$$R_{21} = \tilde{\rho}_{ca}\exp\{-i[(\vec{k}_1 + \vec{k}_2)\vec{r} - (\omega_1 + \omega_2)t]\},\ R_1 = \tilde{\rho}_{aa},\quad R_2 = \tilde{\rho}_{cc},\quad \Lambda = \mathcal{E}_1\mathcal{E}_2^*\Pi_{ca}(\omega_1)/\hbar,$$
$$\Delta = \omega_1 + \omega_2 - \omega_{ca} - (E_c^{St} - E_a^{St})/\hbar,\ N_1 = \rho_a^0,\quad N_2 = \rho_c^0.$$

In case of Raman resonance $\omega_1 - \omega_2 \approx \omega_{ca}$ (D)

$$\left(\frac{\partial}{\partial z}+\frac{1}{c}\frac{\partial}{\partial t}\right)\mathcal{E}_1 = i2\pi k_1\left\{R_{21}\mathcal{E}_2\,\Pi^*_{ca}(-\omega_2) - \mathcal{E}_1\left(\Pi_a(\omega_1)R_1 + \Pi_c(\omega_1)R_2\right)\right\}, \quad (1.3.21a)$$

$$\left(\frac{\partial}{\partial z}+\frac{1}{c}\frac{\partial}{\partial t}\right)\mathcal{E}_2 = i2\pi k_2\left\{R^*_{21}\mathcal{E}_1\Pi_{ca}(\omega_1) - \mathcal{E}_2\left(\Pi_a(\omega_2)R_1 + \Pi_c(\omega_2)R_2\right)\right\}; \quad (1.3.21b)$$

$$R_{21} = \widetilde{\rho}_{ca}\exp\{-i[(\vec{k}_1-\vec{k}_2)\vec{r} - (\omega_1-\omega_2)t]\},\ R_1 = \widetilde{\rho}_{aa},\quad R_2 = \widetilde{\rho}_{cc},\quad \Lambda = \mathcal{E}_1\mathcal{E}^*_2\,\Pi_{ca}(\omega_1)/\hbar,$$
$$\Delta = \omega_1 - \omega_2 - \omega_{ca} - (E^{St}_c - E^{St}_a)/\hbar,\ N_1 = \rho^0_a,\quad N_2 = \rho^0_c.$$

Here, we suppose that all resonant waves propagate along z-axis

The initial conditions at $t = 0$ for the excitation of gas medium in thermodynamic equilibrium for all the considered cases are given by

$$R_{21}\big|_{t=0} = 0,\quad R_1\big|_{t=0} = N_1,\quad R_2\big|_{t=0} = N_2.$$

The results (1.3.21) for Raman resonance follow from (1.3.20) after the replacement $\mathcal{E}_2 \to \mathcal{E}^*_2,\ \omega_2 \to -\omega_2,\ \vec{k}_2 \to -\vec{k}_2$. In (1.3.17)-(1.3.21) the influence of generated harmonics on resonant fields is neglected.

In practice the spectral line of resonant emission/absorption of the medium is often inhomogeneously broadened, e.g. the frequencies ω_{ba} (or ω_{ca}) of different resonant atoms of the same sort are not equal but lie around in the vicinity of some value ω_0. The physical nature reasons of inhomogeneous broadening are different in gas and solid state.

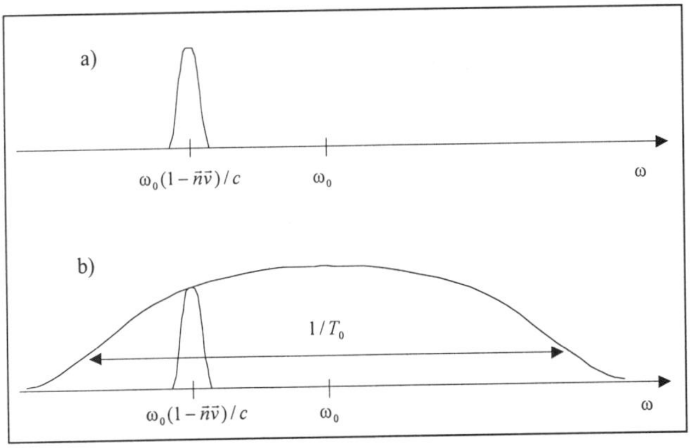

Fig.1.3.1. The spectral line of a single atom moving with a speed $\vec{v}$ (a) and the spectral line of a gas of such atoms (b).

In a gas the resonance atoms move chaotically. The spectral line of radiation of individual atom $\psi(\omega_0)$ moving with a speed $\vec{v}$, is Doppler shifted in dependence of projections of atom velocity on the direction of observation $\vec{n}$, $\psi_{\vec{v}}(\omega_0 - \vec{n}\vec{v}\omega_0/c)$. Here ω_0 is transition frequency of two-level atom in its centre of mass system. At a thermal equilibrium the relative part of atoms $f(\vec{v})\,d\vec{v}$, having a velocity within a narrow range from $\vec{v}$ to $\vec{v}+d\vec{v}$, is defined by Maxwell distribution

$$f(v) = \left(\frac{1}{\sqrt{\pi}u}\right)^3 \exp\left(-\frac{v^2}{u^2}\right),$$

where u is the most probable velocity of atoms. As a result the summarised spectral line of an ensemble of two-level atoms $\int \psi_{\vec{v}} f(v) d\vec{v}$ represents a symmetric contour with the central frequency ω_0 and the width $1/T_0 = \omega_0 u/c$ exceeding noticeably the width of $\psi(\omega_0)$ (see Fig. 1.3.1).

The dynamics of a group of two-level atoms, with the speeds from $\vec{v}$ to $\vec{v}+d\vec{v}$ can be described by Bloch equations (1.1.17). Then, however, it is necessary to replace $\partial/\partial t$ on $\partial/\partial t + \vec{v}\nabla$, N_i with $N_i f(v)$ and to fix Δ in the form

$$\omega - \tilde{\omega}_0 - \vec{k}\vec{v}, \quad 2\omega - \tilde{\omega}_0 - 2\vec{k}\vec{v}, \quad \omega_1 + \omega_2 - \tilde{\omega}_0 - (\vec{k}_1 + \vec{k}_2)\vec{v}$$
$$\omega_1 - \omega_2 - \tilde{\omega}_0 - (\vec{k}_1 - \vec{k}_2)\vec{v}$$

for the cases A $(\omega \approx \omega_0)$, B $(2\omega \approx \omega_0)$, C $(\omega_1 + \omega_2 \approx \omega_0)$ and D $(\omega_1 - \omega_2 \approx \omega_0)$ of resonant interaction accordingly. The sign ~ over the frequency ω_0 denotes the account for the Stark shift - $\tilde{\omega}_0 = \omega_0 + (E_b^{St} - E_a^{St})/\hbar$ or $\tilde{\omega}_0 = \omega_0 + (E_c^{St} - E_a^{St})/\hbar$. One should note that the frequency ω_0 in case A is the frequency of optically allowed resonant transition whereas in cases B-D it is the frequency of optically forbidden resonant transition.

The polarisation of a medium is created by all groups of atoms, therefore it is necessary to carry out the additional integration of the right hand side of expressions (1.1.18)-(1.1.21) over all velocities. For example, equation (1.1.18) in the case of inhomogeneously broadened spectral line becomes

$$\left(\frac{\partial}{\partial z} + \frac{1}{c}\frac{\partial}{\partial t}\right)\mathcal{E} = i2\pi k \int d\vec{v} \{R_{21} d_{ab} - \mathcal{E}(\Pi_a(\omega)R_1 + \Pi_b(\omega)R_2)\}, \qquad (1.3.18')$$

We will denote the analogous forms of equations (1.1.19)-(1.1.20) as (1.1.19′)-(1.1.20′).

The initial conditions at $t = 0$ for excitation of gas medium in thermodynamic equilibrium for all the considered cases are given by

$$R_{21}\big|_{t=0} = 0, \; R_1\big|_{t=0} = N_1 f(v), \; R_2\big|_{t=0} = N_2 f(v). \qquad (1.3.22)$$

Derivation of the density matrix equation for a gas medium is considered in more detail in Appendix 2. The explicit example of operating with the equation is given in section 2.5.5 for the case of photon echoes produced by two standing wave pulses in resonant gas medium.

The frequencies of resonance impurities in a solid state without current carriers locate near some value ω_0. The inhomogeneous broadening is caused by a casual disposition of impurities in a matrix of crystal and various magnitude of interaction of impurity with atoms of crystal. In many cases the distribution of impurities atoms over resonance frequencies ω_{ba} (or ω_{ca}) is satisfactorily described with the help of Lorentz factor

$$\varphi(\omega_{ba} - \omega_0) = \frac{1/\pi T_0}{(\omega_{ba} - \omega_0)^2 + 1/T_0^2},$$

where $1/T_0$ characterises the width of inhomogeneously broadened spectral line. The impurity dipole momentum can be considered as independent from impurity disposition in crystal matrix. Dynamics of group of impurity atoms, having the detuning from a resonance from the interval Δ to $\Delta + d\Delta$, is described by the equations (1.3.17), where N_i is replaced by $N_i F(\Delta)$, where $F(\Delta) = \varphi(\Delta - \Delta_0)$, and for cases A, B, C, and D the value Δ_0 is equal to $\omega - \widetilde{\omega}_0$, $2\omega - \widetilde{\omega}_0$, $\omega_1 + \omega_2 - \widetilde{\omega}_0$, $\omega_1 - \omega_2 - \widetilde{\omega}_0$ accordingly. We have included the Stark shifts in the frequency $\widetilde{\omega}_0$. In the initial conditions (1.3.22) $f(v)$ must be replaced by $F(\Delta)$. The reduced Maxwell equations are given by the expressions (1.3.18′)-(1.3.21′), in which the integral over $d\vec{v}$ in the right hand side is replaced by the integral over $d\Delta$. The obtained Maxwell-Bloch equations can also be used in the case of resonance excitation of gas by waves, propagating in one direction, if to put $F(\Delta) = T_0/\sqrt{\pi}\exp\left[-(\Delta - \Delta_0)^2 T_0^2\right]$ and to consider Δ to be equal to $\Delta_0 - \vec{k}\vec{v}$, $\Delta_0 - 2\vec{k}\vec{v}$, $\Delta_0 - (\vec{k}_1 + \vec{k}_2)\vec{v}$, and $\Delta_0 - (\vec{k}_1 - \vec{k}_2)\vec{v}$ in cases A, B, C and D.

Sometimes, we will designate the integration over the inhomogeneously broadened spectral line by brackets $\langle\ \rangle$. In such cases we change the normalisation of the density matrix so that expression (1.3.18′) will read

$$\left(\frac{\partial}{\partial z} + \frac{1}{c}\frac{\partial}{\partial t}\right)\mathcal{E} = i2\pi k N \langle \mathcal{R}_{21} d_{ab} - \mathcal{E}\left(\Pi_a(\omega)\mathcal{R}_1 + \Pi_b(\omega)\mathcal{R}_2\right)\rangle, \qquad (1.3.18'')$$

where N is the density of atoms, $R \equiv \mathcal{R}F(\Delta)$ or $R \equiv \mathcal{R}f(v)$, and $Sp\rho = 1$. In the Bloch equations and in the initial conditions, the level populations N_i must be replaced by N_i / N, and the distribution function ($f(v)$ or $F(\Delta)$) must be omitted. Then, we follow the definition

$$\langle \text{expression} \rangle \equiv \int d\Delta F(\Delta)(\text{expression}), \text{ or } \langle \text{expression} \rangle \equiv \int d\vec{v} f(v)(\text{expression}).$$

The case of homogeneously broadened spectral line can be treated as a sharp-line limit of an inhomogeneously broadened spectral line when $F(\Delta)=\delta(\Delta-\Delta_0)$. Then,

$$\langle \text{expression} \rangle \equiv \text{expression}\big|_{\Delta=\Delta_0} .$$

In this case eq.(1.3.18") reduces to the eq.(1.3.18) describing the case of the homogeneously broadened spectral line.

1.4. Polarisation effects and resonant level degeneration

The above stated bases of the theory of interaction of coherent radiation with quantum systems are easy to be generalised to involve any sort of polarisation of light wave and a degeneracy of atomic energy levels. This is one of the commonly used models of levels degeneration. Let atomic levels be characterised by energy E_α, total angular momentum is j_α and its projection on the axis of quantisation is m_α $(-j_\alpha \le m_\alpha \le j_\alpha)$. Other quantum numbers (if there are any) in the optical processes under discussion are assumed to be constant.

1.4.1. GENERALISED MAXWELL-BLOCH EQUATIONS

It is convenient to use a formalism of irreducible tensor operators while considering the resonant interaction of light with matter. An example of irreducible tensor operators of a first rank is a vector value represented in spherical components. The scalar product of irreducible tensor operators coincides with the scalar product of vectors, for example:

$$\vec{E}\vec{d} = \sum_{q=0,\pm1} (-1)^q E^q d^{-q} .$$

Spherical components of any vector (say $\vec{d}$) $d^q, q=0,\pm1$, are defined as

$$d^0 = d_z,\ d^{\pm1} = \mp\frac{1}{\sqrt{2}}\left(d_x \pm i d_y\right). \tag{1.4.1}$$

Here d_x, d_y and d_z – projection of a vector $\vec{d}$ on Cartesian axis (z is the axis of quantisation). The matrix elements of the dipole moment operator $\vec{d}$ are related to the reduced dipole momentum d_{ba} of appropriate optically allowed transitions by the formulae [20]

$$d^q_{m_b m_a} = (-1)^{j_b - m_b}\begin{pmatrix} j_b & 1 & j_a \\ -m_b & q & m_b \end{pmatrix} d_{ba},\quad d^q_{m_a m_b} = (-1)^{j_b - m_a}\begin{pmatrix} j_a & 1 & j_b \\ -m_a & q & m_b \end{pmatrix} d^*_{ba} . \tag{1.4.2}$$

For the sake of simplicity we accept the notation simplification of a designation of matrix elements we use the notation

$$d^q_{m_b m_a} \equiv \left\langle E_b j_b m_b \right| d^q \left| E_a j_a m_a \right\rangle .$$

The density matrix in terms of irreducible tensor operator is given by [2,20,21]:

$$\rho_{m_e m_a} = (-1)^{j_e - m_a}(2j_e+1)^{-\frac{1}{2}}\sum_{\aleph q}(2\aleph+1)\begin{pmatrix} j_e & j_a & \aleph \\ m_e & -m_a & q \end{pmatrix}\chi_q^{(\aleph)},$$

$$\rho_{m_a m_a'} = (-1)^{j_a - m_a'}(2j_a+1)^{-\frac{1}{2}}\sum_{\aleph q}(2\aleph+1)\begin{pmatrix} j_a & j_a & \aleph \\ m_a & -m_a' & q \end{pmatrix}\Phi_q^{(\aleph)}, \tag{1.4.3}$$

The expansion of $\rho_{m_e m_e'}$ can be obtained from the expression for $\rho_{m_a m_a'}$ by substitutions $a \to e$ and $\Phi_q^{(\aleph)} \to F_q^{(\aleph)}$. Here $e=b$ or $e=c$ in dependence on resonant conditions, $\chi_q^{(\aleph)}$, $\Phi_q^{(\aleph)}$ and $F_q^{(\aleph)}$ are irreducible components of the appropriate density matrices. Note, that $\Phi_0^{(0)}$ represents the total population of level E_a, $\Phi_q^{(1)}$ and $\Phi_q^{(2)}$ is thought about as an orientation and an alignment. If we take into account depolarising atomic collisions each component of irreducible tensor operator will decay exponentially with the own relaxation constant (see section 2.7.2).

In order to derive the generalisation of the Bloch equations in the case of arbitrary polarisation of exiting fields and arbitrary multiplicity of degeneration of energy levels due to various orientation of total angular momentum, one should: first obtain the equations for the transformed density matrix by means of unitary transformation, then decompose these equations in terms of irreducible tensor operators [21] and at last turn back to a standard form of density matrix. At this stage the number of sums can be summed up. The result is as follows

$$i\hbar\left(\frac{\partial}{\partial t} - i\Delta'\right)R_{m_e m_a} = V_{m_e m_e'}R_{m_e' m_e} - R_{m_e m_a'}V_{m_a' m_a} + V_{m_e m_a'}R_{m_a' m_a} - R_{m_e m_e'}V_{m_e' m_a},$$

$$i\hbar\frac{\partial}{\partial t}R_{m_e m_e'} = V_{m_e m_e''}R_{m_e'' m_e'} - R_{m_e m_e''}V_{m_e'' m_e'} + V_{m_e m_a}R^*{}_{m_e' m_a} - R_{m_e m_a}V^*{}_{m_e' m_a}, \tag{1.4.4}$$

$$i\hbar\frac{\partial}{\partial t}R_{m_a m_a'} = V_{m_a m_a''}R_{m_a'' m_a'} - R_{m_a m_a''}V_{m_a'' m_a'} + V^*{}_{m_e m_a}R_{m_e m_a'} - R^*{}_{m_e m_a}V_{m_e m_a'}.$$

Here and below the repeating matrix indices imply the summation. The resonant polarisation of the medium is represented in the form

$$\vec{P}^{res} = \widetilde{\rho}_{m_e m_a}\vec{D}_{m_a m_e} + c.c. + \widetilde{\rho}_{m_e m_e'}\vec{D}_{m_e' m_e} + \widetilde{\rho}_{m_a m_a'}\vec{D}_{m_a' m_a}. \tag{1.4.5}$$

The values introduced in (1.4.4) and the matrix elements of the effective dipole moment operator have the following expression:

For one-photon resonance $\omega \approx \omega_{ba}$ (A)

$$e = b,\ \Delta' = \omega - \omega_{ba},\ R_{m_b m_a} = \tilde{\rho}_{m_b m_a} \exp[-i(\vec{k}\vec{r} - \omega t)],\ R_{m_b m'_b} = \tilde{\rho}_{m_b m'_b},\ R_{m_a m'_a} = \tilde{\rho}_{m_a m'_a},$$

$$V_{m_b m_a} = (-1)^{j_b - m_a + 1} d_{ba} \sum_q \begin{pmatrix} j_b & j_a & 1 \\ m_b & -m_a & q \end{pmatrix} \mathcal{E}^q,$$

$$V_{m_b m'_b} = (-1)^{j_b - m_b} \sqrt{2j_b + 1} \sum_{\aleph q s} (2\aleph + 1) \begin{pmatrix} j_b & j_b & \aleph \\ m_b & -m'_b & -q-s \end{pmatrix} \begin{pmatrix} 1 & 1 & \aleph \\ s & q & -q-s \end{pmatrix} \mathcal{E}^{-q} \mathcal{E}^{*-s} \Pi^{(\aleph)}_{b\omega} \hbar^{-1},$$

$$D^q_{m_a m_b} = (-1)^{j_b - m_a} \begin{pmatrix} j_b & j_a & 1 \\ m_b & -m_a & q \end{pmatrix} d^*_{ba},\ D^q_{m_b m'_b} = \overline{D}^q_{m_b m'_b} + (-1)^q \left(\overline{D}^{-q}\right)^*_{m'_b m_b},$$

$$\overline{D}^q_{m_b m'_b} = (-1)^{j_b - m'_b + 1} \cdot$$

$$\cdot \sqrt{2j_b + 1} \sum_{\aleph q s} (-1)^s (2\aleph + 1) \begin{pmatrix} j_b & j_b & \aleph \\ m'_b & -m_b & q+s \end{pmatrix} \begin{pmatrix} 1 & 1 & \aleph \\ -s & -q & q+s \end{pmatrix} \mathcal{E}^{-s} \hbar^{-1} \Pi^{(\aleph)}_{b\omega} \exp[i(\vec{k}\vec{r} - \omega t)]$$

(the values $V_{m_a m'_a}$ and $D^q_{m_a m'_a}$ are obtained from $V_{m_b m'_b}$ and $D^q_{m_b m'_b}$ by replacing $b \to a$);

For two-photon one-wave resonance $2\omega \approx \omega_{ca}$ (B)

$$e = c,\ \Delta' = 2\omega - \omega_{ca},\ R_{m_c m_a} = \tilde{\rho}_{m_c m_a} \exp[-2i(\vec{k}\vec{r} - \omega t)],\ R_{m_c m'_c} = \tilde{\rho}_{m_c m'_c},\ R_{m_a m'_a} = \tilde{\rho}_{m_a m'_a},$$

$$V_{m_c m_a} = (-1)^{j_c - m_c + 1} \sqrt{2j_a + 1} \sum_{\aleph q s} (2\aleph + 1) \begin{pmatrix} j_c & j_a & \aleph \\ m_c & -m_a & -q-s \end{pmatrix} \begin{pmatrix} 1 & 1 & \aleph \\ s & q & -q-s \end{pmatrix} \mathcal{E}^{-q} \mathcal{E}^{-s} \frac{\Pi^{(\aleph)}_{\omega}}{2\hbar},$$

$$D^q_{m_a m_c} = (-1)^{j_c - m_c} \sqrt{2j_a + 1} \sum_{\aleph s} (-1)^s (2\aleph + 1) \begin{pmatrix} j_c & j_a & \aleph \\ m_c & -m_a & q+s \end{pmatrix} \begin{pmatrix} 1 & 1 & \aleph \\ -s & -q & q+s \end{pmatrix} \cdot$$

$$\cdot \left(\mathcal{E}^{-s} \Pi^{(\aleph)*}_{-\omega} \exp[i(\vec{k}\vec{r} - \omega t)] + \mathcal{E}^{*-s} \Pi^{(\aleph)*}_{\omega} \exp[-i(\vec{k}\vec{r} - \omega t)]\right) \hbar^{-1}$$

(quantities $V_{m_c m'_c}$ and $V_{m_a m'_a}$ are obtained from $V_{m_b m'_b}$ of previous case by the replacement $b \to c$ and $b \to a$, accordingly. Analogously, quantities $D^q_{m_c m'_c}$ and $D^q_{m_a m'_a}$ are obtained from $D^q_{m_b m'_b}$, by the same replacement $b \to c$ and $b \to a$, accordingly);

For two-photon two-wave resonance $\omega_1 + \omega_2 \approx \omega_{ca}$ (C)

$$e = c\,,\ \Delta' = \omega_1 + \omega_2 - \omega_{ca}\,,$$

$$R_{m_c m_a} = \tilde{\rho}_{m_c m_a} \exp\{-i[(\vec{k}_1 + \vec{k}_2)\vec{r} - (\omega_1 + \omega_2)t]\},\ R_{m_c m_c'} = \tilde{\rho}_{m_c m_c'}\,,\ R_{m_a m_a'} = \tilde{\rho}_{m_a m_a'}\,,$$

$$V_{m_c m_a} = (-1)^{j_c - m_c + 1}\sqrt{2j_a + 1}\sum_{\aleph q s}(2\aleph + 1)\begin{pmatrix} j_c & j_a & \aleph \\ m_c & -m_a & -q-s \end{pmatrix}\begin{pmatrix} 1 & 1 & \aleph \\ s & q & -q-s \end{pmatrix}\mathcal{E}_1^{-q}\mathcal{E}_2^{-s}\frac{\left(\Pi_{\omega_1}^{(\aleph)} + (-1)^{\aleph}\Pi_{\omega_2}^{(\aleph)}\right)}{2\hbar},$$

$$V_{m_c m_c'} = (-1)^{j_c - m_c}\sqrt{2j_c + 1}\sum_{\aleph q s}(2\aleph + 1)\begin{pmatrix} j_c & j_c & \aleph \\ m_c & -m_c' & -q-s \end{pmatrix}\begin{pmatrix} 1 & 1 & \aleph \\ s & q & -q-s \end{pmatrix}\sum_{n=1}^{2}\frac{\mathcal{E}_n^{-q}\mathcal{E}_n^{*-s}\Pi_{c\omega_n}^{(\aleph)}}{\hbar},$$

$$D^q = D_1^q + D_2^q\,,$$

$$\left[D_n^q\right]_{m_a m_c} = (-1)^{j_c - m_c}\sqrt{2j_a + 1}\sum_{\aleph s}(-1)^s(2\aleph + 1)\begin{pmatrix} j_c & j_a & \aleph \\ m_c & -m_a & q+s \end{pmatrix}\begin{pmatrix} 1 & 1 & \aleph \\ -s & -q & q+s \end{pmatrix}\cdot$$

$$\cdot(\mathcal{E}_n^{-s}\Pi_{-\omega_n}^{(\aleph)*}\exp[i(\vec{k}_n\vec{r} - \omega_n t)] + \mathcal{E}_n^{*-s}\Pi_{\omega_n}^{(\aleph)*}\exp[-i(\vec{k}_n\vec{r} - \omega_n t)])\hbar^{-1}$$

($V_{m_a m_a'}$ are obtained from $V_{m_c m_c'}$ by substitution of $c \to a$, and $\left[D_n^q\right]_{m_c m'_c}$ and $\left[D_n^q\right]_{m_a m'_a}$ are obtained from $D_{m_b m'_b}^q$ of case A after substitutions $\mathcal{E} \to \mathcal{E}_n$ and accordingly $b \to c$ and $b \to a$).

The Raman resonance (D) is described by the same equations, as the two-photon two-wave resonance case, where the replacements $\mathcal{E}_2 \to \mathcal{E}_2^*$, $\omega_2 \to -\omega_2$, $\vec{k}_2 \to -\vec{k}_2$ have been made.

The operator of resonant interaction V and the effective dipole moment operator $\vec{D}$ are determined by parameters:

$$\Pi_{\omega}^{(\aleph)} = \sum_{\beta}\frac{d_{c\beta}d_{\beta a}}{\sqrt{2j_a + 1}}(-1)^{j_c + j_a}\begin{Bmatrix} 1 & \aleph & 1 \\ j_a & j_\beta & j_c \end{Bmatrix}\left(\frac{1}{\omega_{\beta a} - \omega} + \frac{(-1)^{\aleph}}{\omega_{\beta c} + \omega}\right),$$

$$\Pi_{c\omega}^{(\aleph)} = \sum_{\beta}\frac{\left|d_{c\beta}\right|^2}{\sqrt{2j_c + 1}}(-1)^{j_c + j_\beta}\begin{Bmatrix} 1 & \aleph & 1 \\ j_a & j_\beta & j_c \end{Bmatrix}\left(\frac{1}{\omega_{c\beta} - \omega} + \frac{(-1)^{\aleph}}{\omega_{c\beta} + \omega}\right). \tag{1.4.6}$$

Here index β marks the nonresonant energy levels, $\Pi_{b\omega}^{(\aleph)}$ and $\Pi_{a\omega}^{(\aleph)}$ are given by $\Pi_{c\omega}^{(\aleph)}$ after the substitutions $c \to b$ and $c \to a$ accordingly. The definition of the 6j symbols is standard [20].

The equations (1.4.4) and (1.4.5) are simplified in the cases of small and large values of the total angular momentum. The proper expressions for small values of the total angular momentum will be given in the relevant sections of the book where we will discuss the methods of their solution. To obtain these particular expressions it is sufficient to use the following formulas:

$$\begin{pmatrix} j & j & 0 \\ m & -m & 0 \end{pmatrix} = (-1)^{j-m} \frac{1}{\sqrt{2j+1}},$$

$$\begin{pmatrix} j_1 & j_2 & j_1+j_2 \\ m_1 & m_2 & -m_1-m_2 \end{pmatrix} =$$

$$= (-1)^{j_1-j_2+m_1+m_2} \sqrt{\frac{(2j_1)!(2j_2)!(j_1+j_2+m_1+m_2)!(j_1+j_2-m_1-m_2)!}{(2j_1+2j_2+1)!(j_1+m_1)!(j_1-m_1)!(j_2+m_2)!(j_2-m_2)!}},$$

$$\begin{pmatrix} j_1 & j_2 & j_3 \\ j_1 & -j_1-m_3 & m_3 \end{pmatrix} =$$

$$= (-1)^{-j_1+j_2+m_3} \sqrt{\frac{(2j_1)!(-j_1+j_2+j_3)!(j_3-m_3)!}{(j_1+j_2+j_3+1)!(j_1-j_2+j_3)!(j_1+j_2-j_3)!(-j_1+j_2-m_3)!(j_3+m_3)!}}.$$

Also, the equations(1.4.4) and (1.4.5) can be considerably simplified [22,12] in the case of large values of angular momentum when the quasiclassical description of angular momentum is valid [22]. The main idea of the quasiclassical approach to the description of angular momentum is considered in Appendix 1.

To involve the inhomogeneous broadening into analysis of $\vec{P}^{res}$ one should proceed analogously to nondegenerate case of section 1.3.4 (see the end of section 1.3.4). We consider here only the case of one-photon resonance in a gas medium with no allowance for the Stark shifts. The Maxwell-Bloch equations read as

$$\left(\frac{\partial}{\partial z} + \frac{1}{c}\frac{\partial}{\partial t}\right)\mathcal{E}^q = i\frac{2\pi}{\sqrt{3}} kNd^*_{ba} \sum_{m_a m_b} \left\langle \mathcal{R}_{m_b m_a} \right\rangle \mathcal{G}^q_{m_b m_a}, \tag{1.4.7a}$$

$$\left(\frac{\partial}{\partial t} + i(\vec{k}\vec{v} - \Delta')\right)\mathcal{R}_{m_b m_a} = \frac{i}{\hbar\sqrt{3}} \sum_q \mathcal{E}^q d_{ba} \left(\sum_{m'_a} \mathcal{G}^q_{m_b m'_a} \mathcal{R}_{m'_a m_a} - \sum_{m'_b} \mathcal{R}_{m_b m'_b} \mathcal{G}^q_{m'_b m_a} \right), \tag{1.4.7b}$$

$$\frac{\partial}{\partial t}\mathcal{R}_{m_a m'_a} = \frac{i}{\hbar\sqrt{3}} \sum_{q m_b} \left(\mathcal{E}^{q*} d^*_{ba} \mathcal{G}^q_{m_b m_a} \mathcal{R}_{m_b m'_a} - \mathcal{E}^q d_{ba} \mathcal{R}_{m_a m_b} \mathcal{G}^q_{m_b m'_a} \right), \tag{1.4.7c}$$

$$\frac{\partial}{\partial t}\mathcal{R}_{m_b m'_b} = \frac{i}{\hbar\sqrt{3}} \sum_{q m_a} \left(\mathcal{E}^q d_{ba} \mathcal{G}^q_{m_b m_a} \mathcal{R}_{m_a m'_b} - \mathcal{E}^{q*} d^*_{ba} \mathcal{R}_{m_b m_a} \mathcal{G}^q_{m'_b m_a} \right), \tag{1.4.7d}$$

where

$$\mathcal{G}^q_{m_b m_a} = (-1)^{j_b - m_b} \begin{pmatrix} j_b & 1 & j_a \\ -m_b & q & m_b \end{pmatrix} \sqrt{3} .$$

In the above notation we have introduced the coefficient $\sqrt{3}$ to simplify expressions for $\mathcal{G}^q_{m_b m_a}$ in the cases of small angular momenta:

$$\mathcal{G}^q_{m_b m_a} = (-1)^{m_b} \delta_{q,-m_b} , \; j_b = j_a + 1 = 1 ; \; \mathcal{G}^q_{m_b m_a} = -\delta_{q,m_a} , \; j_b = j_a - 1 = 0 .$$

1.4.2. NONLINEAR PARABOLIC EQUATION

There is no need to apply the techniques of the total angular momentum and irreducible tensor operators to include polarisation effects in wave equations in nonresonant cases. These complicated methods allow to determine the relationships of optical susceptibility parameters with microscopic parameters of atoms. They also define the symmetry properties of optical susceptibilities. But the latter can be obtained from the more general reasoning.

In isotropic media with the symmetry centre (gas or liquid) the third order optical susceptibility $\chi^{(3)}_{i,klm}(-\omega;\omega,-\omega,\omega)$ has the form

$$\chi^{(3)}_{i,klm}(-\omega;\omega,-\omega,\omega) = \tfrac{1}{6}\alpha(\omega)(\delta_{ik}\delta_{lm} + \delta_{im}\delta_{lk}) + \tfrac{1}{3}\beta(\omega)\delta_{il}\delta_{km} .$$

Thus, assuming the independence of

$$\frac{4\pi(\overline{\omega}+\nu)^2}{c^2} P_i^{NL}(\vec{\bar{k}} + \vec{\kappa}, \overline{\omega} + \nu)$$

in (1.3.5) from ν and $\chi^{(3)}_{ii_1i_2i_3}(-\overline{\omega}-\nu; \overline{\omega}+\nu_1 - \overline{\omega} - \nu_2, \overline{\omega}+\nu_3)$ from ν_1, ν_2, ν_3 and ν and carrying out the reverse Fourier transformation of the nonlinear polarisation

$$P_i^{NL}(\vec{\bar{k}} + \vec{\kappa}, \overline{\omega}+\nu) = 3 \sum_{i_1 i_2 i_3} \int \frac{d\vec{\kappa}_1 d\vec{\kappa}_2 d\vec{\kappa}_3 d\nu_1 d\nu_2 d\nu_3}{(2\pi)^8} \chi^{(3)}_{i,i_1i_2i_3}(-\overline{\omega}-\nu; \overline{\omega}+\nu_1, -\overline{\omega}-\nu_2, \overline{\omega}+\nu_3) \cdot$$
$$\cdot \delta(\vec{\kappa} - \vec{\kappa}_1 - \vec{\kappa}_2 - \vec{\kappa}_3) \delta(\nu - \nu_1 - \nu_2 - \nu_3) \varepsilon_{i_1}(\vec{\kappa}_1 \nu_1) \varepsilon^*_{i_2}(\vec{\kappa}_2 \nu_2) \varepsilon_{i_3}(\vec{\kappa}_3 \nu_3)$$

we can obtain from (1.3.5) and (1.3.9) that

$$\left[-2i\bar{k}\left(\frac{\partial}{\partial z} + \frac{1}{v_g}\frac{\partial}{\partial t} \right) - \Delta_\perp + \frac{1 - \bar{k}v_g'}{v_g^2}\frac{\partial^2}{\partial t^2} \right] \vec{\varepsilon}(\vec{r},t) = \alpha(\omega)\left|\vec{\varepsilon}\right|^2 \vec{\varepsilon} + \beta(\omega)\vec{\varepsilon}^2\vec{\varepsilon}^* . \qquad (1.4.8)$$

It is convenient to consider parameters $\alpha(\omega)$ and $\beta(\omega)$ as phenomenological ones and to determine their values from experimental data. Unlike the resonant situations it is unnecessary to establish the dependence of these parameters from microscopic characteristics of the medium. These parameters are more adequate characteristics of the matter than microscopic ones in view of the exceedingly great number of microscopic parameters for nonresonant wave propagation.

1.5. Conclusion

To understand further development and application of the method of unitary transformations it is useful to consider resonant absorption as the elementary event of many particle collision. For example, two-photon absorption is the collision of one two-level atom of sort *II* with two photons of energies $\hbar\omega_1$ and $\hbar\omega_2$ satisfying the resonance condition $\hbar\omega_1 + \hbar\omega_2 \approx E_c^{II} - E_a^{II}$. Here, the upper index *II* denotes the sort of atoms and we suppose that in this sort of atoms the energy levels are coupled by two-quantum (optically forbidden) transition. Now we replace one photon, say $\hbar\omega_2$, with the two-level atom of another sort *I* with the energy levels coupled by optically allowed transition $E_b^I \to E_a^I$. In a three particle collision of one photon $\hbar\omega_1$ with excited atom *I* and atom *II* on the ground state the resonant absorption of the photon $\hbar\omega_1$ with simultaneous excitation of atom *II* and deactivation of atom *I* is possible. Thus, in external electromagnetic field the excitation energy of atom *I* is transferred to atom *II*. This process is named radiative atomic collision [23] and is very effective if $\hbar\omega_1 + E_b^I - E_a^I \approx E_c^{II} - E_a^{II}$ (see Fig.1.5.1). The method of unitary transformation has turned out to be very powerful for simple and successful derivation of main equations describing radiative atomic collisions [10,15].

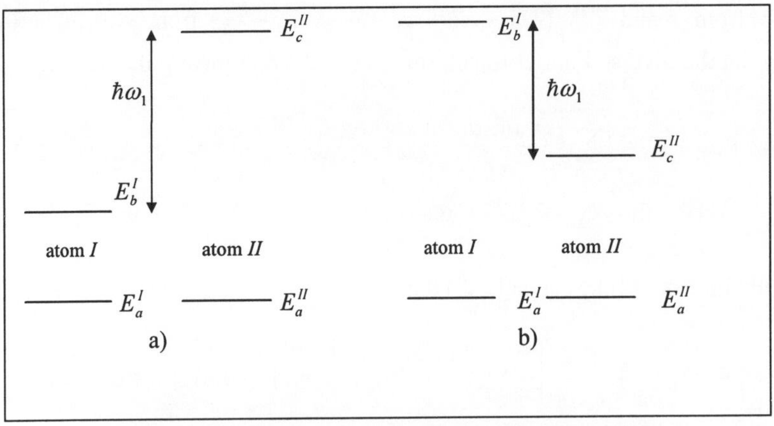

Fig.1.5.1. Level configuration of atoms undergoing radiative collisions.

One can specify in more detail two photons in collision processes with atom *II*. So far we have described electromagnetic fields as classic fields. If we treat one electromagnetic field as classic and another field as quantum, we will obtain the possibility of simple consideration of "spontaneous stimulated" two-photon emission of excited atom *II* as well as stimulated two-photon relaxation which is additional to the one-photon relaxation process. This process is one of the hierarchy of multiphoton spontaneous processes in the external field. The transition in an atom, leading to additional relaxation, occurs with the absorption of one or several photons from the classic coherent field simultaneously with emission (or absorption) of a vacuum photon [13]. Owing to the intensity of an external field, such processes may be very significant. The method of unitary transformation allows to construct the effective Hamiltonian for such processes and to obtain the whole hierarchy of relaxation constants (see Appendix 1).

A complicated picture of two-photon resonance arises in a thin film of width much less than the wavelength of an incident field. If we take into account the reverse influence of film on exciting wave we should simultaneously allow for the Raman scattering of third harmonic. The method of unitary transformation gives a systematic way for the treatment of such processes (see section 9.3).

Another field of applicability of the method of unitary transformation is the construction of Lax pair for integrable models in nonlinear optics [14]. In such problems the method of unitary transformation is more physical and has been used to obtain exactly integrable polarisation models of two-photon resonance (see Appendix 4).

References

1. Allen, L., and Eberly, J.H.: *Optical resonance and two-level atoms,* Wiley-Interscience, New York, 1975.
2. Blum, K.: *Density matrix. Theory and applications,* Plenum Press, New York, 1981.
3. Shen, Y.R.: *The principle of nonlinear optics,* Wiley, New York, 1984.
4. Gardiner, C.W.: *Quantum noise.* Springer, Berlin, 1991.
5. Butcher, P.N., Cotter, D.: The elements of nonlinear optics, Cambridge Univ.Press, 1990.
6. Bjorken, J.D., and Drell, S.D.: *Relativistic quantum mechanics*, McGraw-Hill, New York, 1965
7. Lichtenberg, A.J., and Lieberman, M.A.: *Regular and stochastic motion*, Springer-Verlag, New York, 1983.
8. Takatsuji, M.: Two-level approximation for optical pulse propagation in resonant media, *Physica* **51** (1971), 265-272; Theory of coherent two-photon resonance, *Phys.Rev. A* **11** (1975), 619-624.
9. Grischkowsky, D., Loy, M.M.T., and Liao, P.F.: Adiabatic following model for two-photon transitions: nonlinear mixing and pulse propagation, *Phys.Rev.A* **12** (1975), 2514-2533.
10. Ivanova, A.V., and Melikyan, G.G.: Effective interaction in non-linear optics, *J.Phys. B* **21** (1988), 3017-3031.
11. Basharov, A.M., Maimistov, A.I., and Manykin, E.A.: Polarization features of coherent transition phenomena in two-photon resonance, *Zh.Eksp.Teor.Fiz.* **84** (1983), 487-501 [*Sov.Phys. JETP* **57** (1983), 282-289].
12. Basharov, A.M., and Ivanov, A.Yu.: Spectroscopy of two-quantum transitions on the base of optical nutation, *Optika i spektroskopiya* **66** (1989), 805-811.
13. Basharov, A.M.: Two-photon atomic relaxation in the field of nonresonant electromagnetic wave,

Zh.Eksp.Teor.Fiz. **102** (1992), 1126-1139 [*Sov.Phys. JETP* **75** (1992), 611-618].

14. Basharov, A.M.: Relation between exactly integrable models in resonance optics, *Zh.Eksp.Teor.Fiz.* **97** (1990), 169-178 [*Sov.Phys. JETP* **70** (1990), 94-99].
15. Basharov, A.M.: Photonics. *The method of unitary transformation in nonlinear optics*, MEPI, Moscow, 1990.
16. Butylkin, V.S., Kaplan, A.E., Khromopulo, Yu.G., and Yakubovich, E.I.: *Resonant nonlinear interaction of light with matter*, Springer, Berlin, 1989.
17. Butcher, P.: *Nonlinear optical phenomena*, Ohio State University, Columbus, 1965.
18. Hanna, D.C., Yuratich, M.A., and Cotter D.: *Nonlinear optics of free atoms and molecules,* Springer, New York, 1979.
19. Gibbon, J.D., McGuiness, M.J.: Amplitude equations at the critical points of unstable dispersive physical systems, *Proc.R.Soc.Lond.* A **377** (1981), 185-219.
20. Edmonds, A.R.: *Angular momentum in quantum mechanics*, Princeton Univ.Press, Princeton, 1960.
21. Omont, A.: Irreducible components of the density matrix. Application to optical pumping, *Prog.Quantum Electronics* **5** (1981), 69-138.
22. Nasyrov, K.A., and Shalagin, A.M.: Interaction between intense radiation and atoms or molecules experiencing classical rotary motion, *Zh.Eksp.Teor.Fiz.* **81** (1981), 1649-1663 [*Sov.Phys. JETP* **54** (1981), 877].
23. Yakovlenko, S.I.: Laser induced radiative collisions (the review), Quantum Electronics (Moscow) **5** (1978), 259-269.

CHAPTER 2

COHERENT TRANSIENT PHENOMENA

Three main processes determine the features of optical phenomena. The first is the dynamics of elementary radiators (atoms, molecules, excitons etc.) interacting with light. The second is the emission of light by elementary radiators. The third is the evolution of light waves within the medium under the influence of dispersion, diffraction, nonlinear compression and parametric wave interaction. Essential features of radiator dynamics and emission of light are brightly displayed in coherent optical transients such as free induction decay, optical nutation and various echo phenomena. The favourable condition of their formation is a resonant propagation of an ultrashort light pulses through such thin medium where dispersion and other processes inherent to wave propagation are inessential. In this case coherent transients exhibit the properties directly connected with medium structure. This circumstance is very attractive for spectroscopy especially for investigations of relaxation. The theoretical analysis of coherent transients becomes fairly simple for optically thin medium when the framework of a given field approximation in Maxwell-Bloch equations is valid. This approximation ignores the reaction of the medium to the passing light pulses. In spite of this simplicity coherent transients represent an important large class of nonlinear optical phenomena. This chapter is devoted to the description of the underlying main principles of coherent transients.

Firstly, in section 2.1 we qualitatively consider general characteristics of coherent transients. In section 2.2 we state the given field approximation and general solution of Maxwell-Bloch equations in this approximation. Optical nutation and free induction decay are discussed in sections 2.3 and 2.4. They serve as basis for photon echo formation in resonant interactions with two levels or three levels of atom that we describe in sections 2.5 and 2.6. Polarisation properties of coherent transients we illustrate in section 2.7 on examples of optical nutation by phase switching method and of the method for investigation of atomic relaxation by photon echoes in the three-level systems.

2.1 General characteristics of coherent nonlinear processes

Before developing the theory of behaviour of radiator system (atoms, molecules, impurities in a solid-state crystal matrix etc.) in a field of a resonant or nonresonant electromagnetic wave it is helpful to discuss qualitatively the basic phenomena arising in these processes. It is useful to know the characteristic times determining the basic marks on a time scale where the processes of interaction of radiation with medium take place.

We shall consider a set of atoms without motion and interaction among themselves which are in a resonant field of an electromagnetic wave. Natural width γ_0 of a resonant line of a separate atom gives the life-time of the resonant excited level $1/\gamma_0$ and the time of relaxation of resonant polarisation $2/\gamma_0$. The thermal motion of atoms of the given ensemble (neglecting the processes of collisions and other interatomic interactions) will not change the time $1/\gamma_0$ of population relaxation of levels, but will affect the width of a line of radiation of such an ensemble. Due to the Doppler effect the various groups of atoms, moving in various directions and with various speeds, will have different frequencies of resonant transition. Therefore the total width of an absorption line or emission line consists of separate lines of width γ_0, but their centres are displaced to the value of Doppler shift with respect to the resonant frequency ω_0 of a motionless atom. The total width of a line becomes equal to $1/T_0$. Such broadening of a spectral line is called inhomogeneous, and not only the Doppler effect results in inhomogeneous broadening (but now it is not essential).

One of the results of the interaction between the atoms will be a phase shift of an individual atom without the transition between levels (though in the general case the relaxation speed of the energy levels increases). Besides, the interaction of atoms with the surrounding non-resonant atoms can cause inhomogeneous broadening of a spectral line (for example, in a solid state). Thus, the life-time of an excited level (or relaxation time of the population differences between resonant levels) and relaxation time of resonant polarisation are not wholly determined by the value γ_0. For a group of the atoms having given speed $\vec{v}$, the relaxation time of polarisation and population difference we shall designate $1/\gamma_{21}$ and $1/\gamma$, and the speeds of relaxation γ_{21} and γ can be regarded as independent of the value $\vec{v}$. The inequalities $T_0 < 1/\gamma_{21} < 1/\gamma < 1/\gamma_0$ are usual. Then the time T_0 determines the decay time of the polarisation of atom ensemble and is referred to as dephasing time or time of reversible relaxation. The mechanism of this decay can be explained by the following analogy [1].

Let us consider the start of runners. In the initial moment of time all of them are on a line of start. After the start fast runners escape forward and slow ones are left behind, the others sprawling between them. During the race the runner density per unit of length of the track decreases. The runner density essentially decreases compared with that of at the start during the time T_0. Here the runner density is the analogue of macroscopic polarisation, each runner is the analogue of radiator, the time of the runner's being able to run gives us the concept of $1/\gamma_{21}$ and $1/\gamma$.

Let us involve the time T i.e. the duration of a pulse of an electromagnetic wave, and if the interaction of resonant atoms is "switched on" or "switched off" at any moment of time with the stationary electromagnetic radiation, we designate T the duration of front of the switching process. Let us neglect radiation fluctuations assuming that the external electromagnetic field represents either a monochromatic wave or a set of such waves. In this case the behaviour of ensemble of atoms will depend to a great extent on the ratio between characteristic sizes T, T_0, $1/\gamma_{21}$ and $1/\gamma$.

If $T << 1/\gamma_{21}$, $1/\gamma$, for an interval of time of about several T all atoms behave as being non-interacting. When the external field causes transitions between resonant levels of separate atoms, these transitions appear synchronised i.e. the behaviour of separate atoms is correlated by the common field of an electromagnetic wave. In a system of atoms coherence takes place, the response of this system to external influence is displayed in radiation with the intensity proportional to the square of the number of atoms per unit volume.

Let $T >> 1/\gamma_{21}$, $1/\gamma$. In this case coherence in the system of atoms is kept during the time of the order $1/\gamma_{21}$, and the intensity of radiation response decreases with the time, reaching stationary quantity proportional to the first degree of atomic concentration. If inhomogeneous broadening is present, the loss of coherence in the radiator system occurs more quickly.

Thus, under the resonant conditions the stationary and non-stationary modes of interaction of radiation with medium will differ radically.

The non-resonant situation is characterised by a condition $\Delta T >> 1$, where Δ is the detuning from an exact resonance. During $t_\Delta = \Delta^{-1}$ the polarisation of medium reaches stationary quantity. The population difference does not essentially change. Any coherence displays in the radiator system can happen only before the stationary regime of macroscopic polarisation sets up. The time describing the process of interaction should be less than t_Δ. On the time scale of the order T there will not be any coherence displays, as the condition of non-resonance requires $T >> t_\Delta$.

We shall note that in the cases mentioned above, the role of a monochromatic field can be compared to a role of the conductor of chorus. The conductor's being absent, each singer will sing on his own, the chorus will quickly turn to a rustling crowd. The time of this transformation is also the characteristic time of the process coherence. Each time the conductor tunes up the singers, as a result all of them follow the same plan. If the part is sung "from a sheet" and any pages of the score are absent, we receive an example of a phase shift. We offer the reader to think of other analogies.

2.1.1. NON-STATIONARY AND COHERENT TRANSIENTS

Coherent transients arise with fast perturbation, “switch on” or “off” of interaction of the radiator system with a monochromatic resonant electromagnetic field, for example, due to the abrupt switching of the laser source frequency, the drastic displacement of energy levels, use of ultrashort electromagnetic pulses. The words “fast”, “abrupt”, “ultrashort” mean that the time of the process is far less than that of the irreversible relaxation of polarisation $1/\gamma_{21}$ and population of resonant energy levels $1/\gamma$. During this time the ensemble of radiators can be identified to ideal gas of two-level atoms. Such atoms being affected by an external field do not interact to each other. In optically thin medium all atoms interact with one common field, so processes of the stimulated absorption and emission of photons are synchronised. Medium polarisation is of the order of polarisation of one atom multiplied by atomic density. The intensity of radiation caused by this polarisation will be proportional to a square of the number of atoms N^2. Such dependence of the response on the concentration of atoms displays coherence in the radiator system and is a characteristic feature of coherent transients. Sometimes the coherent response is referred to as superradiation.

The effect of photon echo holds interest among all the coherent transients. The photon echo is the coherent response developing as a result of influence of two or more ultra short pulses of light on resonant system. For example, if two exciting light pulses separated by a time interval τ_{12} pass through a resonant medium, then after an approximate time interval τ_{12} following the second exciting pulse, spontaneous coherent emission is produced by the excited atoms of the medium, and is called two-pulse photon echo [2,3] (Fig.2.1.1).

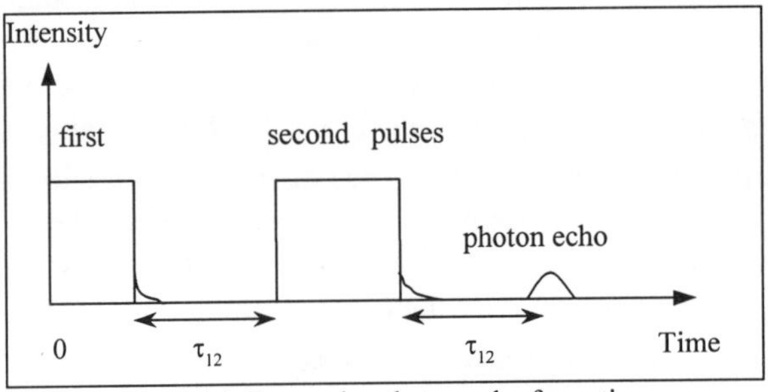

Fig.2.1.1. Two-pulse photon echo formation

The term “two-pulse” is used to distinguish this kind of echo from the others. Also, for the purpose of classification, this echo is named as photon echo on the basis of free induction (decay). The signal “free induction” represents the emission from the medium after the passage of backward edge of exciting pulse. Details can be found in sections 2.4 and 2.5. One should note that the photon echoes, free induction and some others coherent transients are optical analogues of transients in spin systems [4].

The effect of photon echo on the basis of free induction takes place in the presence of inhomogeneous broadening of a resonant absorption line in the system of two-level atoms (and in a general case any radiators). The attenuation of macroscopic polarisation induced by an exciting pulse of an electromagnetic field can occur due to rather fast dephasing of separate radiators because of its different frequency. The specific feature of this relaxation due to inhomogeneous broadening is its reversibility. To demonstrate this statement, we shall consider the idealised system, in which the times $1/\gamma_{21}$ and $1/\gamma$ are especially great, and the duration of exciting pulses T_1 and T_2 are so small that in their action inhomogeneous broadening is insignificant. Let the interval between stimulating pulses τ_{12} considerably surpass the dephasing time T_0, so T_1, $T_2 << T_0 << \tau_{12} << 1/\gamma_{21}$,

$1/\gamma$. Let us turn to the analogy with runners. The first exciting pulse gives start to runners. By the moment of the second stimulating pulse runners will sprawl along a racing track (macroscopic polarisation of medium will disappear or atomic free induction will decay). It appears (and it is proved only by the analysis of the solutions of the corresponding equations), that the action of the second pulse changes a phase of atomic oscillator to the opposite. For the runners it means that the second pulse serves as a signal to turn round and to run back. As a result all the runners simultaneously reach the line of start. After an interval of time τ_{12} after the action of the second exciting pulse there is a macroscopic polarisation in the medium again, the result of which will be the radiation of the signal, which has received the name of photon echo. The analogy involved allows to understand the role of irreversible relaxation in the effects such as an echo. If the interval of time τ_{12} is comparable with $1/\gamma_{21}$, i.e. before the runners turn back, they will strain themselves to the breaking point, then obviously only some of them will either reach the line of start or (as the speeds of runners will change) the runners will cross the line of start in the various moments of time. In any case the effect of an echo will be weakened, and as τ_{12} increases the echo will disappear at all. An interesting classical model of photon echo can be found in ref.[5].

If the length of resonant medium greatly exceeds the length of absorption of electromagnetic radiation, the approximation of the given field is broken. Thus the abnormal weak absorption of energy of radiation is possible, and moreover as the pulse amplitude increases, the moment arises, when the absorption is actually absent. This phenomenon has received the name of a self-induced transparency [6].

The simple explanation of the effect consists in the fact that the absorption of electromagnetic radiation from forward front of a pulse is replaced by stimulated radiation with its back front (after the system of two-level atoms will appear inverted). Such a pulse of a self induced transparency in the process of distribution gets the stationary form which is not dependent on density of resonant atoms, and is spread in medium with the group speed which is appreciably different from the speed of light. The theory of a self induced transparency gives a rare example in theoretical physics: the system of the equations describing evolution of an initial pulse of ultra short duration is quite integrable system. In other words, there is an infinite sequence of the high laws of conservation or integrals of a movement, which are in involution. As a consequence of this the pulse of self induced transparency appears to be a soliton, i.e. a lonely wave keeping the individuality in colliding with another lonely wave. The more powerful pulses break up to some amount of solitons.

2.1.2. COHERENT RADIATION OF RESONANT ATOMS IN THE SPATIALLY SEPARATED LIGHT BEAMS

The specific mechanism of reversible relaxation in gas due to thermal motion of atoms and Doppler effect, allows to realise original spatial analogue of photon echo, which can take place both in pulse, and in stationary light fields [7,8]. We shall look at the formation of usual two-pulse photon echoes in gas in the following way. Let us isolate a group of atoms having a projection of speed onto an axis x being equal to v_x. At the moment of time $t = 0$ they are in the vicinity of a point $x = 0$, in which the stationary light beam spreading along an axis z and having the cross size $0 \le x \le b_1$ is located. The beam exerts pulse influence on the atoms, the duration of which is of the order of the atomic passage through the cross-section of a beam, $T_1 = b_1/v_x$. For the given group of atoms

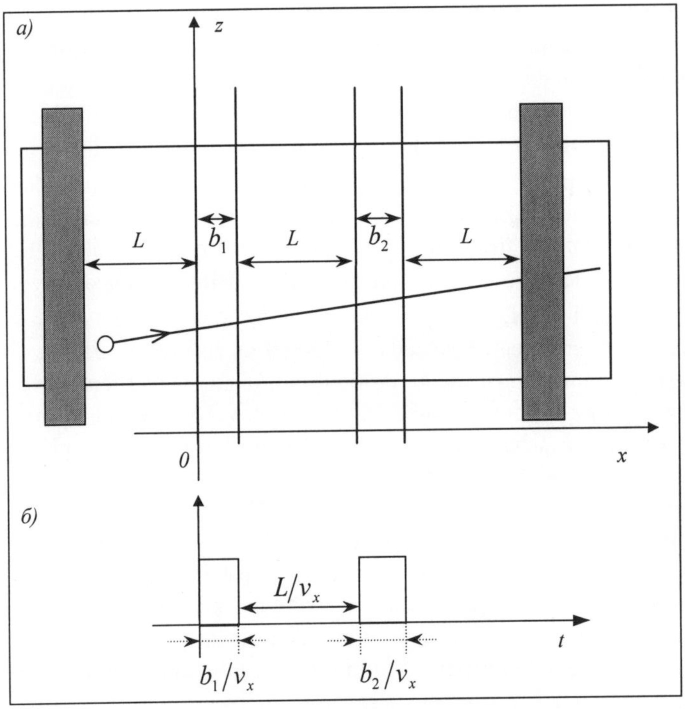

Fig. 2.1.2. Coherent radiation of gas (shaded areas) in the spatially separated light beams (a) and pulse influence on a particle crossing beams (b)

to take part in the formation of photon echo, at the moment of time $t = T_1 + \tau_{12}$ it should undergo the action of the second pulse. However by this moment the atoms will already be at the distance of $v_x\tau_{12}$ from the first beam. Let us assume that at the same distance $L = v_x\tau_{12}$ from the first beam, the second light beam parallel to it with the cross sizes $L + b_1 \le x \le L + b_1 + b_2$ is located. Then the specified atoms undergo the influence of second pulse of duration $T_2 = b_2/v_x$ and "are ready" to radiate at the moment of time $t = 2\tau_{12} + T_1 + T_2$. They radiate a signal of photon echo in the direction of an axis z. The dephasing of atomic radiator necessary for the photon echo formation is provided with disorder of the projections of atomic velocities onto an axis z within the allocated group. The disorder of projections of speeds of atoms onto an axis x will eliminate the display of the echo signal at a glance. However it is not so. To prove it we shall again consider the passage of the allocated group of atoms of two light beams propagating along an axis z and located in areas $0 \le x \le b_1$ and $b_1 + L \le x \le b_1 + b_2 + L$, where distance L between beams does not depend on v_x. Atoms undergo pulse influence in intervals of time $0 \le t \le b_1 / v_x$ and $(b_1 + L)/v_x \le t \le (b_1 + b_2 + L)/v_x$ and radiate a signal of a photon echo in the moment $t_e = (b_1 + b_2 + 2L)/v_x$. Thus they will be in the vicinity of a point with coordinate $b_1 + b_2 + 2L$, that is at the distance L from the second light beam. Let us emphasise that it does not depend on v_x. If to take into account that the atoms move not only in the positive direction of an axis x, the general picture of influence of two parallel light beams located at the distance L from each other and propagating in one direction, will be the following. On both sides from the beams at the distance L from them coherent radiation occurs in gas as the beams of light parallel to exciting beams and propagating in the same direction. Figure. 2.1.2 summarises the above mentioned idea.

For the formation of coherent radiation in separated light fields it is necessary that the size $2L + b_1 + b_2$ should not surpass length λ of free run of atoms.

It seems evident, that all types of echo on the basis of free induction decay have the spatial analogues described.

2.1.3. SUPERFLUORESCENCE

So far we have discussed situations when the external electromagnetic field carried out phase synchronisation of all radiators - two-level atoms. Coherent response of the media to such perturbation was caused by a macroscopic polarisation arising as a result of preparation of radiators in correlated states. The intensity of such response is proportional to a square of the number of the excited atoms N^2, that was the basis to name it as superradiation, opposite to the spontaneous radiation, the intensity of which is proportional to N.

Considering the spontaneous emission of photons by the system of two-level atoms with the size of about the length of a wave, Dicke in 1954 [9] showed that the intensity of spontaneous radiation can be proportional to N^2, despite the obviously absent external correlating influence. Dicke designed a quantum-mechanical state of ensemble of two-level atoms, which is characterised by two quantum numbers: the cooperation number r and half-difference population of the upper and lower energy levels of all the system m. In photon emission the number r is constant and characterises a degree of correlation of a separate radiator, whereas m varies in time. By assuming that the system initially is completely inverted $m(0) = N/2$ and the cooperation number has a maximum value $r = N/2$, Dicke found that the intensity of spontaneous radiation I_N, is proportional to N soon after its beginning. Later number m becomes equal to zero (the case of equal level population), and at this moment of time $I_N \approx N^2$. Besides, if $1/\gamma_0$ is a life time of the excited state of the isolated two-level atom, practically all the energy reserved by system will emit during the time interval of $1/\gamma_0 N$.

Thus, the possibility of spontaneous transformation of atomic system due to usual fluorescence to the highly correlated system was demonstrated. This correlated system is characterised by a high speed of photon emission and by intensity of radiation proportional to N^2. This process has received the name of superfluorescence (or Dicke superradiation). It is necessary to emphasise, that the Dicke state is not the product of eigenstates of the Hamiltonians of separate atoms, and the presence of some a priori correlation between initially states of separate radiator is reflected in it. For realisation of superfluorescence it is necessary for the ensemble of radiators to be prepared in Dicke state, but for it is not enough to invert all atoms.

The realisation of Dicke reasoning in optics led to the principal difficulties. The sizes of atomic system here are greater than the wavelengths of a resonant radiation. The propagation effects of superradiation pulse in the Dicke theory are not considered. It is not clear what is meant by the cooperation number of Dicke state. In any way it is not the number of atoms N.

Being guided by semiclassical reasons, Rehler and Eberly [10] have considered the spontaneous emission of the extended multiatomic system and have found the intensity of superfluorescence in any direction and at any moment of time. It has been shown that the dynamics of intensity spontaneous photon emission reflects qualitatively basic elements of the superfluorescence process. Soon after the excitation the radiation occurs in all directions and its intensity is proportional to N. In the certain time t_D the pulse of superfluorescence is spontaneously formed, which direction is determined by geometry of a sample and direction of an exciting system pulse. In this direction we have $I_N \approx N^2$. The duration of a superfluorescence pulse $t_R = 1/\gamma_0(1+\mu N)$, where μ is a form-factor dependent on the form a sample. Besides, $t_D = t_R \ln(\mu N)$.

Within the limits of a small system volume, when the linear size of a system of atoms about the length of a radiation wave, the Rehler-Eberly theory gives Dicke results. The maximal value of the cooperation number $r_{\max}$ has appeared to be equal to the number of the atoms, which are not participating in induced processes of reemission. The superfluorescence is a purely spontaneous process, if the length of a core containing two-level atoms, is far less than the free run of photon. Such mode is named as pure (one-pulse) superfluorescence. If the specified condition is not carried out, there is oscillator superfluorescence.

For the last years there have been different updatings of the theory of superfluorescence taking into account the relaxation and inhomogeneous broadening of a resonant line of absorption. The satisfactory description of experimental results on supervision of superfluorescence is achieved by introducing several parameters determined basically by the initial (pure quantum) stage of photon emission by the excited system of radiators. The existing problems in the theory of this phenomenon are connected to the quantum description of the occurrence of correlation in an initial condition of radiating system. Let us note, that in the semiclassical theories this condition is simply postulated.

In summary it would be necessary to emphasise, that the bright feature of superfluorescence is the large time of a delay t_D, which can exceed the pulse duration t_R on some orders. The simple explanation is the following. The decay of an initial state of a two-level system begins with spontaneous non-correlated emission of photons by a separate radiator. The arising field of radiation results in the moment t_D to the maximal correlation of these radiators, behaving now as one atom, but with the dipole moment of transition in N time larger than at one radiator. The intensity for this reason becomes proportional to N^2, and the speed of photon emission is increased almost in N time. The curious analogy of such spontaneous occurrence of coherence in radiator system takes a place in human community. It is a storm of applause carried over to an ovation, for example.

2.2. The given field approximation

For the sake of simplicity we consider the case of the one-photon resonance with nondegenerate atomic transition and we neglect any polarisation effects. We will discuss coherent transients in resonant propagation of electromagnetic field

$$E = \mathcal{E} \exp[i(kz - \omega t)] + c.c.. \tag{2.2.1}$$

We write the Maxwell-Bloch equations in the form

$$\left(\frac{\partial}{\partial z} + \frac{1}{c}\frac{\partial}{\partial t}\right)\mathcal{E} = i2\pi k d_{ba} \int R_{21} d\Delta , \tag{2.2.2}$$

$$\left.\begin{aligned}
&\left(\frac{\partial}{\partial t} - i\Delta\right) R_{21} = i\Lambda\left(R_1 - R_2\right),\\
&\frac{\partial}{\partial t} R_1 = i\left(\Lambda^* R_{21} - \Lambda R_{21}^*\right),\\
&\frac{\partial}{\partial t} R_2 = -i\left(\Lambda^* R_{21} - \Lambda R_{21}^*\right)
\end{aligned}\right\} \tag{2.2.3}$$

The spectral line of resonant transition is inhomogeneously broadened so the values R_{21}, R_1 and R_2 represent slowly varying function describing optical coherence and density at lower and upper energy levels of atoms whose detuning from resonance lies within the interval $(\Delta, \Delta + d\Delta)$. The value $\Lambda = \mathcal{E} d_{ba} / \hbar$ is independent from Δ. For details see in item 1.3.4.

It is convenient to introduce new independent variables $z' = z$, $t' = t - z/c$ and rewrite the Maxwell-Bloch equations as

$$\frac{\partial}{\partial z'}\mathcal{E} = i2\pi k d_{ba}\int R_{21} d\Delta \tag{2.2.2'}$$

$$\left.\begin{aligned} &\left(\frac{\partial}{\partial t'} - i\Delta\right)R_{21} = i\Lambda(R_1 - R_2), \\ &\frac{\partial}{\partial t'}R_1 = i\left(\Lambda^* R_{21} - \Lambda R_{21}^*\right), \\ &\frac{\partial}{\partial t'}R_2 = -i\left(\Lambda^* R_{21} - \Lambda R_{21}^*\right). \end{aligned}\right\} \tag{2.2.3'}$$

The electric field at point z in the medium consists of two terms. The first one is the electric field of the exciting wave in the absence of the medium and the second represents the induced field of quantum reemission due to the resonant interaction. Assuming the electric field at point z=0 of entrance to the medium to be as

$$\mathcal{E} = ae^{-i\varphi}, \tag{2.2.4}$$

we write the total field in the form

$$\mathcal{E} = ae^{-i\varphi} + \varepsilon. \tag{2.2.5}$$

The induced field is proportional to the amplitude of dipole emission and the number of emitters:

$$\varepsilon \sim \left(zN|\Lambda|T_0\right)\frac{\omega|d_{ba}|}{c}. \tag{2.2.6}$$

Here, we have taken into account one-dimensional geometry of our consideration and coherent character of resonant interaction. Only a part of order $|\Lambda|T_0 < 1$ of resonant atoms within inhomogeneously broadened spectral line efficiently interacts with an exciting field. We denote by N the density of resonant atoms.

If the reaction of resonant medium of length L to propagating pulse is small

$$|\varepsilon| << a, \tag{2.2.7}$$

i.e.

$$LN|d_{ba}|^2 T_0\,\omega/\hbar c << 1, \tag{2.2.8}$$

the value Λ in eq.(2.2.3) may be treated as following constant

$$\Lambda = ae^{-i\varphi}\, d_{ba}/\hbar .$$

Then eqs.(2.2.3') turn to the system of three ordinary linear differential equations with constant factors, which solution can be easily found by standard methods. The general solution of eqs.(2.2.3') for any initial conditions at the moment of time $t' = t'_0$

$$R_{21}\big|_{t'=t'_0} = R^0_{21},\ R_1\big|_{t'=t'_0} = R^0_1,\ R_2\big|_{t'=t'_0} = R^0_2$$

it is convenient to present in a matrix form

$$R(t') = U(t',t'_0)\, R(t'_0)\, U^{+}(t',t'_0) \tag{2.2.9}$$

where $R(t')$, $R(t'_0)$ and $U(t',t'_0)$ - matrix of the order 2x2:

$$R(t') = \begin{pmatrix} R_1(t') & R^*_{21}(t') \\ R_{21}(t') & R_2(t') \end{pmatrix},\quad R(t'_0) = \begin{pmatrix} R^0_1 & R^{0*}_{21} \\ R^0_{21} & R^0_2 \end{pmatrix},$$

$$U(t',t'_0) = \begin{pmatrix} \cos\dfrac{\Omega(t'-t'_0)}{2} - i\dfrac{\Delta}{\Omega}\sin\dfrac{\Omega(t'-t'_0)}{2} & i\dfrac{2\Lambda^*}{\Omega}\sin\dfrac{\Omega(t'-t'_0)}{2} \\ i\dfrac{2\Lambda}{\Omega}\sin\dfrac{\Omega(t'-t'_0)}{2} & \cos\dfrac{\Omega(t'-t'_0)}{2} + i\dfrac{\Delta}{\Omega}\sin\dfrac{\Omega(t'-t'_0)}{2} \end{pmatrix}, \tag{2.2.10}$$

$\Omega = \sqrt{\Delta^2 + 4|\Lambda|^2}$, and Λ is determined above.

After the solution of the eqs.(2.2.3') has been found, the amplitude of an electrical field caused by reemission of quantums by resonant atoms is given by direct integration of the equation (2.2.2'):

$$\varepsilon(z',t') = i2\pi kz' \int R_{21}(t')d_{ab}d\Delta . \tag{2.2.11}$$

Outside resonant medium $z \geq L$ from (2.2.11) follows

$$\varepsilon(t - z/c) = i2\pi kL \int R_{21}(t - z/c)d_{ab}d\Delta . \tag{2.2.12}$$

The requirement (2.2.7) for (2.2.12) with $|\Lambda|T_0 < 1$ results in a condition (2.2.8). In the case $|\Lambda|T_0 > 1$ the inequality (2.2.8) can be weakened and replaced by

$$\omega N|d_{ab}|\, L/ac << 1 . \tag{2.2.13}$$

Resonant medium which parameters for each ultra short pulse satisfy a condition either (2.2.8) or (2.2.13) is referred to as optically thin, and the stated approach to the solution of the Maxwell-Bloch equations is referred to as the given field (or rotating wave) approximation.

Let us emphasise that though the Maxwell-Bloch equations in the given field approximation become linear, the optical effects described by them are essentially nonlinear, as polarisation of medium depends nonlinearly on intensity of an electrical field (see eqs.(1.1.16), (2.2.9) and (2.2.10)).

2.3. Optical nutation

Let a pulse of rectangular form

$$\mathcal{E} = \begin{cases} a\exp i\varphi, & t - z/c > 0 \\ 0, & t - z/c < 0 \end{cases}$$

enter an optically thin resonant medium. Its intensity $I = c|\mathcal{E}|^2/2\pi$, averaged over a period $2\pi/\omega$ of fast oscillations, at the exit from medium with the account (2.2.7) can be presented as

$$I = \frac{ca^2}{2\pi}\left(1 + 2\tilde{\varepsilon}'(t - z/c)/a\right), \tag{2.3.1}$$

where $\tilde{\varepsilon} = \varepsilon\exp(i\varphi)$, and the prime here designates a real part of $\tilde{\varepsilon}$:

$$\tilde{\varepsilon}'(t) = -E_0 \int_{-\infty}^{\infty} F(\Delta)\frac{\sin\sqrt{\Delta^2 + 4|\Lambda|^2}\,t}{T_0(\Delta^2 + 4|\Lambda|^2)^{1/2}}d\Delta, \quad 0 \le t \le T\,, \tag{2.3.2}$$

$$E_0 = 2\pi(N_1 - N_2)La|d_{ba}|^2\,\omega T_0/\hbar c\,.$$

The expression (2.3.2) describes optical nutation, and the size E_0 characterises a scale of amplitude of nutation oscillations. As it was expected, the period T_n of nutation oscillations is determined by energy of wave interaction with resonant atom (Rabbi frequency). In the case of exact resonance $\Delta_0 = 0$ the approximated formula

$$T_n = 2\pi\Big/ 2|\Lambda| = \frac{\pi}{a}\sqrt{\frac{\omega^3\hbar}{3c^3\gamma}}$$

is valid for any parameter $|\Lambda|T_0$. Here, γ is the probability of spontaneous emission of resonant quantum by an isolated atom, which is connected to oscillator force f of resonant transition by a relation $\gamma = 2e^2\omega^2 f/mc^3$ (e and m are the charge and weight of electron). Certainly, it is possible to speak about nutation oscillations of light pulse only in the case when the duration of a pulse surpasses T_n, and the time of increase of forward pulse front is far less than T_n.

The optical nutation is the analogue of a known phenomenon in spin systems [11] and was first observed by Hocker and Tang [12].

We consider limiting cases of (2.3.2). Let us notice, that the values $F(\Delta)$ and $\Phi(\Delta) = \sin\sqrt{\Delta^2 + 4|\Lambda|^2}\, t \Big/ \sqrt{\Delta^2 + 4|\Lambda|^2}$ are the sharp functions of a variable Δ with the width $1/T_0$ and $2|\Lambda|$, accordingly. With a high-energy mode of a laser irradiation $|\Lambda|T_0 >> 1$ it is possible to neglect inhomogeneous broadening, to take $\Phi(\Delta)$ in a point of a maximum $F(\Delta)$, and to take out for a sign of integral. As a result we shall receive that the amplitude (2.3.2) exhibits harmonic oscillations:

$$\tilde{\varepsilon}' = -E_0 \frac{\sin\sqrt{\Delta_0^2 + 4|\Lambda|^2}\, t}{T_0\sqrt{\Delta_0^2 + 4|\Lambda|^2}}. \tag{2.3.3}$$

In another limiting case $|\Lambda|T_0 << 1$ of low-energy mode of laser irradiation, inhomogeneous broadening is rather essential and strongly has an effect on a picture of nutation oscillations. Instead of harmonic oscillations (2.3.3) optical nutation in the absence of relaxation (!) represents damping oscillations, which are described by Bessel function of the zero order:

$$\tilde{\varepsilon}' = -\sqrt{\pi}\, E_0 \left[J_0(2|\Lambda|t) + \chi(t)\right]. \tag{2.3.4}$$

Here function $\chi(t)$ changes from -1 up to 0 for a small (in a considered case) interval of time T_0. For example, in gaseous medium $\chi(t) = \Phi(t/T_0) - 1$, where $\Phi(\xi)$ is an error function

$$\Phi(\xi) = \frac{2}{\sqrt{\pi}} \int_0^{\xi} \exp(-t^2)dt.$$

Let us emphasise, that though nutation oscillations (2.3.4) is damping, the atoms of medium still continue reemission of resonant quantums, passing from the upper resonant level to the low one and vice versa. However, different groups of atoms inside an inhomogeneous broadened contour of a spectral line emit electromagnetic waves with different frequencies $\omega_0 + \Delta$ and phases $(\omega_0 + \Delta)(t - z/c)$. As the result of such dephasing electromagnetic waves emitted by atoms of different groups are not coherent and weaken each other. Therefore an observable total electrical field (2.3.4) decays. Specified dephasing of separate radiators usually is regarded as display of reversible relaxation, which is caused here by the displacement of atomic frequency of ω_0 owing to either the interaction of resonant atoms with a crystal matrix or the Doppler effect.

Along with the mentioned above mechanism of reversible relaxation of nutation oscillations in molecular gases other mechanism determined by degeneracy of resonant energy levels on the projections of the total angular moment take place also. Its physics is connected to the various width of field broadening of different sublevels of resonant levels, that in the end also results in dephasing of radiation.

The presence in the medium of any from the specified mechanisms of reversible relaxation results in emergence of a nutation echo under certain conditions (see further sections 2.5.3 and 2.7.1).

2.4. Free induction decay

After the ultra short pulse (USP for short) has been passed through an optically thin medium, the excited resonant atoms radiate an electromagnetic wave

$$E = \varepsilon(t - z/c)\exp[i(kz - \omega t - \varphi)] + k.c., \quad T \le t - z/c,$$
$$\varepsilon(t) = -E_0 \int F(\Delta)\frac{1}{\Omega T_0}\left[\sin\Omega t + i\frac{\Delta}{\Omega}(1 - \cos\Omega T)\right]\exp[i\Delta(t - T)d\Delta. \tag{2.4.1}$$

This wave is named as an optical induction [13]. Eq. (2.4.1) describes free (USP has passed) decay of an optical induction. Its intensity averaged for a period $2\pi/\omega$ of fast oscillations, is equal $I = c|\varepsilon|^2/2\pi$.

In discussions of an optical induction and others coherent effects on their basis it is useful to distinguish between two limiting experimental situations, in which the resonant atoms participate in coherent phenomena differently (Fig. 2.4.1). Let us name a spectral line of radiation of free atoms "narrow" in the relation to a power spectrum of USP, if it lays inside the spectral distribution of this pulse energy

$$\frac{1}{T_0} << \frac{1}{T}. \tag{2.4.2}$$

In this case USP excites all atoms inside an inhomogeneously broadened contour of a spectral line. On the contrary, a spectral line of radiation of free atoms we shall name "wide", if the spectral distribution of exciting pulse lies within an inhomogeneously broadened spectral line

$$\frac{1}{T_0} > \frac{1}{T}. \tag{2.4.3}$$

Under condition (2.4.3) only a part of atoms inside the inhomogeneously broadened contour of spectral line is excited.

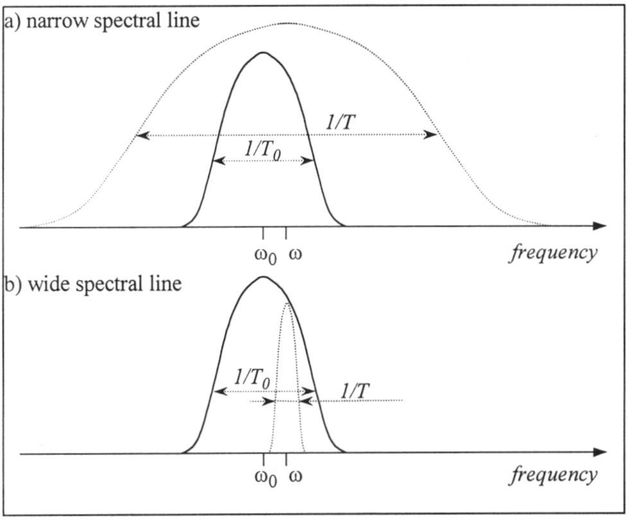

Fig.2.4.1. Relations between inhomogeneously broadened spectral line (continuous curve) and spectral structure of resonant USP in cases of narrow (a) and wide (b) spectral lines.

Frequently one uses the definition of the area of USP as $\theta = 2|\Lambda|T$. In coherent processes pulses with the area $\theta = m\pi$ where $m \sim 1$ usually participate. In this case inequality (2.4.3) imposes restrictions on USP amplitude that corresponds to a low energy mode of laser radiation $|\Lambda|T_0 << 1$. While the condition (2.4.2) corresponds to a high-energy mode $|\Lambda|T_0 >> 1$ for which complicated formulas become far simpler.

As in the case of optical nutation, the free induction decay in the absence of the irreversible relaxation results in attenuation of amplitude of an electrical field (2.4.1). For a narrow spectral line the damping involved is defined by the simple formula

$$\varepsilon(t) = -E_0 \frac{\sin 2|\Lambda|T}{2|\Lambda|T_0} A(t-T)\exp[i\Delta_0(t-T)], \quad t \geq T.$$

For a gaseous resonant medium $A(\xi) = \exp(-\xi^2/4T_0)$ and for a resonant impurity in a crystal $A(\xi) = \exp(-|\xi|/T_0)$. It is visible that the time of reversible relaxation of an optical induction in the case of a narrow spectral line is equal T_0, and the carrying frequency coincides with the centre ω_0 of an inhomogeneously broadened spectral line. In the case of gaseous resonant medium the time T_0 of reversible relaxation is also named as Doppler dephasing time.

In the case of a wide spectral line $1/T_0 >> 1/T$ only the atoms whose frequencies lie in an interval from $\omega - 1/T$ up to $\omega + 1/T$ are effectively excited. For this group of atoms the dephasing time is determined by the duration of pulse T. From here it is possible to conclude that the signal of an optical induction will be radiated with the frequency ω and under the action of reversible relaxation will decay during the time-period of an approximate USP duration. The numerical simulation of (2.4.1) confirms this conclusion. The reversible decay of an optical induction is characterised also by the following formula derived from (2.4.1) in a limiting case $|\Lambda| T_0 << 1$ and $\Delta_0 = 0$:

$$\varepsilon(t) = -\sqrt{\pi} E_0 \left[J_0(x) - 1 + y \int_0^x (\xi^2 + y^2)^{-1/2} J_1(\xi) d\xi + \chi(t) - \chi(t-T) \right], \quad T \le t \le 2T,$$

$$\varepsilon(t) = -\sqrt{\pi} E_0 [\chi(t) - \chi(t-T)], \quad 2T < t,$$

where $J_1(\xi)$ is the Bessel function of the first order, $\chi(\xi)$ for a case of a gaseous resonant medium has been determined in the previous item, and $x = 2|\Lambda|\sqrt{T^2 - (t-T)^2}$, $y = 2|\Lambda|(t-T)$.

If to take into account the degeneration of resonant energy levels, reversible relaxation of an optical induction (2.4.1) is determined extremely by an inhomogeneous broadening of a spectral line. The effects of level degeneration can play a role only in presence of rather powerful non-resonant (without changing an essence of an optical induction) external fields, for example a constant magnetic field.

The mechanism of reversible relaxation of an optical induction underlies a large number of original effects of an echo, the main ones being considered in sections 2.5.1, 2.5.2 and 2.5.5.

2.5. Photon echoes in two-level systems

The propagation of USP in an optically thin resonant medium previously excited by interaction with another USP is also accompanied by the effects of optical nutation and a free induction decay, however their properties are so nontrivial and are unexpected at first sight, that they are spoken about as new coherent nonlinear phenomena, which have received the name an echo-phenomenon or photon echoes (Fig. 2.5.1). Among them are a wide variety of different photon echoes on the basis of an optical induction. Many papers are devoted to theoretical and experimental investigations of echo-phenomena and practical using of echoes in spectroscopy, data processing, optical memory etc. We will discuss only basic ideas underlying physics of echo formation referring the readers to books [8,14] where technical details and numerous references can be found. In sections 2.5.1 and 2.5.2 we shall consider two-pulse and stimulated photon echoes, in section

2.5.3 we shall state the theory photon echo on the basis of optical nutation, in section 2.5.4 we shall generalise the received results in the case of two-photon resonance. Section 2.5.5 is devoted to photon echoes produced by standing wave pulses in gases.

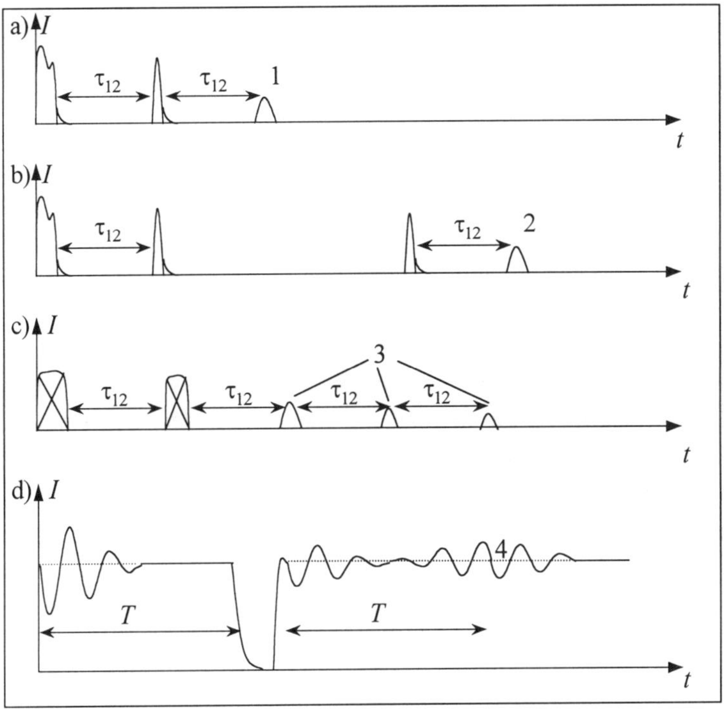

Fig. 2.5.1. The echo phenomena in an optically thin resonant medium under the excitation by pulses of travelling (a), (b), (d) and standing (c) waves:
1 - two-pulse photon echo,
2 - stimulated photon echo,
3 - echoes in standing waves,
4 - nutation echo.

2.5.1. TWO-PULSE PHOTON ECHO

We are discussing now the photon echo formation due to resonant propagation of two USPs of duration T_1 and T_2 separated by an interval of time τ_{12}. For simplicity the pulses propagate along an axis z in one direction and have the following electric field strength

$$E_1 = 2a_1 \cos(\omega t - kz + \varphi_1), \quad 0 \le t - z/c \le T_1, \tag{2.5.1}$$

$$E_2 = 2a_2 \cos(\omega t - kz + \varphi_2), \quad T_1 + \tau_{12} \le t - z/c \le T_1 + T_2 + \tau_{12}, \tag{2.5.2}$$

Here, a_n and φ_n $(n = 1,2)$ are constant, and the USP carrying frequency $\omega = kc$ is resonant to the optically allowed atomic transition $E_b \to E_a$. The frequencies of these transitions are inside an inhomogeneously broadened contour of a spectral line with the central frequency ω_0.

As excited pulses (2.5.1) and (2.5.2) are considered as ultra short $T_1 << 1/\gamma_{21}$, $T_2 << 1/\gamma_{21}$, the irreversible relaxation during the interaction of pulses with medium can be neglected. The basic equations here will be (2.2.2') and (2.2.3'). However, as we shall see below, for the photon echo formation the value τ_{12} should be considerably greater than the USP duration $\tau_{12} >> T_1, \tau_{12} >> T_2$. Therefore in the time interval between the exciting pulses (2.5.1) and (2.5.2) and after them, relaxation must be taken into account. We use (2.2.3') with $\Lambda = 0$. For simplicity of reasoning first we shall be confined to a case (2.4.2) of narrow spectral lines

$$\frac{1}{T_0} << \frac{1}{T_n}, \quad n = 1,2, \tag{2.5.3}$$

and a high-energy mode of a laser irradiation

$$2a_n |d_{ba}| T_0/\hbar >> 1, \quad n = 1,2 .$$

In the time interval $0 \le t - z/c \le T_1$, the intensity of USP (2.5.1) is determined by the formulas (2.3.1) and (2.3.2), and the state of resonant medium - by the density matrix:

$$R_{21}(t') = \frac{id_{ba}}{2|d_{ba}|} e^{-i\varphi_1} (N_1 - N_2) F(\Delta) \sin 2\Lambda_1 t', \tag{2.5.4}$$

$$\left. \begin{aligned} R_1(t') &= \left(N_1 \cos^2 \Lambda_1 t' + N_2 \sin^2 \Lambda_1 t'\right) F(\Delta), \\ R_2(t') &= \left(N_1 \sin^2 \Lambda_1 t' + N_2 \cos^2 \Lambda_1 t'\right) F(\Delta), \end{aligned} \right\} \tag{2.5.5}$$

$$\Lambda_1 = a_1 |d_{ba}|/\hbar, \quad t' = t - z/c .$$

After the passage of the first USP, the free induction decay in the time interval $T_1 \le t - z/c \le T_1 + \tau_{12}$ on account of the relaxation processes follows from

$$R_{21}(t') = R_{21}(T_1) \exp[(i\Delta - \gamma_{21})(t' - T_1)]. \tag{2.5.6}$$

The resonant level populations vary under the law

$$R_1(t') = [R_1(T_1) - N_1 F(\Delta)] \exp[-\gamma_1 (t' - T_1)] + N_1 F(\Delta); \tag{2.5.7}$$

$$R_2(t') = [R_2(T_1) - N_2 F(\Delta)] \exp[-\gamma_2 (t' - T_1)] + N_2 F(\Delta). \tag{2.5.8}$$

Here $R_{21}(T_1)$, $R_1(T_1)$ and $R_2(T_1)$ are the meanings of functions (2.5.4) and (2.5.5) at $t' = T_1$.

The expressions (2.5.6)-(2.5.8), taken at $t' = T_1 + \tau_{12}$, serve as the initial conditions for the solution of the Maxwell-Bloch equations (2.2.3') describing resonant propagation of second USP. Using (2.2.9) and (2.2.10) in the time interval $T_1 + \tau_{12} \le t - z/c \le T_1 + T_2 + \tau_{12}$, we shall obtain

$$R_{21}(t') = \frac{id_{ba}}{2|d_{ba}|} e^{-i\varphi_2} N_0(T_1 + \tau_{12}) F(\Delta) \sin 2\Lambda_2 (t' - T_1 - \tau_{12}) + R_{21}(T_1) e^{(i\Delta - \gamma_{21})\tau_{12}} \times$$
$$\times \cos^2 \Lambda_2 (t' - T_1 - \tau_{12}) + \frac{d_{ba}^2}{|d_{ba}|^2} e^{-2i\varphi_2} e^{(i\Delta - \gamma_{21})\tau_{12}} R_{21}^*(T_1) \sin^2 \Lambda_2 (t' - T_1 - \tau_{12}), \qquad (2.5.9)$$

$$R_1(t') = R_1(T_1 + \tau_{12}) \cos^2 \Lambda_2 (t' - T_1 - \tau_{12}) + R_2(T_1 + \tau_{12}) \sin^2 \Lambda_2 (t' - T_1 - \tau_{12}) -$$
$$- \frac{ie^{-\gamma_{21}\tau_{12}}}{2|d_{ba}|} \left\{ d_{ba} e^{-i\varphi_2} R_{21}^*(T_1) e^{-i\Delta\tau_{12}} - d_{ba}^* e^{i\varphi_2} R_{21}(T_1) e^{i\Delta\tau_{12}} \right\} \sin 2\Lambda_2 (t' - T_1 - \tau_{12}), \qquad (2.5.10)$$

$$R_2(t') = R_1(T_1 + \tau_{12}) \sin^2 \Lambda_2 (t' - T_1 - \tau_{12}) + R_2(T_1 + \tau_{12}) \cos^2 \Lambda_2 (t' - T_1 - \tau_{12}) +$$
$$+ \frac{ie^{-\gamma_{21}\tau_{12}}}{2|d_{ba}|} \left\{ d_{ba} e^{-i\varphi_2} R_{21}^*(T_1) e^{-i\Delta\tau_{12}} - d_{ba}^* e^{i\varphi_2} R_{21}(T_1) e^{i\Delta\tau_{12}} \right\} \sin 2\Lambda_2 (t' - T_1 - \tau_{12}), \qquad (2.5.11)$$

$$N_0(T_1 + \tau_{12}) = \left[R_1(T_1 + \tau_{12}) - R_2(T_1 + \tau_{12}) \right] / F(\Delta) = N_1 - N_2 +$$
$$+ \left[N_1 \left(\cos^2 \Lambda_1 T_1 - 1 \right) + N_2 \sin^2 \Lambda_1 T_1 \right] e^{-\gamma_1 \tau_{12}} - \left[N_1 \sin^2 \Lambda_1 T_1 + N_2 \left(\cos^2 \Lambda_1 T_1 - 1 \right) \right] e^{-\gamma_2 \tau_{12}},$$
$$\Lambda_2 = a_2 |d_{ba}| / \hbar.$$

On the basis of the given formulas we can present the intensity of an electrical field of an optical induction in the time interval $t - z/c \ge T_1 + T_2 + \tau_{12}$ behind a second USP as

$$E = \left(\varepsilon_0(t - z/c) + \varepsilon_i(t - z/c) + \varepsilon_e(t - z/c) \right) \exp\left[i(kz - \omega t) \right] + k.c., \qquad (2.5.12)$$

$$\varepsilon_0(t) = -\widetilde{E}_0 e^{-i\varphi_1} \sin 2\Lambda_1 T_1 \cos^2 \Lambda_2 T_2 \, A(t - T_1 - T_2) \exp\left[(i\Delta_0 - \gamma_{21})(t - T_1 - T_2) \right],$$
$$\varepsilon_i(t) = -\widetilde{E}_0 e^{-i\varphi_2} \frac{N_0(T_1 + \tau_{12})}{N_1 - N_2} \sin 2\Lambda_2 T_2 \, A(t - T_1 - T_2) \exp\left[(i\Delta_0 - \gamma_{21})(t - T_1 - T_2 - \tau_{12}) \right],$$
$$\varepsilon_e(t) = \widetilde{E}_0 e^{i(\varphi_1 - 2\varphi_2)} \sin 2\Lambda_1 T_1 \sin^2 \Lambda_2 T_2 \, A(t - T_1 - T_2 - 2\tau_{12}) \times$$
$$\times \exp\left[i\Delta_0 (t - 2\tau_{12} - T_1 - T_2) - \gamma_{21}(t - T_1 - T_2) \right], \qquad (2.5.13)$$
$$\widetilde{E}_0 = 2\pi |d_{ba}| L (N_1 - N_2) \omega / c.$$

Let us remind that the function $A(\xi)$ is different from zero only in a narrow area close to $\xi = 0$ with the width of about the dephasing time T_0, and for a gaseous medium $A(\xi) = \exp(-\xi^2 / 4T_0^2)$, and for a solid state $A(\xi) = \exp(-|\xi| / T_0)$.

Let the interval of time τ_{12} between USPs exceed T_0:

$$\tau_{12} >> T_0 \tag{2.5.14}$$

but does not exceed the time $1/\gamma_{21}$ of irreversible relaxation, so $A(\tau_{12}) \approx 0$. In these conditions the optical induction after the first USP, before the second USP passes, completely decays due to atomic dephasing. As $t' \geq T_1 + T_2 + \tau_{12}$, $\varepsilon_0 \approx 0$ in all the moments of time. The term ε_i in amplitude of an optical induction represents exactly the same induction, as well as in the case of propagation of only one USP in a resonant medium, in which the equilibrium population is equal to $N_0(T_1 + \tau_{12})$. This term quickly decays during the order T_0 at once after the passage of the second USP. Last term ε_e is everywhere equal to zero except for the area with the sizes of the order T_0 near the moment of time $t' = 2\tau_{12} + T_1 + T_2$, i.e. after the same interval of time after the second exciting pulse, to which it will be separated from the first pulse. It is natural to consider this term as the original phenomenon of an echo in an optical induction, which is named as a two-pulse photon echo, or in brief a photon echo. At the present time there is a great number of experiments performed in from nanoseconds to femtoseconds time scales [14].

We pay attention to how one can simply identify the term R_{21}^e describing photon echo in the expression for a density matrix R_{21}. After the second USP $R_{21} = R_{21}(T_1 + T_2 + \tau_{12})\exp(i\Delta\tilde{t}')$, where $\tilde{t}' = t' - T_1 - T_2 - \tau_{12} \geq 0$. In $R_{21}(T_1 + T_2 + \tau_{12})$ according to (2.5.8) there are terms with factors $e^{i\Delta\tau_{12}}$ and $e^{-i\Delta\tau_{12}}$ and which do not contain them. The term with $e^{-i\Delta\tau_{12}}$ sometimes is spoken about as reversed in time: it enters $R^*{}_{21}$. It is this term that determines echo-phenomenon,

$$\int d\Delta F(\Delta)\exp[i\Delta(\tilde{t}'-\tau_{12})] = A(t'-2\tau_{12} - T_1 - T_2).$$

Finally, we shall write R_{21}^e in a general form

$$\left.\begin{aligned}
&R_{21}^e(t) = U_{21}(t_2^b, t_2^f)R_{21}(t_2^f)U_{21}^*(t_2^b, t_2^f)\exp[(i\Delta - \gamma_{21})(t - t_2^b)],\\
&R_{12}(t_2^f) = R_{12}(t_1^b)\exp[(-i\Delta - \gamma_{21})\tau_{12}],\\
&R_{12}(t_1^b) = R_{21}^*(t_1^b) = U_{21}^*(t_1^b, t_1^f)F(\Delta)(N_1 - N_2)U_{11}(t_1^b, t_1^f),
\end{aligned}\right\} \tag{2.5.15}$$

Where t_n^f and t_n^b are designated as time of an input to a resonant medium of *forward* and *backward* fronts of $n-th$ of an exciting pulse of the rectangular form.

From the quantum-mechanical point of view echo phenomenon represents spontaneous coherent emission of the excited resonant atoms at the moment of time $t' = 2\tau_{12} + T_1 + T_2$. Only the cooperative radiation of a group of the excited atoms results in that remarkable fact that the echo emission (i.e. radiation process) occurs during the time smaller than the time of spontaneous radiation of the isolated excited atom.

The photon echo formation on a broad spectral line

$$\frac{1}{T_n} < \frac{1}{T_0}, \quad n = 1,2, \tag{2.5.16}$$

differs from the considered case only by complicated formulas and value of the reversible relaxation time of an optical induction, which is defined by the duration of exciting pulses. As a result of similar calculations, the intensity of an electrical field after the passing of pulses (2.5.1) and (2.5.2) also can be given by the form (2.5.12). Now, however, the necessary condition to isolate a term $\varepsilon_e(t)$ from two others $\varepsilon_0(t)$ and $\varepsilon_i(t)$ is (instead of (2.5.14)) the inequality

$$\tau >> T_n, \quad n = 1,2. \tag{2.5.17}$$

Then ε_0 and ε_i quickly decay during the time of pulse exciting duration soon after the second exciting pulse, and only after a significant interval of time τ_{12} a photon echo $\varepsilon_e(t)$ will arise from "nothing". Using (2.5.15) it is easy to derive the following expression for amplitude of photon echo:

$$\varepsilon_e(t) = \tilde{E}_0 \exp[i(\varphi_1 - 2\varphi_2) - \gamma_{21}t] \int F(\Delta) \frac{4\Lambda_1\Lambda_2^2}{\Omega_1\Omega_2^2} [\sin\Omega_1 T_1 -$$

$$- i\frac{\Delta}{\Omega_1}(1 - \cos\Omega_1 T_1)](1 - \cos\Omega_2 T_2)\exp[i\Delta(t - t_e)]d\Delta, \tag{2.5.18}$$

$$\Omega_n = \sqrt{\Delta^2 + 4\Lambda_n^2}, \quad n = 1,2, \quad t_e = 2\tau_{12} + T_1 + T_2.$$

The terms $\gamma_{21}T_1$ and $\gamma_{21}T_2$ due to the accuracy excess are neglected here.

The formula (2.5.18) involves the considered above case (2.5.13) as well, which follows from (2.5.18) with $\Lambda_n T_0 >> 1$ and $\Delta T_n << 1$. Let us discuss the basic properties of a two-pulse photon echo (2.5.18).

The duration of a photon echo (2.5.18) in region (2.5.16) approximately coincides with the duration of the most extended exciting pulse, and in case (2.5.3) is equal to $2T_0$. The maximum of amplitude (2.5.18) is reached at the moment of time $t = t_e$ for narrow and later for wide spectral lines. In the case of a narrow spectral line the amplitude of photon echo is maximal if to use exciting pulses with the areas $\theta_1 = 2\Lambda_1 T_1 = \pi/2$ and $\theta_2 = 2\Lambda_2 T_2 = \pi$ (see (2.5.13)). The same sequence $\pi/2$ - and π - pulses is optimal for photon echo formation on a wide spectral line. The carrying frequency of photon echo is equal ω_0 for (2.5.3) and ω in case (2.5.16).

In the absence of irreversible relaxation $\gamma_{21}t_e << 1$ the structure of a photon echo amplitude does not depend on an interval of time τ. Irreversible relaxation breaks conditions of coherent interaction of two USPs and medium by destroying "memory" of influence of the first pulse during the interval of time between pulses. As a result, with increasing of time interval τ_{12} between exciting pulses the photon echo amplitude decreases under the law $\exp(-2\gamma_{21}\tau_{12})$. With $\tau_{12} >> 1/\gamma_{21}$ atomic radiators by the moment of action of the second pulse are completely dephased due to irreversible relaxation and are in a thermodynamic equilibrium. Therefore there is no echo phenomenon and the propagation of the second pulse (2.5.2) is completely similar to the propagation of the first pulse (2.5.1).

A remarkable feature of photon echo as the process caused by "memory" of medium is the effect of correlation of the form of photon echo with structures of exciting pulses [15-19]. The Fig. 2.5.2 evidently demonstrates this property - structure of photon echo reproduces the form of the first exciting pulse. The basic results are obtained here by numerical simulation.

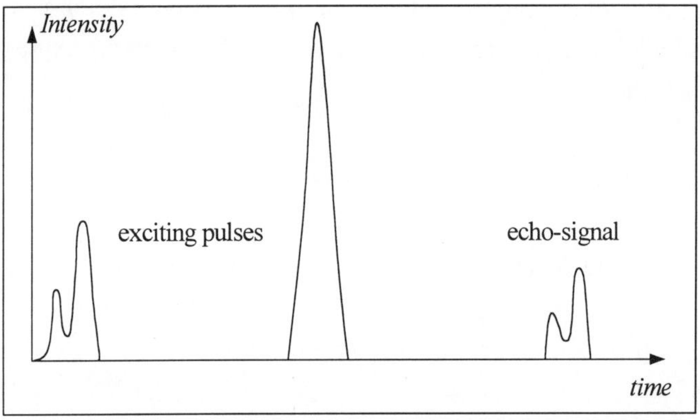

Fig.2.5.2. Correlation of a form of two-pulse photon echo with the form of the first exciting pulse.

Let us consider the conditions of spatial synchronism for a two-pulse photon echo formed by USP with non-collinear wave vectors $\vec{k}_1$ and $\vec{k}_2$:

$$E_1 = 2a_1 \cos(\omega t - \vec{k}_1\vec{r} + \varphi_1), \quad 0 \le t - \frac{\vec{k}_1\vec{r}}{\omega} \le T_1,$$

$$E_2 = 2a_2 \cos(\omega t - \vec{k}_2\vec{r} + \varphi_2), \quad 0 \le t - \frac{\vec{k}_2\vec{r}}{\omega} \le T_1 + T_2 + \tau_{12}.$$

Let us neglect the delay of a signal within the resonant medium in the formulation of initial conditions $L/c << 1/\Lambda_n, T_0$.

In the case of a solid state where the shift of atomic frequency inside the contour of an inhomogeneously broadened spectral line does not depend on the direction of pulse propagation, we do as follows. Let us choose an axis z along a wave vector $\vec{k}_2$. Then in the initial condition before the propagation of the second pulse R_{21}^0 should be taken as

$$R_{21}^0 = R_{21}(T_1)\exp[i(\vec{k}_1\vec{r} - k_2 z)]\exp[(i\Delta - \gamma_{21})\tau_{12}].$$

Here $R_{21}(T_1)$ is given (for a narrow spectral line) by expression (2.5.4) taken at $t' = T_1$. One can see that the phase φ_1 is supplemented by an additional term $k_2 z - \vec{k}_1\vec{r} = \vec{k}_2\vec{r} - \vec{k}_1\vec{r}$. Thus, describing the echo phenomenon in $t - z/c \geq T_1 + T_2 + \tau_{12}$, density matrix R_{21}^e differs from the one similar (2.5.15) in the case (2.5.1) and (2.5.2) by presence of additional phase factor $\exp[i(\vec{k}_2 - \vec{k}_1)\vec{r}]$. As a result the total oscillating phase of $p_{21}^e = R_{21}^e \exp[i(\vec{k}_2\vec{r} - \omega t)]$ is equal to $\exp[i(\vec{k}_e\vec{r} - \omega t + \varphi_1 - 2\varphi_2)]$, $\vec{k}_e = 2\vec{k}_2 - \vec{k}_1$. The signal of photon echo will be emitted from the excited medium provided that the wave vector $\vec{k}_e = 2\vec{k}_2 - \vec{k}_1$ of echo is connected to the frequency ω by the dispersion equation $\vec{k}_e^2 = \omega^2/c^2$. In the case of a two-pulse photon echo it is possible only with $\vec{k}_2 = \vec{k}_1$. If, nevertheless, $\vec{k}_2$ slightly differs from $\vec{k}_1$, a signal of photon echo is emitted in the direction $2\vec{k}_2 - \vec{k}_1$, however its amplitude in comparison with (2.5.18) contains an addition a small factor [20]

$$\frac{\exp[i(k_e - k)L] - 1}{i(k_e - k)L}, \tag{2.5.19}$$

which arises as a result of integration (2.2.3').

In the case of gas a situation is slightly different other because of the dependence of a shift of frequency inside an inhomogeneously broadened contour of a spectral line on the direction of radiation. However for two-pulse excitation the result completely coincides with a case of solid-state resonant medium.

2.5.2 STIMULATED PHOTON ECHO

Propagation through the resonant medium of three USPs of duration T_1, T_2 and T_3, separated in time by intervals accordingly τ_{12} and τ_{23}:

$$E_1 = 2a_1\cos(\omega t - \vec{k}_1\vec{r} + \varphi_1), \quad 0 \leq t - \frac{\vec{k}_1\vec{r}}{\omega} \leq T_1, \tag{2.5.20}$$

$$E_2 = 2a_2\cos(\omega t - \vec{k}_2\vec{r} + \varphi_2), \quad T_1 + \tau_{12} \leq t - \frac{\vec{k}_2\vec{r}}{\omega} \leq T_1 + T_2 + \tau_{12}, \tag{2.5.21}$$

$$E_3 = 2a_3\cos(\omega t - \vec{k}_3\vec{r} + \varphi_3), \; T_1 + T_2 + \tau_{12} + \tau_{23} \leq t - \frac{\vec{k}_3\vec{r}}{\omega} \leq T_1 + T_2 + T_3 + \tau_{12} + \tau_{23}, \tag{2.5.22}$$

is accompanied by a several echo-phenomenon in an optical induction. Among them the so-called stimulated photon echo holds the most interest. It is emitted by the excited medium at the moment of time $2\tau_{12}+\tau_{23}+T_1+T_2+T_3$. It is possible to comment on the stages of its formation as follows. The first pulse polarises medium (see. (2.5.4) and (2.5.6)), that changes dynamics of resonant level population in a field of second USP ((2.5.10) and (2.5.11)). "The memory" in an interval of time between the second and third stimulating pulses is stored about it, since there is no relaxation of level populations in the absence of a field. The third pulse again polarises medium. However, as a result of the previous dephasing processes, atomic radiators appear in phase and coherently emit in the different moments of time. In the moment of time $2\tau_{12}+\tau_{23}+T_1+T_2+T_3$ the "memory" of influence of the first pulse "stored" in an interval of time between second and third USP in populations of resonant energy levels is displayed. As a result in this moment there is an echo signal i.e. a stimulated photon echo.

The necessary conditions of stimulated photon echo formation are described by inequalities

$$2\gamma_{21}\tau_{12}+\gamma_m\tau_{23}<<1, \quad m=1,2, \quad \tau_{12}>>T_n, \quad \tau_{23}>>T_n.$$

The first of them provides coherent interaction of three pulses and medium, including preservation of memory of medium about the influence of each exciting pulse. Other inequalities are necessary for distinct separation of signals of an optical induction, emitted in the different moments of time. It is possible not to separate the moment of time $2\tau_{12}+\tau_{23}+T_1+T_2+T_3$ of stimulated photon echo formation from the moment $\tau_{12}+2\tau_{23}+T_1+T_2+T_3$ of a two-pulse (from (2.5.21) and (2.5.22)) photon echo formation, as we shall see below, because of the various conditions of spatial synchronism for stimulated and two-pulse echoes. The latter means that the wave vectors $\vec{k}_1$, $\vec{k}_2$ and $\vec{k}_3$ of exciting pulses can be chosen so that even with the concurrence of the specified moments of time (i.e. for $\tau_{12}=\tau_{23}$) only a stimulated photon echo is emitted.

Before discussing the properties of a stimulated photon echo, we write the term R^e_{21} of a density matrix R_{21}, responsible for the considered echo-phenomenon. In the case of a solid-state resonant medium we have

$$\rho^e_{21}=R^e_{21}\exp[i(\vec{k}_3\vec{r}-\omega t)],$$
$$R^e_{21}(t)=U_{21}(t^b_3,t^f_3)\,[R^e_1(t^f_3)-R^e_2(t^f_3)]\,U^*_{11}(t^b_3,t^f_3)\exp[(i\Delta-\gamma_{21})(t-t^b_3)],$$
$$R^e_1(t^f_3)-R^e_2(t^f_3)=U_{11}(t^b_2,t^f_2)\,R_{12}(t^f_2)\,U^*_{12}(t^b_2,t^f_2)\,[\exp(-\gamma_1\tau_{23})+\exp(-\gamma_2\tau_{23})],$$
$$R_{12}(t^f_2)=R^*_{21}(t^f_2)=$$
$$=U^*_{21}(t^b_1,t^f_1)\,F(\Delta)(N_1-N_2)\,U_{11}(t^b_1,t^f_1)\exp[-i(\vec{k}_1-\vec{k}_2)\vec{r}]\exp[(-i\Delta-\gamma_{21})\tau_{12}].$$

In these formulas all stages of stimulated echo formation are distinctly traced. In the case of gaseous resonant medium the shift of frequency Δ inside an inhomogeneously broadened contour of a spectral line depends on the direction of wave propagation $\Delta = \Delta_n = \Delta_0 - \vec{k}_n\vec{v}$. If vectors $\vec{k}_1$ and $\vec{k}_3$ are not collinear, the time of emission of a stimulated photon echo changes, and the amplitude of an echo contains an additional small factor. We shall exclude such a situation from the consideration and we shall assume vectors $\vec{k}_1$ and $\vec{k}_3$ to be collinear. If thus $\vec{k}_1 = \vec{k}_3$, the value R^e_{21} coincides with the received above. If $\vec{k}_1 = -\vec{k}_3$, the expression for R^e_{21} is given by:

$$R^e_{21}(t) = U_{21}(t^b_3, t^f_3)\,[R^e_1(t^f_3) - R^e_2(t^f_3)]\; U^*_{11}(t^b_3, t^f_3)\exp\,[(i\Delta_3 - \gamma_{21})(t - t^b_3)],$$
$$R^e_1(t^f_3) - R^e_2(t^f_3) = U_{12}(t^b_2, t^f_2)\,R_{21}(t^f_2)U^*_{11}(t^b_2, t^f_2)\;[\exp(-\gamma_1\tau_{23}) + \exp(-\gamma_2\tau_{23})],$$
$$R^*_{21}(t^f_2) = U^*_{21}(t^b_1, t^f_1)f(v)(N_1 - N_2)\;U_{11}(t^b_1, t^f_1)\exp[-i(\vec{k}_1 - \vec{k}_2)\vec{r}]\exp[(-i\Delta_1 - \gamma_{21})\tau_{12}].$$

Here $f(v) = (1/\sqrt{\pi}u)^3\exp(-v^2/u^2)$ is the Maxwell distribution with the most probable velocity u, and during the action of each of three stimulating pulses the value Δ in the formulas for matrix elements U is equal accordingly $\Delta_n = \Delta_0 - \vec{k}_n\vec{v}, \quad n = 1,2,3$.

Now it is easy to write the intensity of an electrical field of a stimulated photon echo in the form:

$$E = \varepsilon_e(t - \vec{k}_e\vec{r}/\omega)\exp[i(\vec{k}_e\vec{r} - \omega t - \varphi_e)] + c.c., \tag{2.5.23}$$

where

$$\varepsilon_e(t) = 16\tilde{E}_0\int d\Delta F(\Delta)\frac{\Lambda_1\Lambda_2\Lambda_3}{\Omega_1\Omega_2\Omega_3}X_1X_2X^*_3\exp[i\Delta(t - t_e)]\cdot \tag{2.5.24}$$
$$\cdot\exp(-2\gamma_{21}\tau_{12})[\exp(-\gamma_1\tau_{23}) + \exp(-\gamma_2\tau_{23})],$$
$$X_n = \left[\cos\frac{\Omega_nT_n}{2} - i\frac{\Delta}{\Omega_n}\sin\frac{\Omega_nT_n}{2}\right]\sin\frac{\Omega_nT_n}{2},$$
$$t_e = 2\tau_{12} + \tau_{23} + T_1 + T_2 + T_3,\; \varphi_e = \varphi_3 + \varphi_2 - \varphi_1,$$
$$\vec{k}_e = \vec{k}_3 + \vec{k}_2 - \vec{k}_1. \tag{2.5.25}$$

This result directly concerns a solid-state resonant medium. For gaseous resonant medium it is necessary to replace $d\Delta F(\Delta)$ in (2.5.24) on $d\vec{v}f(v)$, and $\exp[i\Delta(t - t_e)]$ with $\exp[i(\Delta_0 - \vec{k}_3\vec{v})(t - t_e)]$. The value Δ in X_n should be replaced by $\Delta_0 - \vec{k}_n\vec{v}$. The received result will concern the case $\vec{k}_1 = \vec{k}_3$. To describe another case $\vec{k}_1 = -\vec{k}_3$, it is necessary that the terms X_1 and X_2 be replaced with complex conjugate values and be put

$$\varphi_e = \varphi_3 + \varphi_1 - \varphi_2, \qquad (2.5.26)$$
$$\vec{k}_e = \vec{k}_3 + \vec{k}_1 - \vec{k}_2.$$

The requirement $k_e^2 = \omega^2/c^2$ determines conditions of spatial synchronism for a stimulated echo emission. The insignificant deviation from this requirement results in occurrence in (2.5.24) in addition to small factor (2.5.19).

Thus, for the stimulated echo formation in a solid state medium it is necessary that the wave vectors of exciting pulses and echo formed figures be represented in a Fig. 2.5.3, a-c.

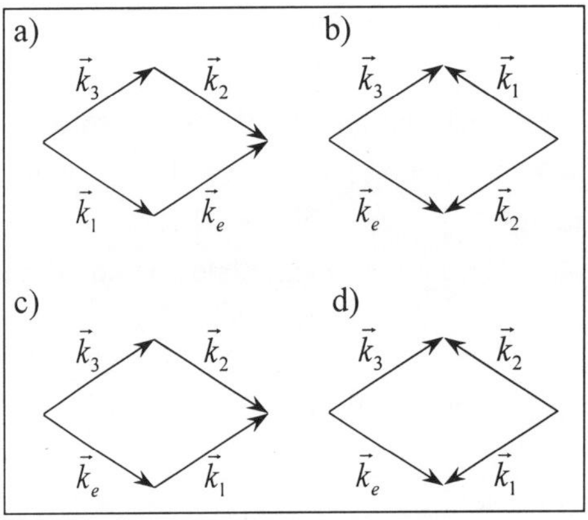

Fig. 2.5.3. Relations between wave vectors of exciting pulses and echo for the echo formation in solid states and in gases.

In a gaseous resonant medium a stimulated photon echo is emitted at the moment of time $t = t_e$, if the wave vectors have a mutual disposition as in Fig. 5.3, c-d. The figure evidently emphasises distinction in a stimulated echo formation in gas and in a solid state due to various physical reasons of inhomogeneous broadening. Though this distinction was displayed only in wave vectors (2.5.25) and (2.5.26) of stimulated echo. It vanishes, if all wave vectors are identical. In the following paragraph the example of an echo-phenomenon wholly dependent from Doppler mechanism of inhomogeneously broadening and absent in a solid state will be considered.

From (2.5.24) it follows that the pulses (2.5.20)-(2.5.22) with the area $\pi/2$ produce the stimulated echo (2.5.23) with maximal intensity.

In absence of the irreversible relaxation the structure of an echo (2.5.24) does not depend on the intervals of time between the exciting pulses. However under the relaxation processes the amplitude of stimulated echo with the increase of τ_{12} decreases under the law $\exp(-2\gamma_{21}\tau_{12})$, and with a variation of τ_{23} changes according to the factor

$[\exp(-\gamma_1\tau_{23}) + \exp(-\gamma_2\tau_{23})]$. Such behaviour reflects the mentioned above stages of the memory storage of medium about the influence of the first pulse (2.5.20).

In the case of excitation of stimulated echo with $\vec{k}_1 = \vec{k}_2 = \vec{k}_3$ and wide spectral width of the first and third exciting pulses and narrow spectral width of the second pulse with respect to atomic spectral line

$$\frac{1}{T_1} >> \frac{1}{T_0}, \quad \frac{1}{T_3} >> \frac{1}{T_0}, \quad \frac{1}{T_2} << \frac{1}{T_0} \tag{2.5.27}$$

there is a distinct correlation of the form of an echo with a structure of the second exciting pulse. This effect in a gas has a quite evident interpretation. It is based on the general picture of the stimulated echo formation and on the following known fact, that monochromatic wave of frequency ω interacts only with those atoms inside an inhomogeneously broadened spectral line, which have velocity projections $v_{\vec{k}}$ onto the direction of the wave propagation in an interval

$$(\omega - \omega_0)/k - \frac{\max\{\gamma_{21}, 2\Lambda\}}{k} \le v_{\vec{k}} \le (\omega - \omega_0)/k + \frac{\max\{\gamma_{21}, 2\Lambda\}}{k}. \tag{2.5.28}$$

Here for definiteness we shall be limited only to the case of gas, and $\vec{k}$ and Λ are a wave vector and Rabbi frequency, accordingly. The above circumstance results in that fact that the distribution R_1 of atoms on v_k at the lower resonant energy level shows the lack of particles with velocities satisfying (2.5.28), and at the upper level in the distribution R_2 the excess of atoms with the same velocities (Fig. 2.5.4) will be formed. As a result on Doppler contour there is a hole, which depth (in linear approximation) is proportional to the wave intensity

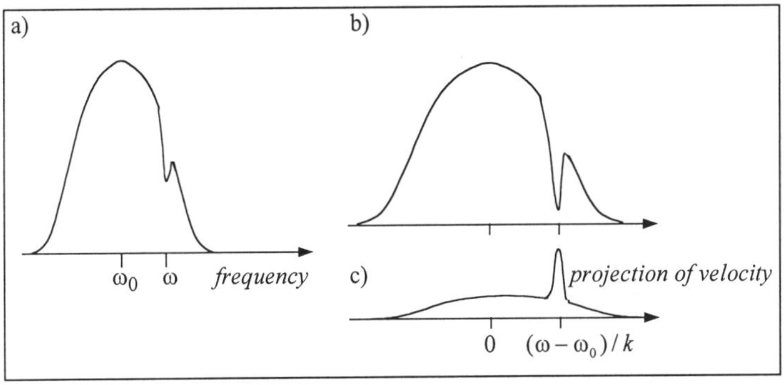

Fig. 2.5.4. a) Hole burning in an inhomogeneously broadened contour of a spectral line. Distribution of atoms at the upper (b) and lower (c) resonant levels on velocity projections to a direction of wave propagation

of a wave. Let us take now a weak pulse of duration $T >> T_0$ instead of a monochromatic wave. Its Fourier spectrum has width $1/T$ smaller than the inhomogeneous width

$1/T_0$. The pulse "burns" "holes" in Doppler contour by their own spectral component, which depth is proportional to the intensity of this spectral component. Thus, in Doppler contour the Fourier image of this pulse is burned (Fig. 2.5.5). The received picture will be stored in the Doppler contour during the life-time $1/\gamma_2$ of the excited level.

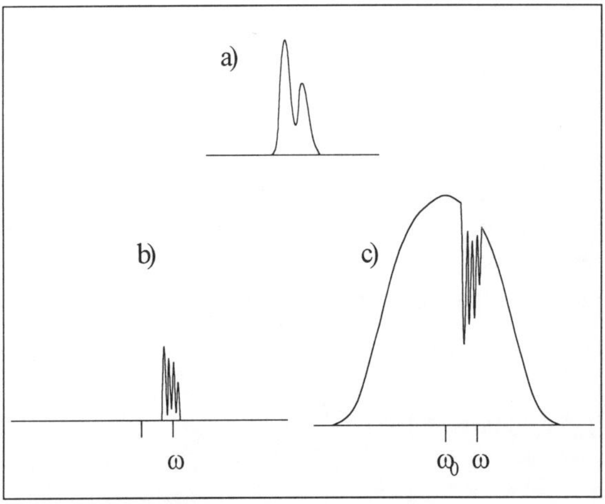

Fig. 2.5.5. a) A structure of exciting USP, b) its Fourier-image and c) "hole burning" of a Fourier-image USP on an inhomogeneously broadened contour of a spectral line.

Now we shall recollect stages of the stimulated echo formation and we shall take into account (2.5.27). The first pulse (2.5.20) excites in the medium optical coherence. The second pulse (2.5.21) "writes down" the Fourier-image on Doppler contour. Thus the phase memory of influence of the first pulse is contained in atomic distributions on velocities. The third pulse (2.5.22) with a wide spectral structure causes the emission of a signal of an echo. Because of the specified distortion of atomic distribution on velocities, the number of atoms in each group of atomic radiators is determined by a Fourier-image of the second pulse. It means that the third pulse makes the reverse Fourier transformation, as a result of which the echo pulse reproduces the form of the second exciting pulse. Similarly transformation properties of photon echoes and their use in optical data processing are discussed in refs. [21-23]. Another important application of stimulated photon echoes is the optical memory devices [24,25].

Other properties of a stimulated photon echo such as its duration, the correlation of the form with the first or third stimulating pulse as a whole are similar to properties of a two-pulse photon echo. They can be easily taken from (2.5.24) or received by numerical simulations.

There exist other methods of photon echo formation by multiple excitation of resonant medium by a set of ultrashort pulses including pulses of standing waves [26-29]. The sequence of Carr and Purcell [30] allows to investigate the relaxation processes with a high accuracy [31].

2.5.3. PHOTON ECHO ON THE BASIS OF OPTICAL NUTATION

In the previous paragraphs it was shown that the USP propagation in a resonance medium previously excited by interaction with another USP has changed the character of reversible relaxation on the basis of optical induction. The influence of each pulse caused the change of a phase of atomic oscillator, therefore they appeared correlated in phases not only soon after the USP, but also in a number of other moments of time dependent from intervals of time between the exciting pulses. Just in these moments the signals of the optical induction which have received the name a photon echo also were radiated. It is possible to speak about similar change of a phase of atomic oscillators as about the reverse of relaxation of electromagnetic oscillations. Really, the change of a phase of oscillations after an interval of time τ after the beginning of their attenuation owing to dephasing, results in that fact that by the moment of time 2τ the oscillations grow again as an echo - phenomenon. Such behaviour is characteristic not only for an optical induction but also for any emission processes, in which the phase $\omega_i t - \vec{k}_i \vec{r} + \varphi_i$ of separate radiator does not have random perturbation, and phase mismatch of radiators (the mechanism of reversible relaxation) is caused by disorder of frequencies ω_i of radiator and/or by the dispersion law $\omega_i = \omega_i(\vec{k}_i)$.

Optical nutation, which properties were studied in section 2.3, belongs to such radiation processes. Let us discuss now the possibility of the echo formation in optical nutation (so-called nutation echo or photon echo on the basis of optical nutation [32-37]). We note that the process of optical nutation formation considered in section 2.3 are not only possible. If the nutation period is less than the time of irreversible relaxation in medium, nutation oscillations always arise after a short-term perturbation of amplitude and/or phase of a light wave, or after the influence of a pulse of a constant electrical field varying a conditions of resonant interaction on resonant medium. For the formation of nutation echo it is necessary, that the interval of time between two consecutive perturbation did not surpass time of irreversible relaxation, but exceeded a nutation period, that nutation oscillations could attenuate owing to atomic dephasing by the moment of influence of the "reverse" perturbation and thus the condition of coherent interaction of a wave with medium did not break.

In the case of USP of the rectangular form the factor "reverse" nutation relaxation is a short-term perturbation of a phase and/or amplitude of USP on distance τ from its forward front. Let us prove that performing the necessary conditions the nutation oscillations of USP will increase after an interval of time τ after the influence of perturbation as a nutation echo. For simplicity we shall be limited to the consideration of an "instant" phase shift of USP. It means, that at an entrance to a resonant medium the amplitude $\mathcal{E}(z,t)$ of USP

$$E = \mathcal{E}(z,t)\exp[i(kz - \omega t - \varphi_0)] + k.c.$$

takes the form

$$\mathcal{E}(z=0,t)=\begin{cases}0, t-z/c<0;\\ a\exp(-i\varphi), \ 0\le t-z/c<\tau;\\ a, \tau\le t-z/c,\end{cases} \qquad (2.5.29)$$

where φ and φ_0 are constant shifts of a phase.

The technique of calculation of USP intensity

$$I=\frac{ca^2}{2\pi}\left(1+2\varepsilon'(t-z/c)/a\right) \qquad (2.5.30)$$

at an exit from resonant medium is stated in section 2.2. In an interval of time $0\le t-z/c<\tau$ the intensity (2.5.30) coincides with (2.3.1) and (2.3.2). After the instant shift of a phase in region $\tau\le t-z/c$ the amplitude of nutation oscillation can be presented as

$$\varepsilon'(t)=\varepsilon_n(t)+\varepsilon_{ne}(t), \qquad (2.5.31)$$

$$\varepsilon_n(t)=-E_0\int d\Delta\frac{F(\Delta)}{\Omega T_0}\{B_1\sin\Omega t+B_2\sin\Omega(t-\tau)\},\ \varepsilon_{ne}(t)=-E_0\int d\Delta\frac{F(\Delta)}{\Omega T_0}B_{ne}\sin\Omega(t-2\tau),$$

$$B_1=\frac{1}{2}\left\{1+\cos\varphi-\frac{\Delta^2}{\Omega^2}(1-\cos\varphi)\right\},\ B_2=\frac{\Delta^2}{\Omega^2}(1-\cos\varphi),\ B_{ne}=(1-\cos\varphi)\frac{2\Lambda^2}{\Omega^2},$$

$$\Lambda=a|d_{ba}|/\hbar,\ \Omega=\left(\Delta^2+4\Lambda^2\right)^{1/2},\ E_0=2\pi(N_1-N_2)La|d_{ba}|^2\omega T_0/\hbar c.$$

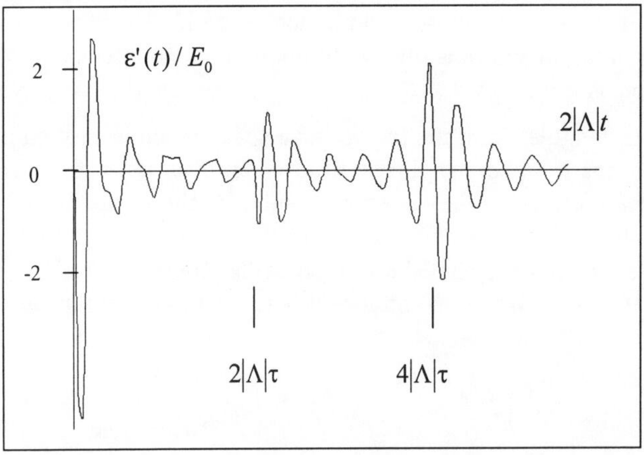

Fig.2.5.6. Optical nutation and nutation echo at $\Delta T_0=0.4$, $\varphi=\pi$, $2|\Lambda|T_0=0.2$, $2|\Lambda|\tau=29$.

In Fig. 2.5.6 a plot of nutation oscillations (2.5.31) is presented. According to the taken parameters nutation the oscillations near the front edge completely decay by the moment of time $t - z/c = \tau$ due to the dephasing of atomic radiators. One can see, that at first nutation oscillations also attenuate, then to the moment of time $t - z/c = 2\tau$ they grow again, then again attenuate. The burst of nutation oscillations near the moment of time $t - z/c = 2\tau$ also represents a nutation echo. Its amplitude is described by the expression $\varepsilon_{ne}(t)$ and is maximal with the shift of a phase on π. It is interesting to note that the form of a nutation echo reproduces the character of nutation oscillations near the forward front of USP. In this sense it is possible to speak about the effect of correlation of the form of a nutation echo and nutation oscillations in the first time interval $0 \le t - z/c < \tau$.

We have already spoken enough about the reasons of occurrence of the phenomena of an echo, therefore only we shall emphasise the basic differences of nutation echo from a photon echo on the basis of an optical induction. First of all we shall note the distinction in the radiation by atoms of signals of an echo. The nutation echo is due to the induced emissions of quantums by atoms which are in the field of a strong electromagnetic wave. For its description it is enough to be limited to the submitted semiclassical approach. Meanwhile, the echo in an induction is formed originally by the coherent spontaneous radiation of free atoms, which, generally speaking, requires consecutive quantum-mechanical consideration. But even in the stated semiclassical approach the nutation echo and photon echo in an optical induction are due to different mechanisms of reversible relaxation. A nutation echo is caused by two equivalent mechanisms determined by inhomogeneous broadening and degeneration of resonant energy levels due to various orientation of the total angular momentum. It enables to form a nutation echo in the conditions when a photon echo in an optical induction is absent (for example, in the case of homogeneously broadened resonant atomic transition between energy levels which are highly degenerate on projections of the total angular moment [32,33,35]).

It is necessary also to specify the following. In a nutation echo the "memory" about the nutation oscillation is displayed not only in polarisation of medium, but also in populations of resonant energy levels. Under the action of irreversible relaxation the specified memory decays differently. It results in more complicated dependence of amplitude from relaxation parameters $\gamma_{21}, \gamma_1, \gamma_2$ and interval of time τ. However in the elementary case, when the constants of polarisation and population relaxation are equal $\gamma_{21} = \gamma_1 = \gamma_2 = \gamma$, an amplitude of nutation echo (2.5.32) with the increase of an interval of time τ occurs an additional factor $\exp(-2\gamma\tau)$, which determines attenuation of an echo owing to relaxation processes.

In the conclusion we shall note, that the effect of optical nutation occurs and with the influence of a pulse of a standing wave on resonant medium. Therefore with the performance of necessary conditions nutation echo can be generated with the short-term perturbation of a phase and/or amplitude of one or two running waves forming a standing wave. Thus the signal of nutation echo will be radiated both in the direct and in the opposite direction axes z. In comparison with nutation echo (2.5.32) it will be distinguished by more fast reversible attenuation of nutation oscillations, as the resonant atom, absorbed quantum from one wave, can it emitted in other wave, and it is the additional mechanism of wave dephasing emitted by different groups of atoms.

2.5.4. PHOTON ECHOES UNDER TWO-PHOTON RESONANCE

According to section 2.2, the investigation of resonant pulse propagation through the optically thin resonant media is divided into two separate problems of dynamics of resonant atoms in the given field and of emission of excited atoms. The first problem consists in solution of one and same systems of Bloch equations (1.1.12), which defines the similarity of atom behaviour in different resonant conditions. The difference here is related to the necessity to take into account the dynamic Stark effect. All features of coherent phenomena under two-photon resonance are in emission problem. The point is that the radiating transition between levels E_a and E_c is forbidden in electric dipole approximation. To make the excited atoms radiate the presence of an external electromagnetic field is necessary (in agreement with (1.1.20)-(1.1.27)). If this field

$$E' = 2a'\cos(\omega' t - \vec{k}'\vec{r} + \varphi') \tag{2.5.32}$$

is nonresonant, the coherent emission of two-photon excited atoms arises both at Stokes $\omega' - \omega_{ca}$, and anti Stokes $\omega' + \omega_{ca}$ frequencies. In the case of running waves this circumstance influences the conditions of spatial synchronism and changes the order of the size of coherent radiation [39-45]. To be more specific, let us consider the excitation of a two-photon absorbing solid-state medium of pair of ultrashort light pulses

$$E_1 = 2a_1\cos(\omega t - \vec{k}_1\vec{r} + \varphi_1), \quad 0 \le t \le T_1,$$
$$E_2 = 2a_2\cos(\omega t - \vec{k}_2\vec{r} + \varphi_2), \quad T_1 + \tau \le t \le T_1 + T_2 + \tau_{12},$$

with carrying frequency in two-photon resonance with the transition $E_c \to E_a$, i.e. $2\omega \approx \omega_0$, where ω_0 is the central frequency of an inhomogeneously broadened spectral line.

In the area $T_1 + T_2 + \tau_{12} < t$ after the passage of USPs the macroscopic polarisation arises near the moment of time $t = 2\tau_{12} + T_1 + T_2$ and is described by a density matrix

$$\rho_{ca} = R_{21}^e \exp\{2i[(2\vec{k}_2 - \vec{k}_1)\vec{r} - \omega t]\}, \tag{2.5.33}$$

in which R_{21}^e is given by expression (2.5.15) on account of a two-photon character of process.

In a field of a probe wave (2.5.32) polarisation (5.33) is displayed as signals of an echo $E_{\pm}$ to anti Stokes (top mark) and Stokes frequencies

$$E_{\pm} = \varepsilon_{\pm}\exp\left\{\pm i\left[\left(\vec{k}'\pm 2(2\vec{k}_2 - \vec{k}_1)\right)\vec{r} - (\omega'\pm 2\omega)t - \varphi_0\right]\right\} + k.c., \qquad (2.5.34)$$

$$\varepsilon_{\pm} = \pm i2\pi\left|\vec{k}'\pm 2(2\vec{k}_2 - \vec{k}_1)\right| La_0\Pi^*_{\pm\omega_p} \int R^e_{21}\, d\Delta. \qquad (2.5.35)$$

The requirement $\omega'\pm 2\omega = \left|\vec{k}'\pm 2(2\vec{k}_2 - \vec{k}_1)\right| c\left[\varepsilon(\omega'\pm 2\omega)\right]^{-1/2}$ determines the conditions of spatial synchronism.

It is necessary to note that under two-photon resonance the amount of possible variants of excitation of medium and coherent radiation as a various echo-phenomenon sharply grows. Thus a picture of coherent radiation essentially varies, for example, with the use of standing waves in gas medium.

The resonant interaction of powerful light waves with a two-photon absorbing medium is accompanied by the processes of parametric transformation of frequency of a wave, therefore resonant medium radiates harmonics with frequencies $3\omega, 2\omega_2 \pm \omega_1, \omega_2 \pm 2\omega_1$. Owing to nutation oscillations (if they are) of the population of resonant levels, the intensity of harmonics is broken into a number of spikes, and in the moments when a photon echo takes place in optical nutation, the similar phenomenon arises and in harmonics (Fig. 2.5.7) [45].

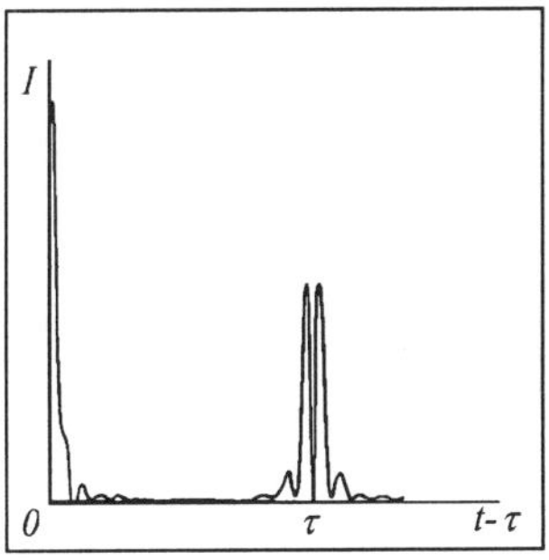

Fig2.5.7. The echo-phenomenon in the third harmonic generation

2.5.5. PHOTON ECHOES PRODUCED BY STANDING-WAVE PULSE

If a short travelling-wave pulse passing through an optically thin medium of length L is reflected from a perpendicular mirror, and propagates in the opposite direction, it produces a standing wave in this medium over a short time interval

$$E = 2a\cos(kz + \varphi)\exp[-i(\omega t + \Phi)] + c.c., \qquad (2.5.36)$$

where $\omega = kc$ is the wave frequency, a is the real amplitude, and φ and Φ are certain phase shifts. The time $2L/c$ of the motion of the leading front in the medium in the forward and backward directions was small compared with the characteristic times. This makes it possible to assume hereafter that the standing wave pulse (2.5.36) is excited simultaneously over the entire length of the medium. Of course, there are also other methods of producing the standing-wave pulse (2.5.36).

The interaction of the pulse (2.5.36) with the resonant atoms is different in gas and solid state media. The features are displayed more brightly in gas medium. Below we will briefly review the stages of the description of resonant interaction of pulse (2.5.36) with optically allowed transition $E_b \to E_a$ in gas medium [46-48].

We start as usual with the equations (1.01), (1.02) and (1.0.3):

$$\left(\frac{\partial^2}{\partial z^2} - \frac{1}{c^2}\frac{\partial^2}{\partial t^2}\right)E = \frac{4\pi}{c^2}\frac{\partial^2}{\partial t^2}\int dv_z Sp(\rho d), \tag{2.5.37}$$

$$\left(\frac{\partial}{\partial t} + v_z\frac{\partial}{\partial z}\right)\rho = \frac{i}{\hbar}[\rho(H_0 - Ed) - (H_0 - Ed)\rho] - \hat{\Gamma}\rho, \tag{2.5.38}$$

$$\hat{\Gamma}\rho_{aa} = \gamma_1\rho_{aa} - \gamma_1 N_1 f(v_z),\ \hat{\Gamma}\rho_{bb} = \gamma_2\rho_{bb} - \gamma_2 N_2 f(v_z),\ \hat{\Gamma}\rho_{ba} = \gamma_{21}\rho_{ba},$$

in which v_z is the projection of atom velocity along the z-axis, $f(v) = (\pi^{1/2}u)^{-1}\exp(-v^2/u^2)$ is the Maxwell distribution function, γ_1 , γ_2 and γ_{21} are the relaxation constants. In resonance approximation equations (2.5.38) become

$$\left(\frac{\partial}{\partial t} + v_z\frac{\partial}{\partial z} - i\Delta + \gamma_{21}\right)R_{21} = i2ad_{ba}\cos(kz+\varphi)\hbar^{-1}(R_1 - R_2),$$

$$\left(\frac{\partial}{\partial t} + v_z\frac{\partial}{\partial z} + \gamma_1\right)R_1 = i2a\cos(kz+\varphi)\hbar^{-1}\left(d_{ba}{}^{*}R_{21} - d_{ba}R_{21}^{*}\right) + \gamma_1 N_1 f(v_z), \tag{2.5.39}$$

$$\left(\frac{\partial}{\partial t} + v_z\frac{\partial}{\partial z} + \gamma_2\right)R_2 = -i2a\cos(kz+\varphi)\hbar^{-1}\left(d_{ba}{}^{*}R_{21} - d_{ba}R_{21}^{*}\right) + \gamma_2 N_2 f(v_z),$$

$$R_{21} = \rho_{ba}\exp(i\omega t),\quad R_1 = \rho_{aa},\quad R_2 = \rho_{bb},\quad \Delta = \omega - \omega_{ba}.$$

The initial condition for these equations at the instant of time $t=0$ prior to excitation by the standing wave are in the form

$$R_{21}(z,t)\big|_{t=0} = 0,\quad R_1(z,t)\big|_{t=0} = N_1 f(v_z),\quad R_2(z,t)\big|_{t=0} = N_2 f(v_z).$$

We change over in (2.5.39) from the variables z and t to two independent variables ξ and t', where $\xi = z - v_z t$ and $t' = t$. Also, we determine the polarisation amplitude

$$p = R_{21}d_{ab}.$$

At the initial instant $t=0$ the sought functions do not depend on ξ, and in the region $0<t$ we have

$$p=p(\xi,t),\quad R_1=R_1(\xi,t),\quad R_2=R_2(\xi,t),$$

where t will no longer be primed $t'\to t$.

The general solution of eqs.(2.3.59) for $\gamma_{21}=\gamma_1=\gamma_2=\Delta=0$ takes the form

$$\begin{aligned}
p(\xi,t)&=p'(\xi,t_0)+ip''(\xi,t_0)\cos\Lambda F+i\tfrac{1}{2}|d_{ba}|(R_1(\xi,t_0)-R_2(\xi,t_0))\sin\Lambda F,\\
R_1(\xi,t)&=\tfrac{1}{2}(R_1(\xi,t_0)+R_2(\xi,t_0))+\tfrac{1}{2}(R_1(\xi,t_0)-R_2(\xi,t_0))\cos\Lambda F-\tfrac{1}{|d_{ba}|}p''(\xi,t_0)\sin\Lambda F,\\
R_2(\xi,t)&=\tfrac{1}{2}(R_1(\xi,t_0)+R_2(\xi,t_0))-\tfrac{1}{2}(R_1(\xi,t_0)-R_2(\xi,t_0))\cos\Lambda F+\tfrac{1}{|d_{ba}|}p''(\xi,t_0)\sin\Lambda F,
\end{aligned}\tag{2.5.40}$$

$$F=\frac{2}{kv_z}\sin\frac{kv_z(t-t_0)}{2}\cos[k\xi+\varphi+\tfrac{1}{2}kv_z(t+t_0)],\quad \Lambda=4a|d_{ba}|/\hbar,$$

where t_0 is the initial instant of time, and the prime and double prime denote the real and imaginary parts of the function $p(\xi,t)$, respectively.

To find the field E_i induced by the excited medium, we use the Fourier expansion

$$p(z-v_zt,t)=\sum_{n=-\infty}^{\infty}p_ne^{inkz},$$

and seek the solution of eq.(2.5.37) in the interior of the medium in the form

$$E_i=\left[\varepsilon_0+\sum_{n=1}^{\infty}\left(\varepsilon_n^{(+)}e^{inkz}+\varepsilon_n^{(-)}e^{-inkz}\right)\right]e^{-i(\omega t+\Phi)}+c.c.,$$

where $\varepsilon_0=\varepsilon_0(z,t)$ and $\varepsilon_0^{(\pm)}=\varepsilon_0^{(\pm)}(z,t)$ are slow functions compared with the exponential $\exp[i(nkz\pm\omega t)]$. Equating the factors of the equal exponentials in both sides of (2.5.37), we get

$$\left[\frac{1}{c}\frac{\partial}{\partial t}\pm n\frac{\partial}{\partial z}+\frac{i\omega}{2c}(n^2-1)\right]\varepsilon_n^{(\pm)}=i\frac{2\pi\omega}{c}\int p_{\pm n}dv_z.$$

As a result of the integration of this equation we find that the quantities ε_0 and $\varepsilon_n^{(\pm)}$ at $n>1$ are small compared with $\varepsilon^{(\pm)}\equiv\varepsilon_1^{(\pm)}$ in the ratio $c/\omega L(n^2-1)<<1$, and can therefore be omitted. Thus, the induced field E_i is a superposition of two wave of frequency ω, propagating in opposite directions:

$$E_i = \varepsilon^{(+)} e^{i(kz-\omega t-\Phi)} + \varepsilon^{(-)} e^{-i(kz+\omega t+\Phi)} + c.c. \tag{2.5.41}$$

This means that the excited medium radiates in the positive and negative directions of the z-axis. The amplitudes $\varepsilon^{(+)}$ and $\varepsilon^{(-)}$ of the forward and backward waves that have emerged to the outside resonant medium are given by

$$\varepsilon^{(\pm)} = iL\frac{\omega^2}{c^2}\int_{-\infty}^{\infty} dv_z e^{\mp ikv_z(t\mp z/c)} \int_{-\pi c/\omega}^{\pi c/\omega} p(\xi, t\mp z/c) e^{\mp ik\xi} d\xi \tag{2.5.42}$$

The derived equations (2.5.40)-(2.5.42) are needed to determine the echo produced after successive excitation of a medium by several pulses of standing waves.

Now let two standing-wave pulses

$$E = 2a_1\cos(kz+\varphi_1)e^{-i(\omega t+\Phi_1)} + c.c., \quad 0 \le t \le T_1,$$
$$E = 2a_2\cos(kz+\varphi_2)e^{-i(\omega t+\Phi_2)} + c.c., \quad T_1+\tau_{12} \le t \le T_1+T_2+\tau_{12},$$

be successively excited in the medium. Here, $a_1, a_2, \varphi_1, \varphi_2, \Phi_1$ and Φ_2 are real constants. The durations T_1 and T_2 of the pulse are so short that the irreversible relaxation and detuning during these time intervals can be neglected in (2.5.39). The interval τ_{12} between the pulses, however, is large so that the irreversible relaxation and detuning during an interval of order τ_{12} must be taken into account.

When the first or second standing wave pulse is turned on, the state of the medium is described by expressions (2.5.40) with the appropriate boundary conditions, while in the absence of these pulses we solve equations (2.5.39) again with the allowance for relaxation and pumping at $a=0$. In the region $T_1+T_1+\tau \le t$ we determine only p. Next, using (2.5.42), we obtain the amplitudes of the forward and backward waves of the induced field (2.5.41). We retain only the terms that contribute to the photon echo. Finally, we obtain

$$\varepsilon^{(\pm)} = \varepsilon_0 T_0 \int_0^{\infty} e^{-(\eta T_0)^2} \sum_{m=2}^{\infty} G_m^{(\pm)} \left[\frac{1+(-1)^m}{2} I_m + \frac{1-(-1)^m}{2} K_m\right] d\eta,$$

$$G_m^{(\pm)} = J_{m-1}\left(\tfrac{2\Lambda_1}{\eta}\sin\tfrac{\eta T_1}{2}\right)\cos\left[\eta\left(t - \tfrac{(2\tau_{12}+T_1+T_2)m+T_1}{2}\right)\right] e^{\pm i[m(\varphi_2-\varphi_1)+\varphi_1]} \exp[(i\Delta-\gamma_{21})(t-\tau_{12})], \tag{2.5.43}$$

$$I_m = J_m\left(\tfrac{2\Lambda_2}{\eta}\sin\tfrac{\eta T_2}{2}\right)\cos(\Delta\tau_{12}+\Phi_2-\Phi_1)\exp(-\gamma_{21}\tau_{12}),$$

$$K_m = -\tfrac{1}{2}J_m\left(\tfrac{2\Lambda_2}{\eta}\sin\tfrac{\eta T_2}{2}\right)[\exp(-\gamma_1\tau_{12})+\exp(-\gamma_2\tau_{12})],$$

$$\varepsilon_0 = \sqrt{4\pi L N_0}\,|d_{ba}|\,\omega/c, \quad T_0 = 1/ku, \quad \Lambda_n = 4a_n|d_{ba}|/\hbar, \quad n=1,2,$$

where $J_m(\vartheta)$ is a Bessel function of order m and of argument ϑ, and the terms $\pm z/c$, which take into account the delay of the signals of the forward and backward waves, are left out for brevity ($t \mp z/c \to t$). When (2.5.43) is substituted in (2.5.41), it is necessary to make the reverse substitutions $t \to t \mp z/c$, and we must put $\Phi = \Phi_2$ in (2.5.41).

It follows from (2.5.43) that following the action of two standing wave-pulses the excited medium emits spontaneously, at approximate instants of time $t = m\tau_{12}$, where $m=2,3,...$, electromagnetic pulses in the form of two travelling waves that propagate in opposite directions (photon echoes). The properties of photon echoes with even numbers $m=2,4,...$ (even echoes) differ substantially from the case of odd echoes with the number $m=3,5,...$. In final analysis this is due to the fact that the even echoes are the result of the phase memory concerning the states of the medium in the interval $T_1 \le t \le \tau_{12} + T_1$, a memory stored in the matrix elements that make up the medium polarisation p. At the same time, the odd echoes are the result of the phase memory stored in the matrix elements which describe the populations of the upper and lower levels in the indicated time interval.

The amplitude of the even echoes contains a characteristic factor $\cos(\Delta\tau_{12} + \Phi_2 - \Phi_1)$ that causes narrow resonances of width $1/2\tau_{12}$ to appear in the radiation intensity when the frequency ω is scanned, if $\Phi_2 - \Phi_1$ is constant. The law of echo decay with a variance of the delay τ_{12} is determined only by relaxation parameter γ_{21}.

The odd echoes do not display narrow resonances and their decay laws depend also on relaxation parameters γ_1 and γ_2.

It is important to emphasise that the appearance of a number of even and odd echoes is due to atomic thermal motion and Doppler dephasing processes. In the case of solid state resonant medium the action of two standing-wave pulses produces echo signal only at moment of time $t = 2\tau_{12}$, since the mechanisms of inhomogeneous broadening of the spectral line work as in the case of travelling waves.

2.6. Photon echoes in three-level systems

The simultaneous or separate in time action on the resonant medium of two and more ultra short light pulses

$$E_1 = a_1 \exp[i(\vec{k}_1\vec{r} - \omega_1 t - \varphi_1)] + c.c. \tag{2.6.1}$$

$$E_2 = a_2 \exp[i(\vec{k}_2\vec{r} - \omega_2 t - \varphi_2)] + c.c. \tag{2.6.2}$$

whose carrier frequencies ω_1 and ω_2 are close to frequencies ω_{ab} and ω_{cb} of adjacent optically allowed transitions $E_a \to E_b$ and $E_c \to E_b$ is accompanied by a variety of optical transients such as various types of three-level echoes. All new types of photon echoes are based on the peculiarities of excitation of one of the transitions among three energy levels E_a, E_b and E_c while the other transitions keep optical memory about preceding excitations. This memory consists of occurrence of optical coherence on transitions not excited at once by the considered travelling pulse. In section 2.6.1 we discuss the mathematics of the optical coherence transfer from one transition to another by resonant excitation of third transition in three-levels of V-, Λ- and Θ-configuration (Fig.2.6.1, the Θ-configuration is also named as the cascade configuration). The peculiar kind of a photon echo in scattering of nonresonant probe wave by three-level systems excited by resonant pulses (2.6.1) and (2.6.2) separated in time is considered in section 2.6.2. The mechanism of optical memory discussed in section 2.6.2 and the related mathematics is applied in section 2.6.3 to describe the nontrivial three-level echoes originally named as tri-level echoes [49,50].

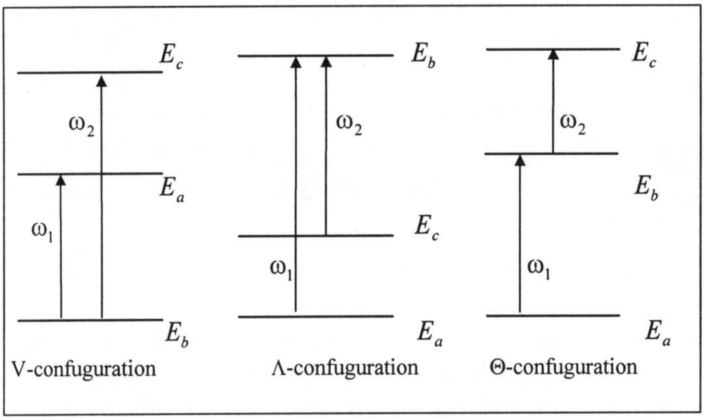

Fig.2.6.1. Configuration of energy levels in three-level system

2.6.1. BASIC EQUATIONS FOR THREE-LEVEL SYSTEMS IN THE GIVEN FIELD APPROXIMATION

The equations for slowly varying matrix elements of density matrix for three types of three-level configurations are written in the following form

$$\left(\frac{\partial}{\partial t} - i\Delta_1 + \gamma_{10}\right)R_{10} = i\Lambda_1(R_{00} - R_{11}) - i\Lambda_2 R_{12}, \tag{2.6.3a}$$

$$\left(\frac{\partial}{\partial t} - i\Delta_2 + \gamma_{20}\right)R_{20} = i\Lambda_2(R_{00} - R_{22}) - i\Lambda_1 R_{21}, \tag{2.6.3b}$$

$$\left(\frac{\partial}{\partial t} - i(\Delta_2 - \Delta_1) + \gamma_{21}\right)R_{21} = i\Lambda_2 R_{01} - i\Lambda_1^* R_{20}, \tag{2.6.3c}$$

$$\left(\frac{\partial}{\partial t}+\gamma_1\right)R_{11}=i(\Lambda_1 R_{01}-\Lambda_1^* R_{10})+\gamma_1 N_1, \tag{2.6.3d}$$

$$\left(\frac{\partial}{\partial t}+\gamma_2\right)R_{22}=i(\Lambda_2 R_{02}-\Lambda_2^* R_{20})+\gamma_2 N_2, \tag{2.6.3e}$$

$$\left(\frac{\partial}{\partial t}+\gamma_0\right)R_{00}=-i(\Lambda_1 R_{01}-\Lambda_1^* R_{10})-i(\Lambda_2 R_{02}-\Lambda_2^* R_{20})+\gamma_0 N_0. \tag{2.6.3f}$$

The introduced values are related to the parameters of three-level system and electromagnetic waves by the formulas

V-*configuration*

$$R_{10}=R_{01}^*=\rho_{ab}\exp[-i(\vec{k}_1\vec{r}-\omega_1 t-\varphi_1)],\quad R_{20}=R_{02}^*=\rho_{cb}\exp[-i(\vec{k}_2\vec{r}-\omega_2 t-\varphi_2)],$$
$$R_{21}=R_{12}^*=\rho_{ca}\exp\{-i[(\vec{k}_2\vec{r}-\omega_2 t-\varphi_2)-(\vec{k}_1\vec{r}-\omega_1 t-\varphi_1)]\},\quad R_{00}=R_{00}^*=\rho_{bb},$$
$$R_{11}=R_{11}^*=\rho_{aa},\ R_{22}=R_{22}^*=\rho_{cc},\quad \Delta_1=\omega_1-\omega_{ab},\quad \Delta_2=\omega_2-\omega_{cb},\quad \omega_{ab}=(E_a-E_b)/\hbar,$$
$$\omega_{cb}=(E_c-E_b)/\hbar,\quad \Lambda_1=a_1 d_{ab}/\hbar,\quad \Lambda_2=a_2 d_{cb}/\hbar.$$

Λ-*configuration*

$$R_{10}=R_{01}^*=\rho_{ab}\exp[i(\vec{k}_1\vec{r}-\omega_1 t-\varphi_1)],\quad R_{20}=R_{02}^*=\rho_{cb}\exp[i(\vec{k}_2\vec{r}-\omega_2 t-\varphi_2)],$$
$$R_{21}=R_{12}^*=\rho_{ca}\exp\{-i[(\vec{k}_1\vec{r}-\omega_1 t-\varphi_1)-(\vec{k}_2\vec{r}-\omega_2 t-\varphi_2)]\},\quad R_{00}=R_{00}^*=\rho_{bb},$$
$$R_{11}=R_{11}^*=\rho_{aa},\quad R_{22}=R_{22}^*=\rho_{cc},\quad \Delta_1=-(\omega_1-\omega_{ba}),\quad \Delta_2=-(\omega_2-\omega_{bc}),$$
$$\omega_{ba}=(E_b-E_a)/\hbar,\quad \omega_{bc}=(E_b-E_c)/\hbar,\quad \Lambda_1=a_1^* d_{ba}^*/\hbar,\quad \Lambda_2=a_2^* d_{bc}^*/\hbar.$$

Θ-*configuration*

$$R_{10}=R_{01}^*=\rho_{ab}\exp[i(\vec{k}_1\vec{r}-\omega_1 t-\varphi_1)],\quad R_{20}=R_{02}^*=\rho_{cb}\exp[-i(\vec{k}_2\vec{r}-\omega_2 t-\varphi_2)],$$
$$R_{21}=R_{12}^*=\rho_{ca}\exp\{-i[(\vec{k}_1\vec{r}-\omega_1 t-\varphi_1)+(\vec{k}_2\vec{r}-\omega_2 t-\varphi_2)]\},\quad R_{00}=R_{00}^*=\rho_{bb},$$
$$R_{11}=R_{11}^*=\rho_{aa},\quad R_{22}=R_{22}^*=\rho_{cc},\quad \Delta_1=-(\omega_1-\omega_{ba}),\quad \Delta_2=\omega_2-\omega_{cb},\quad \omega_{cb}=(E_c-E_b)/\hbar,$$
$$\omega_{ba}=(E_b-E_a)/\hbar,\quad \Lambda_1=a_1^* d_{ba}^*/\hbar,\quad \Lambda_2=a_2 d_{cb}/\hbar.$$

The other parameters in eqs. (2.6.3) are the relaxation constants γ_0, γ_1, γ_2, γ_{10}, γ_{20} and γ_{21} and stationary populations N_0, N_1 and N_2 of energy levels in the absence of external fields. We will use the following matrix form of R:

$$R=\begin{pmatrix} R_{00} & R_{01} & R_{02} \\ R_{10} & R_{11} & R_{12} \\ R_{20} & R_{21} & R_{22} \end{pmatrix}.$$

In accordance with the consideration of two level systems it is convenient to write the general solution of (2.6.3) in the given field approximation as

$$R(t) = U(t,t_0)R(t_0)U^+(t,t_0) \tag{2.6.4}$$

where for ultrashort pulses (2.6.1) and (2.6.2) the evolution operator obeys the equation

$$\frac{\partial}{\partial t}U(t,t_0) = -iH_{eff}U(t,t_0) \tag{2.6.5}$$

with initial condition

$$U(t,t_0) = I$$

and with effective slowly varying Hamiltonian

$$H_{eff} = \begin{pmatrix} (\Delta_1+\Delta_2)/4 & -\Lambda_1^* & -\Lambda_2^* \\ -\Lambda_1 & (\Delta_2-3\Delta_1)4 & 0 \\ -\Lambda_2 & 0 & (\Delta_1-3\Delta_2)/4 \end{pmatrix}.$$

The solution of (2.6.5) becomes very simple for constant values of field amplitudes and $\Delta_1 = \Delta_2$:

$$U(t,t_0) = \tag{2.6.6}$$

$$= \begin{pmatrix} \cos\Xi - i\dfrac{\Delta}{\Omega}\sin\Xi & \dfrac{2i\Lambda_1^*}{\Omega}\sin\Xi & \dfrac{2i\Lambda_2^*}{\Omega}\sin\Xi \\ \dfrac{2i\Lambda_1}{\Omega}\sin\Xi & \dfrac{|\Lambda_2|^2 e^{i\zeta} + |\Lambda_1|^2\left(\cos\Xi + i\dfrac{\Delta}{\Omega}\sin\Xi\right)}{|\Lambda_1|^2+|\Lambda_2|^2} & \dfrac{\Lambda_1\Lambda_2^*\left(-e^{i\zeta}+\cos\Xi + i\dfrac{\Delta}{\Omega}\sin\Xi\right)}{|\Lambda_1|^2+|\Lambda_2|^2} \\ \dfrac{2i\Lambda_2}{\Omega}\sin\Xi & \dfrac{\Lambda_2\Lambda_1^*\left(-e^{i\zeta}+\cos\Xi + i\dfrac{\Delta}{\Omega}\sin\Xi\right)}{|\Lambda_1|^2+|\Lambda_2|^2} & \dfrac{|\Lambda_1|^2 e^{i\zeta} + |\Lambda_2|^2\left(\cos\Xi + i\dfrac{\Delta}{\Omega}\sin\Xi\right)}{|\Lambda_1|^2+|\Lambda_2|^2} \end{pmatrix}.$$

Here, we have introduced the values

$$\Delta = \Delta_1 = \Delta_2, \quad \Omega = \sqrt{\Delta^2 + 4|\Lambda_1|^2 + 4|\Lambda_2|^2}, \quad \Xi = \frac{\Omega(t-t_0)}{2}, \quad \zeta = \frac{\Delta(t-t_0)}{2}.$$

Another important case of general solution of (2.6.5) is written for arbitrary detunings Δ_1 and Δ_2 but for exciting field at only one carrier frequency, e.g. $\Lambda_1 = 0$:

$$U(t,t_0) = \exp[i(\Delta_2 - \Delta_1)(t-t_0)/4] \cdot$$
$$\cdot \begin{pmatrix} \cos\Xi_2 - i\dfrac{\Delta_2}{\Omega_2}\sin\Xi_2 & 0 & \dfrac{2i\Lambda_2^*}{\Omega_2}\sin\Xi_2 \\ 0 & \exp[i(\Delta_1 - \Delta_2/2)(t-t_0)] & 0 \\ \dfrac{2i\Lambda_2}{\Omega_2}\sin\Xi_2 & . \ 0 & \cos\Xi_2 + i\dfrac{\Delta_2}{\Omega_2}\sin\Xi_2 \end{pmatrix}. \qquad (2.6.7a)$$

Here,

$$\Omega_2 = \sqrt{\Delta_2^2 + 4|\Lambda_2|^2}, \quad \Xi_2 = \frac{\Omega_2(t-t_0)}{2}.$$

Solution of (2.6.3) in the form (2.6.4) for the case $\Lambda_1 \neq 0$ but $\Lambda_2 = 0$ can be obtained from (2.6.7a) by simple substitution of indices:

$$U(t,t_0) = \exp[i(\Delta_1 - \Delta_2)(t-t_0)/4] \cdot$$
$$\cdot \begin{pmatrix} \cos\Xi_1 - i\dfrac{\Delta_1}{\Omega_1}\sin\Xi_1 & \dfrac{2i\Lambda_1^*}{\Omega_1}\sin\Xi_1 & 0 \\ \dfrac{2i\Lambda_1}{\Omega_1}\sin\Xi_1 & \cos\Xi_1 + i\dfrac{\Delta_1}{\Omega_1}\sin\Xi_1 & 0 \\ 0 & 0 & \exp[i(\Delta_2 - \Delta_1/2)(t-t_0)] \end{pmatrix} \qquad (2.6.7b)$$

with

$$\Omega_1 = \sqrt{\Delta_2^2 + 4|\Lambda_1|^2}, \quad \Xi_1 = \frac{\Omega_1(t-t_0)}{2}.$$

The common phase factor in eqs.(2.6.7) is inessential for using in the formula (2.6.4).

The main result of separate in time action of pulses (2.6.1) and (2.6.2) under condition of negligible relaxation is considered in the case of V-configuration of resonant levels and the following sequence of pulse excitation:

$$E = \begin{cases} a_1 \exp[i(k_1 z - \omega_1 t)] + c.c., & 0 \le t - z/c \le T_1 \\ 0, & T_1 < t - z/c < T_1 + \tau_{12} \\ a_2 \exp[i(k_2 z - \omega_2 t - \varphi_2)] + c.c., & T_1 + \tau_{12} \le t - z/c \end{cases}, \qquad (2.6.8)$$

where E is the electric field strength acting on resonant atoms.

We write the solution for the matrix R in the region $T_1+\tau_{12}<t$ as

$$R(t)=\begin{pmatrix} \widetilde{R}_{00}\cos^2\Xi_2'+\widetilde{R}_{22}\sin^2\Xi_2' & \widetilde{R}_{01}\cos\Xi_2' & -i\dfrac{\widetilde{R}_{00}-\widetilde{R}_{22}}{2}\sin 2\Xi_2' \\ \widetilde{R}_{10}\cos\Xi_2' & \widetilde{R}_{11} & -i\widetilde{R}_{10}\sin\Xi_2' \\ i\dfrac{\widetilde{R}_{00}-\widetilde{R}_{22}}{2}\sin 2\Xi_2' & i\widetilde{R}_{01}\sin\Xi_2' & \widetilde{R}_{00}\sin^2\Xi_2'+\widetilde{R}_{22}\cos^2\Xi_2' \end{pmatrix}.$$

We denote the initial value of density matrix $R(\tau_{12}+T_1)$ before the excitation by the second pulse as $\widetilde{R}$ and $\Xi_2'=\Lambda_2(t-T_1-\tau_{12})$. We take for simplicity $\Delta_1=\Delta_2=\Delta=0$ and $\Lambda_2=\Lambda_2^*$.

Before the excitation, three-level atoms populate energy levels E_b, E_a and E_c, so the only matrix elements R_{00}, R_{11} and R_{22} are distinct from zero. The interaction of the first exciting pulses in time interval $0\le t\le T_1$ with resonant transition $E_a\to E_b$ induces the optical coherence in this transition $R_{01}=R_{10}^*\neq 0$. The other transitions as adjacent optically allowed transition $E_c\to E_b$ and optically forbidden transition $E_c\to E_a$ remain unperturbed. Therefore, before the action of the second exciting pulse at $t=T_1+\tau_{12}$ only the mentioned above matrix elements can be nonzero. If the action of the second exciting pulse take place after a long time interval τ_{12} compared with the relaxation times, only optical coherence on the resonant transition $E_c\to E_b$ will be induced whereas the optical coherence on $E_a\to E_b$ and $E_c\to E_a$ remains zero. This case corresponds to the independent action of excitation pulses. For a short time interval τ_{12} between the exciting pulses the action of the second pulse induces not only optical coherence on the resonant $E_c\to E_b$ -transition but also excites the optically forbidden two-photon transition $E_c\to E_a$:

$$R_{21}(t)=iR_{01}(\tau+T_1)\sin\Lambda_2(t-\tau_{12}-T_1). \tag{2.6.9}$$

Thus, only optical coherence of optically forbidden transition $E_c\to E_a$ contains the memory about optical coherence of $E_a\to E_b$ -transition at the excitation (2.6.8). This circumstance determines all new coherent phenomena produced by the excitation of adjacent optically allowed transition in three-level system by a sequence of resonant pulses.

Another feature of separated in time excitation of three-level systems consists in the formation of an optical nutation echo of the same type as in the section 2.5.3.

2.6.2. RAMAN SCATTERING BY TWO-PULSE EXCITATION OF THREE-LEVEL SYSTEMS

The results of a previous section allow us to obtain in detail the description of the Raman scattering of a probe nonresonant wave

$$E' = a'\exp[i(\vec{k}'\vec{r} - \omega' t)] + c.c. \tag{2.6.10}$$

in the medium excited by two ultrashort pulses

$$E_1 = a_1\exp[i(k_1 z - \omega_1 t - \varphi_1)] + c.c., \qquad 0 \le t - z/c \le T_1, \tag{2.6.11}$$

$$E_2 = a_2\exp[i(k_2 z - \omega_2 t - \varphi_2)] + c.c., \qquad T_1 + \tau_{12} \le t - z/c \le T_1 + T_2 + \tau_{12} \tag{2.6.12}$$

which are resonant with the two adjacent optically allowed transitions. Below we will consider the case of V-configuration of these resonant transitions: $E_a \to E_b$ ($\omega_1 \approx (E_a - E_b)/\hbar$) and $E_c \to E_b$ ($\omega_2 \approx (E_c - E_b)/\hbar$). For simplicity, the propagation of exciting pulses is taken along z-axis and polarisation effects will be neglected.

The induced optical coherence

$$\rho_{ca} = R_{ca}\exp\{i[(k_2 - k_1)z - (\omega_2 - \omega_1)t - (\varphi_2 - \varphi_1)]\}$$

of two-photon transition $E_c \to E_a$ in the time interval $\tau_{12} + T_1 + T_2 \le t - z/c$ determines in accordance with (1.1.25) the additional polarisation of the medium at frequencies $|\omega' \pm (\omega_2 - \omega_1)|$

$$P' = \int dv_z \rho_{ca} D_{ac}' + c.c., \tag{2.6.13}$$

where for definiteness we consider the resonant medium as a gas and the density matrix ρ_{ca} describes atoms with the velocity projections at z-axis from v_z to $v_z + dv_z$. This matrix obeys equations (2.6.3) with substitutions $\Delta_1 \to \Delta_1 - k_1 v_z$ and $\Delta_2 \to \Delta_2 - k_2 v_z$. The matrix element D_{ac}' of the effective dipole momentum is given by

$$D_{ac}' = a'\Pi^*_{ca}(-\omega')\exp[i(\vec{k}'\vec{r} - \omega' t - \varphi')] + a'^*\,\Pi^*_{ca}(\omega')\exp[-i(\vec{k}'\vec{r} - \omega' t - \varphi')],$$

and the parameter $\Pi_{ca}(-\omega')$ of two-photon transition is defined in (1.1.11).

We rewrite (2.6.13) as

$$\begin{aligned} P' = {} & p_+\exp\{i[(\vec{k}_2 - \vec{k}_1 + \vec{k}')\vec{r} - (\omega_2 - \omega_1 + \omega')t - (\varphi_2 - \varphi_1 + \varphi')]\} + \\ & + p_-\exp\{i[(\vec{k}_2 - \vec{k}_1 - \vec{k}')\vec{r} - (\omega_2 - \omega_1 - \omega')t - (\varphi_2 - \varphi_1 - \varphi')]\} \end{aligned} \tag{2.6.14}$$

where

$$p_+ = \sin(\Lambda_1 T_1)\sin(\Lambda_2 T_2)a'\Pi^*_{ca}(-\omega')f(t'),\ p_- = \sin(\Lambda_1 T_1)\sin(\Lambda_2 T_2)a'^*\Pi^*_{ca}(\omega')f(t'),$$

$$f(t) = N_0 \exp[i((\Delta_2 - \Delta_1)(t-\tau_{12}) - \Delta_1\tau_{12})]\exp[-(k_2-k_1)^2u^2(t-t_e)^2/4],$$

$$t' = t - z/c,\ t_e = k_2\tau_{12}/(k_2-k_1).$$

We omit all the questions related with the phase synchronism and we assume a probe wave (2.6.10) propagating along z-axis as exciting pulses (2.6.11) and (2.6.12). The result is obtained for narrow spectral lines [51].

Polarisation (2.6.14) determines the Stokes and anti-Stokes scattering of a nonresonant wave (2.6.10). The signals of the Stokes and anti-Stokes scattered waves appear at the moment of time $t=t_e$ .The unusual time t_e of the appearance of scattered wave points to the similarity of the discussed phenomenon with the photon echo. The other echo phenomena in the three-level system will be reviewed in the next section.

2.6.3. TRI-LEVEL ECHOES

The further use of induced coherence on optically forbidden transition (2.6.9) can lead to new interesting phenomena of echo-type. Many type of these echoes may be produced only in gas medium [50] since the necessary mechanism of reversible relaxation is due to atomic thermal motion. Below we will discuss only the case of resonant gas medium.

If we consider the action of third pulse after two pulses (2.6.11) and (2.6.12) of the same frequency as the second

$$E_3 = a_3 \exp[i(k_2 z - \omega_2 t - \varphi_3)] + c.c.,$$
$$T_1 + T_2 + \tau_{12} + \tau_{23} \le t - z/c \le T_1 + T_2 + T_3 + \tau_{12} + \tau_{23}$$

we find that the phase memory about optical coherence of $E_c \to E_a$ -transition is displayed in polarisation of optically allowed transition $E_a \to E_b$. That provides the formation of echo signal at carrier frequency ω_1. The moment and the condition of the echo formation are

$$t_e = T_1 + T_2 + T_3 + \frac{\omega_{cb}}{\omega_{ab}}\tau_{23} > T_1 + T_2 + T_3 + \tau_{12} + \tau_{23}.$$

Thus the condition of tri-level echo formation is determined not only by usual assumption about correlation between characteristic time intervals but and by the ratio of transition frequencies. The electric field of this echo is given by

$$E = \varepsilon_e(t - z/c)\exp[i(k_1 z - \omega_1 t - \varphi_e)] + c.c., \tag{2.6.15}$$

where

$$\varepsilon_e(t) = 8\widetilde{E}_0 \int d\vec{v} f(v) \frac{\Lambda_1\Lambda_2\Lambda_3}{\Omega_1\Omega_2\Omega_3} X_1^* X_2 X_3 \exp[-ik_1 v_z (t - t'_e)] \cdot$$

$$\cdot \exp[\Delta_1 (t - \tau_{23}\frac{\Delta_2}{\Delta_1} - T_1) - \Delta_2(T_2 + T_3)/2] \exp[-\gamma_{10}\tau_{23}(\frac{\omega_{cb}}{\omega_{ab}} - 1)] \exp(-\gamma_{21}\tau_{23}),$$

$$X_1 = \left[\cos\frac{\Omega_1 T_1}{2} - i\frac{\Delta_1 - k_1 v_z}{\Omega_1}\sin\frac{\Omega_1 T_1}{2}\right]\sin\frac{\Omega_1 T_1}{2}, \quad X_n = \frac{1}{\Omega_n}\sin\frac{\Omega_n T_n}{2}, \; n = 2,3,$$

$$\Omega_3 = \sqrt{\Delta_2^2 + 4|\Lambda_3|^2}, \quad \Lambda_3 = a_3 d_{cb}/\hbar, \quad \widetilde{E}_0 = 2\pi |d_{ba}| L(N_b - N_a)\omega_{ab}/c,$$

$$t'_e = T_3 + \frac{\omega_{cb}}{\omega_{ab}}\tau_{23} - \frac{\omega_{cb}}{2\omega_{ab}}(T_1 + T_2), \quad \varphi_e = \varphi_3 - \varphi_2 + \varphi_1,$$

We can represent graphically the formation of the echo as it is done in Fig. 2.6.2 a). The arrows indicate the directions of the propagation of exciting pulses, the rectangles show the pulse and corresponding resonant transition.

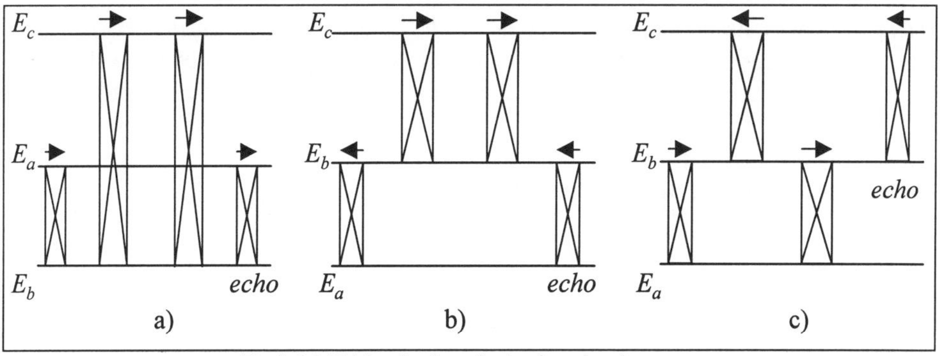

Fig.2.6.2. Tri-level echoes in the three-level systems

In three-level systems with θ-configuration of resonant levels it is convenient to produce photon echoes by pulses that propagate in opposite directions as is shown in Fig. 2.6.2 b) and c). When exciting pulses are given by

$$E_1 = a_1\exp[-i(k_1 z + \omega_1 t - \varphi_1)] + c.c., \quad 0 \le t - z/c \le T_1,$$

$$E_2 = a_2\exp[i(k_2 z - \omega_2 t - \varphi_2)] + c.c., \quad T_1 + \tau_{12} \le t - z/c \le T_1 + T_2 + \tau_{12}$$

$$E_3 = a_3\exp[i(k_2 z - \omega_2 t - \varphi_3)] + c.c., T_1 + T_2 + \tau_{12} + \tau_{23} \le t - z/c \le T_1 + T_2 + T_3 + \tau_{12} + \tau_{23}$$

the tri-level echo is formed at instant

$$t_e = T_1 + T_2 + T_3 + \frac{\omega_{cb}}{\omega_{ba}}\tau_{23}$$

and propagates at the same carrier frequency and in the same direction as the first exciting pulse. In the case of a large spectral line the tri-level echo appears later than the moment t_e at a value of order of pulse duration.

Another set of exciting pulses (Fig. 2.6.2 c))

$$E_1 = a_1\exp[i(k_1 z - \omega_1 t - \varphi_1)] + c.c., \quad 0 \le t - z/c \le T_1,$$

$$E_2 = a_2\exp[-i(k_2 z + \omega_2 t - \varphi_2)] + c.c., \quad T_1 + \tau_{12} \le t - z/c \le T_1 + T_2 + \tau_{12}$$

$$E_3 = a_3\exp[i(k_1 z - \omega_1 t - \varphi_3)] + c.c.,$$

$$T_1 + T_2 + \tau_{12} + \tau_{23} \le t - z/c \le T_1 + T_2 + T_3 + \tau_{12} + \tau_{23}$$

produces tri-level echo at time instant

$$t_e = T_1 + T_2 + T_3 + \tau_{12} + \frac{\omega_{ba}}{\omega_{cb}}(\tau_{12} + \tau_{23})$$

for narrow spectral lines and a little later for large spectral lines. The direction and carrier frequency of its propagation coincides with the direction and carrier frequency of propagation of the second exciting pulse. We discuss the properties of this echo more detail in item 2.7.2 on account of level degeneracy and polarisation of exciting pulses.

2.7. Polarisation properties of coherent transients

The real experiments on coherent transients were made in the media with resonant levels, which are degenerate on the various orientation of the total angular momentum [3,13,33,34,37,42,44,49,52-58]. The examples of resonant media are molecular gas SF_6, atomic vapours of alkaline elements, impurity crystals with rare earth elements and others. The resonant levels of SF_6 are highly degenerate, whereas in other mentioned cases the level degeneracy is defined by small values (1/2 and 3/2) of the total angular momentum. Their correct description is based on the formalism developed in item 1.4. This formalism allows also investigating theoretically polarisation properties of coherent transients. When exciting pulses are different polarisation states, the polarisation of echo phenomena displays new features. These features are useful both in the experimental technique for isolating echo signals from exciting pulses and in spectroscopic purposes for identifying atomic and molecular transitions, the investigation of depolarising atomic collisions and other relaxation properties. Interesting possibilities are being opened for studying fine and superfine interactions by the photon echo technique. Another way in the polarisation studies represents the formation of echo phenomena in external magnetic field. In section 2.7.1 we discuss polarisation properties of optical nutation and nutation echo produced by the phase switching method. In section 2.7.2 we describe polarisation of echo phenomena in a three-level gas medium on account of depolarising atomic collisions.

2.7.1. POLARISATION FEATURES OF OPTICAL NUTATION BY PHASE SWITCHING METHOD

Many investigations are shown that the main polarisation features of optical nutation are independent from the method of their formation. Below we will consider coherent transients produced by a phase switching method. According to this method an additional term $\varphi(t)$ appears in the light wave phase which in time interval $0 \le t \le \tau_1$ equals $\delta_1 t$, in the adjacent region $\tau_1 \le t \le \tau$ it is constant $\varphi(t) = \delta_1\tau_1$, then it linearly decreases $\varphi(t) = \delta_1\tau_1 + \delta_2(t - \tau)$. Simultaneously with the increase and decrease of the light wave phase, instantaneous light wave frequency shifts $\omega + d\varphi / dt$ occur in opposite directions with respect to its initial value ω. Therefore, there exist three atomic groups within the Doppler-broadened atomic line shape (see Fig. 2.7.1). One of these includes atoms with velocities $v \sim (\omega - \omega_0)c / \omega$ (atomic group v) which are at first in resonance with a light wave of frequency ω. Here, ω_0 is the frequency of resonant atomic transition and v is the projection of the atomic velocity along the direction of the light wave propagation. After the light wave frequency switches $\omega \to \omega_1 = \omega + \delta_1$ in the time interval $0 \le t \le \tau_1$, the atomic group v goes off resonance. Simultaneously the second group of atoms with velocities $v_1 \sim (\omega_1 - \omega_0)c / \omega_1$ (atomic group v_1) comes into resonance with the light wave of the new frequency ω_1 . In the time interval $\tau_1 \le t \le \tau$ the atomic group v interacts again with a light wave owing to the frequency switch $\omega_1 \to \omega$. After the repeated light wave frequency switching $\omega \to \omega_2 = \omega + \delta_2$ ($\delta_2 \ne \delta_1$) the new group of atoms with velocities $v_2 \sim (\omega_2 - \omega_0)c / \omega_2$ (atomic group v_2) turns out to be in resonance with a light wave of instantaneous frequency ω_2. In the time interval $\tau + \tau_2 \le t$ both the phase and frequency of the light wave takes their initial values. Thus, the atomic group v is in resonance with a light wave of frequency ω which is turned off for two short time intervals. After each turning on (turning off) of the light wave the atomic group v exhibits the nutation (free induction) effect, and after the repeated turning on (turning off) of the light wave the atomic group v emits, under certain conditions, a nutation echo (induction echo). On the other hand, the atomic group v_1 (atomic group v_2) is irradiated by one ultrashort light pulse of frequency ω_1 (ω_2) and of duration τ_1 (τ_2). Therefore, atomic groups v_1 and v_2 emit no echo signals and, under certain conditions, they do not contribute to the echo signals accompanying the phase shift of a light wave. Thus, echo signals produced by the method of light wave phase switching are due to the radiation of atomic group v only, that is why they are of a simple nature.

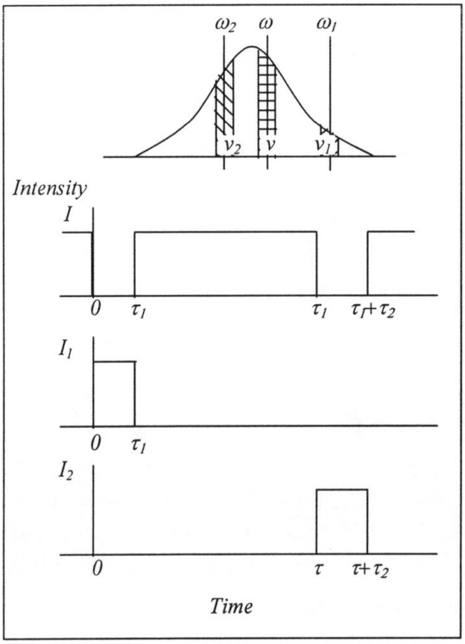

Fig.2.7.1. Interaction of atomic gas with light wave (1). Upper curve: the three atomic groups within a Doppler-broadened atomic line shape. The three lower curves show the intensities I, I_1 and I_2 of the light waves of frequencies ω, $\omega_1=\omega+\delta_1$ and $\omega_2=\omega+\delta_2$ interacting with atomic group v, v_1 and v_2, respectively. Here, and are frequency shift accompanying the phase shift of a light wave.

We assume that the gaseous medium is situated within the boundaries $z_0 \le z \le z_0 + L$ and that it contains two-level atoms (molecules) having upper and lower degenerate levels with angular momenta j_b and j_a and energies E_b and E_a, respectively. In this medium the incident light wave is changed due to resonant interaction with atoms, and a complicated electric field

$$\vec{E} = \vec{l}\mathcal{E}\exp[i(kz - \omega t - \varphi(t - z/c))] + c.c. \tag{2.7.1}$$

arises, where $\vec{l} = \vec{e}_x$ for linear polarisation, $\vec{l} = 2^{-1/2}(\vec{e}_x + i\vec{e}_y)$ for right-handed and $\vec{l} = -2^{-1/2}(\vec{e}_x - i\vec{e}_y)$ for left-handed circular polarisations, $\vec{e}_x$ and $\vec{e}_y$ are unit vectors along the indicated Cartesian axes. A slowly varying complex function $\mathcal{E} = \mathcal{E}(z - z_0, t - z/c)$ in the resonance approximation obeys the equations

$$\left(\frac{\partial}{\partial z} + \frac{1}{c}\frac{\partial}{\partial t}\right)\mathcal{E} = 2\pi i \int \sum_{\mu m} R_{\mu m} \vec{d}_{m\mu} \vec{l}^{\,*} d\eta, \tag{2.7.2}$$

$$\left(\frac{\partial}{\partial t}+i(\eta-\Delta)+\gamma_{21}\right)R_{\mu m}=\frac{i}{\hbar}\mathcal{E}\vec{l}\,(\sum_{m'}\vec{d}_{\mu m'}\rho_{m'm}-\sum_{\mu'}\rho_{\mu\mu'}\vec{d}_{\mu' m}),$$

$$\left(\frac{\partial}{\partial t}+\gamma_1\right)R_{mm'}=\frac{i}{\hbar}(\sum_{\mu}\vec{d}_{m\mu}R_{\mu m'}\mathcal{E}^*\vec{l}^*-R_{m\mu}\vec{d}_{\mu m'}\mathcal{E}\vec{l})+\frac{\gamma_1 N_a}{2j_a+1}f(\eta)\delta_{mm'}, \qquad (2.7.3)$$

$$\left(\frac{\partial}{\partial t}+\gamma_2\right)R_{\mu\mu'}=\frac{i}{\hbar}(\sum_{m}\vec{d}_{\mu m}R_{m\mu'}\mathcal{E}\vec{l}-R_{\mu m}\vec{d}_{m\mu'}\mathcal{E}^*\vec{l}^*)+\frac{\gamma_2 N_b}{2j_b+1}f(\eta)\delta_{\mu\mu'},$$

where

$$R_{\mu m}=\rho_{\mu m}\exp[-i(kz-\omega t-\varphi(t-z/c))],$$

$$f(\eta)=\frac{1}{\sqrt{\pi}u}\exp(-\eta^2T_0^2),\quad \eta=kv,\quad T_0=1/ku,$$

$$\Delta=\omega-\omega_0+d\varphi/dt,\quad \omega_0=(E_b-E_a)/\hbar.$$

We write the solution of equations (2.7.2) and (2.7.3) as

$$\mathcal{E}=\mathcal{E}_0+\varepsilon,\quad R_{\mu m}=R^0_{\mu m}+r_{\mu m},\quad \rho_{mm'}=\rho^0_{mm'}+r_{mm'},\quad \rho_{\mu\mu'}=\rho^0_{\mu\mu'}+r_{\mu\mu'}$$

where the first terms describe the partial solutions of equations (2.7.2) and (2.7.3) in the stationary regime, and the last terms represent the deviation from this stationary regime.

If we take the quantisation axis parallel to $\vec{l}$ in the case of linear polarisation and along the direction of propagation of wave (2.7.1) in the case of circular polarisation, the stationary regime for a constant value of the frequency shift will be

$$R^0_{\mu m}=(-1)^{j_b-m}\begin{pmatrix} j_b & 1 & j_a\\ -\mu & q & m\end{pmatrix}\frac{\mathcal{E}_0 d_{ba}}{\hbar}N_0 f(\eta)\frac{\eta-\Delta+i\gamma_{21}}{(\eta-\Delta)^2+\gamma_{21}^2+\Omega_0^2},$$

$$\rho^0_{mm'}=\frac{N_a f(\eta)}{2j_a+1}\delta_{mm'}-N_0 f(\eta)\frac{\gamma_2}{\gamma_1+\gamma_2}\frac{\Omega_0^2\delta_{mm'}}{(\eta-\Delta)^2+\gamma_{21}^2+\Omega_0^2},$$

$$\rho^0_{\mu\mu'}=\frac{N_b f(\eta)}{2j_b+1}\delta_{\mu\mu'}+N_0 f(\eta)\frac{\gamma_1}{\gamma_1+\gamma_2}\frac{\Omega_0^2\delta_{\mu\mu'}}{(\eta-\Delta)^2+\gamma_{21}^2+\Omega_0^2},$$

$$N_0=\frac{N_a}{2j_a+1}-\frac{N_b}{2j_b+1},\quad \Omega_0^2=2\begin{pmatrix} j_b & 1 & j_a\\ -\mu & q & m\end{pmatrix}^2\left|\frac{\mathcal{E}_0 d_{ba}}{\hbar}\right|^2\frac{\gamma_{21}(\gamma_1+\gamma_2)}{\gamma_1\gamma_2},$$

where $q=0$ for linear polarisation, $q=1$ for right-handed and $q=-1$ for left-handed circular polarisations.

The stationary amplitude $\mathcal{E}_0 = \mathcal{E}_0(z - z_0)$ is the solution of the equation

$$\frac{d\mathcal{E}_0}{dz} = 2\pi i \sum_{\mu m} \int R^0_{\mu m} \vec{d}_{m\mu} \vec{l}^{\,*} d\eta$$

with boundary condition $\mathcal{E}_0(0) = a$ at the point $z = z_0$ of entry into the resonant gaseous medium. Here a is the real constant amplitude of the incident light wave.

We shall consider small deviations from the stationary regime assuming the following inequality

$$|N_0| L T_0 |d_{ba}|^2 \omega / \hbar c << 1$$

to be satisfied, where L is the length of an optically thin resonant medium. In this case the stationary intensity

$$I_0 = c|\mathcal{E}_0(z - z_0)|^2 / 2\pi$$

changes little over the length L and the deviation matrix $r=r(t)$ can be obtained as

$$r(t) = U(t,t_0) r(t_0) U^+(t,t_0)\,. \tag{2.7.4}$$

In this expression, t_0 is an initial moment of time and the matrix elements of the evolution operator $U(t,t_0)$ are given by [36,59,60]

$$\begin{gathered} U_{mm'}(t,t_0) = A_m \delta_{mm'}, \quad U_{\mu\mu'}(t,t_0) = A^*_\mu \delta_{\mu\mu'}, \\ U_{\mu m}(t,t_0) = i(-1)^{j_b - m} \frac{d_{ba}}{|d_{ba}|} B_m \delta_{\mu,m+q}, \quad U_{m\mu}(t,t_0) = i(-1)^{j_b - m} \frac{d^*_{ba}}{|d_{ba}|} B_\mu \delta_{\mu,m+q}, \end{gathered} \tag{2.7.5}$$

where

$$A_\sigma = \cos\tfrac{1}{2}\Omega_\sigma(t - t_0) + i\frac{\eta - \Delta}{\Omega_\sigma}\sin\tfrac{1}{2}\Omega_\sigma(t - t_0), \; B_\sigma = (\chi_\sigma / \Omega_\sigma)\sin\tfrac{1}{2}\Omega_\sigma(t - t_0),$$

$$\Omega_\sigma = \sqrt{(\eta - \Delta)^2 + \chi_\sigma^2}, \quad \sigma = \mu, m,$$

$$\chi_m = \frac{2a|d_{ba}|}{\hbar}\begin{pmatrix} j_b & j_a & 1 \\ m+q & -m & -q \end{pmatrix}, \; \chi_\mu = \frac{2a|d_{ba}|}{\hbar}\begin{pmatrix} j_b & j_a & 1 \\ \mu & q-\mu & -q \end{pmatrix}$$

The intensity I of wave (2.7.1) after its emergence from the resonant medium is

$$I = \frac{c}{2\pi}\left|\mathcal{E}_0(L)\right|^2\left(1 + \frac{2\varepsilon'(t - z/c)}{a}\right) \tag{2.7.6}$$

where $\varepsilon'(t)$ is the real part of the amplitude ε

$$\varepsilon'(t) = \mathrm{Re}\left(2\pi Li\sum_{\mu m}\int r_{\mu m}(t)\vec{d}_{m\mu}\vec{l}^{\,*}d\eta\right). \tag{2.7.7}$$

Assuming the inequalities

$$\gamma_{21}(\tau + \tau_2) << 1, \quad \gamma_1(\tau + \tau_2) << 1, \quad \gamma_2(\tau + \tau_2) << 1, \quad \gamma_{21} << 2\Lambda$$

to be satisfied, we shall henceforth be interested mainly in coherent optical transient phenomena which are within a small time interval compared with the time of irreversible relaxation. For simplicity we shall consider the instantaneous frequency $\omega + d\varphi/dt$ of wave (2.7.1) to be leaving a Doppler-broadened atomic line shape during the time intervals $0 \le t \le \tau_1$ and $\tau \le t \le \tau + \tau_2$,

$$\left|\Delta_n\right| >> 1/T_0, \quad \left|\Delta_n\right| >> 2\Lambda, \quad \Delta_n = \omega + \delta_n - \omega_0, \quad n = 1, 2.$$

The value 2Λ characterises the order of the nutation frequency

$$\Lambda = \frac{a\left|d_{ba}\right|}{\hbar\sqrt{2j_b + 1}} = \frac{a}{2}\left(\frac{3\gamma c^3}{\hbar\omega_0^3}\right)^{1/2}.$$

Thus, we think of the atomic groups v_1 and v_2 as being absent, while atomic group v turns out to be irradiated by light wave (2.7.1) which is turned off in the time intervals $0 \le t - z/c \le \tau_1$ and $\tau \le t - z/c \le \tau + \tau_2$. In these time intervals an atomic group v radiates free induction signals. In other time intervals, a wave (2.7.1) interacts resonantly with gas atoms and the intensity (2.7.6) exhibits nutation oscillations which are due to re-radiation of photons by resonant atoms. The amplitude (2.7.7) of this optical nutation in the region $0 \le t \le \tau_1$ after a single phase shift is given by

$$\varepsilon'(t) = \varepsilon_0\sum_{\mu=-j_b}^{j_b}\mathrm{X}_\mu\int_{-\infty}^{\infty}\exp[-(\eta T_0)^2]G_\mu \cdot$$
$$\cdot\left(\sin(\eta - \Delta_1)\tau_1\cos\Omega(t - \tau_1) - \frac{\eta - \Delta_0}{\Omega}[1 - \cos(\eta - \Delta_1)\tau_1]\sin\Omega(t - \tau_1)\right)d\eta, \tag{2.7.8}$$

$$\varepsilon_0 = 2\sqrt{\pi}LN_0T_0\left|d_{ba}\right|^2 a\omega/\hbar c,$$

$$G_\mu = (\eta - \Delta_0)[(\eta - \Delta_0)^2 + (2j_b + 1)\mathrm{X}_\mu 2\Lambda^2\gamma_{21}(\gamma_1 + \gamma_2)/\gamma_1\gamma_2]^{-1},$$

$$\Omega^2 = (\eta - \Delta_0)^2 + (2j_b + 1)\mathrm{X}_\mu 4\Lambda^2, \ \mathrm{X}_\mu = \begin{pmatrix} j_b & 1 & j_a \\ -\mu & q & m \end{pmatrix}^2, \ \Delta_0 = \omega - \omega_0 .$$

Amplitude (2.7.8) contains arbitrary detuning Δ_0 and τ_1. Their values affect the numerical value of $\varepsilon'(t)$ but do not change the following general features of the optical nutation [36,61]. Below we will indicate the $E_b \to E_a$ transition as $j_b \to j_a$.

The amplitude (2.7.8) of the linear polarised optical nutation (*q=0*) on an atomic transition $j \to j$ with large momenta *j>>1* is less and it decays more rapidly than it does on atomic transitions $j \leftrightarrow j+1$ (see Fig. 2.7.2). The other way round, in the case of a circular polarised optical nutation (*q=1* for right-handed and *q=-1* for left-handed circular polarisations), the amplitude (2.7.8) for the atomic transitions $j \leftrightarrow j+1$ is less and it decays more rapidly than it does for the atomic transition $j \to j$. This relationship makes it possible to distinguish experimentally between the $j \to j$ and $j \leftrightarrow j+1$ transitions. For each of the $j \to j$ and $j \leftrightarrow j+1$ transitions and a specific polarisation of wave (2.7.1) the function $\varepsilon'(t)/\varepsilon_0$ is independent of *j* if *j>>1*.

Analysis has shown that the period of the linearly polarised optical nutation (2.7.8) in the case $\Lambda\tau_1 \le 1$ and *j>>1* is equal $2\pi/2\Lambda$ for the $j \to j$ transition and to $2\sqrt{2}\pi/2\Lambda$ for the $j \leftrightarrow j+1$ transitions. In the case of circular polarisation, the nutation period is equal to $2\sqrt{2}\pi/2\Lambda$ for the $j \to j$ transition and to $2\pi/2\Lambda$ for the $j \leftrightarrow j+1$ transitions. Thus, the nutation period depends only on the amplitude *a* of an incident wave and γ for a given polarisation of the incident wave. This makes it possible to determine the probability γ of spontaneous emission from the nutation period found experimentally. One should note that a small detuning $|\omega - \omega_0| \le 1/T_0$ does not affect the nutation period.

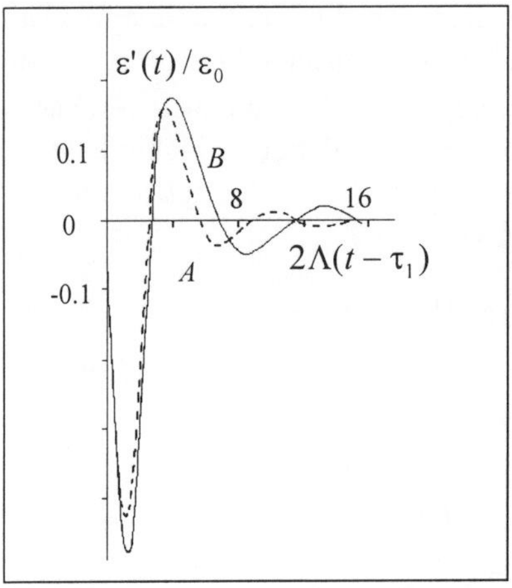

Fig.2.7.2. The optical nutation after a single phase shift. For the $j \to j$ (j>>1) transition, curves A and B correspond to linear and circular polarisation respectively. Conversely, for the $j \leftrightarrow j+1$ (j>>1) transitions, the linear and circular polarisations correspond to curves B and A. It is assumed that $2\Lambda T_0 = 1$, $2\Lambda\tau_1 = \pi/10$, $\Delta_0 T_0 = 0.5$, $\Delta_1 T_0 = 10.5$ and $\gamma_{21} = \gamma_1 = \gamma_2 << 2\Lambda$.

In the case of small values of j the order of magnitude of the nutation period is

$2\pi/\Lambda$. However, in the case of a $j \to j$ transition with *j>1* the period of the linearly polarised optical nutation (2.7.8) is always less than the period of the circular polarised optical nutation. The reverse is true for $j \leftrightarrow j+1$ transitions with *j>0*. The $1 \leftrightarrow 0$ and $1 \leftrightarrow 1$ transitions are characterised by the coincidence of the curves for linearly and circular polarised optical nutations, and for the $\frac{1}{2} \to \frac{1}{2}$ transition the period of the linearly polarised optical nutation is $\sqrt{2}$ times greater than that of a circularly polarised optical nutation.

The above conclusions remain true for the period and curve positions of optical nutation in the region $\tau \le \tau + \tau_2$. There is, however, a specific feature. Nutation oscillations in the region $\tau \le \tau + \tau_2$ first decrease in a similar fashion to (2.7.8), they then increases up to the instant $t = 2\tau + \tau_2 - \tau_1$ in the form of the nutation echo (Fig. 2.7.3). The echo terms in the amplitude of optical nutation in the region $\tau \le \tau + \tau_2$ are given by

$$\varepsilon'_{ne}(t) = -\frac{\varepsilon_0}{2} \sum_{\mu=-j_b}^{j_b} (2j_b+1) X_\mu^2 \int_{-\infty}^{\infty} \exp[-(\eta T_0)^2] G_\mu \left(\frac{2\Lambda}{\Omega}\right)^2 [1-\cos(\eta-\Delta_2)\tau_2] \cdot$$

$$\cdot \left(\sin(\eta-\Delta_1)\tau_1 \cos\Omega(t-2\tau-\tau_2+\tau_1) - \frac{\eta-\Delta_0}{\Omega}[1-\cos(\eta-\Delta_1)\tau_1] \sin\Omega(t-2\tau-\tau_2+\tau_1) \right) d\eta .$$

The inequality $T_0 << 1/\Lambda << \tau - \tau_1$ represents the optimum condition for observing the nutation echo.

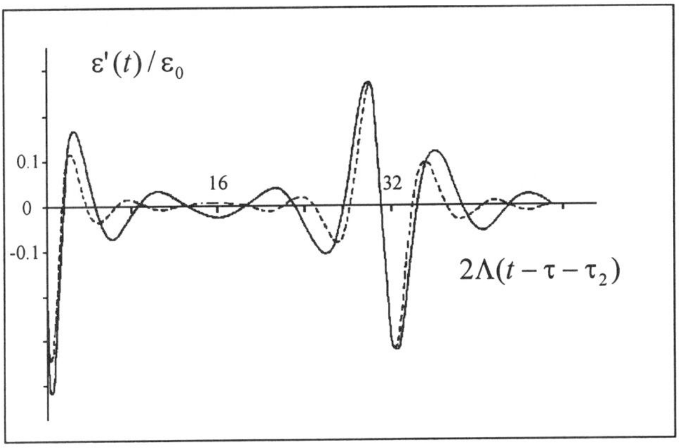

Fig.2.7.3. The optical nutation and nutation echoes. Curves A and B correspond to linear and circular polarisations for the $j \to j$ (j>>1) transition and to circular and linear polarisations for the $j \leftrightarrow j+1$ (j>>1) transitions. It is assumed that $2\Lambda T_0 = 0.5$, $2\Lambda\tau_1 = 2\Lambda\tau_2 = \pi/20$, $\Delta_0 T_0 = 0.5$, $\Delta_1 T_0 = 10.5$ and $\Delta_2 T_0 = -4.5$, $\gamma_{21} = \gamma_1 = \gamma_2 << 2\Lambda$.

2.7.2. POLARISATION OF PHOTON ECHOES IN THE THREE-LEVEL SYSTEMS AND INVESTIGATION OF ATOMIC RELAXATION

To demonstrate polarisation features of echo phenomena based on free induction decay and to illustrate the possibility of their use in spectroscopy we describe one of the tri-level echoes formed by the following set of exciting pulses [62,63]

$$\vec{E}_1 = (\vec{e}_z \cos\psi_1 + \vec{e}_x \sin\psi_1) a_1 \exp[i(k_1 y - \omega_1 t - \varphi_1)] + c.c.,\ 0 \le t - y/c \le T_1,$$

$$\vec{E}_2 = (\vec{e}_z \cos\psi_2 + \vec{e}_x \sin\psi_2) a_2 \exp[-i(k_2 y + \omega_2 t + \varphi_2)] + c.c.,$$
$$T_1 + \tau_{12} \le t + (y - L)/c \le T_1 + T_2 + \tau_{12},$$

$$\vec{E}_1 = \vec{e}_z a_3 \exp[i(k_3 y - \omega_3 t - \varphi_3)] + c.c.,$$
$$T_1 + T_2 + \tau_{12} + \tau_{23} \le t - y/c \le T_1 + T_2 + T_3 + \tau_{12} + \tau_{23},$$

where first and third pulses are resonant to optically allowed transition $E_b \to E_a$ whereas the second pulse is in resonance with adjacent optically allowed transition $E_c \to E_b$ (Fig. 2.6.2 c))

$$\omega_1 = \omega_3 \equiv \omega \approx \omega_{ba} = (E_b - E_a)/\hbar,\quad k_1 = k_3,\quad \omega_2 \equiv \overline{\omega} \approx \omega_{cb} = (E_c - E_b)/\hbar.$$

We have chosen the y-axis along the direction of pulse propagation keeping the z-axis for the designation of the quantisation axis. This change of axes notation allow us to use standard transformation rules for density matrices induced by the rotation of quantisation axis (see eq. (2.7.15), Fig. 2.7.4 and ref.[64,65]).

The resonant energy levels are degenerate due to various orientation of the total angular momentum. We suppose the validity of the theory of depolarising atomic collisions [65-67]. This theory describes atomic collisions of resonant atoms with impurity ones in gaseous medium at lower pressure so that the resonant levels and transitions between them are characterised by a large set of the relaxation parameters [65-67]

$$\gamma_a^{(\kappa)} = \gamma_a + \Gamma_a^{(\kappa)},\quad \gamma_b^{(\kappa)} = \gamma_b + \Gamma_b^{(\kappa)},\quad \gamma_c^{(\kappa)} = \gamma_c + \Gamma_c^{(\kappa)},$$

$$\gamma_{ba}^{(\kappa)} = (\gamma_b + \gamma_a)/2 + \Gamma_{ba}^{(\kappa)},\quad \mathcal{G}_{ba}^{(\kappa)} = \Gamma_{ba}^{(\kappa)} + i\Delta_{ba}^{(\kappa)},$$

$$\gamma_{cb}^{(\kappa)} = (\gamma_c + \gamma_b)/2 + \Gamma_{cb}^{(\kappa)},\quad \mathcal{G}_{cb}^{(\kappa)} = \Gamma_{cb}^{(\kappa)} + i\Delta_{cb}^{(\kappa)},$$

$$\gamma_{ca}^{(\kappa)} = (\gamma_c + \gamma_a)/2 + \Gamma_{ca}^{(\kappa)},\quad \mathcal{G}_{ca}^{(\kappa)} = \Gamma_{ca}^{(\kappa)} + i\Delta_{ca}^{(\kappa)},$$

In this notation the values $\hbar\gamma_a$, $\hbar\gamma_b$ and $\hbar\gamma_c$ represent the radiative widths of the levels E_a, E_b and E_c. The real values $\Gamma_a^{(\kappa)}$ $(0 \le \kappa \le 2j_a), \Gamma_b^{(\kappa)}$ $(0 \le \kappa \le 2j_b)$ and $\Gamma_c^{(\kappa)}$ $(0 \le \kappa \le 2j_c)$ are the contributions of depolarising atomic collisions to the levels broadening. The complex parameters $\mathcal{G}_{ca}^{(\kappa)}$ $(|j_b - j_a| \le \kappa \le j_b + j_a)$ describe the collision relaxation of the optical coherence matrix of the $j_b \to j_a$ transition. The other parameters characterise the $j_c \to j_b$ and $j_c \to j_a$ transitions. These relaxation parameters enter the main equation (1.0.3) through the relaxation operator $\hat{\Gamma}\rho$:

$$(\hat{\Gamma}\rho)_{\mu m} = (\gamma_b + \gamma_a)\rho_{\mu m}/2 + \sum_{\kappa q \mu' m'} (-1)^{2j_b+\mu+\mu'}(2\kappa+1)\mathcal{G}_{ba}^{(\kappa)} \begin{pmatrix} j_b & j_a & \kappa \\ \mu & -m & q \end{pmatrix} \begin{pmatrix} j_b & j_a & \kappa \\ \mu' & -m' & q \end{pmatrix} \rho_{\mu' m'},$$

(2.7.9)

$$(\hat{\Gamma}\rho)_{\mu\mu'} = \gamma_b \rho_{\mu\mu'} + \sum_{\kappa q \mu \mu_1'} (-1)^{2j_b+\mu+\mu_1}(2\kappa+1)\Gamma_b^{(\kappa)} \begin{pmatrix} j_b & j_b & \kappa \\ \mu & -\mu' & q \end{pmatrix} \begin{pmatrix} j_b & j_b & \kappa \\ \mu_1 & -\mu_1' & q \end{pmatrix} \rho_{\mu_1 \mu_1'} - \frac{\gamma_b N_b f(v) \delta_{\mu\mu'}}{2j_b+1}.$$

The expressions for $(\hat{\Gamma}\rho)_{\nu\mu}$, $(\hat{\Gamma}\rho)_{\nu m}$, $(\hat{\Gamma}\rho)_{mm'}$ and others are obtained from the above formulas with appropriate substitutions of indices. In addition to previous agreements the indices $\nu, \nu', ...$ give the projection of the total angular momentum of the state E_c on quantisation axis. We emphasise that the density matrix ρ describes here the group of atoms moving with velocities whose projection on y-axis is v.

The description of the discussed tri-level echoes is based on the following equations generalising eqs. (2.6.3) on account of level degeneracy and pulse polarisation:

$$\left(\frac{\partial}{\partial t} - i(\bar{\eta}+\bar{\Delta}) + \hat{\Gamma}\right) R_{\nu\mu} = \frac{i\bar{a}}{\hbar}\left(\sum_{\mu'} \vec{\bar{l}}\vec{d}_{\nu\mu'} R_{\mu'\mu} - \sum_{\nu'} R_{\nu\nu'} \vec{\bar{l}}\vec{d}_{\nu'\mu}\right),$$

$$\left(\frac{\partial}{\partial t} + i(\eta-\Delta) + \hat{\Gamma}\right) R_{\mu m} = \frac{ia}{\hbar}\left(\sum_{m'} \vec{l}\vec{d}_{\mu m'} R_{m'm} - \sum_{\mu'} R_{\mu\mu'} \vec{l}\vec{d}_{\mu' m}\right),$$

$$\left(\frac{\partial}{\partial t} + i(\eta-\bar{\eta}-\Delta-\bar{\Delta}) + \hat{\Gamma}\right) R_{\nu m} = \sum_{\mu}\left(\frac{i\bar{a}}{\hbar}\vec{\bar{l}}\vec{d}_{\nu\mu} R_{\mu m} - \frac{ia}{\hbar} R_{\nu\mu} \vec{l}\vec{d}_{\mu m}\right),$$

(2.7.10)

$$\left(\frac{\partial}{\partial t} + \hat{\Gamma}\right) R_{\nu\nu'} = \sum_{\mu} \frac{i\bar{a}}{\hbar}\left(\vec{\bar{l}}\vec{d}_{\nu\mu} R_{\mu\nu'} - R_{\nu\mu} \vec{\bar{l}}^{*}\vec{d}_{\mu\nu'}\right),$$

$$\left(\frac{\partial}{\partial t} + \hat{\Gamma}\right) R_{mm'} = \sum_{\mu} \frac{ia}{\hbar}\left(\vec{l}^{*}\vec{d}_{m\mu} R_{\mu m'} - R_{m\mu} \vec{l}\vec{d}_{\mu m'}\right),$$

$$\left(\frac{\partial}{\partial t} + \hat{\Gamma}\right) R_{\mu\mu'} = \sum_{\nu} \frac{i\bar{a}}{\hbar}\left(\vec{\bar{l}}^{*}\vec{d}_{\mu\nu} R_{\nu\mu'} - R_{\mu\nu} \vec{\bar{l}}\vec{d}_{\nu\mu'}\right) + \sum_{m} \frac{ia}{\hbar}\left(\vec{l}\vec{d}_{\mu m} R_{m\mu'} - R_{\mu m} \vec{l}^{*}\vec{d}_{m\mu'}\right).$$

The sought quantities are the following slowly varying functions

$$R_{\nu\mu} = \rho_{\nu\mu} \exp[i(\bar{k}y + \bar{\omega}t + \bar{\varphi})], \quad R_{\mu m} = \rho_{\mu m} \exp[-i(ky - \omega t - \varphi)],$$

$$R_{\nu m} = \rho_{\nu m} \exp\{-i[(k-\bar{k})y - (\omega+\bar{\omega})t - \varphi - \bar{\varphi}]\}, \qquad (2.11)$$

$$R_{mm'} = \rho_{mm'}, \quad R_{\mu\mu'} = \rho_{\mu\mu'}, \quad R_{\nu\nu'} = \rho_{\nu\nu'}.$$

We have written the exciting pulses in the form

$$\vec{\bar{E}} = \vec{\bar{l}}\,\bar{a}\exp[-i(\bar{k}y + \bar{\omega}t + \bar{\varphi})] + c.c.$$
$$\vec{E} = \vec{l}\,a\exp[i(ky - \omega t - \varphi)] + c.c..$$

using the dash to distinguish the parameters of the field $\vec{\bar{E}}$, which are resonant to the transition $j_c \to j_b$, from the parameters of the field $\vec{E}$, which are in resonance with the transition $j_b \to j_a$.

We write the solutions of the system (2.7.10) for separated in time action of the exciting fields $\vec{E}$ or $\vec{\bar{E}}$ only, assuming linear polarisation of these fields. If we take the quantisation axis parallel to polarisation vector, the required solution in the case $\hat{\Gamma} = 0$ will be in the standard form

$$R(t) = U(t,t_0)R(t_0)U^+(t,t_0). \tag{2.7.12}$$

In the case of resonant field $\vec{\bar{E}}$ we have

$$\begin{gathered} U_{\mu\mu'}(t,t_0) = A_\mu \delta_{\mu\mu'},\quad U_{\nu\nu'}(t,t_0) = A_\nu^* \delta_{\nu\nu'}, \\ U_{mm'}(t,t_0) = \exp\{i(t-t_0)[\eta - \Delta - (\bar{\eta} + \bar{\Delta})/2]\}\delta_{mm'}, \\ U_{\nu\mu}(t,t_0) = i(-1)^{j_c-\mu}\frac{d_{cb}}{|d_{cb}|}B_\mu \delta_{\nu\mu},\quad U_{\mu\nu}(t,t_0) = i(-1)^{j_c-\mu}\frac{d_{cb}^*}{|d_{cb}|}B_\mu \delta_{\nu\mu}, \\ U_{\nu m}(t,t_0) = U_{\nu m}(t,t_0) = U_{\mu m}(t,t_0) = U_{m\mu}(t,t_0) = 0, \end{gathered} \tag{2.7.13}$$

where

$$A_\sigma = \cos\tfrac{1}{2}\Omega_\sigma(t-t_0) - i\frac{\bar{\eta}+\bar{\Delta}}{\Omega_\sigma}\sin\tfrac{1}{2}\Omega_\sigma(t-t_0),\ B_\sigma = (\chi_\sigma/\Omega_\sigma)\sin\tfrac{1}{2}\Omega_\sigma(t-t_0),$$

$$\Omega_\sigma = \sqrt{(\bar{\eta}+\bar{\Delta})^2 + \chi_\sigma^2},\ \chi_\sigma = \frac{2\bar{a}|d_{cb}|}{\hbar}\begin{pmatrix} j_c & j_b & 1 \\ \sigma & -\sigma & 0 \end{pmatrix},\quad \sigma = \nu,\mu.$$

In the case of resonant field $\vec{E}$ we have

$$\begin{gathered} U_{mm'}(t,t_0) = A_m \delta_{mm'},\quad U_{\mu\mu'}(t,t_0) = A_\mu^* \delta_{\mu\mu'}, \\ U_{\nu\nu'}(t,t_0) = \exp\{i(t-t_0)[\bar{\eta} + \bar{\Delta} - (\eta - \Delta)/2]\}\delta_{\nu\nu'}, \\ U_{\mu m}(t,t_0) = i(-1)^{j_b-m}\frac{d_{ba}}{|d_{ba}|}B_m \delta_{\mu m},\quad U_{m\mu}(t,t_0) = i(-1)^{j_b-m}\frac{d_{ba}^*}{|d_{ba}|}B_\mu \delta_{\mu m}, \\ U_{\mu\nu}(t,t_0) = U_{\nu\mu}(t,t_0) = U_{\nu m}(t,t_0) = U_{m\nu}(t,t_0) = 0, \end{gathered} \tag{2.7.14}$$

where

$$A_\sigma = \cos\tfrac{1}{2}\Omega_\sigma(t-t_0) + i\frac{\eta-\Delta}{\Omega_\sigma}\sin\tfrac{1}{2}\Omega_\sigma(t-t_0),\ B_\sigma = (\chi_\sigma/\Omega_\sigma)\sin\tfrac{1}{2}\Omega_\sigma(t-t_0),$$

$$\Omega_\sigma = \sqrt{(\eta-\Delta)^2+\chi_\sigma^2},\ \chi_\sigma = \frac{2a|d_{ba}|}{\hbar}\begin{pmatrix} j_b & j_a & 1 \\ \sigma & -\sigma & 0\end{pmatrix},\quad \sigma = \mu, m.$$

In both cases

$$\eta = kv,\quad \bar{\eta} = \bar{k}v,\quad \Delta = \omega - \omega_{ba},\quad \bar{\Delta} = \bar{\omega} - \omega_{cb}.$$

The density matrix ρ of the three-level atoms $E_a < E_c < E_b$ interacting with the field $\vec{E}$ or $\vec{\bar{E}}$ of carrier frequencies $\omega \approx \omega_{ba}$ and $\bar{\omega} \approx \omega_{bc} = (E_b - E_c)/\hbar$ is obtained from eqs.(2.7.11)-(2.7.14) by substitutions $\bar{k} \to -\bar{k}$, $\bar{\omega} \to -\bar{\omega}$, $\bar{\varphi} \to -\bar{\varphi}$, $\bar{\eta} + \bar{\Delta} \to -\bar{\eta} - (\bar{\omega} - \omega_{bc})$ and $d_{cb} \to (-1)^{j_c - j_b} d^*_{bc}$. For the other position of levels $E_b < E_a < E_c$ when the carrier frequencies ω and $\bar{\omega}$ of waves $\vec{E}$ and $\vec{\bar{E}}$ are close to atomic frequencies $\omega_{ab} = (E_a - E_b)/\hbar$ and ω_{cb}, the density matrix is given by eqs. (2.7.11)-(2.7.14) with the following substitutions $k \to -k$, $\omega \to -\omega$, $\varphi \to -\varphi$, $\eta - \Delta \to -\eta + (\omega - \omega_{ab})$ and $d_{ba} \to (-1)^{j_b - j_a} d^*_{ab}$.

We must know the transformation laws for density matrices defined by the change of the coordinate system and the rotation of quantisation axis. This knowledge allows us to formulate the initial conditions for eqs. (2.7.11)-(2.7.14) for describing the interaction of the medium with a set of exciting pulses $\vec{E}$ and $\vec{\bar{E}}$ with different polarisations $\vec{l}$ or $\vec{\bar{l}}$. Let Euler angles α, β and γ describe the rotation of the coordinate system x', y', z' to the system x, y, z where axes z and z' are the quantisation axes (Fig. 2.7.4). We have the following transformation laws:

$$R_{\nu\mu} = \sum_{\nu'\mu'q'q\kappa} (-1)^{\nu'-\nu}(2\kappa+1)\begin{pmatrix} j_b & j_c & \kappa \\ \mu' & -\nu' & q'\end{pmatrix}\mathfrak{D}^{\kappa *}_{q'q}(\alpha,\beta,\gamma)\begin{pmatrix} j_b & j_c & \kappa \\ \mu & -\nu & q\end{pmatrix}R_{\nu'\mu'}. \quad (2.7.15)$$

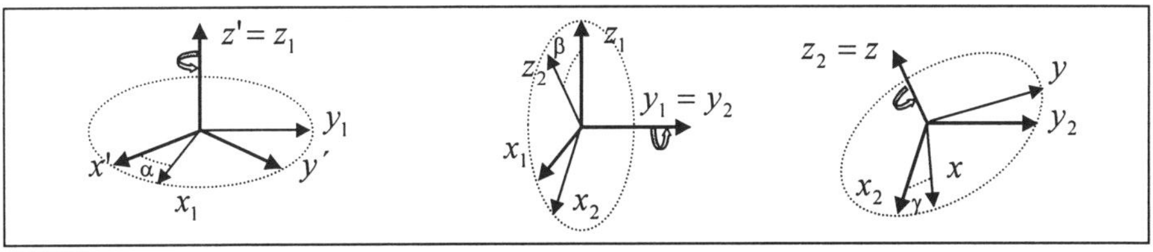

Fig. 2.7.4. Euler angles describing the rotation of the coordinate system x', y', z' to the system x, y, z.

Here the matrix $R_{\nu'\mu'}$ represents the atomic density matrix related to coordinate system x', y', z'. Quantities ν' and μ' are the projections of the total angular momentum on axis z'. The values $R_{\nu\mu}$, ν and μ refer to the coordinate system x, y, z. Function $\mathcal{D}^{\kappa}_{q'q}(\alpha,\beta,\gamma)$ is the Wigner D-function [64]. The proof can be done on the basis of ref.[65].

The electric field $\vec{E}_e$ of the three-level echo of frequency $\overline{\omega}$ produced by exciting pulses $\vec{E}_1$, $\vec{E}_2$ and $\vec{E}_3$ can be obtained using eqs. (2.7.11)-(2.7.14) in the form

$$\vec{E}_e = \vec{\varepsilon}(t + y/c)\exp[-i(\bar{k}y + \overline{\omega}t + \varphi_e)] + c.c., \tag{2.7.16}$$

where the amplitude $\vec{\varepsilon}(t)$ and the phase shift φ_e are given by

$$\vec{\varepsilon}(t) = -\varepsilon_0 \int dv f(v)(\vec{e}_z W_0 + \sqrt{2}\vec{e}_x W_1)\exp[i(\overline{\eta} + \overline{\Delta})(t - t_e) - (\gamma^{(1)}_{cb} + i\Delta^{(1)}_{cb})(t - \tau_{12} - \tau_{23})], \tag{2.7.17}$$

$$\varphi_e = \varphi_1 + \varphi_2 - \varphi_3 - (\Delta + \overline{\Delta}\omega/\overline{\omega})(\tau_{12} + \tau_{23} + T_2 + T_3/2),$$
$$t_e = \tau_{12} + (\tau_{12} + \tau_{23} + T_2 + T_3/2)\omega/\overline{\omega} + T_1 + T_2/2.$$

We have introduced the notation

$$\varepsilon_0 = 2\pi L|d_{cb}|(\overline{\omega}/c)(\frac{N_a}{2j_a+1} - \frac{N_b}{2j_b+1}),$$

$$W_p = \sum_{\kappa\kappa'} G^*_\kappa S_{\kappa\kappa' p} Q_{\kappa' p} \exp[-(\gamma^{(\kappa)}_{ba} + i\Delta^{(\kappa)}_{ba})\tau_{12} - (\gamma^{(\kappa')}_{ca} + i\Delta^{(\kappa')}_{ca})\tau_{23}], \quad p = 0, 1,$$

$$S_{\kappa\kappa' p} = (2\kappa+1)\sum_{\nu m q}(-1)^{j_b-\nu}\begin{pmatrix} j_b & j_a & \kappa \\ \nu & -m & q \end{pmatrix}\begin{pmatrix} j_c & j_a & \kappa' \\ \nu & -m & q \end{pmatrix}\overline{B}_{2\nu} d^{\kappa}_{0q}(\psi_1 - \psi_2) d^{\kappa'}_{qp}(\psi_2),$$

$$Q_{\kappa p} = (2\kappa+1)\sum_{\nu m}(-1)^{j_b-m}\begin{pmatrix} j_c & j_a & \kappa \\ \nu & -m & p \end{pmatrix}\begin{pmatrix} j_c & j_b & 1 \\ \nu & -m & p \end{pmatrix} B_{3m},$$

$$G_\kappa = \sum_{\mu}\begin{pmatrix} j_b & j_a & \kappa \\ \mu & -\mu & 0 \end{pmatrix} A_{1\mu} B_{1\mu},$$

$$A_{n\sigma} = \cos\tfrac{1}{2}\Omega_{n\sigma}T_n + i\frac{\eta - \Delta}{\Omega_{n\sigma}}\sin\tfrac{1}{2}\Omega_{n\sigma}T_n, \quad B_{n\sigma} = (\chi_{n\sigma}/\Omega_{n\sigma})\sin\tfrac{1}{2}\Omega_{n\sigma}T_n,$$

$$\Omega_{n\sigma} = \sqrt{(\eta - \Delta)^2 + \chi^2_{n\sigma}}, \quad \chi_{n\sigma} = \frac{2a|d_{ba}|}{\hbar}\begin{pmatrix} j_b & j_a & 1 \\ \sigma & -\sigma & 0 \end{pmatrix}, \quad n = 1, 2, 3, \quad \sigma = \nu, \mu, m.$$

Here, L/c is small compared with τ_n $(n = 1,2,3)$, and $d^{\kappa}_{qp}(\beta)$ is the Wigner D-function

$\mathcal{D}^{\kappa}_{qp}(\alpha,\beta,\gamma)$ at $\alpha=\gamma=0$. The value $\overline{B}_{2\nu}$ is obtained from $B_{2\sigma}$ by substitutions $\sigma\to\nu$, $j_a\to j_b$, $j_b\to j_c$, $d_{ba}\to d_{cb}$, $\Delta\to\overline{\Delta}$ and $\eta\to-\overline{\eta}$. Eqs. (2.7.16) and (2.7.17) describe also the three-level echo produced by the excitation of the three-level systems $E_a<E_c<E_b$ by pulses $\vec{E}_1$, $\vec{E}_2$ (with $L=0$ and y replaced by $-y$) and $\vec{E}_3$. In this case in eq.(2.7.15) one should substitute $\overline{\omega}$ and $\overline{\varepsilon}(t+y/c)$ by $-\overline{\omega}$ and $-\overline{\varepsilon}(t-y/c)$, whereas in eq.(2.7.16) $\Delta^{(1)}_{cb}$, $\overline{\Delta}$ and φ_2 must be replaced by $-\Delta^{(1)}_{bc}$, $-(\overline{\omega}-\omega_{bc})$ and $-\varphi_2$.

The amplitude (2.7.17) contains a specific term $\vec{e}_zW_0+\sqrt{2}\vec{e}_xW_1$ which defines polarisation properties and decay laws of three-level echo (2.7.16). In the case of small angular momenta 1/2 and 3/2 this term has a simple form

$$\vec{e}_zW_0+\sqrt{2}\vec{e}_xW_1=\vec{F}A^*_{1\frac{1}{2}}B_{1\frac{1}{2}}\overline{B}_{2\frac{1}{2}}B_{3\frac{1}{2}}\exp[-(\gamma^{(1)}_{ba}+i\Delta^{(1)}_{ba})\tau_{12}],$$

where the vector $\vec{F}$ is given by the following equations:

1) for $j_a=j_b=j_c=1/2$

$$\vec{F}=\frac{2}{\sqrt{6}}\{\vec{e}_z\cos(\psi_1-\psi_2)\exp[-(\gamma^{(0)}_{ca}+i\Delta^{(0)}_{ca})\tau_{23}]-\vec{e}_x\sin(\psi_1-\psi_2)\exp[-(\gamma^{(1)}_{ca}+i\Delta^{(1)}_{ca})\tau_{23}]\},$$

2) for $j_a=j_b=1/2$ and for $j_c=3/2$

$$\vec{F}=\frac{1}{2\sqrt{6}}\{\vec{e}_x[\sin(\psi_1-\psi_2)\exp[-(\gamma^{(1)}_{ca}+i\Delta^{(1)}_{ca})\tau_{23}]-3\sin(\psi_1+\psi_2)\exp[-(\gamma^{(2)}_{ca}+i\Delta^{(2)}_{ca})\tau_{23}]]-$$
$$-\vec{e}_z[\cos(\psi_1-\psi_2)+3\cos(\psi_1+\psi_2)]\exp[-(\gamma^{(2)}_{ca}+i\Delta^{(2)}_{ca})\tau_{23}]\},$$

3) for $j_a=j_c=1/2$ and for $j_b=3/2$

$$\vec{F}=\frac{1}{2\sqrt{6}}\{\vec{e}_x\sin(\psi_1-\psi_2)\exp[-(\gamma^{(1)}_{ca}+i\Delta^{(1)}_{ca})\tau_{23}]-\vec{e}_z4\cos(\psi_1-\psi_2)\exp[-(\gamma^{(0)}_{ca}+i\Delta^{(0)}_{ca})\tau_{23}].$$

We demonstrate the approach for using the polarisation feature of echo decay on the example of the $j_c=3/2\to j_b=1/2$ and $j_b=1/2\to j_a=1/2$ transitions from general formula (2.7.17). In this case the echo intensity as a function of τ_{12} at $t=t_e$ decays as $\exp[-2(\gamma^{(1)}_{ba}+g\gamma^{(1)}_{cb})\tau_{12}]$, where $g=\omega/\overline{\omega}$. By choosing the appropriate values of the angles ψ_1 and ψ_2 between polarisation planes of the first and third and the second and third pulses we can obtain the simple exponential law for decay of echo intensity, corresponding to the projections of echo electric field.

If $\psi_1 = -\psi_2 = \pi/4$, the intensity, corresponding to the projection on x-axis of the echo electric field decay with varying of τ_{23} as $\exp\{-2[\gamma_{ca}^{(1)} + (g-1)\gamma_{cb}^{(1)}]\tau_{23}\}$. The echo intensity as a function of τ_{23} at $\psi_1 = \psi_2$ is $\exp\{-2[\gamma_{ca}^{(2)} + (g-1)\gamma_{cb}^{(1)}]\tau_{23}\}$.

One should note that the model of depolarising atomic collisions gives $\gamma_{ca}^{(0)} = \gamma_{ca}^{(1)}$ and $\Delta_{ca}^{(0)} = \Delta_{ca}^{(1)}$ for the $j_c = 1/2 \rightarrow j_a = 1/2$ transition and $\gamma_{ca}^{(1)} = \gamma_{ca}^{(2)}$ and $\Delta_{ca}^{(1)} = \Delta_{ca}^{(2)}$ for the $j_c = 3/2 \rightarrow j_a = 1/2$ transition (in the case of Van der Waals interaction between colliding atoms). In other case of angular momenta the relaxation parameters $\gamma_{ca}^{(0)}$, $\gamma_{ca}^{(1)}$, $\gamma_{ca}^{(2)}$ differ from each other by 10-20 per cent.

To simplify the general formula for arbitrary values of angular momenta it is convenient to use exciting pulses $\vec{E}_1$, $\vec{E}_2$ and $\vec{E}_3$ with small areas. Then the amplitude (2.7.17) has the polarisation structure similar to one in the case of small angular momenta:

$$\begin{aligned}\vec{\varepsilon}(t) = b\{\vec{e}_z[2L_0\cos(\psi_1-\psi_2) + L_2(\cos(\psi_1-\psi_2) + 3\cos(\psi_1+\psi_2))] - \\ - 3\vec{e}_x[L_1\sin(\psi_1-\psi_2) - L_2\sin(\psi_1+\psi_2)]\}\cdot \qquad (2.7.18)\\ \cdot\exp\{-(\gamma_{ba}^{(1)} + i\Delta_{ba}^{(1)})\tau_{12} - i\Delta_{cb}^{(1)}[(\omega/\overline{\omega})\tau_{12} + (\omega/\overline{\omega}-1)\tau_{23}] + i(\overline{\Delta}\omega/\overline{\omega} + \Delta)\tau_1/2\},\end{aligned}$$

$$\begin{aligned}b = -\frac{\varepsilon_0}{6}\int dv f(v)\frac{\Lambda_1^0\Lambda_2^0\Lambda_3^0}{(\eta-\Delta)^2(\overline{\eta}+\overline{\Delta})}\sin\frac{(\eta-\Delta)\tau_1}{2}\sin\frac{(\overline{\eta}+\overline{\Delta})\tau_2}{2}\sin\frac{(\eta-\Delta)\tau_3}{2}\cdot\\ \cdot\exp[i(\overline{\eta}+\overline{\Delta}-\Delta_{cb}^{(1)})(t-t_e-\omega\tau_1/2\overline{\omega}) - \gamma_{cb}^{(1)}(t-\tau_{12}-\tau_{23})],\end{aligned}$$

$$L_\kappa = \begin{Bmatrix}1 & \kappa & 1\\ j_a & j_b & j_c\end{Bmatrix}^2 \exp[-(\gamma_{ca}^{(\kappa)} + i\Delta_{ca}^{(\kappa)})\tau_{23}], \quad \kappa = 0, 1, 2,$$

$$\Lambda_2^0 = 2a_2|d_{cb}|/\hbar, \quad \Lambda_m^0 = a_m|d_{ba}|/\hbar, \quad m = 1, 3.$$

Now, we analyse the position of polarisation plane of echo (2.7.16) and (2.7.18) with respect to polarisation planes of exciting pulses. This position is determined by the type of $j_c \rightarrow j_a$ transition. For example, the $j \leftrightarrow j+2$ transitions are characterised by the following properties. If $\psi_1 = -\psi_2$, the echo polarisation plane always coincides with polarisation plane of third exciting pulse. In case $\psi_1 = \psi_2 \neq 0$ echo polarisation plane is situated within blunt angle formed by polarisation planes of the exciting pulses.

The polarisation plane of echo (2.7.18) at $\psi_1 = \psi_2 \neq 0$ in the case of $j \leftrightarrow j+1$ $(j \geq 1/2)$ optically forbidden transitions lies within the same blunt angle. But if $\psi_1 = -\psi_2$ the echo polarisation plane essentially differs from the polarisation plane of the third exciting pulse (contrary to $j \leftrightarrow j+2$ transitions). The angle ψ_e of inclination of the echo polarisation plane to z-axis at $\psi_1 = -\psi_2$ is $\tan\psi_e = -5\sin 2\psi_1 /(3+\cos 2\psi_1)$, where we have supposed for simplicity that the relaxation processes are inessential and the three-level system with $j \leftrightarrow j+1$ optically forbidden transition is characterised by large values of angular momentum $j >> 1$.

The three-level echo (2.7.17) in the case of $j \to j$ $(j \geq 3/2)$ optically forbidden transitions at $\psi_1 = \psi_2 \neq 0$ has always the polarisation plane within the acute angle formed by polarisation planes of exciting pulses (contrary to $j \leftrightarrow j+1$ and $j \leftrightarrow j+2$ transitions).

Instructive studies of photon echo polarisation can be found in ref.[45,68-72].

2.8. Conclusion

Free induction decay, nutation oscillations as well as photon echoes on their basis can be regarded as separate well distinguished phenomena only in optically thin media where they serve as basic elements of the analysis of coherent transients. In extended medium these phenomena are distorted by dispersion broadening, diffraction, nonlinear compression and other processes inherent to nonlinear propagation [73]. Therefore the nonlinear analysis of pulse propagation needs its own basic elements of the language for describing wave evolution in an extended medium. The main element is the notion of soliton directly connected with the inverse scattering transform method. This notion will be discussed in detail in next chapters. Other new elements characterising nonlinear regime of pulse interaction with medium arise in the description of wave transmission through thin films of resonant atoms. The notions of optical bistability and self-pulsation along with the above mentioned ones allow us to analyse different regimes of the resonant reflection and refraction by thin films (see Chapter 9).

References

1. Allen, L., and Eberly, J.H.: *Optical resonance and two-level atoms,* Wiley-Interscience, New York, 1975.
2. Kopvillem, U.Kh., and Nagibarov, V.R.: Light echo on paramagnetic crystals, *Fiz.Metallov i Metallovedenie* **15** (1963), 313-315.
3. Kurnit, N.A., Abella, I.D., and Hartmann, S.R.: Observation of a Photon Echo, *Phys.Rev.Lett.* **13** (1964), 567-568; Photon echoes, *Phys.Rev.* **141** (1966), 391-406.
4. Hahn, E.L.: *Phys.Rev.* **80** (1950), 580.
5. Chebotaev, V.P., and Dubetski, B.Ya.: A classical model of photon echo, *Appl.Phys. B* **31** (1983), 45-52.
6. McCall, S.L., and Hahn, E.L.: Self-induced transparency by pulsed coherent light, *Phys.Rev.Lett.* **18** (1967), 908-911; Self-induced transparency, *Phys.Rev.* **183** (1969), 457-485.
7. Chebotaev, V.P.: *Appl.Phys.* **15** (1978), 219-223.
8. Letokhov, V.S., and Chebotaev, V.P.: Nonlinear laser spectroscopy of super-high resolution, Nauka, Moscow, 1984.

9. Dicke, R.H.: Coherence in spontaneous radiation processes, *Phys.Rev.* **93** (1954), 99-110.
10. Rehler, N.E., and Eberly, J.H.: Superradiance, *Phys.Rev. A* **3** (1971), 1735-1751.
11. Torrey, H.C.: Transient nutation in nuclear magnetic resonance, *Phys.Rev.* **76** (1949), 1059-1069.
12. Hocker, G.B., and Tang, C.L.: Observation of the optical transient nutation effect, *Phys.Rev.Lett.* **21** (1968), 592-594.
13. Brewer, R.G., and Shoemaker, R.L.: Photon echoes and optical induction in molecules, *Phys.Rev.Lett.,* **27** (1971), 631-635.
14. Manykin, E.A., and Samartzev, V.V.: *Optical echo-spectroscopy*, Nauka, Moscow, 1984.
15. Elyutin, S.O., Zakharov, S.M., and Manykin, E.A.: On the features of the photon-echo form, *Optika i spektroskopiya* **42** (1977), 1005-1007.
16. Elyutin, S.O., Zakharov, S.M., and Manykin, E.A.: Theory of the photon-echo formation, *Zh.Eksp.Teor.Fiz.* **76** (1979), 835-845.
17. Zuikov, V.A., Samartzev, V.V., and Usmanov, R.G.: The correlation of the photon-echo form with the form of exciting pulses, *Pis'ma Zh.Eksp.Teor.Fiz.* **32** (1980), 293-297.
18. Carlson, N.W., Babbitt, W.R., Bai, Y.S., and Mossberg, T.W.: Field-inhibited optical dephasing and shape locking of photon echoes, *Opt.Lett.* **9** (1984), 232-234.
19. Rebane, A.K., Kaarli, R.K., and Saari, P.M.: Dynamical picosecond holography by photochemical hole burning, *Pis'ma Zh.Eksp.Teor.Fiz.* **38** (1983), 320-323.
20. Kaarli, R.K., Saari, P.M., and Sonajalg, H.R.: Storage and reproduction of an ultrafast optical signal with arbitrary time dependent wavefront and polarization, *Opt.Commun.* **65** (1988), 170-174.
21. Manykin, E.A.: Spatial synchronism in non-stationary processes of echo-type, *Pis'ma Zh.Eksp.Teor.Fiz.* **7** (1968), 345-348.
22. Heer, C.V., and McManamon, P.F.: Wavefront correction with photon echoes, *Opt.Commun.* **23** (1977), 49-50.
23. Zakharov, S.M., and Manykin, E.A.: Temporary and correlation characteristics of echo-signals in two- and three-level systems with inhomogeneously broadened resonant energy levels, *Zh.Eksp.Teor.Fiz.* **91** (1986), 1289-1301.
24. Zakharov, S.M., and Manykin, E.A.: Scaled transformations of non-stationary pictures by photon echoes, *Zh.Eksp.Teor.Fiz.* **95** (1989), 1587-1597.
25. Mossberg, T.W.: Time-domain frequency-selective optical data storage, *Opt.Lett.* **7** (1982), 77-79.
26. Akhmediev, N.N., Borisov, B.S., Zuikov, V.A., et al: Observation of multiple long-living photon echoes, *Pis'ma Zh.Eksp.Teor.Fiz.* **48** (1988), 585-587.
27. Ershov, G.M., and Kopvillem, U.Kh.: Theory of multiple-pulses excitation of echo-type signals, *Zh.Eksp.Teor.Fiz.* **72** (1963), 279-289.
28. Zuikov, V.A., Samartzev, V.V., and Turianski, E.A.: Excitation of photon echoes by a set of running and standing waves, *Zh.Eksp.Teor.Fiz.* **81** (1981), 653-663.
29. Zuikov, V.A., and Samartzev, V.V.: Reverse photon echo as a method of investigation of resonant medium parameters, *Phys.Status Solidi* A **73** (1982), 625-632.
30. Schenzle, A., DeVoe, R.G., and Brewer, R.G.: Cumulative two-pulse photon echoes, *Phys.Rev. A* **30** (1984), 1866-1872.
31. Car, N.Y., and Purcell, T.M.: Effects of diffusion on free precession in nuclear magnetic resonance experiments, *Phys.Rev.* **94** (1954), 46-61.
32. Schmidt, J., Berman, P.R., and Brewer, R.G.: Coherent transient study of velocity-changing collisions, *Phys.Rev.Lett.* **31** (1973), 1103-1105.
33. Nurmikko, A. V., and Schwartz, S. E.: Optical nutation echo in degenerate resonant systems, *Opt.Communs* **2** (1971), 416-418.
34. Rohart, F., Glorieux., P., and Makke, B.: "Rotary" photon echoes, *J.Phys. B* 10 (1977), 3835-3848.
35. Alekseev, A.I., and Basharov, A.M.: Modulation oscillations of a light wave in the Stark-pulse technique, *Zh.Eksp.Teor.Fiz.* **77** (1979), 537-547 [*Sov.Phys.JETP* **50** (1979), 272-277].
36. Alekseev, A.I., and Basharov, A.M.: Optical nutation and photon echo by phase shift of a light wave, *Phys.Lett. A* **77** (1980), 123-125; Optical nutation and photon echo by phase or amplitude shift of light

wave, *Opt.Commun.* **36** (1981), 291-296; Optical nutation and free induction in degenerate systems, *J.Phys. B* **15** (1982),4269-4282.

37. Wong, N.C., Kano, S.S., and Brewer, R.G.: Optical rotary echoes, *Phys.Rev. A* **21** (1980), 260-267.
38. Muramoto, T., Nakanishi, S., Tamura, 0., and Hashi, T.: Optical nutation echoes in ruby, *Jap.J.Appl.Phys.* **19** (1980), L211-L214.
39. Brewer, R.G., and Hahn, E.L.: Coherent two-photon processes: transient and steady-state cases, *Phys.Rev. A* **11** (1975), 1641-1649.
40. Makhviladze, T.M., and Sarychev, M.E.: Light echo in nonlinear systems, *Zh.Eksp.Teor.Fiz.* **69** (1975), 1594-1600 [*Sov.Phys.JETP* **42** (1975), 812-816].
41. Loy, M. M. T.: Observation of two-photon nutation and free-induction decay, *Phys.Rev.Lett.* **36** (1976), 1454-1457.
42. Flusberg, A., Mossberg, T., Kachru, R., and Hartmann, S.R.: Observation and relaxation of the two-photon echo in Na vapor, *Phys.Rev.Lett.*, **41** (1978), 305-308.
43. Maimistov, A.I., and Manykin, E.A.: Stimulated photon echo at two-quantum resonance, *Optika i spektroskopiya* **46** (1982), 958-960.
44. Bruckner, V., Bente, E.A.J.M., Langelaar, J., Bebelaar, D., and van Voorst, J.D.W.: Raman Echo on a Picosecond Timescale in Nitrogen Gas, *Opt.Commun.* **51** (1984), 49-52.
45. Basharov, A.M., Maimistov, A.I., and Manykin, E.A.: Polarisation features of coherent transition phenomena in two-photon resonance, *Zh.Eksp.Teor.Fiz.* **84** (1983), 487-501 [*Sov.Phys. JETP* **57** (1983), 282-289].
46. Le Gouet, J.-L., and Berman, P.R.: Photon echoes in standing-wave pulse: time separation of spatial harmonics, *Phys.Rev. A* **20** (1979), 1105-1115.
47. Baklanov, E.V., Dubetsky, B.Ya., and Semibalamut, V.M.: Theory of stimulated coherent emission from atoms in spatially separate optical fields, *Zh.Eksp.Teor.Fiz.* **76** (1979), 482-504 [*Sov.Phys. JETP* **49** (1979), 244-254].
48. Alekseev, A.I., Basharov, A.M., and Beloborodov, V.N.: Polarisation of photon echo produced by standing-wave pulses, *Zh.Eksp.Teor.Fiz.* **79** (1980), 787-796 [*Sov.Phys.JETP* **52** (1980), 401-405].
49. Mossberg, T., Flusberg, A., Kachru, R., and Hartmann, S.R.: Tri-level echoes, *Phys.Rev.Lett.* **19** (1979), 1523-1526.
50. Mossberg, T.W., Kachru, R., Hartmann, S.R., and Flusberg, A.: Echoes in gaseous media: a generalized theory of rephasing phenomena, *Phys.Rev.* A **20** (1979), 1976-1996.
51. Basharov, A.M.: Raman scattering by coherent excitation of three-level systems, *Optika i spektroskopiya* **57** (1984), 961-962.
52. Patel, C.K.N., and Slusher, R.E.: Photon echoes in gases, *Phys.Rev.Lett.*, **20** (1968), 1087-1089.
53. Gordon, J. P., Wang, C.H., Patel, C.K.N. et al.: Photon echo in gases, *Phys.Rev.* **179** (1969), 294-309.
54. Alimpiev, S.S., and Karlov, N.V.: Photon echoes in SF_6 and BCl_3 gases, *Zh.Eksp.Teor.Fiz.* **63** (1972), 482-490.
55. Alimpiev, S.S., and Karlov, N.V.: Nutation effect in molecular gases BCl_3 and SF_6, *Zh.Eksp.Teor.Fiz.* **66** (1974), 542-551 [*Sov.Phys.JETP* **39** (1974), 260-264].
56. Baer, T., and Abella, I.D.: Polarisation rotation of photon echoes in cesium vapour in magnetic field, *Phys.Rev. A* **16** (1977), 2093-2100.
57. Kachru, R., Mossberg, T. W., and Hartmann, S. R.: Stimulated photon echo study of $Na(3^2S_{1/2})$-CO velocity-changing collisions, *Opt.Communs.* **30** (1979), 57-62.
58. Vasilenko, L.S., Rubtsova, N.N., and Khvorostov, E.B.: Investigation of collision-induced decay of population, orientation and alignment by stimulated photon echoes in molecular gas, *Zh.Eksp.Teor.Fiz.* **113** (1998), 826-834.
59. Wang, C.H..: Effects of mixing collisions on photon echoes in gases, *Phys.Rev. B* **1** (1970), 156-183.
60. Alekseev, A.I., and Basharov, A.M.: Coherent emission by atoms induced by a travelling or standing wave, *Zh.Eksp.Teor.Fiz.* **80** (1981), 1361-1370 [*Sov.Phys.JETP* **53** (1981), 694-698].
61. Alekseev, A.I., Basharov, A.M., and Khabakkpashev, M.A.: Some features of optical nutation in a gas, *Zh.Eksp.Teor.Fiz.* **75** (1978), 2122-2131 [*Sov.Phys.JETP* **48** (1978), 1069-1073].

62. Yevseyev, I. V., and Yermachenko, V. M.: Polarisation properties of the tri-level echoes, *Phys.Lett. A* **90** (1982), 37-40.
63. Alekseev, A.I., and Basharov, A.M.: Investigation of atomic relaxation by the three-level echoes, *Opt.Commun.* **45** (1983), 171-178.
64. Edmonds, A.R.: *Angular momentum in quantum mechanics*, Princeton Univ.Press, Princeton, 1960.
65. Omont, A.: Irreducible components of the density matrix. Application to optical pumping, *Prog.Quantum Electronics* **5** (1981), 69-138.
66. Matskevich, V.K.: Depolarising atomic collisions and spectral line broadening, *Optika i spektroskopiya* **37** (1974), 411-416.
67. Bakaev, D.S., Evseev, I.V., and Ermachenko, V.M.: Effect of depolarizing collisions on the photon echo in a magnetic field, *Zh.Eksp.Teor.Fiz.* **76** (1979), 1212-1225 [*Sov.Phys.JETP* **49** (1979), 615-621].
68. Alekseev, A.I., and Evseev, I. V.: Polarisation of photon echo in gas medium, , *Zh.Eksp.Teor.Fiz.* **56** (1969), 2118-2128.
69. Heer, C. V., and Nordstrom, P. J.: Polarisation of photon echoes from SF_6 molecules, *Phys.Rev. A* **11** (1975), 536-548.
70. Alekseev, A.I., and Basharov, A.M.: On ultrahigh resolution spectroscopy based on photon echo, *Zh.Eksp.Teor.Fiz.* **74** (1978), 1988-1998.
71. Alekseev, A.I., Basharov, A.M., and Beloborodov, V.N.: Photon-echo quantum beats in a magnetic field, *J.Phys. B* **16** (1983), 4697-4715.
72. Evseev, I.V., Ermachenko, V.M., and Reshetov, V.A.: On the possibility of measuring relaxation times of population, orientation and alignment by the photon echo method, *Zh.Eksp.Teor.Fiz.* **78** (1980), 2213-2221.
73. Zakharov, S.M., and Manykin, E.A.: Optical nutation effects in multiple photon echo, *Pis'ma Zh.Eksp.Teor.Fiz.* **17** (1973), 431-434.

CHAPTER 3

INVERSE SCATTERING TRANSFORM METHOD

Let us consider the Fourier transform method for solving the linear partial differential equation. It is known that the Fourier transformation converts this equation into a linear ordinary differential equation, which can be easily integrated. Thus we obtain the evolution of Fourier components of the initial data in the problem under consideration. Inverse Fourier transformation yields the solution of initial equation in the integral form. It would be more correct to say that the Fourier transforms method permits to solve the Cauchy problem of a linear partial differential equation. This is the main reason why this method is attractive to apply in mathematical physics.

The inverse scattering transform method is an extension of the Fourier transform method when we consider scattering data as the combination of the generalised Fourier components and spectral parameter. It is interesting to note that one of the first survey article by Ablowitz, Kaup, Newell and Segur was titled "The Inverse Scattering Transform - Fourier Analysis for Nonlinear Problems" [1].

There are the Hamiltonian systems in mechanics and in the field theory. There are cases when the canonical transformation makes the equations of motion trivially integrable after the conversion to new variables. Then they say that Hamiltonian system admits the action-angle variables. Thus, if the action-angle variables exist, then the Hamiltonian system is *completely integrable*. By using the IST method we can now state that some equations, which admit zero-curvature representation, are completely integrable Hamiltonian equations [2].

In practical problems the systems of equations may not be completely integrable due to dissipation, dispersion or an external force. In this situation the IST method leads to some sorts of perturbation theories and another approximation schemes to analyse such problems.

Nowadays the IST method is extended to cover many problems involving quantum models and quantum statistical physics. According to Wadati's statement the IST method is one of the most important inventions in the modern mathematical physics. It is a unique method for solving the initial value problem of non-linear evolution equations. A number of excellent books could be recommended as guides on application of the IST method to various non-linear evolution equations [3-15]. Below we shall develop basic elements of the IST method necessary for understanding the mathematics underlying the theory of nonlinear optical waves.

3.1. Basic principle of the inverse scattering transform

Let us consider the pair of linear operators

$$\hat{\mathbf{X}}(\lambda) = \partial_x - \hat{U}(\lambda, \{q\}), \quad \hat{\mathbf{T}}(\lambda) = \partial_t - \hat{V}(\lambda, \{q\}),$$

where $\hat{U}$ and $\hat{V}$ are $N \times N$ matrices depending on a set $\{q\} \equiv \{q, q_x, q_{xx}, \ldots\}$ and the parameter λ, $q(x,t)$ is, in general, a matrix-valued function of x and t. The integrability condition for the pair of equations

$$\hat{\mathbf{X}}(\lambda)\psi = 0, \quad \hat{\mathbf{T}}(\lambda)\psi = 0, \tag{3.1.1}$$

means that the commutator of these operators vanishes,

$$[\hat{\mathbf{X}}(\lambda), \hat{\mathbf{T}}(\lambda)] = 0$$

or in matrix form

$$\frac{\partial \hat{U}}{\partial t} - \frac{\partial \hat{V}}{\partial x} + [\hat{U}, \hat{V}] = 0. \tag{3.1.2}$$

Equation (3.1.2.) must hold for all λ. If so, this equation determines a set of partial differential (generally non-linear) equations for the components of the potential $q(x,t)$,

$$\frac{\partial q}{\partial t} = \hat{\mathbf{N}}[q].$$

Here $\hat{\mathbf{N}}[\,]$ denotes a non-linear operator acting in space of the potentials.

The presentation of non-linear evolution equations of this kind in the form (3.1.2) was named a zero-curvature representation of these equations and the matrices $\hat{U}$ and $\hat{V}$ make up the U-V-pair.

The differential operators in the equations (3.1.1) and (3.1.2) may be considered as the operators of covariant differentiation

$$\hat{D}_1 = \hat{\mathbf{X}}(\lambda), \quad \hat{D}_2 = \hat{\mathbf{T}}(\lambda),$$

so matrices $\hat{U}$ and $\hat{V}$ are understood as local coefficients of connectivity in fibre bundle $R^2 \times C^N$, where Euclidean space-time R^2 is a base manifold, and $\psi(x,t;\lambda)$ has its values in C^N, which is a N-dimension complex plane. Equation (3.1.2) means, therefore, that this connection has zero curvature [2,16].

One of the equations (3.1.1) or (3.1.2), for example

$$\hat{\mathbf{X}}(\lambda)\psi = 0\,,$$

can be interpreted as a scattering problem. More precisely, it describes the scattering of the wave $\psi(x,t;\lambda)$ on the potential $q(x,t)$ (or potentials $\{q(x,t), q_x(x,t),\dots\}$), while the second equation of (3.1.1) describes the variation of the scattering matrix with time t. In the same words, this equation describes the variation of the scattering matrix with the deformation of potential $q(x,t)$ governed by the non-linear evolution equation or system of equations. The determination of the scattering matrix, the eigenvalues and eigenfunctions of the operator $\hat{\mathbf{X}}(\lambda)$ is a direct spectral problem of scattering theory. The reconstruction of the potential $q(x,t)$ by using the solution of the spectral problem is the inverse problem of scattering theory. Let us denominate the scattering matrix, the eigenfunctions and the spectrum of the operator $\hat{\mathbf{X}}(\lambda)$ by scattering data and denote this set by the symbol $\mathbf{S}(t)$, bearing in mind that it may vary in time. It is known that the scattering data is that one needs to solve the inverse scattering problem.

By solving the direct scattering problem, we obtain a map of the potential $q(x,t)$ into scattering data, i.e. there is a mapping

$$\text{Scattering Transform: } q(x,t) \to \mathbf{S}(t)\,.$$

By solving the inverse scattering problem, we obtain a mapping

$$\text{Inverse Scattering Transform: } \mathbf{S}(t) \to q(x,t)\,.$$

of the scattering data into potential. Now we are able to use these transforms in order to solve the Cauchy problem for non-linear evolution equation (or the system of equations). Let $q(x,t=0) = q_0(x)$ is the initial condition for some non-linear evolution equation, whereas $\lim\limits_{x\to\pm\infty} q(x,t) = q^{\pm}$. Taking $q_0(x)$ as a potential in the direct scattering problem we may find $\mathbf{S}(t=0)$. Using the second equation of (3.1.1) at the next step we are able to reconstruct $\mathbf{S}(t)$ in terms of $\mathbf{S}(t=0)$. Solution of the inverse scattering problem yields $q(x,t)$, which is the solution of a given Cauchy problem. The steps mentioned above can be schematically represented as

$$\begin{array}{ccc} q_0(x) \to \mathbf{S}(t=0) & & \\ & \downarrow & \\ & & \mathbf{S}(t) \to q(x,t) \end{array}$$

Every step in this chain results in the solution of a linear problem. This is the principal advantage of this method, named inverse scattering problem or inverse scattering transform (IST).

3.2. The Zakharov-Shabat spectral problem

The IST method was developed at the same time by Zakharov-Shabat (ZS) [17] and Ablowitz, Kaup, Newell, Segur (AKNS) [18,19]. The particular cases of the spectral problem of IST method appeared in the analysis of the SIT equations and the non-linear Schrödinger equation. The more general case of the ZS spectral problem is presented by the following linear system of equations

$$\frac{\partial \psi_1}{\partial x} = -i\lambda\psi_1 + q(x)\psi_2 \, , \tag{3.2.1a}$$

$$\frac{\partial \psi_2}{\partial x} = i\lambda\psi_2 + r(x)\psi_1 \, . \tag{3.2.1b}$$

Complex number λ plays the role of a spectral parameter, functions $q(x,t)$ and $r(x,t)$ are the solutions of non-linear evolution equation. It is expected that under condition $|x|\to\infty$ $q(x,t)$ and $r(x,t)$ disappear rapidly enough.

The *Jost functions* are chosen as the solutions of equations (3.2.1), that with real λ satisfy the following boundary conditions

$$\text{as } x\to-\infty \quad \Phi(x,\lambda)\to\begin{pmatrix}1\\0\end{pmatrix}\exp(-i\lambda x), \quad \overline{\Phi}(x,\lambda)\to\begin{pmatrix}0\\-1\end{pmatrix}\exp(+i\lambda x) \, ,$$

$$\text{as } x\to+\infty \quad \Psi(x,\lambda)\to\begin{pmatrix}0\\1\end{pmatrix}\exp(+i\lambda x), \quad \overline{\Psi}(x,\lambda)\to\begin{pmatrix}1\\0\end{pmatrix}\exp(-i\lambda x) \, .$$

The pair of functions Φ and $\overline{\Phi}$ represent the fundamental solutions of equations(3.2.1), i.e., the Wronskian

$$W[\Phi,\overline{\Phi}] \equiv \Phi_1\overline{\Phi}_2 - \Phi_2\overline{\Phi}_1$$

does not equal to zero. The same is true for other pair Ψ and $\overline{\Psi}$. By using (3.2.1), it is possible to demonstrate that Wronskians $W[\Phi,\overline{\Phi}]$ and $W[\Psi,\overline{\Psi}]$ are independent from co-ordinate x. Its magnitudes can be calculated at $|x|\to\infty$ in correspondence with asymptotic expressions

$$W[\Phi,\overline{\Phi}] = -1, \qquad W[\Psi,\overline{\Psi}] = +1. \tag{3.2.2}$$

So far as Ψ and $\overline{\Psi}$ are linear independent functions, then the any solution of equations (3.2.1) can be expressed as their linear combination, including functions Φ and $\overline{\Phi}$ themselves

$$\begin{aligned} \Phi(x,\lambda) &= a(\lambda)\overline{\Psi}(x,\lambda) + b(\lambda)\Psi(x,\lambda), \\ \overline{\Phi}(x,\lambda) &= -\overline{a}(\lambda)\Psi(x,\lambda) + \overline{b}(\lambda)\overline{\Psi}(x,\lambda). \end{aligned} \tag{3.2.3a}$$

These expressions define the *matrix of transition* (or *transfer matrix*) $\hat{T}(\lambda)$

$$T_{11}(\lambda) = a(\lambda), \quad T_{12}(\lambda) = \overline{b}(\lambda),$$
$$T_{21}(\lambda) = b(\lambda), \quad T_{22}(\lambda) = -\overline{a}(\lambda).$$

Matrix $\hat{T}(\lambda)$ can be written in terms of Wronskians of the four Jost functions introduced above

$$\hat{T}(\lambda) = \begin{pmatrix} W[\Phi,\Psi] & W[\overline{\Phi},\Psi] \\ W[\overline{\Psi},\Phi] & W[\overline{\Psi},\overline{\Phi}] \end{pmatrix}. \tag{3.2.4}$$

Using (3.2.2) and (3.2.3a) we can find a unimodularity condition of transition matrix $\hat{T}(\lambda)$

$$a(\lambda)\overline{a}(\lambda) + b(\lambda)\overline{b}(\lambda) = 1$$

In the points of discrete spectrum λ_n and $\overline{\lambda}_n$, for which $a(\lambda_n) = 0$ and $\overline{a}(\overline{\lambda}_n) = 0$ ($\operatorname{Im}\lambda_n > 0, \operatorname{Im}\overline{\lambda}_n < 0$), one has

$$\begin{aligned} \Phi(x,\lambda_n) &= b_n\Psi(x,\lambda_n), \\ \overline{\Phi}(x,\lambda_n) &= \overline{b}_n\overline{\Psi}(x,\overline{\lambda}_n). \end{aligned} \tag{3.2.3b}$$

By using an asymptotic formulas for Φ and $\overline{\Phi}$, Ψ and $\overline{\Psi}$, one can show that both $\Phi(x,\lambda_n)$ and $\overline{\Phi}(x,\overline{\lambda}_n)$, are exponentially decrease under condition $|x| \to \infty$.

The analytic properties of Jost functions are very important in solving the inverse spectral problem. These properties may be formulated in the form of theorem:

If $q(x)$ and $r(x)$ are a complex functions satisfying

$$\int_{-\infty}^{\infty} |q(x)|\,dx < \infty, \quad \int_{-\infty}^{\infty} |r(x)|\,dx < \infty$$

then $\Phi(x,\lambda)\exp(i\lambda x)$ and $\Psi(x,\lambda)\exp(-i\lambda x)$ are analytical functions of λ in the upper half-plane $\operatorname{Im}\lambda > 0$, whereas $\overline{\Phi}(x,\lambda)\exp(-i\lambda x)$ and

$\overline{\Psi}(x,\lambda)\exp(i\lambda x)$ *are analytical functions of* λ *in the lower-plane* $\operatorname{Im}\lambda < 0$. *Under condition* $\operatorname{Im}\lambda = 0$ *all these functions are limited.*

From this theorems and (3.2.4) it follows that with these restrictions on the potentials of spectral problem (3.2.1) function $a(\lambda)$ is analytical under condition $\operatorname{Im}\lambda > 0$, whereas $\overline{a}(\lambda)$ is an analytical function when $\operatorname{Im}\lambda < 0$.

There are following integral representation (the triangular presentation) of the Jost functions:

$$\Psi(x,\lambda) = \begin{pmatrix} 0 \\ 1 \end{pmatrix} \exp(i\lambda x) + \int_x^{\infty} K(x,y)\exp(i\lambda y)dy \ , \tag{3.2.5a}$$

$$\overline{\Psi}(x,\lambda) = \begin{pmatrix} 1 \\ 0 \end{pmatrix} \exp(-i\lambda x) + \int_x^{\infty} \overline{K}(x,y)\exp(-i\lambda y)dy \ , \tag{3.2.5b}$$

$$\Phi(x,\lambda) = \begin{pmatrix} 1 \\ 0 \end{pmatrix} \exp(-i\lambda x) - \int_{-\infty}^{x} M(x,y)\exp(-i\lambda y)dy \ , \tag{3.2.5c}$$

$$\overline{\Phi}(x,\lambda) = \begin{pmatrix} 0 \\ -1 \end{pmatrix} \exp(i\lambda x) - \int_{-\infty}^{x} \overline{M}(x,y)\exp(i\lambda y)dy \ . \tag{3.2.5d}$$

Here $K(x,y)$, $\overline{K}(x,y)$, $M(x,y)$, $\overline{M}(x,y)$ are two-component column-matrices, and moreover

$$K(x,y) = \overline{K}(x,y) = 0 \quad \text{under conditions } y > x \text{ and } y \to \infty$$
$$M(x,y) = \overline{M}(x,y) = 0 \quad \text{under conditions } y < x \text{ and } y \to -\infty$$

Potentials of the spectral problem (3.2.1) $q(x,t)$ and $r(x,t)$ are expressed in terms of components of column- matrixes $K(x,y)$ and $\overline{K}(x,y)$. In order to find these expressions one should proceed the following routine. Let us differentiate a triangular presentation $\Psi(x,\lambda)$ (3.2.5a) with respect to *x* by using the Leibniz rule

$$\frac{d}{dx}\int_b^{a(x)} f(x,y)dy = \int_b^{a(x)} \frac{\partial}{\partial x} f(x,y)dy + f(x,a(x))\frac{da}{dx} \ .$$

Thereby, we have

$$\frac{\partial}{\partial x}\Psi(x,\lambda) = \begin{pmatrix} 0 \\ 1 \end{pmatrix} i\lambda\exp(i\lambda x) - K(x,x)\exp(i\lambda x) + \int_x^{\infty} \frac{\partial}{\partial x} K(x,y)\exp(i\lambda y)dy \ . \tag{3.2.6}$$

Integration of $\Psi(x,\lambda)$ in (3.2.5a) by parts yields

$$\Psi(x,\lambda)=\begin{pmatrix}0\\1\end{pmatrix}\exp(i\lambda x)+\frac{i}{\lambda}K(x,x)\exp(i\lambda x)+\frac{i}{\lambda}\int_x^{\infty}\frac{\partial}{\partial y}K(x,y)\exp(i\lambda y)dy\,. \quad (3.2.7)$$

The equations (3.2.1) can be transcribed in matrix form as

$$\frac{\partial}{\partial x}\Psi=-i\lambda\hat{\sigma}_3\Psi+\hat{Q}(x)\Psi\ , \quad (3.2.8)$$

where

$$\hat{\sigma}_3=\begin{pmatrix}1&0\\0&-1\end{pmatrix},\quad \hat{Q}(x)=\begin{pmatrix}0&q(x)\\r(x)&0\end{pmatrix}.$$

Let $\Psi(x,\lambda)$ in the first summand in the right hand side of (3.2.8) be changed according to (3.2.7), the function $\Psi(x,\lambda)$ in the second summand be changed according to (3.2.5a) and the derivation with respect to x from in (3.2.8) is written in the form (3.2.6). The result is

$$\left\{\hat{\sigma}_3K(x,x)+\hat{Q}(x)\begin{pmatrix}0\\1\end{pmatrix}+K(x,x)\right\}\exp(i\lambda x)=$$
$$=\int_x^{\infty}\left\{-\frac{\partial}{\partial x}K(x,y)+\hat{\sigma}_3\frac{\partial}{\partial y}K(x,y)+\hat{Q}(x)K(x,y)\right\}\exp(i\lambda y)dy \quad (3.2.9)$$

Equating the coefficients before $\exp(i\lambda x)$ and $\exp(i\lambda y)$ in (3.2.9) to zero we get the system of equations

$$\hat{\sigma}_3\frac{\partial}{\partial y}K(x,y)+\hat{Q}(x)K(x,y)-\frac{\partial}{\partial x}K(x,y)=0\ ,\ \text{where } y>x \quad (3.2.10a)$$

with a boundary conditions at $y=x$

$$\hat{\sigma}_3K(x,x)+K(x,x)+\hat{Q}(x)\begin{pmatrix}0\\1\end{pmatrix}=0\,. \quad (3.2.10b)$$

The differential equation (3.2.10) with boundary conditions (3.2.10b) is a Goursat problem, for which it is known that its solution exists and is unique. Besides that, from

(3.2.10b) it follows a necessary relation to allow to terminate a solution of the inverse scattering problem:

$$q(x) = -2K_1(x,x). \tag{3.2.11a}$$

With the use of the expressions for $\overline{\Psi}(x,\lambda)$ (3.2.5b) after some algebra we can obtain

$$r(x) = -2\overline{K}_2(x,x), \tag{3.2.11b}$$

and for $\overline{K}(x,y)$ vanishing as $y \to \infty$, we have a similar Goursat problem

$$\hat{\sigma}_3 \frac{\partial}{\partial y}\overline{K}(x,y) - \hat{Q}(x)\overline{K}(x,y) + \frac{\partial}{\partial x}\overline{K}(x,y) = 0, \text{ when } y > x \tag{3.2.12a}$$

with the boundary conditions at $y = x$:

$$\hat{\sigma}_3\overline{K}(x,x) - \overline{K}(x,x) - \hat{Q}(x)\begin{pmatrix}1\\0\end{pmatrix} = 0 . \tag{3.2.12b}$$

Whence it follows the existence and uniqueness of $\overline{K}(x,y)$ and the formula (3.2.11b).

Equations (3.2.10) give rise to a useful formula

$$2K_2(x,x) = \int_x^\infty r(x')q(x')dx' \tag{3.2.13}$$

Functions $K(x,y)$ and $\overline{K}(x,y)$ satisfy the linear system of integral equations, which are named by Gelfand-Levitan-Marchenko equations (GLM)

$$\overline{K}(x,z) + \begin{pmatrix}0\\1\end{pmatrix}F(x+z) + \int_x^\infty K(x,y)F(y+z)dy = 0 , \tag{3.2.14a}$$

$$K(x,z) - \begin{pmatrix}1\\0\end{pmatrix}\overline{F}(x+z) - \int_x^\infty \overline{K}(x,y)\overline{F}(y+z)dy = 0 , \tag{3.2.14b}$$

where $F(z)$ and $\overline{F}(z)$ are defined by the scattering data as

$$\begin{aligned} F(z) &= -i\sum_{n=1}^{N} C_n \exp(i\lambda_n z) + \frac{1}{2\pi}\int_{-\infty}^{\infty}\rho(\xi)\exp(i\xi z)d\xi , \\ \overline{F}(z) &= i\sum_{n=1}^{\overline{N}} \overline{C}_n \exp(-i\overline{\lambda}_n z) + \frac{1}{2\pi}\int_{-\infty}^{\infty}\overline{\rho}(\xi)\exp(-i\xi z)d\xi . \end{aligned} \tag{3.2.15}$$

Here N and $\bar{N}$ are the number of eigenvalues $\{\lambda_n\}$ and $\{\bar{\lambda}_n\}$ accordingly,

$$\rho(\xi) = b(\xi)/a(\xi), \; \bar{\rho}(\xi) = \bar{b}(\xi)/\bar{a}(\xi), \; C_k = ib_k / a'_k, \text{ where } a'_k = \frac{da}{d\lambda}\Big|_{\lambda=\lambda_k}$$

Thereby, the inverse spectral problem consists in the determination of $q(x,t)$ and $r(x,t)$ by their relation with $K(x,y)$ and $\bar{K}(x,y)$, which are the solution of integral equations (3.2.14) under given kernels $F(z)$ and $\bar{F}(z)$.

One should note some characteristics of this spectral problem (3.2.1) that appears as a result from the additional relationships between $q(x,t)$ and $r(x,t)$ named reduction.

In the case of

$$r(x,t) = \alpha q(x,t), \tag{3.2.16a}$$

where α is an arbitrary complex number, we have

$$\bar{\Psi}(x,\lambda) = \hat{R}\Psi(x,-\lambda), \quad \bar{\Phi}(x,\lambda) = -\alpha^{-1}\hat{R}\Phi(x,-\lambda),$$

with $\hat{R} = \begin{pmatrix} 0 & 1 \\ \alpha & 0 \end{pmatrix}$.

This result can be directly checked. Thence it follows from (3.2.5) that

$$\bar{a}(\lambda) = a(-\lambda), \quad \bar{b}(\lambda) = -\alpha^{-1}b(-\lambda),$$

and the number of zeroes of functions $a(\lambda)$ and $\bar{a}(\lambda)$ is the same. Besides, $\bar{\lambda}_n = -\lambda_n$, where $n = 1, 2, \ldots, N = \bar{N}$.

If we have

$$r(x,t) = \alpha q^*(x,t), \tag{3.2.16b}$$

where α is an arbitrary real nonzero number, then the relations

$$\bar{\Psi}(x,\lambda) = \hat{R}\Psi^*(x,\lambda^*), \quad \bar{\Phi}(x,\lambda) = -\alpha^{-1}\hat{R}\Phi^*(x,\lambda^*)$$

are equitable. Thence it follows

$$\bar{a}(\lambda) = a^*(\lambda^*), \quad \bar{b}(\lambda) = -\alpha^{-1}b^*(\lambda^*)$$

Functions $a(\lambda)$ and $\overline{a}(\lambda)$ have the same number of zeroes and $\overline{\lambda}_n = \lambda_n{}^*$, where $n = 1, 2, \ldots, N = \overline{N}$. Under such reduction it is often $\alpha = \pm 1$. Then for the kernels and solutions of integral GLM equation $F(z)$ and $\overline{F}(z)$, $K(x,y)$ and $\overline{K}(x,y)$ the relations

$$\overline{F}(z) = \mp F^*(z)$$

and

$$\overline{K}_1(x,y) = K_2^*(x,y), \quad \overline{K}_2(x,y) = \pm K_1^*(x,y)$$

are correct.

Finally, if $r(x,t) = \pm q(x,t)$ are real functions, then both reductions considered above take place simultaneously. That leads to

$$\overline{a}(\lambda) = a^*(\lambda^*) = a(-\lambda),$$

whence it follows $\lambda_n{}^* = -\lambda_n$. This condition means that either λ_n is an imaginary quantity or there is a pair of the different eigenvalues λ_n and λ_{n+1} such that

$$\operatorname{Im}\lambda_n = \operatorname{Im}\lambda_{n+1}, \quad \operatorname{Re}\lambda_n = -\operatorname{Re}\lambda_{n+1}.$$

In this case both $F(z)$ and $K(x,y)$ are real functions.

3.3. Solution of the Gelfand-Levitan-Marchenko equations for reflectionless potentials

Now it is useful to find the soliton solutions of those nonlinear equations, which have zero-curvature representation with the Zakharov-Shabat's *U-V*-pair (or AKNS equations). These solutions correspond to reflectionless potentials of the spectral problem (3.2.1). The spectrum, related to this case, is discrete that implies the GLM equations to be exactly solvable.

Denoting the two-component column-matrices $K(x,y)$ and $\overline{K}(x,y)$ as $K(x,y) = \mathrm{colon}\big(K_1(x,y),\, K_2(x,y)\big)$ and $\overline{K}(x,y) = \mathrm{colon}\big(\overline{K}_1(x,y),\, \overline{K}_2(x,y)\big)$, equations (3.2.14) can be written in a scalar form

$$\overline{K}_1(x,z) + \int_x^\infty K_1(x,y)F(y+z)dy = 0, \tag{3.3.1a}$$

$$\overline{K}_2(x,z) + F(x+z) + \int_x^\infty K_2(x,y)F(y+z)dy = 0 \ , \tag{3.3.1b}$$

$$K_1(x,z) - \overline{F}(x+z) - \int_x^\infty \overline{K}_1(x,y)\overline{F}(y+z)dy = 0 \ , \tag{3.3.1c}$$

$$K_2(x,z) - \int_x^\infty \overline{K}_2(x,y)\overline{F}(y+z)dy = 0 \ , \tag{3.3.1d}$$

Since

$$q(x) = -2K_1(x,x),$$
$$|q(x)|^2 = 2\frac{\partial}{\partial x}K_2(x,x) = 2\frac{\partial}{\partial x}\overline{K}_1^*(x,x),$$

we may choose from this systems only those equations which define $K_1(x,x)$. Thereby, it is necessary to deal with only the pair of equations (3.3.1a) and (3.3.1c).

Let us consider one-soliton potential, i.e., $N = \overline{N} = 1$ and let

$$F(z) = -iC_1 \exp(i\lambda_1 z), \quad \overline{F}(z) = i\overline{C}_1 \exp(-i\overline{\lambda}_1 z) \ . \tag{3.3.2}$$

Substitution of these expressions into (3.3.1a) and (3.3.1c) allows to notice that $K(x,y)$ and $\overline{K}(x,y)$ must depend on y in the same way as $F(y)$ and $\overline{F}(y)$ accordingly do:

$$K_1(x,y) = h_1(x)\exp(-i\overline{\lambda}_1 y), \quad \overline{K}_1(x,y) = \overline{h}_1(x)\exp(i\lambda_1 y) \ . \tag{3.3.3}$$

Then the unknown functions $h_1(x)$ and $h_2(x)$ can be determined from the linear algebraic system of equation

$$\overline{h}_1(x) + C_1 \exp[ix(\lambda_1 - \overline{\lambda}_1)](\lambda_1 - \overline{\lambda}_1)^{-1} h_1(x) = 0,$$
$$h_1(x) + \overline{C}_1 \exp[ix(\lambda_1 - \overline{\lambda}_1)](\lambda_1 - \overline{\lambda}_1)^{-1} \overline{h}_1(x) = i\overline{C}_1 \exp(-ix\overline{\lambda}_1) \ .$$

Thence follows that

$$h_1(x) = \frac{i\overline{C}_1 \exp(-i\overline{\lambda}_1 x)}{1 - C_1\overline{C}_1(\lambda_1 - \overline{\lambda}_1)^{-2}\exp[2i(\lambda_1 - \overline{\lambda}_1)x]}$$

and the solution of the GLM equations under condition $N = \overline{N} = 1$ has the form

$$K_1(x,y) = \frac{i\overline{C}_1 \exp[-i\overline{\lambda}_1(x+y)]}{1 - C_1\overline{C}_1(\lambda_1 - \overline{\lambda}_1)^{-2}\exp[2i(\lambda_1 - \overline{\lambda}_1)x]} \ . \tag{3.3.4}$$

Now we can turn to the case of the arbitrary $N = \overline{N}$, when

$$F(z) = -i\sum_{n=1}^{N} C_n \exp(i\lambda_n z)\,, \quad \overline{F}(z) = i\sum_{n=1}^{N} \overline{C}_n \exp(-i\overline{\lambda}_n z)\,. \tag{3.3.5}$$

With taking into account that $r(x,t) = -q^*(x,t)$, we can rewrite the relevant system of GLM equations in the following form

$$K_1(x,z) - F^*(x+z) - \int_x^\infty K_2^*(x,y)F^*(y+z)dy = 0\,, \tag{3.3.6a}$$

$$K_2^*(x,z) + \int_x^\infty K_1(x,y)F(y+z)dy = 0 \tag{3.3.6b}$$

Let us introduce the vectors

$$\phi_n(x) = \sqrt{C_n}\exp(i\lambda_n x)\,.$$

Then we can write the kernels of the integral equations (3.3.6) in the factorized forms:

$$F(y+z) = -i\sum_{n=1}^{N}\phi_n(y)\phi_n(z), \quad F^*(y+z) = i\sum_{n=1}^{N}\phi_n^*(y)\phi_n^*(z)\,.$$

So we may try to search for the solutions of equations (3.3.6) in the generalized form of expressions (3.3.3):

$$K_1(x,z) = \sum_{n=1}^{N} h_n(x)\phi_n^*(z), \quad K_2^*(x,z) = \sum_{n=1}^{N}\overline{h}_n(x)\phi_n(z)\,. \tag{3.3.7}$$

By substituting the expansions (3.3.7) into (3.3.6a) and (3.3.6b), we will show that the resulting expressions can be represented in the form of algebraic sum, where the n-th term is proportional to either $\phi_n(y)$ or $\phi_n^*(y)$. Integrating of these expressions with respect to y and collecting similar members we again obtain the linear system of equations

$$\overline{h}_n(x) - i\sum_{m=1}^{N}\widetilde{V}_{nm}(x)h_m(x) = 0\,, \tag{3.3.8a}$$

$$h_n(x) - i\sum_{m=1}^{N} V_{nm}(x)\overline{h}_m(x) = i\phi_n^*(x) \tag{3.3.8b}$$

where matrices $\hat{V}(x)$ and $\hat{\tilde{V}}(x) = \hat{V}^*(x)$ are introduced as

$$V_{nm}(x) = \int_x^\infty \phi_n^*(y)\phi_m(y)dy\,, \quad \tilde{V}_{nm}(x) = \int_x^\infty \phi_n(y)\phi_m^*(y)dy = V_{mn}(x) = V_{nm}^*(x)\,.$$

One can represents the matrix elements of $\hat{V}(x)$ in the explicit form

$$V_{nm}(x) = \frac{i\sqrt{C_n^* C_m}\,\exp[ix(\lambda_m - \lambda_n^*)]}{(\lambda_m - \lambda_n^*)}\,.$$

When the eigenvalues of the spectral problem of IST method are imaginary, i.e., $\lambda_n = i\eta_n$. These matrix elements have the following form:

$$V_{nm}(x) = \frac{\sqrt{C_n^* C_m}\,\exp[-x(\eta_m + \eta_n)]}{(\eta_m + \eta_n)}\,.$$

In this case coefficients C_n are pure real or imaginary and matrix elements of the $\hat{V}(x)$ matrix are real.

The linear system of equations for $h_n(x)$ follows from (3.3.8)

$$\sum_{m=1}^{N}\left\{\delta_{nm} + (\hat{V}\hat{V}^*)_{nm}\right\}h_m(x) = i\phi_n^*(x)\,. \tag{3.3.9}$$

Let $\hat{D} = \hat{\text{I}} + \hat{V}\hat{V}^*$. Solution of the linear algebraic system (3.3.9) can be written as

$$h_n(x) = i\sum_{m=1}^{N}(\hat{D}^{-1})_{nm}\phi_m^*(x)\,.$$

With the account of this expression, equation (3.3.8a) results in

$$h_n(x) = -\sum_{m=1}^{N}(\hat{V}^*\hat{D}^{-1})_{nm}\phi_m^*(x)\,.$$

In the accord to (3.3.7) solution of GLM equations (3.3.6) writes in the form

$$K_1(x,y) = i\sum_{n=1}^{N}\sum_{m=1}^{N}(\hat{D}^{-1}(x))_{nm}\phi_m^*(x)\phi_n^*(y)\,.$$

$$K_2^*(x,y) = -\sum_{n=1}^{N}\sum_{m=1}^{N}(\hat{V}^*(x)\hat{D}^{-1}(x))_{nm}\phi_m^*(x)\phi_n(y)\,..$$

Let us consider the latter expression at $y = x$. The definition of matrix $\hat{V}(x)$ yields

$$\phi_n^*(x)\phi_m(x) = -\frac{\partial}{\partial x}V_{nm}(x).$$

Hence

$$K_2^*(x,x) = \sum_{n=1}^{N}\sum_{m=1}^{N}(\hat{V}^*(x)\hat{D}^{-1}(x))_{nm}\frac{\partial}{\partial x}V_{mn}(x) = \mathrm{tr}\left(\hat{V}^*(x)\hat{D}^{-1}(x)\frac{\partial \hat{V}(x)}{\partial x}\right).$$

If we take into account the symmetry properties of matrix $\hat{V}(x)$, i.e., $V_{mn}(x) = V_{nm}^*(x)$, then

$$K_2^*(x,x) = \mathrm{tr}\left(\frac{\partial \hat{V}^*(x)}{\partial x}\hat{D}^{-1}(x)\hat{V}(x)\right).$$

Thus,

$$K_2^*(x,x) = \frac{1}{2}\mathrm{tr}\left(\hat{V}^*(x)\hat{D}^{-1}(x)\frac{\partial \hat{V}(x)}{\partial x}\right) + \frac{1}{2}\mathrm{tr}\left(\frac{\partial \hat{V}^*(x)}{\partial x}\hat{D}^{-1}(x)\hat{V}(x)\right) =$$

$$= \frac{1}{2}\mathrm{tr}\left(\hat{D}^{-1}(x)\frac{\partial \hat{V}(x)}{\partial x}\hat{V}^*(x)\right) + \frac{1}{2}\mathrm{tr}\left(\hat{D}^{-1}(x)\hat{V}(x)\frac{\partial \hat{V}^*(x)}{\partial x}\right) =$$

$$= \frac{1}{2}\mathrm{tr}\left(\hat{D}^{-1}(x)\frac{\partial}{\partial x}\{\hat{V}(x)\hat{V}^*(x)\}\right) = \frac{1}{2}\frac{\partial}{\partial x}\mathrm{tr}\left(\ln\{\hat{\mathrm{I}} + \hat{V}(x)\hat{V}^*(x)\}\right).$$

Note, that for any non-degenerate matrix $\hat{A}$:

$$\mathrm{tr}\log\hat{A} = \log\det\hat{A}.$$

Then we can write

$$K_2^*(x,x) = \frac{1}{2}\frac{\partial}{\partial x}\ln\left(\det\{\hat{\mathrm{I}} + \hat{V}(x)\hat{V}^*(x)\}\right). \tag{3.3.10}$$

We obtain the well-known expression for multi-soliton solution of equation (or systems of equations), which is solvable by the IST method:

$$|q(x)|^2 = \frac{\partial^2}{\partial x^2}\ln\det\{\hat{\mathrm{I}} + \hat{V}^*(x)\hat{V}(x)\}. \tag{3.3.11}$$

In the cases of reductions (3.2.16) or for the real functions $r(x,t) = \pm q(x,t)$, expression for $q(x)$ can be written in more compact form [4, 20]:

$$q(x) = -2i\frac{\partial}{\partial x}\arctan\left\{\frac{\operatorname{Im}\det\{\hat{\mathrm{I}} + \hat{V}(x)\}}{\operatorname{Re}\det\{\hat{\mathrm{I}} + \hat{V}(x)\}}\right\}. \tag{3.3.12}$$

This is the famous solution of the evolution equation which include N_1 solitons and $N_2 = (N - N_1)/2$ breathers. It was named by single-pole soliton solution.

3.4. Some examples of the solution of spectral problem

Let us consider the potentials of the spectral problem (3.2.1) in the form

$$r(x) = -q^*(x), \quad q(x) = 2i\eta\,\operatorname{sech}(2\eta x)$$

Instead of Jost function $\Phi(x,\lambda)$, it is convenient to introduce the following function $f(x) = \Phi(x,\lambda)\exp(i\lambda x)$ with the asymptotic as $x \to -\infty$

$$f(x) \to \begin{pmatrix} 1 \\ 0 \end{pmatrix},$$

Equations (3.2.1) yield for the components of this function

$$\frac{\partial f_1}{\partial x} = 2i\eta\,\operatorname{sech}(2\eta x) f_2, \tag{3.4.1a}$$

$$\frac{\partial f_2}{\partial x} = 2i\lambda f_2 + 2i\eta\,\operatorname{sech}(2\eta x) f_1 \tag{3.4.1b}$$

If $f_1(x)$ and $f_2(x)$ are chosen in the form

$$f_1(x) = a + b\tanh(2\eta x), \quad f_2(x) = c\,\operatorname{sech}(2\eta x),$$

then substitution of these expressions in (3.4.1) and equating the coefficients at the functions $\operatorname{sech}(2\eta x)$, $\tanh(2\eta x)\operatorname{sech}(2\eta x)$ and $\operatorname{sech}^2(2\eta x)$ results in the relations between constants a, b and c:

$$a = -(\lambda/\eta)c, \quad b = ic.$$

Constant c is defined from asymptotic $f_1(x)$ as $x \to -\infty$: it is $c = -\eta(\lambda + i\eta)^{-1}$. Thus, Jost function $f(x)$ or Jost function $\Phi(x,\lambda)$ are found to be:

$$\Phi(x,\lambda) = \frac{\exp(-i\lambda x)}{\lambda + i\eta}\begin{pmatrix} \lambda - i\eta\tanh(2\eta x) \\ -\eta\,\mathrm{sech}(2\eta x) \end{pmatrix} \tag{3.4.2}$$

In order to find the second Jost function $\overline{\Phi}(x,\lambda)$ it is suitable to define an auxiliary function $\bar{f}(x) = \overline{\Phi}(x,\lambda)\exp(-i\lambda x)$, which has an asymptotic as $x \to -\infty$

$$\bar{f}(x) \to \begin{pmatrix} 0 \\ -1 \end{pmatrix} .$$

The equations for the component of this functions follow from (3.2.1)

$$\frac{\partial \bar{f}_1}{\partial x} = -2i\lambda\bar{f}_1 + 2i\eta\,\mathrm{sech}(2\eta x)\bar{f}_2, \tag{3.4.3a}$$

$$\frac{\partial \bar{f}_2}{\partial x} = 2i\eta\,\mathrm{sech}(2\eta x)\bar{f}_1 . \tag{3.4.3b}$$

Like for $f(x)$ with taking into account its behavior as $x \to -\infty$, here we can search for components $\bar{f}_1(x)$ and $\bar{f}_2(x)$ in the form:

$$\bar{f}_1(x) = c\,\mathrm{sech}(2\eta x), \quad \bar{f}_2(x) = a + b\tanh(2\eta x)$$

Substitution of these expressions in (3.4.3) allows to find the relationships between the constants a, b and c. The condition $\bar{f}_2(x) \to -1$ as $x \to -\infty$ fixes the magnitude of the constant c. The final result for $\overline{\Phi}(x,\lambda)$ is

$$\overline{\Phi}(x,\lambda) = \frac{\exp(i\lambda x)}{\lambda - i\eta}\begin{pmatrix} \eta\,\mathrm{sech}(2\eta x) \\ -\lambda - i\eta\tanh(2\eta x) \end{pmatrix} \tag{3.4.4}$$

Similarly we can find another Jost functions $\Psi(x,\lambda)$ and $\overline{\Psi}(x,\lambda)$:

$$\Psi(x,\lambda) = \frac{\exp(i\lambda x)}{\lambda + i\eta}\begin{pmatrix} \eta\,\mathrm{sech}(2\eta x) \\ \lambda + i\eta\tanh(2\eta x) \end{pmatrix} \tag{3.4.5}$$

and

$$\overline{\Psi}(x,\lambda) = \frac{\exp(-i\lambda x)}{\lambda - i\eta}\begin{pmatrix} \lambda - i\eta\tanh(2\eta x) \\ -\eta\,\mathrm{sech}(2\eta x) \end{pmatrix}. \tag{3.4.6}$$

The consideration of the asymptotics of $\Phi(x,\lambda)$ and $\overline{\Phi}(x,\lambda)$ as $x \to +\infty$ provides the matrix elements of the transition matrix:

$$T_{11}(\lambda) = \frac{\lambda - i\eta}{\lambda + i\eta}, \quad T_{12}(\lambda) = 0,$$
$$T_{21}(\lambda) = 0, \quad T_{22}(\lambda) = .\frac{\lambda + i\eta}{\lambda - i\eta}. \tag{3.4.7}$$

As far as $\rho(\lambda) = T_{21}(\lambda) / T_{11}(\lambda) = 0$, the potential considered here was named the reflectionless one. The single eigenvalues of the Zakharov-Shabat inverse spectral problem (3.2.1) are $\lambda_1 = i\eta$ and $\overline{\lambda}_1 = -i\eta$. The eigenfunction, which corresponds with $\lambda_1 = i\eta$, can be found exactly by the same way as it was done for Jost functions. The result is:

$$\Phi_1(x) = \operatorname{sech}(2\eta x)\begin{pmatrix} i^{-1/2} \exp(-\eta x) \\ i^{1/2} \exp(+\eta x) \end{pmatrix}, \tag{3.4.8}$$

Let us consider now the more general form of potential in equations (3.2.1) [21,22]

$$r(x) = -q^*(x), \quad q(x) = i\tilde{q}(x) = iA\operatorname{sech}(2\eta x) \tag{3.4.9}$$

By changing the variable x one can go over to the case of $\eta = 1/2$. Without losing the generality, we can consider only this particular case.

First of all it is necessary to obtain the one equation of second order for ψ_1 on the base of system (3.2.1).

$$\frac{d^2\psi_1}{dx^2} - \left(\frac{1}{\tilde{q}}\frac{d\tilde{q}}{dx}\right)\frac{d\psi_1}{dx} + \left[-i\lambda\left(\frac{1}{\tilde{q}}\frac{d\tilde{q}}{dx}\right) + \lambda^2 + \tilde{q}^2\right]\psi_1 = 0.$$

If to introduce new variable s

$$s = \frac{1}{2}(1 + \tanh x),$$

then this equation writes

$$s(1-s)\frac{d^2\psi_1}{ds^2} - \left(\frac{1}{2} - s\right)\frac{d\psi_1}{ds} + \left[A^2 + \frac{\lambda[\lambda - i(1-2s)]}{4s(1-s)}\right]\psi_1 = 0. \tag{3.4.10}$$

Now the new function $w(s)$ can be defined as

$$\psi_1(x) = s^{\alpha}(1-s)^{\beta} w(s),$$

[21,23]. Parameters α and β should be selected to transform equation (3.4.10) into the hypergeometric equation for $w(s)$:

$$s(1-s)\frac{d^2 w}{ds^2} - \left[c-(1+a+b)s\right]\frac{dw}{ds} - abw = 0 , \qquad (3.4.11)$$

where $c = 2\alpha + 1/2$, $a+b = 2(\alpha+\beta)$. The parameters α and β are fixed by the requirement that the multiplier ab is constant. Thus,

$$\alpha_{\pm} = \left[1 \pm (1+2i\lambda)\right]/4 , \quad \beta_{\pm} = \left[1 \pm (1-2i\lambda)\right]/4 .$$

The asymptotic behavior of the function ψ_1 as $x \to -\infty$ determines by the choice of the parameters α and β

Let ψ_1 is a first component of the Jost functions $\Phi(x,\lambda)$ that behaves as $\exp(-i\lambda x)$ as $x \to -\infty$. The two linear independent solution of (3.4.11) are expressed in terms of hypergeometric functions [24] as follows

$$w^{(1)}(s) = F(a, b, c; s),$$
$$w^{(2)}(s) = s^{1-c} F(a-c+1, b-c+1, 2-c; s) .$$

Thus

$$\Phi_1(x,\lambda) = s^{\alpha}(1-s)^{\beta}\left[B_1 w^{(1)}(s) + B_2 w^{(2)}(s)\right] .$$

The constants B_1 and B_2, α and β should be chosen to make these functions have a correct asymptotic at $x \to -\infty$. With taking this condition into account, we can find that

$$B_1 = 1, B_2 = 0 , \alpha = \alpha_- = -i\lambda/2 , \beta = \beta_- = i\lambda/2 .$$

Herewith $a = -b = A$, $c = 1/2 - i\lambda$ and we have

$$\Phi_1(x,\lambda) = s^{-i\lambda/2}(1-s)^{i\lambda/2} F(A, -A, 1/2 - i\lambda; s) . \qquad (3.4.12)$$

It is possible to exclude ψ_1 from equations (3.2.1) and then equation for $\psi_2 = \Phi_2(x,\lambda)$ gets similar to equation (3.4.10) under the replacing $\lambda \to -\lambda$. Consequently, the expression for $\Phi_2(x,\lambda)$ reads

$$\Phi_2(x,\lambda) = s^{\alpha}(1-s)^{\beta}\left[B_1' w^{(1)}(s) + B_2' w^{(2)}(s)\right],$$

where $\lambda \to -\lambda$. Hence, as $x \to -\infty$ we have

$$\Phi_2(x,\lambda) \to s^{\alpha} B_1' + s^{1/2-\alpha} B_2' = B_1' \exp(i\lambda x) + B_2' \exp(x - i\lambda x).$$

As far as in this limit $\Phi_2(x,\lambda) \to 0$, we need to put $B_1' = 0$ and for $\Phi_2(x,\lambda)$ we have the expression:

$$\Phi_2(x,\lambda) = B_2' s^{1/2+i\lambda/2}(1-s)^{-i\lambda/2} F(1/2 - i\lambda + A, 1/2 - i\lambda - A, 3/2 - i\lambda; s). \quad (3.4.13)$$

The relationship between $\Phi_2(x,\lambda)$ and $\Phi_1(x,\lambda)$ (i.e., equations (3.2.1)) can be used to define the constant B_2'

$$\Phi_2(x,\lambda) = -A^{-1}\cosh(x)\left(\frac{d\Phi_1}{dx} + i\lambda\Phi_1\right). \quad (3.4.14)$$

At $x \to -\infty$ (that is $s \to 0$) the hypergeometric functions has the following asymptotic

$$F(a,b,c;s) \approx 1 + \frac{ab}{c}s.$$

By using the expressions (3.4.12), (3.4.14) and this formula we find that as $x \to -\infty$

$$\Phi_2(x,\lambda) \to B_2' \exp(x - i\lambda x), \quad \Phi_1(x,\lambda) \to \exp(-i\lambda x)\left[1 - \frac{A^2 \exp(2x)}{1/2 - i\lambda}\right].$$

On the other hand, according to (3.4.14), one has

$$\Phi_2(x,\lambda) \to \frac{iA}{(1/2 - i\lambda)}\exp(x - i\lambda x).$$

Thereby $B_2' = iA(1/2 - i\lambda)^{-1}$. And finally

$$\Phi_2(x,\lambda) = \frac{iA}{(1/2 - i\lambda)} s^{1/2+i\lambda/2}(1-s)^{-i\lambda/2} F\left(\frac{1}{2} - i\lambda + A, \frac{1}{2} - i\lambda - A, \frac{3}{2} - i\lambda; s\right). \quad (3.4.15)$$

Now we are able to find the elements of the transition matrix $T_{11}(\lambda)$ and $T_{21}(\lambda)$. Considering the limit $x \to +\infty$, when $s \to 1 - \exp(-x)$, and using the formula

$$F(a,b,c;1) = \frac{\Gamma(c)\Gamma(c-a-b)}{\Gamma(c-a)\Gamma(c-b)},$$

we get

$$T_{11}(\lambda) = a(\lambda) = \frac{\Gamma(1/2 - i\lambda)\Gamma(1/2 - i\lambda)}{\Gamma(1/2 - i\lambda - A)\Gamma(1/2 - i\lambda + A)}, \tag{3.4.16}$$
$$T_{21}(\lambda) = b(\lambda) = \frac{iA\Gamma(3/2 - i\lambda)\Gamma(1/2 - i\lambda)}{(1/2 - i\lambda)\Gamma(1-A)\Gamma(1+A)}.$$

By using the following formulas for Γ-function [24]:

$$\Gamma(1+z) = z\Gamma(z), \quad \Gamma(1-z)\Gamma(z) = \frac{\pi}{\sin(\pi z)}, \quad \Gamma(\frac{1}{2} - z)\Gamma(\frac{1}{2} + z) = \frac{\pi}{\cos(\pi z)},$$

we can obtain for $T_{21}(\lambda)$

$$T_{21}(\lambda) = b(\lambda) = \frac{i\sin(\pi A)}{\cosh(\pi\lambda)}. \tag{3.4.17}$$

The points of discrete spectrum, which corresponds to the zeroes of functions $T_{11}(\lambda)$ at $\operatorname{Im}\lambda > 0$, are defined by the poles of Γ-functions in (3.4.16) , whence it follows that

$$\lambda_n = i(A - n + 1/2), \tag{3.4.18}$$

where $n = 1, 2, \ldots, N$, and N is determined from the condition of $\operatorname{Im}\lambda_n > 0$.

The coefficients C_n, which are introduced in the definition of the kernel of GLM equations, are determined by the residues of the function $b(\lambda)/a(\lambda)$ at the points of discrete spectrum. Let A in (3.4.9) be not an integer number. It follows from (3.4.17) that $b(\lambda_n) = i(-1)^{n+1}$. If A is an integer number, this result can be obtained as well but in more accurate manner. Thereby,

$$C_n = \operatorname{Res}\frac{b(\lambda)}{a(\lambda)} = \frac{i(-1)^{n+1}\Gamma(1/2 + A - i\lambda_n)}{\Gamma^2(1/2 - i\lambda_n)} \operatorname{Res}\Gamma(1/2 - A - i\lambda_n)\Big|_{\lambda=\lambda_n}.$$

Utilizing the formula

$$\mathrm{Res}\,\Gamma(z)\Big|_{z=(1-n)} = \frac{(-1)^{n-1}}{(n-1)!},$$

we obtain

$$C_n = \frac{i\Gamma(2A+1-n)}{(n-1)!\,\Gamma^2(A+1-n)}. \tag{3.4.19}$$

Other pair of the Jost functions $\Psi(x,\lambda)$ and $\overline{\Psi}(x,\lambda)$ can be found by repeating the routine used to derive $\Phi(x,\lambda)$. But in this time it is needed to enter new variable $s' = 1 - s$, so as $x \to +\infty$ we can analyze the hypergeometric functions near the point $s' = 0$. However, the alternative approach can be offered. From (6.3a) follows that

$$\begin{aligned} \Phi_1(x,\lambda) &= a(\lambda)\overline{\Psi}_1(x,\lambda) + b(\lambda)\Psi_1(x,\lambda), \\ \Phi_2(x,\lambda) &= a(\lambda)\overline{\Psi}_2(x,\lambda) + b(\lambda)\Psi_2(x,\lambda). \end{aligned} \tag{3.4.20}$$

As far as $\Phi(x,\lambda)$, $b(\lambda)$ and $a(\lambda)$ are already known, one can find $\Psi(x,\lambda)$ and $\overline{\Psi}(x,\lambda)$, by expressing $\Phi(x,\lambda)$ in terms of $s' = 1 - s$. The resulting expression should be compared with (3.4.20). By using the formula [24]

$$\begin{aligned} F(a,b,c;s) &= A_1 F(a,b,\, a+b-c+1;\, 1-s) + \\ &+ A_2(1-s)^{c-a-b} F(c-a, c-b,\, c-a-b+1;\, 1-s), \end{aligned}$$

where

$$A_1 = \frac{\Gamma(c)\Gamma(c-a-b)}{\Gamma(c-a)\Gamma(c-b)}, \quad A_2 = \frac{\Gamma(c)\Gamma(a+b-c)}{\Gamma(a)\Gamma(b)},$$

after a series of manipulates with Γ-functions, we can get

$$\begin{aligned} \Phi_1(x,\lambda) &= a(\lambda) s^{+i\lambda/2}(1-s)^{i\lambda/2} F\left(A, -A, \frac{1}{2} + i\lambda; 1-s\right) + \\ &+ b(\lambda) s^{+i\lambda/2}(1-s)^{1/2+i\lambda/2} \frac{(-iA)}{(1/2 - i\lambda)} F\left(\frac{1}{2} - i\lambda - A, \frac{1}{2} - i\lambda + A, \frac{3}{2} - i\lambda; 1-s\right) \end{aligned}$$

Hence it follows the desired results

$$\overline{\Psi}_1(x,\lambda) = (s')^{i\lambda/2}(1-s')^{-i\lambda/2} F\left(A, -A, \frac{1}{2} + i\lambda; s'\right), \tag{3.4.21}$$

$$\Psi_1(x,\lambda) = \frac{-iA}{(1/2 - i\lambda)}(s')^{1/2 - i\lambda/2}(1-s')^{-i\lambda/2} F\left(\frac{1}{2} - i\lambda - A, \frac{1}{2} - i\lambda + A, \frac{3}{2} - i\lambda; s'\right).$$

The similar approach leads to the expression for $\Phi_2(x,\lambda)$:

$$\Phi_2(x,\lambda)=a(\lambda)\frac{(-iA)}{(1/2+i\lambda)}s^{1/2+i\lambda/2}(1-s)^{1/2+i\lambda/2}F\left(1-A,1+A,\frac{3}{2}+i\lambda;1-s\right)+$$

$$+b(\lambda)s^{1/2-i\lambda/2}(1-s)^{-i\lambda/2}F\left(\frac{1}{2}-i\lambda+A,\frac{1}{2}-i\lambda-A,\frac{1}{2}-i\lambda;1-s\right).$$

Both summands in this expression should be converted according to the formula

$$F(a,b,c;s)=(1-s)^{c-a-b}F(c-a,c-b,c;s) \ .$$

Thus, we obtain

$$\overline{\Psi}_2(x,\lambda)=\frac{-iA}{(1/2+i\lambda)}(s')^{1/2+i\lambda/2}(1-s')^{i\lambda/2}F\left(\frac{1}{2}+i\lambda+A,\frac{1}{2}+i\lambda-A,\frac{3}{2}+i\lambda;s'\right)$$

$$\Psi_2(x,\lambda)=(s')^{-i\lambda/2}(1-s')^{i\lambda/2}F\left(A,-A,\frac{1}{2}-i\lambda;s'\right).$$

These results can be checked by means of symmetry relation, which follows from the reduction $r=-q^*$ when $\lambda=\lambda^*$:

$$\overline{\Psi}(x,\lambda)=\begin{pmatrix}0 & 1\\ -1 & 0\end{pmatrix}\Psi^*(x,\lambda) \ .$$

3.5. Evolution of scattering data

In order to solve the non-linear evolution equation it needs to find the linear equation governing the variation of the scattering data. We begin by considering the linear equations of IST method (3.1.1)

$$\frac{\partial}{\partial x}\hat{G}=\hat{U}\hat{G}, \qquad (3.5.1a)$$

$$\frac{\partial}{\partial t}\hat{G}=\hat{V}\hat{G}, \qquad (3.5.1b)$$

where $\hat{G}$ is matrix function. We consider the equation (3.5.1a) as the spectral problem.

Let U-V-matrices are such that

$$\lim_{x\to\pm\infty} \hat{U}(\lambda,x) = \hat{U}^{(\pm)}(\lambda), \quad \lim_{x\to\pm\infty} \hat{V}(\lambda,x) = \hat{V}^{(\pm)}(\lambda),$$

where $\hat{U}^{(\pm)}(\lambda)$ and $\hat{V}^{(\pm)}(\lambda)$ are the diagonal constant matrices. Solution of the (3.5.1a) represents the Jost matrices either $\hat{\Phi}$ or $\hat{\Psi}$ for which

$$\hat{\Psi}(x,\lambda) \to \hat{\Psi}^{(+)}(x,\lambda) = \exp[\hat{U}^{(+)}(\lambda)x], \quad \text{as} \quad x \to +\infty,$$
$$\hat{\Phi}(x,\lambda) \to \hat{\Psi}^{(-)}(x,\lambda) = \exp[\hat{U}^{(-)}(\lambda)x], \quad \text{as} \quad x \to -\infty.$$

These Jost matrices define the transfer matrix $\hat{T}(\lambda)$ as

$$\hat{T}(\lambda) = \hat{\Psi}^{-1}\hat{\Phi}.$$

Taking (3.5.1a) into account we can show that $\partial\hat{T}(\lambda)/\partial x = 0$.

Differentiating (3.5.1a) with respect to t, we have

$$\frac{\partial}{\partial t}\left(\frac{\partial}{\partial x}\hat{\Phi}\right) = \left(\frac{\partial}{\partial t}\hat{U}\right)\hat{\Phi} + \hat{U}\left(\frac{\partial}{\partial t}\hat{\Phi}\right),$$

where it takes $\hat{G} = \hat{\Phi}$. It can be rewritten as

$$\frac{\partial}{\partial x}\left(\hat{\Phi}^{-1}\frac{\partial}{\partial t}\hat{\Phi}\right) = \hat{\Phi}^{-1}\left(\frac{\partial}{\partial t}\hat{U}\right)\hat{\Phi}.$$

Integrating this expression with respect to x from $-\infty$ to x, we have

$$\hat{\Phi}^{-1}\frac{\partial\hat{\Phi}}{\partial t}\Bigg|_{-\infty}^{x} = \int_{-\infty}^{x}\hat{\Phi}^{-1}\frac{\partial\hat{U}}{\partial t}\hat{\Phi}dx'.$$

It is convenient to use

$$\frac{\partial}{\partial t}\hat{\Phi} = \hat{V}\hat{\Phi} - \hat{\Phi}\hat{V}^{(-)} \tag{3.5.2}$$

as the second linear equations of IST method associated with spectral problem (3.5.1a). The integrability condition for the pair equations (3.5.1a) (where it takes $\hat{G} = \hat{\Phi}$) and (3.5.2) results in the same zero-curvature representation (3.1.2):

$$\frac{\partial\hat{U}}{\partial t} - \frac{\partial\hat{V}}{\partial x} + [\hat{U}, \hat{V}] = 0. \tag{3.5.3}$$

From (3.5.2) it follows

$$\frac{\partial \hat{\Phi}}{\partial t} \to 0 \text{ as } x \to -\infty ,$$

thus

$$\frac{\partial \hat{\Phi}}{\partial t} = \hat{\Phi} \int_{-\infty}^{x} \hat{\Phi}^{-1} \frac{\partial \hat{U}}{\partial t} \hat{\Phi} dx' . \tag{3.5.4}$$

Now, we can take the $x \to +\infty$ and obtain the equation for transfer matrix. By using

$$\frac{\partial}{\partial t} \hat{\Psi} = \hat{V} \hat{\Psi} - \hat{\Psi} \hat{V}^{(+)}$$

as equation that determine time variation of Jost matrix $\hat{\Psi}$, we can find

$$\frac{\partial \hat{\Psi}}{\partial t} \to 0 \quad \text{as} \quad x \to +\infty .$$

Thus, with taking $\det \hat{\Psi}^{(+)} \neq 0$ into account the equation (3.5.4) leads to

$$\frac{\partial \hat{T}}{\partial t} = \hat{T} \int_{-\infty}^{+\infty} \hat{\Phi}^{-1} \frac{\partial \hat{U}}{\partial t} \hat{\Phi} dx' . \tag{3.5.5}$$

From (3.5.3) and (3.5.5) we have

$$\frac{\partial \hat{T}}{\partial t} = \hat{T} \int_{-\infty}^{+\infty} \hat{\Phi}^{-1} \left(\frac{\partial \hat{V}}{\partial x} + [\hat{V}, \hat{U}] \right) \hat{\Phi} dx' . \tag{3.5.6}$$

Equation (3.5.2) results in

$$\hat{V} = \frac{\partial \hat{\Phi}}{\partial t} + \hat{\Phi} \hat{V}^{(-)} \hat{\Phi}^{-1} .$$

Differentiating this expression with respect to x and using the (3.5.1a) and (3.5.2) we can find

$$\hat{\Phi}^{-1} \frac{\partial \hat{V}}{\partial x} \hat{\Phi} = \frac{\partial}{\partial x} \left(\hat{\Phi}^{-1} \frac{\partial \hat{\Phi}}{\partial x} \right) + \left[\hat{\Phi}^{-1} \frac{\partial \hat{\Phi}}{\partial x} , \hat{V}^{(-)} \right] ,$$

and

$$\hat{\Phi}^{-1} [\hat{V}, \hat{U}] \hat{\Phi} = - \frac{\partial \hat{\Phi}^{-1}}{\partial t} \frac{\partial \hat{\Phi}}{\partial x} + \frac{\partial \hat{\Phi}^{-1}}{\partial x} \frac{\partial \hat{\Phi}}{\partial t} + \left[\hat{V}^{(-)} , \hat{\Phi}^{-1} \frac{\partial \hat{\Phi}}{\partial x} \right] .$$

Substitution of these expressions into integral in the (3.5.6) result in the following equation

$$\frac{\partial \hat{T}}{\partial t} = \hat{T} \int_{-\infty}^{+\infty} \frac{\partial}{\partial x} \left(\hat{\Phi}^{-1} \frac{\partial \hat{\Phi}}{\partial t} \right) dx' = \hat{T} \left(\hat{\Phi}^{-1} \frac{\partial \hat{\Phi}}{\partial t} \right) \Bigg|_{-\infty}^{+\infty} .$$

With taking into account the following formulae

$$\hat{\Phi}^{(+)} = \hat{\Psi}^{(+)} \hat{T}, \quad \hat{\Phi}^{(+)-1} = \hat{T}^{-1} \hat{\Psi}^{(+)-1}$$

we can obtain the equation for transfer matrix

$$\frac{\partial \hat{T}}{\partial t} = \hat{T} \left(\hat{T}^{-1} \hat{\Psi}^{(+)-1} \hat{V}^{(+)} \hat{\Psi}^{(+)} \hat{T} - \hat{\Phi}^{(-)-1} \hat{V}^{(-)} \hat{\Phi}^{(-)} \right),$$

or

$$\frac{\partial \hat{T}}{\partial t} = \hat{\Psi}^{(+)-1} \hat{V}^{(+)} \hat{\Psi}^{(+)} \hat{T} - \hat{T} \hat{\Phi}^{(-)-1} \hat{V}^{(-)} \hat{\Phi}^{(-)} .$$

So far as $\hat{U}^{(\pm)}(\lambda)$ and $\hat{V}^{(\pm)}(\lambda)$ are the diagonal matrices this equation is reduced to

$$\frac{\partial \hat{T}}{\partial t} = \hat{V}^{(+)} \hat{T} - \hat{T} \hat{V}^{(-)} . \tag{3.5.7}$$

If $\hat{V}^{(+)} = \hat{V}^{(-)}$, then (3.5.7) is converted into simple matrix equation

$$\frac{\partial \hat{T}}{\partial t} = [\hat{V}^{(-)}, \hat{T}] . \tag{3.5.8}$$

Equation (3.5.8) shows that in this case the diagonal elements of transfer matrix are the integrals of motion in considered non-linear evolution equation. More correctly, they are the generating functions of the infinite series of the integrals.

3.6. Conclusion

Both the Fourier integral transform method of solving a linear evolution equations and IST method of solving the non-linear evolution equation are the main technique in modern theoretical investigations of the non-linear optical waves. In this chapter we consider only spectral problem, which was proposed by Zakharov and Schabat, and was developed by Ablowitz, Kaup, Newell and Segur. In most cases it is enough to analyse the formation and propagation of the optical solitons. In the next chapters we shall consider the main examples of that. However, there are other spectral problems. It is noteworthy that non-linear optics provides the examples of the wave equation that can be solved by the IST method based on these spectral problems, too. It is convenient to discuss the relevant variant of the IST methods for the specific instances of the wave phenomena. In any case the principal features of this method observed above are invariable.

References

1. Ablowitz, M.J., Kaup, D.J., Newell, A.C., and Segur, H.: The inverse scattering transform - Fourier analysis for nonlinear problems, *Stud.Appl.Math.* **53** (1974), 249-315.
2. Takhtajan, L.A., and Faddeev L.D.: *Hamiltonian approach to theory of solitons*, Nauka, Moscow, 1986.
3. Ablowitz, M.J., and Segur, H.: *Solitons and the Inverse Scattering Transform*, SIAM, Philadelphia, 1981.
4. Zakharov, V.E., Manakov, S.V., Novikov, S.P., and Pitaevskii, L.P.: *Theory of Solitons: The Inverse Problem Method* [in Russian], Nauka, Moscow, 1980. *Theory of Solitons: The Inverse Scattering Method*, Plenum, New York, 1984.
5. Eilenberger, G.: *Solitons. Mathematical Methods for Physicists*, Springer-Verlag, Berlin, Heidelberg, New York, 1981.
6. Calogero, F., Degasperis, A.: *Spectral Transform and Solitons*. vol.1, Amsterdam: North- Holland, 1982.
7. Calogero F., edit. *Nonlinear Evolution Equations Solvable by Spectral Transform* Pitman 1978.
8 Dodd, R.K., Eilbeck, J.C., Gibbon, J.D., and Morris, H.C.: *Solitons and Nonlinear Waves*, Academic Press, New York, 1982.
9. Newell, A.C.: *Solitons in Mathematics and Physics*, University of Arizona, 1985.
10. Drazin, P.G., and Johnson, R.S.: *Solitons: An Introduction*, Cambridge University Press, Cambridge, 1989
11. Lamb, G.L., Jr.: *Elements of Soliton Theory*, Wiley, New York, 1980.
12. Hasegawa, A.: *Optical Solitons in Fibers*, Springer-Verlag, Berlin, 1989.
13. Miura, R.M., ed., *Backlund Transformations, the Inverse Scattering Method, Solitons and Their Applications*, (Lect.Notes in Math. 515), Springer-Verlag, Berlin, 1976
14. Bishop, A.R. and Schneider, T., eds., *Solitons and Condensed Matter Physics*, Springer, Berlin, Heidelberg, New York, 1978
15. Bullough, R.K. and Caudrey, P.J., eds., *Solitons,* Springer, Berlin, Heidelberg, New York, 1980.
16. Monastyrsky M.I.: *Topology of gauge fields and condensed mater*, PAIMS, Moscow, 1995.
17. Zakharov, V.E., and Faddeev, L.D.: The Korteweg-de Vries equations: A completely integrable Hamiltonian system, *Funct. Analysis Appl.* **5** (1971), 280-287.
18. Ablowitz, M.J., Kaup, D.J., Newell, A.C., and Segur, H.: Method for solution of the Sine-Gordon equation, *Phys.Rev.Lett.* **30** (1973), 1262-1264.
19. Ablowitz, M.J., Kaup, D.J., Newell, A.C., and Segur, H.: Nonlinear evolution equations of physical significance, *Phys.Rev.Lett.* **31** (1973), 125-127.
20. Wadati, M., and Ohkuma K.: *J.Phys.Soc.Japan,* **51** (1982), 2029-2035.
21. Satsuma J., and Yajima N.: Initial value problems of one-dimensional self-modulation of non-linear waves in dispersive media. *Progr.Theor.Phys.Suppl.*, №55 (1974), 284-306.
22. Michalska-Trautman R., Exact Solution of the Initial Eigenvalue Problem for Coherent Pulse Propagation. Phys. Rev. A23, №1, 352-359 (1981)
23. Michalska-Trautman R.: Exact solution of the initial eigenvalue problem for coherent pulse propagation, *Phys.Rev.* **A 23** (1981), 352-359.
24. *Higher Transcendental Functions*, vol. 1, eds. H.Bateman and A.Erdelyi, McGraw-Hill, New York, 1955.

CHAPTER 4

SELF-INDUCED TRANSPARENCY

A self-induced transparency (SIT) phenomenon consists in the propagation of a powerful ultrashort pulse (USP) of light through a resonance medium without the distortion and energy loss of this pulse [1-4]. This phenomenon is characterised by the continuous absorption and re-emission of electromagnetic radiation by resonant atoms of medium in such a manner that steady-state optical pulse propagates. In the ideal case the energy dissipation of the USP is invisible, and the state of the resonant medium is not varying. It means that the medium is transparent. The group velocity of a such steady-state pulse, called 2π-pulse or soliton of SIT, is less than the phase speed of light in a medium. The group velocity depends on a 2π-pulse duration: the shorter is the duration, the higher is its speed [2-5]. When two pulses of the different velocities spread in the medium, the second pulse may overtake the first and a collision will take place. After the collision, the solitons keep their shape and velocity (but in general all other parameters of solitons may alter). This fundamental property of the SIT solitons has been studied many times both theoretically and experimentally [3,6,7].

From the mathematical point of view this property is a consequence of the complete integrability of the reduced Maxwell-Bloch equations, describing the SIT in the two-level media with non-degenerated levels [8-13]. The 2π-pulses correspond to the single-soliton solutions of these equations, and the process of "collision" reflects the evolution of the double-soliton solution – its asymptotically transformation into a pair of solitons under certain conditions (see, for example, [6,9,14-16]).

The simplest theory describing the self-induced transparency phenomenon was developed by McCall and Hahn. Later the more complicated theories appeared, however the McCall-Hahn theory is quite instructive one. The base of the McCall-Hahn theory is represented in section 4.1. We shall consider the steady-state solution both of the total Maxwell-Bloch equations and the McCall-Hahn's system of the equations. These equations may be represented as the condition of integrability of some linear equations that provides the solution of these equations be the IST method. Generation of the SIT theory in frame of the IST method will be considered in section 4.2. Coherent pulse propagation along the interface of dielectric medium and SIT for surface polaritons are the subject of consideration in section 4.3. Non-trivial features of the self-induced transparency are discussed in section 4.4 as an example of the pulse propagation under two-photon resonance. This chapter is finished by the analysis of the relations between several field-theoretical models and the SIT theories.

4.1. Theory of SIT phenomena by McCall-Hahn

In general, the theory of the interaction of radiation with an ensemble of two-level atoms is based on the Bloch equations for atoms and the Maxwell equations for the classical electromagnetic field (for the sake of details see section 1.1.2. and 1.3.4.). In an isotropic dielectric the set of Maxwell equations reduced to one equation for the electric field $\vec{E} = E \cdot \vec{l}$. For a plane wave with constant polarisation vector $\vec{l}$ one can obtain the following system of total Maxwell-Bloch (MB) equations;

$$\frac{\partial^2 E}{\partial z^2} - \frac{1}{c^2}\frac{\partial^2 E}{\partial t^2} = \left(\frac{4\pi n_A d}{c^2}\right)\left\langle \frac{\partial^2 r_1}{\partial t^2} \right\rangle, \tag{4.1.1a}$$

$$\frac{\partial r_1}{\partial t} = -\omega_a r_2, \quad \frac{\partial r_2}{\partial t} = \omega_a r_1 + \frac{2d}{\hbar} E r_3, \quad \frac{\partial r_3}{\partial t} = -\frac{2d}{\hbar} E r_2, \tag{4.1.1b}$$

where d is the projection of a matrix element of the dipole operator on the direction of $\vec{l}$, n_A is the concentration of resonant atoms. It should be noted that the components of Bloch vector r_1, r_2, and r_3 depend on the atomic resonance frequency ω_a. Hereafter the angular brackets represent summation over all the atoms characterised by the frequency ω_a.

The Bloch equations contain products of the field E and the polarisations r_2 and r_3 responsible for interference between the opposite propagated waves. It has been shown [58,60], however, that if the density of resonant atoms is small enough to make the parameter $4\pi n_A d^2 / \hbar\omega_a$ less than unity, interference may be neglected. It was found that for a typical value of $d \sim 1$ Debye, $\omega_0 \sim 10^{15}$ s^{-1} and $n_A \ll 10^{23}$ cm^{-3} one may not take into account the backward wave generation by a forward running pulse. Thus, the MB equations convert into the simple reduced Maxwell-Bloch (RMB) system of equations.

Following [60] we introduce an auxiliary function $B(z,t)$, satisfying the equations

$$\frac{\partial B}{\partial t} = c\frac{\partial E}{\partial z},$$

so that the equation (4.1.1a) can be presented in the equivalent form

$$\frac{\partial B}{\partial t} - c\frac{\partial E}{\partial z} = 0,$$

$$\frac{\partial E}{\partial t} - c\frac{\partial B}{\partial z} = -4\pi n_A d\left\langle \frac{\partial r_1}{\partial t} \right\rangle. \tag{4.1.2}$$

After simple algebra one obtains

$$\left(\frac{\partial B}{\partial t}+c\frac{\partial B}{\partial z}\right)-\left(\frac{\partial E}{\partial t}+c\frac{\partial E}{\partial z}\right)=4\pi n_A d\left\langle\frac{\partial r_1}{\partial t}\right\rangle, \tag{4.1.3a}$$

$$\left(\frac{\partial B}{\partial t}-c\frac{\partial B}{\partial z}\right)+\left(\frac{\partial E}{\partial t}-c\frac{\partial E}{\partial z}\right)=-4\pi n_A d\left\langle\frac{\partial r_1}{\partial t}\right\rangle. \tag{4.1.3b}$$

The equations of characteristics are $\xi=t+z/c$, $\eta=t-z/c$. In terms of ξ and η equations (4.1.3) look more compact

$$\frac{\partial B}{\partial\eta}+\frac{\partial E}{\partial\eta}=-2\pi n_A d\left\langle\frac{\partial r_1}{\partial t}\right\rangle,\quad \frac{\partial B}{\partial\xi}-\frac{\partial E}{\partial\xi}=2\pi n_A d\left\langle\frac{\partial r_1}{\partial t}\right\rangle. \tag{4.1.4}$$

One may assume now, that E, B and r_1 are the waves moving preferably in one dimension, for instance along characteristic $\eta=t-z/c$. If there were, no response atoms in a sample than all quantities in (4.1.4) would depend only on η. In the case under consideration a back scattering wave occurs. Set $2\pi n_A d=\varepsilon$ and expand E, B and r_1 into a series over degrees of ε considering this parameter to be small

$$E=E^{(0)}(\eta)+\varepsilon E^{(1)}(\eta,\xi)+\varepsilon^2E^{(2)}(\eta,\xi)+\ldots,$$
$$B=B^{(0)}(\eta)+\varepsilon B^{(1)}(\eta,\xi)+\varepsilon^2B^{(2)}(\eta,\xi)+\ldots,$$
$$r_1=r_1^{(0)}(\eta)+\varepsilon r_1^{(1)}(\eta,\xi)+\varepsilon^2r_1^{(2)}(\eta,\xi)+\ldots\,.$$

Substitution of the expansion of E, B and r_1 in the first equation of (4.1.4) gives in the first order of ε:

$$\frac{\partial}{\partial\eta}\left(B^{(0)}+\varepsilon B^{(1)}\right)+\frac{\partial}{\partial\eta}\left(E^{(0)}+\varepsilon E^{(1)}\right)=-\varepsilon\left\langle\frac{\partial r_1^{(0)}}{\partial\eta}\right\rangle,$$

or

$$B+E=-\varepsilon\left\langle r_1^{(0)}\right\rangle.$$

In the equations above we have assumed that $r_1^{(0)}$ does not depend on ξ and that field disappears at some point along with the polarisation of the medium. For example, electric field and polarisation vanish at $t\to\pm\infty$.

Substituting the above expression in the second equation of (4.1.4) and holding the terms up to the first order of ε in the right hand side one obtains

$$\frac{\partial E}{\partial \xi} = -\frac{\varepsilon}{2}\frac{\partial r_1^{(0)}}{\partial \eta}.$$

In terms of initial z and t variables the above equation reads

$$\frac{\partial E}{\partial z} + \frac{1}{c}\frac{\partial E}{\partial t} = -\left(\frac{2\pi n_A d}{c}\right)\left\langle\frac{\partial r_1}{\partial t}\right\rangle, \tag{4.1.5a}$$

This equation together with the Bloch equations in the same order of ε

$$\frac{\partial r_1}{\partial t} = -\omega_a r_2, \quad \frac{\partial r_2}{\partial t} = \omega_a r_1 + \frac{2d}{\hbar}Er_3, \quad \frac{\partial r_3}{\partial t} = -\frac{2d}{\hbar}Er_2, \tag{4.1.5b}$$

yields the RMB equation. In (4.1.5) the upper script of all the values is omitted for brevity. It should be emphasised that in both the MB and RMB equations symbol E denotes the real value of the electric field strength. Representation of the electromagnetic wave as a quasi-monochromatic one is

$$E(z,t) = 2A(z,t)\cos[k_0 z - \omega_0 t + \varphi(z,t)] = \mathcal{E}(z,t)\exp[i(k_0 z - \omega_0 t)] + c.c.$$

where ω_0 is the radiation frequency, k_0 is the wave number, and the real envelope $A(z,t)$ and the phase $\varphi(z,t)$ are slowly varying functions of z and t. This is an approximation which means that the envelope and the phase obey the inequalities

$$\left|\frac{\partial A}{\partial t}\right| << \omega_0 |A|, \quad \left|\frac{\partial A}{\partial z}\right| << k_0 |A|, \quad \left|\frac{\partial \varphi}{\partial t}\right| << \omega_0 |\varphi|, \quad \left|\frac{\partial \varphi}{\partial z}\right| << k_0 |\varphi|.$$

Besides, the envelope amplitudes are usually so weak that Rabi frequency turns out to be much less than the resonance transition frequency. The resulting system of equations was obtained and discussed in Chapter 1 (section 1.3.4). Here we represent resulting equations

$$\frac{\partial \tilde{E}}{\partial z} + \frac{1}{c}\frac{\partial \tilde{E}}{\partial t} = -\alpha'\langle P\rangle, \quad \tilde{E}\left(\frac{\partial \varphi}{\partial z} + \frac{1}{c}\frac{\partial \varphi}{\partial t}\right) = \alpha'\langle Q\rangle, \tag{4.1.6a}$$

$$\frac{\partial Q}{\partial t} = \left(\Delta\omega + \frac{\partial \varphi}{\partial t}\right)P, \quad \frac{\partial P}{\partial t} = -\left(\Delta\omega + \frac{\partial \varphi}{\partial t}\right)Q + \tilde{E}R_3, \quad \frac{\partial R_3}{\partial t} = -\tilde{E}P, \tag{4.1.6b}$$

where $\Delta\omega = (\omega_a - \omega_0)$, $\alpha' = 2\pi\omega n_A d^2 / \hbar c$, $\tilde{E} = dA / 2\hbar$ is a normalised slowly varying pulse envelope and P, Q and R_3 are connected with initial Bloch vector components by the relations

$$r_1 = -P(z,t)\sin[k_0 z - \omega_0 t + \varphi(z,t)] + Q(z,t)\cos[k_0 z - \omega_0 t + \varphi(z,t)],$$
$$r_3 = -R_3(z,t).$$

Alternatively, the slowly varying functions P, Q and R_3 are defined in terms of variables from section 1.3.4 as

$$R_{21} = (1/2)(Q + iP)\exp[i\varphi(z,t)], \quad R_1 - R_2 = R_3 .$$

To describe the SIT phenomenon McCall and Hahn used the system (4.1.6). Equations (4.1.6) will be referred to as the general SIT equations.

Furthermore, if we confine the SIT analysis to situations when the input optical pulse does not carry any phase modulation, i.e. $\partial\varphi / \partial z = \partial\varphi / \partial t = 0$ at $z = 0$ and the form factor of the inhomogeneous line is a symmetrical function of frequency detuning $\Delta\omega$, then equations (4.1.6) yield $\partial\varphi / \partial z = \partial\varphi / \partial t = 0$ at any z and t. In this case equations (4.1.6) reduce to the system of SIT equations

$$\frac{\partial \tilde{E}}{\partial z} + \frac{1}{c}\frac{\partial \tilde{E}}{\partial t} = -\alpha' \langle P \rangle , \qquad (4.1.7a)$$

$$\frac{\partial Q}{\partial t} = \Delta\omega P , \quad \frac{\partial P}{\partial t} = -\Delta\omega Q + \tilde{E} R_3 , \quad \frac{\partial R_3}{\partial t} = -\tilde{E} P . \qquad (4.1.7b)$$

We have now set up all the systems of equations required to consider the ultra-short optical pulse propagation in a resonant medium.

4.1.1. STATIONARY SOLUTION OF SIT EQUATIONS

Before that moment when it became clear that SIT equations could be solved by the inverse scattering transform method, a unique analytical solution of these equations had been given by McCall and Hahn describing the stationary pulse propagation (steady-state solution). Besides the solutions in the form of the solitary wave there were periodic stationary solutions of the SIT equations which we call *cnoidal waves* [61-63]. These solutions do not satisfy the conditions of the applicability of SIT theory as far as their duration exceeds the polarisation and inversion relaxation times of resonant atoms. But together with solitary waves cnoidal waves play a certain role in understanding the processes of the optical pulses non-linear propagation in a resonant environment.. Cnoidal waves were discussed and investigated in literature. They should be considered like illustrations of the development of SIT theory.

So, let us denote a unique independent variable for all dependent variables as $\tau = (t - z/V)$, where V is a velocity of a steady state pulse respectively, which have not been determined yet. In terms of this notation SIT equations can be written as:

$$\left(1 - \frac{c}{V}\right)\frac{d\tilde{E}}{d\tau} = -c\alpha'\langle P\rangle, \tag{4.1.8a}$$

$$\left(1 - \frac{c}{V}\right)\tilde{E}\frac{d\varphi}{d\tau} = +c\alpha'\langle Q\rangle, \tag{4.1.8b}$$

$$\frac{dQ}{d\tau} = \left(\Delta\omega + \frac{d\varphi}{d\tau}\right)P , \tag{4.1.8c}$$

$$\frac{dP}{d\tau} = -\left(\Delta\omega + \frac{d\varphi}{d\tau}\right)Q + \tilde{E}R_3, \tag{4.1.8d}$$

$$\frac{dR_3}{d\tau} = -\tilde{E}P . \tag{4.1.8e}$$

Some additional assumptions are made:

1. the frequency of the carrier waves ω_0 is tuned to the centre of the symmetric inhomogeneously broadened line of absorption : $F(\Delta\omega) = F(-\Delta\omega)$;
2. $Q(\Delta\omega) = -Q(-\Delta\omega)$;
3. as $\tau \to \infty$ we state $d\varphi/d\tau = 0$, i.e. there is no phase modulation.

It follows from assumptions 2 and 3 and the phase equations that $d\varphi/d\tau = 0$ and $\varphi = 0$ for all τ. Then from (4.1.8c) and equations for envelope (4.1.8a) we have:

$$\left(1 - \frac{c}{V}\right)\tilde{E} = -\alpha' c\left\langle \frac{Q}{\Delta\omega}\right\rangle,$$

where we take into account that both $\tilde{E}$ and Q vanish simultaneously. Then

$$\left(1 - \frac{c}{V}\right) = -c\alpha'\int_{-\infty}^{\infty}\left(\frac{F(\Delta\omega)}{\Delta\omega}\right)\frac{Q(\tau,\Delta\omega)}{E(\tau)}d\Delta\omega . \tag{4.1.9}$$

Having differentiated both parts of (4.1.9) with respect to τ and taking into account that $F(\Delta\omega)$ is an arbitrary symmetric function we can show that

$$\frac{\partial}{\partial\tau}\left[\frac{Q(\tau,\Delta\omega)}{E(\tau)}\right] = 0,$$

Expression in square brackets is a function of only frequency detuning $\Delta\omega$, which function hereinafter will be referred to as $\chi(\Delta\omega)$:

$$Q(\tau,\Delta\omega) = \chi(\Delta\omega)\widetilde{E}(\tau). \tag{4.1.10}$$

Providing (4.1.10), equations (4.1.8d) and (4.1.8e) give rise to an integral of motion

$$P^2 + R_2^3 - 2\Delta\omega\chi(\Delta\omega)R_3 = I_1. \tag{4.1.11}$$

It is worth reminding that Bloch equations (4.1.8c) - (4.1.8e) have a integral of motion, which is the consequence of normalisation of the density matrix for a two-level system. This integral of motion does not depend on the assumption made before,

$$P^2 + R_3^2 + Q^2 = 1 . \tag{4.1.12}$$

Taking into account these two integrals and relation (4.1.10) we can express $R_3(\Delta\omega,\tau)$ as

$$R_3(\Delta\omega,\tau) = -\frac{I_1 - 1}{2\Delta\omega\chi(\Delta\omega)} - \frac{\chi(\Delta\omega)}{2\Delta\omega}\widetilde{E}^2(\tau) . \tag{4.1.13}$$

Equation (4.1.8c) and relation (4.1.10) allow to express $P(\Delta\omega,\tau)$ in terms of the functions depending only on $\Delta\omega$ or τ:

$$P(\Delta\omega,\tau) = \frac{\chi(\Delta\omega)}{\Delta\omega}\left(\frac{d\widetilde{E}}{d\tau}\right). \tag{4.1.14}$$

If now to substitute Bloch vector components (4.1.10), (4.1.13) and (4.1.14) in (4.1.12), we obtain the equation with respect to an ultrashort pulse envelope $\widetilde{E}(\tau)$. Now it is useful to fix different boundary conditions defining the integrating constant and the value of the integral of motion in (4.1.11). As a result different types of solution of the equation for $\widetilde{E}(\tau)$ will be obtained.

Solitary waves

Let us now assume that if the envelope of the optical pulse vanishes, then the all atoms be in the ground states. Formally, that means that,

$$\text{if } \widetilde{E} = 0 \text{ then } P = Q = 0,\ R_3 = +1.$$

Having calculated the value $I_1 = 1 - 2\Delta\omega\, \chi(\Delta\omega)$, one can write

$$R_3(\Delta\omega, \tau) = 1 - \frac{\chi(\Delta\omega)}{2\Delta\omega} \widetilde{E}^2(\tau),$$

and the substitution of all components of Bloch vector to (4.1.12) yields

$$\left(\frac{d\widetilde{E}}{d\tau}\right)^2 = \left(\frac{\Delta\omega}{\chi(\Delta\omega)} - \Delta\omega^2\right)\widetilde{E}^2 - \frac{1}{4}\widetilde{E}^4 . \tag{4.1.15}$$

The factor in brackets in the right hand side of equation (4.1.15) must be positive otherwise, this equation has no solutions. Denote this factor as

$$\left(\frac{\Delta\omega}{\chi(\Delta\omega)} - \Delta\omega^2\right) = \frac{1}{t_p^2}$$

so, function $\chi(\Delta\omega)$ can be explicitly written as:

$$\chi(\Delta\omega) = \frac{t_p^2 \Delta\omega}{1 + t_p^2 \Delta\omega^2} .$$

Now equation (4.1.15) writes

$$\frac{d\widetilde{E}}{d\tau} = \pm \frac{\widetilde{E}}{t_p}\left(1 - \frac{t_p^2}{4}\widetilde{E}^2\right)^{1/2} .$$

Its solution is

$$\widetilde{E}(\tau) = \frac{2}{t_p} \operatorname{sech}\left(\frac{\tau - \tau_0}{t_p}\right). \tag{4.1.16}$$

The pulse duration t_p stays a free parameter, which can be fixed by solving the Cauchy problem for SIT equations. One can say that information about initial profile of a pulse entering resonant medium is hidden in parameter t_p. Another parameter τ_0 appears here as an integrating constant for equation (4.1.15). It is possible to set it equal to zero. This is equivalent to choosing a reference point on the scale of time or co-ordinate.

The area shaped by the envelope $E(\tau)$ is equal to 2π. This pulse is the 2π-pulse. The components of the Bloch vector vary by formulae:

$$Q(\tau,\Delta\omega)=\frac{2t_p\Delta\omega\,\mathrm{sech}(\tau/t_p)}{1+t_p^2\Delta\omega^2};$$

$$P(\tau,\Delta\omega)=\frac{-2\tanh(\tau/t_p)\,\mathrm{sech}(\tau/t_p)}{1+t_p^2\Delta\omega^2};$$

$$R_3(\tau,\Delta\omega)=1-\frac{2\,\mathrm{sech}^2(\tau/t_p)}{1+t_p^2\Delta\omega^2}.$$

It is worth noting that all Bloch vectors for any $\Delta\omega$ rotate through an angle of 2π about the individual direction of the $\Delta\omega$ depending on an efficient field.

Equation (4.1.9) yields an expression for the group velocity of 2π-pulse:

$$V^{-1}=c^{-1}+\alpha' t_p^2\left\langle(1+t_p^2\Delta\omega^2)^{-1}\right\rangle. \tag{4.1.17}$$

The 2π-pulse velocity may depend on a duration in a very sophisticated way. If, for instance, to chose the form of inhomogeneously broadened absorption line as:

$$F(\Delta\omega)=\frac{T_2^*}{2\sqrt{\pi}}\exp\left\{-\left(\frac{\Delta\omega T_2^*}{2}\right)^2\right\},$$

then

$$V^{-1}=c^{-1}+2\alpha' t_p^2\left(\frac{T_2^*}{2t_p}\right)\exp\left[\left(\frac{T_2^*}{2t_p}\right)^2\right]\mathrm{erfc}\left(\frac{T_2^*}{2t_p}\right),$$

where $\mathrm{erfc}(y)=\int_y^\infty \exp(-t^2)\,dt$ is the complementary probability integral. In a sharp line limit, when $t_p \ll T_2^*$ V takes the minimum value:

$$V^{-1}=c^{-1}+\alpha' t_p^2.$$

In the opposite limit case ($t_p \gg T_2^*$) only a small part of resonant atoms, which frequencies are located close to the centre of an absorption line, interacts with the electromagnetic wave, and so $V\approx c$.

Cnoidal Wave

Let us rewrite expressions (4.1.10), (4.1.13), (4.1.14) for real component of Bloch vector in the form

$$Q(\tau,\Delta\omega)=\chi(\Delta\omega)\widetilde{E}(\tau), \tag{4.1.18a}$$

$$P(\Delta\omega,\tau)=\frac{\chi(\Delta\omega)}{\Delta\omega}\left(\frac{d\widetilde{E}}{d\tau}\right), \tag{4.1.18b}$$

$$R_3(\Delta\omega,\tau)=R_0(\Delta\omega)-\frac{\chi(\Delta\omega)}{2\Delta\omega}\widetilde{E}^2(\tau), \tag{4.1.18c}$$

Above the new notation for an integrating constant in the formula (4.1.13) has been introduced. By substitution of these expressions in (4.1.12), it is possible to obtain an equation for the electric field envelope

$$\left(\frac{d\widetilde{E}}{d\tau}\right)^2=a+b\widetilde{E}^2-\frac{1}{4}\widetilde{E}^4\ , \tag{4.1.19}$$

where parameters a and b should be constants, as soon as the left side of this equation does not depend on the frequency detuning:

$$a=\frac{\Delta\omega^2}{\chi^2(\Delta\omega)}\left[1-R_0^2(\Delta\omega)\right],\quad b=\left(\frac{R_0(\Delta\omega)\Delta\omega}{\chi(\Delta\omega)}-\Delta\omega^2\right). \tag{4.1.20}$$

Equation (4.1.19) can be written as

$$\left(\frac{d\widetilde{E}}{d\tau}\right)^2=\frac{1}{4}\left(\alpha_1^2-\widetilde{E}^2\right)\left(\alpha_2^2+\widetilde{E}^2\right), \tag{4.1.21}$$

where $\alpha_1^2=2(D+b)$, $\alpha_2^2=2(D-b)$, $D^2=a+b^2$.

Consider the case of $\alpha_1^2>\alpha_2^2>0$. Substitution $E=\alpha_1\cos\phi$ transforms equation (4.1.21) into

$$\left(\frac{d\phi}{d\tau}\right)^2=\frac{1}{4}(\alpha_1^2+\alpha_2^2)\left(1-m^2\sin^2\phi\right),$$

where $m=\alpha_1(\alpha_1^2+\alpha_2^2)^{-1/2}$.

Solution of this equation is expressed in terms of the Jacobian elliptic functions [64]. We can write in particular that

$$\widetilde{E}(\tau) = \alpha_1 \,\mathrm{cn}\left(\frac{\tau}{2}\sqrt{\alpha_1^2 + \alpha_2^2},\, \frac{\alpha_1}{\sqrt{\alpha_1^2 + \alpha_2^2}} \right)$$

or

$$\widetilde{E}(\tau) = \sqrt{2(D+b)}\,\mathrm{cn}\left(\tau\sqrt{D},\, \sqrt{\frac{D+b}{2D}} \right) .$$

Now it is suitable to introduce new parameters that can be treated as a certain amplitude and duration:

$$\sqrt{D} = t_p^{-1} , \quad \sqrt{2(D+b)} = E_0$$

Relations (4.1.20) allow to express $\chi(\Delta\omega)$ and $R_0(\Delta\omega)$:

$$\chi(\Delta\omega) = \frac{\Delta\omega^2}{\sqrt{a + (b + \Delta\omega^2)^2}} , \quad R_0(\Delta\omega) = \frac{b + \Delta\omega^2}{\sqrt{a + (b + \Delta\omega^2)^2}} ,$$

where constants a and b are expressed in terms of physically significant constants:

$$a = E_0^2\left(t_p^{-2} - \frac{1}{4}E_0^2 \right), \quad b = t_p^{-2}\left(\frac{1}{2}t_p^2 E_0^2 - 1 \right) .$$

The same constants allow to write the modulus of elliptic cosine as $m = E_0 t_p / 2$. Thus, it is possible to obtain the solution of the reduced Maxwell-Bloch equations (4.1.8) in the case of $m = E_0 t_p / 2 < 1$ in the form

$$\widetilde{E}(\tau) = E_0 \,\mathrm{cn}\left(\tau / t_p ,\, E_0 t_p / 2 \right) , \tag{4.1.22a}$$

$$Q(\tau, \Delta\omega) = \frac{2t_p \Delta\omega\, m\, \mathrm{cn}(\tau / t_p , m)}{\sqrt{(1 - t_p^2\Delta\omega^2)^2 + 4m^2 t_p^2 \Delta\omega^2}} , \tag{4.1.22b}$$

$$P(\tau, \Delta\omega) = \frac{2m\, \mathrm{sn}(\tau / t_p , m)\, \mathrm{dn}(\tau / t_p , m)}{\sqrt{(1 - t_p^2\Delta\omega^2)^2 + 4m^2 t_p^2 \Delta\omega^2}} , \tag{4.1.22c}$$

$$R_3(\tau, \Delta\omega) = \frac{-\left(1 + t_p^2\Delta\omega^2\right) + 2\,\mathrm{dn}^2(\tau / t_p , m)}{(1 - t_p^2\Delta\omega^2)^2 + 4m^2 t_p^2 \Delta\omega^2} . \tag{4.1.22d}$$

To obtain these expressions it is sufficient to engage formulae (4.1.18a) - (4.1.18c) and the relations

$$\chi(\Delta\omega) = \frac{\Delta\omega t_p^2}{\sqrt{(1-t_p^2\Delta\omega^2)^2 + 4m^2 t_p^2 \Delta\omega^2}} \ , \quad R_0(\Delta\omega) = \frac{-1+2m^2+(\Delta\omega t_p)^2}{\sqrt{(1-t_p^2\Delta\omega^2)^2 + 4m^2 t_p^2 \Delta\omega^2}}$$

Expression for inversion of resonance energy levels populations (4.1.22d) was obtained after some algebra by using the properties of Jacobian elliptic functions.

Cnoidal wave under consideration describes an infinite sequence of short pulses with duration t_p, spread in resonant medium free of any dissipation. Period of this sequence is determined by the period of the Jacobian elliptic function, which is expressed in terms of the complete elliptic integral of the first genus:

$$T_c = 4t_p \mathbf{K}\left(\frac{1}{2}E_0 t_p\right).$$

This period becomes infinite and a cnoidal wave transforms into a solitary wave (4.1.16) when the argument of complete elliptic integral becomes unity.

Notice that condition $\alpha_1^2 > \alpha_2^2 > 0$ means $m = E_0 t_p / 2 < 1$. For solitary wave, relation $\alpha_1^2 > \alpha_2^2 = 0$ holds. Formally, it is possible to consider the case $\alpha_1^2 > 0, \alpha_2^2 < 0$. That means that $m = E_0 t_p / 2 > 1$. In this case, it is possible to find a solution of the equation (4.1.21). But it is simpler to get the solution by transforming the found solutions by the known formulae of transformation from elliptic cosine with modulus exceeding unity to elliptic functions with modulus less than unity. Using the expression [64]

$$\mathrm{cn}\left(kz, 1/k\right) = \mathrm{dn}(z, k) \ ,$$

solution of equation (4.1.21) can be written as

$$\widetilde{E}(\tau) = E_0 \,\mathrm{dn}\left(\frac{\tau E_0}{2}, \frac{2}{E_0 t_p}\right).$$

Expressions for a component of the Bloch vector can be found from (4.1.22b)–(4.1.22d) in which it is necessary to change parameters and arguments of the elliptic functions according to the rules of transformation of a modulus from $m = E_0 t_p / 2$ to $\widetilde{m} = 2 / E_0 t_p$.

This solution describes waves with amplitude modulation, and the electric field strength does not vanish anywhere. The period of such a wave is

$$T_c = 2t_p \mathbf{K}\left(\frac{2}{E_0 t_p}\right)$$

So far as its amplitude does not vanish there is a nonzero polarisation in a medium where this wave propagates. A realistic process of polarisation and inversion relaxation in two-level medium make such waves unrealised.

4.1.2. STEADY-STATE SOLUTIONS OF THE MAXWELL-BLOCH EQUATIONS

In a scalar form the Maxwell-Bloch (MB) equations without slowly varying envelope and phase approximation are [7, 58, 65]

$$\frac{\partial^2 E}{\partial z^2} - \frac{1}{c^2}\frac{\partial^2 E}{\partial t^2} = \frac{4\pi n_A d}{c^2}\frac{\partial^2 r_1}{\partial t^2}, \tag{4.1.23a}$$

$$\frac{\partial r_1}{\partial t} = -\omega_a r_2, \quad \frac{\partial r_2}{\partial t} = \omega_a r_1 + \frac{2d}{\hbar} E r_3, \quad \frac{\partial r_3}{\partial t} = -\frac{2d}{\hbar} E r_2 . \tag{4.1.23b}$$

If we introduce new dimensionless variables

$$\tau = \omega_a t, \quad \xi = \omega_a z / c, \quad q(\tau, \xi) = \frac{2dE}{\hbar\omega_a},$$

these equations can be represented in the form

$$\frac{\partial^2 q}{\partial \xi^2} - \frac{\partial^2 q}{\partial \tau^2} = \alpha \frac{\partial^2 r_1}{\partial \tau^2}, \tag{4.1.24a}$$

$$\frac{\partial r_1}{\partial \tau} = -r_2, \quad \frac{\partial r_2}{\partial \tau} = r_1 + q r_3, \quad \frac{\partial r_3}{\partial \tau} = -q r_2, \tag{4.1.24b}$$

where $\alpha = 8\pi n_A d^2 (\hbar\omega_a)^{-1}$. The integral of motion follows from the Bloch equations (4.1.24b) in the form

$$r_1^2 + r_2^2 + r_3^2 = r_0^2 = 1. \tag{4.1.25}$$

for absorbing medium $r_0 = -1$. Parameter α can be expressed in terms of characteristic time of a two-level system [66]

$$t_c^{-1} = 4\pi n_A d^2 / \hbar,$$

so $\alpha = (2/t_c\omega_a)$.

To obtain the equations describing the propagation of stationary USP one should suppose that the components of the Bloch vector and the normalised pulse envelope depend on only one variable $t \pm x/V$ or $\eta = \omega_a(t \pm x/V)$ in dimensionless form. That means that a stationary (steady state) wave propagates only in one direction. Under this assumption the system (4.1.24) transforms into the system of ordinary differential equations

$$\left(\frac{c^2}{V^2}-1\right)\frac{d^2q}{d\eta^2} = \alpha\frac{d^2r_1}{d\eta^2}, \tag{4.1.26a}$$

$$\frac{dr_1}{d\eta} = -r_2, \tag{4.1.26b}$$

$$\frac{dr_2}{d\eta} = r_1 + qr_3, \tag{4.1.26c}$$

$$\frac{dr_3}{d\eta} = -qr_2. \tag{4.1.26d}$$

Solitary waves

Let us consider the solution of equations (4.1.26) describing a solitary steady-state wave. Boundary conditions at $|\eta| \to \infty$ must look like the following

$$\frac{dq}{d\eta} = q = 0, \quad r_1 = r_2 = 0, \quad r_3 = -1\ . \tag{4.1.27}$$

By integrating equation (4.1.26a) and taking (4.1.27) into account we obtain

$$\left(\frac{c^2}{V^2}-1\right)\frac{dq}{d\eta} = \alpha\frac{dr_1}{d\eta}, \tag{4.1.28}$$

or, if one uses equations (4.1.26b),

$$\left(\frac{c^2}{V^2}-1\right)\frac{dq}{d\eta} = -\alpha r_2. \tag{4.1.29}$$

The above equations and (4.1.26d) give

$$r_3 = -1 + \frac{1}{2\alpha}\left(\frac{c^2}{V^2}-1\right)q^2. \tag{4.1.30}$$

So far as $r_3 \geq -1$, the second term in the right side of (4.1.30) must be positive. That means that the velocity of a stationary USP is less than the velocity of light in the medium. This is a typical result for coherent propagation of a stationary optical pulse in the medium of two-level atoms. The following parameter is convenient

$$q_0^2 = \frac{4\alpha V^2}{c^2 - V^2}.$$

Then

$$r_3 = -1 + 2(q/q_0)^2 = -1 + 2w^2 .$$

Equation (4.1.29) yields

$$r_2 = -\frac{4}{q_0}\frac{dw}{d\eta}.$$

Integrating of equation (4.1.28) under boundary conditions provides $r_1 = 4w/q_0$. The substitution of the Bloch vector components, founded in (4.1.25), leads at once to the equation to determine $w(\eta)$:

$$\left(\frac{dw}{d\eta}\right)^2 = \left(\frac{q_0^2}{4} - 1\right)w^2 - \frac{q_0^2}{4}w^4 .$$

This equation has a solution in the form of a solitary wave if the factor at the first term is positive. Let us denote it as

$$\frac{q_0^2}{4} - 1 = \theta^{-2} > 0,$$

and rewrite the resulting equation in the form

$$\frac{dw}{d\eta} = \pm\frac{w}{\theta}\left(1 - \frac{q_0^2\theta^2 w^2}{4}\right)^{1/2} . \tag{4.1.31}$$

The solution of this equation is

$$w(\eta) = \frac{2}{\theta\ \cosh[\eta/\theta]}$$

or

$$q(z,t) = \frac{2}{\theta}\,\mathrm{sech}\left[\frac{\omega_a}{\theta}\left(t \pm \frac{z}{V}\right)\right]. \tag{4.1.32}$$

This expression demonstrates that the duration of stationary USP can be determined as

$$t_p = \theta / \omega_a$$

The expression for an electric field strength follows from (4.1.32)

$$E(z,t) = E_0\,\mathrm{sech}\left[\frac{dE_0}{\hbar}\left(t \pm \frac{z}{V}\right)\right], \tag{4.1.33}$$

where pulse amplitude is $E_0 = \hbar t_p^{-1} d^{-1}$. This expression for stationary pulse correlates with the one found in [58, 65].

Expression for the stationary USP velocity can be obtained from definitions of pulse duration and amplitude. As in SIT theory by McCall-Hahn, these values are related to each other, so the pulse is able to invert a two-level system and then return it to the initial state while the pulse lasts. After some algebra one may find

$$\frac{1}{V^2} = \frac{1}{c^2}\left(1 + \frac{\alpha\theta^2}{1+\theta^2}\right) = \frac{1}{c^2}\left(1 + \frac{4(t_p/t_c)^2}{1+(t_p\omega_a)^2}\right)$$

or

$$\frac{1}{V^2} = \frac{1}{c^2}\left(1 + \frac{8\pi n_A d^2 \hbar\omega_a}{(dE_0)^2 + (\hbar\omega_a)^2}\right), \tag{4.1.34}$$

that corresponds to what was found in [58].

Cnoidal waves

There is another class of stationary waves – cnoidal waves amongst the solutions of Maxwell-Bloch equations (4.1.23) or (4.1.26). These are periodic continuous waves different from solitary waves observed above. Since the equations (4.1.23) are valid when the duration of the wave less or much less than relaxation times of the atomic subsystem, the cnoidal waves seem to present a mathematical object lying beyond the physical meaning of the source equations. However, it is worth to discuss this class of stationary solutions as an illustration of the properties of the models under consideration.

Let us again turn to equation (4.1.26). We will not fix boundary conditions for a while and we will consider only the waves with the limited amplitudes.

If now to integrate equation (4.1.26a) with arbitrary boundary conditions at $|\eta|\to\infty$, then we obtain the following expression

$$\frac{dq}{d\eta}=\frac{1}{4}q_0^2\frac{dr_1}{d\eta}+C_1' ,$$

where C_1' is the integrating constant. The repeated integrating of this expression results in the relationship

$$q(\eta)=\frac{1}{4}q_0^2 r_1(\eta)+C_1'\eta+C_1'' ,$$

where C_1'' is a new integrating constant. The second term in the right side of this expression displays an unlimited growing of the electromagnetic wave amplitudes. To avoid it the integrating constant above should be zero.

Equation (4.1.29) can be rewritten as

$$\frac{dq}{d\eta}=-\frac{1}{4}q_0^2 r_2(\eta),$$

and being combined with (4.1.26d) it provides

$$q^2(\eta)=\frac{1}{2}q_0^2 r_3(\eta)+C_3' .$$

This equation is a generalisation of equation (4.1.30). For the further analysis it is convenient to collect all expressions found above in one list:

$$r_1=\frac{4}{q_0^2}q+C_1 , \tag{4.1.35a}$$

$$r_2=-\frac{4}{q_0^2}\frac{dq}{d\eta} , \tag{4.1.35b}$$

$$r_3=C_3+\frac{2}{q_0^2}q^2(\eta) . \tag{4.1.35c}$$

The constants introduced here are defined from boundary conditions, which were not determined before this moment. These conditions should be chosen from the physical point of view. For instance, if to adopt that in the absence of an electromagnetic field all atoms are in the ground state, then constants should be taken in the form

$C_1 = 0, C_3 = -1$. It relates to a situation considered earlier, when a steady-state pulse obeys equation (4.1.31) and is described by (4.1.32) or by (4.1.33).

Let the field have a zero amplitude at a certain point $\eta = \eta_0$, the atoms are in the "superradiance condition" [3, 67, 68].

$$r_1 = 0, r_2 = \pm 1, r_3 = 0 . \qquad (4.1.36)$$

Then from (4.1.2.13) it is possible to define $C_1 = 0, C_3 = 0$. From the relation (4.1.25) and (4.1.2.13) equation for $w = q / q_0$ arises

$$\left(\frac{dw}{d\eta}\right)^2 = \frac{q_0^2}{16}\left(1 - \frac{16}{q_0^2} w^2 - 4w^4\right).$$

This equation can be transformed into the form that is more convenient

$$\left(\frac{dw}{d\xi}\right)^2 = \left[\left(\alpha_1^2 - w^2\right)\left(\alpha_2^2 + w^2\right)\right], \qquad (4.1.37)$$

where

$$\xi = q_0\eta/2, \quad \alpha_1^2 = (2q_0^{-2})\left[\sqrt{1 + q_0^4/16} - 1\right], \quad \alpha_2^2 = (2q_0^{-2})\left[\sqrt{1 + q_0^4/16} + 1\right].$$

Substituting $w = \alpha_1 \cos\phi$ into equation (4.1.37) leads to

$$\left(\frac{d\phi}{d\xi}\right)^2 = (\alpha_1^2 + \alpha_2^2)\left(1 - m^2 \sin^2\phi\right),$$

where $m = \alpha_1(\alpha_1^2 + \alpha_2^2)^{-1/2}$. Solution of this equation can be written in terms of the elliptic integral of the first genus [64]:

$$(\xi - \xi_0) = \frac{F(\phi, m)}{\sqrt{\alpha_1^2 + \alpha_2^2}} .$$

Then, using the Jacobian elliptic functions it is possible to obtain an explicit expression for $w(\xi)$:

$$w(\xi) = \alpha_1 \,\mathrm{cn}\left[\alpha_1(\xi - \xi_0)m^{-1}, m\right] . \qquad (4.1.38)$$

In the initial variables, this result takes the following form

$$E(z,t) = E_0 \operatorname{cn}\left[\frac{q_0\omega_a\sqrt{\alpha_1^2+\alpha_2^2}}{2}\left(t \pm z/V - t_0\right), \frac{\alpha_1}{\sqrt{\alpha_1^2+\alpha_2^2}}\right], \qquad (4.1.39)$$

where $E_0 = \hbar\omega_a q_0 \alpha_1 / 2d$.

Usually a period of a harmonic wave is defined as the time, during which the argument changes to 2π. In so far as the period of Jacobian elliptic functions is equal to $4\mathbf{K}(m)$ (where $\mathbf{K}(m)$ is a total elliptic integral of the first genus), the period T_p of non-linear (cnoidal *ad hoc*) waves can be introduced by the following expression

$$T_p = \frac{8m\mathbf{K}(m)}{q_0\omega_a\alpha_1}.$$

One more time parameter naturally appears in (4.1.39).

$$t_p = \frac{2}{q_0\omega_a\sqrt{\alpha_1^2+\alpha_2^2}},$$

Besides, it is possible to write $q_0^2/4 = \left[(1/t_p\omega_a)^4 - 1\right]^{1/2}$. Then the restriction on duration of the "splash of a cnoidal wave" is $\omega_a t_p \le 1$, that is the frequency of oscillation of the steady-state continuous non-linear wave does not exceed optical transition frequencies ω_a.

Formula for phase velocity of this wave comes from the determination of parameter q_0

$$\left(\frac{c^2}{V^2} - 1\right) = \frac{\alpha\omega_a^2 t_p^2}{\sqrt{1-(\omega_a t_p)^4}}.$$

The amplitude of the cnoidal wave and the modulus of elliptic functions can be expressed in terms of this parameter:

$$E_0^2 = \left(\frac{\hbar\omega_a}{2d}\right)^2 \frac{1-\left(\omega_a t_p\right)^2}{2\left(\omega_a t_p\right)^2}, \qquad m = \frac{1}{\sqrt{2}}\sqrt{1-\left(\omega_a t_p\right)^2}.$$

Let us consider the system (4.1.2.13) and choose new boundary conditions $r_1 = 0, r_2 = 0, r_3 = r_0$. This choice corresponds to the excited medium, but there is no macroscopic polarisation in the initial moment. In this case, the following equation will replace (4.1.36):

$$\left(\frac{dw}{d\eta}\right)^2 = \frac{q_0^2}{16}\left((1-r_0^2)-(\frac{16}{q_0^2}-4r_0)w^2-4w^4\right). \quad (4.1.40)$$

This equation can be transformed to equation (4.1.37) with substitution

$$\alpha_1^2 = \frac{1}{2}\left[\sqrt{(1-r_0^2)+\left(\frac{4}{q_0^2}-r_0\right)^2}-\left(\frac{4}{q_0^2}-r_0\right)\right],$$

$$\alpha_2^2 = \frac{1}{2}\left[\sqrt{(1-r_0^2)+\left(\frac{4}{q_0^2}-r_0\right)^2}+\left(\frac{4}{q_0^2}-r_0\right)\right]$$

With the choice of $r_0 < 4q_0^{-2}$ and $w = \alpha_1 \cos\phi$, the integrating of equation (4.1.40) yields

$$w(\xi) = \alpha_1 \operatorname{cn}\left[\alpha_1(\xi-\xi_0)m^{-1}, m\right],$$

where ξ_0 is a constant of integrating. Parameter m is the modulus of an elliptic integral as it is in (4.1.38), but with the different α_1^2 and α_2^2. It is possible to introduce physical parameters E_0 and t_p instead of parameters q_0 and r_0:

$$E_0^2 = \left(\frac{\hbar\omega_a}{2d}\right)^2 q_0\alpha_1, \quad t_p^{-1} = (q_0\omega_a/2)\sqrt{\alpha_1^2+\alpha_2^2}$$

In terms of these physical variables the solution (4.1.40) can be written as

$$E(z,t) = E_0 \operatorname{cn}\left[\frac{t \pm z/V - t_0}{t_p}, \frac{t_p dE_0}{\hbar}\right]. \quad (4.1.41)$$

The period of oscillation of the electric field strength of these cnoidal waves is

$$T_C = 4t_p\mathbf{K}(dE_0t_p\hbar^{-1}). \quad (4.1.42)$$

Notice that infinite period corresponds to the unit value of the elliptic cosine modulus in (4.1.41) and then

$$E(z,t) = E_0 \operatorname{sech}\left[\frac{t \pm z/V - t_0}{t_p}\right], \quad E_0 = \hbar(t_p d)^{-1},$$

that corresponds to the expression (4.1.33). For any cnoidal wave, inequality $E_0 < \hbar(t_p d)^{-1}$ should be held.

4.1.3. INVERSE SCATTERING TRANSFORM METHOD FOR THE SIT THEORY

In order to illustrate one of the ways to find zero-curvature representation of the equations of the SIT theory, let us consider at first the simplified version of the theory. If an absorption line is homogeneously broadened (this is a sharp-line limit of the absorption line) and exact resonance condition holds, then the SIT-equations reduce to the well-known Sine-Gordon equation or SG-equation:

$$\frac{\partial^2 \theta}{\partial z \partial t} = -\sin\theta \ ,$$

which appears in the different fields of physics and often attracts attention as the classical example of the equation having soliton solutions.

If we abandon the condition of exact resonance and consider an inhomogeneously broadened absorption line, then SIT-equations are not already reducible to SG-equation. However, a zero-curvature representation of the SIT-equations can be found as well.

Denote A as a real slowly varying in the space and time envelope of the electric field strength of an optical wave. The SIT-equations (4.1.7) in the case of $\Delta\omega = 0$ write as

$$\frac{\partial q}{\partial \zeta} = -P, \quad \frac{\partial P}{\partial \tau} = qR_3 \ , \quad \frac{\partial R_3}{\partial \tau} = -qP \ , \qquad (4.1.43a)$$

where $q = dAt_{p0}/2\hbar$ is the real normalised pulse envelope, $\zeta = \alpha' t_{p0} z = L_a^{-1} z$ is the normalised space co-ordinate, $\tau = (t - z/c)/t_{p0}$ is the retarded time, c is a velocity of light pulse in the medium in linear approximation, d is the projection of a matrix element of the dipole operator on x-axis.

In addition to (4.1.43a), initial and boundary conditions are chosen in the form

$$q(\zeta = 0, \tau) = q_0(\tau), \quad \lim_{|\tau| \to \infty} q(\zeta, \tau) = 0,$$
$$R_3(\zeta, \tau \to \pm\infty) = 1, \quad P(\zeta, \tau \to \pm\infty) = 0. \qquad (4.1.43b)$$

Matrix $\hat{U}(\lambda)$ of the IST method can be taken in the simplest form

$$\hat{U}(\lambda) = \begin{pmatrix} -i\lambda & w \\ \overline{w} & i\lambda \end{pmatrix} . \tag{4.1.44a}$$

The spectral problem with this matrix has been well studied, and this is one of the reasons to choose it for the equations under consideration. Matrix $\hat{V}(\lambda)$ can be chosen in a general form:

$$\hat{V}(\lambda) = \begin{pmatrix} A & B \\ C & -A \end{pmatrix} . \tag{4.1.44b}$$

The dependency of the factors *A*, *B* and *C* on P, R_3, q and λ is unknown in advance. It must be chosen to match the zero-curvature condition (3.1.2) with (4.1.43a). Substitution of the equations (4.1.44a) and (4.1.44b) in (3.1.2) results in the system of equation:

$$\frac{\partial A}{\partial \tau} = wC - \overline{w}B, \tag{4.1.45a}$$

$$\frac{\partial B}{\partial \tau} + 2i\lambda B + 2wA = \frac{\partial w}{\partial \zeta}, \tag{4.1.45b}$$

$$\frac{\partial C}{\partial \tau} - 2i\lambda C - 2\overline{w}A = \frac{\partial \overline{w}}{\partial \zeta} . \tag{4.1.45c}$$

The equation (4.1.43a) contains only partial derivative $\partial q / \partial \zeta$, the equation (4.1.45) contains partial derivatives $\partial w / \partial \zeta$ and $\partial \overline{w} / \partial \zeta$. Consequently, it is possible to expect that

$$w = \alpha_1 q, \quad \overline{w} = \alpha_2 q . \tag{4.1.46a}$$

Constants α_1 and α_2 have not been determined yet. The equations (4.1.43a) contain partial derivatives of *P* and R_3 with respect to τ whereas in equations (4.1.45) there are only partial derivatives of *A*, *B* and *C* with respect to τ. Consequently, *A*, *B*, and *C* are linear combinations of *P* and R_3:

$$\begin{aligned} A(\tau,\zeta;\lambda) &= a_1(\lambda)P(\tau,\zeta) + a_2(\lambda)R_3(\tau,\zeta), \\ B(\tau,\zeta;\lambda) &= b_1(\lambda)P(\tau,\zeta) + b_2(\lambda)R_3(\tau,\zeta), \\ C(\tau,\zeta;\lambda) &= c_1(\lambda)P(\tau,\zeta) + c_2(\lambda)R_3(\tau,\zeta). \end{aligned} \tag{4.1.46b}$$

Substitution of (4.1.46) in the first equation of (4.1.45) yields

$$a_1 \frac{\partial P}{\partial \tau} + a_2 \frac{\partial R_3}{\partial \tau} + \alpha_2 q b_1 P + \alpha_2 q b_2 R_3 - \alpha_1 q c_1 P - \alpha_1 q c_2 R_3 = 0 \ .$$

Using (4.1.43a) it is necessary to exclude partial derivatives of P and R_3 with respect to τ from the above and comparing the factors before the equal degrees of qP and qR_3, one can find the relations:

$$\alpha_1 c_2 = a_1 + \alpha_2 b_2, \quad \alpha_1 c_1 = -a_2 + \alpha_2 b_1 \ , \tag{4.1.47a}$$

Now it is necessary to repeat the same procedure, substituting equations (4.1.46b) into (4.1.45b). It will give

$$2i\lambda b_1 = -\alpha_1, \quad 2i\lambda b_2 = 0, \quad 2\alpha_1 a_1 = b_2, \quad 2\alpha_1 a_2 = -b_1 \tag{4.1.47b}$$

Substitution of (4.1.46b) into (4.1.45c), eliminating the partial derivatives of P and R_3 with respect to τ by (4.1.43a) from resulting expressions and reduction of similar terms bring the rest of the necessary relations:

$$2i\lambda c_1 = \alpha_2, \quad 2i\lambda c_2 = 0, \quad 2\alpha_2 a_1 = c_2, \quad 2\alpha_2 a_2 = c_1 \ . \tag{4.1.47c}$$

From (4.1.47b,c) it is seen that $c_2 = b_2 = a_1 = 0$. Other relations permit to define b_1, c_1 and a_2:

$$c_1 = \alpha_2 (2i\lambda)^{-1}, \quad b_1 = -\alpha_1 (2i\lambda)^{-1}, \tag{4.1.48a}$$

$$a_2 = (4i\lambda)^{-1} \ . \tag{4.1.48b}$$

If one substitutes $b_1(\lambda)$ and $c_1(\lambda)$ into equation (4.1.47a), then one gets another expression for $a_2(\lambda)$:

$$a_2 = -\alpha_1 \alpha_2 (i\lambda)^{-1}. \tag{4.1.48c}$$

By comparing (4.1.48c) with (4.1.48b), we find that free constants α_1 and α_2 are connected by the relation

$$\alpha_1 \alpha_2 = -1/4, \tag{4.1.49}$$

but for the rest they may be arbitrary numbers. For instance, it is possible to choose:

$$\alpha_1 = i/2, \quad \alpha_2 = i/2$$

or

$$\alpha_1 = \pm 1/2, \quad \alpha_2 = \mp 1/2.$$

So the *U*-*V*-pair can be chosen ambiguously: this is a single-parameter family of *U*-*V*-pairs. It is easy to be convinced of it if one takes α_1 and α_2 in the forms

$$\alpha_1 = i(\cosh\mu - \sinh\mu)/2, \quad \alpha_2 = i(\cosh\mu + \sinh\mu)/2,$$

here μ enumerates all *U*-*V*-pairs for system (4.1.43a) just as parameter λ does.

Let us now write a *U*-*V*-pair for equations (4.1.43a) in an explicit form and point out the evolution law for scattering data. So

$$\hat{U}(\lambda) = i\begin{pmatrix} -\lambda & q/2 \\ q/2 & \lambda \end{pmatrix}, \quad \hat{V}(\lambda) = \frac{1}{4i\lambda}\begin{pmatrix} R_3 & -iP \\ iP & -R_3 \end{pmatrix}. \tag{4.1.50}$$

Due to boundary conditions for the given problem we have

$$\hat{V}^{(+)} = \hat{V}^{(-)} = \frac{1}{4i\lambda}\begin{pmatrix} 1 & 0 \\ 0 & -1 \end{pmatrix}.$$

The evolution of the transfer matrix $\hat{T}(\lambda;\zeta)$ will be assigned by the equation (3.5.8):

$$\frac{\partial \hat{T}}{\partial \zeta} = \left[\hat{V}^{(-)}, \hat{T}\right].$$

Then it follows that

$$T_{12}(\lambda;\zeta) = T_{12}(\lambda;0)\exp\{\zeta/2i\lambda\}, \quad T_{21}(\lambda;\zeta) = T_{21}(\lambda;0)\exp\{-\zeta/2i\lambda\},$$
$$T_{11}(\lambda;\zeta) = T_{11}(\lambda;0), \quad T_{22}(\lambda;\zeta) = T_{22}(\lambda;0).$$

Initial values of the transfer matrix elements at $\zeta = 0$ are defined here by the solution of the spectral problem for initial optical pulse with envelope $q_0(\tau)$.

It is seen from (4.1.3.1) that for new functions

$$\theta(\tau,\zeta) = \int_{-\infty}^{\tau} q(\tau',\zeta)d\tau'$$

the equation

$$\frac{\partial^2\theta}{\partial\tau\partial\zeta} = -\sin\theta$$

can be found. Here $P = \sin\theta$, $R = \cos\theta$. The U-V-pair for SG-equations can be derived from (4.1.50):

$$\hat{U}(\lambda) = i\begin{pmatrix} -\lambda & (1/2)\partial\theta/\partial\tau \\ (1/2)\partial\theta/\partial\tau & \lambda \end{pmatrix}, \quad \hat{V}(\lambda) = \frac{1}{4i\lambda}\begin{pmatrix} \cos\theta & -i\sin\theta \\ i\sin\theta & -\cos\theta \end{pmatrix}.$$

The most general form of the SIT-equations, accounting the propagation of the ultrashort pulse of the plane wave in the medium with an inhomogeneously broadened absorption line, is [2,3,6]:

$$\frac{\partial q}{\partial\zeta} = i\langle p\rangle, \quad \frac{\partial p}{\partial\tau} = i\delta p + iqR_3, \quad \frac{\partial R_3}{\partial\tau} = \frac{i}{2}(q^* p - qp^*), \tag{4.1.51}$$

where now one makes substitutions:

$$q\exp(+i\varphi) = q' \to q, \quad (Q + iP)\exp(+i\varphi) = p' \to p,$$
$$\text{and} \quad \delta = (\omega_0 - \omega_a)t_{p0}.$$

To find zero-curvature representation for equations (4.1.51) we shall follow the procedure discussed above. Let $w = \alpha_1 q$, $\overline{w} = \alpha_2 q^*$ as it was assumed in (4.1.46a). By comparing equations (4.1.45) and (4.1.51) one can notice that in (4.1.51) partial derivatives of p and R_3 with respect to τ are taken, but in (4.1.45) partial derivatives of A, B and C are taken. For this reason it is naturally to expect that A, B and C are a linear combination of only p, p^* and R_3. Let us assume

$$B(\tau,\zeta;\lambda) = \langle b(\lambda)p(\tau,\zeta)\rangle, \quad C(\tau,\zeta;\lambda) = \langle c(\lambda)p^*(\tau,\zeta)\rangle, \tag{4.1.52}$$

where $b(\lambda)$ and $c(\lambda)$ should be chosen in the form for eqs. (4.1.45) to coincide with eqs. (4.1.51). Of course expressions (4.1.52) are not the most general form, but the experience in searching for U-V-pair in the case of equations (4.1.3.1) prompts that it is possible to start with the assumption (4.1.52). If an attempt to select A, B and C in this case is not a success, then it is necessary to present these quantities as the general linear combination of p, p^* and R_3. This is recommended to do as an exercise.

By substituting (4.1.52) into equation (4.1.45) and eliminating the terms with partial derivatives with respect to τ and ζ by means of (4.1.51), we obtain two equations

$$b(\lambda)\left(i\delta p + iqR_3\right) + 2i\lambda\, b(\lambda)p = i\alpha_1 p - 2\alpha_1 qA\,,$$
$$c(\lambda)\left(i\delta p^* + iq^* R_3\right) + 2i\lambda\, c(\lambda)p^* = i\alpha_2 p^* - 2\alpha_2 q^* A\,.$$

By equating the factors at p, p^*, and q, q^* in the right and left sides of these expressions, factors $b(\lambda)$ and $c(\lambda)$ as well as expression for $A(\tau,\zeta;\lambda)$ can be found:

$$b(\lambda) = \alpha_1(\delta + 2\lambda)^{-1}, \quad c(\lambda) = \alpha_2(\delta + 2\lambda)^{-1}\,,$$
$$A(\tau,\zeta;\lambda) = (-i/2)\left\langle R_3(\tau,\zeta)(\delta + 2\lambda)^{-1}\right\rangle\,.$$

If one substitutes now $A(\tau,\zeta;\lambda)$ and the found $B(\tau,\zeta;\lambda)$ and $C(\tau,\zeta;\lambda)$ into the equation (4.1.45a), then a restriction, imposed on parameters α_1 and α_2, will be obtained: $\alpha_1\alpha_2 = -1/4$. This is the same condition as in (4.1.49). It is convenient to choose $\alpha_1 = i/2$ and $\alpha_2 = i/2$.

Thereby, the matrices $\hat{U}(\lambda)$ and $\hat{V}(\lambda)$, assigning a zero-curvature representation for the system of equations (4.1.51) are found. The linear equations of IST method can be written as

$$\frac{\partial\psi_1}{\partial\tau} = -i\lambda\psi_1 + i\frac{q}{2}\psi_2, \quad \frac{\partial\psi_2}{\partial\tau} = i\lambda\psi_2 + i\frac{q^*}{2}\psi_1\,, \tag{4.1.53a}$$

$$\begin{aligned} \frac{\partial\psi_1}{\partial\zeta} &= -\frac{i}{2}\left\langle\frac{R_3}{\delta + 2\lambda}\right\rangle\psi_1 + \frac{i}{2}\left\langle\frac{p}{\delta + 2\lambda}\right\rangle\psi_2, \\ \frac{\partial\psi_2}{\partial\zeta} &= \frac{i}{2}\left\langle\frac{p^*}{\delta + 2\lambda}\right\rangle\psi_1 + \frac{i}{2}\left\langle\frac{R_3}{\delta + 2\lambda}\right\rangle\psi_2. \end{aligned} \tag{4.1.53b}$$

The matrix elements of the transfer matrix $\hat{T}(\lambda)$ are varied with ζ by the laws:

$$\begin{aligned} T_{12}(\lambda;\zeta) &= T_{12}(\lambda;0)\exp\{-i\zeta < (\delta + 2\lambda)^{-1} >\}\,, \\ T_{21}(\lambda;\zeta) &= T_{21}(\lambda;0)\exp\{+i\zeta < (\delta + 2\lambda)^{-1} >\}\,, \\ T_{11}(\lambda;\zeta) &= T_{11}(\lambda;0), \qquad T_{22}(\lambda;\zeta) = T_{22}(\lambda;0)\,, \end{aligned} \tag{4.1.54}$$

where the boundary conditions (4.1.43b) were accepted.

The zero-curvature representation of SIT equations (4.1.51) allows to obtain the solutions related to McCall-Hahn's 2π-pulses and, furthermore, to the $N\pi$-pulses, which describe the interaction of the 2π-pulses. We may use the results from section 3.3.

Let us consider the single soliton solution of the non-linear equations (4.1.51). Soliton solution is related to the discrete spectrum of a single point. Let $\lambda_1 = \xi_1 + i\eta_1$, $\overline{\lambda}_1 = \xi_1 - i\eta_1$. The equations (4.1.54) lead to the evolution law for the kernel coefficients of GLM equation:

$$C_1(\zeta) = C_1(0)\exp[\mathcal{K}_1(\lambda_1)\zeta + i\mathcal{K}_2(\lambda_1)\zeta], \quad \overline{C}_1(\zeta) = C_1^*(\zeta),$$

where

$$\mathcal{K}(\lambda) = \frac{i}{\delta + 2\lambda} = \mathcal{K}_1(\lambda) + i\mathcal{K}_2(\lambda).$$

It is convenient to define two parameters τ_0 and φ_0 as $C_1(0) = 2\eta_1 \exp(2\eta_1\tau_0 + i\varphi_0)$. Then it follows from the relation (3.3.4) that

$$K_1(\tau,\tau;\zeta) = \frac{2i\eta_1 \exp(-\vartheta - i\widetilde{\Phi})}{1+\exp(-2\vartheta)} = i\eta_1 \operatorname{sech}(\vartheta)\exp(-i\widetilde{\Phi}),$$

where

$$\vartheta = 2\eta_1\left(\tau - \tau_0 - \frac{1}{2\eta_1}\mathcal{K}_1\zeta\right), \quad \widetilde{\Phi} = \varphi_0 + \mathcal{K}_2\zeta + 2\xi_1\tau.$$

Thereby, the corresponding solitary wave $q_S(\tau,\zeta)$ is a soliton with the duration $\tau_S = (2\eta_1)^{-1}$. The constant phase shift φ_0, the position of the soliton centre τ_0 and the soliton amplitude proportional to η_1 are determined by the initial envelope profile and expressed in terms of scattering data

$$\varphi_0 = \arg[C_1(0)/2\eta_1], \quad \tau_0 = (2\eta_1)^{-1}\ln[C_1(0)/2\eta_1].$$

We can see that the real envelope of an electric field of a light pulse does not experience any distortion while the pulse propagates in a resonant absorbing medium. It is

$$q(t,z) = \frac{2}{t_S}\operatorname{sech}\left(\frac{t - z/V_g}{t_S}\right)$$

where the group velocity V_g is related to the pulse duration $t_S = t_{p0}\tau_S$ and the resonant absorption coefficient L_a^{-1} by the formula

$$V_g^{-1} = c^{-1}\left(1 + ct_S^2 L_a^{-1}\right)$$

This is a well-known result for the coherent propagation of light pulses, where the speed of 2π-pulse is less than the speed of light.

The multi-soliton, in particular *N*-soliton solutions of the SIT equation can be obtained by considering the case of a spectrum consisting of only *N* discrete points. Let $\lambda_n = i\eta_n$, $n = 1,2,\dots,N$. Expression (3.3.11) from section 3.3. allows to do it. The compact form of *N*-soliton solution for SIT equation has been found in [7,16]. It is

$$|q|^2 = \frac{\partial^2}{\partial\tau^2} \ln\det \hat{M}(\tau,\zeta), \tag{4.1.55}$$

where $N \times N$ matrix $\hat{M}(\tau,\zeta)$ is defined by the matrix elements

$$M_{nm} = \frac{8\eta_n\eta_m}{\eta_n + \eta_m}\left\{\exp[\vartheta_n] + (-1)^{m+n}\exp[-\vartheta_m]\right\},$$

where

$$\vartheta_n = 2\eta_n\left(\tau + \tau_n - \mathcal{K}_1(\eta_n)\zeta/2\eta_n\right).$$

Here functions $\mathcal{K}_1(\eta_n)$ and $\mathcal{K}_2(\eta_n)$ are defined as

$$\mathcal{K}_1(\eta_n) = \left\langle \frac{2\eta_n}{4\eta_n^2 + \delta^2} \right\rangle, \quad \mathcal{K}_2(\eta_n) = \left\langle \frac{\delta}{4\eta_n^2 + \delta^2} \right\rangle.$$

If the form factor of the inhomogeneously broadened absorption line is the symmetric function, then $\mathcal{K}_2(\eta_n)$ is equal to zero.

4.1.4. DERIVATION OF BÄCKLUND TRANSFORMATIONS

One of the important properties of the equations, which allows a zero-curvature representation, is a possibility to specify a certain relation for any pair of its solutions, which converts one solution into another. This relation for Sine-Gordon equation was found by Bäcklund in the previous century and now it is known as *Bäcklund transformation* (BT). Here it will be shown how it is possible to find BT for the whole class of a non-linear evolution equations solved by IST method with Zakharov-Schabat (or AKNS) spectral problem under the reduction of $\overline{w} = -w^* = w$, $C = -B$. The RMB equations will serve as an illustration of the method.

If one introduces a new dependent variable $\Gamma = \psi_2 / i\psi_1$, then the linear equations of IST method can be presented in the form of the Riccati equations:

$$\frac{\partial\Gamma}{\partial\tau} = 2i\lambda\Gamma + W(1+\Gamma^2), \tag{4.1.56a}$$

$$\frac{\partial\Gamma}{\partial\zeta} = -2A\Gamma + iB(1-\Gamma^2), \tag{4.1.4.2b}$$

where $W = q/2$ is a real quantity.

Following [69], the pair of equations (4.1.56) can been considered as a mapping of the solution of the RMB equations onto the space of functions Γ and complex numbers λ, denoted as (λ, Γ). The further procedure is general. The linear equations of the AKNS spectral problem are chosen only for illustrations. Let us denote this mapping as $\mathfrak{R}$ and display it schematically by a diagram

$$\mathfrak{R}: (W, A, B) \to (\lambda, \Gamma) \ . \tag{4.1.57}$$

This mapping can be inverted as it is seen from (4.1.56a) provided A and B are expressed in terms of W and spatial derivatives of W. Now in the space (λ, Γ) one can make some transformation $\mathfrak{I}$:

$$\mathfrak{I}: (\lambda, \Gamma) \to (\lambda', \Gamma') \ . \tag{4.1.58}$$

As a result of this transformation Γ' and λ' will satisfy in the general case the equations different from (4.1.56) by form. One should select only those transformations $\mathfrak{I}$, which keep the form of equations (4.1.56). Herewith A, B and W perhaps have to be replaced by A', B' and W'. Thus, new Riccati equations result and they define the mapping

$$\mathfrak{R}^{-1}: (\lambda', \Gamma') \to (W', A', B') \ . \tag{4.1.59}$$

The relationship of A', B' and W' with A, B and W is the result identified as the *Bäcklund transformation* (Fig.4.1.1.)

$$\begin{array}{l} \mathfrak{R}:(W, A, B) \to (\lambda, \Gamma) \\ \qquad\qquad\quad \mathfrak{I} \downarrow \\ \qquad\qquad\quad \mathfrak{R}:(\lambda', \Gamma') \to (W', A', B') \end{array}$$

Fig. 4.1.1 Scheme of Bäcklund transformation

To illustrate the Bäcklund transformation algorithm, let us choose a concrete transformation $\mathfrak{I}$:

$$\mathfrak{I}: (\lambda, \Gamma) \to (\lambda', \Gamma') = (\lambda, -\Gamma^{-1}) \ . \tag{4.1.60}$$

In primed variables Riccati equations can be written in the same form as (4.1.56):

$$\frac{\partial\Gamma'}{\partial\tau} = 2i\lambda'\Gamma' + W'(1+\Gamma'^2), \tag{4.1.61a}$$

$$\frac{\partial\Gamma'}{\partial\zeta} = -2A'\Gamma' + iB'(1-\Gamma'^2). \tag{4.1.61b}$$

By taking (4.1.60) into account, equation (4.1.61a) can be rewritten as

$$\frac{\partial\Gamma}{\partial\tau} = -2i\lambda\Gamma + W'(1+\Gamma^2) \ . \tag{4.1.62}$$

The equations (4.1.62) and (4.1.4.1a) give rise to

$$4i\lambda\Gamma + (1+\Gamma^2)(W-W') = 0 \ , \tag{4.1.63}$$

$$2\frac{\partial\Gamma}{\partial\tau} = (1+\Gamma^2)(W+W') \ . \tag{4.1.64}$$

Thence

$$2(1+\Gamma^2)^{-1}\frac{\partial\Gamma}{\partial\tau} = 2\frac{\partial \arctan\ \Gamma}{\partial\tau} = (W+W') \ , \tag{4.1.65}$$

$$\frac{4i\lambda\Gamma}{(1+\Gamma^2)} = (W'-W). \tag{4.1.66}$$

Let $q = \partial u/\partial\tau$, then equation (4.1.65) yields

$$\Gamma = \tan\left(\frac{u+u'}{4}\right). \tag{4.1.67}$$

One of the BT (say τ-part of this transformation) follows from equations (4.1.56) and (4.1.66):

$$\frac{1}{2}\frac{\partial(u-u')}{\partial\tau} = 2i\lambda\sin\left(\frac{u'+u}{2}\right). \tag{4.1.68}$$

In order to obtain the second part of BT (i.e., ζ-part of BT) it is necessary to produce similar calculations for (4.1.56b) and (4.1.61b). Nevertheless, there is a shorter way. The transformation law for W (4.1.65)

$$W + W' = 2\frac{\partial \arctan\ \Gamma}{\partial\tau} \tag{4.1.69}$$

leads to

$$u' = u + 2\arctan\Gamma . \tag{4.1.70}$$

Differentiating of the right and left sides of (4.1.70) with respect to ζ with the account of (4.1.56b) yields

$$\frac{1}{2}\frac{\partial(u'+u)}{\partial\zeta} = -4A(u)\frac{\Gamma}{1+\Gamma^2} + 2iB(u)\frac{1-\Gamma^2}{1+\Gamma^2} .$$

It is needed to eliminate Γ by use of (4.1.70):

$$\frac{1}{2}\frac{\partial(u'+u)}{\partial\zeta} = 2\left\{-A(u)\sin\left(\frac{u'+u}{2}\right) + iB(u)\cos\left(\frac{u'+u}{2}\right)\right\} . \tag{4.1.71}$$

In this expression $A(u)$ and $B(u)$ are not specified. Therefore, expressions (4.1.68) and (4.1.71) define BT for the whole class of non-linear partial differential equations, which solutions can be found by IST method with Zakharov-Shabat (AKNS) spectral problem.

The RMB equations are equivalent to Sine-Gordon equation in the case of exact resonance and an infinitely narrow absorption line. In fact, if $R_3 = \cos u$, $P = \sin u$ and $q = \partial u / \partial \tau$, then the equations (4.1.3.1) transform into

$$\frac{\partial^2 u}{\partial\tau\partial\zeta} + \sin u = 0 . \tag{4.1.72}$$

As it follows from (3.1.53b)

$$A = \frac{\cos u}{4i\lambda} , \quad B = -\frac{\sin u}{4\lambda}$$

and expressions (4.1.68) and (4.1.71) yield well known form of BT for SG-equation:

$$\frac{1}{2}\frac{\partial(u'-u)}{\partial\tau} = a\sin\left(\frac{u+u'}{2}\right), \quad \frac{1}{2}\frac{\partial(u'+u)}{\partial\zeta} = \frac{1}{a}\sin\left(\frac{u-u'}{2}\right), \tag{4.1.73}$$

where $a = (2i\lambda)$ is a parameter of BT .

It is worth noting that the τ-part of BT (4.1.68) is due to the spectral problem only. That means that (4.1.68) is a general result for the whole class of non-linear equations solvable by the IST method with the given spectral problem. If it is possible to derive some relationship only from (4.1.68) without using (4.1.71), then this relationship will be applicable for the same class of non-linear equations.

Let u_0, u_1, u_2 and u_3 be solutions of the equation (4.1.72), $q_k = \partial u_k / \partial \tau$ ($k = 0, 1, 2, 3$) are solutions of RMB equations under current assumption. Besides let solutions u_1 and u_3 be related accordingly to solutions u_0 and u_2 by Bäcklund transformation with parameter a_1, but u_3 and u_2 be related accordingly to solutions u_1 and u_0 by BT with parameter a_2. These requirements are formally expressed by the relations (4.1.68):

$$\frac{\partial(u_1 - u_0)}{\partial \tau} = 2a_1 \sin\left(\frac{u_1 + u_0}{2}\right), \qquad \frac{\partial(u_2 - u_0)}{\partial \tau} = 2a_2 \sin\left(\frac{u_2 + u_0}{2}\right),$$

$$\frac{\partial(u_3 - u_1)}{\partial \tau} = 2a_2 \sin\left(\frac{u_3 + u_1}{2}\right), \qquad \frac{\partial(u_3 - u_2)}{\partial \tau} = 2a_1 \sin\left(\frac{u_3 + u_2}{2}\right).$$

All derivatives should be excluded from these expressions. The result is an identity, which is valid for only SG-equation solutions:

$$\tan\left(\frac{u_3 - u_0}{4}\right) = \left(\frac{a_1 + a_2}{a_1 - a_2}\right) \tan\left(\frac{u_1 - u_2}{4}\right). \tag{4.1.74}$$

This is an exceedingly important result known as a *formula of tangents*. Firstly, the relation (4.1.74) is true for the infinite number of the non-linear evolution equations. Secondly, this relation allows constructing a new solution u_3 of the corresponding equations in terms of the three already known solutions u_0, u_1 and u_2, leaning only on algebraic transformations (no integration is involved). Before the IST method was applied to the equations of SIT theory, the formula of tangents (4.1.74) was a unique means to obtain the multi-soliton solutions. The 0π-pulses ("breathers") were found in this way too and, in particular, the decay of the 4π-a pulse on the pair of the 2π-pulses was described [70]. This period of the SIT theory development has been summed up in the survey [6].

4.1.5. BÄCKLUND TRANSFORMATION IN SIT THEORIES

Bäcklund transformation (4.1.73) allows to find soliton solutions of self-induced transparency equations in the case of the exact resonance, when inhomogeneous broadening of the absorbing lines is either absent or neglected. More general BT or the formula of tangents (4.1.74) are applicable in a more general case of SIT theory. Here this simple case is chosen only for the sake of illustration of the BT method. In this limiting case the system of SIT equations (4.1.7) is reduced to SG-equation (4.1.72), for which the BT are expressed by the formulae (4.1.73).

2π-pulse

Let us choose the trivial solution of the SIT equations $u_0 = 0$ as a starting-point. Then the solutions $u_a(\tau,\zeta)$ related to the trivial one by BT with parameter a, can be defined by the differential first-order equation

$$\frac{1}{2}\frac{\partial u_a}{\partial \tau} = a\sin\left(\frac{u_a}{2}\right), \quad \frac{1}{2}\frac{\partial u_a}{\partial \zeta} = -\frac{1}{a}\sin\left(\frac{u_a}{2}\right). \qquad (4.1.75)$$

Therefore

$$\frac{1}{a}\frac{\partial u_a}{\partial \tau} + a\frac{\partial u_a}{\partial \zeta} = 0.$$

That means that sought-for solution depends on one variable $\eta = a\tau - \zeta / a$. It describes a simple wave moving at the velocity a^{-2} in the τ and ζ frame of reference. Now both equations (4.1.75) are reduced to one

$$du_a / d\eta = \sin(u_a / 2).$$

The solution of this equation is

$$u_a(\eta) = 4\arctan\{\exp(\eta - \eta_0)\}, \qquad (4.1.76)$$

where η_0 is an integrating constant. The envelope of the electrical field strength of the ultrashort light pulse follows from the above solution in the form

$$q(\tau,\zeta) = \frac{\partial u_a}{\partial \tau} = 2a\operatorname{sech}(\eta - \eta_0). \qquad (4.1.77a)$$

The integrating constant may be chosen to locate the maximum of the pulse at $\eta = 0$, so $\eta_0 = 0$. One may be convinced that pulse area

$$\theta = \int_{-\infty}^{\infty} q(\tau,\zeta)d\tau ,$$

equals to 2π and for this reason this pulse is named *2π-pulse*. Expression (4.1.77a) can be re-written in the form

$$q(\tau,\zeta) = 2a\operatorname{sech}(a\tau - \zeta / a),$$

or

$$\tilde{E}(\tau,\zeta) = \frac{2}{t_p}\operatorname{sech}\left(\frac{\tau - \zeta / a^2}{t_p}\right)$$

where the introduced parameter $t_p = t_{p0}a^{-1}$ has an obvious meaning of the 2π-pulse duration. If to restore initial coordinates z and t, it is seen that the group velocity of 2π-pulse V_g is different from the light velocity in a linear medium.

$$V_g^{-1} = c^{-1}\left(1 + c\alpha' t_p^2\right),$$

where $\alpha' = 2\pi n_a \omega_0 d^2 / \hbar c$. This expression shows that 2π-pulses of greater duration (and smaller amplitude $\tilde{E}_{\max} = 2/t_p$) propagate in the medium with smaller velocities, which means that solitons tend to stay in line in the height order.

4π-pulses

Let we have a trivial solution $u_0 = 0$ and two soliton solutions u_1 and u_2 (4.1.76). Now we can use relationship (4.1.74) to find more complex solutions of equation (4.1.72). Let

$$u_{1,2} = 4\arctan[\exp(v_{1,2})],$$

where $v_{1,2} = a_{1,2}\tau - \zeta / a_{1,2}$. Then (4.1.74) provides a new unknown solution $u_3(\tau,\zeta)$

$$u_3(\tau,\zeta) = 4\arctan\left[\left(\frac{a_1 + a_2}{a_1 - a_2}\right)\tan\left(\frac{v_1 - v_2}{2}\right)\right]$$

or

$$u_3(\tau,\zeta) = 4\arctan\left[\left(\frac{a_1 + a_2}{a_1 - a_2}\right)\left(\frac{\sinh[(v_1 - v_2)/2]}{\cosh[(v_1 + v_2)/2]}\right)\right]. \tag{4.1.78a}$$

The pulse area θ in a general case is $\theta = u(+\infty,\zeta) - u(-\infty,\zeta)$. In the case under consideration we can simply obtain from (4.1.78a) that $u_3(\pm\infty,\zeta) = \pm 2\pi$. In so far as the area $\theta = 4\pi$ corresponds to this pulse, the envelope of USP corresponds to the 4π-pulse of SIT.

Depending on a_1 and a_2, this expression provides different two-soliton solutions of the considered equation.

Let $a_1 > 0$, but $a_2 < 0$. If we introduce parameters $t_{p1} = t_{p0}a_1^{-1}$ and $t_{p2} = t_{p0} \mid a_2 \mid^{-1}$, then from (4.1.78a) we obtain $\widetilde{E}(t,z)$ in the form:

$$\widetilde{E}(t,z) = A\frac{(2/t_{p1})\operatorname{sech} Y_1 + (2/t_{p2})\operatorname{sech} Y_2}{1 - B(\tanh Y_1 \tanh Y_2 - \operatorname{sech} Y_1 \operatorname{sech} Y_2)}, \qquad (4.1.79a)$$

where

$$A = \frac{t_{p2}^2 - t_{p1}^2}{t_{p2}^2 + t_{p1}^2}, \quad B = \frac{2t_{p1}t_{p2}}{t_{p2}^2 + t_{p1}^2}, \qquad Y_1 = \left(t - z/V_1\right)t_{p1}^{-1}, \quad Y_2 = \left(t - z/V_2\right)t_{p2}^{-1}.$$

This 4π-pulse is not a stationary solution of (4.1.72) and it describes collision of two solitons (of 2π-pulses) which propagate in the same direction with the different velocities V_1 and V_2 where

$$V_k^{-1} = c^{-1}\left(1 + c\alpha' t_{pk}^2\right), \quad k = 1,2.$$

As $\zeta \to +\infty$ expression (4.1.79a) yields the approximate formula for the envelope of this 4π-pulse. It describes two solitons separated in space.

$$\widetilde{E}(t,z) = (2/t_{p1})\operatorname{sech}(Y_1 \pm \beta) + (2/t_{p2})\operatorname{sech}(Y_2 \mp \beta), \qquad (4.1.78b)$$

Whereas at $\zeta \to -\infty$, $\widetilde{E}(\tau,\zeta)$ is given by the same expression but with $\beta = 0$. Parameter $\beta = \operatorname{artanh} B$ in formula (4.1.78b) is a phase shift appearing under the two-soliton collision. The upper sign before β in the arguments in (4.1.78b) corresponds to the condition $t_{p1} < t_{p2}$ and the lower one corresponds to $t_{p1} > t_{p2}$. It is worthy to note that character of the soliton collision depends on the ratio of the pulse durations t_{p1}/t_{p2}. Figures 4.1.2 and 4.1.3 show different evolution of these 4π-pulses. The condition of the formation of 4π-pulse profile peaked at $t = z = 0$ is $\partial^2\widetilde{E}/\partial t^2 < 0$ and $\widetilde{E} > 0$ in this point. So, we can obtain the following inequality

$$\left(1 - \frac{t_{p1}}{t_{p2}}\right)\left[1 + \left(\frac{t_{p1}}{t_{p2}}\right)^2 - 3\frac{t_{p1}}{t_{p2}}\right] > 0.$$

Thus, the solitons collision results in overlapping if ratio t_{p1}/t_{p2} is obeyed to inequality

$$t_{p1}/t_{p2} < (3 - \sqrt{5})/2 \approx 0.382.$$

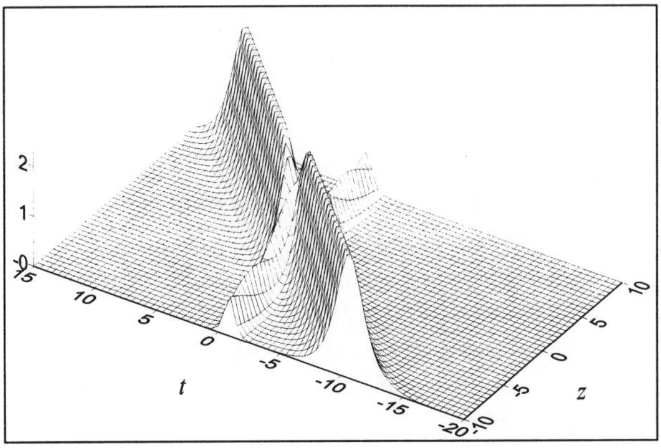

Fig.4.1.2.

Crossing collision of two solitons

This kind of collision is referred to as crossing one. Fig. 4.1.2 shows the evolution of the 4π-pulse for which t_{p1}/t_{p2} is 1/3. In the opposite case when $t_{p1}/t_{p2} > 0.382$, the picture of the soliton collision alike with an elastic ball interaction.

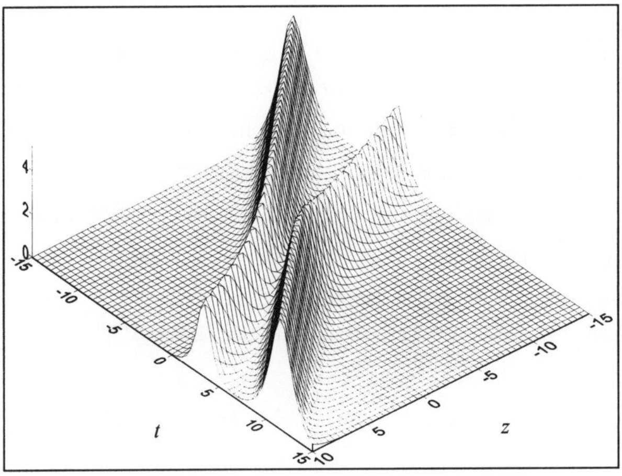

Fig.4.1.3.

Repulsive collision of two solitons

The evolution of the 4π-pulse in the case of $t_{p1}/t_{p2} = 2/3$ is shown in Fig.4.1.3.

The terms of "crossing collision" and "repulsive one" are illustrated well by the Fig.4.1.4.

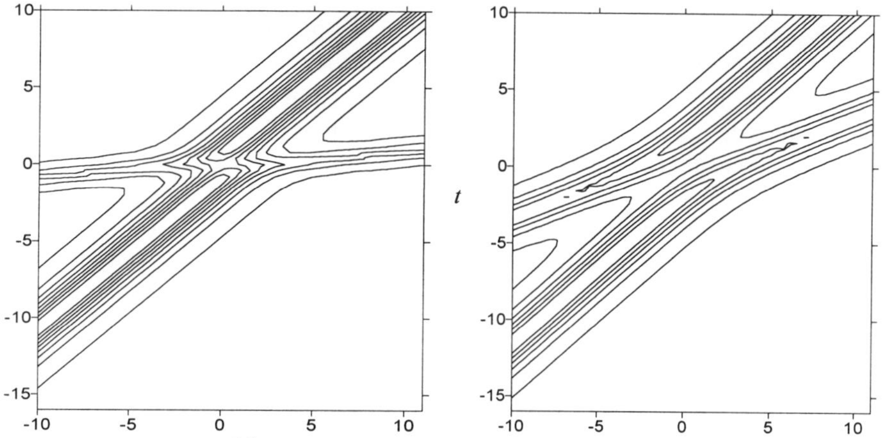

(a) (b)

Fig. 4.1.4. Contour plots of crossing (a) and repulsive (b) collision of two solitons

Let $a_1 > 0$ and $a_2 > 0$. In this case $u_3(\pm\infty,\zeta) = 2\pi$ and the pulse area θ is zero. It is natural to call such pulse as 0π-pulse, but this term doesn't prove to be proper first. It is useful to find the envelope of USP from (4.1.78a) for positive parameters of Bäcklund transformations. By introducing pulse duration $t_{p1} = t_{p0}a_1^{-1}$ and $t_{p2} = t_{p0}a_2^{-1}$, and differentiating $u_3(\tau,\zeta)$ with respect to τ, we can find

$$\widetilde{E}(t,z) = A\frac{(2/t_{p1})\operatorname{sech}Y_1 - (2/t_{p2})\operatorname{sech}Y_2}{1 - B(\tanh Y_1 \tanh Y_2 + \operatorname{sech}Y_1 \operatorname{sech}Y_2)}. \tag{4.1.79b}$$

Fig. 4.1.5.

Collision of soliton with anti-soliton

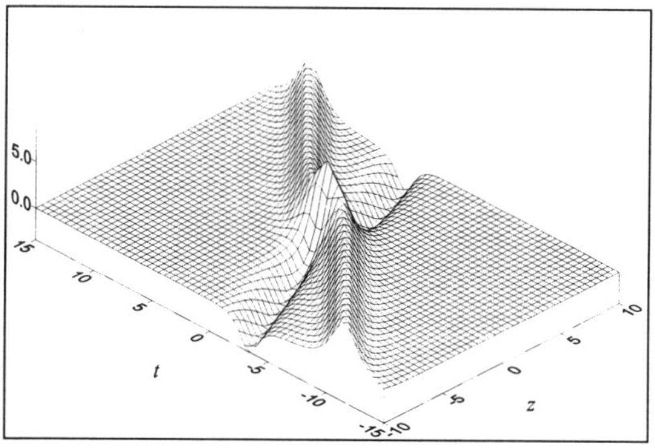

Under condition $\zeta \to +\infty$ this expression approximately equals the sum of two solitons of the (4.1.77a) type with the opposite phases (Fig. 4.1.5). Pulse areas of both solitons are equal, but they are opposite by the sign. Indeed, such pulses are the "sum" of 2π-pulse and anti-2π-pulse. The word "sum" has been taken in the quotation marks, as far as such 4π-pulse decays into two 2π-pulses only asymptotically and, strictly speaking, is not a linear superposition of the two real solitons.

Breathers and degenerated 0π-pulses

There are no restrictions imposed on the parameter a of BT in the above method of solving the equation (4.1.72). Thus a can be considered as a complex number $a = \alpha + i\beta$, $\alpha > 0$. As a result the 2π-pulse obtains the phase modulation. Formula of tangents (4.1.74) is correct in this case of complex parameters a_1 and a_2 too. It is particularly interesting to consider the case when $a_1 = a_2^* = \alpha + i\beta$, $\alpha > 0$. Then the formula of tangents yields

$$u_3(\tau,\zeta) = 4\arctan\left[\left(\frac{\alpha}{\beta}\right)\frac{\sin\overline{q}}{\cosh\overline{p}}\right], \tag{4.1.80}$$

where the pulse duration can be defined as $t_p = t_{p0}\alpha^{-1}$. The $\overline{p} = \alpha(\tau - \zeta/|a_1|^2)$, $\overline{q} = \beta(\tau + \zeta/|a_1|^2)$ are introduced for the sake of simplicity. The envelope of the USP obtained from (4.1.80) can be written as

$$\widetilde{E}(\tau,\zeta) = \frac{4}{t_p}\sinh(\overline{p})\left[\frac{\cos(\overline{q}) - (\alpha/\beta)\sin(\overline{q})\tanh(\overline{p})}{1 + (\alpha/\beta)^2\sin^2(\overline{q})\operatorname{sech}^2(\overline{p})}\right]. \tag{4.1.81}$$

Fig.4.1.6.

Breather or 0π-pulse

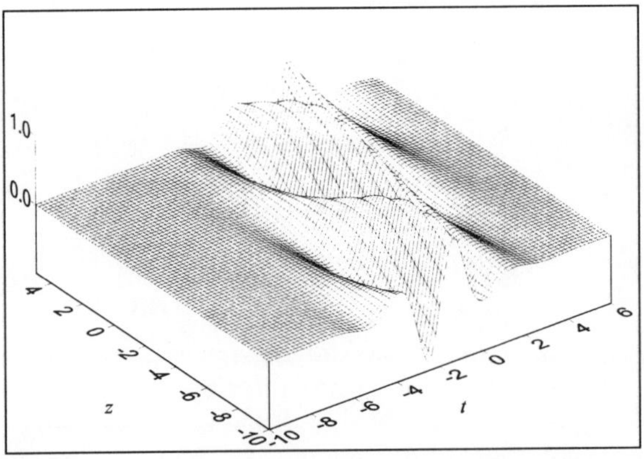

This pulse does not decay into two 2π-pulses as it happened with the 4π-pulse, considered above, but it remains as the indivisible object being a non-steady-state solitary wave (Fig.4.1.6). The pulse area θ of this USP is zero, so such a pulse should be named as 0π-pulse. But they often call it *breather*, emphasizing the periodic changing of its shape. Breather is a specific case of two-soliton solutions of SIT equations or of SG-equation. It is interpreted as (nonlinear) superposition of two solitons, which have equal group velocities but different phase velocities.

If parameter β approaches to zero in the above formulae, then we can find solution of SIT equations under condition $a_1 = a_2$ ($\mathrm{Im}\, a_{1,2} = 0$). It is impossible to find this solution from the formula of tangents. In this degenerated case the two-soliton USP the envelope takes the form:

$$\widetilde{E}(\tau,\zeta) = \frac{4}{t_p}\operatorname{sech}(v)\frac{1 - u\tanh(v)}{1 + u^2\operatorname{sech}^2(v)}, \tag{4.1.82}$$

where $u = a\tau + \zeta / a$ and $v = a\tau - \zeta / a$. This degenerated 0π-pulse exhibits a pair of 2π- and anti-2π-pulses propagating as shown in (Fig.4.1.7).

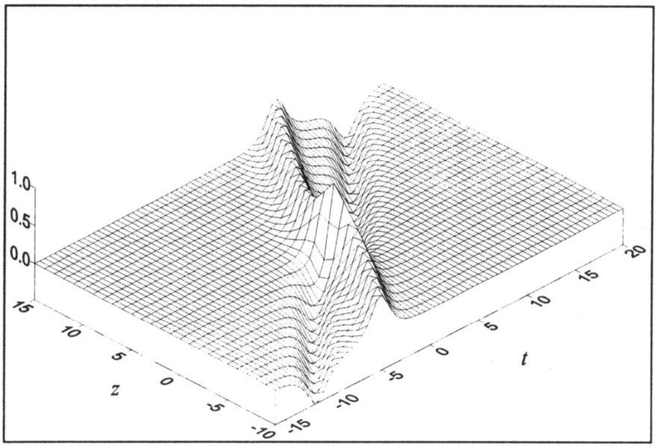

Fig.4.1.7.

Collision of two solitons corresponded with degenerated 0π-pulse

All these expressions for two-soliton solutions of the SIT equation can be obtained in the framework of the inverse scattering transform method too.

4.1.6. LINEAR POLARISED ULTRASHORT PULSE PROPAGATION

Let us dwell on the known limiting version of the equations, which describes the polarised ultrashort pulse propagation in a resonant medium with energy levels degenerated over the orientations of the total angular momentum (1.4.7). It is easy to see that for a USP linearly polarised along the same axis, the initial equations (1.4.7) for resonance transitions $1/2 \rightarrow 1/2$, $J \rightarrow J$, $1 \rightarrow 1$ and $3/2 \leftrightarrow 1/2$ reduce to the conventional SIT equations

$$\frac{\partial q}{\partial \zeta} = i\langle p \rangle, \quad \frac{\partial p}{\partial \tau} = i\delta p + iqR_3, \quad \frac{\partial R_3}{\partial \tau} = -\frac{i}{2}(qp^* - q^* p), \tag{4.1.83}$$

with the initial and boundary conditions

$$q(\tau, \zeta = 0) = q_{in}(\tau), \quad \lim_{|\tau| \to \infty} q = \lim_{|\tau| \to \infty} p = 0, \quad \lim_{|\tau| \to \infty} R_3 = 1,$$

where R_3 is the inversion of the lower and upper levels for one pair of Zeeman sublevels. The quantisation axis is chosen along x-axis, q is related to the field $\mathcal{E}_x$ by the formulae:

- for transitions $1 \leftrightarrow 0$: $q = q^{(x)} = d\mathcal{E}_x t_{p0} / \hbar$
- for transitions $1/2 \rightarrow 1/2$, $1 \rightarrow 1$ and $3/2 \leftrightarrow 1/2$: $q^{(x)} = d\mathcal{E}_x t_{p0} / \hbar\sqrt{2}$.

Equations (4.1.83) can also be derived for the circularly polarised waves and the transitions $1/2 \rightarrow 1/2$, $1 \leftrightarrow 0$ and $1 \rightarrow 1$, as well as for a fixed (identical for all USPs) elliptical polarisations for the transitions $1 \leftrightarrow 0$ and $1 \rightarrow 1$. Thus, all those versions of the USP propagation do not provide any new aspects to the theory of SIT.

A simplification of the problem can be achieved in the sharp-line limit, the perfect resonance and ignoring the phase chirp of the input USPs with the identical linear or circular polarisation. Then, assuming field $\mathcal{E}_j$ to be given, equations (1.4.7) can be integrated easily and the substitution of the result in equation (1.47a) yields

$$\frac{\partial^2 \theta}{\partial \zeta \partial \tau} = -\frac{1}{2} \sum_m \mathcal{G}^J_{m,m+J} \sin\left(2\theta \mathcal{G}^J_{m,m+J}\right), \tag{4.1.84}$$

where we introduce the following definitions:

- for linear polarisation ($j = 0$) along the x-axis

$$\theta = \theta(\tau, \zeta) = \int_{-\infty}^{\tau} q^{(x)}(\tau', \zeta) d\tau';$$

- for circular polarisation ($j = \pm 1$)

$$\theta = \theta(\tau, \zeta) = \int_{-\infty}^{\tau} q^{(\pm)}(\tau', \zeta) d\tau'.$$

The index (x) or $(\pm)$ labels the spherical components of vector $\vec{q}$ For linear polarisation the quantisation axis is chosen along x-axis, and for circular polarisation it is supposed to be parallel to the z-axis.

For the transitions $1/2 \to 1/2$, $1 \leftrightarrow 0$, $1 \to 1$ and either linear or circular polarisation of the light wave as well as for the transition $3/2 \leftrightarrow 1/2$ and linear polarisation of the light wave, equation (4.1.84) provides the famous Sine-Gordon equation

$$\frac{\partial^2 \theta}{\partial \zeta \partial \tau} = -\sin\theta \, .$$

In the case of plane polarisation and $2 \to 2$ transition, equation (4.1.84) gives Double-Sine-Gordon equation (DSG-equation)

$$\frac{\partial^2 \theta}{\partial \zeta \partial \tau} = -\left(\sin\theta + \frac{1}{2}\sin\frac{\theta}{2}\right), \tag{4.1.85}$$

where

$$\theta = \theta(\tau, \zeta) = \sqrt{8/5} \int_{-\infty}^{\tau} q^{(x)}(\tau', \zeta) d\tau' \, .$$

Independent variables ζ and τ are in a factor of $\sqrt{5}/2$ larger than the old variables.

The DSG-equation is a representative of the family of the non-linear evolution equations. It was named the "Multiple-Sine-Gordon equation"

$$\frac{\partial^2 \theta}{\partial \zeta \partial \tau} = \sum_{m=1}^{J} \frac{m}{J} \sin\frac{m}{J}\theta \, . \tag{4.1.86}$$

This equation was engaged to describe the USP propagation in a degenerate medium with the $Q(J)$ vibration-rotational transitions. The selection rules for the $Q(J)$ branch transition in the wave of linear polarised light are $J \to J$, $\Delta m = 0$. Thus equation (4.1.84) can be reduced to (4.1.86). In [6,21] other transitions ($J \to J \pm 1$) were observed and other families of the generalised Sine-Gordon equations were discussed. In [21] these families of equations were written as

$$\frac{\partial^2 \theta}{\partial \zeta \partial \tau} = \frac{1}{2J+1} \sum_{m=-J}^{J} \frac{\sqrt{J^2 - m^2}}{J} \sin\frac{\sqrt{J^2 - m^2}}{J}\theta$$

for transitions ($J \to J-1$), and

$$\frac{\partial^2 \theta}{\partial \zeta \partial \tau} = \frac{1}{2J+1} \sum_{m=-J}^{J} \frac{\sqrt{(J+1)^2 - m^2}}{J+1} \sin\frac{\sqrt{(J+1)^2 - m^2}}{J+1}\theta$$

for transitions ($J \to J+1$).

However, only DSG-equation attracts a noticeable attention due to its relations with physical problems. For example, the DSG-equation is the model equation for the charge-density waves phenomenon [71,72]. This equation describes the spin wave dynamics in the B-phase of superfluid ^{3}He [73-75]. The DSG equation governs the stable 2π-kinks in the theory of commensurate-incommensurate phase transitions [76,77]. The non-linear internal dynamics of dislocations in Frenkel-Kontorova model is described by this equation [78] too. More detailed surveys may be found in [79].

It is known [80] that the DSG equation is not completely integrable. So it does not have soliton solutions. However, there are steady-state solutions. The change of the independent variables in (4.1.85) yields the standard form of DSG equation

$$\frac{\partial^2\theta}{\partial\xi^2} - \frac{\partial^2\theta}{\partial\eta^2} = \sin\theta + \frac{1}{2}\sin\frac{\theta}{2}, \qquad (4.1.87)$$

where $\xi = \zeta - \tau$, $\eta = \zeta + \tau$. The steady-state solution of (4.1.87) [79] can be written as

$$\begin{aligned}\theta(\vartheta) = \theta_0 &+ 4\arctan\left[\cosh\delta_3 \exp(\vartheta - \delta_1) + \sinh\delta_3\right] + \\ &+ 4\arctan\left[\cosh\delta_3 \exp(\vartheta - \delta_2) - \sinh\delta_3\right],\end{aligned} \qquad (4.1.88)$$

where

$$\vartheta = \frac{c\Omega}{2V}\left[\left(\frac{2V}{c} - 1\right)\eta - \xi\right] + \vartheta_0 = \frac{\sqrt{5}\Omega}{2t_{p0}}\left(t - \frac{z}{V}\right) + \vartheta_0,$$

and δ_1, δ_2, δ_3, ϑ_0, Ω, V and θ_0 are free parameters. Of particular interest are the next four specific solutions of the DSG equation:

I. This is the so-called 4π-kink [79-82]

$$\theta(\vartheta) = 4\arctan(\exp[\vartheta - \Delta]) + 4\arctan(\exp[\vartheta + \Delta]), \qquad (4.1.89)$$

which follows from (4.1.88) under condition $\theta_0 = 0$, $\delta_1 = -\delta_2 = \Delta = \ln(2 + \sqrt{5})$, $\delta_3 = 0$. The pulse envelope can be then written in the form

$$q(\tau,\zeta) = 2\Omega\,\mathrm{sech}(\vartheta - \Delta) + 2\Omega\,\mathrm{sech}(\vartheta + \Delta), \qquad (4.1.90)$$

$$V = \frac{4\Omega^2}{5 + 4\Omega^2}c.$$

This steady-state pulse is shown in Fig. 4.1.8.

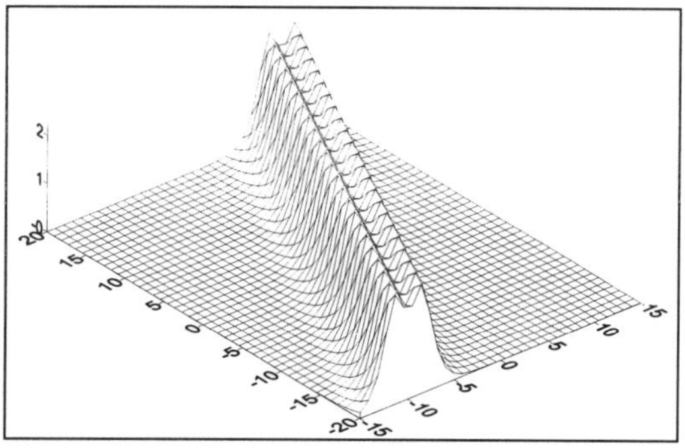

Fig.4.1.8.

4π -kink of the DSG equation

II. It is the case of solitary wave named 0π -kink

$$\theta(\vartheta) = 2\pi + 4\arctan\left(\exp[\vartheta - \Delta]\right) - 4\arctan\left(\exp[\vartheta + \Delta]\right). \qquad (4.1.91)$$

It follows from (4.1.88) under conditions $\theta_0 = 2\pi$, $\delta_1 = i\pi - \delta_2 = \Delta = \ln\left(2+\sqrt{3}\right)$ and $\delta_3 = 0$. The pulse envelope of the 0π -kink is

$$q(\tau,\zeta) = 2\Omega\,\mathrm{sech}(\vartheta - \Delta) - 2\Omega\,\mathrm{sech}(\vartheta + \Delta), \qquad (4.1.92)$$

$$V = 4\Omega^2\left(3 + 4\Omega^2\right)^{-1} c.$$

III. This solution was named the $4\pi - 2\delta$ -kink:

$$\theta(\vartheta) = 2\pi - \delta + 4\arctan\left(\frac{4}{\sqrt{15}}\exp[\vartheta] + \frac{1}{15}\right), \qquad (4.1.93)$$

It can be obtained from (4.1.88) under conditions $\theta_0 = \delta - 2\pi$, $\delta_1 = 0$, $\delta_2 = \infty$, and $\delta_3 = \ln(5/3)/2$ with the pulse envelope in the form

$$q(\tau,\zeta) = \frac{2\;\;15\Omega}{1 + 4\cosh\vartheta}, \qquad (4.1.94)$$

$$V = 16\Omega^2\left(15 + 16\Omega^2\right)^{-1} c.$$

This solution corresponds to the rotation of the Bloch vector from $4\pi - \delta$ at $z \to -\infty$ to $+\delta$ as $z \to +\infty$, so the total angle of rotation is $4\pi - 2\delta$.

IV. This solution was named the 2δ-kink:

$$\theta(\vartheta) = \delta - 2\pi + 4\arctan\left(\frac{4}{\sqrt{15}}\exp[\vartheta] - \frac{1}{15}\right). \qquad (4.1.95)$$

It can be obtained from the expression (4.1.88) under conditions that $\theta_0 = 2\pi - \delta$, $\delta_1 = 0, \delta_2 = -\infty$, and $\delta_3 = -\ln(5/3)/2$. The corresponding pulse envelope is

$$q(\tau, \zeta) = \frac{2\sqrt{15}\Omega}{4\cosh\vartheta - 1}, \qquad (4.1.96)$$

$$V = 16\Omega^2\left(15 + 16\Omega^2\right)^{-1} c.$$

This solution corresponds to the rotation of the Bloch vector from $+\delta$ at $z \to -\infty$ to $-\delta$ as $z \to +\infty$, so the total angle of rotation is $4\pi - 2\delta$.

These expressions allow to interpret the physical meaning of Ω and V. So V is the group velocity of a steady-state pulse and Ω determines the pulse duration by the formula

$$t_p = \frac{2}{\sqrt{5}\Omega} t_{p0}.$$

The numerical solutions for $J = 2$ and $J = 3$, obtained in [80], demonstrate all characteristic features of solitons. For example, when $J = 2$ the initial condition $\theta_0 = 0$ corresponds to an initially unexcited medium of the five-fold degenerated two-level atoms. The unique pulses, which propagate without distortion, are the double-humped pulses with pulse area 4π (4.1.90). The decay of the pulses converts them into the train of such 4π-pulses of different amplitudes. Two 4π-pulses can collide and pass through each other without a change of their shapes and velocities, but with the phase shifts, which are small.

As it is pointed out in [83], solutions, corresponding to 0π-kink, $4\pi - 2\delta$-kink and 2δ-kink, are steady-state pulses in a partially inverted medium. Analysis of the integral curves related to the DSG equation on a phase plane yields four points of the rest, i.e. zeros of function $2\sin\theta + \sin(\theta/2)$. These zeros are 0π, 2π, $4\pi - \delta$, and δ, where $\cos(\delta/2) = -1/4$. These values of the parameter δ label the steady-state solutions under consideration.

4.2. SIT -theory in frame of inverse scattering method. Polarised solitons

The development of the SIT theory is characterised by its going beyond the framework of two-level approximation and by the spectral composition of the USP field becoming more complex. The latter means that the resonant medium interacts with radiation containing several carrier frequencies. Besides, such factors as the direct interaction between resonant atoms, non-linear properties of the dielectric doped with resonant atoms and polarisation of the electromagnetic field should be taken into account. We shall consider now some of these further generations of the coherent propagation theory.

4.2.1. SIMULTONS

In a number of works [22-28,91] the ultrashort pulses propagation in three-level medium was studied. In the simplest case of this model the resonance levels have V, Λ and "cascade" configurations (see Fig.4.2.1.). It was determined [23-25] that, if the oscillator forces for every transition in a V or a Λ configuration are equal, then a two-frequency pulse (characterised by two different frequencies of the carrier wave) is able to propagate in such a medium without the envelope distortion. An ultrashort pulse of the kind was called *simulton* [22].

Fig. 4.2.1.

Energy level configurations:
(a) Λ, (b) V and (c)"cascade"

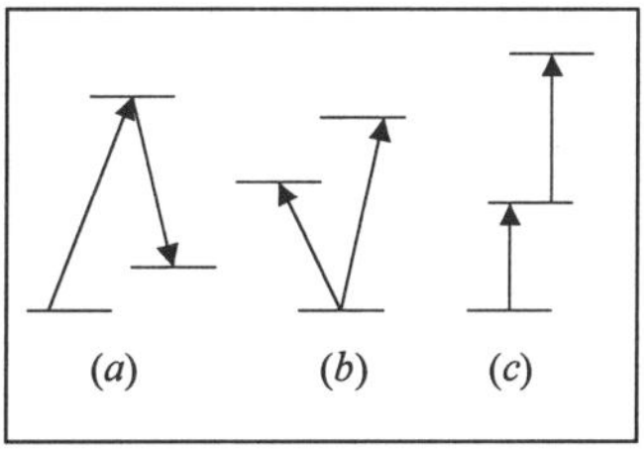

Further studies [26,27] demonstrated that the system of equation governing the evolution of USP in the case of V and Λ configurations are completely integrable. The solutions of these equations are solitons. The theory of simultons includes the infinite series of higher conservation laws, Bäcklund transformation and non-linear superposition principle. The simultons are single-soliton solutions of the above equations. At the same time solutions occur which are responsible for the propagation and collisions of simultons. The oscillating simultons (colour breathers) are the two-frequency generalisations of 0π -pulse by McCall-Hahn. It is worth noting at this point that a simulton is generally unstable with respect to transformation into one-frequency 2π -pulse [92] and may remain as two-frequency pulse only for a special choice of resonance level populations.

In the case of "cascade" configuration the corresponding equations are unlikely to be completely integrable and to have soliton solutions. Nevertheless, numerical simulations

[24,93] displayed a solitary wave differing very little from a soliton. These solitary waves propagate without any significant distortion.

The propagation of three-frequency simultons under triple resonance with simultaneous non-resonant parametric three-wave interaction was studied by Zabolotskii [94]. This model can serve as a generalisation of a three-wave interaction.

Dynamics of simultons under multiple resonance conditions was studied in [95-97]. We know both this model and the multifrequency simulton model do not provide completely integrable system of equations. Numerical simulations free of any restrictions showed that a USP with a simulton-like envelope keeps its form steady long enough. So one may recognise it as nonrigorous SIT, though there are no soliton solutions of the corresponding evolution equations.

Let the resonant medium consist of there-level atoms with non-degenerated resonance states. We shall consider the collinear propagation of two ultrashort electromagnetic pulses

$$E_{1,2} = \mathcal{E}_{1,2} \exp[i(k_{1,2}z - \omega_{1,2}t)] + c.c.,$$

resonant to adjacent optically permitted transitions $|1> \to |3>$ and $|2> \to |3>$ in the Λ configuration $\omega_1 \approx \omega_{31}$, $\omega_2 \approx \omega_{32}$. The generalised Maxwell-Bloch equations under slowly varying complex envelope approximation can be written as

$$\frac{\partial \mathcal{E}_1}{\partial z} + \frac{1}{c}\frac{\partial \mathcal{E}_1}{\partial t} = i\frac{2\pi\omega_1 n_a}{cn(\omega_1)}\langle d_{13}\rho_{31}\rangle, \qquad \frac{\partial \mathcal{E}_2}{\partial z} + \frac{1}{c}\frac{\partial \mathcal{E}_2}{\partial t} = i\frac{2\pi\omega_2 n_a}{cn(\omega_2)}\langle d_{23}\rho_{32}\rangle,$$

$$i\hbar\frac{\partial \rho_{31}}{\partial t} = \hbar\,\Delta\omega_1\rho_{31} + d_{31}(\rho_{33} - \rho_{11})\mathcal{E}_1 - d_{32}\rho_{21}\mathcal{E}_2,$$

$$i\hbar\frac{\partial \rho_{32}}{\partial t} = \hbar\,\Delta\omega_2\rho_{32} + d_{32}(\rho_{33} - \rho_{22})\mathcal{E}_2 - d_{31}\rho_{21}^*\mathcal{E}_1$$

$$i\hbar\frac{\partial \rho_{21}}{\partial t} = \hbar(\Delta\omega_1 - \Delta\omega_2)\rho_{21} + d_{31}\rho_{32}^*\mathcal{E}_1 - d_{32}^*\rho_{31}\mathcal{E}_2^*$$

$$\hbar\frac{\partial(\rho_{33} - \rho_{11})}{\partial t} = -2\,\mathrm{Im}(d_{32}^*\rho_{32}\mathcal{E}_2^*) - 4\,\mathrm{Im}(d_{31}^*\rho_{31}\mathcal{E}_1^*),$$

$$\hbar\frac{\partial(\rho_{33} - \rho_{22})}{\partial t} = -4\,\mathrm{Im}(d_{32}^*\rho_{32}\mathcal{E}_2^*) - 2\,\mathrm{Im}(d_{31}^*\rho_{31}\mathcal{E}_1^*).$$

Here as usually ρ_{ij} is slowly varying amplitudes of matrix elements of the density matrix of a three-level atom. Besides, $\Delta\omega_1 = \omega_{31} - \omega_1$ and $\Delta\omega_2 = \omega_{32} - \omega_2$. The velocities of weak pulses in matter are supposed to be equal.

When the USP is in resonance with optically permitted transitions $|1\rangle \to |3\rangle$ and $|1\rangle \to |2\rangle$ in V configuration $\omega_1 \approx \omega_{21}$, $\omega_2 \approx \omega_{31}$, the basic equations can be written as following

$$\frac{\partial \mathcal{E}_1}{\partial z} + \frac{1}{c}\frac{\partial \mathcal{E}_1}{\partial t} = i\frac{2\pi\omega_1 n_a}{cn(\omega_1)}\langle d_{12}\rho_{21}\rangle, \qquad \frac{\partial \mathcal{E}_2}{\partial z} + \frac{1}{c}\frac{\partial \mathcal{E}_2}{\partial t} = i\frac{2\pi\omega_2 n_a}{cn(\omega_2)}\langle d_{13}\rho_{31}\rangle,$$

$$i\hbar\frac{\partial \rho_{21}}{\partial t} = \hbar\,\Delta\omega_1\rho_{21} + d_{21}(\rho_{11} - \rho_{22})\mathcal{E}_1 - d_{31}\rho_{32}^*\mathcal{E}_2,$$

$$i\hbar\frac{\partial \rho_{31}}{\partial t} = \hbar\,\Delta\omega_2\rho_{31} + d_{31}(\rho_{11} - \rho_{33})\mathcal{E}_2 - d_{21}\rho_{32}\mathcal{E}_1,$$

$$i\hbar\frac{\partial \rho_{32}}{\partial t} = \hbar(\Delta\omega_2 - \Delta\omega_1)\rho_{32} + d_{31}\rho_{21}^*\mathcal{E}_2 - d_{21}^*\rho_{31}\mathcal{E}_1^*,$$

$$\hbar\frac{\partial(\rho_{22} - \rho_{11})}{\partial t} = -4\,\mathrm{Im}(d_{21}^*\rho_{21}\mathcal{E}_1^*) - 2\,\mathrm{Im}(d_{31}^*\rho_{31}\mathcal{E}_2^*),$$

$$\hbar\frac{\partial(\rho_{33} - \rho_{11})}{\partial t} = -2\,\mathrm{Im}(d_{21}^*\rho_{21}\mathcal{E}_1^*) - 4\,\mathrm{Im}(d_{31}^*\rho_{31}\mathcal{E}_1^*),$$

where $\Delta\omega_1 = \omega_{21} - \omega_1$ and $\Delta\omega_2 = \omega_{31} - \omega_2$.

When the USP propagates in the medium under double resonance condition with the optically permitted transitions $|1\rangle \to |2\rangle$ and $|2\rangle \to |3\rangle$ in the "cascade" configuration $\omega_1 \approx \omega_{21}$, $\omega_2 \approx \omega_{32}$, evolution of the slowly varying complex envelopes and matrix elements ρ_{ij} is governed by the following system of equations

$$\frac{\partial \mathcal{E}_1}{\partial z} + \frac{1}{c}\frac{\partial \mathcal{E}_1}{\partial t} = i\frac{2\pi\omega_1 n_a}{cn(\omega_1)}\langle d_{12}\rho_{21}\rangle, \qquad \frac{\partial \mathcal{E}_2}{\partial z} + \frac{1}{c}\frac{\partial \mathcal{E}_2}{\partial t} = i\frac{2\pi\omega_2 n_a}{cn(\omega_2)}\langle d_{23}\rho_{32}\rangle,$$

$$i\hbar\frac{\partial \rho_{21}}{\partial t} = \hbar\,\Delta\omega_1\rho_{21} + d_{21}(\rho_{22} - \rho_{11})\mathcal{E}_1 - d_{32}^*\rho_{31}\mathcal{E}_2^*,$$

$$i\hbar\frac{\partial \rho_{32}}{\partial t} = \hbar\,\Delta\omega_2\rho_{32} - d_{32}(\rho_{22} - \rho_{33})\mathcal{E}_2 + d_{21}^*\rho_{31}\mathcal{E}_1^*,$$

$$i\hbar\frac{\partial \rho_{31}}{\partial t} = \hbar(\Delta\omega_2 + \Delta\omega_1)\rho_{31} + d_{21}\rho_{32}\mathcal{E}_1 - d_{32}\rho_{21}\mathcal{E}_2,$$

$$\hbar\frac{\partial(\rho_{22} - \rho_{11})}{\partial t} = 4\,\mathrm{Im}(d_{21}^*\rho_{21}\mathcal{E}_1^*) + 2\,\mathrm{Im}(d_{32}\rho_{32}^*\mathcal{E}_2),$$

$$\hbar\frac{\partial(\rho_{22} - \rho_{33})}{\partial t} = 2\,\mathrm{Im}(d_{21}^*\rho_{21}\mathcal{E}_1^*) + 4\,\mathrm{Im}(d_{32}\rho_{32}^*\mathcal{E}_2),$$

where $\Delta\omega_1 = \omega_{21} - \omega_1$ and $\Delta\omega_2 = \omega_{32} - \omega_2$.

Now we introduce the new normalisation variables to unify these systems. Here they are for V configuration

$$q_1 = -d_{21}\mathcal{E}_1 t_{p0}\hbar^{-1}, \quad q_2 = -d_{31}\mathcal{E}_2 t_{p0}\hbar^{-1},$$
$$u = \rho_{21}, v = \rho_{31}, w = -\rho_{32}, \; n_1 = \rho_{11} - \rho_{22}, \; n_2 = \rho_{11} - \rho_{33},$$

and for Λ configuration

$$q_1 = -d_{31}\mathcal{E}_1 t_{p0}\hbar^{-1}, \quad q_2 = -d_{32}\mathcal{E}_2 t_{p0}\hbar^{-1},$$
$$u = \rho_{31}, v = \rho_{32}, w = \rho_{12}, \; n_1 = \rho_{11} - \rho_{33}, \; n_2 = \rho_{22} - \rho_{33}.$$

Besides, we introduce $\tau = (t - z/c)t_{p0}^{-1}$ and $\zeta = z/L$, where L is a normalised factor, which is related to resonant absorption of optical radiation at frequency ω_1. In terms of these variables, we can write the unified form of evolution equations for these two cases

$$\frac{\partial q_1}{\partial \zeta} = iu, \quad \frac{\partial q_2}{\partial \zeta} = i\gamma v, \tag{4.2.1a}$$

$$\frac{\partial u}{\partial \tau} = i\delta u + iq_1 n_1 + iq_2 w^*, \quad \frac{\partial v}{\partial \tau} = i\delta v + iq_2 n_2 + iq_1 w, \tag{4.2.1b}$$

$$\frac{\partial w}{\partial \tau} = iq_1^* v - iq_2 u^*, \tag{4.2.1c}$$

$$\frac{\partial n_1}{\partial \tau} = 2i(q_1^* u - q_1 u^*) + i(q_2^* v - q_2 v^*), \tag{4.2.1d}$$

$$\frac{\partial n_2}{\partial \tau} = i(q_1^* u - q_1 u^*) + 2i(q_2^* v - q_2 v^*). \tag{4.2.1e}$$

Here we assume that, for partial carrier frequencies ω_1 and ω_2, the conditions $\delta = (\omega_1 - \omega_{21})t_{p0} = (\omega_2 - \omega_{31})t_{p0}$ and $\delta = (\omega_1 - \omega_{31})t_{p0} = (\omega_2 - \omega_{32})t_{p0}$ are satisfied. The parameter γ is the ratio of oscillator forces

$$\gamma = \frac{\omega_2 |d_{13}|^2 \, n(\omega_1)}{\omega_1 |d_{12}|^2 \, n(\omega_2)}, \quad \text{for V configuration}$$

and

$$\gamma = \frac{\omega_2 |d_{23}|^2 \, n(\omega_1)}{\omega_1 |d_{13}|^2 \, n(\omega_2)}, \quad \text{for } \Lambda \text{ configuration.}$$

We can also rewrite the evolution equations for the case of "cascade" configuration. Let

$$q_1 = -d_{21}\mathcal{E}_1 t_{p0}\hbar^{-1}, \quad q_2 = -d_{32}\mathcal{E}_2 t_{p0}\hbar^{-1},$$
$$u = \rho_{21}, v = \rho_{32}, w = \rho_{31}, \ n_1 = \rho_{11} - \rho_{22}, \ n_2 = \rho_{33} - \rho_{22}.$$
$$\delta_1 = (\omega_{21} - \omega_1)t_{p0}, \quad \delta_2 = (\omega_{32} - \omega_2)t_{p0}.$$

Then the normalised equations are

$$\frac{\partial q_1}{\partial \zeta} = i\langle u\rangle, \quad \frac{\partial q_2}{\partial \zeta} = i\gamma\langle v\rangle, \tag{4.2.2a}$$

$$\frac{\partial u}{\partial \tau} = i\delta_1 u + iq_1 n_1 - iq_2^* w, \quad \frac{\partial v}{\partial \tau} = i\delta_2 v - iq_2 n_2 + iq_1^* w, \tag{4.2.2b}$$

$$\frac{\partial w}{\partial \tau} = i(\delta_1 + \delta_2)w + iq_1 v - iq_2 u, \tag{4.2.2c}$$

$$\frac{\partial n_1}{\partial \tau} = 2i(q_1^* u - q_1 u^*) + i(q_2^* v - q_2 v^*), \tag{4.2.2d}$$

$$\frac{\partial n_2}{\partial \tau} = i(q_1^* u - q_1 u^*) + 2i(q_2^* v - q_2 v^*). \tag{4.2.2e}$$

We can see that for the "cascade" configuration Bloch equations differ from the equations (4.2.1.).

Generally speaking, system (4.2.1) cannot be integrated by IST method. The reason is clear from the following. There are two lengths of non-linear compression or resonant absorption in three-level systems due to interaction of the double-frequency USP with two different transitions. The equality of these lengths is necessary for solitary steady-state wave to exist. This condition can be written in the form $\gamma = 1$. At the same time, system of equations (4.2.2) is not known to be integrable under any conditions.

The integrable version of unified system (4.2.1) coincides with the generalised Maxwell-Bloch equations in the certain case of USP propagation in the resonant medium with degenerated energy levels. Now we turn to these instances.

4.2.2. THE POLARISED SOLITONS

In this section the equations describing the ultrashort pulse propagation in resonant medium with the degeneration of energy levels will be introduced. The resulting equations are to be rewritten in the homogeneous form, so that we will be able to observe their solutions and properties from the common point of view by considering vector and matrix solitons.

The case of one-photon resonance

Let us consider the resonant medium as an ensemble of two-level atoms embedded in the linear host medium with the index $n(\omega)$. Let the optical pulse propagate along axis z and Cartesian components of the electric field strength vector be presented in the form

$$E^a(t,z) = \mathcal{E}^a(t,z)\exp\left(-i\omega_0 t + ik_0 z\right) + c.c.$$

The carrier frequency $\omega_0 = k_0 c$ is in resonance with the frequency $\omega_{21} = (E_2 - E_1)/\hbar$ of an optically permitted atomic transition $j_1 \to j_2$ between energy levels E_1 and E_2 degenerated over the projections m and l of the total angular momenta j_1 and j_2.

In a general case the evolution of the envelope of the ultrashort pulse and states of the resonant medium are described by the system of equations

$$\begin{aligned}
\left(\frac{\partial}{\partial z} + \frac{1}{c}\frac{\partial}{\partial t}\right)\mathcal{E}^a &= i\left(2\pi\omega_0 n_A \,|\, d \,|\, / c\right)\sum_{l,m} \mathcal{G}^a_{lm}\langle \rho_{lm}\rangle, \\
\hbar\left(\frac{\partial}{\partial t} - i\Delta\omega\right)\rho_{lm} &= i\sum_a d\mathcal{E}^a\left(\sum_j \mathcal{G}^a_{lj}\rho_{jm} - \sum_m \mathcal{G}^a_{lm}\rho_{mn}\right), \\
\hbar\frac{\partial}{\partial t}\rho_{mn} &= i\sum_a\sum_l\left(d\mathcal{E}^{a*}\mathcal{G}^a_{lm}\rho_{ln} - d\mathcal{E}^a\mathcal{G}^a_{ln}\rho_{ml}\right), \qquad (4.2.3)\\
\hbar\frac{\partial}{\partial t}\rho_{lk} &= i\sum_a\sum_m\left(d\mathcal{E}^a\mathcal{G}^a_{lm}\rho_{mk} - d\mathcal{E}^{a*}\mathcal{G}^a_{km}\rho_{lm}\right)
\end{aligned}$$

In these equations d is the reduced dipole moment of atomic transition defining the matrix element of the dipole operator

$$d^a_{ml} = \mathcal{G}^a_{ml} d = (-1)^{j_2 - m}\begin{pmatrix} j_1 & 1 & j_2 \\ -m & a & l \end{pmatrix} d,$$

written by means of the $3j$ Wigner symbol. Index $a = \pm 1, 0$ labels the spherical components of the vectors under consideration. In the case accepted here $d^0 = d_z$, $d^{\pm 1} = \mp(d_x \pm id_y)/\sqrt{2}\,d$. Indices x, y and z denote the Cartesian components of vectors, $\Delta\omega = (\omega_0 - \omega_{21})$ is the frequency detuning from the centre of the absorbing line. Angular brackets mean a summation over all frequency detuning within the inhomogeneously broadened line. The slowly varying envelopes of matrix elements of the density matrix $\hat{\rho}$ are determined as

$$\rho_{ml} = \langle j_1, m|\hat{\rho}|\, j_2, l\rangle,\ \ \rho_{lk} = \langle j_2, l|\hat{\rho}|\, j_2, k\rangle,\ \ \rho_{mm'} = \langle j_1, m|\hat{\rho}|\, j_1, m'\rangle \quad ,$$

It is convenient to introduce the independent variables $\tau = (t - z/c)t_{p0}^{-1}$, $\zeta = z / L_a$, and the normalised pulse envelopes $q_a = d\mathcal{E}^a t_{p0}\hbar^{-1}$, where L_a is defined by

$$L_a = \left(2\pi\omega_0 d^2 n_A t_{p0} / cn(\omega)\hbar\right)^{-1}.$$

Hereafter n_A is a density of resonant atoms. The system of equations (4.2.3) with the initial condition

$$\mathcal{E}^a(t, z=0) = \mathcal{E}_0^a(t)$$

and the boundary conditions

$$\mathcal{E}^a(t \to \pm\infty, z) = \rho_{lm}(t \to \pm\infty, z) = 0,$$
$$\rho_{ll'}(t \to \pm\infty, z) = 0, \quad \rho_{mm'}(t \to \pm\infty, z) = \delta_{mm'},$$

form the boundary problem for which the exact solution is not known in the case of arbitrary values j_1 and j_2. But the certain choice of transitions $j_1 = 0 \leftrightarrow j_2 = 1$, $j_1 = 1 \to j_2 = 1$, and $j_1 = 1/2 \to j_2 = 1/2$ makes this system of equations exactly integrable. Its solution can be obtained by the inverse scattering transform method [16], as it was shown in [30,31,33].

Generalised reduced self-induced transparency equations (GSIT equations), for transitions $j_1 = 0 \leftrightarrow j_2 = 1, j_1 = 1 \to j_2 = 1$, can be represented in the homogeneous form [30,31]:

$$\frac{\partial q_j}{\partial \zeta} = -i\sum_{a=1,2} \beta_a \langle P_j^{(a)} \rangle,$$
$$\left(\frac{\partial}{\partial \tau} - i\Delta\omega t_{p0}\right)P_j^{(a)} = -i\sum_l \left(q_l M_{lj}^{(a)} - q_j N^{(a)}\right), \tag{4.2.4}$$
$$\frac{\partial M_{jl}^{(a)}}{\partial \tau} = -i\left(q_j^* P_l^{(a)} - q_l P_j^{(a)*}\right), \quad \frac{\partial N^{(a)}}{\partial \tau} = i \sum_j \left(q_j^* P_j^{(a)} - q_j P_j^{(a)*}\right).$$

If we consider transition $j_1 = 0 \to j_2 = 1$, then in (4.2.4) we assign

$$q_j = d\mathcal{E}^j t_{p0} / \hbar, \quad \beta_1 = 1, \quad \beta_2 = 0, \tag{4.2.5a}$$

for transition $j_1 = 1 \to j_2 = 0$ we have

$$q_j = d\mathcal{E}^j t_{p0} / \hbar, \quad \beta_1 = 0, \quad \beta_2 = 1, \tag{4.2.5b}$$

and for transition $j_1 = 1 \rightarrow j_2 = 1$ we have

$$q_j = jd\mathcal{E}^j t_{p0} / \hbar\sqrt{2}, \quad \beta_1 = \beta_2 = 1/2, \tag{4.2.5c}$$

Everywhere here the sub-index and upper index take the values $j = \pm 1$, and the determinations are used as following

$$\begin{aligned} P_j^{(1)} &= \langle j_2, 0|\hat{\rho}|\, j_1, j\rangle, \quad P_j^{(2)} = \langle j_2, -j|\hat{\rho}|\, j_1, 0\rangle, \\ N^{(1)} &= \langle j_2, 0|\hat{\rho}|\, j_2, 0\rangle, \quad N^{(2)} = -\langle j_1, 0|\hat{\rho}|\, j_1, 0\rangle, \\ M_{jl}^{(1)} &= \langle j_1, j|\hat{\rho}|\, j_1, l\rangle, \quad M_{jl}^{(2)} = \langle j_2, -l|\hat{\rho}|\, j_2, -j\rangle. \end{aligned} \tag{4.2.6}$$

The case of double resonance

Let the double-frequency USPs with the electric field strength

$$E^a(t,z) = \mathcal{E}_1^a \exp(-i\omega_1 t + ik_1 z) + \mathcal{E}_2^a \exp(-i\omega_2 t + ik_2 z) + c.c.$$

propagate in the linear medium containing resonant impurity atoms. The carrier frequencies are satisfied to the double resonance condition

$$\omega_1 \approx |E_a - E_b| / \hbar, \quad \omega_2 \approx |E_b - E_c| / \hbar,$$

where E_a, E_b and E_c are the energy of the resonant levels characterised by the total angular moments j_a, j_b, j_c and their projections m_a, m_b, m_c on the axis of quantisation.

Depending on the positions of E_a, E_b, and E_c on the scale of energy we will distinguish three configurations, namely

- V-configuration: $E_b < E_a < E_c$;
- Λ-configuration: $E_a < E_c < E_b$;
- Cascade configuration: $E_a < E_b < E_c$.

In the case of arbitrary values of the angular momenta j_a, j_b, j_c, the equations, describing the evolution of the USP and the states of resonant medium present the cumbersome system of equation [32]. Here we shall consider only the case of $j_a = j_c = 0, j_b = 1$ and $j_a = j_b = 1$, $j_c = 0$ as the examples.

Transition $j_a = j_c = 0$, $j_b = 1$

If we introduce new variables $\tau = (t - z/c)t_{p0}^{-1}$, $\zeta = z / L_a$ and define the normalised vector pulse envelope $\vec{q}_{1,2} = d\vec{\mathcal{E}}_{1,2} t_{p0} \hbar^{-1}$, then the GSIT-equations will look like the following [32,59]:

$$
\begin{gathered}
i\frac{\partial q_j^l}{\partial \zeta} = \langle P_j^l \rangle, \quad j,l = 1,2, \\
\left(\frac{\partial}{\partial \tau} - i\Delta\omega t_{p0}\right) P_1^l = -i\left[\sum_k q_1^k M_{lk} - q_1^l N_1 - q_2^l R\right], \\
\left(\frac{\partial}{\partial \tau} - i\Delta\omega t_{p0}\right) P_2^l = -i\left[\sum_k q_2^k M_{lk} - q_2^l N_2 - q_1^l R\right], \\
\frac{\partial N_j}{\partial \tau} = -i\sum_k \left(q_j^k P_j^{k*} - q_j^{k*} P_j^k\right), \quad \frac{\partial M_{kl}}{\partial \tau} = -i\sum_j \left(q_j^{k*} P_j^l - q_j^l P_j^{k*}\right), \\
\frac{\partial R}{\partial \tau} = -i\sum_k \left(q_1^k P_2^{k*} - q_2^{k*} P_1^k\right)
\end{gathered}
\tag{4.2.7}
$$

In these equations we have

$$
\begin{gathered}
\vec{q}_1 = d_{ab}\vec{\mathcal{E}}_1 t_{p0}/\hbar, \quad \vec{q}_2 = d_{cb}\vec{\mathcal{E}}_2 t_{p0}/\hbar, \\
P_1^k = \langle j_a,0|\hat{\rho}|j_b,k\rangle, \quad P_2^k = \langle j_c,0|\hat{\rho}|j_b,k\rangle, \quad M_{kl} = \langle j_b,k|\hat{\rho}|j_b,l\rangle, \\
N_1 = \langle j_a,0|\hat{\rho}|j_a,0\rangle, \quad N_2 = \langle j_c,0|\hat{\rho}|j_c,0\rangle, \quad R = \langle j_a,0|\hat{\rho}|j_c,0\rangle
\end{gathered}
\tag{4.2.8}
$$

for V-configuration, and

$$
\begin{gathered}
\vec{q}_1 = d_{ba}\vec{\mathcal{E}}_1 t_{p0}/\hbar, \quad \vec{q}_2 = d_{bc}\vec{\mathcal{E}}_2 t_{p0}/\hbar, \\
P_1^k = \langle j_b,-k|\hat{\rho}|j_a,0\rangle, \quad P_2^k = \langle j_c,-k|\hat{\rho}|j_b,0\rangle, \quad M_{kl} = -\langle j_b,-l|\hat{\rho}|j_b,-k\rangle, \\
N_1 = -\langle j_a,0|\hat{\rho}|j_a,0\rangle, \quad N_2 = -\langle j_c,0|\hat{\rho}|j_c,0\rangle, \quad R = -\langle j_a,0|\hat{\rho}|j_c,0\rangle
\end{gathered}
\tag{4.2.9}
$$

for Λ - configuration.

The following expressions serve as the boundary and the initial conditions for the equations (4.2.7)

$$\vec{q}_j(\tau, \zeta = 0) = \vec{q}_{j0}(\tau),$$

and at $|\tau| \to \infty$

$$
\begin{gathered}
\vec{q}_j(\tau,\zeta) \to 0, \quad \vec{P}_j(\tau,\zeta) \to 0, \quad R_j(\tau,\zeta) \to 0, \\
M_{kl}(\tau,\zeta) \to m_0\delta_{ml}, \quad N_j(\tau,\zeta) \to n_{j0}.
\end{gathered}
$$

In (4.2.7) the length of absorption is determined by the formula

$$L_a = \left(2\pi\omega_1 \mid d_{ab} \mid^2 n_A t_{p0} / cn(\omega)\hbar\right)^{-1},$$

and we expect the oscillator forces to be equal, i.e., $\omega_1 |d_{ab}|^2 = \omega_2 |d_{cb}|^2$.

Transition $j_a = j_b = 1$, $j_c = 0$

For this case the system of generalised self-induced transparency equations can also be written in the homogeneous form [37, 59]:

$$\begin{gathered}
i\frac{\partial q_j^l}{\partial\zeta} = \langle P_j^l \rangle, \quad j,l = 1,2, \\
\left(\frac{\partial}{\partial\tau} - i\Delta\omega t_{p0}\right) P_1^k = -i\left[\sum_l \left(q_1^l M_{lk} + q_2^l R_{kl}^*\right) - q_1^k N\right], \\
\left(\frac{\partial}{\partial\tau} - i\Delta\omega t_{p0}\right) P_2^k = -i\left[\sum_l \left(q_2^l M_{lk} + q_1^l R_{kl}\right) - q_2^k N\right], \\
\frac{\partial M_{kl}}{\partial\tau} = -i\left(q_1^{k*} P_1^l - q_1^l P_1^{k*}\right), \quad \frac{\partial K_{kl}}{\partial\tau} = -i\left(q_2^{k*} P_2^l - q_2^l P_2^{k*}\right), \\
\frac{\partial R_{kl}}{\partial\tau} = -i\left(q_1^{k*} P_2^l - q_2^l P_1^{k*}\right), \quad \frac{\partial N}{\partial\tau} = -i\sum_k\sum_l \left(q_j^k P_j^{k*} - q_j^{k*} P_j^k\right),
\end{gathered} \tag{4.2.10}$$

In these equations indices k,j,l take only two values -1 and +1.

For V-configuration we have

$$\begin{aligned}
&\vec{q}_1 = d_{ab}\vec{\mathcal{E}}_1 t_{p0} / \hbar\sqrt{3}, && \vec{q}_2 = d_{cb}\vec{\mathcal{E}}_2 t_{p0} / \hbar\sqrt{3}, \\
&P_1^k = \langle j_a, -k |\hat{\rho}| j_b, 0\rangle, && P_2^k = \langle j_c, -k |\hat{\rho}| j_b, 0\rangle, \\
&M_{kl} = -\langle j_a, -l |\hat{\rho}| j_a, -k\rangle, && K_{kl} = -\langle j_c, -l |\hat{\rho}| j_c, -k\rangle, \\
&R_{kl} = -\langle j_a, -l |\hat{\rho}| j_c, -k\rangle, && N = -\langle j_b, 0 |\hat{\rho}| j_b, 0\rangle
\end{aligned} \tag{4.2.11a}$$

For Λ -configuration we have

$$\begin{aligned}
&\vec{q}_1 = d_{ba}\vec{\mathcal{E}}_1 t_{p0} / \hbar\sqrt{3}, && \vec{q}_2 = d_{bc}\vec{\mathcal{E}}_2 t_{p0} / \hbar\sqrt{3} \\
&P_1^k = \langle j_b, 0 |\hat{\rho}| j_a, k\rangle, && P_2^k = \langle j_b, 0 |\hat{\rho}| j_c, k\rangle, \\
&M_{kl} = \langle j_a, k |\hat{\rho}| j_a, l\rangle, && K_{kl} = \langle j_c, k |\hat{\rho}| j_c, l\rangle, \\
&R_{kl} = \langle j_c, l |\hat{\rho}| j_c, k\rangle, && N = \langle j_b, 0 |\hat{\rho}| j_b, 0\rangle,
\end{aligned} \tag{4.2.11b}$$

Everywhere above $\hat{\rho}$ is understood as a slowly varying amplitude of the density matrixes (speaking more correctly, this is a density matrix in the interaction picture). The length of absorption is determined by the formula

$$L_a = \left(2\pi\omega_1 \mid d_{ab} \mid^2 t_{p0} n_A / 3cn(\omega)\hbar\right)^{-1}.$$

We also assign the equality of the oscillator forces

$$\omega_1 |d_{ab}|^2 = \omega_2 |d_{cb}|^2 . \tag{4.2.12}$$

The boundary and initial conditions are the same as in the preceding case.

4.2.3. ZERO-CURVATURE REPRESENTATION FOR THE GSIT EQUATIONS

In order to use the inverse scattering transform method for investigation of the solutions of the considered systems of GSIT-equations, it is necessary to represent these equations in the form of the condition of integrability for a set of linear equations [98-101]

$$\partial\Psi / \partial\tau = \hat{U}\Psi \ , \quad \partial\Psi / \partial\zeta = \hat{V}\Psi, \tag{4.2.13a,b}$$

or in a matrix notation

$$\frac{\partial\hat{U}}{\partial\zeta} = \frac{\partial\hat{V}}{\partial\tau} + \left[\hat{V}, \hat{U}\right]. \tag{4.2.14}$$

In equations (4.2.13) and (4.2.14) $\Psi = \text{colon}(\psi_1, \psi_2, \ldots, \psi_N)$, functions $\hat{U}(\lambda)$ and $\hat{V}(\lambda)$ are ($N\times N$) matrices, depending on some constant complex parameter λ (it is spectral parameter or eigenvalue of the spectral problem under consideration), These matrices also depend on both the dependent variables of the non-linear equation under consideration, and their derivatives.

The important problem now is to find matrixes $\hat{U}(\lambda)$ and $\hat{V}(\lambda)$ in order to obtain the considered system of the GSIT-equations from (4.2.14) for any λ. For (4.2.4), (4.2.7) and (4.2.10) this problem was solved in [30-33, 37] by using previous experience and some prompts. But now it is possible to derive the results of these works, i.e. the zero-curvature representation of the GSIT-equations in a more formal way.

Notice that usual SIT-equations (or reduced Maxwell-Bloch equations) have zero-curvature representation with the *U*-*V*-matrixes, which belong to AKNS hierarchies [98, 99]. The electric field of USP in this case is described by the scalar function $q(\zeta, \tau)$. In the case considered here we deal with the vector value which describes the envelope of the USP $q_j(\zeta,\tau)$. It is natural to find the zero-curvature representation of (4.2.4) in the vector expansion of the AKNS hierarchy. On the other hand as we can see from equations (4.2.7) and (4.2.10) the electric field of USP is characterised by matrix $q_j^k(\zeta,\tau)$. Consequently, it is necessary to pay attention to the matrix expansion of the AKNS hierarchy.

Vector variant of the GSIT-equations

Let us consider the vector expansion of the AKNS hierarchy. Let $\hat{U}(\lambda)$ and $\hat{V}(\lambda)$ be in the form

$$\hat{U}(\lambda)=\begin{pmatrix} -i\lambda & \vec{R} \\ \vec{Q} & i\lambda\hat{\mathrm{I}} \end{pmatrix}, \quad \hat{V}(\lambda)=\begin{pmatrix} A & \vec{B} \\ \vec{C} & \hat{D} \end{pmatrix},$$

where $\vec{Q}$, $\vec{R}$, $\vec{B}$, $\vec{C}$ are the N-components vectors, $\hat{\mathrm{I}}$ is a matrix unity, $\hat{D}$ is the ($N\times N$) matrix. The AKNS-like system of equations follows from (4.2.14)

$$\begin{gathered} \frac{\partial A}{\partial\tau}=\vec{Q}\cdot\vec{C}-\vec{R}\cdot\vec{B}, \quad \frac{\partial\hat{D}}{\partial\tau}=\vec{R}\otimes\vec{B}-\vec{C}\otimes\vec{Q}, \\ \frac{\partial\vec{B}}{\partial\tau}+2i\lambda\vec{B}+A\vec{Q}-\vec{Q}\cdot\hat{D}=\frac{\partial\vec{Q}}{\partial\zeta}, \\ \frac{\partial\vec{C}}{\partial\tau}-2i\lambda\vec{C}-A\vec{R}+\hat{D}\cdot\vec{R}=\frac{\partial\vec{R}}{\partial\zeta}. \end{gathered} \tag{4.2.15}$$

where dots mark the internal (scalar) multiplying, i.e., $(\vec{a}\cdot\hat{B})_j=a_kB_{kj}$, $\vec{a}\cdot\vec{b}=a_kb_k$, and symbol $\otimes$ serves for tensor multiplying, i.e., $(\vec{a}\otimes\vec{b})_{kn}=a_kb_n$

The system of equations (4.2.4) can be rewritten by using vector notation

$$\begin{gathered} \frac{\partial\vec{q}}{\partial\zeta}=-i\sum_a\beta_a\langle\vec{P}^{(a)}\rangle, \\ \frac{\partial\vec{P}^{(a)}}{\partial T}=i\Delta\omega\vec{P}^{(a)}-i\vec{q}\cdot\hat{M}^{(a)}+i\vec{q}N^{(a)}, \\ \frac{\partial\hat{M}^{(a)}}{\partial T}=-i\left(\vec{q}^{\,*}\otimes\vec{P}^{(a)}-\vec{P}^{(a)*}\otimes\vec{q}\right), \\ \frac{\partial N^{(a)}}{\partial T}=-i\left(\vec{q}\cdot\vec{P}^{(a)*}-\vec{q}^{\,*}\cdot\vec{P}^{(a)}\right). \end{gathered} \tag{4.2.16}$$

Let $\vec{Q} = \gamma_1 \vec{q}$, $\vec{R} = \gamma_2 \vec{q}^*$ and it is assigned that

$$\vec{B} = \left\langle \sum_a b_a(\lambda) \vec{P}^{(a)} \right\rangle, \qquad \vec{C} = \left\langle \sum_a c_a(\lambda) \vec{P}^{(a)*} \right\rangle.$$

Utilising the last two equations from (4.2.15) and the equation for the envelope of the USP one can find that

$$b_a(\lambda) = -\gamma_1 \beta_a (2\lambda + \Delta\omega)^{-1}, \quad c_a(\lambda) = -\gamma_2 \beta_a (2\lambda + \Delta\omega)^{-1}, \tag{4.2.17}$$

and

$$A = i\left\langle\left\langle \sum_a \beta_a N^{(a)} \right\rangle\right\rangle, \qquad \hat{D} = i\left\langle\left\langle \sum_a \beta_a \hat{M}^{(a)} \right\rangle\right\rangle, \tag{4.2.18}$$

where the shortened designation is used $\langle\langle ... \rangle\rangle = \langle ... (2\lambda + \Delta\omega)^{-1} \rangle$.

Taking $\vec{B}$ and $\vec{C}$ into account, one can find from the first two equations of system (4.2.15) that

$$A = -i\gamma_1\gamma_2 \left\langle\left\langle \sum_a \beta_a N^{(a)} \right\rangle\right\rangle, \qquad \hat{D} = -i\gamma_1\gamma_2 \left\langle\left\langle \sum_a \beta_a \hat{M}^{(a)} \right\rangle\right\rangle. \tag{4.2.19}$$

Comparison of the expressions (4.2.18) and (4.2.19) makes it possible to find parameters γ_1 and γ_2 which satisfy the condition: $\gamma_1\gamma_2 = -1$. For the sake of certainty let $\gamma_1 = \gamma_2 = -i$. Thereby, we obtain zero-curvature representation or *U*-*V*-pair for the system of equation (4.2.4) or (4.2.16):

$$\hat{U}(\lambda) = \begin{pmatrix} -i\lambda & -i\vec{q} \\ -i\vec{q}^* & i\lambda\hat{I} \end{pmatrix}, \tag{4.2.20a}$$

$$\hat{V}(\lambda) = i\left\langle\left\langle \sum_a \beta_a \hat{V}(\lambda)^{(a)} \right\rangle\right\rangle, \quad \hat{V}(\lambda)^{(a)} = \begin{pmatrix} N^{(a)} & \vec{P}^{(a)} \\ \vec{P}^{(a)*} & \hat{M}^{(a)} \end{pmatrix}. \tag{4.2.20a}$$

Matrix variant of the GSIT-equations

Now the matrix expansion of the AKNS hierarchy will be considered. Let $\hat{U}(\lambda)$ and $\hat{V}(\lambda)$ have the form

$$\hat{U}(\lambda) = \begin{pmatrix} -i\lambda\hat{I} & \hat{R} \\ \hat{Q} & i\lambda\hat{I} \end{pmatrix}, \qquad \hat{V}(\lambda) = \begin{pmatrix} \hat{A} & \hat{B} \\ \hat{C} & \hat{D} \end{pmatrix},$$

where $\hat{Q}$, $\hat{R}$, $\hat{A}$, $\hat{B}$, $\hat{C}$, $\hat{D}$ are matrices of $(N\times N)$ size and $N>3$. That is a particular case of the more general approach considered in a number of works [102-106].The system, named matrix AKNS, follows from (4.2.14):

$$\frac{\partial\hat{A}}{\partial\tau}=\hat{Q}\hat{C}-\hat{B}\hat{R},\quad \frac{\partial\hat{D}}{\partial\tau}=\hat{R}\hat{B}-\hat{C}\hat{Q},$$
$$\frac{\partial\hat{B}}{\partial\tau}+2i\lambda\hat{B}+\hat{A}\hat{Q}-\hat{Q}\hat{D}=\frac{\partial\hat{Q}}{\partial\zeta}, \tag{4.2.21}$$
$$\frac{\partial\hat{C}}{\partial\tau}-2i\lambda\hat{C}-\hat{R}\hat{A}+\hat{D}\hat{R}=\frac{\partial\hat{R}}{\partial\zeta}.$$

System (4.2.7) can be transcribed in the matrix form by introducing the (2×2) matrices

$$\hat{Q}(\lambda)=\begin{pmatrix} q_1^{-1} & q_1^{+1}\\ q_2^{-1} & q_2^{+1}\end{pmatrix},\quad \hat{P}(\lambda)=\begin{pmatrix} P_1^{-1} & P_1^{+1}\\ P_2^{-1} & P_2^{+1}\end{pmatrix},$$
$$\hat{N}(\lambda)=\begin{pmatrix} N_1 & R\\ R^* & N_2\end{pmatrix},\quad \hat{M}(\lambda)=\begin{pmatrix} M_{-1,-1} & M_{-1,+1}\\ M_{+1,-1} & M_{+1,+1}\end{pmatrix}.$$

Now equations (4.2.7) can be rewritten in the form

$$\frac{\partial\hat{Q}}{\partial\zeta}=-i\langle\hat{P}\rangle,$$
$$\frac{\partial\hat{P}}{\partial\tau}=i\Delta\omega t_{p0}\hat{P}+i\left(\hat{N}\hat{Q}-\hat{Q}\hat{M}\right),$$
$$\frac{\partial\hat{M}}{\partial\tau}=i\left(\hat{P}^{+}\hat{Q}-\hat{Q}^{+}\hat{P}\right),\quad \frac{\partial\hat{N}}{\partial\tau}=i\left(\hat{P}\hat{Q}^{+}-\hat{Q}\hat{P}^{+}\right). \tag{4.2.22}$$

By referring to (4.2.8), (4.2.9), it is worth to note that $\hat{N}$ and $\hat{M}$ are Hermitian matrices by their determination, i.e., $\hat{N}^{+}=\hat{N}$, $\hat{M}^{+}=\hat{M}$.

Alike the preceding case, we expect that

$$\hat{Q}=\gamma_1\hat{Q}(\zeta,\tau),\quad \hat{R}=\gamma_2\hat{Q}^{+}(\zeta,\tau),$$
$$\hat{B}=\langle b(\lambda)\hat{P}(\zeta,\tau)\rangle,\quad \hat{C}=\langle c(\lambda)\hat{P}^{+}(\zeta,\tau)\rangle. \tag{4.2.23}$$

Taking into account equations (4.2.22), substitution of (4.2.23) into the last two equations of the system (4.2.21) yields $b(\lambda)=-\gamma_1(2\lambda+\Delta\omega)^{-1}$, $c(\lambda)=-\gamma_2(2\lambda+\Delta\omega)^{-1}$,

and

$$\hat{B} = -\gamma_1 \langle\langle \hat{P} \rangle\rangle, \qquad \hat{C} = -\gamma_2 \langle\langle \hat{P}^+ \rangle\rangle, \tag{4.2.24a}$$

$$i\langle b(\lambda)(\hat{N}\hat{Q} - \hat{Q}\hat{M})\rangle + \gamma_1(\hat{A}\hat{Q} - \hat{Q}\hat{D}) = 0, \tag{4.2.24b}$$

$$i\langle c(\lambda)(\hat{Q}^+\hat{N} - \hat{M}\hat{Q}^+)\rangle + \gamma_2(\hat{D}\hat{Q}^+ - \hat{Q}^+\hat{A}) = 0. \tag{4.2.24c}$$

It follows from the first two equations of (4.2.21) and (4.2.24a) that

$$\hat{A} = \hat{A}_0 - i\gamma_1\gamma_2 \langle\langle \hat{N} \rangle\rangle, \qquad \hat{D} = \hat{D}_0 - i\gamma_1\gamma_2 \langle\langle \hat{M} \rangle\rangle,$$

where $\hat{A}_0$ and $\hat{D}_0$ are integrating constants. These results are in agreement with (4.2.24b) and (4.2.24c), i.e. these relations must be identities. It is reached if $\hat{A}_0 = \hat{D}_0 = 0$ and $\gamma_1\gamma_2 = -1$. By choosing $\gamma_1 = \gamma_2 = -i$, we are able to write *U-V*-pair explicitly

$$\hat{U}(\lambda) = \begin{pmatrix} -i\lambda\hat{I} & -i\hat{Q} \\ -i\hat{Q}^+ & i\lambda\hat{I} \end{pmatrix}, \qquad \hat{V}(\lambda) = \begin{pmatrix} i\langle\langle \hat{N} \rangle\rangle & i\langle\langle \hat{P} \rangle\rangle \\ i\langle\langle \hat{P}^+ \rangle\rangle & i\langle\langle \hat{M} \rangle\rangle \end{pmatrix}. \tag{4.2.25}$$

These matrices constitute the zero-curvature representation of the GSIT-equations (4.2.7).

The case of double resonance with $j_a = j_c = 1$ and $j_b = 0$ described by the system of equation (4.2.10), does not provide the new type of the AKNS equations. In order to show it, equations (4.2.10) should be rewritten by introducing the vector designations which correspond to the space of polarisations $\vec{q}_j = \{q_j^{-1}, q_j^{+1}\}$, $\vec{P}_j = \{P_j^{-1}, P_j^{+1}\}$. Then equations (4.2.10) will be re-written as

$$\frac{\partial \vec{q}_1}{\partial \zeta} = -i\langle \vec{P}_1 \rangle, \qquad \frac{\partial \vec{q}_2}{\partial \zeta} = -i\langle \vec{P}_2 \rangle,$$

$$\frac{\partial \vec{P}_1}{\partial \tau} = i\Delta\omega t_{p0}\vec{P}_1 - i\vec{q}_1 \cdot M - i\vec{q}_2 \cdot \hat{R}^+ + i\vec{q}_1 \cdot \hat{N},$$

$$\frac{\partial \vec{P}_2}{\partial \tau} = i\Delta\omega t_{p0}\vec{P}_2 - i\vec{q}_2 \cdot \hat{K} - i\vec{q}_1 \cdot \hat{R} + i\vec{q}_2 \cdot \hat{N}, \tag{4.2.26}$$

$$\frac{\partial \hat{M}}{\partial \tau} = -i(\vec{q}_1^* \otimes \vec{P}_1 - \vec{P}_1^* \otimes \vec{q}_1), \qquad \frac{\partial \hat{K}}{\partial \tau} = -i(\vec{q}_2^* \otimes \vec{P}_2 - \vec{P}_2^* \otimes \vec{q}_2),$$

$$\frac{\partial \hat{R}}{\partial \tau} = -i(\vec{q}_1^* \otimes \vec{P}_2 - \vec{P}_1^* \otimes \vec{q}_2), \qquad \frac{\partial \hat{N}}{\partial \tau} = -i\sum_k (\vec{q}_k \otimes \vec{P}_k^* - \vec{q}_k^* \otimes \vec{P}_k).$$

Hereinafter, one can define the vectors (and matrices) in the space, which presents the direct sum of the space of polarisations and the spaces of "colour"

$$\vec{q} = \vec{q}_1 \oplus \vec{q}_2 = \{q_1^{-1}, q_1^{+1}, q_2^{-1}, q_2^{+1}\}, \quad \vec{P} = \vec{P}_1 \oplus \vec{P}_2 = \{P_1^{-1}, P_1^{+1}, P_2^{-1}, P_2^{+1}\},$$
$$\widehat{M}_{11} = \widehat{M}, \quad \widehat{M}_{12} = \widehat{R}, \quad \widehat{M}_{21} = \widehat{R}^{+}, \quad \widehat{M}_{22} = \widehat{K}.$$

In terms of these designations equations (4.2.26) will take the form

$$\frac{\partial \vec{q}}{\partial \zeta} = -i\langle \vec{P} \rangle,$$
$$\frac{\partial \vec{P}}{\partial \tau} = i\Delta\omega t_{p0}\vec{P} - i\vec{q}\cdot\widehat{M} + i\vec{q}N, \tag{4.2.27}$$
$$\frac{\partial \widehat{M}}{\partial \tau} = -i\left(\vec{q}^{\,*} \otimes \vec{P} - \vec{P}^{*} \otimes \vec{q}\right), \quad \frac{\partial N}{\partial \tau} = -i\left(\vec{q}\cdot\vec{P}^{*} - \vec{q}^{\,*}\cdot\vec{P}\right),$$

where internal (i.e. scalar) and direct (i.e. tensor) products are extended to the case of the direct sum of the vector spaces.

It is seen now that the system of equations (4.2.27) complies with the (4.2.16) by the form. Consequently, it is possible to write at once the *U-V*-pair for the system of equations (4.2.10) in the given designations.

$$\widehat{U}(\lambda) = \begin{pmatrix} -i\lambda & -i\vec{q} \\ -i\vec{q}^{\,*} & i\lambda\widehat{I} \end{pmatrix}, \quad \widehat{V}(\lambda) = \begin{pmatrix} i\langle\langle N \rangle\rangle & i\langle\langle \vec{P} \rangle\rangle \\ i\langle\langle \vec{P}^{*} \rangle\rangle & i\langle\langle \widehat{M} \rangle\rangle \end{pmatrix}. \tag{4.2.28}$$

This result was presented in [32, 37].

4.2.4. SOLITON SOLUTIONS, INTEGRALS OF MOTION, AND BÄCKLUND TRANSFORMATION

If we know a zero-curvature representation of the GSIT equations, then the large number of the exact results referring to these equations can be found. So multiple-soliton solution can be obtained by solving linear algebraic equations following from the system of Gelfand-Levitan-Marchenko equations. This approach was presented in [37]. It bases on the results from the article by Manakov [107]. Equation (4.2.13a) with $\hat{U}(\lambda)$ in the form (4.2.20a) is the spectral problem considered in this work in detail. Equation (4.2.13a) with $\hat{U}(\lambda)$ from expression (4.2.29) is the same problem, but with the matrices of a greater size, i.e. it is a (5×5) matrix. The alternative approach will be presented here based on both the representation (4.2.13) in the form of Riccati equations and the Bäcklund transformations following from representations [27, 32].

Bäcklund Transformation

Let us determine vector $\vec{\Gamma}$ as $\vec{\Gamma} = \mathrm{colon}(\psi_2/\psi_1, \psi_3/\psi_1, \ldots, \psi_N/\psi_1)$, where it is supposed that $\psi_1 \neq 0$. Then from (4.2.13) we can get the system of equations

$$\partial\Gamma_a/\partial\tau = R_a + 2i\lambda\Gamma_a - (\vec{Q}\cdot\vec{\Gamma})\Gamma_a, \tag{4.2.29a}$$

$$\partial\Gamma_a/\partial\zeta = C_a + (D_{ab} - A\delta_{ab})\Gamma_b - (\vec{B}\cdot\vec{\Gamma})\Gamma_a. \tag{4.2.29b}$$

On the other hand, if the condition

$$\frac{\partial}{\partial\tau}\left(\frac{\partial\vec{\Gamma}}{\partial\zeta}\right) = \frac{\partial}{\partial\zeta}\left(\frac{\partial\vec{\Gamma}}{\partial\tau}\right)$$

is valid, then system of equations (4.2.29) yields the AKNS-system (4.2.13) so, that we may consider (4.2.29) as an alternative to (4.2.13) and forget about the assumption, that $\psi_1 \neq 0$.

Now we may consider equation (4.2.29a) as the rule to transfer any "potentials" $\vec{R} = -\vec{Q}^*$ and $\vec{Q}$ into "pseudo-potentials" $\vec{\Gamma}$ and parameter λ. The potentials $\vec{Q}$ forms a set of solutions of the considered system (4.2.15) if we use both equation (4.2.29b) and equation (4.2.29a). In the space, formed by pseudo-potentials $\vec{\Gamma}$ and λ, one can assign a transformation (map) $(\vec{\Gamma},\lambda) \to (\vec{\Gamma}',\lambda')$, so that equations (4.2.29) keeps their form. It is possible when the transformation of potentials is done simultaneously with the above mentioned. This transformation was named the Bäcklund transformation (adding "in the pseudo-potential form") [96,108,109].

Let $\vec{\Gamma} \to \vec{\Gamma}'$, then in accordance with (4.2.29a) one has

$$\partial\vec{\Gamma}'/\partial\tau = \vec{R}' + 2i\lambda'\vec{\Gamma}' - (\vec{Q}'\cdot\vec{\Gamma}')\vec{\Gamma}'.$$

If we expect that $\vec{\Gamma}' = \vec{\Gamma}$, $\lambda' = \lambda^*$, then this equation will take the form

$$\partial\vec{\Gamma}/\partial\tau = \vec{R}' + 2i\lambda^*\vec{\Gamma} - (\vec{Q}'\cdot\vec{\Gamma})\vec{\Gamma}. \tag{4.2.30}$$

By taking into account $\vec{R} = -\vec{Q}^*$, from equations (4.2.29a) and (4.2.30) we obtain that $\Delta\vec{Q} = \vec{Q}' - \vec{Q}$ expresses as $\Delta\vec{Q}^* = -2i(\lambda - \lambda^*)\vec{\Gamma} - (\Delta\vec{Q}\cdot\vec{\Gamma})$.

By assuming $\Delta\vec{Q} = f(|\vec{\Gamma}|)\vec{\Gamma}$, where $f(|\vec{\Gamma}|) = f^*(|\vec{\Gamma}|)$, we can define the function $f(|\vec{\Gamma}|)$ and the final result has the form

$$\vec{Q}' = \vec{Q} - 2i(\lambda - \lambda^*)\vec{\Gamma}\left(1 + |\vec{\Gamma}|^2\right)^{-1}. \tag{4.2.31}$$

Now let us consider the matrix AKNS-equations (4.2.21). One may be convinced that the equations alternative to equations (4.2.13) in this case are the matrix Riccati equations

$$\partial\hat{F}/\partial\tau = 2i\hat{F} + \hat{R} - \hat{F}\hat{Q}\hat{F}, \qquad (4.2.32a)$$

$$\partial\hat{F}/\partial\zeta = \hat{C} + \hat{D}\hat{F} - \hat{F}\hat{A} - \hat{F}\hat{B}\hat{F}. \qquad (4.2.32b)$$

From the condition

$$\frac{\partial}{\partial\tau}\left(\frac{\partial\hat{F}}{\partial\zeta}\right) = \frac{\partial}{\partial\zeta}\left(\frac{\partial\hat{F}}{\partial\tau}\right)$$

the system of equations (4.2.21) follows.

If we make a transformation $\hat{F} \to \hat{F}'$, $\lambda \to \lambda'$ then the conservation of the form of equations (4.2.32a) must be ensured by the additional transformation $\hat{Q} \to \hat{Q}'$, $\hat{R} \to \hat{R}'$. Let

$$\hat{Q}' = \hat{Q}^*, \quad \lambda' = \lambda^*.$$

It can be obtained from the equation for $\hat{F}'$ with the account of (4.2.30a) that

$$0 = 2i(\lambda - \lambda^*)\hat{F} + \Delta\hat{Q}^+ + \hat{F}\Delta\hat{Q}\hat{F}, \qquad (4.2.33)$$

where $\Delta\hat{Q} = \hat{Q}' - \hat{Q}$. Let us assume that $\Delta\hat{Q} = f(\hat{F})\hat{F}$, $\det\hat{F} \neq 0$, and $f(\hat{F})$ is a real function, which is necessary to be defined now. From expression (4.2.33) it follows

$$0 = 2i(\lambda - \lambda^*) + f(\hat{F}) + f(\hat{F})\hat{F}^+\hat{F}.$$

Thence we can find $f(\hat{F})$ and formulate the transformation law for $\hat{Q}$

$$\hat{Q}' = \hat{Q} - 2i(\lambda - \lambda^*)\hat{F}^+\left[1 + d^{-1}\,\mathrm{tr}(\hat{F}^+\hat{F})\right]^{-1}, \qquad (4.2.34)$$

where d is the dimensionality of the space which $\hat{Q}$ and $\hat{F}$ act to, i.e. here $d = 2$.

Vector and matrix solitons

The generalised SIT equations, called here as GSIT-equations, present the USP propagating in the resonant medium and keeping its form. In order to find a soliton solution of these equations, formulae (4.2.31) and (4.2.34) may be used. But, beforehand the pseudo-potentials $\vec{\Gamma}$ or $\hat{F}$ should be determined accordingly. If one is interested in single-soliton solution, then the pseudo-potentials for the trivial case, when $\vec{Q} = 0$, $\hat{Q} = 0$

(i.e. when the USP is absent) should be found. Then $\vec{Q}'$ and $\hat{Q}'$ accordingly from the (4.2.31) and (4.2.34) will be the single-soliton solutions.

Consider firstly the vector USP, i.e. the relationship (4.2.31). From equation (4.2.29) it follows

$$\partial\Gamma_a/\partial\tau = 2i\lambda\Gamma_a, \quad \partial\Gamma_a/\partial\zeta = -(A\delta_{ab} - D_{ab})\Gamma_b.$$

The boundary conditions for (4.2.4) yields diagonal matrix $\hat{D} = (D_{ab})$, A is simply the constant

$$\hat{D} = i\langle\langle\beta_1\rangle\rangle\hat{\mathrm{I}}, \qquad A = -i\langle\langle\beta_2\rangle\rangle,$$

so that

$$\partial\Gamma_a/\partial\zeta = \Re(\lambda)\Gamma_a,$$

where

$$\Re(\lambda) = i\left\langle\frac{(\beta_1 + \beta_2)}{(2\lambda + \Delta\omega t_{p0})}\right\rangle.$$

Thereby, it is possible to find $\vec{\Gamma}$

$$\Gamma_a = C_{0a}\exp[2i\lambda\tau + \Re(\lambda)\zeta], \tag{4.2.35}$$

where C_{0a} is an integrating constant, which has been chosen as a real vector for the sake of simplicity. Substitution of (4.2.35) into (4.2.31) with the account of $\vec{Q} = 0$, $\vec{Q}' = -i\vec{q}$ provides a solution of equations (4.2.4)

$$q_a(\tau,\zeta) = \frac{4\eta C_{0a}\exp[-2\eta\tau + \Re'\zeta]\exp(i\varphi)}{1 + |\vec{C}_0|^2\exp[-4\eta\tau + 2\Re'\zeta]},$$

where $\varphi = -2\xi\tau + \Re''\zeta$, $\lambda = \xi + i\eta$, and $\Re(\lambda) = \Re' + i\Re''$.

After introducing the unit vector $\vec{l} = \vec{C}_0/|\vec{C}_0|$ and a new constant $\tau_0 = \ln|\vec{C}_0|/2\eta$ [27,30], $q_a(\tau,\zeta)$ can be expressed by the following:

$$q_a(\tau,\zeta) = l_a 2\eta\,\mathrm{sech}[2\eta(\tau - \tau_0) - \Re'\zeta]\exp\{i\varphi(\tau,\zeta)\}. \tag{4.2.36}$$

This formula gives the soliton solution of the system of equations (4.2.4), which corresponds to the USP of self-induced transparency or the polarised 2π-pulse.

We can get a matrix generalisation of the 2π-pulse, which is the two-frequency polarised 2π-pulse in the similar way. Let us consider the relation (4.2.32), in which

$\hat{Q}$=0, and let $\hat{F}$ be a solution of (4.2.30) under $\hat{Q}$=0. When USP is absent, the medium is in the ground state, so that

$$\hat{M} = \begin{pmatrix} m_0 & 0 \\ 0 & m_0 \end{pmatrix}, \qquad \hat{N} = \begin{pmatrix} n_1 & 0 \\ 0 & n_2 \end{pmatrix},$$

consequently equations (4.2.32) will take the following form

$$\partial\hat{F} / \partial\tau = 2i\lambda\hat{F} \ , \quad \partial\hat{F} / \partial\zeta = \hat{K}\hat{F} \ , \tag{4.2.37}$$

where $\hat{K} = i\left((2\lambda + \Delta\omega t_{p0})^{-1}\right)\mathrm{diag}(m_0 - n_1,\, m_0 - n_2)$. Solution of these equations can be written in the form

$$F_{ab} = \sqrt{1/2}\ \ C_a^{(1)} C_b^{(2)} \exp[2i\lambda\tau + K_b(\lambda)\zeta], \tag{4.2.38}$$

where $K_b(\lambda) = i\left((2\lambda + \Delta\omega t_{p0})^{-1}\right)(m_0 - n_b)$, and the integrating constant is presented in the form of a direct multiplication of a vector from the space of polarisations $\vec{C}^{(1)}$ and vector from the space of "colours" (or frequencies) $\vec{C}^{(2)}$. Alike in the preceding case of the vector solitons, these vectors may be chosen real (that means that we ignore the constant phase of the USP envelope). Substitution of the relation (4.2.38) into (4.2.34) with the setting $\hat{Q}' = -i\hat{Q}$ allows to write the following expression for the *j*th component of the envelope of USP

$$Q_{kj}(\tau,\zeta) = \frac{2\sqrt{2}i\eta C_k^{(1)} C_j^{(2)} \exp\left[-2\eta\tau + K_j'\zeta - i\varphi_j\right]}{1 + |\vec{C}^{(1)}|^2 \sum_i C_i^{(2)2} \exp\left[-4\eta\tau + 2K_i'\zeta\right]}, \tag{4.2.39}$$

where the polarisation is marked by the index *k*, $\lambda = \xi + i\eta$, $\varphi_j = 2\xi\tau + K_j''\zeta$ and $K_j = K_j' + iK_j''$. If to introduce the orthogonal unit vectors in spaces of polarisations $\vec{l}^{(1)}$ and "colour" $\vec{l}^{(2)}$, so that $C_k^{(1)} C_j^{(2)} = l_k^{(1)} l_j^{(2)} |\vec{C}^{(1)}||\vec{C}^{(2)}|$, and to define parameter τ_0 by the formula $\tau_0 = \ln\left(|\vec{C}^{(1)}||\vec{C}^{(2)}|\right)/2\eta$, then the expression (4.2.39) can be transformed into the following form

$$Q_{kj}(\tau,\zeta) = \frac{2\sqrt{2}i\eta l_k^{(1)} l_j^{(2)} \exp\left[-2\eta(\tau - \tau_0) + K_j'\zeta - i\varphi_j\right]}{1 + \sum_i l_i^{(2)2} \exp\left[-4\eta(\tau - \tau_0) + 2K_i'\zeta\right]}. \tag{4.2.40}$$

If only the lower resonant atom level was populated, then

$$n_j = 0, \quad m_0 = 1, \quad K_1 = K_2 = K = i(2\lambda + \Delta\omega t_{p0})^{-1} .$$

The expression for $Q_{kj}(\tau, \zeta)$ will take the form of the well known 2π-pulse:

$$Q_{kj}(\tau,\zeta) = 2\sqrt{2}i\eta l_k^{(1)} l_j^{(2)} \operatorname{sech}[2\eta(\tau - \tau_0) - K'\zeta] \exp[-i\varphi(\tau,\zeta)] .$$

Conservation laws and integrals of motion

The fact that generalised self-induced equations (GSIT-equations) allow zero-curvature representation leads to the existence of the infinite number of relations in the form of the equations of continuity

$$\frac{\partial \Im_n}{\partial \zeta} + \frac{\partial \Re_n}{\partial \tau} = 0, \quad n = 1,2,3,\ldots \tag{4.2.41}$$

where $\Im_n$ and $\Re_n$ are the functions of the variables from equations (4.2.4), (4.2.7), (4.2.10), and their space derivations. Expressions (4.2.41) are named the (local) conservation laws. If the boundary conditions on the T -axis is $\Re_n^+ = \Re_n^-$, where $\Re_n{}^{\pm} = \lim \Re_n$, at $\tau \to \pm\infty$, then

$$\frac{\partial}{\partial \zeta} \int_{-\infty}^{+\infty} \Im_n d\tau = 0 \text{ and } \overline{\Im}_n = \int_{-\infty}^{+\infty} \Im_n d\tau .$$

The quantities $\overline{\Im}_n$ can be understood as the integrals of motion.

It is necessary to emphasise that the structure of $\Im_n$ and $\Re_n$ was defined by the linear equations of IST method (4.2.29), or by their alternative Riccati form versions (4.2.32).

Let us consider the case of vector expansion of AKNS-equations. Let $\overline{O} = (\vec{\Gamma} \cdot \vec{Q})$ and, by using the (4.2.15) and (4.2.29b), one may find that derivative $\overline{O}$ with respect to ζ has the form

$$\frac{\partial \overline{O}}{\partial \zeta} = \frac{\partial A}{\partial \tau} + \frac{\partial (\vec{\Gamma} \cdot \vec{B})}{\partial \tau} .$$

Then

$$\frac{\partial (\vec{\Gamma} \cdot \vec{Q})}{\partial \zeta} = \frac{\partial}{\partial \tau} \left(A + \vec{\Gamma} \cdot \vec{B} \right) . \tag{4.2.42}$$

Here $\vec{\Gamma} = \vec{\Gamma}(\vec{Q}, \lambda)$. The conservation laws must not contain parameter λ. Following the standard procedure from [99-101], the pseudopotential $\vec{\Gamma}$ can be expanded in a power series of $(2i\lambda)^{-1}$:

$$\vec{\Gamma}(\vec{Q},\lambda) = \sum_{n=1}^{\infty}(2i\lambda)^{-n}\vec{\Gamma}^{(n)}(\vec{Q}).$$

The (4.2.29a) relation results in

$$2i\lambda(\vec{\Gamma}\cdot\vec{Q}) = (\vec{\Gamma}\cdot\vec{Q})^2 + (\vec{Q}\cdot\vec{Q}^*) + (\vec{Q}\cdot\partial\vec{\Gamma}/\partial\tau).$$

The substitution of the series for $\vec{\Gamma}(\vec{Q},\lambda)$ into this relation and equating the factors at the same degrees of $(2i\lambda)^{-1}$ give rise to the recursion relationships for the determination of $\vec{\Gamma}^{(n)}(\vec{Q})$

$$\begin{aligned} \vec{\Gamma}^{(n+1)} &= \frac{\partial\vec{\Gamma}^{(n)}}{\partial\tau} + \sum_{m=1}^{n-1}\vec{\Gamma}^{(m)}(\vec{\Gamma}^{(n-m)}\cdot\vec{Q}), \quad n \geq 3 \\ \vec{\Gamma}^{(1)} &= \vec{Q}^*, \quad \vec{\Gamma}^{(2)} = \frac{\partial\vec{Q}^*}{\partial\tau}. \end{aligned} \tag{4.2.43a}$$

We can obtain now the set of the conservation laws from equation (4.2.40) after the substitution of (4.2.43a) and expanding of A and $\vec{B}$ in a power series of $(2i\lambda)^{-1}$.

If $\lim\vec{B} = 0$, $\lim A = A^{\pm}$ at $T \to \pm\infty$ and $A^+ = A^-$, then the quantities

$$\Im_n = \int_{-\infty}^{+\infty}(\vec{\Gamma}^{(n)}\cdot\vec{Q})d\tau, \quad n = 1,2,3,\ldots$$

form an infinite set of the integrals of motion. As an example, it is useful to demonstrate here the first three integrals

$$\Im_1 = \int_{-\infty}^{+\infty}(\vec{Q}\cdot\vec{Q}^*)d\tau, \quad \Im_2 = \int_{-\infty}^{+\infty}(\vec{Q}\cdot\frac{\partial\vec{Q}^*}{\partial\tau})d\tau, \quad \Im_3 = \int_{-\infty}^{+\infty}\left(\vec{Q}\cdot\frac{\partial^2\vec{Q}^*}{\partial\tau^2} + (\vec{Q}\cdot\vec{Q}^*)^2\right)d\tau.$$

Let us consider now the matrix form of AKNS-system (4.2.30). By using the equations (4.2.32a) and (4.2.21), we can get for $\hat{O} = \hat{Q}\hat{F}$ the following expression

$$\frac{\partial\hat{O}}{\partial\zeta} = \frac{\partial\hat{B}\hat{F}}{\partial\tau} + \frac{\partial\hat{A}}{\partial\tau} + [\hat{A}, \hat{Q}\hat{F}] + [\hat{B}\hat{F}, \hat{Q}\hat{F}].$$

Thence we can see that after calculating the *trace* of all matrices in this expression the following relation holds

$$\frac{\partial}{\partial\zeta}\mathrm{tr}(\hat{Q}\hat{F}) = \frac{\partial}{\partial\tau}\mathrm{tr}(\hat{A}+\hat{B}\hat{F}).$$

As in the preceding case, $\hat{F}$ is expanded in a power series of $(2i\lambda)^{-1}$

$$\hat{F}(\hat{Q},\lambda) = \sum_{n=1}^{\infty}(2i\lambda)^{-n}\hat{F}^{(n)}(\hat{Q}).$$

Equation (4.2.32a) results in

$$2i\lambda(\hat{Q}\hat{F}) = (\hat{Q}\hat{F})^2 + (\hat{Q}\hat{Q}^+) + (\hat{Q}\partial\hat{F}/\partial\tau).$$

Substitution of the power series of $\hat{F}$ in this equation and equating the factors at the same degrees of $(2i\lambda)^{-1}$ yield the recursion relationships for $\hat{F}^{(n)}(\hat{Q})$

$$\begin{aligned}\hat{F}^{(n+1)} &= \frac{\partial\hat{F}^{(n)}}{\partial\tau} + \sum_{m=1}^{n-1}\hat{F}^{(m)}(\hat{F}^{(n-m)}\hat{Q}), \quad n\geq 3\\ \hat{F}^{(1)} &= \hat{Q}^+, \quad \hat{F}^{(2)} = \frac{\partial\hat{Q}^+}{\partial\tau}.\end{aligned} \tag{4.2.43b}$$

If $\lim\hat{B}=0$, $\lim\hat{A}=\hat{A}^{(\pm)}$ and $\hat{A}^{(+)}=\hat{A}^{(-)}$ as $\tau\to\pm\infty$, then the integrals of motion are in the form

$$\Im_n = \int_{-\infty}^{+\infty}\mathrm{tr}(\hat{F}^{(n)}\hat{Q})d\tau, \quad n=1,2,3,\ldots$$

For instance,

$$\Im_1 = \int_{-\infty}^{+\infty}\mathrm{tr}(\hat{Q}\hat{Q}^+)d\tau, \quad \Im_2 = \int_{-\infty}^{+\infty}\mathrm{tr}(\hat{Q}\frac{\partial\hat{Q}^+}{\partial\tau})d\tau, \quad \Im_3 = \int_{-\infty}^{+\infty}\mathrm{tr}\left(\hat{Q}\frac{\partial^2\hat{Q}^+}{\partial\tau^2} + (\hat{Q}\hat{Q}^+)^2\right)d\tau.$$

We can see that the form of these integrals is the same as considered above.

4.2.5. INTERACTION OF POLARIZED SOLITONS

The solitons (both vector or polarised 2π-pulses (4.2.34), and matrix or polarised two-frequency pulses (4.2.38)) obtained here, differ a little as a separate USP from 2π-pulses by McCall-Hahn [1,2], which are the scalar solitons of the SIT equations. The polarisation of solitons, as an additional degree of freedom, manifests itself in an interaction between them. Such interaction is described by multiple-soliton solutions of the GSIT-equations, which we can find by the IST method, basing on (4.2.18), or (4.2.23), or (4.2.26) and by using standard procedures of the soliton theory [100,101].

Let us consider a collision of two vector (polarised) solitons as an example of their interaction. Instead of searching for the two-soliton solution it is possible to use more simple procedure, described in [100] for the case of scalar solitons. When $\zeta \to \pm\infty$ two-soliton solution transforms to a pair of space and time diverged solitons. If one considers that as $\zeta \to -\infty$ there are two single solitons before the interaction, then as $\zeta \to +\infty$ these solitons should be understood as solitons after interactions. It is important to know how their parameters have changed with respect to the initial data.

For further purposes some information on the spectral problem of the IST method (4.2.13a) with matrix $\hat{U}(\lambda)$ in the form (4.2.18) is necessary. The spectral problem with such a sort of $\hat{U}(\lambda)$ was first reported by Manakov [107] to describe the self-focusing of the polarised light beams.

Manakov's spectral problem

As it has been noticed above, the advantage of a representation of non-linear evolution equations as a condition for zero curvature consists in the possibility to solve the Cauchy problem by means of the auxiliary linear equations (4.2.13a). One of them serves to pass from the functions desired to scattering data of the $\hat{U}(\lambda)$ operator. Let us write the equations of Manakov's spectral problem in the form

$$\partial\psi_1/\partial\tau + i\lambda\psi_1 + iq_1\psi_2 + iq_2\psi_3 = 0,$$
$$\partial\psi_2/\partial\tau - i\lambda\psi_2 + iq_1^*\psi_1 = 0, \quad \partial\psi_3/\partial\tau - i\lambda\psi_3 + iq_2^*\psi_1 = 0. \tag{4.2.13c}$$

If the potentials decrease rapidly enough as $\tau \to \pm\infty$, then the solutions of this system of equations is uniquely determined by one of its asymptotic forms for $\tau \to \pm\infty$.

There are two sets of special solutions known as Jost functions for the spectral problem (4.2.13c). $\Phi^{(i)}(\tau,\lambda)$ and $\Psi^{(i)}(\tau,\lambda)$ $(i = 1,2,3)$ are defined as the fundamental solutions of (4.2.13c) with the following asymptotic at real $\lambda = \xi$, as $\tau \to -\infty$,

$$\Phi^{(1)} \to \begin{pmatrix}1\\0\\0\end{pmatrix}\exp(-i\xi\tau), \quad \Phi^{(2)} \to \begin{pmatrix}0\\1\\0\end{pmatrix}\exp(+i\xi\tau), \quad \Phi^{(3)} \to \begin{pmatrix}0\\0\\1\end{pmatrix}\exp(+i\xi\tau),$$

and, as $\tau \to +\infty$, then

$$\Psi^{(1)} \to \begin{pmatrix}1\\0\\0\end{pmatrix}\exp(-i\xi\tau), \quad \Psi^{(2)} \to \begin{pmatrix}0\\1\\0\end{pmatrix}\exp(+i\xi\tau), \quad \Psi^{(3)} \to \begin{pmatrix}0\\0\\1\end{pmatrix}\exp(+i\xi\tau),$$

The transfer (or scattering) matrix $\hat{T}$ in frame of the Jost functions is defined as

$$\Phi^{(i)}(\tau,\zeta,\xi) = \sum_{j=1}^{3} T_{ij}(\xi)\Psi^{(j)}(\tau,\zeta,\xi). \tag{4.2.44a}$$

The eigenvalues of the spectral problem (4.2.13c) are found as the solutions of the equation $T_{11}(\lambda) = 0$.

If $\vec{q}(\tau,\zeta)$ decreases sufficiently quickly as $\tau \to \pm\infty$, then $T_{11}(\xi)$, $\Phi^{(1)}(\xi)$, $\Psi^{(3)}(\xi)$ and $\Psi^{(2)}(\xi)$ are analytically continued in the upper half-plane of the complex variable λ, and $\Phi^{(2)}(\xi), \Phi^{(3)}(\xi), \Psi^{(1)}(\xi)$ are analytic for $\operatorname{Im}\lambda \le 0$.

Let λ_n be the zeroes of T_{11}, i.e. $T_{11}(\lambda_n) = 0$, $n = 1,2,3,\dots,N$ ($\operatorname{Im}\lambda_n > 0$). Then

$$\Phi^{(1)}(\tau,\lambda_n) = C_{12}^{(n)}\Psi^{(2)}(\tau,\lambda_n) + C_{13}^{(n)}\Psi^{(3)}(\tau,\lambda_n). \tag{4.2.44b}$$

The set of quantities $T_{ij}(\xi)$, $C_{12}^{(n)}$, $C_{13}^{(n)}$ and the zeros of T_{11} establish scattering data which are unambiguously determined by the potentials of operator in (4.2.13a) (variable ζ plays the role of the "renumbering" parameter for all potentials corresponding to these scattering data). The solution of the inverse scattering problem may be obtained on the base of the singular integral equations from Manakov's article [107]. It is possible to use an alternative approach based on the system of Gelfand-Levitan-Marchenko integral equations, which were presented in the article by Kaup [131] and in [101]. These equations have the following form

$$-\vec{K}^{(1)}(x,y) = \sum_{b=2,3} F_{1b}(x+y)\vec{g}^{(b)} + \sum_{b=2,3}\int_x^{\infty} \vec{K}^{(b)}(x,z)F_{1b}(z+y)dz,$$

$$\vec{K}^{(b)}(x,y) = F_{1b}^{*}(x+y)\vec{g}^{(1)} + \int_x^{\infty} \vec{K}^{(1)}(x,z)F_{1b}^{*}(z+y)dz.$$

$$F_{1b}(y) = -i\sum_{n=1}^{N} \frac{C_{1b}^{(n)}}{T_{11}'(\lambda_n)}\exp(i\lambda_n y) + \int_{-\infty}^{\infty}\frac{T_{1b}(\xi)\exp\{i\xi\ y\}}{2\pi T_{11}(\xi)}d\xi,$$

$$T_{11}'(\lambda_n) = (dT_{11}/d\lambda)(\lambda_n),$$

where $\vec{K}^{(1)}, \vec{K}^{(b)}$ and $\vec{g}^{(i)}$ are the three-component vectors, $(\vec{g}^{(i)})_j = \delta_{ij}$, b =2, 3. Potentials $\vec{q}(\tau,\zeta)$ are expressed in terms of solutions of these equations under the limiting transition $\tau' \to \tau + 0$

$$iq_1(\tau,\zeta) = -2K_1^{(2)}(\tau,\tau') = 2K_2^{(1)*}(\tau,\tau'),$$
$$iq_2(\tau,\zeta) = -2K_1^{(3)}(\tau,\tau') = 2K_3^{(1)*}(\tau,\tau').$$

Here the lower index of $K_j^{(1)}, K_j^{(b)}$ numbers the elements j of the column vectors $\vec{K}^{(1)}, \vec{K}^{(b)}$. The system of Gelfand-Levitan-Marchenko equations can be reduced to one integral equation

$$-\vec{K}^{(1)}(x,y) = \sum_{b=2,3} F_{1b}(x+y)\vec{g}^{(b)} + \mathcal{R}(x;x,y)\vec{g}^{(1)} + \int_x^{\infty} \vec{K}^{(1)}(x,z)\mathcal{R}(x;z,y)dz\ ,$$

where

$$\mathcal{R}(x;x',y) = \sum_{a=2,3} \int_x^{\infty} F_{1a}^*(x'+z)F_{1a}(z+y)dz\ .$$

It is convenient to obtain the quantities $K_a^{(1)}$, $a = 2,3$, from the following equations

$$-K_a^{(1)}(x,y) = F_{1a}(x+y) + \int_x^{\infty} K_a^{(1)}(x,z)\mathcal{R}(x;z,y)dz$$

There are specific solutions of the GSIT equations, which scattering data of the spectral problem (4.2.13a) at $\zeta = 0$ are nontrivial only for a discrete spectrum, which means they are characterised by a transfer matrix with zero non-diagonal matrix elements on a real axis $T_{1b}(\xi) = T_{b1}(\xi) = 0$. That is a reflectionless potential. Solutions of this kind are usually called N-soliton solutions, according to the N points of the discrete spectrum. They correspond to the functions $F_{1b}(x)$ defining the kernel of the integral Gelfand-Levitan-Marchenko equations,

$$F_{1b}(y) = -i\sum_{n=1}^{N} \frac{C_{1b}^{(n)}}{T_{11}'(\lambda_n)} \exp(i\lambda_n y) = \sum_{n=1}^{N} C_b^{(n)} \exp(i\lambda_n y)\ .$$

Note that the coefficients $C_b^{(n)}$ are the functions of ζ.

Single soliton solution

If there is only one eigenvalue $\lambda_1 = i\eta$ and there is no continuous spectrum of the differential operator in (4.2.13a), then the spectral data can be written in the form

$$F_{1b}(y) = -iC_b\, e^{-\eta y}, \quad T_{11}(\lambda, \lambda_1) = (\lambda - i\eta)/(\lambda + i\eta), \quad T_{1b}(\lambda, \lambda_1) = T_{b1}(\lambda, \lambda_1) = 0,$$
$$T_{ab}(\lambda, \lambda_1) = \delta_{ab} + \frac{2i\eta C_a^* C_b}{(\lambda - i\eta)(|C_1|^2 + |C_2|^2)}.$$

In this case the solution of Gelfand-Levitan-Marchenko equations yields

$$q_{-1}(\tau,\zeta) = -2iC_2^*(\zeta)\mathrm{H}(\tau,\zeta), \quad q_{+1}(\tau,\zeta) = -2iC_3^*(\zeta)\mathrm{H}(\tau,\zeta),$$
$$\mathrm{H}(\tau,\zeta) = \left[\exp(2\eta\tau) + \left(|C_2|^2 + |C_3|^2\right)\exp(-2\eta\tau)/4\eta^2\right]^{-1},$$
$$C_2^*(\zeta) = 2\eta l_{-1}\exp\{2\eta\tau_0'(\zeta)\}, \quad C_3^*(\zeta) = 2\eta l_{+1}\exp\{2\eta\tau_0'(\zeta)\}.$$

These expressions present an analogue of the 2π-pulse by McCall-Hahn

$$\vec{q}(\tau,\zeta) = -2i\eta\vec{l}(\zeta)\,\mathrm{sech}\left[2\eta\{\tau - \tau_0'(\zeta)\}\right].$$

The duration of the pulse is $1/2\eta$ and the polarisation vector $\vec{l}$ that obeys expression (4.2.34) arisen from Bäcklund transformations.

The second equation from (4.2.13) sets an evolution of the scattering data (see section 3.5), where V- matrix is given under the limits $\zeta \to \pm\infty$, i.e. $\hat{V}^{(\pm)}$,

$$\partial\hat{T}/\partial\zeta = \hat{V}^{(+)}\hat{T} - \hat{T}\hat{V}^{(-)},$$

and the eigenvalues $\{\lambda_n\}$ are stable. In the case under consideration we can find that

$$\tau_0' = \tau_0 - (2\eta)^{-1}\mathrm{Re}\,\Re(\lambda_1)\zeta, \quad \vec{l}(\zeta) = \vec{l}_0\exp\{i\,\mathrm{Im}\,\Re(\lambda_1)\zeta\},$$

where $\Re(\lambda) = i\langle(\beta_1 + \beta_2)/(2\lambda + \Delta\omega t_{p0})\rangle$.

Double-soliton solution

Let two points λ_1 and λ_2 from the discrete spectrum of the Manakov problem be on the imaginary axis, i.e. $\lambda_1 = i\eta_1$ and $\lambda_2 = i\eta_2$. Herewith we have

$$F_{1b}(y) = C_b^{(1)}\exp(-\eta_1 y) + C_b^{(2)}\exp(-\eta_2 y), \quad b = 2,3.$$

It is suitable to seek for a solution in the form

$$K_b^{(1)}(x,y) = A_b^{(1)}(x)\exp(-\eta_1 y) + A_b^{(2)}(x)\exp(-\eta_2 y).$$

After some algebra it can be expressed in the form

$$K_a^{(1)}(\tau,\tau) = \frac{i}{\Delta}\Big\{2\eta_1 B_a^{(1)*}\, e^{\rho_1+\varphi_1}\Big(1+e^{2\rho_2} + \frac{4\eta_1\eta_2}{(\eta_1+\eta_2)^2}\sum_b B_b^{(1)} B_b^{(2)*} e^{\rho_1+\rho_2+\varphi_2-\varphi_1}\Big) +$$

$$+ 2\eta_2 B_a^{(2)*}\, e^{\rho_2+\varphi_2}\Big(1+e^{2\rho_1} + \frac{4\eta_1\eta_2}{(\eta_1+\eta_2)^2}\sum_b B_b^{(2)} B_b^{(1)*} e^{\rho_1+\rho_2-\varphi_2+\varphi_1}\Big) -$$

$$- \frac{4\eta_1\eta_2}{(\eta_1+\eta_2)}\, e^{\rho_1+\rho_2}\Big[B_b^{(1)*}\Big(e^{\rho_2+\varphi_1} + \sum_b B_b^{(1)} B_b^{(2)*} e^{\rho_1+\varphi_2}\Big) + B_b^{(2)*}\Big(e^{\rho_1+\varphi_2} + \sum_b B_b^{(2)} B_b^{(1)*} e^{\rho_2+\varphi_1}\Big)\Big]\Big\}.$$

where

$$\rho_n = \mathrm{Re}\,\Re(\lambda_n)\zeta - 2\eta_n(\tau-\tau_n), \quad \varphi_n = i\,\mathrm{Im}\,\Re(\lambda_n)\zeta\,,$$
$$C_b^{(n)} = -2\lambda_n B_b^{(n)*}\exp[4\eta_n\tau_n + \Re(\lambda_n)\zeta], \quad \sum_b |B_b^{(n)}|^2 = 1,$$
$$b = 1,2; \quad n = 1,2;$$
$$\Delta = 1 + e^{2\rho 1} + e^{2\rho 2} + e^{2\rho 1 + 2\rho 2 + A_1} + e^{2\rho 1 + 2\rho 2 + A_2}\,,$$

$$\exp(A_1) = \left(\frac{\eta_1-\eta_2}{\eta_1+\eta_2}\right)^2\left[1 - \left|\sum_b B_b^{(2)*} B_b^{(1)}\right|^2 \frac{4\eta_1\eta_2}{(\eta_1+\eta_2)^2}\right],$$
$$\exp(A_2) = \frac{4\eta_1\eta_2}{(\eta_1+\eta_2)^2}\left(\sum_b B_b^{(2)} B_b^{(1)*}\, e^{\varphi 1-\varphi 2} + c.c.\right).$$

Assume that $\tau_2 - \tau_1 >> (2\eta_1)^{-1}, \quad (2\eta_2)^{-1}$ and $\eta_2 > \eta_1$. Then at the point $\zeta = 0$ on the characteristic $\rho_1 = \mathrm{const} \approx 1$, we have $\rho_2 >> 1$ and

$$K_b^{(1)}(\tau,\tau) = \frac{2i\eta_1\exp(\rho_1+\varphi_1)}{1+\exp(2\rho_1+A_1)}\left(\frac{\eta_1-\eta_2}{\eta_1+\eta_2}\right)\left[B_b^{(1)*} - B_b^{(2)*}\frac{2\eta_2}{\eta_1+\eta_2}\sum_a B_a^{(1)*} B_a^{(2)}\right],$$

but on the characteristic $\rho_2 = \mathrm{const} \approx 1$ we have $\rho_1 << -1$ and

$$K_b^{(1)}(\tau,\tau) = \frac{2i\eta_2\exp(\rho_2+\varphi_2)}{1+\exp(2\rho_2)} B_b^{(2)*}.$$

As $\zeta \to \infty$ on the characteristic $\rho_1 = \text{const} \approx 1$, then $\rho_2 \to -\infty$ and

$$K_b^{(1)}(\tau,\tau) = \frac{2i\eta_1 \exp(\rho_1 + \varphi_1)}{1 + \exp(2\rho_1)} B_b^{(1)*}.$$

On the other characteristic $\rho_2 = \text{const} \approx 1$, we have $\rho_1 \to +\infty$ and expression for $K_b^{(1)}(\tau,\tau)$ therefor gives

$$K_b^{(1)}(\tau,\tau) = \frac{2i\eta_2 \exp(\rho_2 + \varphi_2)}{1 + \exp(2\rho_2 + A_1)} \left(\frac{\eta_2 - \eta_1}{\eta_1 + \eta_2} \right) \left[B_b^{(2)*} - B_b^{(1)*} \frac{2\eta_1}{\eta_1 + \eta_2} \sum_a B_a^{(2)*} B_a^{(1)} \right].$$

Thus the double-soliton solution of GSIT equations splits at $\zeta = 0$ and $\zeta \to \infty$ into two single-soliton solutions. This result was interpreted as the collision of two polarised solitons.

Collision of vector (polarised) soliton

Assume two differently polarised soliton with duration $1/2\eta_1$ and $1/2\eta_2$ and unit polarisation vectors $\vec{l}_1$ and $\vec{l}_2$ enter a resonant medium at $\zeta = 0$. Let them be separated in time by the interval $\tau_2 - \tau_1$,

$$\vec{q}(\tau,0) = -2i\eta_1\vec{l}_1 \operatorname{sech}[2\eta_1(\tau - \tau_1)] - 2i\eta_2\vec{l}_2 \operatorname{sech}[2\eta_2(\tau - \tau_2)]$$

On the τ axis the first soliton is at the left of the second one and the inequality $\tau_2 - \tau_1$ >> $1/2\eta_1$, $1/2\eta_2$ holds. If $\eta_2 > \eta_1$, then the velocity of second soliton exceeds the velocity of first one. Therefore the second soliton overtakes the first one in the course of propagation and after the collision continues its course, so that at $\zeta \to \infty$ the second pulse occurs at the left of the first one on τ-axis

$$\vec{q}(\tau,\zeta) = -2i\eta_1\vec{l}_1{}' \operatorname{sech}[2\eta_1(\tau - \tau_1')] - 2i\eta_2\vec{l}_2{}' \operatorname{sech}[2\eta_2(\tau - \tau_2')].$$

The possible phase shifts are not taken into account here, but they can be taken into consideration by expecting that vectors $\vec{l}_a{}'$ include phase shifts in their determination. During the interaction of solitons the positions of their centres (position of the maximum of each of them) change that was taken into account in the parameters τ_n'.

Instead of the analysis of two-soliton formulae it is possible to use the method of "passing" by Zakharov and Shabat [101]. When $\zeta = 0$, the Jost functions in the field $\tau \to -\infty$ have the form

$$\Phi^{(1)}(\tau, i\eta_1) = \vec{g}^{(1)} \exp(\eta_1\tau), \quad \Phi^{(1)}(\tau, i\eta_2) = \vec{g}^{(1)} \exp(\eta_2\tau).$$

In the field between solitons the Jost functions write as

$$\Phi^{(1)}(\tau, i\eta_1) = \left(C_{12}^{(1)}\vec{g}^{(2)} + C_{13}^{(2)}\vec{g}^{(3)}\right)\exp(-\eta_1\tau),$$
$$\Phi^{(1)}(\tau, i\eta_2) = T_{11}(i\eta_2, i\eta_1)\vec{g}^{(1)}\exp(\eta_2\tau).$$

Here the upper indices of the factors $C_{12}^{(n)}$ and $C_{13}^{(n)}$ indicate the number of solitons. At the passing of the Jost functions through the second soliton, vectors from the right parts of these expressions $\vec{g}^{(2)}\exp(-\eta_1\tau)$ and $\vec{g}^{(3)}\exp(-\eta_1\tau)$ transform by (33.2.44a), but vector $\vec{g}^{(1)}\exp(\eta_2\tau)$ transforms by (4.2.44b). Thereby, at $\tau >> \tau_2$ we have

$$\Phi^{(1)}(\tau, i\eta_1) = \sum_{b,a} T_{ab}(i\eta_1, i\eta_2)C_{1a}^{(1)}\vec{g}^{(b)}\exp(-\eta_1\tau),$$
$$\Phi^{(1)}(\tau, i\eta_2) = T_{11}(i\eta_2, i\eta_1)\sum_{b} C_{1b}^{(2)}\vec{g}^{(b)}\exp(-\eta_2\tau). \qquad (4.2.45)$$

If $\zeta \to \infty$, then the order of soliton changes to the opposite and for the Jost functions we can similarly find the expressions as $\tau \to \infty$

$$\Phi^{(1)}(\tau, i\eta_1) = T_{11}(i\eta_1, i\eta_2)\sum_{b} C_{1b}'^{(1)}\vec{g}^{(b)}\exp(-\eta_1\tau),$$
$$\Phi^{(1)}(\tau, i\eta_2) = \sum_{b,a} T'_{ab}(i\eta_2, i\eta_1)\, C_{1a}'^{(2)}\vec{g}^{(b)}\exp(-\eta_2\tau). \qquad (4.2.46)$$

The quantities $T'_{ab}(i\eta_1, i\eta_2)$ and $C_{1b}'^{(1,2)}$ were calculated after soliton collisions.

Comparing expressions (4.2.45) and (4.2.46) and using the evident forms T_{ab} and $C_{1b}'^{(n)}$, we can obtain a relationship between polarisation vectors $\vec{l}_a$ (before) and $\vec{l}_a'$ (after) the collision for each soliton

$$\vec{l}_1' = \tilde{f}\left[-\vec{l}_1 + \frac{2\eta_2}{(\eta_2 - \eta_1)}\vec{l}_2(\vec{l}_1 \cdot \vec{l}_2^*)\right], \quad \vec{l}_2' = \tilde{f}\left[-\vec{l}_2 + \frac{2\eta_1}{(\eta_2 - \eta_1)}\vec{l}_1(\vec{l}_2 \cdot \vec{l}_1^*)\right], \qquad (4.2.47)$$

$$\tilde{f} = \left[1_1 + \frac{4\eta_1\eta_2}{(\eta_2 - \eta_1)^2}|\vec{l}_1 \cdot \vec{l}_2^*|^2\right]^{-1/2},$$

and $(\vec{l}_1' \cdot \vec{l}_2'^*) = (\vec{l}_1 \cdot \vec{l}_2^*)$.

Shifts of the position of the n-th soliton centre are given by the known formulae, which can be obtained within this approach:

$$\Delta\tau_n = (-1)^n (2\eta_n)^{-1} \ln\left[(\eta_1 + \eta_2)(\eta_2 - \eta_1)^{-1} \tilde{f}^{-1}\right]. \tag{4.2.48}$$

Formula (4.2.47) means that if the USPs were linearly polarised before the collision, they remain linearly polarised after it. The polarisation vectors of these solitons $\vec{l}_a$ rotate by different angles θ_1 and θ_2

$$(\vec{l}_1' \cdot \vec{l}_1) = \cos\theta_1 = -\left(1 + B_{12}\cos^2\theta\right)\left(1 - B_{12}B_{21}\cos^2\theta\right)^{-1/2},$$
$$(\vec{l}_2' \cdot \vec{l}_2) = \cos\theta_2 = -\left(1 + B_{21}\cos^2\theta\right)\left(1 - B_{12}B_{21}\cos^2\theta\right)^{-1/2}, \tag{4.2.49}$$
$$B_{12} = \frac{2\eta_2}{\eta_1 - \eta_2}, \quad B_{21} = \frac{2\eta_1}{\eta_2 - \eta_1}, \quad \cos\theta = (\vec{l}_1 \cdot \vec{l}_2).$$

It is interesting to observe that the angle between the polarisation vectors of the soliton does not change. It is equal to θ. It follows from (4.2.49) that a rotation of the polarisation plane does not occur if $\theta = 0$ and if $\theta = \pi/2$, i.e., if the initial pulses are polarised collinearly or perpendicularly. A collision of circularly polarised solitons does not change the polarisation either.

A collision of a linearly polarised soliton with a circularly polarised one results in two elliptically polarised solitons. One can obtain the parameters of the ellipses from (4.2.45) where they are determined by the pulse durations and by the value of the reduced dipole moment of the resonance transition. The pulses with any polarisation behave as repulsive particles. The solitons being overtaken are given a positive increment in the co-ordinate $\Delta\tau_2 > 0$, while the surpassed soliton is given a negative one, $\Delta\tau_1 < 0$.

The result (4.2.47) can also be obtained using the Bäcklund transformation of GSIT [111]. Also, this result can be obtained by analysing the solutions of Gelfand-Levitan-Marchenko equations provided there are only two eigenvalues of the spectral problem $\lambda_1 = i\eta_1$ and $\lambda_2 = i\eta_2$.

It is noteworthy that the law of variation of polarisation (4.2.47) is defined under the analysis of the spectral problem based only on the *U*-matrix in the form (4.2.18). Therefore the interaction of polarised soliton in the case of transition $j_1 = 1 \rightarrow j_2 = 1$ gives rise to the same results. Furthermore, the features given above do not depend on either the degree of an inhomogeneous broadening or of detuning. The latter can only affect the velocity of the solitons within the validity of the resonance approximation.

The interaction of the scalar two-frequency USPs (i.e., simultons [22-25]) has the same nature [27], since simulton propagation is described by the equations (4.2.4). The partial amplitudes of the waves with different frequencies [27,92] correspond to the polarisation states. The collision of polarised simultons can be considered similarly [32,37].

Alternative solution of spectral problem by Manakov

There is a different approach to the solution of the inverse spectral problem for the system of equations (4.2.13c). It was developed by Manakov [107]. This approach bases on the system of linear singular integral equations [107, 110]

$$\vec{f}^{(1)}(\tau;\lambda) = \vec{g}^{(1)} + \frac{1}{2i\pi}\oint_C \frac{\exp(2i\lambda'\tau)}{\lambda' - \lambda}\big[\widetilde{\rho}_1(\lambda')\vec{f}^{(2)}(\tau;\lambda') + \\ + \widetilde{\rho}_3(\lambda')\vec{f}^{(3)}(\tau;\lambda')\big], \tag{4.2.50a}$$

$$\vec{f}^{(2)}(\tau;\lambda) = \vec{g}^{(2)} + \frac{1}{2i\pi}\oint_{\bar{C}} \frac{\exp(-2i\lambda'\tau)}{\lambda' - \lambda}\widetilde{\rho}_1^*(\lambda')\vec{f}^{(1)}(\tau;\lambda'), \tag{4.2.50b}$$

$$\vec{f}^{(3)}(\tau;\lambda) = \vec{g}^{(3)} + \frac{1}{2i\pi}\oint_{\bar{C}} \frac{\exp(-2i\lambda'\tau)}{\lambda' - \lambda}\widetilde{\rho}_3^*(\lambda')\vec{f}^{(3)}(\tau;\lambda'), \tag{4.2.50c}$$

where the following functions are introduced $\widetilde{\rho}_b(\lambda) = T_{1b}(\lambda)/T_{11}(\lambda)$.

The C is the contour in the complex plane above all zeros of $T_{11}(\lambda)$ (i.e., the poles of $\widetilde{\rho}_b(\lambda)$) while $\overline{C}$ is its mirror image in the lower half complex plane. The potentials $q_1(\tau)$ and $q_2(\tau)$ can be recovered from the following asymptotic limit for large λ

$$q_1(\tau) = -2\lambda \lim_{\lambda\to\infty}(\vec{g}^{(1)}\cdot\vec{f}^{(2)}), \quad q_2(\tau) = -2\lambda \lim_{\lambda\to\infty}(\vec{g}^{(1)}\cdot\vec{f}^{(3)}).$$

We are often interested only in soliton solutions of the evolution equations, which are determined by the discrete component of the spectrum. In this case only the poles contribute to equations (4.2.50), which are reduced to set of linear non-homogeneous algebraic equations

$$\vec{f}^{(1)}(\tau;\lambda) = \vec{g}^{(1)} - \sum_{n=1}^{N}\frac{\exp(2i\lambda_n\tau)}{\lambda_n - \lambda}\big[\widetilde{C}_{2n}\vec{f}^{(2)}(\tau,\lambda_n) + \widetilde{C}_{3n}\vec{f}^{(3)}(\tau,\lambda_n)\big],$$

$$\vec{f}^{(2)}(\tau;\lambda) = \vec{g}^{(2)} + \sum_{n=1}^{N}\frac{\exp(-2i\lambda_n^*\tau)}{\lambda_n^* - \lambda}\widetilde{C}_{2n}^*\vec{f}^{(1)}(\tau,\lambda_n^*),$$

$$\vec{f}^{(3)}(\tau;\lambda) = \vec{g}^{(3)} + \sum_{n=1}^{N}\frac{\exp(-2i\lambda_n^*\tau)}{\lambda_n^* - \lambda}C_{3n}^*\vec{f}^{(3)}(\tau,\lambda_n^*),$$

where the following notations are introduced

$$\widetilde{C}_{bn} = T_{1b}(\lambda_n)/T'_{11}(\lambda_n) = iC_b^{(n)}.$$

4.2.6. POLARIZED SIMULTONS

Let us discuss the propagation and interaction of differently polarised two-frequency USPs in a resonant medium consisting of three-level atoms under the conditions of double resonance

$$\omega_1 \approx |E_a - E_b| / \hbar, \quad \omega_2 \approx |E_b - E_c| / \hbar,$$

as it was considered in section 3.2.3. We shall focus our attention on a specific choice of $j_a = j_c = 0$, $j_b = 1$. In section 4.2.5. the single soliton solution of the GSIT equations has been obtained by Bäcklund transform method (4.2.39). This solution is

$$Q_{kj}(\tau,\zeta) = \frac{2\sqrt{2i\eta l_k^{(1)} l_j^{(2)}} \exp[-2\eta(\tau-\tau_0) + K_j'\zeta - i\varphi_j]}{1+\sum_i l_i^{(2)2} \exp[-4\eta(\tau-\tau_0) + 2K_i'\zeta]}$$

or

$$Q_{kj}(\tau,\zeta) = 2\sqrt{2i\eta l_k^{(1)} l_j^{(2)}} \operatorname{sech}[2\eta(\tau-\tau_0) - K_j'\zeta] \exp[-i\varphi_j(\tau,\zeta)] \qquad (4.2.51)$$

in the case of purely absorption medium .

Let us now determine two linearly dependent sets of Jost functions of the spectral problem (4.2.13a) $\Phi^{(i)}(\tau,\lambda)$ and $\Psi^{(i)}(\tau,\lambda)$, (i = 1,2,3,4) which are defined as the fundamental solutions of (4.2.13a) with the U-matrix in the form (4.2.25) and with the following asymptotic at real $\lambda = \xi$

$$\Phi^{(1)}(\tau,\lambda) \to \vec{g}^{(1)} \exp(-i\lambda\tau), \quad \Phi^{(2)}(\tau,\lambda) \to \vec{g}^{(2)} \exp(-i\lambda\tau),$$
$$\Phi^{(3)}(\tau,\lambda) \to \vec{g}^{(3)} \exp(+i\lambda\tau), \quad \Phi^{(4)}(\tau,\lambda) \to \vec{g}^{(4)} \exp(+i\lambda\tau),$$

as $\tau \to -\infty$, and

$$\Psi^{(1)}(\tau,\lambda) \to \vec{g}^{(1)} \exp(-i\lambda\tau), \quad \Psi^{(2)}(\tau,\lambda) \to \vec{g}^{(2)} \exp(-i\lambda\tau),$$
$$\Psi^{(3)}(\tau,\lambda) \to \vec{g}^{(3)} \exp(+i\lambda\tau), \quad \Psi^{(4)}(\tau,\lambda) \to \vec{g}^{(4)} \exp(+i\lambda\tau),$$

as $\tau \to +\infty$. Since

$$\frac{d}{dx}\operatorname{sech} x = -\tanh x \operatorname{sech} x, \quad \frac{d}{dx}\tanh x = \operatorname{sech}^2 x,$$

it is natural to seek for a solution of the equations of the spectral problem (4.2.13a) in the form

$$\psi_k = \{\alpha_k \operatorname{sech}[2\eta(\tau-\tau_0)] + \beta_k \tanh[2\eta(\tau-\tau_0)] + \gamma_k\} \exp(i\delta\lambda\tau),$$

where $k = 1,2,3,4$ and constants $\delta, \alpha_k, \beta_k$ and γ_k are different for every Jost function. After algebra we obtain that:

- for $\Phi^{(1)}(\tau,\lambda)$,

$$\delta=-1,\ \beta_1=\gamma_1-1=-\frac{i\eta}{\lambda+i\eta}|l_{-1}^{(1)}|^2,\quad \beta_2=\gamma_2=-\frac{i\eta}{\lambda+i\eta}l_{-1}^{(1)*}l_{+1}^{(1)},$$

$$\alpha_3=\frac{\eta}{\lambda+i\eta}l_{-1}^{(1)*}l_1^{(2)*},\quad \alpha_4=\frac{\eta}{\lambda+i\eta}l_{-1}^{(1)*}l_2^{(2)*},\quad \alpha_1=\alpha_2=\beta_3=\beta_4=\gamma_3=\gamma_4=0;$$

- for $\Phi^{(2)}(\tau,\lambda)$,

$$\delta=-1,\ \beta_1=\gamma_1=-\frac{i\eta}{\lambda+i\eta}l_{-1}^{(1)}l_{+1}^{(1)*},\quad \beta_2=\gamma_2-1=-\frac{i\eta}{\lambda+i\eta}|l_{+1}^{(1)}|^2,$$

$$\alpha_3=\frac{\eta}{\lambda+i\eta}l_{+1}^{(1)*}l_1^{(2)*},\quad \alpha_4=\frac{\eta}{\lambda+i\eta}l_{+1}^{(1)*}l_2^{(2)*},\quad \alpha_1=\alpha_2=\beta_3=\beta_4=\gamma_3=\gamma_4=0;$$

- for $\Phi^{(3)}(\tau,\lambda)$

$$\delta=+1,\ \beta_4=\gamma_4=\frac{i\eta}{\lambda-i\eta}l_1^{(2)}l_2^{(2)*},\quad \beta_3=\gamma_3-1=\frac{i\eta}{\lambda-i\eta}|l_1^{(2)}|^2,$$

$$\alpha_1=\frac{-\eta}{\lambda-i\eta}l_{-1}^{(1)}l_1^{(2)},\quad \alpha_2=\frac{-\eta}{\lambda-i\eta}l_{+1}^{(1)}l_1^{(2)},\quad \alpha_3=\alpha_4=\beta_1=\beta_2=\gamma_1=\gamma_2=0;$$

- for $\Phi^{(4)}(\tau,\lambda)$,

$$\delta=+1,\ \beta_3=\gamma_3=\frac{i\eta}{\lambda-i\eta}l_2^{(2)}l_1^{(2)*},\quad \beta_4=\gamma_4-1=\frac{i\eta}{\lambda-i\eta}|l_2^{(2)}|^2,$$

$$\alpha_1=\frac{-\eta}{\lambda-i\eta}l_{-1}^{(1)}l_2^{(2)},\quad \alpha_2=\frac{-\eta}{\lambda-i\eta}l_{+1}^{(1)}l_2^{(2)},\quad \alpha_3=\alpha_4=\beta_1=\beta_2=\gamma_1=\gamma_2=0;$$

- for $\Psi^{(1)}(\tau,\lambda)$,

$$\delta=-1,\ \beta_1=\gamma_1-1=\frac{-i\eta}{\lambda-i\eta}|l_{-1}^{(1)}|^2,\quad \beta_2=-\gamma_2=\frac{-i\eta}{\lambda-i\eta}l_{-1}^{(1)*}l_{+1}^{(1)},$$

$$\alpha_3=\frac{\eta}{\lambda-i\eta}l_{-1}^{(1)*}l_1^{(2)*},\quad \alpha_4=\frac{\eta}{\lambda-i\eta}l_{-1}^{(1)*}l_2^{(2)*},\quad \alpha_1=\alpha_2=\beta_3=\beta_4=\gamma_3=\gamma_4=0;$$

- for $\Psi^{(2)}(\tau,\lambda)$,

$$\delta=-1,\ \beta_1=-\gamma_1=\frac{-i\eta}{\lambda-i\eta}l_{-1}^{(1)}l_{+1}^{(1)*},\quad \beta_2=\gamma_2-1=-\frac{i\eta}{\lambda+i\eta}|l_{+1}^{(1)}|^2,$$

$$\alpha_3=\frac{\eta}{\lambda-i\eta}l_{+1}^{(1)*}l_1^{(2)*},\quad \alpha_4=\frac{\eta}{\lambda-i\eta}l_{+1}^{(1)*}l_2^{(2)*},\quad \alpha_1=\alpha_2=\beta_3=\beta_4=\gamma_3=\gamma_4=0;$$

- for $\Psi^{(3)}(\tau,\lambda)$,

$$\delta=+1,\ \beta_4=-\gamma_4=\frac{i\eta}{\lambda+i\eta}l_1^{(2)}l_2^{(2)*},\quad \beta_3=1-\gamma_3=\frac{i\eta}{\lambda+i\eta}|l_1^{(2)}|^2,$$

$$\alpha_1=\frac{-\eta}{\lambda+i\eta}l_{-1}^{(1)}l_1^{(2)},\quad \alpha_2=\frac{-\eta}{\lambda+i\eta}l_{+1}^{(1)}l_1^{(2)},\quad \alpha_3=\alpha_4=\beta_1=\beta_2=\gamma_1=\gamma_2=0;$$

- for $\Psi^{(4)}(\tau,\lambda)$,

$$\delta=+1,\ \beta_3=-\gamma_3=\frac{i\eta}{\lambda+i\eta}l_2^{(2)}l_1^{(2)*},\quad \beta_4=1-\gamma_4=\frac{i\eta}{\lambda+i\eta}|l_2^{(2)}|^2,$$

$$\alpha_1=\frac{-\eta}{\lambda+i\eta}l_{-1}^{(1)}l_2^{(2)},\quad \alpha_2=\frac{-\eta}{\lambda+i\eta}l_{+1}^{(1)}l_2^{(2)},\quad \alpha_3=\alpha_4=\beta_1=\beta_2=\gamma_1=\gamma_2=0.$$

These formulae clearly demonstrate the analyticity of the Jost functions and determine the transfer matrix for simulton (4.2.51):

$$\hat{T}(\lambda)=\begin{pmatrix} 1-\dfrac{2i\eta}{\lambda+i\eta}|l_{-1}^{(1)}|^2 & \dfrac{-2i\eta}{\lambda+i\eta}l_{-1}^{(1)*}l_{+1}^{(1)} & 0 & 0 \\ \dfrac{-2i\eta}{\lambda+i\eta}l_{+1}^{(1)*}l_{-1}^{(1)} & 1-\dfrac{2i\eta}{\lambda+i\eta}|l_{+1}^{(1)}|^2 & 0 & 0 \\ 0 & 0 & 1+\dfrac{2i\eta}{\lambda-i\eta}|l_1^{(2)}|^2 & \dfrac{2i\eta}{\lambda-i\eta}l_2^{(2)*}l_1^{(2)} \\ 0 & 0 & \dfrac{2i\eta}{\lambda-i\eta}l_1^{(2)*}l_2^{(2)} & 1+\dfrac{2i\eta}{\lambda-i\eta}|l_2^{(2)}|^2 \end{pmatrix}.$$

As $\tau\to-\infty$ only $\Phi^{(1)}(\tau,\lambda)$ and $\Phi^{(2)}(\tau,\lambda)$ functions are bounded at the point $\lambda=i\eta$ and as $\tau\to+\infty$ only $\Psi^{(3)}(\tau,\lambda)$ and $\Psi^{(4)}(\tau,\lambda)$ functions are also bounded. Therefore the bounded solution $\Phi(\tau,\lambda)$ can be written in the form

$$\begin{aligned}\Phi(\tau,\lambda)&=l_{-1}^{(1)}\Phi^{(1)}(\tau,i\eta)+l_{+1}^{(1)}\Phi^{(2)}(\tau,i\eta)=\\&=-\left[l_1^{(2)*}\Psi^{(3)}(\tau,i\eta)+l_2^{(2)*}\Psi^{(4)}(\tau,i\eta)\right]\exp(2\eta\tau_0).\end{aligned}\qquad(4.2.52)$$

where $\lambda = i\eta$ is a root of the equation

$$\det\begin{pmatrix} T_{11}(\lambda) & T_{12}(\lambda) \\ T_{21}(\lambda) & T_{22}(\lambda) \end{pmatrix} = 0. \tag{4.2.53}$$

For arbitrary potentials Q_{kj} roots of equation (4.2.53) determine, in general, a discrete spectrum of a spectral problem under consideration. When (4.2.53) is satisfied, we can eliminate $\Psi^{(1)}(\tau,\lambda)$ and $\Psi^{(2)}(\tau,\lambda)$ from the equations

$$\Phi^{(1)}(\tau,\lambda) = \sum_{a=1}^{4} T_{1a}(\lambda)\Psi^{(a)}(\tau,\lambda), \qquad \Phi^{(2)}(\tau,\lambda) = \sum_{a=1}^{4} T_{2a}(\lambda)\Psi^{(a)}(\tau,\lambda),$$

and a bounded solution $\Phi(\tau,\lambda)$ can be constructed as a linear superposition of $\Phi^{(1)}(\tau,\lambda)$ and $\Phi^{(2)}(\tau,\lambda)$ (or $\Psi^{(3)}(\tau,\lambda)$ and $\Psi^{(4)}(\tau,\lambda)$) as follows

$$\Phi(\tau,\lambda) = -T_{21}(\lambda)\Phi^{(1)}(\tau,\lambda) + T_{11}(\lambda)\Phi^{(2)}(\tau,\lambda).$$

This expression differs from (4.2.52) by the factor $l_{+1}^{(1)*}$. It is important to emphasise that this definition does not depend on the co-ordinate ζ since, if we assume that the resonant medium is in the ground state as $\tau \to \pm\infty$, then T_{21} and T_{11} (as well as T_{12}, T_{22}, T_{33}, T_{34}, T_{43} and T_{44}) do not depend on ζ

Below we will need the two-simulton transfer matrix for the "potential"

$$Q_{kj}(\tau,0) = 2\sqrt{2}i\eta_1 l_{1k}^{(1)} l_{1j}^{(2)} \operatorname{sech}[2\eta_1(\tau-\tau_1)] + 2\sqrt{2}i\eta_2 l_{2k}^{(1)} l_{2j}^{(2)} \operatorname{sech}[2\eta_2(\tau-\tau_2)],$$

for $\eta_1 \neq \eta_2$. Instead of being directly solved the equation of the spectral problem, we assume that the simultons are separated by a large time interval, i.e., the inequality $\tau_2 - \tau_1 \gg 1/2\eta_1,\ 1/2\eta_2$ holds. Now consider the Jost function $\Phi^{(1)}(\tau,\lambda)$ and $\Phi^{(2)}(\tau,\lambda)$.

As $\tau \to -\infty$, we have $\Phi^{(1)}(\tau,\lambda) = \vec{g}^{(1)} \exp(-i\lambda\tau)$, $\Phi^{(2)}(\tau,\lambda) = \vec{g}^{(2)} \exp(-i\lambda\tau)$. For $\tau_1 \ll \tau \ll \tau_2$, we have

$$\Phi^{(i)}(\tau,\lambda) = \sum_{k=1,2} T_{ik}^{(1)}(\lambda)\vec{g}^{(k)} \exp(-i\lambda\tau).$$

So that, for $\tau \to +\infty$ we obtain

$$\Phi^{(i)}(\tau,\lambda) = \sum_{k=1,2} T_{ik}^{(1)}(\lambda) T_{km}^{(2)}(\lambda)\vec{g}^{(m)} \exp(-i\lambda\tau).$$

The superscripts 1 and 2 on T label the one-simulton transfer matrix for $\vec{l}_1^{(1)}$, $\vec{l}_1^{(2)}$, η_1 and $\vec{l}_2^{(1)}$, $\vec{l}_2^{(2)}$, η_2, respectively. Thereby, the two-simulton transfer matrix can be expressed in terms of one-simulton scattering matrices

$$T_{im}(\lambda) = \sum_{k=1,2} T_{ik}^{(1)}(\lambda) T_{km}^{(2)}(\lambda).$$

For $\lambda = i\eta_1$ and $\lambda = i\eta_2$ the matrix elements of the above matrix can be transformed into the following form

$$\begin{aligned} T_{11}(i\eta_1) &= l_{1,+1}^{(1)} L_{+1}^{(1)*} \frac{\eta_1 - \eta_2}{\eta_1 + \eta_2}, \quad T_{11}(i\eta_2) = l_{2,+1}^{(1)*} L_{+1}^{(2)} \frac{\eta_2 - \eta_1}{\eta_1 + \eta_2}, \\ T_{21}(i\eta_1) &= -l_{1,-1}^{(1)} L_{+1}^{(1)*} \frac{\eta_1 - \eta_2}{\eta_1 + \eta_2}, \quad T_{21}(i\eta_2) = -l_{2,+1}^{(1)*} L_{-1}^{(1)} \frac{\eta_1 - \eta_2}{\eta_1 + \eta_2}, \end{aligned} \tag{4.2.54}$$

where $L_q^{(n)}$ are the spherical components of the vectors $\vec{L}^{(n)}$, $n = 1, 2$, which are determined by the following relations

$$\vec{L}^{(1)} = \vec{l}_1^{(1)} - \frac{2\eta_2}{\eta_2 - \eta_1} (\vec{l}_1^{(1)} \cdot \vec{l}_2^{(1)*}) \vec{l}_2^{(1)}, \quad \vec{L}^{(2)} = \vec{l}_2^{(1)} + \frac{2\eta_1}{\eta_2 - \eta_1} (\vec{l}_2^{(1)} \cdot \vec{l}_1^{(1)*}) \vec{l}_1^{(1)}.$$

Let us now consider the collision of polarised simultons. Suppose that two differently polarised simultons be incident on a three-level medium:

$$\hat{Q}(\tau,0) = 2i\eta_1 \vec{l}_1^{(1)} \otimes \vec{l}_1^{(2)} \operatorname{sech}[2\eta_1(\tau - \tau_1)] + 2i\eta_2 \vec{l}_2^{(1)} \otimes \vec{l}_2^{(2)} \operatorname{sech}[2\eta_2(\tau - \tau_2)].$$

Here the normalisation of the pulse envelops is modified. Let $\tau_2 > \tau_1$ and the velocity of the second polarised simulton be higher than the velocity of the first one. During its course the second simulton overtakes, interacts with and passes first simulton. That means that, as $\zeta \to +\infty$, the first simulton will have the position to the right of the second one on the τ axis. Generally simultons will be characterised by new polarisation vectors $\vec{l}_n'^{(1)}$, colour-state vectors $\vec{l}_n'^{(2)}$, the phase shifts and the soliton centre positions τ_n'. In order to obtain these parameters one may use the "passing" (or "transporting") method by Zakharov and Schabat [112]. We shall use the procedure from [32, 59].

Let us assume that, for both $\zeta = 0$ and $\zeta \to +\infty$ the simultons are sufficiently far from each other, i.e., $\tau_2 - \tau_1 >> (2\eta_1)^{-1}$, $(2\eta_2)^{-1}$ and $\tau_2' - \tau_1' >> (2\eta_1)^{-1}$, $(2\eta_2)^{-1}$. Consider a linear combination of Jost functions $\Phi^{(1)}(\tau,\lambda)$ and $\Phi^{(2)}(\tau,\lambda)$:

$$\Phi(\tau,\zeta,\lambda) = -T_{21}(\lambda)\Phi^{(1)}(\tau,\zeta,\lambda) + T_{11}(\lambda)\Phi^{(2)}(\tau,\zeta,\lambda).$$

As $\tau \to -\infty$, this function is independent from ζ:

$$\Phi(\tau,\zeta,\lambda) \to -T_{21}(\lambda)\vec{g}^{(1)} \exp(-i\lambda\tau) + T_{11}(\lambda)\vec{g}^{(2)} \exp(-i\lambda\tau).$$

One can calculate $\Phi(\tau,\zeta,\lambda)$ for $\tau \to +\infty$ and two values of λ, namely, $\lambda = i\eta_1$ and $\lambda = i\eta_2$, by "transporting" the Jost functions through the both simultons. The results of $\zeta = 0$ and $\zeta \to +\infty$ must coincide if we take the evolution of the scattering data into account.

By utilising (4.2.52) and (4.2.54), for $\zeta = 0$ we have: as $\tau \to -\infty$

$$\Phi(\lambda_1) = \left[l_{1,-1}^{(1)}\vec{g}^{(1)} \exp(-i\lambda_1\tau) + l_{1,+1}^{(1)}\vec{g}^{(2)} \exp(-i\lambda_1\tau)\right]L_{+1}^{(1)*} \frac{\eta_1 - \eta_2}{\eta_1 + \eta_2},$$

$$\Phi(\lambda_2) = \left[L_{-1}^{(2)}\vec{g}^{(1)} \exp(-i\lambda_2\tau) + L_{+1}^{(2)}\vec{g}^{(2)} \exp(-i\lambda_2\tau)\right]l_{2,+1}^{(1)*} \frac{\eta_2 - \eta_1}{\eta_1 + \eta_2};$$

- for $\tau_1 << \tau << \tau_2$,

$$\Phi(\lambda_1) = -\left[l_{1,1}^{(2)*}\vec{g}^{(3)} \exp(+i\lambda_1\tau) + l_{1,2}^{(2)*}\vec{g}^{(4)} \exp(+i\lambda_1\tau)\right]L_{+1}^{(1)*} \frac{\eta_1 - \eta_2}{\eta_1 + \eta_2} \exp(2\eta_1\tau_1),$$

$$\Phi(\lambda_2) = \left[l_{2,-1}^{(1)}\vec{g}^{(1)} \exp(-i\lambda_2\tau) + l_{2,+1}^{(1)}\vec{g}^{(2)} \exp(-i\lambda_2\tau)\right]l_{2,+1}^{(1)*} \frac{\eta_2 - \eta_1}{\eta_1 + \eta_2};$$

- as $\tau \to +\infty$

$$\Phi(\lambda_1) = -\left\{\left[l_{1,1}^{(2)*}T_{33}^{(2)}(\lambda_1) + l_{1,2}^{(2)*}T_{43}^{(2)}(\lambda_1)\right]\vec{g}^{(3)} \exp(i\lambda_1\tau) + \right.$$
$$\left. + \left[l_{1,1}^{(2)*}T_{43}^{(2)}(\lambda_1) + l_{1,2}^{(2)*}T_{44}^{(2)}(\lambda_1)\right]\vec{g}^{(4)} \exp(i\lambda_1\tau)\right\}L_{+1}^{(1)*} \frac{\eta_1 - \eta_2}{\eta_1 + \eta_2} \exp(2\eta_1\tau_1)$$

$$\Phi(\lambda_2) = -\left[l_{2,1}^{(2)*}\vec{g}^{(3)} \exp(+i\lambda_2\tau) + l_{2,2}^{(2)*}\vec{g}^{(4)} \exp(+i\lambda_2\tau)\right]l_{2,+1}^{(1)*} \frac{\eta_2 - \eta_1}{\eta_1 + \eta_2} \exp(2\eta_2\tau_2).$$

To simplify the cumbersome form of the formulae below, the evolutionary factors and primes on the quantities $\vec{l}_n^{\,\prime(1)}$, $\vec{l}_n^{\,\prime(2)}$, τ_n' and $T_{ab}^{(n)}$ will be omitted, so that one should remember that these parameters correspond to the simultons after interaction. For $\zeta \to +\infty$, the situation after the collision of the simultons is as follows

- as $\tau \to -\infty$

$$\Phi(\lambda_1) = \left[L_{-1}^{(1)}\vec{g}^{(1)} \exp(-i\lambda_1\tau) + L_{+1}^{(1)}\vec{g}^{(2)} \exp(-i\lambda_1\tau)\right]l_{2,+1}^{(1)*} \frac{\eta_1 - \eta_2}{\eta_1 + \eta_2},$$

$$\Phi(\lambda_2) = \left[l_{2,-1}^{(1)}\vec{g}^{(1)} \exp(-i\lambda_2\tau) + l_{2,+1}^{(1)}\vec{g}^{(2)} \exp(-i\lambda_2\tau)\right]L_{+1}^{(2)*} \frac{\eta_2 - \eta_1}{\eta_1 + \eta_2};$$

- for $\tau_1 << \tau << \tau_2$,

$$\Phi(\lambda_1) = \left[l_{1,-1}^{(1)}\vec{g}^{(1)} \exp(-i\lambda_1\tau) + l_{1,+1}^{(1)}\vec{g}^{(2)} \exp(-i\lambda_1\tau)\right] l_{1,+1}^{(1)*} \frac{\eta_1 - \eta_2}{\eta_1 + \eta_2},$$

$$\Phi(\lambda_2) = -\left[l_{2,1}^{(2)*}\vec{g}^{(3)} \exp(+i\lambda_2\tau) + l_{2,2}^{(2)*}\vec{g}^{(4)} \exp(+i\lambda_2\tau)\right] L_{+1}^{(2)*} \frac{\eta_2 - \eta_1}{\eta_1 + \eta_2} \exp(2\eta_2\tau_2);$$

- as $\tau \to +\infty$

$$\Phi(\lambda_1) = -\left[l_{1,1}^{(2)*}\vec{g}^{(3)} \exp(+i\lambda_1\tau) + l_{1,2}^{(2)*}\vec{g}^{(4)} \exp(+i\lambda_1\tau)\right] l_{1,+1}^{(1)*} \frac{\eta_1 - \eta_2}{\eta_1 + \eta_2} \exp(2\eta_1\tau_1),$$

$$\Phi(\lambda_2) = -\left\{\left[l_{2,1}^{(2)*}T_{33}^{(2)}(\lambda_2) + l_{2,2}^{(2)*}T_{43}^{(2)}(\lambda_2)\right]\vec{g}^{(3)} \exp(i\lambda_2\tau) + \right.$$

$$\left. + \left[l_{2,1}^{(2)*}T_{43}^{(2)}(\lambda_2) + l_{2,2}^{(2)*}T_{44}^{(2)}(\lambda_2)\right]\vec{g}^{(4)} \exp(i\lambda_2\tau)\right\} L_{+1}^{(2)*} \frac{\eta_2 - \eta_1}{\eta_1 + \eta_2} \exp(2\eta_2\tau_2).$$

Comparison of the formulae for $\Phi(\lambda_2)$ as $\tau \to +\infty$ at $\zeta = 0$ and $\zeta \to +\infty$ yields

$$\vec{l}_2^{(2)} = \frac{\exp\{2\eta_2(\tau_2' - \tau_2)\}}{l_{2,+1}^{(1)}} L_{+1}'^{(2)} \left(\vec{l}_2'^{(2)} + \frac{2\eta_1}{\eta_2 - \eta_1} (\vec{l}_2'^{(2)} \cdot \vec{l}_1'^{(2)*}) \vec{l}_1'^{(2)} \right),$$

$$\vec{l}_1'^{(2)} = \frac{\exp\{2\eta_1(\tau_1' - \tau_1)\}}{l_{1,+1}'^{(1)}} L_{+1}^{(1)} \left(\vec{l}_1^{(2)} - \frac{2\eta_2}{\eta_2 - \eta_1} (\vec{l}_1^{(2)} \cdot \vec{l}_2^{(2)*}) \vec{l}_2^{(2)} \right).$$

We note that the equations of spectral problem are invariant under the transformations

$$\psi_1 \to \psi_3^*, \quad \psi_2 \to \psi_4^*, \quad \psi_3 \to \psi_1^*, \psi_4 \to \psi_2^*,$$

$$q_1^{-1} \to -q_1^{-1}, \quad q_2^{-1} \to -q_1^{+1}, \quad q_1^{+1} \to -q_2^{-1}, \quad q_2^{+1} \to -q_2^{+1}.$$

Therefore we can write

$$\vec{l}_2^{(1)} = \frac{1}{l_{2,2}^{(2)}} \vec{L}'^{(2)} \left(l_{2,2}^{(2)} + \frac{2\eta_1}{\eta_2 - \eta_1} (\vec{l}_2'^{(2)} \cdot \vec{l}_1'^{(2)*}) l_{1,2}'^{(2)} \right),$$

$$\vec{l}_1'^{(1)} = \frac{1}{l_{1,2}'^{(2)}} \vec{L}^{(2)} \left(l_{1,2}^{(2)} - \frac{2\eta_2}{\eta_2 - \eta_1} (\vec{l}_1^{(2)} \cdot \vec{l}_2^{(2)*}) l_{2,2}^{(2)} \right).$$

These expressions result in the following transformation laws for the polarisation vector

$\vec{l}_n^{(1)}$ and the colour-state vectors $\vec{l}_n^{(2)}$:

$$\vec{l}_1{}'^{(1)} = \frac{1}{B^{(1)}}\left(-\vec{l}_1^{(1)} + \frac{2\eta_2}{\eta_2-\eta_1}(\vec{l}_1^{(1)}\cdot\vec{l}_2^{(1)*})\vec{l}_2^{(1)}\right), \tag{4.2.55a}$$

$$\vec{l}_2{}'^{(1)} = \frac{1}{B^{(1)}}\left(\vec{l}_2^{(1)} + \frac{2\eta_1}{\eta_2-\eta_1}(\vec{l}_2^{(1)}\cdot\vec{l}_1^{(1)*})\vec{l}_1^{(1)}\right), \tag{4.2.55b}$$

and

$$\vec{l}_1{}'^{(2)} = \frac{1}{B^{(2)}}\left(-\vec{l}_1^{(2)} + \frac{2\eta_2}{\eta_2-\eta_1}(\vec{l}_1^{(2)}\cdot\vec{l}_2^{(2)*})\vec{l}_2^{(2)}\right), \tag{4.2.56a}$$

$$\vec{l}_2{}'^{(2)} = \frac{1}{B^{(2)}}\left(\vec{l}_2^{(2)} + \frac{2\eta_1}{\eta_2-\eta_1}(\vec{l}_2^{(2)}\cdot\vec{l}_1^{(2)*})\vec{l}_1^{(2)}\right), \tag{4.2.56b}$$

where

$$B^{(1)} = \left(1+\frac{4\eta_1\eta_2}{(\eta_2-\eta_1)^2}|\vec{l}_1^{(1)}\cdot\vec{l}_2^{(1)*}|\right)^{1/2}, \quad B^{(2)} = \left(1+\frac{4\eta_1\eta_2}{(\eta_2-\eta_1)^2}|\vec{l}_1^{(2)}\cdot\vec{l}_2^{(2)*}|\right)^{1/2}.$$

It is seen that the collision of the polarised simultons is accompanied by the rotation of the polarisation vectors independently from each other and by redistribution of the energy between the partial pulses forming the simultons. These processes are governed by the relations (4.2.55) and (4.2.56), i.e., this law is identical to the transformation law for the polarisation vectors of pulses of single-frequency SIT involving the transitions $j_a = 1 \leftrightarrow j_b = 0$, $j_a = 1 \rightarrow j_b = 1$ [30,31,33], but differs from the properties of the interaction of two-frequency waves in a resonant Kerr medium investigated in [113].

The phase shifts of simultons that accompany their collisions are found to depended both on polarisation and on the colour-state of USPs (i.e., on the energy distribution between them). These quantities are defined as

$$\tau_1' - \tau_1 = \ln(B^{(2)}B^{(1)})/2\eta_1, \quad \tau_2' - \tau_2 = -\ln(B^{(2)}B^{(1)})/2\eta_2.$$

Whatever the polarisation, the simultons behave as if they repel one another: the overtaken simulton receives a positive increase in its co-ordinate $\tau_1' - \tau_1 > 0$, and the overtaking simulton receives a negative co-ordinate increase $\tau_2' - \tau_2 < 0$.

The special case resonant transition $j_a = j_c = 0$, $j_b = 1$ is not unique example of exactly integrable systems. Another one is provided by the tree-level system of Λ- or V-configuration with $j_a = j_b = j_c = 1/2$. Here the left- and the right-polarised circular components of the USP with the identical carrier frequency propagate independently of one another, so that the interaction between arbitrary polarised simultons for this sort of level configuration is described by two independent sets of non-linear evolution equations [26,27]. It is readily shown that, in contrast to (4.2.56), the linearly polarised simultons preserve their states of polarisation in a collision in the three-level medium with $j_a = j_b = j_c = 1/2$, but the collision between a linearly polarised simulton and a circular polarised simulton results in the formation of three circular polarised ones. The latter result is a consequence of the separation of the left- and right-polarised components of the linearly polarised simulton due to the interaction between one of them and the circular polarised simulton.

4.3. SIT for surface polaritons

Propagating of an ultrashort pulse of an electromagnetic wave along a thin film of resonant atoms (thickness of which is less than the light wave length) can be considered on the basis of the Maxwell and Bloch equations, as it was done in the fundamental work by Agranovich et al. [38]. The simplest case is a two-layer system, when a thin film of the resonant atoms is located on the interface of the two linear dielectric media. Let wave propagate along a planar interface of two linear dielectric media in the z-direction. The normal to the surface is fixed to lie in the x-direction while the y-direction lies in the interface plane, giving a right-handed Cartesian frame. Since in the frame chosen all y-co-ordinates are equivalent, the electric field strength as well as the magnetic field and electric induction vectors do not depend on y. This planar symmetry leads to splitting the Maxwell equations into two group of equations describing two kinds of surface waves. These are the transverse electric wave (TE wave) and the transverse magnetic field wave (TM wave).The single non-zero component of the electric field of the TE wave is E^y, while the magnetic field of the TM wave has only H^y as a non-zero component.

4.3.1. SURFACE POLARITON

Let us consider the TM waves propagation. It is convenient to consider the wave equation for the vector of magnetic field strength of light wave instead of the wave equation for $\mathbf{E}$ in the case of TE waves. Indeed, the boundary conditions at the interface use a single nonzero component of a magnetic field strength vector in TM waves.

$$E^z(0-) - E^z(0+) = 0, \quad D^x(0-) - D^x(0+) = 0, \quad H^y(0-) - H^y(0+) = \frac{4\pi}{c}\frac{\partial P_S^z}{\partial t}. \tag{4.3.1}$$

This component $H = H^y(x,z,t)$ satisfies the equations

$$\frac{\partial H}{\partial z} = -\frac{1}{c}\frac{\partial D^y}{\partial t}, \quad \frac{\partial H}{\partial x} = \frac{1}{c}\frac{\partial D^z}{\partial t}, \quad \frac{\partial E^x}{\partial z} - \frac{\partial E^z}{\partial x} = -\frac{1}{c}\frac{\partial H}{\partial t}.$$

It can be found that for the isotropic dielectric medium at $x < 0$ and $x > 0$ the Fourier component of the magnetic field strength $\widetilde{H}(x,z,\omega)$ satisfies wave equation

$$\partial^2 \widetilde{H} / \partial z^2 + \partial^2 \widetilde{H} / \partial x^2 + k_0^2 \varepsilon_i \widetilde{H} = 0,$$

where $\varepsilon_1 = \varepsilon(x > 0)$, $\varepsilon_2 = \varepsilon(x < 0)$, $k = \omega / c$. Let $\widetilde{H}(x,z,\omega)$ is given by

$$\widetilde{H}(x,z,\omega) = \int_{-\infty}^{\infty} \frac{d\beta}{2\pi} h(x,\beta,\omega) \exp[i\beta(\omega)z] \quad ,$$

then for $h(x,\beta,\omega)$ we obtain

$$\partial^2 h / \partial x^2 + [k_0^2 \varepsilon_i - \beta^2] h = 0 .$$

For the solution of this equation to correspond to a surface wave, it is necessary that k_i^2 be less than β^2 $(k_i^2 = k_0^2 \varepsilon_1)$. Then $h(x,\beta,\omega)$ is a decreasing function of $|x|$,

$$h(x,\beta,\omega) = \begin{cases} h^{(1)}(x) = A(\beta,\omega) \exp(-qx) \,, x > 0 \\ h^{(2)}(x) = B(\beta,\omega) \exp(px) \,, x < 0 \end{cases}$$

where $q^2 = \beta^2 - k_1^2$, $p^2 = \beta^2 - k_2^2$ and $p > 0, q > 0$ are chosen. The boundary conditions yield the relation between $h^{(1)}(0)$ and $h^{(2)}(0)$

$$h^{(1)}(0,\beta,\omega) - h^{(2)}(0,\beta,\omega) = ik_0 4\pi \, p_S(\beta,\omega) , \tag{4.3.2}$$

where

$$p_S(\beta,\omega) = \int_{-\infty}^{\infty} dt dz P_S^z \exp[i\omega t - i\beta(\omega)z] .$$

Using the equation $\operatorname{rot} \vec{H} = c^{-1} \partial \vec{D} / \partial t$ and continuity of z-components of an electrical field of the light wave one can find that

$$A(\beta,\omega) = -ik_0(\varepsilon_1 / q)\widetilde{E}_z^{(1)}(0,\beta,\omega), \;\; B(\beta,\omega) = +ik_0(\varepsilon_2 / p)\widetilde{E}_z^{(2)}(0,\beta,\omega) , \tag{4.3.3}$$

and also $\widetilde{E}^{z(1)}(0,\beta,\omega) = \widetilde{E}^{z(2)}(0,\beta,\omega) = \widetilde{E}(\beta,\omega)$, where

$$\widetilde{E}^{z(i)}(x,\beta,\omega) = \int_{-\infty}^{\infty} dt \, dz \widetilde{E}^{z(i)}(x,z,t) \exp[i\omega t - i\beta(\omega)z] .$$

Substitution of (4.3.3) into (4.3.2) yields the equation to determine $\widetilde{E}(\beta,\omega)$,

$$\left(\varepsilon_1 q^{-1} + \varepsilon_2 p^{-1}\right)\widetilde{E}(\beta,\omega) = -4\pi\, p_S(\beta,\omega), \tag{4.3.4}$$

or in space-temporary representation

$$-4\pi P_S^z = \int_{-\infty}^{\infty} \frac{d\omega d\beta}{(2\pi)^2} f(\beta,\omega)\widetilde{E}(\beta,\omega), \tag{4.3.4a}$$

where

$$f(\beta,\omega) = q^{-1}\varepsilon_1 + p^{-1}\varepsilon_2 .$$

If there are no resonant atoms on the interface, expression (4.3.4) yields the dispersion relation for the linear surface wave of TM type. Besides, if the TE wave had been considered, equation

$$4\pi k_0^2 P_S^y = \int_{-\infty}^{\infty} \frac{d\omega d\beta}{(2\pi)^2} f(\beta,\omega)\widetilde{E}(\beta,\omega),$$

where $f(\beta,\omega) = q + p$, instead of (4.3.4) would have been received, where p and q are determined in the same way, as they are here. $\widetilde{E}(\beta,\omega)$ is Fourier component of the electrical field strength of light wave $E^y(x,z,t)$ at $x = 0$.

No assumptions have been made here that a thin film is formed by resonant atoms located on the interface. Equation (4.3.4) is the basis of the theory of light pulse propagating along the interface of two linear dielectric materials. The film is made of linear or non-linear material. It is important only that the thickness of the film is less than the wavelength. Such film serves as a guiding structure like an optical waveguide does. However, if the guiding ability of a waveguide can be explained in terms of total internal reflection from its boundaries this interpretation is not valid in the case of film. Besides that, the film is a single mode guiding structure, i.e. a single-mode waveguide.

Let the light pulse be described by a complex envelope which is slowly varying in space and time, namely

$$E(x,z,t) = \mathcal{E}(z,t)\Phi(x,\omega)\exp[-i\omega_0 t + i\beta_0(\omega)z]$$

Function $\Phi(x,\omega)$ can be related to the function $h(x,\beta,\omega)$ found above, but it is not necessary here.

If $\widetilde{E}(\beta,\omega)$ is a Fourier component of a surface wave envelope, then one can show that $\mathcal{E}(\beta,\omega) = \widetilde{E}(\beta_0 + \beta,\, \omega_0 + \omega)$. The Fourier component of the slowly varying surface polarisation envelope is defined analogously as $\tilde{p}(\beta,\omega) = p_S(\beta_0+\beta,\omega_0+\omega)$. Equation for $\tilde{e}(\beta,\omega)$ and $\tilde{p}(\beta,\omega)$ follows from (4.3.4)

$$-4\pi\tilde{p}(\beta,\omega) = f(\beta_0+\beta,\omega_0+\omega)\tilde{e}(\beta,\omega). \tag{4.3.5}$$

Since $e(z,t)$ and $p(z,t)$ are slowly varying functions in space and in time, the quantities $\tilde{e}(\beta,\omega)$ and $\tilde{p}(\beta,\omega)$ contribute essentially only in a small interval of β and ω around zero, i.e., $\beta \ll \beta_0, \omega \ll \omega_0$. Therefore, we may make a power series expansion in (4.3.5). Let

$$f(\beta_0+\beta,\omega_0+\omega) = f(\beta_0,\omega_0) + (\partial f/\partial\omega)_0\omega + (\partial f/\partial\beta)_0\beta, \tag{4.3.6}$$

where index zero means that the corresponding derivatives are taken in the point (β_0,ω_0). Then (4.3.5) is converted to

$$-4\pi\,\tilde{p}(\beta,\omega) = f(\beta_0,\omega_0)\tilde{e}(\beta,\omega) + (\partial f/\partial\omega)_0\omega\,\tilde{e}(\beta,\omega) + (\partial f/\partial\beta)_0\beta\,\tilde{e}(\beta,\omega),$$

Inverse Fourier transformation provides the differential equation

$$f(\beta_0,\omega_0)e + i(\partial f/\partial\omega)_0\frac{\partial e}{\partial t} - i(\partial f/\partial\beta)_0\frac{\partial e}{\partial z} = -4\pi p. \tag{4.3.7}$$

We assume that ω_0 perfectly matches the resonance frequency of the two-level atoms in the film and the inhomogeneous broadening of the absorption line is negligible.

The atomic system is described by Bloch equations similarly to a bulk medium, but unlike the latter the resonant atoms are affected by the field in the thin film. As it follows from the Bloch equations (3.1.7), the polarisation is purely imaginary when $\Delta\omega = 0$ Atoms absorb electromagnetic radiation and reradiate it due to induced emission. One can easily find from the Bloch equations (1.3.17) or (4.1.43)

$$p(z,t) = in_A d\sin\Psi(z,t),$$

where

$$\Psi(z,t) = (d/\hbar)\int_{-\infty}^{t} e(z,t')dt'.$$

Substituting $p(z,t)$ in (4.3.7) and separating the real and imaginary parts of the equation (in solution of (3.1.7) it was already implied that $e(z,t)$ is a real function), one obtains

$$f(\beta_0,\omega_0) = 0, \tag{4.3.8a}$$

$$\partial e/\partial t + v_g\partial e/\partial z = -4\pi n_a d(\partial f/\partial\omega)_0^{-1}\sin\Psi, \tag{4.3.8b}$$

where $v_g = (d\omega / d\beta)_0$ is the group velocity. Equation (4.3.8a) gives a relationship between β and ω or dispersion relation, from which the group velocity can be obtained. If we introduce new independent variables $\tau = (t - z/v_g)/t_{p0}$, $\zeta = z/L_a$, equation (4.3.8b) reduces to the well known SG-equation [100,101] for $\Psi(\tau,\zeta)$

$$\partial^2\Psi / \partial\tau\partial\zeta = -\eta' \sin\Psi , \qquad (4.3.9)$$

where $\eta' = \mathrm{sgn}(\partial f / \partial\omega)_0$ and $L_a^{-1} = 4\pi n_A d^2 t_{p0}[\hbar v_g \,|(\partial f / \partial\omega)_0|]^{-1}$. If $\eta' = -1$, then equation (4.3.9) has soliton solutions, which determine the electrical field strength of a surface solitary wave moving along the film. Non-steady-state solitary solutions (breathers) [100] serve as another example of non-linear lossless ultrashort pulse of the non-linear surface wave. On the base of these results the phenomenon of *self-induced transparency of surface polaritons* was predicted in [38] (see section 4.1. also)

Attention should be drawn to the equation (4.3.8a). As it is well known, this equation has no solutions in the case of TE waves. That means that SIT for TE waves does not exist, at least in the framework of the assumptions made. This has been one of the reasons to pay major attention to a TM wave, for which equation (4.3.8a) can have a solution.

4.3.2. POLARITONS IN THE THREE-LAYER SYSTEM

We now consider the propagation of an optical pulse in a planar waveguide with both sides coated by a thin film of resonant atoms. The planar waveguide is assumed to be a layer of dielectric material with a permittivity ε_2 and a thickness l placed on a substrate ($x < 0$) with a permittivity ε_3. The permittivity of the covering medium ($x > l$) is ε_1. It is known [114] that if $\varepsilon_2 > \varepsilon_1, \varepsilon_3$ a three-layer system can guide the propagation of electromagnetic TE and TM waves, setting up a discrete series of waveguide modes. The propagating constant β for every mode is determined by the respective branch of the dispersion relation.

The presence of thin films of polarisable matter brings new qualitative aspects into this picture. First of all the characteristics of the waveguide modes are changed. It would be more correct to call them modified modes. In addition, since both interfaces (at $x = 0$ and at $x = l$) are now able to support surface polariton waves, in the guided wave spectrum a modified polariton wave appears in the form of two bounded surface polariton waves. In the linear waveguide theory the surface TE waves are absent in the absence of a thin polarisable layer but TM waves occur when both ε_1 or ε_3 are negative.

Let us consider the case of the TE waves. Maxwell equations yield

$$\partial^2 e / \partial x^2 + [k_0^2 \varepsilon(x) - \beta^2] e = 0 , \qquad (4.3.10)$$

where

$$E(x,z,t)=\int_{-\infty}^{\infty}\frac{d\beta}{2\pi}\frac{d\omega}{2\pi}e(x,\beta,\omega)\exp[-i\omega t+i\beta z]\,,$$

and

$$\varepsilon(x)=\{\varepsilon_1,\ x>l;\ \varepsilon_2,\ 0<x<l;\ \varepsilon_3,\ x<0\}\ .$$

The boundary and continuity conditions are

$$\lim_{|x|\to\infty}e(x,\beta,\omega)=0,\quad e(0+)-e(0-)=0,\quad e(l+0)-e(l-0)=0\ ,$$
$$e,_x(0+)-e,_x(0-)=4\pi k_0^2p_S(0),\quad e,_x(l+0)-e,_x(l-0)=4\pi k_0^2p_S(l)\,.$$

Taking into account these conditions the solution of equation (4.3.10) can be written as

$$e(x,\beta,\omega)=\begin{cases}e^{(1)}(x,\beta,\omega)=A_1\exp(-q_1x)\,, & x>l\\ e^{(2)}(x,\beta,\omega)=A_2\exp(q_2x)+B_2\exp(-q_2x)\,, & 0<x<l\\ e^{(3)}(x,\beta,\omega)=A_3\exp(q_3x)\,, & x<0\end{cases}$$

where $q_1^2=\beta^2-k_0^2\varepsilon_c$, $q_2^2=\beta^2-k_0^2\varepsilon_f$, $q_3^2=\beta^2-k_0^2\varepsilon_s$ and $q_1^2>0, q_3^2>0$. For modified surface polariton waves $q_2^2>0$, for modified modes $q_2^2<0$. The coefficients A_1, A_2, A_3 and B_2 are coupled by continuity conditions at the interfaces x=0 and x=l,

$$A_1\exp(-q_1l)=A_2\exp(q_2l)+B_2\exp(-q_2l) \tag{4.3.11a}$$
$$q_2\left[A_2\exp(q_2l)-B_2\exp(-q_2l)\right]+q_1A_1\exp(-q_1l)=-4\pi k_0^2p_S(l) \tag{4.3.11b}$$
$$A_3=A_2+B_2 \tag{4.3.11c}$$
$$q_3A_3-q_2(A_2-B_2)=-4\pi k_0^2p_S(0) \tag{4.3.11d}$$

Expressions (4.3.11a) and (4.3.11c) yield A_2 and B_2

$$\begin{aligned}A_2&=\frac{1}{2}\sinh^{-1}(q_2l)\left[e(l)-e(0)\exp(-q_2l)\right],\\ B_2&=\frac{1}{2}\sinh^{-1}(q_2l)\left[e(0)\exp(-q_2l)-e(l)\right],\end{aligned} \tag{4.3.12}$$

where the new notations are introduced

$$e(l)=A_1\exp(-q_1l)\,,\quad e(0)=A_3$$

The equations (4.3.11b), (4.3.11d) and (4.3.12) give the following expressions

$$\begin{aligned}&\left[q_3 + q_2 \tanh^{-1}(q_2 l)\right]e(0) - q_2 \sinh^{-1}(q_2 l)e(l) = -4\pi k_0^2 p_S(0)\,,\\&\left[q_1 + q_2 \tanh^{-1}(q_2 l)\right]e(l) - q_2 \sinh^{-1}(q_2 l)e(0) = -4\pi k_0^2 p_S(l)\,.\end{aligned} \tag{4.3.13}$$

If we introduce the notations

$$f_1(\beta,\omega) = q_1 + q_2 \tanh^{-1}(q_2 l)\,, \quad f_2(\beta,\omega) = q_2 \sinh^{-1}(q_2 l)\,,$$
$$f_3(\beta,\omega) = q_3 + q_2 \tanh^{-1}(q_2 l)\,,$$

equations (4.3.13) can be rewritten as

$$\begin{aligned}&\left(f_1 f_3 - f_2^2\right)e(0) = -4\pi k_0^2\left[f_1 p_S(0) + f_2 p_S(l)\right],\\&\left(f_1 f_3 - f_2^2\right)e(l) = -4\pi k_0^2\left[f_2 p_S(0) + f_3 p_S(l)\right].\end{aligned} \tag{4.3.14}$$

Following the paper [39] we shall consider here the particular case of a symmetric waveguide ($\varepsilon_1 = \varepsilon_3$) and we shall assume that the condition of exact resonance is fulfilled and inhomogeneous broadening of the absorbing lines is absent. In this special case $f_1(\beta,\omega) = f_3(\beta,\omega)$ and $e(x=0,\beta,\omega) = \pm e(x=l,\beta,\omega)$. Here the positive sign corresponds to a symmetric solution of the wave equation, the negative sign corresponds to an anti-symmetric one. It results from (4.3.14) that

$$F_\pm(\beta,\omega)e(l) = -4\pi k_0^2 \bar{p}_S(\beta,\omega)\,, \tag{4.3.15}$$

where

$$F_\pm(\beta,\omega) = f_1(\beta,\omega) \pm f_2(\beta,\omega)\,, \quad \bar{p}_S(\beta,\omega) = p_S(l) = \pm p_S(0)\,.$$

Equation (4.3.15) is written for the Fourier components of the observable field strength and surface polarisations. As in the previous section we must transfer to slowly varying in time and space envelopes $e(z,t)$ and $p(z,t)$. The Fourier components of these envelopes $\widetilde{E}(\beta,\omega)$ and $p_S(\beta,\omega)$ are related to $\tilde{e}(\beta,\omega)$ and $\tilde{p}(\beta,\omega)$ by the following relations

$$\tilde{e}(\beta,\omega) = \widetilde{E}(\beta_0 + \beta,\, \omega_0 + \omega)\,, \qquad \tilde{p}(\beta,\omega) = p_S(\beta_0 + \beta, \omega_0 + \omega)\,,$$

and they satisfy the equation

$$-4\pi k_0^2 \tilde{p}(\beta,\omega) = F_\pm(\beta_0 + \beta, \omega_0 + \omega)\tilde{e}(\beta,\omega)\,. \tag{4.3.16}$$

Analogous to surface polariton wave spreading along only one interface, a reduced equation for slowly varying envelope $e(z,t)$ results from (4.3.16)

$$\partial e / \partial t + v_g \partial e / \partial z = 4\pi i k_0^2 r_S^{\pm} p ,$$

where v_g is the group velocity of a modified surface wave light pulse and

$$r_S^{\pm} = (\partial F_{\pm} / \partial \omega)_0^{-1} .$$

The polarisation $p(z,t)$ can be found from the Bloch equation (3.1.7b). Finally, for the value $\Psi(\tau,\zeta)$ we obtain the SG-equation

$$\partial^2 \Psi / \partial \tau \partial \zeta = -\eta'' \sin \Psi ,$$

where $\tau = (t - z / v_g) t_{p0}^{-1}$, $\zeta = z / L_a$, $\eta'' = \mathrm{sgn}(r_S^{\pm})$, $L_a^{-1} = 4\pi d^2 n_A t_{p0} (\hbar v_g \mid r_S^{\pm} \mid)^{-1}$. That means that the problem is reduced to a known theory SIT by McCall-Hahn in the limit of exact resonance and a narrow absorption line, the sharp-line approximation.

Unlike the surface polariton waves considered above, the modified surface polariton waves can be even and odd with respect to the sign in the expression for $F_{\pm}(\beta,\omega)$ (4.3.15). The dispersion law for modified surface polariton waves is given by expression

$$F_{\pm}(\beta,\omega) = q_1 + q_2 \tanh^{-1}(q_2 l) \pm q_2 \sinh^{-1}(q_2 l) = 0 . \tag{4.3.17}$$

Note that a SIT phenomenon is possible only for the positive sign of $r_S^{\pm}$.

To find modified modes of the planar waveguide all derivations must be redone substituting $q_2^2 = k_0^2 \varepsilon_2 - \beta^2$ for all q_2^2. However, it is simpler to change q_2 for $i\kappa = i(k_0^2 \varepsilon_2 - \beta^2)^{1/2}$ in all expressions. Then the dispersion relation for the modified modes can be shown to be

$$F_{\pm}(\beta,\omega) = q_1 + \kappa \tan^{-1}(\kappa l) \pm \kappa \sin^{-1}(\kappa l) = 0 . \tag{4.3.18}$$

This expression may be rewritten in a different form. Relation (4.3.18) yields

$$q_1 \sin(\kappa l) + \kappa \cos(\kappa l) \pm \kappa = 0 .$$

Using the known formulae $\sin\alpha = \tan\alpha\left(1 + \tan^2\alpha\right)^{-1/2}$, $\cos\alpha = \left(1 + \tan^2\alpha\right)^{-1/2}$ we obtain that

$$\tan(\kappa l) = \frac{2\kappa q_1}{\kappa^2 - q_1^2} .$$

This is a well-known expression for the dispersion relation of the guided modes of a planar symmetric waveguide. Since $\tan(\kappa l)$ is a multivalued function the solutions of this equation (which implicates the relation between β and ω) give a set of dispersion relation branches for modified modes.

Surface waves of TM type can be examined in exactly the same way as stated above. The analogue of equation (4.3.13) here is equations

$$\begin{aligned}&\left[\varepsilon_s q_3^{-1}+\varepsilon_f q_2^{-1}\tanh^{-1}(q_2 l)\right]e(l)-\varepsilon_f q_2^{-1}\sinh^{-1}(q_2 l)e(0)=-4\pi p_S(0),\\&\left[\varepsilon_c q_1^{-1}+\varepsilon_f q_2^{-1}\tanh^{-1}(q_2 l)\right]e(0)-\varepsilon_f q_2^{-1}\sinh^{-1}(q_2 l)e(l)=-4\pi p_S(l).\end{aligned}\tag{4.3.19}$$

where $e(x;\beta,\omega)$ is the Fourier component of z-projection of the electric field strength $E^z(x,z,t)$. Instead of equation (4.3.14) we obtain the following system

$$\begin{aligned}&\left(f_1 f_3-f_2^2\right)e(0)=-4\pi\left[f_1 p_S(0)+f_2 p_S(l)\right],\\&\left(f_1 f_3-f_2^2\right)e(l)=-4\pi\left[f_2 p_S(0)+f_3 p_S(l)\right]\end{aligned}\tag{4.3.20}$$

where

$$f_1(\beta,\omega)=\varepsilon_c q_1^{-1}+\varepsilon_f q_2^{-1}\tanh^{-1}(q_2 l),\quad f_2(\beta,\omega)=\varepsilon_f q_2^{-1}\sinh^{-1}(q_2 l),$$
$$f_3(\beta,\omega)=\varepsilon_s q_3^{-1}+\varepsilon_f q_2^{-1}\tanh^{-1}(q_2 l).$$

For a symmetric waveguide $f_1(\beta,\omega)=f_3(\beta,\omega)$, the equation for the slowly varying pulse envelope is similar to (4.3.16): $-4\pi\tilde{p}(\beta,\omega)=F_\pm(\beta_0+\beta,\omega_0+\omega)\tilde{e}(\beta,\omega)$. One can reduce it to the SG-equation. The dispersion relation for the modified surface polariton of TM type can be written in the form:

$$\varepsilon_1 q_1^{-1}+\varepsilon_2 q_2^{-1}\tanh^{-1}(q_2 l)\pm\varepsilon_2 q_2^{-1}\sinh^{-1}(q_2 l)=0,$$

and for the guided modes it has the form

$$\varepsilon_1 q_1^{-1}+\varepsilon_2\kappa^{-1}\tan^{-1}(\kappa l)\pm\varepsilon_2\kappa^{-1}\sin^{-1}(\kappa l)=0.$$

As in the case of TE waves, dispersion properties of TM waves are determined completely by the linear characteristics of the three-layer medium.

The propagation of a surface wave USP along a thin layer of the resonant atoms lying near the interface was considered by Agranovich, Rupasov and Chernyak in the work [39]. The layer was assumed to be polarised normal with respect to the interface. It was also assumed that the layer of the atoms is completely embedded in one of the medium, for instance, in the one with permittivity ε_1, at a depth much less than a wavelength. The difference from the cases considered so far consists in the boundary conditions, which now read

$$E^z(x=0+)-E^z(x=0-)=-4\pi\varepsilon_1^{-1}\partial P/\partial z, \quad H^z(x=0+)-H^z(x=0-)=0,$$
$$E^y(x=0+)-E^y(x=0-)=-4\pi\varepsilon_1^{-1}\partial P/\partial y, \quad H^y(x=0+)-H^y(x=0-)=0\,.$$

where $P=P_S(z,t)$ is the thin film polarisation per units of surface.

Let us consider a planar waveguide with a thin layer of resonant atoms located near the interface at $x = l$. Assume that the layer was polarised normal with respect to the interface. We can obtain the equations for the surface TM waves by analogue with (4.3.19). Then the equation for the electric field strength $E^z(x=l)$ will have the same form, as (4.3.20). The transition to the slowly varying envelope $e(z,t)$ is made by shifting arguments of function $\widetilde{E}(\beta,\omega)$ $\beta\to\beta_0+\beta$ and $\omega\to\omega_0+\omega$, as it has been done for TE waves. The difference is in the concrete type of functions $F(\beta,\omega)$. In the present case

$$F(\beta,\omega)=f_1(\beta,\omega)-f_2^2(\beta,\omega)/f_3(\beta,\omega),$$

where

$$f_1(\beta,\omega)=\frac{\varepsilon_c^2}{\beta^2}\left(\frac{q_1}{\varepsilon_c}+\frac{q_2}{\varepsilon_f}\tanh^{-1}(q_2l)\right),$$

$$f_2(\beta,\omega)=\frac{\varepsilon_c\varepsilon_s}{\varepsilon_f\beta^2}\frac{q_2}{\sinh(q_2l)},$$

$$f_3(\beta,\omega)=\frac{\varepsilon_s^2}{\beta^2}\left(\frac{q_3}{\varepsilon_s}+\frac{q_2}{\varepsilon_f}\tanh^{-1}(q_2l)\right).$$

In the sharp-line approximation and under exact resonance the Fourier component $\widetilde{e}(\beta,\omega)$ is governed by the equation similar to the equation for TE waves, with substitution $r_S^{\pm}\to r_S=(\partial F/\partial\omega)_0$. Then the Bloch angle defined as

$$\Psi(z,t)=(d/\hbar)\int_{-\infty}^{t}e(z,t')dt'$$

satisfies the SG-equation.

Theory by Agranovich, Rupasov and Chernyak can be generalised by including the cubic non-linear non-resonant susceptibility of a thin film, three-level atoms and double resonance and two-photon absorption by two-level atoms. A somewhat more complicated subject is the IST theory under the condition of degeneration of the energy levels of resonant atoms of a film.

4.4. SIT under two-photon resonance

So far the one-photon resonance picture has been discussed. Progress in laser physics has made possible the generation of intense ultrashort pulses to produce the multi-photon processes. The simplest one develops when the resonant matter interacts with a pair of USPs of different carrier frequencies ω_1 and ω_2 so that $\omega_1 \pm \omega_2$ coincides with a transition frequency ω_{ca}. The nonlinear processes are two-photon absorption and Raman scattering, respectively. The coherent pulse propagation takes place under both resonance conditions when the pulse duration is much less then relaxation times. This process is similar to SIT and it was referred to as two-photon self-induced transparency (TPSIT). In the next sections we shall consider the theory of TPSIT.

4.4.1. BASIC EQUATIONS OF TWO-PHOTON SELF-INDUCED TRANSPARENCY THEORY

The generalised Maxwell-Bloch equations, which describe the joint evolution of two differently polarised ultrashort pulses, (a = 1, 2 are vector indices)

$$E_{1a} = \mathcal{E}_{1a} \exp[i(k_1 z - \omega_1 t)] + \text{c.c.}, \quad E_{2a} = \mathcal{E}_{2a} \exp[i(k_2 z - \omega_2 t)] + \text{c.c} \qquad (4.4.1)$$

under the conditions of Raman resonance (RR),

$$\omega_1 - \omega_2 \approx \omega_{ca}, \qquad (4.4.2)$$

or two-photon resonance (TPR),

$$\omega_1 + \omega_2 \approx \omega_{ca}, \qquad (4.4.3)$$

with a double-quantum (optically forbidden) transition $j_c \to j_a$, between atomic (or molecular) energy levels E_c, and E_a, $E_c - E_a = \hbar\omega_{ca}$ are extensively described in sections 1.1.2, 1.13 and 1.3.4. There the degeneracy due to different orientations is taken into account. Here we consider the simplest case of zero momentum, i.e. $j_c = j_a = 0$. Let us assume that group velocities of different waves are the same, i.e., $c_1 = c_2 = c$. The following definitions are convenient to be used:

$$\Pi_\pm(\omega) = [\Pi_a(\omega) \pm \Pi_c(\omega)],$$
$$R_3 = R_1 - R_2, \quad R_0 = R_1 + R_2.$$

Now the systems of equations (1.317), (1.3.20), and (1.3.21) can be exploited.

Raman resonance $(\omega_1 - \omega_2 \approx \omega_{ca})$

The generalised Maxwell-Bloch equations read:

$$i\left(\frac{\partial \mathcal{E}_1}{\partial z} + \frac{1}{c}\frac{\partial \mathcal{E}_1}{\partial t}\right) = -2\pi n_A k_1 \{\Pi^*_{ca}(-\omega_2) < R_{21} > \mathcal{E}_2 - \\ - \Pi_+(\omega_1) < R_0 > \mathcal{E}_1 - \Pi_-(\omega_1) < R_3 > \mathcal{E}_1\}, \tag{4.4.4a}$$

$$i\left(\frac{\partial \mathcal{E}_2}{\partial z} + \frac{1}{c}\frac{\partial \mathcal{E}_2}{\partial t}\right) = -2\pi n_A k_2 \{\Pi^*_{ca}(-\omega_1) < R^*_{21} > \mathcal{E}_1 - \\ - \Pi_+(\omega_2) < R_0 > \mathcal{E}_2 - \Pi_-(\omega_2) < R_3 > \mathcal{E}_2\}, \tag{4.4.4b}$$

$$\frac{\partial R_{21}}{\partial t} = i(\Delta\omega + \Delta\omega_{St})R_{21} + i\hbar^{-1}\Pi_{ca}(\omega_1)R_3\mathcal{E}_1\mathcal{E}_2^*, \tag{4.4.4c}$$

$$\frac{\partial R_3}{\partial t} = -2i\left(\hbar^{-1}\Pi_{ca}(\omega_1)R^*_{21}\mathcal{E}_1\mathcal{E}_2^* - c.c.\right), \tag{4.4.4d}$$

where $\Delta\omega = \omega_1 - \omega_2 - \omega_{ca}$ is frequency detuning and the high-frequency Stark shift determined by the formula $\Delta\omega_{St} = 2\Pi_-(\omega_1) | \mathcal{E}_1 |^2 + 2\Pi_-(\omega_2) | \mathcal{E}_2 |^2$. As usual, the corner brackets denotes the summation over all the resonant atoms. Let

$$\Pi_{ca}(\omega_1) = \Pi_{ca}(-\omega_2) = | \Pi_{ca}(\omega_1) | \exp(i\psi), \quad R_{21} = (1/2)p\exp[i\psi + i\Delta k_1 z - i\Delta k_2 z].$$

If one introduces $\tau = (t - z/c)t_0^{-1}$, $\zeta = z/L_0$, and $q_{1,2} = \sqrt{2}A^{-1}_{01,2}\mathcal{E}_{1,2}\exp[-i\Delta k_{1,2}z]$, then the system of equation (4.4.4) reduced to normalisation one

$$i\partial q_1/\partial\zeta = b_1 \langle R_3 \rangle q_1 - (1/2)\langle p \rangle q_2, \tag{4.4.5a}$$

$$i\partial q_2/\partial\zeta = b_2 \langle R_3 \rangle q_2 - (1/2)\langle p^* \rangle q_1, \tag{4.4.5b}$$

$$\partial p/\partial\tau = i(\delta + \delta_{St})p + iR_3 q_1 q_2^*, \tag{4.4.5c}$$

$$\partial R_3/\partial\tau = (i/2)\left(pq_1^* q_2 - p^* q_1 q_2^*\right), \tag{4.4.5d}$$

where $\delta_{St} = 2b_1 | q_1 |^2 + 2b_2 | q_2 |^2$ and $\delta = \Delta\omega t_0$. These equations were obtained under proposition that

$$\Delta k_1 = -2\pi k_1 n_A \Pi_+(\omega_1) < R_0 >, \ \Delta k_2 = -2\pi k_2 n_A \Pi_+(\omega_2) < R_0 >$$

$$A^2_{01} = k_1 | \Pi_{ca}(-\omega_2) | A_0^2, \qquad A^2_{02} = k_2 | \Pi_{ca}(-\omega_1) | A_0^2,$$

$$t_0 = \frac{\hbar}{| \Pi_{ca}(\omega_1) |^2 \quad k_1 k_2 A_0^2}, \quad L_0^{-1} = 2\pi n_A | \Pi_{ca}(\omega_2) | \quad k_1 k_2 .$$

Note that $|\Pi_{ca}(\omega_2)| = |\Pi_{ca}(\omega_1)|$. Normalised Stark-shift parameters b_1 and b_2 are defined as

$$b_1 = \sqrt{\frac{k_1}{k_2}}\frac{\Pi_-(\omega_1)}{|\Pi_{ca}(\omega_1)|}, \quad b_2 = \sqrt{\frac{k_2}{k_1}}\frac{\Pi_-(\omega_2)}{|\Pi_{ca}(\omega_2)|}$$

Two-photon resonance $(\omega_1 + \omega_2 \approx \omega_{ca})$

The generalised Maxwell-Bloch equations (1.3.20) and (1.3.17) in this case read

$$i\left(\frac{\partial \mathcal{E}_1}{\partial z} + \frac{1}{c}\frac{\partial \mathcal{E}_1}{\partial t}\right) = -2\pi n_A k_1 \{\Pi^*_{ca}(\omega_2) < R_{21} > \mathcal{E}^*_2 - \quad \text{(4.4.6a)}$$
$$- \Pi_+(\omega_1) < R_0 > \mathcal{E}_1 - \Pi_-(\omega_1) < R_3 > \mathcal{E}_1\},$$

$$i\left(\frac{\partial \mathcal{E}_2}{\partial z} + \frac{1}{c}\frac{\partial \mathcal{E}_2}{\partial t}\right) = -2\pi n_A k_2 \{\Pi^*_{ca}(\omega_1) < R_{21} > \mathcal{E}^*_1 - \quad \text{(4.4.6b)}$$
$$- \Pi_+(\omega_2) < R_0 > \mathcal{E}_2 - \Pi_-(\omega_2) < R_3 > \mathcal{E}_2\},$$

$$\frac{\partial R_{21}}{\partial t} = i(\Delta\omega + \Delta\omega_{St})R_{21} + i\hbar^{-1}\Pi_{ca}(\omega_1)R_3\mathcal{E}_1\mathcal{E}_2, \quad \text{(4.4.6c)}$$

$$\frac{\partial R_3}{\partial t} = -2i\left(\hbar^{-1}\Pi_{ca}(\omega_1)R^*_{21}\mathcal{E}_1\mathcal{E}_2 - c.c.\right), \quad \text{(4.4.6d)}$$

Let again $R_{21} = (1/2)p\exp[i\psi + i\Delta k_1 z + i\Delta k_2 z]$, and all other parameters are the same. The system of equations (4.4.6) becomes

$$i\partial q_1 / \partial\zeta = b_1\langle R_3\rangle q_1 - (1/2)\langle p\rangle q^*_2, \quad \text{(4.4.7a)}$$
$$i\partial q_2 / \partial\zeta = b_2\langle R_3\rangle q_2 - (1/2)\langle p\rangle q^*_1 \quad \text{(4.4.7b)}$$
$$\partial p / \partial\tau = i(\delta + \delta_{St})p + iR_3 q_1 q_2, \quad \text{(4.4.7c)}$$
$$\partial R_3 / \partial\tau = (i/2)\left(p q^*_1 q^*_2 - p^* q_1 q_2\right), \quad \text{(4.4.7d)}$$

where $\Delta\omega = \omega_1 + \omega_2 - \omega_{ca}$, and $\delta = \Delta\omega t_0$.

Steudel-Kaup equation

The Bloch equations under a two-photon resonance condition can be interpreted as usual Bloch equations in which an effective field is more complicated complex. The equations for this effective field can be obtained taking into account the reduced Maxwell equations.

Let us consider the case of Raman resonance at first. From equations (4.4.5a,b) we can find

$$\frac{\partial}{\partial \zeta}|q_1|^2 = \frac{i}{4}\left(\langle p\rangle S^* - \langle p^*\rangle S\right), \quad \frac{\partial}{\partial \zeta}|q_2|^2 = -\frac{i}{4}\left(\langle p\rangle S^* - \langle p^*\rangle S\right),$$

where $S = 2q_1 q_2^*$. Hence,

$$\frac{\partial}{\partial \zeta}\left(|q_1|^2 + |q_2|^2\right) = 0, \quad \frac{\partial}{\partial \zeta}\left(|q_1|^2 - |q_2|^2\right) = \frac{i}{2}\left(\langle p\rangle S^* - \langle p^*\rangle S\right).$$

Thus, if one introduces the new variables $A = \left(|q_1|^2 + |q_2|^2\right)$, $S_3 = \left(|q_1|^2 - |q_2|^2\right)$ then the system of equations (4.4.5) can be rewritten as

$$\begin{aligned}
&\partial p / \partial \tau = i\delta p + ib_+ Ap + ib_- pS_3 + iR_3 S, \\
&\partial R_3 / \partial \tau = (i/2)\left(pS^* - p^* S\right), \\
&\partial S_3 / \partial \zeta = (i/2)\left(\langle p\rangle S^* - \langle p^*\rangle S\right), \\
&\partial S / \partial \zeta = -i\langle p\rangle S_3 - ib_- \langle R_3\rangle S, \\
&\partial A / \partial \zeta = 0,
\end{aligned}$$

where $b_\pm = b_1 \pm b_2$. Let denote $S = S_1 - iS_2$ and $R = R_1 - iR_2$. Thus, these equations describe the interaction of two real vector fields: $\vec{S}$ and $\vec{R}$.

We can do the same manipulations in the case of two-photon resonance with the system of equations (4.4.7), too. Resulting equations may be compared with above equations that lead to the system of equations for quadratic combinations of the electric field envelope and Bloch vector

$$\partial S / \partial \zeta = -im\langle p\rangle S_3 - ib_- \langle R_3\rangle S, \tag{4.4.8a}$$

$$\partial S_3 / \partial \zeta = (i/2)\left(\langle p\rangle S^* - \langle p^*\rangle S\right), \tag{4.4.8b}$$

$$\frac{\partial p}{\partial \tau} = i\delta p + ib_+ Ap + ib_- pS_3 + iR_3 S, \tag{4.4.8c}$$

$$\partial R_3 / \partial \tau = (i/2)\left(pS^* - p^* S\right). \tag{4.4.8d}$$

In these equations we use the following parameters and values

$$A = \left(|q_1|^2 + m|q_2|^2\right), \quad S_3 = \left(|q_1|^2 - m|q_2|^2\right),$$

$$S = S_1 - iS_2 = (1-m)q_1 q_2 + (1+m)q_1 q_2^*,$$

and $b_{\pm} = b_1 \pm mb_2$, where $m = +1$ for RR and $m = -1$ for TPR.

It is seen from equations (4.4.8) that

$$\frac{\partial}{\partial \tau}\left(R_1^2 + R_2^2 + R_3^2\right) = 0, \quad \frac{\partial}{\partial \zeta}\left(mS_1^2 + mS_2^2 + S_3^2\right) = 0.$$

Besides it follows from equations (4.4.5a,b) and (4.4.7a,b) that $\partial A / \partial \zeta = 0$. If we know the solution of equations (4.4.8) for quadratic combinations S_j, (j = 1, 2, 3) of electric field envelopes, then q_1, and q_2, can be found from equations (4.4.7a,b) for TPR and from equations (4.4.5a,b) for RR.

The quantities $< R >$ and $< p >$ are connected with the quantities S_j by the following expressions:

$$p = -mS_3^{-1}\left(b_- R_3 S - i\partial S / \partial \zeta\right), \qquad \partial R_3 / \partial \tau = \partial S_3 / \partial \zeta. \tag{4.4.9}$$

We shall consider only the easiest case of the sharp-line approximation and $\delta = 0$. Let us make the following transformation:

$$p^* = r_+ \exp\left(-ib_+ \int_0^{\tau} A(\tau')d\tau'\right), \quad R_3 = r_3, \quad S^* = Bs_+ \exp\left(-ib_+ \int_0^{\tau} A(\tau')d\tau'\right), \quad S_3 = Bs_3,$$

where $B(\tau)^2 = m(S_1^2 + S_2^2) + S_3^2$. By using the new independent variable T, where

$$T = \int_0^{\tau} B(\tau')\, d\tau',$$

we can obtain from equations (4.4.8) the Steudel-Kaup equations of a two-photon SIT (TPSIT) theory,

$$\frac{\partial s_+}{\partial \zeta} = imr_+ s_3 + igr_3 s_+, \quad \frac{\partial s_3}{\partial \zeta} = \frac{i}{2}\left(r_+^* s_+ - r_+ s_+^*\right), \tag{4.4.10a}$$

$$\frac{\partial r_+}{\partial T} = igr_+ s_3 + ir_3 s_+, \quad \frac{\partial r_3}{\partial T} = -\frac{i}{2}\left(r_+^* s_+ - r_+ s_+^*\right), \tag{4.4.10b}$$

where we redefine $g = -b_-$. For the real quantities r_n and s_n the following relationships hold:

$$r_1^2 + r_2^2 + r_3^2 = 1, \qquad m(s_1^2 + s_2^2) + s_3^2 = 1. \tag{4.4.11}$$

The generalised reduced Maxwell-Bloch equations for TPSIT on the transitions 1/2→1/2 and 1↔0 may also be transformed to equations (4.4.10) with the certain limitations on the parameters of the two-photon interaction operator. A difference should be noted between the model in question and the one examined by Steudel [53]. Though the outward form of equations (4.4.10) coincides exactly with Steudel-Kaup equations, the field of application of equations (4.4.10) is much wider. It covers the case of differently polarised pump pulses. Formally, this is displayed by a mismatch of the quantities B^2 and A^2 in equation (4.4.10), as happens in the case of plane polarised pump pulses.

The properties of equations (4.4.10) were thoroughly analysed by Steudel [48, 53], Kaup [54], and Meinel [55]. In this section we only underline that the homogeneous solutions of equations (4.4.10), $mr_3 = s_3 = 1$, $r_+ = s_+ = 0$, are stable.

4.4.2. STEADY-STATE SOLITARY WAVES TWO-PHOTON ABSORPTION

When the carrier waves frequencies are equal ($\omega_1 = \omega_2 = \omega$), they say that the *degenerated two-photon absorption* takes place. The normalised system of equations, describing the propagation of USP can be written as:

$$i\partial q/\partial\zeta = b\langle R_3\rangle q - (1/2)\langle p\rangle q^* , \tag{4.4.12a}$$

$$\partial p/\partial\tau = i(\delta + \delta_{St})p + iR_3 q^2 , \tag{4.4.12b}$$

$$\partial R_3/\partial\tau = (i/2)\left(pq^{*2} - p^* q^2\right), \tag{4.4.12c}$$

where $q = A_0^{-1}\mathcal{E}\exp[i2\pi k n_A \Pi_+(\omega_0) < R_0 > z]$ is normalised complex envelope of the USP, $\delta_{St} = 2b|q|^2$ is Stark shift of the resonant levels, and other parameters were defined in the section 4.4.1.

Here we will consider the theory by Belenov-Poluektov [42], where the stationary pulse of light was determined. This pulse is alike to conserving solitary wave with the zero asymptotic, propagating at exact two-photon resonance with the narrow absorption line, i.e., at $\delta = 0$. In the following works [43-52] other solutions were found, amongst which the solution by Belenov-Poluektov is presented as a specific case.

From the equations in complex variable we can change to equations for real variables determined by the following formulae $q = a\exp(i\phi)$, $p = (u_1 + iu_2)\exp(2i\phi)$, and $R_3 = u_3$. We assume that $\delta = 0$ an the inhomogeneous broadening is absent. System (4.4.12) results in the following system of the real equations

$$\partial a/\partial\zeta = -(1/2)au_2, \tag{4.4.13a}$$

$$\partial\phi/\partial\zeta = -bu_3 + (1/2)u_1, \tag{4.4.13b}$$

$$\partial u_1/\partial\tau = \left(2\partial\phi/\partial\tau - 2ba^2\right)u_2, \tag{4.4.13b}$$

$$\partial u_2/\partial\tau = -\left(2\partial\phi/\partial\tau - 2ba^2\right)u_1 + u_3 a^2, \tag{4.4.13c}$$

$$\partial u_3/\partial\tau = -u_2 a^2. \tag{4.4.13d}$$

By differentiating equation (4.4.13b) with respect to τ and using equations (4.4.13b) and (4.4.13d) for "instant frequency" $w = (\partial\phi / \partial\tau)$, we get

$$\partial w / \partial\zeta = wu_2 + ba^2 u_2 .$$

Accounting equation (4.4.13a), we can obtain then:

$$dw / da^2 = -wa^{-2} + b .$$

This equation has a general solution $w(a^2) = C_0 a^{-2} + ba^2 / 2$. Constant of integrating must be chosen by the assumption that the instant frequency is limited for all values of the USP envelope amplitude including those when $a \to 0$. So that

$$\partial\phi / \partial\tau = ba^2 / 2 . \tag{4.4.14}$$

Let us substitute the expression (4.4.14) in the Bloch equations. In terms of the angle variable ϑ determined as

$$\vartheta = \int_{-\infty}^{\tau} a^2(\tau')d\tau' .$$

Bloch equations (4.4.13c-3.5.13d) can be written in the form of the system of ordinary equations

$$du_1 / d\vartheta = -bu_2, \quad du_2 / d\vartheta = bu_1 + u_3, \quad du_3 / d\vartheta = -u_2,$$

with the initial conditions $u_1(\vartheta = 0) = u_2(\vartheta = 0) = 0,\ u_3(\vartheta = 0) = 1$ that correspond to the absorbing medium. Solution of this system can be found by standard methods in the following form

$$u_1(\theta) = -b(1+b^2)^{-1}[1-\cos(\theta)],\ u_2(\theta) = \sqrt{1+b^2}\sin(\theta)\ \ u_3(\theta) = (1+b^2)^{-1}[b^2 + \cos(\theta)],$$

where $\theta(\tau,\zeta) = \sqrt{1+b^2}\,\vartheta(\tau,\zeta)$ is the Bloch angle in the two-photon resonance case. Now the equation (4.4.13a) leads to an equation for this principal variable

$$\partial^2\theta / \partial\zeta\partial\tau = -(1+b^2)^{-1/2}\sin(\theta)\partial\theta / \partial\tau . \tag{4.4.15}$$

By integrating equation (4.4.15), we can obtain

$$\partial\theta / \partial\zeta = -(1+b^2)^{-1/2}[1-\cos(\theta)], \tag{4.4.16}$$

where the integrating constant was determined from the condition of the vanishing the USP envelope as $t \to -\infty$. If to determine Ψ, named a pulse area [42,47-51] (it is more precise to speak about energy), by the expression

$$\Theta(\zeta) = \sqrt{1+b^2} \int_{-\infty}^{+\infty} a^2(\tau,\zeta) d\tau ,$$

then equation (4.4.16) leads to an expression which is similar to the area theorem by McCall-Hahn in the one-photon SIT theory.

$$d\Theta / d\zeta = -(1+b^2)^{-1/2} [1 - \cos\Theta]. \tag{4.4.17}$$

This area theorem for two-photon resonance shows that pulses may be steady-state when their area Θ is an integer multiple of 2π, but unlike one-photon SIT, these pulses are generally unstable. A small distortion of the $2\pi N$-pulse causing the decrease of its area converts it into $2\pi(N\text{-}1)$-pulse.

By integrating equation (4.4.16), one can obtain

$$(1+b^2)^{-1/2} \zeta = \cot(\theta/2) - \cot(\theta_0/2),$$

where $\theta_0 = \theta(\tau, \zeta = 0)$. Hence, we get the exact solution of the equation (4.4.15).

$$\theta(\tau,\varsigma) = 2 \operatorname{arccot}\left[(1+b^2)^{-1/2} \zeta + \cot(\theta_0/2)\right]. \tag{4.4.18}$$

Turning back to the initial variables, it is possible to get a square of envelope amplitude of the USP from (4.4.18):

$$a^2(\tau,\zeta) = \frac{a^2(\tau,\zeta=0)}{\sin^2(\theta_0(\tau)/2) + \left[\cos(\theta_0(\tau)/2) + (1+b^2)^{-1/2} \zeta \sin(\theta_0(\tau)/2)\right]^2} . \tag{4.4.19}$$

The steady-state pulse can be derived either by solving equations (4.4.13), having assumed that all variables depend only on $\eta = \zeta - u_0 \tau$, or by using the relation (4.4.18). To do this one should change the argument of the function in the right part of (4.4.18) as

$$\theta(\tau,\varsigma) = 2 \operatorname{arccot}\left[(1+b^2)^{-1/2} (\varsigma - u_0 \tau)\right].$$

It is possible if

$$\cot(\theta_0(\tau)/2) = -\left((1+b^2)^{-1/2} u_0\right)\tau .$$

Hence, it follows from the initial value $\theta_0 = \theta_0(\tau)$, which provides propagation of the USP without altering its form:

$$a^2(\tau,\zeta=0) = \frac{2u_0}{(1+b^2)\left[1+\left(u_0\tau/\sqrt{1+b^2}\right)^2\right]}. \tag{4.4.20}$$

So if the "Lorentzian pulse" (4.4.20), under the conditions considered here, enter non-linear medium it does not change its Lorentzian form in the course of his further propagation in the medium.

$$a^2(\tau,\zeta) = \frac{2u_0}{(1+b^2)\left[1+\left((\zeta-u_0\tau)/\sqrt{1+b^2}\right)^2\right]}. \tag{4.4.21}$$

Having substituted (4.4.21) in the integral defining the pulse area Θ, we can find that Θ is equal to 2π. So they call this Lorentzian USP a 2π-pulse, too. Under the action of this USP a Bloch vector rotates through 2π, i.e., the state of a resonant medium after the passing of such a pulse returns into the initial one. Numerical study of the collisions of 2π-pulses [115] showed that, though the total area of pulses stayed unaltered, the shapes of pulses after the collision changed. For instance, one of the pulses experienced auto-compression. It should be pointed out that the formula (4.4.19) gives an analytical description of this process.

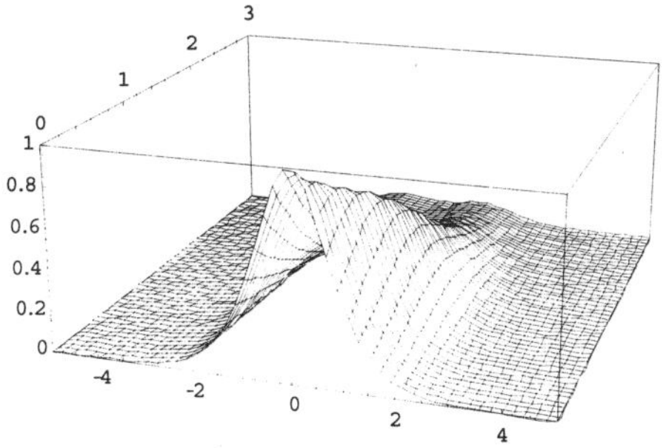

Fig.4.4.1.

Evolution of the 0.8 π-pulse.

The illustration of the pulse propagation under the THR condition is shown in Figs.4.4.1-4.4.3. We assume that the initial pulse is characterised by the sech-envelope

$$a(\tau,\zeta=0) = a_0\mathrm{sech}(\tau/\tau_0).$$

In this case we have

$$\theta_0(\tau) = a_0^2\tau_0\sqrt{1+b^2}\,(1+\tanh(\tau/\tau_0)).$$

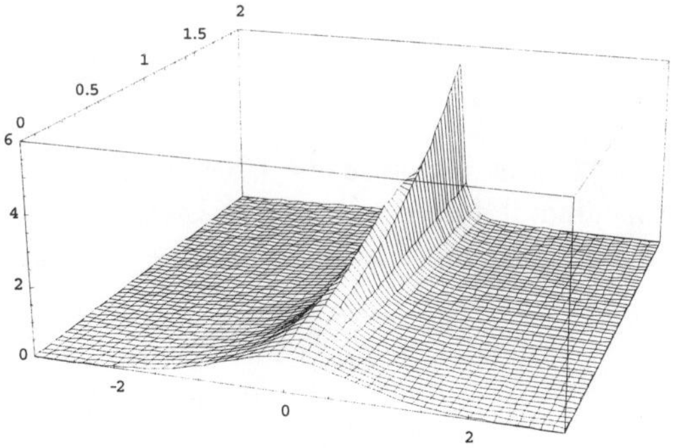

Fig.4.4.2.

Evolution of the 2.5 π-pulse

Let $x = \tau/\tau_0$, $y = \zeta(1+b^2)^{-1/2}$, $\vartheta_0 = a_0^2\tau_0\sqrt{1+b^2}$, and $w(x,y) = a^2(\tau,\zeta)/a_0^2$. Then

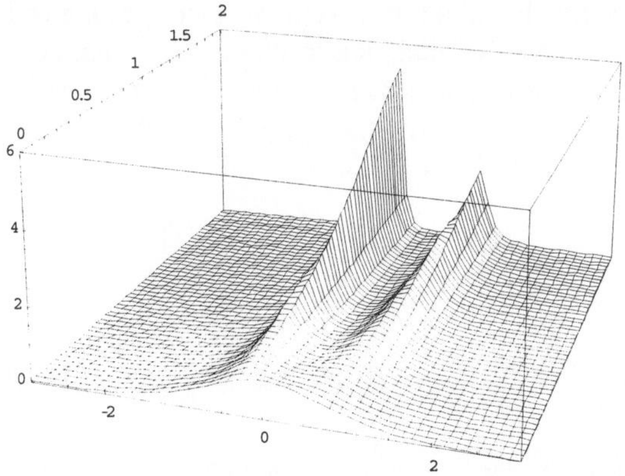

Fig.4.4.3.

Evolution of the 4.1 π-pulse

the expression (4.4.10) can be re-written as

$$w(x,y) = \frac{\operatorname{sech}^2(x)}{\sin^2(\vartheta_0(1+\tanh x)/2) + [\cos(\vartheta_0(1+\tanh x)/2) + y\sin(\vartheta_0(1+\tanh x)/2)]^2}.$$

The Fig.4.4.2 show that the USP decreases along co-ordinate y at $\vartheta < \pi$. The initial pulse collapses when the parameter $2\pi < \vartheta < 4\pi$, i.e. the pulse duration decreases USP propagates in the medium (see Fig.4.4.2). The initial USP transforms into two-peaked pulse if $4\pi < \vartheta < 6\pi$ as it is shown in Fig.4.4.3.

4.4.3. ZERO-CURVATURE REPRESENTATION OF THE TPSIT EQUATIONS

Of course, one may obtain the zero-curvature representation of the TPSIT equations by means of the empirical McLaughlin-Corones method [116]. But in order to show the capability of the Wahlquist-Estabrook method [108,109,117-119] we shall try to find the form of the zero-curvature representation of the equations (4.4.10) following Wahlquist and Estabrook.

Let us consider the pair of linear equations of the IST method,

$$\partial\psi / \partial T - \hat{U}\psi = 0, \quad \partial\psi / \partial\zeta - \hat{V}\psi = 0\,, \tag{4.4.22}$$

The condition of integrability for (4.4.22) gives

$$\partial\hat{U} / \partial\zeta - \partial\hat{V} / \partial T + [\hat{U}, \hat{V}] = 0\,. \tag{4.4.23}$$

We shall take the matrices $\hat{U}$ and $\hat{V}$ in the form

$$\hat{V} = \hat{\mathbf{X}}_0 + r_+\hat{\mathbf{X}}_+ + r_-\hat{\mathbf{X}}_- + r_3\hat{\mathbf{X}}_3, \quad \hat{U} = \hat{\mathbf{Y}}_0 + s_+\hat{\mathbf{Y}}_+ + s_-\hat{\mathbf{Y}}_- + s_3\hat{\mathbf{Y}}_3\,, \tag{4.4.24}$$

where the $\hat{\mathbf{X}}$'s and $\hat{\mathbf{Y}}$'s are constant matrices.

Substituting (4.4.24) into (4.4.23) and equating the coefficients of dependent variables one can obtain the commutation relations for the $\hat{\mathbf{X}}$'s and $\hat{\mathbf{Y}}$'s,

$$\begin{gathered}
[\hat{\mathbf{X}}_0, \hat{\mathbf{Y}}_0] = [\hat{\mathbf{X}}_0, \hat{\mathbf{Y}}_+] = [\hat{\mathbf{X}}_0, \hat{\mathbf{Y}}_-] = [\hat{\mathbf{X}}_0, \hat{\mathbf{Y}}_3] = [\hat{\mathbf{X}}_+, \hat{\mathbf{Y}}_0] = [\hat{\mathbf{X}}_+, \hat{\mathbf{Y}}_+] = 0, \\
[\hat{\mathbf{X}}_+, \hat{\mathbf{Y}}_-] = i(\hat{\mathbf{X}}_3 + \hat{\mathbf{Y}}_3)/2, \quad [\hat{\mathbf{X}}_+, \hat{\mathbf{Y}}_3] = -i\theta\hat{\mathbf{Y}}_+ - ib_-\hat{\mathbf{X}}_+\,, \\
[\hat{\mathbf{X}}_-, \hat{\mathbf{Y}}_0] = 0, \quad [\hat{\mathbf{X}}_-, \hat{\mathbf{Y}}_+] = -i(\hat{\mathbf{X}}_3 + \hat{\mathbf{Y}}_3)/2, \quad [\hat{\mathbf{X}}_-, \hat{\mathbf{Y}}_-] = 0, \\
[\hat{\mathbf{X}}_-, \hat{\mathbf{Y}}_3] = i\theta\hat{\mathbf{Y}}_- + b_-\hat{\mathbf{X}}_-\,, \quad [\hat{\mathbf{X}}_3, \hat{\mathbf{Y}}_+] = ib_-\hat{\mathbf{Y}}_+ + i\hat{\mathbf{X}}_+\,, \\
[\hat{\mathbf{X}}_3, \hat{\mathbf{Y}}_0] = 0, \quad [\hat{\mathbf{X}}_3, \hat{\mathbf{Y}}_-] = -ib_-\hat{\mathbf{Y}}_- - i\hat{\mathbf{X}}_-\,, \quad [\hat{\mathbf{X}}_3, \hat{\mathbf{Y}}_3] = 0\,.
\end{gathered} \tag{4.4.25}$$

It is seen that with the assumptions

$$\hat{\mathbf{X}}_0 = \hat{\mathbf{Y}}_0 = 0\,, \quad \hat{\mathbf{X}}_\pm = \alpha_\pm\hat{\mathbf{I}}_\pm\,, \quad \hat{\mathbf{Y}}_\pm = \beta_\pm\hat{\mathbf{I}}_\pm\,, \quad \hat{\mathbf{X}}_3 = \alpha_3\hat{\mathbf{I}}_3\,, \quad \hat{\mathbf{Y}}_3 = \beta_3\hat{\mathbf{I}}_3$$

and requiring

$$\begin{aligned}
&i\alpha_+\beta_- = -(\alpha_3 + \beta_3)/2\,, && i\alpha_+\beta_3 = -(b_-\alpha_+ + \theta\beta_+)/2\,, \\
&i\alpha_-\beta_+ = -(\alpha_3 + \beta_3)/2\,, && i\alpha_-\beta_3 = -(b_-\alpha_- + \theta\beta_-)/2\,, \\
&i\alpha_3\beta_+ = -(\alpha_+ + b_-\beta_+)/2\,, && i\alpha_3\beta_- = -(\alpha_- + \theta\beta_-)/2\,,
\end{aligned} \tag{4.4.26}$$

one may obtain from (4.4.25) the structural equations of the sl(2, **R**) Lie algebra,

$$[\hat{\mathbf{I}}_+,\hat{\mathbf{I}}_-]=\hat{\mathbf{I}}_3\,,\quad [\hat{\mathbf{I}}_+,\hat{\mathbf{I}}_3]=-2\hat{\mathbf{I}}_+\,,\quad [\hat{\mathbf{I}}_-,\hat{\mathbf{I}}_3]=2\hat{\mathbf{I}}_-\,.$$

Choosing β_3 as a spectral parameter, i.e. $\beta_3 = i\lambda$, then relations (4.4.26) yield

$$\beta_+ = -(if+\lambda)\,,\quad \beta_- = -m(if-\lambda)\,,$$
$$\alpha_+ = -\frac{m(if+\lambda)}{(2\lambda+g)}\,,\quad \alpha_- = -\frac{(if-\lambda)}{(2\lambda+g)}\,,\quad \alpha_3 = -\frac{i}{2}\left[g-\frac{m}{(2\lambda+g)}\right],$$

where $f^2=(m-g^2)/4$, $g=-b_+$. A two-dimensional representation of sl(2, **R**) is realised by the matrices

$$\hat{\mathbf{I}}_+ = \begin{pmatrix} 0 & 1 \\ 0 & 0 \end{pmatrix},\quad \hat{\mathbf{I}}_- = \begin{pmatrix} 0 & 0 \\ 1 & 0 \end{pmatrix},\quad \hat{\mathbf{I}}_3 = \begin{pmatrix} 1 & 0 \\ 0 & -1 \end{pmatrix}. \tag{4.4.27}$$

Thus we can obtain the linear equations of the IST method,

$$\frac{\partial\psi_1}{\partial T} = -i\lambda s_3\psi_1 + (if+\lambda)s_+\psi_2,\quad \frac{\partial\psi_2}{\partial T} = m(if-\lambda)s_-\psi_1 + i\lambda s_3\psi_2, \tag{4.4.28a}$$

and

$$\begin{aligned} \frac{\partial\psi_1}{\partial\zeta} &= \frac{i}{2}\left(g-\frac{m}{2\lambda+g}\right)r_3\psi_1 + \frac{m(if+\lambda)}{2\lambda+g}r_+\psi_2, \\ \frac{\partial\psi_2}{\partial\zeta} &= \frac{(if-\lambda)}{2\lambda+g}r_-\psi_1 - \frac{i}{2}\left(g-\frac{m}{2\lambda+g}\right)r_3\psi_2, \end{aligned} \tag{4.4.28b}$$

Equations (4.4.28) provide a zero-curvature representation of the TPSIT equations (4.4.10).

It is noteworthy that the zero-curvature representation (4.4.28) constructed by Steudel cannot be directly reduced to the one found by Maimistov and Manykin [119]. This is the result of the ambiguity of both the introduction of the constraints (4.4.26) and the choice of the spectral parameter. Nevertheless, according to a general result all spectral problems of dimension 2×2 are gauge equivalent to the Zakharov-Shabat spectral problem. With respect to the problem under consideration, this equivalence was demonstrated by Kaup [54].

To reduce the system of the equations (4.4.28) to the Zakharov-Shabat form let us introduce new independent variables

$$s_3\cos\beta,\quad s_+ = \exp(i\alpha)\sin\beta,\quad s_- = m\exp(-i\alpha)\sin\beta, \tag{4.4.29}$$

where β is a real quantity for $m = +1$ (RR) and imaginary for $m = -1$ (TPR). To be definite we shall assume that $0 \le \beta \le \pi$ for $m = +1$ and $i\beta \ge 0$ for $m = -1$.

Transform the variables $\hat{\psi} = \text{colon}(\psi_1, \psi_2)$ of the spectral problem (4.4.28a) by the following rotation:

$$\hat{\psi}' = \hat{\Omega}^{-1}(\alpha)\hat{\psi}, \qquad \hat{\Omega}(\alpha) = \begin{pmatrix} \exp(i\alpha/2) & 0 \\ 0 & \exp(-i\alpha/2) \end{pmatrix} \tag{4.4.30}$$

to eliminate the factor $\exp(\pm i\alpha)$. Then the system (4.4.28a) takes the form

$$\frac{\partial \hat{\psi}'}{\partial T} = -i\lambda \begin{pmatrix} \cos\beta & i\sin\beta \\ -i\sin\beta & -\cos\beta \end{pmatrix} \hat{\psi}' + \begin{pmatrix} -\dfrac{i}{2}\dfrac{\partial\alpha}{\partial T} & if\sin\beta \\ if\sin\beta & \dfrac{i}{2}\dfrac{\partial\alpha}{\partial T} \end{pmatrix} \hat{\psi}'$$

Rotation over the angle β,

$$\hat{\psi}'' = \hat{\Omega}(\beta)\hat{\psi}', \qquad \hat{\Omega}(\beta) = \begin{pmatrix} \cos\beta/2 & i\sin\beta/2 \\ i\sin\beta/2 & \cos\beta/2 \end{pmatrix} \tag{4.4.31}$$

yields

$$\frac{\partial \hat{\psi}''}{\partial T} = \begin{pmatrix} -i\lambda - \dfrac{i}{2}\dfrac{\partial\alpha}{\partial T}\cos\beta & \left(if - \dfrac{1}{2}\dfrac{\partial\alpha}{\partial T}\right)\sin\beta + \dfrac{i}{2}\dfrac{\partial\beta}{\partial T} \\ \left(if + \dfrac{1}{2}\dfrac{\partial\alpha}{\partial T}\right)\sin\beta + \dfrac{i}{2}\dfrac{\partial\beta}{\partial T} & i\lambda + \dfrac{i}{2}\dfrac{\partial\alpha}{\partial T}\cos\beta \end{pmatrix} \hat{\psi}''.$$

A third rotation,

$$\hat{\psi}''' = \hat{\Omega}^{-1}(\gamma)\hat{\psi}'', \qquad \hat{\Omega}(\gamma) = \begin{pmatrix} \exp(i\gamma/2) & 0 \\ 0 & \exp(-i\lambda/2) \end{pmatrix} \tag{4.4.32}$$

reduces the spectral problem to the Zakharov-Shabat form,

$$\frac{\partial \hat{\psi}'''}{\partial T} = \begin{pmatrix} -i\lambda & q \\ r & i\lambda \end{pmatrix} \hat{\psi}''', \tag{4.4.33}$$

where

$$q = \left[\left(if - (1/2)\partial\alpha/\partial T\right)\sin\beta + (i/2)\partial\beta/\partial T\right]\exp(i\gamma), \tag{4.4.34}$$

$$r = \left[\left(if + (1/2)\partial\alpha/\partial T\right)\sin\beta + (i/2)\partial\beta/\partial T\right]\exp(-i\gamma), \tag{4.4.35}$$

if the equality $\partial\gamma / \partial T = (\partial\alpha / \partial T)\cos\beta$ holds. Utilising (4.4.29) and (4.4.10) one can show that

$$\partial\gamma / \partial T = \rho_+ / 2\sin\beta \ , \qquad \rho_\pm = r_- \exp(i\alpha) \pm m r_+ \exp(-i\alpha) \ .$$

Note, that $\partial\beta / \partial\zeta = -(i/2)\rho_-$ and

$$\partial\alpha / \partial\zeta = g r_3 + (1/2)\rho_+ \tan^{-1}\beta, \quad \partial\rho_+ / \partial\zeta = i\rho_-\left(\partial\alpha / \partial T - g\cos\beta\right).$$

In the RR case and when f is real, then $q = -r^*$.

Another auxiliary linear equation following from (4.4.28b) has the form

$$\frac{\partial\hat{\psi}'''}{\partial\zeta} = (Y_1\hat{s}_1 + Y_2\hat{s}_2 + Y_3\hat{s}_3)\hat{\psi}''' = \hat{Y}\hat{\psi}''' ,$$

where

$$Y_1 = -i\sin\gamma\left[\frac{1}{2}\sin\beta\left(\rho_+\tan^{-1}\beta + \frac{m r_3}{2\lambda + g}\right) + \cos\beta\frac{if\rho_- - \lambda\rho_+}{2(2\lambda + g)}\right] + \cos\gamma\frac{if\rho_+ - \lambda\rho_-}{2(2\lambda + g)} ,$$

$$Y_2 = -i\cos\gamma\left[\frac{1}{2}\sin\beta\left(\rho_+\tan^{-1}\beta + \frac{m r_3}{2\lambda + g}\right) + \cos\beta\frac{if\rho_- - \lambda\rho_+}{2(2\lambda + g)}\right] - \sin\gamma\frac{if\rho_+ - \lambda\rho_-}{2(2\lambda + g)} ,$$

$$Y_3 = \frac{i\rho_+}{4\sin\beta} - \frac{i}{2}\cos\beta\left(\rho_+\tan^{-1}\beta + \frac{m r_3}{2\lambda + g}\right) + i\sin\beta\frac{if\rho_- - \lambda\rho_+}{2(2\lambda + g)} ,$$

$\hat{s}_i$ are the Pauli matrices.

We stress that equations (4.4.33) differ from the usual Zakharov-Shabat spectral problem by the fact that the variable T varies on a finite interval from 0 to T_{max}:

$$T_{max} = \int_0^\infty B(\tau')d\tau' .$$

The details of this problem were examined by Kaup [54].

4.4.4. A SINGLE-SOLITON SOLUTION OF THE TPSIT EQUATIONS

In comparison with one-photon and double resonances the TPSIT theory and its polarisation aspects in particular have not been sufficiently elaborated yet. One of the reasons for this is the complex form of the transformation (4.4.29), (4.4.33)-(4.4.35) reducing the initial spectral problem to Zakharov-Shabat's form. The other reason is associated with the bilinearity of the functions desired in (4.4.10) over the field amplitudes (4.4.1). So to obtain the pulse shapes (4.4.1) in TPSIT one should in addition solve equations.

(4.4.4a,b) or (4.4.5a,b) for the transition $0 \to 0$ or analogous equations for the transitions $1 \leftrightarrow 0$ and $1/2 \to 1/2$. It is clear from the very form of the spectral problem (4.4.28a) that the inverse scattering problem has a solution only when S_3 does not vanish when $\tau \to \pm\infty$. Thereby a TPSIT soliton is a USP of one frequency on the background of another frequency. There are evidently no restrictions on the mutual orientations of the polarisation vectors of the pulses. We will not dwell on the TPSIT polarisation effects any longer but just quote the results for the fundamental equations (4.4.10). From our point of view these results are quite sufficient for further considerations following the philosophy of section 4.2. They were obtained by Steudel [53]. A single-soliton solution of the equations (4.4.10) is written in the form [53]:

– RR ($m = +1$) for $g^2 < 1$,

$$(1+s_3)\sigma^{-1} = (1+r_3)\vartheta^{-1} = 2\sin 2W[f(\cosh 4\mu + \cosh 2W)]^{-1},$$

$$\alpha = 2\xi T + (2l-g)\zeta + \tan^{-1}\left(\frac{\sigma - f\tan W}{\xi}\tanh 2\mu\right),$$

$$\nu = 2\xi T + (2l-g)\zeta + \tan^{-1}\left(\frac{\vartheta - f\tan W}{g-2l}2\tanh 2\mu\right) + \frac{\pi}{2}\operatorname{sign}(2lf - \vartheta g\tanh W),$$

$$\tanh 2W = 2f\sigma(\xi^2 + \sigma^2 + f^2)^{-1};$$

– for $g^2 > 1$ ($f = -i\phi$, $\operatorname{Im}\phi = 0$),

$$(1+s_3)\sigma^{-1} = (1+r_3)\vartheta^{-1} = 2\sin 2W[\phi(\cosh 4\mu + \cos 2W)]^{-1},$$

$$\alpha = 2\xi T + (2l-g)\zeta + \tan^{-1}\left(\frac{\sigma + (\xi+\phi)\tan W}{\xi + \phi - \sigma\tan W}\tanh 2\mu\right),$$

$$\nu = 2\xi T + (2l-g)\zeta + \tan^{-1}\left(\frac{(2l-g-2\phi)\tan W - 2\vartheta}{2\vartheta\tan W - 2\phi - g + 2l}2\tanh 2\mu\right) +$$

$$+\frac{\pi}{2}\operatorname{sign}(2l\phi - \vartheta g\tan W),$$

$$\tanh 2W = 2\phi\sigma(\phi^2 - \xi^2 - \sigma^2)^{-1}, \quad 0 \le W \le \pi/2;$$

– TPR ($m = -1, f = -i$),

$$(s_3 - 1)\sigma^{-1} = (1+r_3)\vartheta^{-1} = 2\sin 2W[\phi(\cosh 4\mu + \cos 2W)]^{-1},$$

$$\alpha = -2\xi T - (2l+g)\zeta - \tan^{-1}\left(\frac{\sigma + (\xi-\phi)\tan W}{\xi - \phi - \sigma\tan W}\tanh 2\mu\right),$$

$$\nu = -2\xi T - (2l+g)\zeta - \tan^{-1}\left(\frac{2\vartheta + (2\phi - g - 2l)\tan W}{g - 2\phi + 2l - 2\vartheta \tan W} 2\tanh 2\mu\right) +$$

$$+\frac{\pi}{2}[\text{sign}(2l\phi + \vartheta g \tan W) - 1],$$

$$\tanh 2W = 2\phi\sigma(\phi^2 - \xi^2 - \sigma^2)^{-1}, \quad 0 \le W \le \pi/2\,;$$

Everywhere above the $\tan^{-1}$ branch belongs to the interval $(-\pi/2, \pi/2)$. The quantity s_+ is determined by equation (4.4.29) and $r_+ = i(1 - r_3^2)^{1/2}\exp(i\nu)$. Besides, ξ and σ are the real and imaginary parts of λ, $\lambda = \xi + i\sigma$, from the discrete spectrum. The other notations are

$$\mu = \sigma T - \vartheta\zeta, \quad l = \frac{(2\xi + g)}{2[4\sigma^2 + (g + 2\xi)^2]}, \quad \vartheta = \frac{\sigma}{[4\sigma^2 + (g+2\xi)^2]}.$$

It should be emphasised following [54] that IST method leads to generalised Zakharov and Shabat problem (4.4.33). Typical of this problem is that the bound state eigenvalues will not be stationary. These eigenvalues will be at the poles of the scattering data $\rho(\lambda) = b(\lambda)/a(\lambda)$, and since the magnitude of the $b(\lambda)/a(\lambda)$ is ζ dependent, we may expect any pole to move as ζ varies. It may even be possible for new poles to appear and old poles to disappear.

Remark
Equations describing propagation of the USP under conditions of exact two-photon resonance in the absence of inhomogeneous broadening of the absorbing lines can be transcribed in the form which was named "unified". To do this it is necessary to redenote a complex variable as follows:

$$p = R_1 + iR_2, q^2 = S = S_1 + iS_2, |q|^2 = S_3$$

In these variables the system of equations (4.4.12) looks as

$$\partial p/\partial\tau = iR_3 S + 2ibRS_3, \qquad \partial R_3/\partial\tau = (i/2)(pS^* - p^*S),$$

$$\partial S/\partial\zeta = 2ibR_3 S - pS_3, \qquad \partial S_3/\partial\zeta = (i/2)(pS^* - p^*S).$$

It should be noted that pseudo-spin vector (S_1, S_2, S_3) has a zero length. In the case of non-degenerated two-photon resonance, its length is not a zero. These equations have a structure of equations by Kaup and Steudel and this is the reason to consider the IST method to be applicable here as it is in the general case.

4.5. Relation between SIT-theory and field-theoretical models

One of the fundamental features of the completely integrable evolution equation (or system of equations) is their nontrivial relationship with each other named the gauge equivalence. The origin of this term is connected with the geometric interpretation of the linear equations of IST method. Now it is useful to make some remarks to consider this interpretation [120,121].

The differential operators in the equations (4.2.13) may be considered as the operators of the covariant differentiation, so matrices $\hat{U}$ and $\hat{V}$ are understood as the local coefficients of connectivity in the trivial fiber bundle $R^2 \times C^N$, where Euclidean space-time R^2 is the basic manifold, and $\Psi(\tau,\zeta)$ takes the values in a typical fiber C^N, which is an N-dimension complex plane. Equation (4.2.14) means that this connection has a zero curvature. The transformation of the fiber variables, under which the curvature does not change (in the general case the curvature tensor must transform as the covariant one), was named the *gauge transformation.*

If the local transformation in fiber is executed by means of matrix functions $g(\tau,\zeta)$

$$\Psi(\lambda;\tau,\zeta) \to \Psi'(\lambda;\tau,\zeta) = g^{-1}(\tau,\zeta)\Psi(\lambda;\tau,\zeta),$$

then in order to conserve the equations (4.2.13) in the form

$$\partial\Psi'/\partial T - \hat{U}'\Psi' = 0, \quad \partial\Psi'/\partial\zeta - \hat{V}'\Psi' = 0 \tag{4.5.1}$$

the U-V-matrices should be transformed by the formulae

$$\hat{U}' = g^{-1}\hat{U}g - g^{-1}\partial g/\partial\tau, \quad \hat{V}' = g^{-1}\hat{V}g - g^{-1}\partial g/\partial\zeta. \tag{4.5.2}$$

The direct calculations show that the zero-curvature condition (4.2.14) transforms into the following relation

$$\partial\hat{U}'/\partial\zeta = \partial\hat{V}'/\partial\tau + [\hat{V}', \hat{U}']. \tag{4.5.3}$$

The system of nonlinear evolution equations, which may be solved by the IST method on the base of the linear equations (4.2.13), follows from (4.2.14). Exactly in the same way the general case of another integrable system, for which IST method is based on the (4.5.1), follows from (4.5.3). The obtained systems of nonlinear equations are named *gauge equivalent*.

Zakharov and Takhtajan in [120] discovered the gauge equivalence of the Landau-Lifshitz equation for nonlinear waves of the macroscopic magnetization in an isotropic classical one-dimensional Heisenberg ferromagnetic and nonlinear Schrödinger equation. After them the great progress has been achieved in studying the gauge relation among the various nonlinear evolution equations. The detail review of this subject is given in [122-124]. It is worth pointing to the references [124,125], where the gauge equivalence between the Sin-Gordon equation and the field equation of $O(3)$-nonlinear σ-model was discussed. The first appears, apart of other fields of physics, in the theory of self-induced transparency by McCall-Hahn in the sharp-line limit, the latter plays the main role in the chiral field models.

Function $f(z)$, determined on R^K (where it is often $K = 2$), with the values belonged to some nonlinear manifold M in the most general case is named the *chiral field*. The *principal chiral field* is that for which M complies with the Lie group [126,127]. The chiral fields are remarkable as their Lagrangian is a free field Lagrangian, i.e., it does not contain potential energy at all. But the domain of the values of the field belongs to some nonlinear manifold. For the principal chiral models this manifold is a Lie group, for other models this manifold may be the homogeneous space of a Lie group or coset space. This is the reason why the Euler-Lagrange equations, that is the field equations, are nonlinear. One says that the nonlinearity is induced by the geometry.

4.5.1. SIT UNDER ONE-PHOTON RESONANCE CONDITION

Let us consider the SIT-equations, which has a zero-curvature representation in the form of equations (4.5.20), or (4.5.25), or (4.5.28), under the condition of homogeneous broadening of the resonant absorption line and exact resonance. In all these cases the U-V-pair may be presented in the unification form

$$\hat{U}(\lambda;\tau,\zeta) = \lambda\hat{J} + \hat{E}(\tau,\zeta), \quad \hat{V}(\lambda;\tau,\zeta) = \lambda^{-1}\hat{P}(\tau,\zeta) .$$

The zero-curvature condition (i.e., SIT-equations) in these designations have the universal form

$$\partial\hat{E}/\partial\zeta = [\hat{P}, \hat{J}], \quad \partial\hat{P}/\partial T = [\hat{E}, \hat{P}].$$

The expressions for matrices $\hat{J}, \hat{E}, \hat{P}$ can be easily found by the comparison with (4.5.20), (4.5.25) and (4.5.28). Let $g(\tau,\zeta)$ be determined by a solution of equations

$$\partial g/\partial\tau = (\hat{J} + \hat{E}(\tau,\zeta))g , \quad \partial g/\partial\zeta = \hat{P}(\tau,\zeta)g ,$$

that is $g(\tau,\zeta)$ complies with $\Psi(\lambda = 1;\tau,\zeta)$. From (4.5.2) we obtain that

$$\hat{U}' = (\lambda - 1)g^{-1}\hat{J}g , \qquad \hat{V}' = (\lambda^{-1} - 1)g^{-1}\hat{P}g . \tag{4.5.4}$$

Introducing of the new quantities

$$l_1(T,\zeta) = -2g^{-1}\hat{J}g , \quad l_0(T,\zeta) = -2g^{-1}\hat{P}g , \tag{4.5.5}$$

from (4.5.3) we obtain the system of equations

$$\partial l_1 / \partial\zeta = \partial l_0 / \partial\tau + [l_0, l_1]\,, \quad \partial l_1 / \partial\zeta = (1/2)[l_0, l_1]\,, \quad \partial l_0 / \partial\tau = -(1/2)[l_0, l_1]\,.$$

Whence the equations of chiral field models are

$$\frac{\partial l_1}{\partial\zeta} + \frac{\partial l_0}{\partial\tau} = 0\,, \quad \frac{\partial l_1}{\partial\zeta} - \frac{\partial l_0}{\partial\tau} = [l_0, l_1]\,. \tag{4.5.6}$$

If l_0 and l_1 are the elements of Lie algebra of Lie groups G, defined by *U-V*-pair, then equations (4.5.6) are the equations of the principal chiral field (PCF). For instance, $G = SU(2)$ for the SIT theory by McCall-Hahn, for the SIT theories in the vector case $G = GL(3,C)$, for the SIT at transition $j_1 = 1/2 \to j_2 = 1/2$ [30] $G = SU(2) \times SU(2)$.

If one considers l_0 and l_1 as the left currents $h(\tau,\zeta)$ of the PCF, i.e., $\partial h / \partial\tau = l_1 h$, $\partial h / \partial\zeta = l_0 h$, then the system (4.5.6) results in the standard PCF equation

$$\frac{\partial}{\partial\zeta}\left(\frac{\partial h}{\partial\tau} h^{-1}\right) + \frac{\partial}{\partial\tau}\left(\frac{\partial h}{\partial\zeta} h^{-1}\right) = 0\,,$$

Thus, it has been shown the gauge equivalence of the generalized models of SIT to the PCF models.

It is necessary to note that $h(\tau,\zeta)$ is a function determined on the two-dimensional Minkowski space-time $R^{1,1}$ with values in *G*. That provides obstacles in the definition of the topological charges, self-duality equation and other conceptions inherent to the PCF models in Euclidean space.

The solution of the PCF models can be expressed in terms of the solutions of the spectral problem of IST method, i.e., the Jost matrix function: $\Psi(\lambda;\tau,\zeta)$

$$h(\tau,\zeta) = \Psi^{-1}(\lambda = 1;\tau,\zeta)\Psi(\lambda = -1;\tau,\zeta)\;.$$

On the another hand, the solutions of the given model of PCF generate new solutions of SIT (or GSIT) equations, describing the propagation of the USP in a resonant medium. This is the consequence of the reversibility of the gauge transformations.

4.5.2. SIT UNDER TWO-PHOTON RESONANCE CONDITION

In order to find another example of the hidden relations between different physical models let us consider the theories of the USP propagation in the medium under a two-photon resonance condition. The zero-curvature representation of reduced Maxwell-Bloch equations in the case of such resonance was founded in [53,54,119]. Here we reproduce the *U-V*-pair from [53]:

$$\hat{U}(\lambda)=\begin{pmatrix} -i\lambda s_3 & (\lambda+if)s_+ \\ -m(\lambda-if)s_- & i\lambda s_3 \end{pmatrix},$$

$$\hat{V}(\lambda)=\frac{i}{2}\left(-b_- - \frac{m}{2\lambda-b_-}\right)\begin{pmatrix} r_3 & 0 \\ 0 & -r_3 \end{pmatrix}+\frac{1}{(2\lambda-b_-)}\begin{pmatrix} 0 & m(\lambda+if)r_+ \\ -(\lambda-if)r_- & 0 \end{pmatrix},$$

where $f^2=(m-b_-^2)/2$, $m=+1$ is chosen for stimulated Raman scattering and $m=-1$ holds for two-photon absorption. The quantities s_3, $s_\pm=s_1\pm is_2$ are expressed in terms of the envelopes of electrical fields of the two-frequency USP and the quantities r_3, $r_\pm=r_1\pm ir_2$ are connected with the matrix elements of density matrix of the two-level atoms ensemble (see Section 4.4.1). The zero-curvature condition brings about the system of equations offered by Steudel [53] and Kaup to describe the phenomenon of the two-photon SIT

$$\begin{aligned} &\partial s_+/\partial\zeta=imr_+s_3-ib_-r_3s_+, \quad \partial s_3/\partial\zeta=(i/2)\left(r_+^*s_+-r_+s_+^*\right), \\ &\partial r_+/\partial T=-ib_-r_+s_3+ir_3s_+, \quad \partial r_3/\partial T=(i/2)\left(s_+^*r_+-s_+r_+^*\right) \end{aligned} \tag{4.5.7}$$

Choose the solution of equations (4.2.13) at $\lambda=1$ as $g(\tau,\zeta)$. According to the relation (4.5.2), a new *U-V*-pair may be written in the following form

$$\hat{U}'=(\lambda-1)g^{-1}\hat{S}g, \quad \hat{V}'=(\lambda-1)(2\lambda-b_-)^{-1}g^{-1}\hat{R}g,$$

where matrixes $\hat{S}$ and $\hat{R}$ are introduced as

$$\hat{S}=\begin{pmatrix} -is_3 & s_+ \\ -ms_- & is_3 \end{pmatrix},$$

$$\hat{R}=\left(\frac{im}{2-b_-}\right)\begin{pmatrix} r_3 & 0 \\ 0 & -r_3 \end{pmatrix}+\frac{1}{(2-b_-)}\begin{pmatrix} 0 & -m(b_--2if)R_+ \\ (b_--2if)R_- & 0 \end{pmatrix}.$$

If one defines left currents l_0 and l_1 as $l_1(T,\zeta)=-(2-b_-)g^{-1}\hat{S}g$, $l_0(T,\zeta)=g^{-1}\hat{R}g$, then we get the transformed *U-V*-pair $\hat{U}'$ and $\hat{V}'$ in the following form

$$\hat{U}'=(1-\lambda)(2-b_-)^{-1}l_1(T,\zeta), \quad \hat{V}'=-(1-\lambda)(2\lambda-b_-)^{-1}l_0(T,\zeta).$$

Introducing new spectral parameter $\mu=(2\lambda-b_-)(2-b_-)^{-1}$, we obtain

$$\hat{U}' = (1/2)(1-\mu)l_1(T,\zeta), \qquad \hat{V}' = (1/2)(1-\mu^{-1})l_0(T,\zeta).$$

By comparing these expressions with (4.5.4) and (4.5.5), we may conclude that Steudel-Kaup equations (4.5.7) are gauge equivalent to the equations of the PCF model with $G = SU(2)$.

It is interesting that both McCall-Hahn equations in the sharp-line limit and exact resonance (that is Sin-Gordon equation) and Steudel-Kaup equations can be mapped into the same PCF model. This is the essence of the geometric interpretation of the solution of the Steudel-Kaup equations given by Kaup [54]. Thereby, we obtain the gauge equivalence of the SIT models based on completely integrable equations and integrable PCF models. Another example was presented in [59]. There it has been found that the simplest scalar SIT model is equivalent to the theory of a charged vector field interacting with the Abelian gauge field. In [128] the system of two-photon resonance SIT equations was transformed into two connected Klein-Gordon equations. Recently the authors of [129,130] have investigated the GSIT equations and found the hidden non-Abelian group structure of the these equations in the case of multi-level resonant medium. They have discovered that a non-degenerate two-level system of self-induced transparency is associated with symmetric space $G/H = SU(2)/U(1)$ while three-level V- or Λ-systems are associated with $G/H = SU(3)/U(2)$. The same symmetric space is associated with the degenerate two-level system of SIT in the case of transitions $j_1 = 1 \to j_2 = 0$ and $j_1 = 0 \to j_2 = 1$. When one considers the transition $j_1 = 1/2 \to j_2 = 1/2$, the GSIT equations are associated with $G/H = (SU(3)/U(2))^2$. There are many complex aspects related to the degeneration of the energy levels in a three-level system of SIT. For instance, the transitions between states $j_a = j_c = 0, j_b = 1$ (or $j_a = j_c = 1, j_b = 0$) are associated with the symmetric space $G/H = SU(4)/S(U(2)\times U(2))$ (accordingly $G/H = SU(5)/U(4)$.). If to consider $\hat{U}(\lambda)$ and $\hat{V}(\lambda)$ as the matrix functions of complex variable λ, then one may notice that nature of the singularities of these functions do not change under gauge transformations. This fact could be used to classify the gauge equivalent integrable models from the general point of view.

4.6. Conclusion

In this chapter we have considered the SIT models, which are based on the completely integrable equations. It allows us to use the powerful method of the inverse scattering transform for obtaining the soliton solutions of these equations. However, we have left aside some questions related to such models.

The McCall-Hahn theory of SIT phenomena can be represented in the Hamiltonian formulation [131]. Furthermore, by using the zero-curvature representation of the SIT equations one can show that the Hamiltonian formulation allows the ultralocal Poisson brackets and r-matrix [121]. The theory of SIT without SVEPA (see equation (4.1.5)) permits the Hamiltonian formulation but the Poisson brackets are not ultralocal, which means they contain not only the δ-function, but its derivative as well. Thus we can conclude that there is the fundamental distinction between these two theories of the ultrashort pulse propagation in a resonant two-level medium. However, it has been shown [132] that fundamental Poisson brackets for a transfer matrix can be written in a compact form in accordance with the requirement of the generalised Hamiltonian formulation [121] by using two r-matrices. It opens a new way to the development of the quantum theory of SIT.

There is a hypothesis that completely integrable systems of non-linear evolution equations possess the Painleve property. It means that appropriate transformation of variables converts the non-linear partial equations to ordinary ones without movable critical points in the solutions [101,133-135]. It has been showed [136,137] that SIT equations (4.1.7) stand the Painleve test. This property has also been found for the system of equations describing USP propagation in the cubic non-linear medium containing the two-level resonant impurities [138-140]. As for the variety of the generalised SIT equations accounting for two-photon resonance the Painleve property test has not been carried out yet.

In section 4.1 and 4.2 we discussed the regular soliton solutions which follow from the relevant systems of the equations. However, there is another technique for solving these equations. Some new solutions have been found in [34,35] under consideration polarised USP propagating in medium with the degeneration of the energy levels (section 4.2.2). These solutions were obtained via a generalised formulation of the dressing operator approach of the Riemann-Hilbert problem with the second order poles. It is important to point out that the shape of the USP envelopes are no longer of the simple sech-type, but they tend to zero with increasing the time. By using the $\bar{\partial}$-formalism for rational reflection coefficient of the spectral problem (3.2.1) in [141] solutions of the SIT equations which belong to class of the discontinuous soliton-like solutions of these equations have been obtained.

Analysis of the non-soliton part of the general solution of the SIT equations is the important problem when the amplification of the optical pulse in the laser amplifier is considered. It has been found that the output pulse is always of a quasi self-similar nature, and the parameters of the optical pulse in the amplifier are determined exclusively by the front of the initial pulse [142,143]. The more detailed description of the evolution of the amplified pulse was represented in [144]. An approximate solution of the boundary problem for the SIT equations which describes the superfluorescence has been obtained in [143] as well.

We have considered the SIT theories based on equations in (1+1)-dimensional space-time. The simplest way to increase the space dimension is expressed in the substitution

$$\partial / \partial z \rightarrow \partial / \partial z + D_x \partial^2 / \partial x^2 + D_y \partial^2 / \partial y^2$$

in the equations for the electric field of the USP. It results in a non-integrable system of equations. The analysis of the USP envelope evolution can be done in the frame of the approximate methods or numerical simulation.

We finish this section by noting that the theory of the self-induced transparency is the most fruitful region of non-linear optics where the non-linear wave dynamics is represented in full measure.

References

1. McCall, S.L., and Hahn, E.L.: Self-induced transparency by pulsed coherent light, *Phys.Rev.Letts.* **18** (1967), 908-911.
2. McCall, S.L., and Hahn, E.L.: Self-induced transparency, *Phys.Rev.*\ **183** (1969), 457-485.
3. Allen, L., and Eberly, J.H.: *Optical Resonance and Two-Level Atoms*, Wiley, New York, 1975.
4. Slusher, R.E.: Self-induced transparency, *Progr.Optics*, **12** (1974), 53-100.
5. Courtens, E.: Giant Faradey rotations in self-induced transparency. *Phys.Rev.Letts.* **21** (1968), 3-5.
6. Lamb, G.L., Jr.: Analytical descriptions of ultrashort optical pulse propagation in a resonant medium, *Rev.Mod Phys.* **43** (1971), 99-124.
7. Eilbeck, J.L., Gibbon, J.D., Caudrey, P.J., and Bullough, R.K.,: Solitons in nonlinear optics. I. A more accurate description of the 2π-pulse in self-induced transparency, *J.Phys.* A, **6**, (1973), 1337-1347.
8. Lamb, G.L., Jr.: Coherent optical pulse propagation as inverse problem, *Phys.Rev.* A **9**, (1974), 422-430.
9. Gibbon, J.D., Caudrey, P.J., Bullough, R.K., and Eilbeck, J.L.: An N-soliton solution of a nonlinear optics equation derived by a general inverse method, *Letts.Nuovo Cimento* **8**, (1973), 775-779.
10. Takhtajan, L.A. : Exact theory of propagation of ultrashort optical pulses in two-level media, *Zh.Eksp.Teor.Fiz.* **66** (1974), 476-489 [*Sov.Phys. JETP* **39** (1974), 228-238]
11. Ablowitz, M.J., Kaup, D.J., Newell, A.C., and Segur, H.: Coherent pulse propagation, a dispersive, irreversible phenomenon, *J.Math.Phys.* **15** (1974),1852-2858.
12. Kaup, D.J.: Coherent pulse propagation: A comparison of the complete solution with the McCall-Hahn theory and others, *Phys.Rev.* A **16**, (1965), 240-243.
13. Haus H.A.: Physical interpretation of inverse scattering formalism applied to self-induced transparency, *Rev.Mod.Phys.* **51**, (1979), 331-339.
14. Gibbon, J.D., and Eilbeck, J.L.: A possible N solitons solution for nonlinear optics equation, *J.Phys.* A **5**, (1972), .L122-L124.
15. Caudrey, P.J., Gibbon, J.D., Eilbeck, J.L., and Bullough, R.K.: Exact multisolitons solution of the self-induced transparency and Sine-Gordon equations, *Phys. Rev. Letts.* **30**, (1973), 237-238.
16. Caudrey, P.J., Eilbeck, J.L., Gibbon, J.D., and Bullough, R.K.: Exact multisolitons solution of the inhomogeneously broadened self-induced transparency equations, *J.Phys.* A, **6** (1973), L53-L56.
17. Rhodes, C.K., Szoke, A., and Javan, A.: The influence of level degeneracy on the self-induced transparency effect, *Phys.Rev.Letts.* **21** (1968), 1151-1155.
18. Lu, E.Y.C., and Wood L.E.: Single-peaked self-induced transparency pulses in a degenerate resonant medium, *Phys.Letts.* A, **45** (1973), 373-374.
19. Lu, E.Y.C., and Wood L.E.: Analytic description of distortionless propagation in a resonant absorbing medium with overlapping *Q*(*j*)-transitions, *Opt.Commun.* **10**, (1974), 169-171.
20. Duckworth, S., Bullough, R.K., Caudrey, P.J., and Gibbon, J.D.: Unusualy soliton behavior in the self-induced transparency of *Q*(2) vibration-rotation transitions, *Phys.Letts.* A, **57**, (1976), 19-22 .
21. Gundersen, R.M.: Self-induced transparency with level degeneracy, *J.Phys.* A, **23** (1990) 4237-4247.

22. Konopnicki, M.J., Drummond, P.D., and Eberly J.H.: Simultons: theory of simultaneous propagation of short different-wavelength optical pulses, *Bull.Amer. Phys Soc.* **25** (1980), 1124; *Appl.Phys.* **B 28** (1982), 103.
23. Konopnicki, M.J., Drummond,dP.D., and Eberly J.H.: Theory of lossless propadation of simultaneous ddfferent-wavelength optidal pulses, *Opt.Commun.***3d** (1981), 313-316.
24. Konopnicki, M.J., and Ebedly J.H.: Simultaneous propagation of short different-wavelength optical pulses, *Phys.Rev.* A **24** (1981), 2567-2583.
25. Kujawski, A.: Soliton properties of optical simultons, *Opt.Commun.* **43** (1982), 375-377.
26. Bol'shov, L.A., Likhanskii, V.V., and Persiancev, M.I.: On the theory of coherent interaction of light pulse with a resonant multilevel media, *Zh.Eksp.Teor.Fiz.* **84** (1983), 903-911 [*Sov.Phys. JETP* **57** (1983), 524]
27. Maimistov, A.I.: A rigorous theory of self-induced trancparency under double resonance in a three-level medium, *Kvantovaya Electron., Moskva,* **11** (1984), 567-577 [*Sov.J. Quantum Electron.* **14** (1984), 385
28. Bol'shov, L.A., and Likhanskii, V.V.: Coherent interaction between emission pulses and resonant multilevel media (Review article), *Kvantovaya Electron., Moskva,* **12** (1985), 1339-1364. [*Sov.J. Quantum Electron.* **12** (1985), 889] .
29. Maimistov, A.I.: New examples of exactly solvable problems in nonlinear optics. *Opt.Spektrosk.* **57** (1984), 564-566. [*Opt.Spectrosc.*(*USSR*) **57**, (1984) 340-341.]
30. Basharov,A.M., and Maimistov, A.I.: On self-induced transparency under condition of degeneration of resonant energy levels, *Zh.Eksp.Teor.Fiz.* **87** (1984), 1594-1605 [*Sov.Phys. JETP* **60** (1984), 913]
31. Basharov, A.M., Maimistov, A.I., and Sklyarov Yu.M.: Self-induced transparency on the transition $1\rightarrow 1$ is a exactly solvable polarization model of nonlinear optics, *Opt.Spektrosk.* **63** (1987), 707-709 [*Opt.Spectrosc.*(*USSR*) **62** (1987), 418]
32. Basharov, A.M., and Maimistov, A.I.: Polarized solitons in three-level media, *Zh.Eksp.Teor.Fiz.* **94** (1988), 61-75 [*Sov.Phys. JETP* **67** (1988), 2426-2433]
33. Chernyak, V.Ya., and Rupasov, V.I.: Polarization effects in self-induced transparency theory, *Phys.Letts.* A. **108** (1985), 434-436.
34. Roy Chowdhury,A., and M.De, M.: Riemann-Hilbert problem with order poles for self-induced transparency with degenerate energy levels, *Austr.J.Phys.* **41** (1988), 735-741.
35. Roy Chowdhury,A., and M.De, M.: On a Riemann-Hilbert approach to degenerate SIT: Soliton solutions with time-dependent velocity, *Inverse Probl.* **4** (1988), 901-911.
36. Steudel, H.: N-soliton solutions to degenerate self-induced transparency, *J.Mod. Opt.* **35** (1988), 693-702
37. Basharov, A.M., and Maimistov, A.I.: Interaction of polarized waves in three-level medium, *Opt.Spektrosk.* **68**, (1990), 1112-1117.
38. Agranovich, V.M., Chernyak, V.Y., and Rupasov, V.I.: Self-induced transparency in wave guides, *Opt.Commun.* **37** (1981), 363-365.
39. Agranovich, V.M., Rupasov, V.I., and Chernyak, V.Y.,: Theory of self-induced transparency of the surface and guided wave solutions. *Fiz.Tverd.Tela (Leningrad)* **24** (1982), 2992-2999 [Sov. Phys.Solid State **24** (1982), 1693]
40. Ponath, H.E., and Schubert, M.: Optical soliton-like surface phenomena, *Opt.Acta* **30** (1983), 1139-1149.
41. Basharov, A.M., Maimistov, A.I., and Manykin E.A.: Exactly integrable models of resonance interaction of light with a thin film of three-particle levels, *Zh.Eksp.Teor. Fiz.* **97** (1990), 1530-1543 [*Sov.Phys. JETP* **70** (1990), 864-871]
42. Belenov, E.M., and Poluektov, I.A.: Coherent effects under ultrashort light pulse in medium under condition of two-photon resonance absorption, *Zh.Eksp.Teor.Fiz.* **56** (1969), 1407-1411 [*Sov.Phys. JETP* **29** (1969), 754]
43. Takatsuji, M.: Propagation of a Light Pulse in a Two-Photon Resonant Medium. *Phys.Rev.* **A 4** (1971), 808-810.

44. Takatsuji, M.: Theory of Coherent Two-Photon Resonance, *Phys.Rev.* **A 11** (1975), 619-624.
45. Tan-no, N., and Higuchi, Y.: Solitary Wave Solutions in Coherent Two-Photon Pulse Propagation, *Phys.Rev.* **A16** (1977), 2181-2183 (1977)
46. Tan-no, N., Shirahata, T., and Yokoto, K.: Coherent Transient Effect in Raman Pulse Propagation, *Phys.Rev.* **A 12** (1975), 159-168.
47. Hanamura, E.: Self-Indused Transparency Due to Two-Photon Transition and Area Theorem, *J.Phys.Soc.Japan* **37** (1974), 1598-1605.
48. Steudel, H.: Solitons in stimulated Raman scattering, *Ann.Phys.(DDR)* **34** (1977), 188-202.
49. Poluectov, I.A., Popov, Yu.M., and Roitberg V.S.: Coherent effects under propagation of ultrashort light pulses in medium with resonant two-photon absorption, *Pis'ma Zh. Eksp. Teor.Fiz.* 18 (1973) 638-641 [*Sov.Phys. - JETP Lett.* **18** (1973), 373-376].
50. Poluectov, I.A., Popov, Yu.M., and Roitberg V.S.: Coherent propagation of powerful light pulses through medium under condition of two-photon interaction, *Pis'ma Zh. Eksp. Teor.Fiz.* 20 (1974) 533-537 [*Sov.Phys. - JETP Lett.* **20** (1974), 243-246].
51. Poluectov, I.A., Popov, Yu.M., and Roitberg V.S.: Coherent effects appearing in propagation of ultrashort light pulse in medium under condition of two-photon resonance, *Kvantovaya Electron., Moskva*, **2** (1975), 1147-1152 [*Sov.J. Quantum Electron.* **5** (1975), 620]
52. Dutta, N.: Theory of Coherent Two-Photon Absorption, *Phys.Lett.* **A69** (1978), 21-23.
53. Steudel, H.: Solitons in stimulated Raman scattering and resonant two-photon propagation, *Physica* D **6**, (1983), 155-178.
54. Kaup, D.J.: The method of solution for stimulated Raman scattering and two-photon propagation, *Physica* D **6**, (1983), 143-154.
55. Meinel, R.: Backlund transformation and N-soliton solutions for stimulated Raman scattering and resonant two-photon propagation, *Opt.Commun.* **49**, (1984), 224-228.
56. Maimistov, A.I.: Gauge relation between self-induced transparency theory and principal chiral models, *Phys.Letts.* **A144** (1990), 11-14.
57. Poluectov, I.A., Popov, Yu.M., and Roitberg, V.S.: Effect of self-induced-transparency, *Usp. Fiz. Nauk* **114** (1974), 97-131.
58. Bullough, R.K., Jack, P.M., Kitchenside, P.W., and Saudders, R.: Solitons in laser physics, *Phys.Scr.* **20** (1979), 364-381.
59. Maimistov, A.I., Basharov, A.M., Elyutin, S.O., and Sklyarov, Yu.M.: Present state of self-induced transparency theory, *Phys.Rept*, **C 191** (1990), 1-108.
60. Eilbeck, J.C., and Bullough, R.K.: The method of characteristics in the theory of resonant or nonresonant nonlinear optics, *J.Phys.* **A 5** (1972), 820-829.
61. Arecchi, F.T., DeGiorgio, V., and Someda, C.G.: Self-induced light propagation in a resonant medium, *Phys.Lett.* **A27** (1968), 588-589.
62. Eberly, J.H.: Optical Pulse and Pulse-Train Propagation in a Resonant Medium. *Phys.Rev.Lett.* **22** (1969), 760-762.
63. Crisp, M.D.: Distortionless propagation of light through an optical medium, *Phys.Rev.Lett.* **22** (1969), 820-823.
64. Byrd, P.F., and Friedman M.D.: *Handbook of elliptic integrals for engineers and physicists,* Springer, Berlin, Gottinberg, Heidelberg, 1954.
65. Bullough, R.K., and Ahmad, F.: Exact solutions of the self-induced transparency equations, *Phys.Rev.Lett.* **27** (1971), 330-333.
66. Arecchi, F.T., and Courtens, E.: Cooperative phenomena in resonant electromagnetic propagation, *Phys.Rev.* **A 2** (1970), 1730-
67. Dicke, R.H.: Coherence in spontaneous radiation processes, *Phys.Rev.* **93** (1954), 99-110.
68. Rehler, N.E., Eberly, J.H.: Superradiance, *Phys.Rev.* **A3** (1971), 1735-1751.
69. Chen, H.-H., Relation between Bäcklund transformations and inverse scattering problems, in R.M.Miura (ed.), *Backlund Transformations, the Inverse Scattering Method, Solitons and Their Applications*, (Lect.Notes in Math. 515), Springer-Verlag, Berlin, 1976, pp.241-252

70. Barnard, T.W.: $2N\pi$ ultrashort light pulses, *Phys.Rev.* **A7** (1973), 373-376.
71. Bak, P.: Solitons in incommensurate systems, in A.R.Bishop, T.Schneider (eds.), *Solitons and Condensed Matter Physics*, Springer, Berlin, Heidelberg, New York, 1978, pp.216-233.
72. Rice, M.J.: Charge density wave systems: the ϕ-particle model, in A.R.Bishop, T.Schneider (eds.), *Solitons and Condensed Matter Physics*, Springer, Berlin, Heidelberg, New York, 1978, pp.246-253.
73. Kitchenside, P.W., Bullough, R.K., and Caudrey, P.J.: Creation of spin wave in ^{3}He-B, in A.R.Bishop, T.Schneider (eds.), *Solitons and Condensed Matter Physics*, Springer, Berlin, Heidelberg, New York, 1978, pp.291-296.
74. Gibbon J.: The interaction of spin waves in liquid ^{3}He in several dimensions, in A.R.Bishop, T.Schneider (eds.), *Solitons and Condensed Matter Physics*, Springer, Berlin, Heidelberg, New York, 1978, pp.297-300.
75. Rozhkov, S.S.: Dynamics of order parameter in superfluid phases of helium-3, *Usp.Fiz.Nauk (USSR)* **142** (1986), 325-346.
76. Iwabuchi, Ch.: Commensurate-incommensurate phase transition in double sine-Gordon system, *Progr.Theor.Phys.* **70** (1983), 941-953.
97. Hudak, O.: Double Sine-Gordon equation, A stable 2π-kink and commensurate-incommensurate phase transitions: *Phys.Lett.* **A 82** (1981), 95-96.
78. Burdick, S., El-Batanouny, M., and Willis, C.R.: Nonlinear internal dynamics of the double-sine-Gordon soliton, *Phys.Rev.* **B 34** (1986), 6575-6578.
79. Bullough, R.K., Caudrey, P.J., and Gibbs H.M.: The double-sine-Gordon equations: a physically applicable system of equations, in R.K. Bullough, P.J.Caudrey (eds.), *Solitons,* Springer, Berlin, Heidelberg, New York, 1980, pp.107-147.
80. Dodd, R.K., Bullough, R.K., and Duckworth, S.: Multisoliton solutions of nonlinear dispersive wave equations not soluble by the inverse method, *J.Phys.* **A 8** (1975), L64-L68.
81. Hudak, O.: The double sine-Gordon equation: On the nature of internal oscillation of the 2π-kink, *Phys.Lett.* **A 86** (1981), 208-212.
82. van der Merwe, P. du T.: Phase-locked solutions, *Lett.Nouvo Cimento* **25** (1979), 93-96.
83. Duckworth, S., Bullough, R.K., Caudrey, P.J., and Gibbon J.D.: Unusualy soliton behavior in the self-induced transparency of *Q*(2) vibration-rotation transitions, *Phys.Lett.* **A 57** (1976), 19-22.
84. Salamo, G.J., Gibbs, H.M., and Churchill, G.G.: Effects of degeneracy on self-induced transparency, *Phys.Rev.Lett.* **33** (1974), 273-276
85. Gibbs, H.M., McCall, S.L., and Salamo, G.J.: Near-ideal self-induced-transparency breakup in highly degenerated systems, *Phys.Rev.* **A 12** (1975), 1032-1035.
86. Xu Gan, King, T.A., and Bannister, J.J.: Self-induced transparency on degenerate magnetic dipole transitions, *Phys.Rev.* **A 29** (1984), 3455-3457.
87. Xu Gan, and King, T.A.: Coherent pulse propagation and self-induced transparency on degenerate transitions in atomic iodine, *Phys.Rev.* **A 30** (1984), 354-364.
88. Xu Gan, King, T.A., and Bannister, J.J.: Pulse reshaping in coherent interaction with a resonant absorber and application to measurement of homogeneous relaxation time. *Opt.Acta* **31** (1984), 487-495.
89. Xu Gan, King, T.A., and Bannister, J.J.: Coherent pulse propagation on a *Q*-branch transition in atomic iodine vapour, *Opt.Acta* **32** (1985), 7-15.
90. Bannister, J.J., Baker, H.J., King, T.A., and McNaught, W.G.: Self-induced transparency and resonant self-focusing in atomic iodine vapor, *Phys.Rev.Lett.* **44** (1980), 1062-1065.
91. Drummond, P.D.: Formation and stability of vee simultons, *Opt.Commun.* **49** (1984), 219-223.
92. Maimistov, A.I., and Sklyarov Yu.M.: On coherent interaction light pulses with three-level medium, *Opt.Spektrosk.***59** (1985), 760-763 [*Opt.Spectrosc. (USSR)* **59** (1985), 459]
93. Bol'shov L.A., and Napartovich, A.P.: Coherent interaction of light pulses with three-level systems, *Zh.Eksp.Teor.Fiz.* **68** (1975), 1763-1767

94. Zabolotskii, A.A.: Resonance and parametric interaction of light in a nonlinear medium, *Phys. Lett.* **A 113** (1986), 459-462.
95. Hioe, F.T.: Exact solitary-wave solution of short different-wavelength optical pulses in many-level atomic absorbers, *Phys.Rev.* **A 26** (1982), 1466-1472.
96. Drummond, P.D., and Eberly J.H.: Four-dimensional simulation of multiple-laser pulse propagation, *Bull.Amer.Phys.Soc.*, **20** (1980), 1124.
97. Chelkowski, S., and Bandrauk, A.D.: Coherent propagation of intense ultrashort laser pulses in a molecular multilevel medium. *J.Chem.Phys.* **89** (1988), 3618-3628.
98. Ablowitz, M.J., Kaup, D.J., Newell, A.C., and Segur, H.: Nonlinear evolution equations of physical significance, *Phys.Rev. Letts.* **31**, (1973), 125 -127.
99. Ablowitz, M.J., Kaup, D.J., Newell, A.C., and Segur, H.: The inverse scattering transform - Fourier analysis for nonlinear problems, *Stud.Appl.Math.* **53** (1974) 249-315.
100. Zakharov, V.E., Manakov, S.V., Novikov, S.P., and Pitaevskii, L.P.: *Theory of Solitons: The Inverse Problem Method* [in Russian], Nauka, Moscow, 1980. *Theory of Solitons: The Inverse Scattering Method*, Plenum, New York, 1984.
101. Ablowitz, M.J., and Segur, H.: *Solitons and the Inverse Scattering Transform*, SIAM, Philadelphia, 1981.
102. Ablowitz, M.J., and Huberman, R.: Resonantly coupled nonlinear evolution equations, *J.Math.Phys.* **16** (1975), 2301-2305.
103. Huberman, R.: Note on generating nonlinear evolution equations, *SIAM J. Appl .Math.* **31** (1976), 47-50.
104. Caudrey, P.J.: The inverse problem for a general $N \times N$ spectral equation, *Physica* **D 6** (1982), 51-66.
105. Gerdjikov, V.S., and Kulish, P.P.: The generating operator for the $N \times N$ linear system, *Physica* **D 3** (1981), 549-564.
106. Fordy, A.P., and Kulish, P.P.: Nonlinear Schrodinger equations and simple Lie algebra's, *Commun.Math.Phys.* **89** (1983), 427-443.
107. Manakov, S.V.: On the theory of two-dimensional stationary self-focusing of electromagnetic waves, *Zh.Eksp.Teor.Fiz.* **65** (1973), 505-516 [*Sov.Phys. JETP* **38** (1974), 248]
108. Wahlquist, H.D., and Estabrook, F.B.: Prolongation structures and nonlinear evolution equations, *J.Math.Phys.***16** (1975), 1-7.
109. Wahlquist, H.D., and Estabrook, F.B.: Prolongation structures and nonlinear evolution equations, *J.Math.Phys.***17** (1976), 1293-1297.
110. Kaup, D.J., and Malomed, B.A.: Soliton trapping and daughter waves in the Manakov model, *Phys.Rev.* **A 48** (1993), 599-604.
111. Steudel, H.: *N*-Soliton solutions to degenerate self-induced transparency, *J.Modern Opt.* **35** (1988), 693-702.
112. Zakharov, V.E., and Schabat, A.B.: The exact theory of two-dimensional self-focussing and one-dimensional self-modulating of waves in nonlinear medium, *Zh.Eksp.Teor.Fiz.* **61** (1971), 118-134 [*Sov.Phys. JETP* **34** (1972), 62-69].
113. Tratnik, M.V., Sipe J.E.: Polarization solitons, *Phys.Rev.Lett.* **58** (1987), 1104-1107.
114. *Introduction to Integrated Optics,* M.K.Barnosky, (ed.), Plenum, New York, 1974.
115. Elyutin, S.O., Maimistov, A.I., and Manykin E.A.: On propagation of coherent optical pulses under two-photon resonance condition, *Opt.Spektrosk* **50** (1981), 354-361.
116. McLaughlin, D.W., and Corones, J.: Semiclassical radiation theory and the inverse method, *Phys.Rev.* **A10** (1974), 2051-2062.
117. Kaup D.J.: The Estabrook-Wahlquist method with examples of application, *Physica* **D 1** (1980), 391-411.
118. Dodd, R., and Fordy, A.: The prolongation structures of quasi-polynomial flows, *Proc.Roy.Soc.(London)* **A385** (1983), 389-429

119. Maimistov, A.I., and Manykin, E.A.: Prolongation structure for the reduced Maxwell-Bloch equations describing the two-photon self-induced transparency, *Phys.Lett.* **A95** (1983), 216-218.
120. Zakharov, V.E., and Takhtajan, L.A.: Equivalence of nonlinear Schrodinger equation and equation of Heisenberg ferromagnetic, *Teor.Mat.Fiz.* **38** (1979), 26-35.
121. Takhtajan, L.A., and Faddeev L.D.: *Hamiltonian approach to theory of solitons*, Nauka, Moscow, 1986.
122. Kundu, A.: Gauge equivalence of s-models with non-compact Grassmannian manifolds, *J.Phys.* **A19** (1986), 1303-1313.
123. Kundu, A.: Gauge unification of integrable nonlinear systems, in *Solitons.Introduction and Application*, Springer- Verlag, Berlin, 1989, p.86 - 104.
124. Orfanidis, S.J.: σ -models of nonlinear evolution equations, *Phys.Rev.* **D 21** (1980), 1513-1522.
125. Pohlmayer, K.: Integrable hamiltonian systems and interactions through quadratic constraints, *Commun.Math.Phys.* **46** (1976), 207-221.
126. Perelomov, A.M.: Chiral models: geometrical aspects, *Phys.Rept.* **C 146** (1987), 135-213.
127. Dubrovin, B.A., Novikov S.P., and Fomenko A.T.: *Modern geometry*, Nauka, Moscow, 1979, p. 740-750
128. Steudel, H.: Integrable theories of nonlinearly coupled Klein-Gordon fields, *Phys.Letts.* **A109** (1985), 85-86.
129. Park Q-H., and Shin, H.J.: Field theory for coherent optical pulse propagation, SNUTP 97-110 (1997) August, pp.1-43.
130. Park Q-H., and Shin, H.J.: Field theory for coherent optical pulse propagation, *Phys.Rev.* **A57** (1998), 4621-4643.
131. Aiyer, R.N.: Hamiltonian and recursion operator for the Reduced Maxwell-Bloch Equations, *J.Phys.* **A16** (1983), 1809-1811.
132. Maimistov, A.I., and Elyutin, S.O.: Hamiltonian formulation of self-induced transparency theory without slowly varying envelope approximation, *Chaos, Solitons & Fractals* **8** (1997), 369-376.
133. Ablowitz, M.J., Ramani, A., and Segur, H.: A connection between nonlinear evolution equations and ordinary differential equations of P-type.II., *J.Math.Phys.* **21** (1980), 1006-1015.
134. Ablowitz, M.J., Ramani, A., and Segur, H.: A connection between nonlinear evolution equations and ordinary differential equations of P-type. I, *J.Math.Phys.* **21** (1980), 715-721.
135. Ablowitz, M.J.: Remarks on nonlinear evolution equations and ordinary differential equations of Painleve type, *Physica* **D3** (1981), 129-141.
136. Grauel, A.: The Painleve' test, Backlund transformation and solutions of the Reduced Maxwell-Bloch equations, *J.Phys.* **A 19** (1986), 479-484.
137. Goldstein, P.P.: Testing the Painleve' property of Maxwell-Bloch and Reduuced Maxwell-Bloch equations, *Phys.Lett.* **A 121** (1987), 11-14
138. Sakovich, S.Yu: Painleve analysis and Backlund transformations of Doctorov-Vlasov equations, *J.Phys.* **A27** (1994), L33-L38.
139. Porsezian, K., Nakkeeran, K.: Optical soliton propagation in a coupled system of the non-linear Schrödinger equation and the Maxwell-Bloch equations, *J.Modern Opt.* **42** (1995), 1953-1958.
140. Nakkeeran, K., Porsezian, K.: Solitons in an Erbium-doped non-linear fibre medium with stimulated inelastic scattering, *J.Phys.* **A28** (1995), 3817-3823.
141. Leon, J.J.P.: Discontinuous soliton-like solution to the self-induced transparency equations, *Phys.Lett.* **A131** (1988), 79-84.
142. Manakov, S.V.: Propagation of the ultrashort optical pulse in a two-level laser amplifier, *Zh.Eksp.Teor.Fiz.* **83** (1982), 68-83.
143. Gabitov I.R., Zakharov V.E., and Mikhailov A.V.: Maxwell-Bloch equation and inverse scattering method, *Teor.Mat.Fiz.* **63** (1985), 11-31.
144. Manakov, S.V., and Novokshenov V.Yu.: Complete asymptotic representation of electromagnetic pulse in a long two-level amplifier, *Teor.Mat.Fiz.* **69** (1986), 40-54.

CHAPTER 5

COHERENT PULSE PROPAGATION

Propagation of an ultrashort optical pulse is often associated with the self-induced transparency phenomenon. But not always the optical pulse propagation can be described in the frames of the models based on the systems of completely integrable equations. By assuming, however, that the duration of an optical pulse is less than the characteristic relaxation times, we should admit the synchronous interaction of the resonance atoms with the electromagnetic field. Then they say that the system possesses a phase memory. A number of effects were considered in Chapter 2, where the evolution of an ultrashort pulse in the course of its propagation through the medium was not taken into account. These are the coherent transient processes. The phenomena arise when an ultrashort pulse spreads in a system with the phase memory are called the transient processes. The self-induced transparency, examined in the previous chapter, is a typical representative of the coherent transient processes. In the current chapter we shall consider several examples of the similar phenomena, which theory is based on the sets of non-completely integrable equations. Strictly speaking, there are no true solitons in such systems, but steady pulses have soliton-like properties.

Section 5.1 is devoted to the theory of the propagation of an ultrashort pulse in a medium with the non-resonance non-linearity. Here the optical pulse propagates without noticeable distortions as it takes place in SIT phenomenon [1-7]. Recently this phenomenon has been observed and investigated in [8].

The allowance of interaction between resonance atoms in addition to their resonance interaction with the electromagnetic field can be considered as another generalisation of SIT theory. A SIT in the system of excitons is a typical example of such an approach [9-27]. The simplest model [19] including the interaction between atoms is observed in section 5.2.

The propagation of extremely short optical pulses has been exciting an interest since McCall and Hahn's theory was created. In this theory the model of the two-level atoms is not the most successful model, but the first interesting results were obtained by a simple generalisation of the SIT equations [28-35]. In section 5.3 we consider the theoretical models and their solutions, which were used to analyse this problem. It will be demonstrated that both resonance and non-resonance models of a non-linear medium can describe the propagation of a stationary ultrashort pulse often called as the pulses in one period of the optical wave oscillation. Precisely speaking, such pulses, named video pulses , are not quasi-harmonic waves. They represent a separate (solitary) splash of the electromagnetic field. A similar object in non-linear optics is an optical shock wave, whose brief description completes this chapter.

5.1. SIT in a Kerr-type non-linear mediums

The progress has been achieved recently in the production of the extremely short optical pulses, which durations are about several femtoseconds [36-40]. Such ultrashort pulse is usually characterised by a high electric field strength, so that the non-linear optical effects take place. One of them is solitons generation and propagation. There are intensive investigations of the non-linear pulse propagation in the optical fibres [41]. However, practically all materials used for fibre fabrication contain impurities that contribute to the absorption spectrum of the fibres. The losses due to the resonant absorption decrease if the frequency of the carrier wave is located within the window of transparency of glass fibre. Another means to decrease losses is to make pulse duration shorter than the characteristic relaxation times of the resonant states or in other words to make optical pulses ultrashort. In this case the well known self-induced transparency phenomenon can be expected to arise.

It is known that the Nonlinear Schrödinger equation which is used to describe optical solitons in non-linear monomode optical fibre is completely integrable [42-46]. The reduced Maxwell-Bloch equations or their generalisations considered hereafter as RMB-equations are completely integrable too. The IST method for both the NLS and the RMB-equations enabling to solve some non-linear evolution equations is based on the same spectral problem. The model of the USP propagation in a Kerr-type non-linear medium doped by resonant impurity atoms incorporates both of these systems. But there are no reasons for the resulting system of equations to possess a complete integrability. The search for zero curvature representation of this system of equations allows to give a certain answer here.

5.1.1. THE CASE OF SCALAR SOLITONS

The evolution of the USP propagating in a non-linear monomode optical fibre in z-direction is described by the equations which generalise the Maxwell-Bloch equations [2]. We could name them the Nonlinear Schrödinger and Bloch equations.

$$i\frac{\partial q}{\partial \zeta} + s\frac{\partial^2 q}{\partial \tau^2} + \mu |q|^2 q + a\langle p\rangle = 0, \qquad (5.1.1a)$$

$$\partial p / \partial \tau = i\delta p + if R_3 q , \qquad (5.1.1b)$$

$$\partial R_3 / \partial \tau = (if/2)\left(q^* p - q \ p^*\right), \qquad (5.1.1c)$$

where q is the normalised slowly varying complex envelope of the USP defined by the following expression

$$E(t,x,y,z) = A_0 q(t,z)\Psi(x,y)\exp\{-i\omega_0 t + i\beta_0 z\},$$

$\Psi(x,y)$ is a mode function that determines the transverse distribution of the electric field over the fibre cross-section. The interaction of the radiation with the resonant impurities is characterised by the dimensionless constant $f = \bar{d}A_0 t_{p0}/2\hbar$, where $\bar{d}$ is an effective matrix element of the dipole transition between the resonant states:

$$\bar{d} = \int d(\vec{\varrho})\Psi(\vec{\varrho})\,|\,\Psi(\vec{\varrho})\,|^2\, d\vec{\varrho} \Big/ \int |\,\Psi(\vec{\varrho})\,|^2\, d\vec{\varrho}\,.$$

Here $\zeta = z/L_D$, $\tau = (t - z/v_g)t_{p0}^{-1}$ are normalised independent variables of co-ordinate and time, accordingly, t_{p0} is a pulse duration at $z = 0$ and v_g is the USP propagation group velocity. The term in (5.1.1a) with the second derivative with respect to τ describes the USP dispersion broadening ($s = -1$ for the normal dispersion and $s = +1$ for the anomalous dispersion). The coefficient a is expressed in terms of the dispersion length L_D and the resonant absorption length L_a [47] as $a = L_D L_a^{-1} f^{-1}$, where

$$L_D = 4\beta_0 t_{p0}^2 \left(|\,\partial\beta^2(\omega)/\partial\omega^2\,|\right)^{-1}, \quad L_a = c\hbar n_{eff}\left(2\pi\omega_0 n_A d^2 t_{p0}\right)^{-1}.$$

The effective refractive index n_{eff} is defined as $\beta(\omega) = (\omega/c)n_{eff}$ The third term in (5.1.1a) is responsible for the self-action effect. Coefficient μ is equal to the ratio of the dispersion length L_D to the Kerr length L_K

$$L_K = c^2\beta_0\left(2\pi\omega^2 A_0^2\,|\,\chi_{K,eff}\,|\right)^{-1}.$$

Here β_0 is the propagation constant depending on the frequency of the carrier wave ω_0 and $\chi_{K,eff}$ is the effective non-linear susceptibility responsible for the Kerr effect (These definitions will be discussed in more detail in the next chapter, in section 6.1.1). The angle brackets in (5.1.1a) denote summation over all normalised frequency detuning $\delta = \Delta\omega t_{p0}$ from the centre of the inhomogeneously broadened line, $\Delta\omega$ is the difference between the pulse carrier frequency and the atomic transition.

In contrast to a uniform infinite medium case, where Kerr susceptibility χ_K is constant, the effective non-linear susceptibility $\chi_{K,eff}$ depends on the mode of the wave propagating in fibre. This value can be written as

$$\chi_{K,eff} = \int \chi_K(\vec{\varrho})\,|\,\Psi(\vec{\varrho})\,|^4\, d\vec{\varrho} \Big/ \int |\,\Psi(\vec{\varrho})\,|^2\, d\vec{\varrho},$$

where $\vec{\varrho} = \{x, y\}$.

The dependence on the transverse co-ordinates $\vec{\varrho}$ is extracted from the variables describing the resonant atoms (i.e. impurities) and from the envelope q of the USP,

$$R_{12} = p(z,t)\Psi(\vec{\varrho}), \quad R_{11} - R_{22} = R_3(z,t)\Psi(\vec{\varrho}).$$

Here R_{ij} (i, j = 1,2) are the slowly varying density-matrix elements of the two-level atoms representing the models of resonant impurities.

To define the condition for the existence of non-broadening optical pulse we must find at what ratio of the parameters in (5.1.1) this system permits zero curvature representation.

Let matrices of the U-V-pair be chosen in the form

$$\hat{U} = \begin{pmatrix} -i\lambda & \alpha_1 q \\ \alpha_2 q^* & i\lambda \end{pmatrix}, \quad \hat{V} = \begin{pmatrix} A & B \\ C & -A \end{pmatrix}.$$

Constants a_1, a_2 and A, B, C (which are the functions of q, p, R_3, $\partial q/\partial\tau$) should be so chosen that the condition of zero curvature

$$\frac{\partial \hat{U}}{\partial \zeta} = \frac{\partial \hat{V}}{\partial \tau} + \left[\hat{V}, \hat{U}\right]$$

would coincide with (5.1.1). This expression can be rewritten in the expanded form

$$\frac{\partial A}{\partial \tau} = \alpha_1 q C - \alpha_2 q^* B, \tag{5.1.2a}$$

$$\frac{\partial B}{\partial \tau} + 2i\lambda B = \alpha_1 \frac{\partial q}{\partial \zeta} - 2\alpha_1 q A, \tag{5.1.2b}$$

$$\frac{\partial C}{\partial \tau} - 2i\lambda C = \alpha_2 \frac{\partial q^*}{\partial \zeta} + 2\alpha_2 q^* A, \tag{5.1.2c}$$

Let B and C have the form of the linear combination

$$\begin{aligned} B &= \langle b_1 p \rangle + b_2 q + b_3 \partial q / \partial \tau, \\ C &= \langle c_1 p^* \rangle + c_2 q^* + c_3 \partial q^* / \partial \tau, \end{aligned} \tag{5.1.3}$$

where the unknown coefficients b_j and c_j (j = 1,2,3) depend only on λ. By substituting (5.1.3) into (5.1.2b) and (5.1.2c), equating coefficients of p, p^*, $\partial q/\partial\tau$, $\partial q^*/\partial\tau$, $\partial^2 q/\partial\tau^2$, $\partial^2 q^*/\partial\tau^2$ and taking the systems of equations (5.1.1) into account we can obtain that

$$b_1 = \alpha_1 a/(\delta + 2\lambda), \quad b_2 = 2\lambda s\alpha_1, \quad b_3 = is\alpha_1,$$
$$c_1 = \alpha_2 a/(\delta + 2\lambda), \quad c_2 = 2\lambda s\alpha_2, \quad c_3 = -is\alpha_2,$$

and

$$A = \frac{i}{2}\left\{\mu|q|^2 - 4s\lambda^2 - af\left\langle\frac{R_3}{\delta + 2\lambda}\right\rangle\right\}. \tag{5.1.4a}$$

Thereby

$$B = \alpha_1\left\{2\lambda sq + is\frac{\partial q}{\partial\tau} + a\left\langle\frac{p}{\delta + 2\lambda}\right\rangle\right\}, \tag{5.1.4b}$$

$$C = \alpha_2\left\{2\lambda sq^* - is\frac{\partial q^*}{\partial\tau} + a\left\langle\frac{p^*}{\delta + 2\lambda}\right\rangle\right\}. \tag{5.1.4c}$$

Compatibility of (5.1.4) and (5.1.2a) imposes the constraints on the parameters of the problem, i.e., the following conditions must be satisfied:

$$-\mu = 2\alpha_1\alpha_2 s, \quad f^2 = -\alpha_1\alpha_2 .$$

Hence

$$\mu = 2sf^2 \tag{5.1.5a}$$

Since f is a positive value, equation (5.1.5a) leads to a relation for the signs of the Kerr susceptibility and the constant of the second order group velocities dispersion: $\operatorname{sign}\mu = s$. In terms of dimensioned variables, condition (5.1.5a) takes the form $L_D L_K^{-1} = 2f^2$ or

$$4\pi\left(\frac{\hbar\omega_0}{cd}\right)^2 |\chi_{K,eff}| = \frac{\partial^2\beta^2(\omega)}{\partial\omega^2} \approx \frac{2\lambda_0}{c}|D|, \tag{5.1.5b}$$

where the parameter of group velocities dispersion D is introduced.

The only condition for parameters α_1 and α_2 is the requirement that the spectral problem of the IST method be anti-Hermitian one. Hence it follows that $\alpha_1 = \alpha_2 = if$.

If there are no resonant impurities in a fibre then the existence of soliton would be guaranteed by the following conditions $\alpha_1 = \alpha_2 = i, \quad \mu = 2s$. In this case the system of equations (5.1.1) reduces to the NLS.

The presence of resonant impurities radically changes the situation: 2π-pulse of self-induced transparency should simultaneously be also a soliton of the NLS equation, i.e. the amplitude and duration of the 2π-pulse should precisely be of such values that the corresponding self-action (due to the high-frequency Kerr effect) would lead to complete compensation of the dispersion broadening of the USP.

Let us assume that a dipole moment d and Kerr susceptibility χ_K of a fibre do not depend on transverse co-ordinates, then

$$|\chi_{K,eff}|\bar{d}^{-2} = \Im|\chi_K|d^{-2},$$

where the geometric factor $\Im = \Im_4\Im_2/\Im_3^2$ is explicitly separated. Under condition (5.1.5) the difference between the cases of unbounded inhomogeneous medium and fibre is clearly seen. Here

$$\Im_j = \int|\Psi(\vec{\varrho})|^j\, d\vec{\varrho}, \qquad (j = 1 - 4).$$

The appearance of a geometric factor is typical for the problems of integrated and fibre optics. It is useful to estimate the influence of the geometric factor on the solitons existence condition (5.1.1). For a fundamental mode the Gaussian function of $|\vec{\varrho}|$ serves as good approximation of $\Psi(\vec{\varrho})$ with the half-maximum width equal to the effective thickness of the fibre light-guided core. For the fibre with the parabolic profile of the distribution of the refractive index it is at any rate the exact result. For this particular case the geometric factor is $\Im = 8/9$. Of course for each specific fibre and mode the value of the geometric factor will differ but it is unlikely to deviate greatly from unity.

Thereby, the existence of the optical solitons in a fibre doped with resonant impurities is restricted. It is more reasonable to consider these impurities as a perturbation factor for the optical soliton in picosecond range of duration (soliton of NLS). Therefore we can consider the influence of the resonant centres by means of the perturbation theory. In another limiting case the non-resonant effects of group velocities dispersion and Kerr-type non-linearity could play the role of perturbations for solitons of self-induced transparency phenomena (2π-pulses).

5.1.2. THE CASE OF VECTOR SOLITONS

Now let us consider ultrashort pulses propagation in a Kerr-type dispersive medium when the transition between energy levels of the impurity atoms are degenerated over the orientations of the total angular momentum j_a and j_b. The same system of equations appears when fibre contains three-level impurities atoms, so that formally we could speak about optical vector solitons in a general case.

Let the electric field strength write as

$$E^{(j)}(t,x,y,z) = A_0 q_j(t,z)\Psi(\vec{\varrho})\exp\{-i\omega_0 t + i\beta_0 z\}.$$

The equations for the normalised envelope $q_j(t,z)$ and the variables of the atomic resonant system can be written in a unified form as it was done above

$$i\frac{\partial q_j}{\partial \zeta}+s\frac{\partial^2 q_j}{\partial \tau^2}+\mu|\vec{q}|^2 q_j - a\langle P_j\rangle = 0,$$

$$\frac{\partial P_j^{(a)}}{\partial \tau}-i\delta P_j^{(a)} = -if\left\{\sum_l q_l M_{lj}^{(a)} - q_j N^{(a)}\right\} \qquad (5.1.6)$$

$$\frac{\partial M_{jl}^{(a)}}{\partial \tau} = -if\left\{q_j^* P_l^{(a)} - q_l P_j^{(a)*}\right\}, \quad \frac{\partial N^{(a)}}{\partial \tau} = -if\sum_j\left\{q_j P_j^{(a)*} - q_j^* P_j^{(a)}\right\}$$

where j and l mark the spherical components of the vectors, and $P_j = \sum_a \beta_a P_j^a$. The variables into the Bloch equations in the system (5.1.6) have been determined by the formulae from Chapter 4 (section 4.2.2). Parameter $f = \bar{d}A_0 t_{p0}/\hbar$ characterises the interaction of the USP with resonant atoms. The rest of parameters remained as before.

To find zero curvature representation, matrices $\hat{U}(\lambda)$ and $\hat{V}(\lambda)$ can be taken in the form

$$\hat{U}(\lambda) = \begin{pmatrix} -i\lambda & -i\vec{q} \\ -i\vec{q}^* & i\lambda\hat{I} \end{pmatrix}, \qquad \hat{V}(\lambda) = \begin{pmatrix} A & \vec{B} \\ \vec{C} & \hat{D} \end{pmatrix},$$

and then the matrix elements of matrix $\hat{V}(\lambda)$ should be resolved in the form of the linear combinations of variables characterising impurities and the USP electric field

$$\vec{B} = \sum_a \left\langle b_a(\lambda)\vec{P}^{(a)}\right\rangle + b_3(\lambda)\vec{q} + b_4(\lambda)\partial\vec{q}/\partial\tau,$$

$$\vec{C} = \sum_a \left\langle c_a(\lambda)\vec{P}^{(a)*}\right\rangle + c_3(\lambda)\vec{q}^* + c_4(\lambda)\partial\vec{q}^*/\partial\tau$$

Substitution of these expressions in the two last equations of system (4.2.15) and equating the coefficients at $\vec{P}^{(a)}$, $\vec{P}^{(a)*}$, $\partial\vec{q}/\partial\tau$, $\partial\vec{q}^*/\partial\tau$, $\partial^2\vec{q}/\partial\tau^2$, $\partial^2\vec{q}^*/\partial\tau^2$ provide

$$\vec{B} = -a\gamma_1\sum_a \beta_a\left\langle\left\langle \vec{P}^{(a)}\right\rangle\right\rangle + 2\lambda\gamma_1 s\vec{q} + i\gamma_1 s\partial\vec{q}/\partial\tau,$$

$$\vec{C} = -a\gamma_2\sum_a \beta_a\left\langle\left\langle \vec{P}^{(a)*}\right\rangle\right\rangle + 2\lambda\gamma_2 s\vec{q}^* - i\gamma_2 s\partial\vec{q}^*/\partial\tau,$$

where it is accounted that $\langle \vec{P} \rangle = \sum_a \beta_a \langle \vec{P}^{(a)} \rangle$ and <<...>> = <...(2λ + δ)$^{-1}$>.

Using these expressions together with the two first equations from the system (4.2.15), one can find A and $\hat{D}$ by integrating these equations:

$$A = A_0' - i\gamma_1\gamma_2(a/f)\sum_a \beta_a \langle\langle N^{(a)} \rangle\rangle - i\gamma_1\gamma_2 s(\vec{q}\cdot\vec{q}^*),$$

$$\hat{D} = \hat{D}_0 - i\gamma_1\gamma_2(a/f)\sum_a \beta_a \langle\langle \hat{M}^{(a)} \rangle\rangle + i\gamma_1\gamma_2 s(\vec{q}^*\otimes\vec{q}),$$

where A_0' and matrix $\hat{D}_0$ are the integrating constants. In order to define A_0' and $\hat{D}_0$, γ_1 and γ_2 it is necessary to substitute $\vec{B}$, $\vec{C}$, A and $\hat{D}$ into the two last equations from (4.2.15) and to find out the conditions how to convert these equations into identity. This procedure brings in the following expressions

$$A_0' = -(8/3)is\lambda^2, \quad \hat{D}_0 = (4/3)is\lambda^2\hat{1}$$

and

$$f^2 = -\gamma_1\gamma_2, \quad s\mu = -2\gamma_1\gamma_2.$$

Having chosen $\gamma_1 = \gamma_2 = -if$, we receive the condition of complete integrability of the system of equation (5.1.6) $\mu = 2sf^2$. It should be noted that this result has already been obtained (refer to (5.1.5)) in the analysis of the scalar solitons, with the same interpretation.

5.1.3. VECTOR SOLITON SOLUTIONS

In order to find soliton solutions of equations (5.1.6) by the IST method with the *U-V*-pair found above, we can make use of the fact that herein the well known Manakov spectral problem [48] takes place.

The solutions of equations (5.1.6) can be written as

$$fq_{-1} = 2iK_2^{(1)*}(\tau,\tau), \qquad fq_{+1} = 2iK_3^{(1)*}(\tau,\tau), \tag{5.1.7}$$

where $K_j^{(1)}(x,y)$ is the matrix elements of the three-component vector $\vec{K}^{(1)}(x,y)$ satisfying the equation:

$$-\vec{K}^{(1)}(x,y) = \vec{F}_1(x+y) + \int_x^\infty \vec{K}^{(1)}(x,u)\mathcal{R}(x,u,y)du. \tag{5.1.8}$$

This equation comes from the Gelfand-Levitan-Marchenko equations allowing to solve the inverse problem for Manakov spectral problem. Here

$$\mathcal{R}(x,u,y)=\int_x^{\infty}\sum_{s=2,3}F_{1s}^*(u+w)F_{1s}(w+y)dw$$

and

$$F_{1s}(x)=-i\sum_{j=1}^{N}\frac{C_{1s}^{(j)}}{T'_{1j}}\exp(i\lambda_j x)+\frac{1}{2\pi}\int_{-\infty}^{+\infty}\frac{T_{1s}(\xi)}{T_{11}(\xi)}\exp(i\xi x)d\xi .$$

The scattering data $C_{1s}^{(j)}$, $T_{1s}(\xi)$ and $T'_{1j}=(dT_{11}/d\lambda)(\lambda_j)$ depend on ζ by the formulae [42-46,48]:

$$C_{1s}^{(j)}(\zeta)=C_{1s}^{(j)}(0)\exp\{\Re(\lambda_j)\zeta\},$$
$$T_{1s}(\xi,\zeta)=T_{1s}(\xi,0)\exp\{\Re(\xi)\zeta\},$$

where

$$\Re(\lambda)=4is\lambda^2+i\left(\frac{\beta_1+\beta_2}{2\lambda+\delta}\right),$$

and λ_j are the points of discrete spectrum of the Manakov spectral problem, $T_{11}(\lambda_j)=0$, $\mathrm{Im}\lambda_j>0, j=1,\ldots,N$.

It is useful to consider the case of the discrete spectrum consisting of only one point $\lambda_j=\lambda_1$ (and moreover $\lambda_1=i\eta$) and $T_{1s}(\xi)=0$. This situation is well known [44,45] to correspond to a single-soliton solution of the completely integrable equations. The equation for $\vec{K}^{(1)}(x,y)$ is reduced to the system of linear equation which solution provides the result

$$f\cdot\vec{q}(\zeta,\tau)=2\eta\vec{l}\,\mathrm{sech}(2\eta\tau-\mathrm{Re}\,\Re_1\zeta-2\eta\tau_0)\exp[i\varphi], \tag{5.1.9}$$

where $\tau_0=\ln|\vec{C}(0)|/2\eta$, $\vec{l}=\vec{C}(0)/|\vec{C}(0)|$, vector $\vec{C}(0)=\{C_{11}^{(1)}(0),C_{12}^{(1)}(0)\}$ is formed from the initial conditions by solving spectral problem (1.11.a), $\Re_1=\Re(\lambda_1)$, and $\varphi(\zeta,\tau)$ $=\mathrm{Im}\,\Re_1\zeta+\arg[\vec{C}(0)]$. This is the polarised soliton, which propagates in a Kerr-type medium with the resonant degenerated energy levels impurities. These results are valid for transitions $j_1=0\to j_2=1$ $\beta_1=1,\beta_2=0)$, $j_1=1\to j_2=0$ $(\beta_1=0,\beta_2=1)$, $j_1=1\to j_2=1$ $(\beta_1=\beta_2=1/2)$. The case of resonant impurities complying to the model of three-level atoms of Λ- and *V*-configurations of levels belongs to the class of integrable systems too. These situations should be referred to as "simulton in the non-linear fibre".

The collision (i.e. the interaction) of the polarised solitons in such medium results in the changing of their polarisation vectors (or partial amplitudes of the simultons) $\vec{l}^{(1)}$ and $\vec{l}^{(2)}$. If we recall that the transformation law for $\vec{l}^{(1)}$ and $\vec{l}^{(2)}$ was defined solely by the form of *U*-matrices in a spectral problem (4.2.13a) or (4.2.13c), then the corresponding expressions must coincide with (4.2.47), (4.2.48) and (4.2.49). In the work [49] this statement was checked out by the analysis of double-soliton solution obtained from the solution system (5.1.6) provided the discrete spectrum was formed by two points $\lambda_{1,2} = i\eta_{1,2}$ only and there was no continuum spectrum at all.

5.2. Self-induced transparency in the media with spatial dispersion

The spatial dispersion appears with allowance for intermolecular interaction. It leads to an additional influence upon the process of USP propagation due to phase modulation. We can observe these effects in the frames of the simplest models. We follow here the work [19] where the model of the cubic molecular crystal with one two-level molecule per elementary cell was considered. In Heitler-London approximation the Hamiltonian of such systems of molecules interacting with electromagnetic wave has the form:

$$\hat{H} = \varepsilon\sum_{n} \hat{P}_n^+ \hat{P}_n + \sum_{m,n} \Delta(m-n)\hat{P}_n^+ \hat{P}_m + \hat{H}_{\text{int}},$$
$$\hat{H}_{\text{int}} = -\frac{g}{2}\sum \left\{\hat{P}_n^+ E^{(-)}(n) + \hat{P}_n E^{(+)}(n)\right\}, \tag{5.2.1}$$

where ε is an excitation energy of a molecule in the crystal, $\Delta(m-n)$ characterises the resonance interaction between nodes *m* and *n* of crystalline lattice, $\hat{P}_n$ is an annihilation operator of excitation on *n*th node, g is a constant of dipole interaction of the electromagnetic field with the two-level molecules. The electric field strength is presented as the sum of positive and negative frequency parts

$$E = E^{(+)} + E^{(-)}.$$

Furthermore, the terms responsible for harmonic generations and shifts of levels are omitted in Hamiltonian of the interactions of field with molecules. Such approach named "Rotating Wave Approximation" is often in use. Operators of creation and annihilation of molecular excitations satisfy the Pauli commutative relations.

In work [19] the classical equations for this models for polarisations $P_n(t) = <\hat{P}_n>$ and differences of the population of excited and ground states $n_n(t) = <\hat{P}_n^+ \hat{P}_n>$ were obtained. Here, as usual, the expectation value of the quantum mechanical operator $\hat{O}$ is denoted as $<\hat{O}>$. In one dimensional case and in continuum approximation these equations can be written as:

$$i\hbar\frac{\partial P}{\partial t}-\frac{\hbar^2 n}{2m}\frac{\partial^2 P}{\partial z^2}-\varepsilon P+\Delta_0 P=gnE^{(-)}\ , \qquad (5.2.2a)$$

$$i\hbar\frac{\partial n}{\partial t}+\frac{\hbar^2}{4m}\frac{\partial}{\partial z}\left(P^*\frac{\partial P}{\partial z}-P\frac{\partial P^*}{\partial z}\right)=\frac{g}{2}\left(E^{(+)}P-E^{(-)}P^*\right), \qquad (5.2.2b)$$

$$\frac{\partial^2 E^{(-)}}{\partial z^2}-\frac{1}{c^2}\frac{\partial^2 E^{(-)}}{\partial t^2}=\frac{4\pi n_A g}{c^2}\frac{\partial^2 <P>}{\partial t^2}\ . \qquad (5.2.2c)$$

Here the lattice variables have been changed in the following manner $P_n(t)\to P(z,t)$ and $n_n(t)\to n(z,t)$. The angle brackets in (5.2.2c) denote summation over all excitation energies of the molecules. While deriving the equations (5.2.2), the Fourier component of the resonant intermolecular interaction $\Delta(\vec{k})$ was expanded in a power series of wave vector $\vec{k}$ limited only by the second order:

$$\Delta(\vec{k})=\Delta_0+\frac{\hbar^2}{2m}k^2$$

where m is an efficient exciton mass. Similar equations were used in [20,23].

The next simplification is connected with the transition to slowly varying envelopes of optical pulse and polarisation:

$$E^{(-)}(z,t)=\mathcal{E}(z,t)\exp[-i\omega_0 t+ik_0 z],$$
$$P(z,t)=\mathcal{P}(z,t)\exp[-i\omega_0 t+ik_0 z]$$

For the complex envelope of USP, the polarisation and the population differences equations (5.2.2) lead to following:

$$i\hbar\frac{\partial\mathcal{P}}{\partial t}=(\varepsilon-\hbar\omega_0+\gamma n)\mathcal{P}+\frac{\hbar^2 k_0}{m}n\frac{\partial\mathcal{P}}{\partial z}+gn\mathcal{E}\ , \qquad (5.2.3a)$$

$$i\hbar\frac{\partial n}{\partial t}+i\frac{\hbar^2 k_0}{2m}\frac{\partial}{\partial z}\left(|\mathcal{P}|^2\right)-\frac{\hbar^2}{4m}\frac{\partial}{\partial z}\left(\mathcal{P}^*\frac{\partial\mathcal{P}}{\partial z}-\mathcal{P}\frac{\partial\mathcal{P}^*}{\partial z}\right)=\frac{g}{2}\left(\sigma\mathcal{P}^*-\mathcal{P}^*\mathcal{E}\right), \qquad (5.2.3b)$$

$$\left(\frac{\partial\mathcal{E}}{\partial z}+\frac{1}{c}\frac{\partial\mathcal{E}}{\partial t}\right)=i\frac{2\pi n_A g\omega_0}{c}<\mathcal{P}>\ , \qquad (5.2.3c)$$

where $\gamma=(\Delta_0+k_0^2/2m)$ and $k_0=\omega_0/c$.

If the interaction between molecules were absent, we would obtain a usual system of two-level molecules interacting with the electromagnetic field as it was considered in the theory of SIP by McCall-Hahn. The energy spectrum of this system is presented by a single highly degenerated energy level (multiplicity of degeneration is equal to the number of molecules in this system). Intermolecular interaction removes this degeneration and the energy level broadens into a band of allowed energies with a width of the order of Δ_0. Naturally, the next step to simplify the problem can be a transition to *heavy exciton approximation*. If to discard all terms proportional to m^{-1} in the system of equation (5.2.3), a new system of equations follows:

$$i\hbar\frac{\partial\mathcal{P}}{\partial t} = (\varepsilon - \hbar\omega_0 + \gamma n)\mathcal{P} + gn\mathcal{E} \;, \tag{5.2.4a}$$

$$i\hbar\frac{\partial n}{\partial t} = \frac{g}{2}\left(\mathcal{P}\mathcal{E}^* - \mathcal{P}^*\mathcal{E}\right), \tag{5.2.4b}$$

$$\left(\frac{\partial\mathcal{E}}{\partial z} + \frac{1}{c}\frac{\partial\mathcal{E}}{\partial t}\right) = i\frac{2\pi n_A g\omega_0}{c} < \mathcal{P} >, \tag{5.2.4c}$$

The set of equations of the same form was considered in works [20, 24-26].

Now we begin considering the solitary waves, and the boundary condition as $|t| \to \infty$

$$\mathcal{P} \to 0, \quad n \to n_0 = -1,$$

will be accepted. It means that in absence of the field the molecular system is in the ground state. It is convenient to transfer to real and non-dimensional variables. Let

$$\varepsilon = \hbar\omega_a, \quad \tau = \omega_a t, \quad \xi = \omega_a z / c,$$
$$q(\tau,\xi) = \frac{g\mathcal{E}}{\hbar\omega_a}$$

then the system of equations (5.2.4) reads

$$i\frac{\partial\mathcal{P}}{\partial\tau} = (\delta + \beta n)\mathcal{P} + nq, \tag{5.2.5a}$$

$$\frac{\partial n}{\partial\tau} = \frac{1}{2i}\left(\mathcal{P}q^* - \mathcal{P}^*q\right), \tag{5.2.5b}$$

$$\left(\frac{\partial q}{\partial\xi} + \frac{\partial q}{\partial\tau}\right) = i\alpha < \mathcal{P} >, \tag{5.2.5c}$$

where parameters characterising the intensity of the USP interaction with resonant molecules, the intermolecular interaction and the frequency detuning are introduced

$$\alpha = 2\pi g^2 n_A \omega_0 / \hbar\omega_a^2, \quad \beta = \left(\Delta_0 + k_0^2 / 2m\right) / \hbar\omega_a \approx (\Delta_0 / \varepsilon), \quad \delta = (\varepsilon - \hbar\omega_0) / \varepsilon .$$

Parameter α can be expressed in terms of the characteristic transition time t_c:

$$t_c^{-2} = 2\pi n_A g^2 \omega_a \hbar^{-1} = 2\pi n_A g^2 \varepsilon \hbar^{-2}$$

so $\alpha = (\omega_0 / \omega_a)(t_c \omega_a)^{-2}$.

It is noteworthy that parameter β which characterises intermolecular interaction is equal to the ratio of the exciton band width to photon energy or transition energy. So, it is natural to consider this ratio as a small value.

In order to obtain equations (5.2.5) in real variables, the real envelope and the phase of USP and the real components of the Bloch vector associated with the two-level system should be defined. Let

$$q = a \exp(i\phi), \quad \mathcal{P} = (v + iu)\exp(i\phi) .$$

In terms of these variables the real form of equations (5.2.5) can be written as:

$$\frac{\partial v}{\partial \tau} = \left(\delta + \frac{\partial \phi}{\partial \tau} + \beta n\right) u , \quad \frac{\partial u}{\partial \tau} = -\left(\delta + \frac{\partial \phi}{\partial \tau} + \beta n\right) v - an , \quad \frac{\partial n}{\partial \tau} = au , \qquad (5.2.6a)$$

$$\left(\frac{\partial a}{\partial \tau} + \frac{\partial a}{\partial \xi}\right) = -\alpha < u > , \quad a\left(\frac{\partial \phi}{\partial \tau} + \frac{\partial \phi}{\partial \xi}\right) = \alpha < v > . \qquad (5.2.6b)$$

Assume that there is no inhomogeneous broadening, and all dependent variables are the functions of only one variable $\eta = \omega_a (t - z/V)$, i.e., the characteristic for the wave propagating in one direction. The equations describing stationary propagation of an optical pulse follow from (5.2.6) and have the form

$$dv / d\eta = (\delta + d\phi / d\eta + \beta n) u , \qquad (5.2.7a)$$

$$du / d\eta = -(\delta + d\phi / d\eta + \beta n) v - an , \qquad (5.2.7b)$$

$$dn / d\eta = au , \qquad (5.2.7c)$$

$$(1 - c/V) da / d\eta = -\alpha u , \qquad (5.2.7d)$$

$$a(1 - c/V) d\phi / d\eta = \alpha v . \qquad (5.2.7e)$$

Keeping in mind the boundary conditions, from (5.2.7c) and (5.2.7d) it follows that

$$n(\eta) = n_0 + \frac{1}{2\alpha}\left(\frac{c}{V} - 1\right)a^2(\eta) \ , \tag{5.2.8}$$

and besides this, as with two-level models, the length of the Bloch vector conserves in the absence of relaxation:

$$u^2(\eta) + v^2(\eta) + n^2(\eta) = n_0^2 = 1 . \tag{5.2.9}$$

So far as the absorbing medium is considered, from (5.2.8) the inequality

$$(c/V - 1) > 0$$

comes out. It means that the velocity of propagation of the steady-state pulse is less than velocity of light in a medium. (In the inverted medium everything would be vice versa). It is convenient to re-write the expression (5.2.8) by introducing the parameter

$$a_0^2 = \frac{4\alpha V}{c - V},$$

so that

$$n(\eta) = -1 + 2\frac{a^2(\eta)}{a_0^2} . \tag{5.2.10}$$

Equations (5.2.7d) and (5.2.7e) result in

$$u = \frac{4}{a_0^2}\left(\frac{da}{d\eta}\right), \quad \frac{d\phi}{d\eta} = -\frac{a_0^2}{4}\left(\frac{v}{a}\right).$$

Substitution of these expressions into (5.2.7a) gives rise to the equation

$$\frac{d(av)}{d\eta} = \frac{4}{a_0^2}\left\{(\delta + \beta)a\frac{da}{d\eta} - \frac{2\beta}{a_0^2}a^3\frac{da}{d\eta}\right\}.$$

Hence the integral of the system (5.2.7) follows

$$av - \frac{2}{a_0^2}\left\{(\delta + \beta)a^2 - \frac{\beta}{a_0^2}a^4\right\} = \text{const} .$$

Under the assumption that the field and polarisation vanish simultaneously one can find that this constant is zero. Now the expression for the instant frequency of carrying wave of the USP (or phase modulation) can be written

$$\frac{d\phi}{d\eta} = -\frac{1}{2}\left\{\Delta - \frac{\beta}{a_0^2}a^2\right\}, \tag{5.2.11}$$

where the re-normalised detuning from resonance is introduced by

$$\Delta = \delta + \beta \approx (\varepsilon + \Delta_0 - \hbar\omega_0)/\varepsilon .$$

In order to get an equation for the real envelope of USP, we can make use of the integral of motion (5.2.9) and the expressions for Bloch vector components in terms of re-normalised real envelope of USP $w = (a/a_0)$. It provides the following equation

$$2\left(\frac{dw}{d\eta}\right)^2 = w^2\left[(a_0^2 - \Delta^2) + (2\beta\Delta - e_0^2)w^2 - \beta^2 w^4\right].$$

If one turns now to a variable $f = w^{-2}$, a new equation will be obtained:

$$\frac{df}{d\eta} = -\left[(a_0^2 - \Delta^2)f^2 - (a_0^2 - 2\beta\Delta)f - \beta^2\right]^{1/2}. \tag{5.2.12}$$

The behaviour of the solution of this equation depends on the sign of the factor at the senior degree of polynomial under the radical, namely $(a_0^2 - \Delta^2)$. Under exact (re-normalised) resonance $\Delta = 0$, so it naturally to begin the analysis of the solution of the equation (5.2.12) with small Δ, by supposing that

$$(a_0^2 - \Delta^2) = \theta^{-2} > 0. \tag{5.2.13}$$

In this case the change of variables

$$f = f_0 + f_1 \cosh\tilde{\varphi},$$

where

$$f_0 = \theta^2(a_0^2 - 2\beta\Delta)/2, \quad f_1 = (f_0^2 + \beta^2\theta^2)^{1/2},$$

reduces equation (5.2.12) to

$$d\tilde{\varphi}/d\eta = -\theta^{-1}.$$

Thereby, we can write expression for the real USP envelope

$$a^2(\zeta) = \frac{a_0^2}{f_0 + f_1 \cosh[(\eta - \eta_0)/\theta)}, \tag{5.2.14}$$

where integrating constant η_0 can be chosen arbitrarily so far as it defines a position of a maximum of USP, and the initial equations themselves possess translation symmetry, i.e. they are invariant with respect to Galilean transformations.

Solution of equation system (5.2.7) (or (5.2.6)) founded here, describes a steady-state pulse propagation in the non-linear medium with the spatial dispersion and in this sense it generalises 2π-pulse by McCall-Hahn. But, strictly speaking, this is not a soliton solution because we know nothing of the complete integrability of the equations (5.2.6). Some numerical studies [27] exhibited the unstable behaviour of these pulses in their mutual collisions.

Velocity of the steady-state USP propagation can be found from the expression for the parameter $(a_0^2 - \Delta^2)$:

$$\frac{1}{V} = \frac{1}{c}\left[1 + \frac{4\alpha\theta^2}{1 + \Delta^2\theta^2}\right] = \frac{1}{c}\left[1 + \frac{4(\omega_0/\omega_a)(t_p/t_c)(1/t_c\omega_a)}{1 + (\Delta\omega)^2 t_p{}^2}\right]. \tag{5.2.15}$$

Here the steady-state pulse duration t_p and frequency detuning $\Delta\omega$ are introduced

$$t_p = \theta/\omega_a, \quad \Delta\omega = (\varepsilon + \Delta_0 - \hbar\omega_0)/\hbar.$$

The increasing of normalised detuning leads to equality $(a_0^2 - \Delta^2)$ =0. In this marginal case the solution of equation (5.2.12) exists only if $a_0 < 2\beta$, and it has the form

$$a^2(\eta) = \frac{2a_0^2(2\beta a_0 - a_0^2)}{2\beta^2 - (\eta - \eta_0)^2}. \tag{5.2.16}$$

Velocity of the propagation for such "rational" steady-state USP is given by the formula:

$$\frac{1}{V} = \frac{1}{c}\left[1 + \frac{4\alpha}{1 + \Delta^2}\right] = \frac{1}{c}\left[1 + \frac{4(\omega_0/\omega_a)}{1 + (t_c\Delta\omega)^2}\right].$$

The further increase of this detuning, when the parameter $(a_0^2 - \Delta^2)$ stays negative, makes the existence of solitary wave impossible, but stationary periodic waves may exist.

5.3. Propagation of ultrashort pulses in a non-linear medium

The recent progress in the field of generation of femtosecond pulses [36-40] has made it necessary to revise theoretical models of their propagation in a non-linear dispersive medium. Indeed, the derivation of the evolutionary equations for electromagnetic radiation is often based on the approximation of slowly varying complex envelopes of the optical pulses. In this case Maxwell equations or d'Alembert wave equations are reduced to the equation for envelope. Such approximation is adequate when the pulse duration appreciably exceeds the optical period $T_{opt} = 2\pi / \omega_0$, where ω_0 is the carrier wave frequency. The situation may change in the femtosecond range and the reduced equations may become invalid. It is interesting to find a method to describe an ultrashort pulse (USP) evolution without the use of slowly varying envelope approximation. The simplest way to do this is to combine the wave equation for electromagnetic field with the equations specifying the changes in the state of the medium. This can be done provided the proper model of the medium is chosen. The typical model is an ensemble of N-level atoms. If the pulse duration is much shorter than all relaxation times in the atomic system, the USP propagation is accompanied only by the induced absorption and reemission processes. At $N = 2$ this process develops in the self-induced transparency effect considered above.

Following [50, 51], we can arrange the hierarchy of the different degrees of approximation in the theory of coherent interaction and propagation of optical pulses in resonance media.

(•) *Resonance condition.* The USP spectral width is much smaller than the resonance frequency. The wave equation in this case is complemented by the Bloch equations [51-56]. The system of Maxwell-Bloch equations has a steady-state solution rather than soliton solutions. These solutions were obtained in Chapter 4, section 4.1.1 and 4.1.2. The instability of these solitary steady-state waves under collision was demonstrated by numerical simulation.

(••) *Unidirectional propagation condition.* When the concentration of the resonance atoms is sufficiently low, we can neglect the interaction between the counter-propagating waves [50, 51]. Under this approximation the Maxwell-Bloch equations transform into reduced Maxwell-Bloch (RMB) equations. The procedure of such reduction is demonstrated in section 4.1.1. In the scalar case these equations have soliton solutions [50, 57].

(•••) *Slowly varying envelope approximation.* This means that the complex envelope of the USP obeys the inequalities

$$\left|\frac{\partial \mathcal{E}}{\partial t}\right| << \omega_0 |\mathcal{E}|, \quad \left|\frac{\partial \mathcal{E}}{\partial z}\right| << k_0 |\mathcal{E}|,$$

where ω_0 is a carrier wave frequency and k_0 is a wave number corresponding ω_0. The RMB equations then transform into the system of McCall-Hahn equations (i.e. SIT-equations).

It is noteworthy that Bloch equations are referred to as an example of material equations. We can use any model equations for medium description. There is, for instance, the non-linear oscillator model, the plasma of electrons in metals, conductive electrons in semiconductors, excitons in molecular systems and so on.

In the case of resonant medium we have the resonance transition frequency ω_a as a scale parameter for time. When pulse duration t_p obeys the inequality $t_p\omega_a >> 1$, the slowly varying envelope approximation is adequate to describe the pulse propagation. On the contrary, if $t_p\omega_a << 1$ we can use at least the unidirectional propagation approximation. The ratio $\varepsilon = \omega_R / \omega_a$, where ω_R is the Rabi frequency, provides a new small parameter. So, we can attempt to solve Bloch equations in a certain order of ε, then find the polarisation in the same order of ε, and thus obtain an approximate equations of the USP electric field strength without the assumption of slowly varying envelope. The condition $\varepsilon \approx 1$ means that the amplitude of the electric field strength approximately equals he strength of atomic field. Thus parameter $\varepsilon = \omega_R / \omega_a$ defines the meaning of the term "strong field".

Below we will consider some cases of USP interaction with resonant medium when the slowly varying envelope approximation is not assumed.

5.3.1. SIT WITHOUT THE SLOWLY VARYING ENVELOPES APPROXIMATION

The system of equations, describing the propagation of a scalar ultrashort pulse without the approximation of slowly varying envelopes, was obtained in section 4.1. as the RMB-equations (4.1.5). In terms of new normalised variables

$$\tau = t - z/c, \quad \zeta = (4\pi n_A d^2 / c\hbar) z, \quad q = (2d/\hbar)E,$$

the RMB-equations can be rewritten in the form

$$\begin{gathered} \partial q / \partial \zeta = - \partial r_1 / \partial \tau , \\ \partial r_1 / \partial \tau = -\omega_a r_2 , \quad \partial r_2 / \partial \tau = \omega_a r_1 + q r_3 , \quad \partial r_3 / \partial \tau = -q r_2 . \end{gathered} \tag{5.3.1}$$

The direct verification shows that the *U-V*-matrices in zero-curvature representation of the equations (5.3.1) have the form [57,58]:

$$\hat{U}(\lambda) = \begin{pmatrix} -i\lambda & iq(\tau,\zeta)/2 \\ iq(\tau,\zeta)/2 & i\lambda \end{pmatrix}, \quad \hat{V}(\lambda) = \begin{pmatrix} A(\lambda;\tau,\zeta) & B(\lambda;\tau,\zeta) \\ C(\lambda;\tau,\zeta) & -A(\lambda;\tau,\zeta) \end{pmatrix},$$

where

$$A(\lambda;\tau,\zeta)=\frac{i\lambda\omega_a r_3(\tau,\zeta;\omega_a)}{4\lambda^2-\omega_a^2},$$

$$B(\lambda;\tau,\zeta)=\frac{\lambda\omega_a}{4\lambda^2-\omega_a^2}\left\{r_2(\tau,\zeta;\omega_a)+\frac{i\omega_a}{2\lambda}r_1(\tau,\zeta;\omega_a)\right\},$$

$$C(\lambda;\tau,\zeta)=\frac{-\lambda\omega_a}{4\lambda^2-\omega_a^2}\left\{r_2(\tau,\zeta;\omega_a)-\frac{i\omega_a}{2\lambda}r_1(\tau,\zeta;\omega_a)\right\}.$$

If we assume, that before the arrival of an ultrashort pulse all two-level atoms are in the ground state and after the passing of the USP all atoms recover in the initial states, then boundary conditions are

$$\lim_{|\tau|\to\infty}P_3(\tau,\zeta;\omega_a)=-1,\quad \lim_{|\tau|\to\infty}P_{1,2}(\tau,\zeta;\omega_a)=0.$$

The RMB-equations with these boundary conditions can be solved by the IST method in a regular way.

According to the results of section 3.3 the N-soliton solution of the RMB-equations can be written as

$$q(\tau,\zeta)^2=4\frac{d^2}{d\tau^2}\ln\det(1+\hat{H}^*\hat{H}),\tag{5.3.2}$$

where matrix $\hat{H}$ is defined by its matrix elements as

$$H_{nm}=\frac{(C_nC_m)^{1/2}\exp[i\tau(\lambda_n-\lambda_m^*)]}{\lambda_n-\lambda_m^*}$$

and

$$C_n(\zeta)=C_n(0)\exp\left\{\frac{2i\lambda_n\omega_a}{4\lambda_n^2-\omega_n^2}\zeta\right\}.$$

Now, we shall transform this solution following [57]. Matrix $\hat{J}$ is determined as

$$(\hat{H}\cdot\hat{J})_{nm}=\frac{-i\exp[i(\lambda_n+\lambda_m)\tau-\alpha_n-\alpha_m]}{\lambda_n+\lambda_m},\tag{5.3.3a}$$

$$(\hat{J}\cdot\hat{H}^{-1})_{nm}=\frac{-i\exp[i(\lambda_n+\lambda_m)\tau+\alpha_n+\alpha_m+2(\beta_n+\beta_m)]}{\lambda_n+\lambda_m}.\tag{5.3.3b}$$

Where parameters α_n and β_n are established by expressions

$$iC_n(\tau) = \exp[-2\alpha_n(\tau)], \qquad \frac{\prod_{j}(\lambda_j + \lambda_n)}{\prod_{j \neq n}(\lambda_j - \lambda_n)} = -i\exp[2\beta_n]$$

We have taken into account that $q(x)$ is a real quantity. It implies that λ_m and C_n are either purely imaginary or they are the part of anti-Hermitian pairs

$$\lambda^*_m = -\lambda_n, \qquad C^*_m = -C_n.$$

Further, $\hat{H} \cdot \hat{J} = \hat{J} \cdot \hat{H}^*$ so that equations (5.3.3) yield

$$(\hat{J} \cdot \hat{H}^{-1} + \hat{J} \cdot \hat{H}^*)_{nm} = \exp(\beta_n + \beta_m) M_{nm},$$

where

$$M_{nm} = \frac{\cosh(\vartheta_n + \vartheta_m)}{2i(\lambda_n + \lambda_m)}$$

and

$$\vartheta_n = \frac{1}{4}\left(E_n \tau - \left\langle \frac{4\omega_a E_n}{E_n^2 + \omega_a^2} \right\rangle \zeta + \delta_n \right), \quad E_n = 4i\lambda_n.$$

By taking into account all expressions above we finally obtain an elegant formula

$$q(\tau, \zeta)^2 = 4 \frac{d^2}{d\tau^2} \ln \det(\hat{M}). \tag{5.3.4}$$

So far as λ_m are composed of purely imaginary numbers L_1 and anti-Hermitian pairs L_2 (so that $N = L_1 + 2L_2$), the solution of RMB-equations consists of L_1 single solitons and L_2 breathers (or bions - soliton-antisoliton bounded state). Breather is a solitary wave with internal oscillations. It has the same collision stability as ordinary solitons in both the bion-bion and the bion-soliton collisions. It is worth to note that the Sine-Gordon equations has the same breather solution, which is studied quite well.

The single soliton solution of the RMB-equations follows from (5.3.4)

$$E(\tau, \zeta) = E_1 \operatorname{sech}\left[\frac{1}{2} E_1 (t - z/V_1)\right], \tag{5.3.5}$$

where the pulse group velocity V_1 can be written as

$$\frac{1}{V_1}=\frac{1}{c}\left(1+\left\langle\frac{\alpha'\omega_a}{E_1^2+4\omega_a^2}\right\rangle\right),$$

where $\alpha'=4\pi n_A d^2/\hbar\omega_a$. This is the RMB version of the soliton solution of the complete Maxwell-Bloch equations (4.1.1) [29-32, 59]. This sort of USP has no carrier wave and represents a unipolar spike of electromagnetic radiation. Sometimes these pulses are named the *video pulses*.

Among two-soliton solutions of the RMB-equations there is one which looks like following

$$E(t,z)=\left(\frac{E_1^2-E_2^2}{E_1^2+E_2^2}\right)\frac{E_1\operatorname{sech}\vartheta_1+E_2\operatorname{sech}\vartheta_2}{\{1-B_{12}[\tanh\vartheta_1\tanh\vartheta_2-\operatorname{sech}\vartheta_1\operatorname{sech}\vartheta_2]\}},\tag{5.3.6}$$

where

$$B_{12}=2E_1E_2(E_1^2+E_2^2)^{-1}\text{ and }\vartheta_n=\frac{E_n}{2}\left[t-\frac{z}{c}\left\langle1+\frac{4\alpha'\omega_a}{E_n^2+4\omega_a^2}\right\rangle\right].$$

This solution describes the collision of two video pulses in the same fashion as it was done for two soliton solution of SIT equations by McCall-Hahn. However, the two-soliton solution of the RMB-equations can be used to obtain the generation of McCall-Hahn 2π-pulses. By setting λ_1 and λ_2 or E_1 and E_2 to be a pair of anti-Hermitian complex quantities (i.e., $E_1=-E_2^*$) we obtain the breather, i.e., a real solution in the form of localised pulse with internal oscillations. This is an exact analogue of the McCall-Hahn 0π-pulse.

Let $E_1=-E_2^*=E_0+2i\Omega$ and $\delta_1=-\delta_2^*=\delta'+i\delta''$. Then expression (5.3.6) yields the exact solution of (5.3.1)

$$E_b(t,z)=2E_0\operatorname{sech}\vartheta_R\left(\frac{\cos\vartheta_I-\gamma\sin\vartheta_I\tanh\vartheta_R}{1+\gamma^2\sin^2\vartheta_I\operatorname{sech}^2\vartheta_R}\right),\tag{5.3.7}$$

where $\gamma=E_0/2\Omega$,

$$\vartheta_R=\frac{1}{2}E_0\left(t-\frac{z}{c}\left\langle1+\frac{4\alpha'\omega_a\{E_0^2+4(\omega_a^2+\Omega^2)\}}{E_0^2+8E_0^2(\omega_a^2+\Omega^2)+16(\omega_a^2+\Omega^2)^2}\right\rangle\right)+\delta',$$

$$\vartheta_I=\Omega\left(t-\frac{z}{c}\left\langle1+\frac{4\alpha'\omega_a\{4(\omega_a^2-\Omega^2)-E_0^2\}}{E_0^2+8E_0^2(\omega_a^2+\Omega^2)+16(\omega_a^2+\Omega^2)^2}\right\rangle\right)+\delta''.$$

As in the theory of 0π-pulse by McCall-Hahn, solution (5.3.7) of the RMB *field* equation has a zero pulse area. Let us choose the magnitudes of parameters E_0 and Ω such that $E_0 << \Omega$. Expanding (5.3.7) to the zero order of γ, we have

$$E_b(t,z) \approx 2E_0 \,\mathrm{sech}\, \vartheta_R \cos\vartheta_I .$$

Thus we got the result that the 2π-pulse of McCall-Hahn is the limiting case of the 0π-pulse of the RMB equations. By expanding the expression (5.3.7) in a power series of γ, we can obtain the corrections to the McCall-Hahn steady-state pulse (2π-pulse). In the first order of γ expression (5.3.7) can be written as

$$E_b(t,z) \approx 2E_0 \,\mathrm{sech}\, \vartheta_R \cos\{\vartheta_I + \phi(t,z)\}, \tag{5.3.8}$$

where $\phi(t,z) = \gamma \tanh \vartheta_R$. Formula (5.3.8) represents a 2π-pulse which is "chirped". Defining the chirping as $\Delta\omega_{\mathrm{ch}} = \partial\phi / \partial t$, we find the chirp frequency ratio as

$$\frac{\Delta\omega_{\mathrm{ch}}}{\omega_a} = \gamma^2 \,\mathrm{sech}^2\, \vartheta_R .$$

5.3.2. ULTRASHORT PULSE OF POLARIZED RADIATION IN RESONANT MEDIUM

We now consider the ultrashort electromagnetic pulse propagation in a resonant medium which contains two-level atoms with quantum transitions between the energy levels degenerated with respect to the projections of an angular momenta j_a and j_b [60,61]. Our concern is the transition $j_a = 1 \to j_b = 0$. The following notations:

$$\rho_{12} = <a,-1|\hat{\rho}|a,+1>, \quad \rho_{13} = <a,-1|\hat{\rho}|b>,$$
$$\rho_{23} = <a,+1|\hat{\rho}|b>, \quad \rho_{11} = <a,-1|\hat{\rho}|a,-1>,$$
$$\rho_{22} = <a,+1|\hat{\rho}|a,+1>, \quad \rho_{33} = <b|\hat{\rho}|b>,$$
$$\rho_{kl} = \rho_{lk}^{*}, \quad l,k = 1,2,3$$

are introduced for the elements of the density matrix $\hat{\rho}$, describing the transitions between quantum states

$$|a,m> = |j_a = 1, m = \pm 1>, \qquad |b> = |j_b = 0, m = 0>$$

The generalised system of Maxwell-Bloch equations can be written in the form:

$$\frac{\partial^2 E^{(+1)}}{\partial z^2} - \frac{1}{c^2}\frac{\partial^2 E^{(+1)}}{\partial t^2} = \frac{4\pi n_A}{c^2}\frac{\partial^2}{\partial t^2}\langle d_{13}\rho_{31} + d_{31}\rho_{13}\rangle, \tag{5.3.9a}$$

$$\frac{\partial^2 E^{(-1)}}{\partial z^2} - \frac{1}{c^2}\frac{\partial^2 E^{(-1)}}{\partial t^2} = \frac{4\pi n_A}{c^2}\frac{\partial^2}{\partial t^2}\langle d_{23}\rho_{32} + d_{32}\rho_{23}\rangle, \tag{5.3.9b}$$

$$i\hbar\frac{\partial \rho_{13}}{\partial t} = -\hbar\,\omega_a\rho_{13} + d_{13}(\rho_{33} - \rho_{11})E^{(+1)} - d_{23}\rho_{12}E^{(-1)}, \tag{5.3.10a}$$

$$i\hbar\frac{\partial \rho_{23}}{\partial t} = -\hbar\omega_a\rho_{23} + d_{23}(\rho_{33} - \rho_{22})E^{(-1)} - d_{13}\rho_{21}E^{(+1)}, \tag{5.3.10b}$$

$$i\hbar\frac{\partial \rho_{12}}{\partial t} = d_{13}\rho_{32}E^{(+1)} - d_{32}\rho_{13}E^{(-1)}, \tag{5.3.10c}$$

$$i\hbar\frac{\partial}{\partial t}(\rho_{11} - \rho_{33}) = 2\left(d_{13}\rho_{31} - d_{31}\rho_{13}\right)E^{(+1)} + \left(d_{23}\rho_{32} - d_{32}\rho_{23}\right)E^{(-1)}, \tag{5.3.10d}$$

$$i\hbar\frac{\partial}{\partial t}(\rho_{22} - \rho_{33}) = \left(d_{13}\rho_{31} - d_{31}\rho_{13}\right)E^{(+1)} + 2\left(d_{23}\rho_{32} - d_{32}\rho_{23}\right)E^{(-1)}. \tag{5.3.10e}$$

Where $E^{(l)}$ is a spherical l-component of the electromagnetic wave electric field strength vector, $l = \pm 1$, d_{kn} – are reduced matrix elements of the dipole operator for $j_a = 1 \to j_b = 0$ transition. Hereafter $d_{13} = d_{23} = d_{31}^* = d_{32}^*$. In equations (5.3.9) n_A is the density of resonant atoms, the corner brackets mean a summation over all atoms with the transition frequency ω_a.

Let us consider the sharp-line limit for the resonant medium. It is convenient to transfer to the real variables and the normalised strength of the electric field by the formulae:

$$\begin{gathered} \rho_{13} = r_1 + ir_2, \quad \rho_{23} = s_1 + is_2, \quad \rho_{12} = p_1 + ip_2, \\ \rho_{33} - \rho_{11} = n_1, \quad \rho_{33} - \rho_{22} = n_2, \\ \frac{dE^{(+1)}}{\hbar\omega_a} = q_1, \quad \frac{dE^{(-1)}}{\hbar\omega_a} = q_2 \end{gathered} \tag{5.3.11}$$

The generalised system of Maxwell-Bloch equations can be rewritten in new notations as the following

$$\frac{\partial r_1}{\partial \tau} = -r_2 - q_2 p_2\,, \quad \frac{\partial r_2}{\partial \tau} = r_1 - q_1 n_1 + q_2 p_1\,, \quad \frac{\partial n_1}{\partial \tau} = 4q_1 r_2 + 2q_2 s_2\,,$$
$$\frac{\partial s_1}{\partial \tau} = -s_2 + q_1 p_2\,, \quad \frac{\partial s_2}{\partial \tau} = s_1 - q_2 n_2 + q_1 p_1\,, \quad \frac{\partial n_2}{\partial \tau} = 2q_1 r_2 + 4q_2 s_2\,, \qquad (5.3.12)$$
$$\frac{\partial p_1}{\partial \tau} = -q_1 s_2 - q_2 r_2\,, \quad \frac{\partial p_2}{\partial \tau} = -q_1 s_1 + q_2 r_1\,,$$

$$\left(\frac{\partial^2 q_1}{\partial \xi^2} - \frac{\partial^2 q_1}{\partial \tau^2}\right) = \alpha \frac{\partial^2 r_1}{\partial \tau^2}, \quad \left(\frac{\partial^2 q_2}{\partial \xi^2} - \frac{\partial^2 q_2}{\partial \tau^2}\right) = \alpha \frac{\partial^2 s_1}{\partial \tau^2}, \qquad (5.3.13)$$

where $\tau = \omega_a t,\ \ \xi = \omega_a z c^{-1}$, and $\alpha = 8\pi n_A \,|\, d\,|^2 / \hbar\omega_a$. These equations describe the propagation of an ultrashort pulse of polarised light in resonant medium. It is unlikely to find an exact solution of this system by analytical methods, but in any case we can try to find the stationary solution following the conventional approach used in investigations of non-linear waves.

Assume that the desired solution of the equations (5.3.12) and (5.3.13) is represented by the functions depending on one variable $\eta = \omega_a (t \pm x / V)$. Then all the equations of (5.3.12) and (5.3.13) can be transformed to obtain the following

$$\left(\frac{c^2}{V^2} - 1\right)\frac{d^2 q_1}{d\eta^2} = \alpha \frac{d^2 r_1}{d\eta^2}, \quad \left(\frac{c^2}{V^2} - 1\right)\frac{d^2 q_2}{d\eta^2} = \alpha \frac{d^2 s_1}{d\eta^2}, \qquad (5.3.14)$$

and

$$\frac{dr_1}{d\eta} = -r_2 - q_2 p_2, \qquad (5.3.15a)$$
$$\frac{dr_2}{d\eta} = r_1 - q_1 n_1 + q_2 p_1, \qquad (5.3.15b)$$
$$\frac{ds_1}{d\eta} = -s_2 + q_1 p_2, \qquad (5.3.15c)$$
$$\frac{ds_2}{d\eta} = s_1 - q_2 n_2 + q_1 p_1, \qquad (5.3.15d)$$
$$\frac{dp_1}{d\eta} = -q_1 s_2 - q_2 r_2, \quad \frac{dp_2}{d\eta} = -q_1 s_1 + q_2 r_1, \qquad (5.3.15e)$$
$$\frac{dn_1}{d\eta} = 4q_1 r_2 + 2q_2 s_2, \quad \frac{dn_2}{d\eta} = 2q_1 r_2 + 4q_2 s_2\,. \qquad (5.3.15f)$$

By integrating equations (5.3.14) with the corresponding boundary conditions and utilising equations (5.3.15a) and (5.3.15c), one can find

$$dq_1 / d\eta = -q_0^2 (r_2 + p_2 q_2), \quad dq_2 / d\eta = -q_0^2 (s_2 + p_2 q_1) . \tag{5.3.16}$$

After the second integrating we have

$$q_1 = q_0^2 r_1 , \quad q_2 = q_0^2 s_1 , \tag{5.3.17}$$

where $q_0^2 = \alpha \ V^2 (c^2 - V^2)^{-1}$. On substitution (5.3.17) into the equation for p_2 one can see p_2 be constant value. This constant is equal to zero due to the boundary conditions. Therefore, instead of equations (5.3.16), we have

$$dq_1 / d\eta = -q_0^2 r_2 , \quad dq_2 / d\eta = -q_0^2 s_2 . \tag{5.3.18}$$

If we substitute (5.3.18) into the equations for population differences (5.3.15f), the resulted equations are the complete derivatives and can be integrated:

$$n_1 + q_0^{-2} \left(2q_1^2 + q_2^2 \right) = n_{10} , \qquad n_2 + q_0^{-2} \left(q_1^2 + 2q_2^2 \right) = n_{20} . \tag{5.3.19}$$

Here n_{10} and n_{20} are the integrals of motion for equations (5.3.14) and (5.3.15). Equation for p_1 with the account of (5.3.18) takes the form

$$\frac{dp_1}{d\eta} = q_0^{-2} \left(q_1 \frac{dq_2}{d\eta} + q_2 \frac{dq_1}{d\eta} \right),$$

hence, due to the boundary condition it follows that

$$p_1 = q_0^{-2} q_1 q_2 . \tag{5.3.20}$$

The formulae (5.3.17) - (5.3.20) express all variables of the medium in terms of the field variables q_1 and q_2. Substitution of these expressions into (5.3.15b) and (5.3.15d) results in equations for the normalised electrical field strength of USP:

$$\frac{d^2 q_1}{dy^2} + (q_1^2 + q_2^2) q_1 = \frac{1}{2} (q_0^2 n_{10} - 1) q_1 , \tag{5.3.21a}$$

$$\frac{d^2 q_2}{dy^2} + (q_1^2 + q_2^2) q_2 = \frac{1}{2} (q_0^2 n_{20} - 1) q_2 , \tag{5.3.21b}$$

where $y = \sqrt{2}\eta$. Let us introduce the parameters

$$\Omega = \frac{1}{4}\left[q_0^2\left(n_{10} + n_{20}\right) - 2\right], \quad \mu = \frac{1}{4}\left[q_0^2\left(n_{10} - n_{20}\right)\right].$$

If one assume that all atoms of a resonant medium are not excited in the absence of USP, i.e., $n_{10} = n_{20} = 1$, then μ is zero. This is the case of an *unpolarised resonant medium.* If the resonant atoms are specially prepared, for example, by irradiating with a weak signal of circular polarised resonant radiation (there is no limitation on the duration of a signal), the resonant medium will be polarised before the USP arrival. In this case $n_{10} \neq n_{20}$ and $\mu \neq 0$.

Let us consider the case when the populations of the excited levels are not equal $n_{10} \neq n_{20}$. Denote

$$a_1^2 = \frac{1}{2}\left(q_0^2 n_{10} - 1\right), \quad a_2^2 = \frac{1}{2}\left(q_0^2 n_{20} - 1\right)$$

and suppose that [62, 63] $q_1 = g/f$, $q_2 = h/f$, where

$$q_1^2 + q_2^2 = 2\frac{d^2}{dy^2}\ln f.$$

In terms of Hirota's *D*-operators [64]

$$D(a \cdot b) = \left(\frac{da}{dy}b - a\frac{db}{dy}\right)$$

the equations (5.3.21) can be transformed into bilinear form

$$D^2(g \cdot f) = a_1^2 gf, \quad D^2(h \cdot f) = a_2^2 hf, \quad D^2(f \cdot f) = g^2 + h^2. \quad (5.3.22)$$

Several useful relationships are known for *D*-operators [64, 65]

$$D^m(a \cdot 1) = \frac{d^m a}{dy^m}, \quad D^m(a \cdot b) = (-1)^m D^m(b \cdot a), \quad D^{2m}(a \cdot a) = 2D^{2m-1}\left(\frac{da}{dy} \cdot a\right)$$

$$\frac{d^2 \ln f}{dy^2} = \frac{1}{2f^2} D^2(f \cdot f),$$

$$D^m\left(\exp[k_1 y] \cdot \exp[k_2 y]\right) = (k_1 - k_2)^m \exp[(k_1 + k_2)y]$$

Let functions *g, h, f* be represented by an expansion

$$g = \varepsilon g_1 + \varepsilon^3 g_3\,, \quad h = \varepsilon h_1 + \varepsilon^3 h_3\,, \quad f = 1 + \varepsilon^2 f_2 + \varepsilon^4 f_4\,. \tag{5.3.23}$$

Substitution of the above expressions into equations (5.3.22) and grouping the terms with the equal power of ε provides

$$D^2(g_1 \cdot 1) = a_1^2 g_1, \quad D^2(h_1 \cdot 1) = a_2^2 h_1, \tag{5.3.24a}$$

$$2D^2(f_2 \cdot 1) = g_1^2 + h_1^2\,, \tag{5.3.24b}$$

$$\begin{aligned} D^2(g_3 \cdot 1) &= a_1^2 g_3 + a_1^2 g_1 f_2 - D^2(g_1 \cdot f_2)\,, \\ D^2(h_3 \cdot 1) &= a_2^2 h_3 + a_2^2 h_1 f_2 - D^2(h_1 \cdot f_2)\,, \end{aligned} \tag{5.3.24c}$$

$$2D^2(f_4 \cdot 1) = 2(g_1 g_3 + h_1 h_3) - D^2(f_2 \cdot f_2), \tag{5.3.24d}$$

$$\begin{aligned} D^2(g_3 \cdot f_2) + D^2(g_1 \cdot f_4) &= a_1^2\left(g_1 f_4 + g_3 f_2\right), \\ D^2(h_3 \cdot f_2) + D^2(h_1 \cdot f_4) &= a_2^2\left(h_1 f_4 + h_3 f_2\right), \end{aligned} \tag{5.3.24e}$$

$$2D^2(f_2 \cdot f_4) = g_3^2 + h_3^2 \tag{5.3.24f}$$

$$D^2(g_3 \cdot f_4) = a_1^2 g_3 f_4\,, \quad D^2(h_3 \cdot f_4) = a_2^2 h_3 f_4 \tag{5.3.24g}$$

$$D^2(f_4 \cdot f_4) = 0 \tag{5.3.24h}$$

It follows from (5.3.24a) that the solution of these equations are in the form of exponential functions, and, in the special case we know the constant of integration, we can write down

$$g_1 = 2\sqrt{2}a_1 \exp(\theta_1)\,, \quad h_1 = 2\sqrt{2}a_2 \exp(\theta_2)\,, \tag{5.3.25}$$

where $\theta_{1,2} = a_{1,2}(y - y_{1,2})$. Substitution of this expression into (5.3.24b) results in the equation for f_2, which solution has the form

$$f_2 = \exp(2\theta_1) + \exp(2\theta_2)\,. \tag{5.3.26}$$

The further substitution of (5.3.25) and (5.3.26) into (5.3.24c) with the use of the properties of a *D*-operators leads to the following equations:

$$\frac{d^2 g_3}{dy^2} = a_1^2 g_3 + 8\sqrt{2}a_1 a_2 (a_1 - a_2)\exp(\theta_1 + \theta_2)\,,$$

$$\frac{d^2 h_3}{dy^2} = a_2^2 h_3 - 8\sqrt{2}a_1 a_2 (a_1 - a_2)\exp(\theta_1 + \theta_2)\,.$$

The solutions of these equations can be found by a standard procedure. The particular solutions are

$$g_3 = 2\sqrt{2}a_1\left(\frac{a_1 - a_2}{a_1 + a_2}\right)\exp(\theta_1 + 2\theta_2), \quad h_3 = 2\sqrt{2}a_2\left(\frac{a_2 - a_1}{a_1 + a_2}\right)\exp(\theta_2 + 2\theta_1).$$

By utilising these expressions and also equations (5.3.25) and (5.3.26), we can now find from (5.3.24d) the equation to determine f_4

$$\frac{d^2 f_4}{dy^2} = 4(a_1 - a_2)^2 \exp(2\theta_1 + 2\theta_2),$$

whence one can obtain

$$f_4 = \left(\frac{a_1 - a_2}{a_1 + a_2}\right)^2 \exp(2\theta_1 + 2\theta_2) .$$

All other equations (5.3.24e)-(5.3.24h) are satisfied identically. Thereby, we obtain the particular solutions of the system of equations (5.3.22)

$$g = 2\sqrt{2}a_1 \exp(\theta_1)\{1 + \exp[2\theta_2 + a_{12}]\}, \tag{5.3.27a}$$

$$h = 2\sqrt{2}a_2 \exp(\theta_2)\{1 - \exp[2\theta_1 + a_{12}]\}, \tag{5.3.27b}$$

$$f = 1 + \exp(2\theta_1) + \exp(2\theta_2) + \exp(2\theta_1 + 2\theta_2 + a_{12}), \tag{5.3.27c}$$

where

$$\exp(a_{12}) = \frac{a_1 - a_2}{a_1 + a_2}.$$

The solutions of (5.3.14) and (5.3.15) can be written as

$$q_1(y) = \frac{2\sqrt{2}\exp(\theta_1)\{1 + \exp[2\theta_2 + a_{12}]\}}{1 + \exp(2\theta_1) + \exp(2\theta_2) + \exp(2\theta_1 + 2\theta_2 + a_{12})}, \tag{5.3.28a}$$

$$q_2(y) = \frac{2\sqrt{2}\exp(\theta_2)\{1 - \exp[2\theta_1 + a_{12}]\}}{1 + \exp(2\theta_1) + \exp(2\theta_2) + \exp(2\theta_1 + 2\theta_2 + a_{12})}, \tag{5.3.28b}$$

If we assign $n_{10} = n_{20}$, i.e. the resonant medium is unpolarised, then the expressions (5.3.25) give $g = h = 2\sqrt{2}a_1 \exp(\theta_1)$ and $f = 1 + 2\exp(2\theta_1)$. In such a way we obtain the solution of (5.3.14) and (5.3.15) in this specific case

$$q_1(y) = q_2(y) = a_1 \operatorname{sech}\{a_1(y - y_1)\}$$

Returning to the initial physical variables, we can express the strength of the electrical field of the USP in the form

$$E^{(+)}(t,x) = E^{(-)}(t,x) = e^{(+)} E_0 \operatorname{sech}\left[\frac{dE_0}{\hbar}\left(t \pm x / V - t_0\right)\right], \qquad (5.3.29)$$

where $e^{(\pm)}$ are the components of a unit vector of electromagnetic fields. This is a simple generalisation of the results obtained by Bullough and Ahmad [59, 66] to the case of vector (polarised) USP and specific model of a resonant medium. As in Ref. [59], the pulse duration t_p is expressed in terms of the pulse peak amplitude $t_p = \hbar(dE_0)^{-1}$. The propagation velocity of the USP appears from the definition of the parameters q_0 and Ω:

$$\frac{1}{V^2} = \frac{1}{c^2}\left[1 + \frac{8\pi n_A |d|^2 \hbar\omega_a}{(\hbar\omega_a)^2 + 2(dE_0)^2}\right]. \qquad (5.3.30)$$

Besides the solitary wave solutions, equations (5.3.14) and (5.3.15) have the solutions in the form of cnoidal wave [67]

$$q_1(y) = q_2(y) = \alpha_1 \operatorname{cn}\left[\sqrt{\frac{\alpha_1^2 + \alpha_2^2}{2}}\left(y - y_0\right),\ k\right], \qquad (5.3.31)$$

where $\alpha_1^2 = \sqrt{2a_1 + \Omega^2} + \Omega$, $\alpha_2^2 = \sqrt{2a_1 + \Omega^2} - \Omega$ and $k^2 = \alpha_1^2(\alpha_1^2 + \alpha_2^2)^{-1}$.

The above analysis of the possibility of propagation of a steady-state USP of an electromagnetic field in a resonant medium under conditions of level degeneration shows that the nature of such stationary USP depends on the state of the medium. The scalar solution [59] can be trivially generalised to cover a vector video pulse case under condition $n_{10} = n_{20} = 1$. So we obtain a circular polarised pulse [68] with a duration of a half of a reciprocal atomic transition frequency. The propagation velocity of this pulse does not depend on the conditions of its polarisation and the value of this velocity is in agreement with what was found in [59].

The new solution of the generalised complete Maxwell-Bloch equations arises when initially the resonance medium has an asymmetrical distribution of level populations of the excited states that have different projections of the angular momentum $n_{10} \neq n_{20}$. One of the spherical components of the electric field strength vector behaves as in the scalar case, i.e., it is a unipolar spike of the electrical field. Another component is a variable-sign solitary wave. All stationary solutions like these form a three-parametric family, where each member is defined by the propagation speed V and n_{10}, n_{20} determining the polarisation state of the pulse. In Fig.5.3.1 the spherical components of electric field strength vector of a pulse for two different pairs of parameters n_{10} and n_{20} are depicted as the functions of the retarded time. When $q_0^2 n_{01}$ or $q_0^2 n_{02}$ approaches to unity on the right, the oscillating component of the pulse disappears. Thus, the video pulse becomes a circular polarised as a whole.

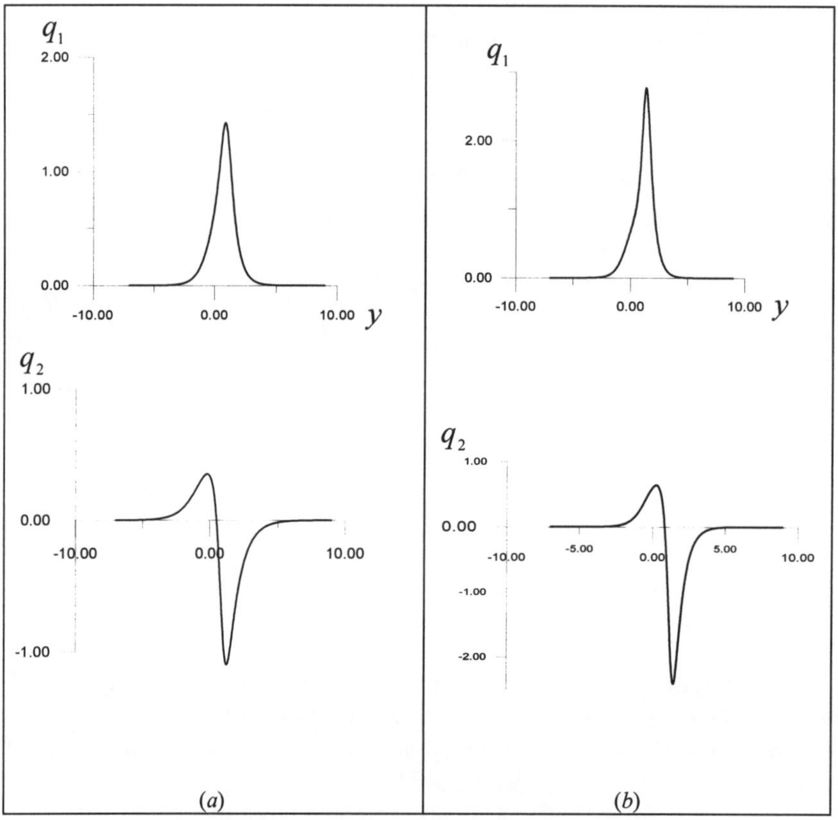

Fig.5.3.1 Spherical components of the video pulse at
(a) $q_0^2 n_{01} = 4$, $q_0^2 n_{02} = 7$, and (b) $q_0^2 n_{01} = 9$, $q_0^2 n_{02} = 8$

By a similar way we can study the propagation of the extremely short pulses in the medium characterised by transitions $j_a = 0 \to j_b = 1$ and $j_a = 1 \to j_b = 1$.

The condition of applicability of the two-level atom model should be considered in detail. Starting from the expressions (5.3.29), we can find the width $\Delta\omega$ of a USP spectrum. If its duration is $\tau_p = 2\hbar(dE_0)^{-1}$, then the Fourier image of that pulse is

$$E(\omega) = \int_{-\infty}^{+\infty} |E^{(q)}(t,x)| \exp(i\omega t)dt = \frac{\pi}{2} E_0 \tau_p \operatorname{sech}\left(\frac{\pi}{4}\omega\tau_p\right) .$$

Hence $\Delta\omega$ is defined by the relationship

$$\Delta\omega\tau_p = \frac{4}{\pi}\ln(2+\sqrt{3}) \approx 1{,}677 . \qquad (5.3.32)$$

The maximum of $E(\omega)$ is located at $\omega = 0$, but with increase in ω, the amplitudes of the Fourier components, which form this wave package, decrease exponentially. If the wave is quasimonochromatic, it has a carrier-wave frequency which is comparable with (or equal to) the frequency of the atomic (or molecular) transition. That permits to select the resonant transition and ignore all the others. In our case the half-width $\Delta\omega$ of the video pulse spectrum is the only quantity that can be used to establish a criterion for choosing just one particular transition. Such criterion is the condition that the spectral half-width $\Delta\omega$ of an ultrashort pulse is less than or of the order of the frequency of transition between the ground state and the nearest excited state, whereas the other energy levels are further away than several $\hbar\Delta\omega$ from the ground state.

The applicability of the two-level models is restricted by neglecting the cascade transitions. Such situation is studied, for instance, in [69].

Another restriction on a video pulse duration is related to the photo-ionisation limitation onto the USP electric field amplitude $E_0 \le E_{\text{atom}} \sim 10^9$ V/cm. As far as duration τ_p is related with the E_0, this inequality causes the restriction $\tau_p \ge \tau_{\text{at}}$, where $\tau_{\text{at}} = 2\hbar(dE_{\text{atom}})^{-1}$, $\tau_{\text{at}} \approx 70$ fs for $d = 1$ debye. Hence the photo-ionisation limit of the spectrum width of a video pulse is $\Delta\omega \le \Delta\omega_{\text{phot}}$, where $\Delta\omega_{\text{phot}} \approx 2{,}25 \cdot 10^{13}$ c^{-1}. Consequently, the steady-state video pulses can propagate in resonant medium in which the energy difference between the ground state and the nearest excited state is more than $10\hbar\Delta\omega_{\text{phot}} \approx 10^{-19} J$. When the video pulse duration becomes shorter than τ_{st}, the amplitude of stationary pulse can exceed an atomic field strength. In this case the perturbation theory is no longer valid for the description of the interaction of atoms with electromagnetic field. For instance, it is suggested in [70] that an atomic system should be considered classically if the electric field strength of a pulse is comparable with the atomic electric field strength.

5.3.3. ULTRA-SHORT PULSE PROPAGATION UNDER QUASI-RESONANCE CONDITION

Let the USP amplitude be of such magnitude that the Rabi frequency ω_R is small compared with the minimum of atomic transition frequency. That means $\varepsilon = \omega_R / \omega_a$ is a small parameter. Then we can attempt to solve the Bloch equations approximately. Let (4.1.5b) be the model equations for a resonant two-level medium.

$$\frac{\partial P_1}{\partial t} = -\omega_a P_2 , \quad \frac{\partial P_2}{\partial t} = \omega_a P_1 + \mu E P_3 , \quad \frac{\partial P_3}{\partial t} = -\mu E P_2 , \tag{5.3.33}$$

where $\mu = 2d\hbar^{-1}$, and $r_{1,2,3}$ are replaced by $P_{1,2,3}$.

In terms of new notations

$$B = P_1 + iP_2 , \quad C = P_1 - iP_2 ,$$
$$q = r = iE / \max | E | ,$$

equations (5.3.33) give rise to

$$\omega_R \frac{\partial B}{\partial T} - i\omega_a B = 2\omega_R q P_3 , \quad \omega_R \frac{\partial C}{\partial T} + i\omega_a C = -2\omega_R r P_3 , \tag{5.3.34a}$$

$$\frac{\partial P_3}{\partial T} = Br - Cq , \tag{5.3.34b}$$

where $T = \omega_R t$, and $\omega_R = d \max|E|/\hbar$ is the Rabi frequency at $E = \max|E|$. Formally from (5.3.34b) we have that

$$P_3 = \sigma + \int_{-\infty}^{T} (Br - Cq)\, dT' ,$$

where $\sigma = -1$ for the absorbing medium. If we introduce two-component vectors

$$\vec{\chi} = \mathrm{colon}(B, C) , \quad \vec{\psi} = \mathrm{colon}(q, r) ,$$

then equations (5.3.34a) can be written as a single vector equation

$$(\omega_R \hat{\mathbf{R}} - i\omega_a)\vec{\chi} = 2\omega_R \sigma \vec{\psi} , \tag{5.3.35}$$

where the matrix operator $\hat{\mathbf{R}}$ is defined by

$$\hat{\mathbf{R}} = \begin{pmatrix} 1 & 0 \\ 0 & -1 \end{pmatrix} \frac{\partial}{\partial T} - 2 \begin{pmatrix} q\int r & -q\int q \\ r\int r & -r\int q \end{pmatrix} .$$

This operator is referred to in the theory of solitons as the recursion operator [45]. The integral operator $u\int v$ entering into definition of $\hat{\mathbf{R}}$ is given by

$$(u\int v) f(T) = u(T) \int_{-\infty}^{T} v(t) f(t) dt , \quad \forall f(t) .$$

Using the resolvent operator $\hat{\mathbf{G}} = (1 + i\varepsilon \hat{\mathbf{R}})^{-1}$ the solution of (5.3.35) can be written as

$$\vec{\chi} = 2i\sigma\varepsilon\hat{\mathbf{G}}\vec{\psi} . \tag{5.3.36}$$

Regarding $\varepsilon = \omega_R / \omega_a$ as a small parameter, the resolvent operator can be expanded as a power series in ε

$$\hat{\mathbf{G}} = (1 + i\varepsilon \hat{\mathbf{R}})^{-1} = 1 - i\varepsilon\hat{\mathbf{R}} - \varepsilon^2 \hat{\mathbf{R}}^2 + \dots . \tag{5.3.37}$$

Employing the relationships

$$\hat{\mathbf{R}} \begin{pmatrix} q \\ r \end{pmatrix} = \begin{pmatrix} \dfrac{\partial q}{\partial T} \\ -\dfrac{\partial r}{\partial T} \end{pmatrix} , \quad \hat{\mathbf{R}}^2 \begin{pmatrix} q \\ r \end{pmatrix} = \begin{pmatrix} \dfrac{\partial^2 q}{\partial T^2} - 2q(rq) \\ \dfrac{\partial^2 r}{\partial T^2} - 2r(qr) \end{pmatrix} ,$$

we obtain at the second order of ε

$$B = 2i\varepsilon\sigma \left(q - i\varepsilon \frac{\partial q}{\partial T} - \varepsilon^2 \frac{\partial^2 q}{\partial T^2} + 2\varepsilon^2 q(qr) \right) ,$$

$$C = 2i\varepsilon\sigma \left(r + i\varepsilon \frac{\partial r}{\partial T} - \varepsilon^2 \frac{\partial^2 r}{\partial T^2} + 2\varepsilon^2 r(rq) \right) .$$

Now, taking into account that $P_1 = (B + C)/2$ and $q = r$, we derive the expression for polarisation of the single atoms

$$P_1 = 2i\varepsilon\sigma\left(q - \varepsilon^3 \frac{\partial^2 q}{\partial T^2} + 2\varepsilon^2 q^3\right).$$

The polarisation of the ensemble of two-level atoms in terms of initial variables is

$$\langle P_1 \rangle = -\left\langle \frac{2d\sigma}{\hbar\omega_a} \right\rangle E + \left\langle \frac{2d\sigma}{\hbar\omega_a^3} \right\rangle \frac{\partial^2 E}{\partial t^2} + \left\langle \frac{4d^3\sigma}{\hbar^3\omega_a^3} \right\rangle E^3. \quad (5.3.38)$$

Thus we have an expression for polarisation in the Maxwell equations and we can consider some following degrees (or levels) of approximation in the theory of non-linear wave propagation.

Let us consider unidirectional waves described by the equations (4.1.5a), i.e.

$$\frac{\partial E}{\partial z} + \frac{1}{c}\frac{\partial E}{\partial t} = -\left(\frac{2\pi n_a d_a}{c}\right)\left\langle \frac{\partial P_1}{\partial t} \right\rangle.$$

Substitution of the expression for polarisation (5.3.38) into this equation yields

$$\frac{\partial E}{\partial z} + \frac{1}{V}\frac{\partial E}{\partial t} + aE^2 \frac{\partial E}{\partial t} + b\frac{\partial^3 E}{\partial t^3} = 0, \quad (5.3.39)$$

where

$$a = \left\langle \frac{24\pi n_a \sigma |d|^4}{c\hbar^3\omega_a} \right\rangle, \quad b = \left\langle \frac{4\pi n_a \sigma |d|^2}{c\hbar\omega_a^3} \right\rangle.$$

The re-normalised velocity of the USP propagation V is defined by

$$\frac{1}{V} = \frac{1}{c}\left[1 - \left\langle \frac{4\pi n_a \sigma |d|^2}{\hbar\omega_a} \right\rangle\right].$$

In terms of new variables $\tau = |b| z$, $\zeta = t - z/V$, $u(\tau,\zeta) = (a/6b)^{1/2} E(z,t)$ the equation (5.3.39) takes the form of the modified Korteweg-de Vries equation (mKdV)

$$\sigma\frac{\partial u}{\partial \tau} + 6u^2 \frac{\partial u}{\partial \zeta} + \frac{\partial^3 u}{\partial \zeta^3} = 0.$$

As it is known [71] this equation is a completely integrable, and its solutions can be found by the IST method [43-45]. The one-soliton solution is

$$u_S(\tau,\zeta) = u_0 \operatorname{sech}\left(u_0^3\tau - \sigma u_0\zeta + \delta_0\right); \tag{5.3.40}$$

where parameters u_0 and δ_0 are determined from the initial conditions by using the IST method. This ultrashort pulse can be interpreted as a video pulse, i.e., a half-period pulse.

The result of a greater interest is a breather solution of mKdV [71]

$$u_B(\tau,\zeta) = -\left(\frac{4u_0}{v_0}\right)\frac{v_0 \cosh\vartheta_1 \sin\vartheta_2 - u_0 \sinh\vartheta_1 \cos\vartheta_2}{\cosh^2\vartheta_1 + (u_0/v_0)\cos^2\vartheta_2}, \tag{5.3.41}$$

where

$$\vartheta_1 = 2u_0\zeta + 8u_0(3v_0^2 - u_0^2)\tau + \delta_1,$$
$$\vartheta_2 = 2v_0\zeta + 8v_0(v_0^2 - 3u_0^2)\tau + \delta_2.$$

Parameters u_0, v_0, δ_1 and δ_2 are derived from the initial condition. If inequality $u_0 << v_0$ holds for this initial condition, then

$$u_B(\tau,\zeta) \approx -4u_0 \operatorname{sech}\vartheta_1 \sin\vartheta_2. \tag{5.3.42}$$

Thus, the expression (5.3.42) describes the USP (Fig.5.3.2) with a sech-envelope and the high-frequency filling, i.e., carrier wave. The breather description of the femtosecond optical

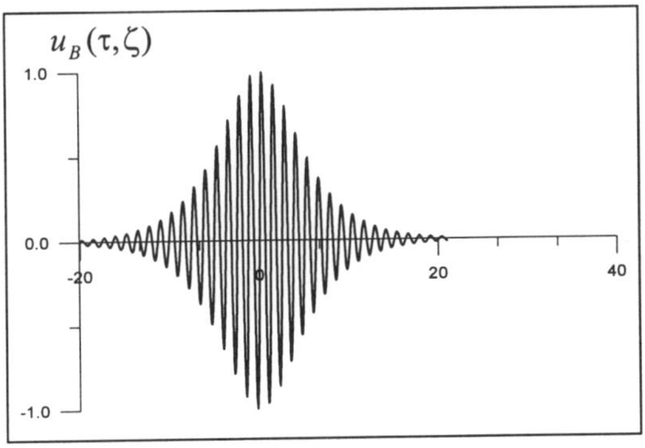

Fig.5.3.2.

High frequency breather

pulses is likely to be more adequate than the soliton described by the NLS or by its modifications.

It should be noted that among the solutions of equation (5.3.39) there are those describing the "dark" solitons have been studied recently in non-linear fibre optics [72, 73]. Among those solutions, there are rational soliton solutions [74, 75], such as, for example,

$$u(\tau,\zeta) = u_0 - \frac{4u_0}{1+4u_0^2(\zeta - 6u_0^2\tau)^2}. \tag{5.3.43}$$

Equation (5.3.39) provides the description of ultra-short optical pulse propagation in a dispersive non-linear medium without the approximation of slowly varying envelopes and phases under condition of unidirectional wave propagation. The same result was obtained in a different way in [76]. This proves the validity of the method proposed here.

We now consider the more correct theory of USP propagation. We shall assume that the condition of unidirectional propagation is violated. Then we must start from the wave equation (4.1.1)

$$\frac{\partial^2 E}{\partial z^2} - \frac{1}{c^2}\frac{\partial^2 E}{\partial t^2} = \left(\frac{4\pi n_a d_a}{c^2}\right)\left\langle\frac{\partial^2 P_1}{\partial t^2}\right\rangle, \tag{5.3.44}$$

Specifying the polarisation by equations (5.3.38), we assume that no restrictions are imposed on the direction of the USP propagation. Hence, one can substitute (5.3.38) into (5.3.44), and thus obtain the non-linear wave equation

$$\frac{\partial^2 E}{\partial z^2} - \frac{1}{V^2}\frac{\partial^2 E}{\partial t^2} = \frac{\partial^2}{\partial t^2}\left(a_1 E^3 + b_1\frac{\partial^2 E}{\partial t^2}\right), \tag{5.3.45}$$

where

$$a_1 = \left\langle\frac{16\pi n_a \sigma |d|^4}{c^2\hbar^3\omega_a^3}\right\rangle, \quad b_1 = \left\langle\frac{8\pi n_a \sigma |d|^2}{c^2\hbar\omega_a^3}\right\rangle.$$

The re-normalised velocity of the USP propagation V is given by the following expression

$$\frac{1}{V^2} = \frac{1}{c^2}\left[1 - \left\langle\frac{8\pi n_a \sigma |d|^2}{\hbar\omega_a}\right\rangle\right].$$

Let us introduce the following new variables $\tau = (|b_1|V^4)^{-1/2} z$, $\zeta = (|b_1|V^2)^{-1/2} t$, $u(\tau,\zeta) = (|a_1|V^2)^{1/2} E(z,t)$. Then the non-linear wave equations (5.3.45.) can be written as

$$\frac{\partial^2 u}{\partial \tau^2} = \frac{\partial^2}{\partial \zeta^2}\left(u + \sigma u^3 + \sigma \frac{\partial^2 u}{\partial \zeta^2} \right), \tag{5.3.46}$$

where $\sigma = \mathrm{sgn}(a_1) = \mathrm{sgn}(b_1)$ is a sign of the populations difference of the resonant states.

This non-linear wave equation describes the propagation of ultra-short pulses in the dispersive non-linear medium. The description is free from limitations of slowly varying approximation. At the same time, other approximations are used, and the most important of them being the formal solution of the Bloch equations.

It is worth to note that the non-linear wave equation (5.3.46) resembles the Boussinesq equation

$$\frac{\partial^2 \overline{u}}{\partial \tau^2} = \frac{\partial^2}{\partial \zeta^2}\left(\overline{u} + \overline{u}^2 + \frac{\partial^2 \overline{u}}{\partial \zeta^2} \right)$$

very much. This equation is completely integrable and it has soliton solutions. Unfortunately, this cannot be said about the non-linear wave equation (5.3.46). However, we are able to find a steady-state solution of this equation. Let u be dependent on a single variable $y = \zeta \pm \alpha \tau$ and

$$u = \partial u / \partial \tau = \partial u / \partial \zeta = \partial^2 u / \partial \tau^2 = \partial^2 u / \partial \zeta^2 = \ldots = 0$$

as $\tau \to \pm\infty$. These conditions can be considered as zero boundary conditions for a solitary steady-state wave. From (5.3.46) it follows

$$d^2 u / dy^2 = \sigma(\alpha^2 - 1)u - u^3 .$$

After multiplying both sides of this equation by (du/dy) and subsequent integrating, we have

$$(du/dy)^2 = p^2 u^2 - 2u^4 . \tag{5.3.47}$$

This equation admits a real solution, provided $p^2 = \sigma(\alpha^2 - 1) > 0$. By integrating (5.3.47) under zero boundary conditions, we obtain

$$u(y) = \frac{p}{\sqrt{2}\cosh[p(y - y_0)]}, \tag{5.3.48a}$$

where y_0 is the integrating constant, which can be set equal to zero. Turning to the initial variables, we obtain

$$u(y) = u_0 \operatorname{sech}\left[\sqrt{2}u_0\left(\zeta \pm \tau\sqrt{1+2\sigma u_0^2}\right)\right]. \quad (5.3.48b)$$

Different signs in this expression correspond to the opposite propagation directions of the solitary steady-state wave. It is noteworthy that for an absorbing medium (i.e., $\sigma = -1$) the maximum of the amplitude is limited: $u_0^2 \le 0{,}5$.

Let us now consider other boundary conditions as $\tau \to \pm\infty$

$$u = u_1, \quad \partial u / \partial\tau = \partial u / \partial\zeta = \partial^2 u / \partial\tau^2 = \partial^2 u / \partial\zeta^2 = \ldots = 0$$

that corresponds to the nonvanishing solitary steady-state wave as $\tau \to \pm\infty$. After double sequential integration of (5.3.46) with taking nonvanishing boundary conditions into account we obtain the equation

$$(du/dy)^2 = u_1^2\left(p^2 - (3/2)u_1^2\right) - 2u_1(p^2 - u_1^2)u + p^2u^2 - (1/2)u^4.$$

If we denote $w(y) = u(y)/u_1$ and $\kappa^2 = p^2/u_0^2$, then we can rewrite this equation as

$$(dw/dy)^2 = (1-w)^2[2(\kappa^2 - 1) - (1+w)^2]. \quad (5.3.49)$$

Thereby, we obtain a simple equation

$$(dr/dy)^2 = (\Theta^2 - 4)(r^2 - r_0^2) \quad (5.3.50)$$

for a new variable $r(y) = 2(\Theta^2 - 4) + 1/w(y)$, where $r_0^2 = \Theta^2(\Theta^2 - 4)^{-2}$ and $\Theta^2 = 2(\kappa^2 - 1)$. As the left-hand side of (5.3.49) is positive, the inequality $\Theta^2 > \max(1+w)^2$ holds. We assume that $w = 1$ at $y \to \pm\infty$, therefore, $\Theta^2 > 4$ and the integration of (5.3.50) yields

$$w(y) = \frac{u(y)}{u_1} = 1 + \frac{\Theta^2 - 4}{2 - \Theta\cosh[y\sqrt{\Theta^2 - 4}]}. \quad (5.3.51)$$

This solution of the non-linear wave equation (5.3.46) describes a stationary dark wave propagation either in forward or backward direction with the amplitude $w_0(2+\Theta)$. When the parameter Θ^2 approaches 4, then one can find either from (5.3.51) or from direct solution of (5.3.49) the algebraic solution

$$u(y) = u_1 - \frac{4u_1}{1 + 4(\zeta \pm \tau\sqrt{1 + 3\sigma u_1^2})^2}, \tag{5.3.52}$$

similar to (3.7.14). The solution (5.3.51) can be formally extrapolated over the parameter Θ^2 into the region $0 < \Theta^2 < 4$, but the result obtained in this way does not satisfy the boundary conditions. This is a periodic solution describing the quasi-harmonic wave

$$u(y) = u_1 - \frac{u_1(4 - \Theta^2)}{2 - \Theta\cos[y\sqrt{4 - \Theta^2}]}.$$

Besides this wave, other periodic solutions of (5.3.46) can be found in terms of elliptic Jacobian functions.

5.3.4. ULTRA-SHORT PULSE PROPAGATION IN NON-RESONANCE MEDIUM

In this paragraph we consider the USP propagation under the unidirectional propagation approximation. Let the medium be a non-linear and non-resonant one. A similar approach was demonstrated in [77] for the self-induced transparency in ionic crystals in the frames of the Duffing's type model. The reduced Maxwell equation is

$$\frac{\partial E}{\partial z} + \frac{1}{c}\frac{\partial E}{\partial t} = -\left(\frac{2\pi n_a d_a}{c}\right)\frac{\partial P_a}{\partial t}. \tag{5.3.53}$$

Scalar Duffing model

We complete this wave equation by the non-linear oscillator equation for the medium

$$\frac{\partial^2 P_a}{\partial t^2} + \omega_a^2 P_a + \beta_a P_a^3 = \frac{e^2}{md_a} E. \tag{5.3.54}$$

It is assumed that all oscillators have a universal vibration frequency ω_a and no relaxation is involved. In terms of the dimensionless variables

$$\zeta = z / ct_p, \quad \tau = (t - x / c)t_p^{-1},$$
$$E = E_0 q(\zeta, \tau), \quad P = 2\pi n_a d_a E_0^{-1} P_a, \quad Q = 2\pi n_a d_a E_0^{-1} \partial P_a / \partial t$$

the system of equations (5.3.53) and (5.3.54) can be written as

$$\frac{\partial q}{\partial \zeta} = -Q, \quad \frac{\partial P}{\partial \tau} = Q,$$
$$\frac{\partial Q}{\partial \tau} + \nu_a^2 Q + 2b_a Q = \alpha_a q, \tag{5.3.55}$$

where $\alpha_a = (2\pi e^2 n_a d_a t_p^2 / m)$, $\nu_a = \omega_a t_p$, $b_a = (\beta_a / 2)(E_0 t_p / 2\pi n_a d_a)^2$. For q, P and Q depending on a composite variable $\eta = \tau - \zeta / V$ (where V is a propagation velocity of stationary pulse) equations (5.3.55) can be converted into the form

$$\frac{\partial q}{\partial \eta} = VQ, \quad \frac{\partial P}{\partial \eta} = Q,$$
$$\frac{\partial Q}{\partial \eta} + \nu_a^2 Q + 2b_a Q = \alpha_a q. \tag{5.3.56}$$

The first pair of these equations gives

$$q(\eta) = VP(\eta). \tag{5.3.57}$$

The relation between $Q(\eta)$ and $P(\eta)$ follows from (5.3.56)

$$Q = (\alpha_a V - \nu_a^2)P^2 - b_a P^4. \tag{5.3.58}$$

In the case when $\alpha_a V < \nu_a^2$ the $Q = P = 0$ is a unique solution of equation (5.3.58) so far as P and Q are real. Hence, the non-trivial solution of (5.3.56) takes place only at $\alpha_a V > \nu_a^2$. That means that the propagation velocity of a stationary pulse V is always greater than the critical velocity V_c, for which we have

$$V_c < \nu_a^2 / \alpha_a = 2\omega_a^2 / \omega_p^2$$

where $\omega_p = \left(4\pi e^2 n_a / m\right)^{1/2}$ is plasma oscillation frequency.

It should be emphasised that in the (x, t) variables the stationary pulse propagates with the velocity V_{st} which is associated with the V by the relationship

$$V_{st} = c\frac{V}{1+V},$$

and satisfies the inequality

$$c > V_{st} \geq c \frac{2\omega_a^2}{\omega_a^2 + \omega_p^2}.$$

The stationary pulse shape is determined from (5.3.56) and (5.3.58)

$$q(\eta) = V\sqrt{(\alpha_a V - \nu_a^2)/b_a}\ \mathrm{sech}\left[\eta\sqrt{\alpha_a V - \nu_a^2}\right]. \tag{5.3.59}$$

This solution described the shortest optical pulse permitted by the model. It has no carrier wave and may be referred to as the video pulse considered in [78]. In a general case optical pulse evolution was investigated numerically in [79].

Vector Duffing model

Let us consider the USP propagation in a non-linear medium, which is again presented by an ensemble of the molecules. However, now let us suppose that the inner degrees of freedom are described by the potential :

$$U(x,y) = \frac{1}{2}\omega_1 x^2 + \frac{1}{2}\omega_2 y^2 + \frac{1}{2}\kappa_2 x^2 y^2 + \frac{1}{4}\kappa_4 (x^4 + y^4). \tag{5.3.60}$$

If the coupling parameter $\kappa_2 = 0$, then this potential describes the scalar Duffing model of anharmonic oscillator, which can oscillate independently in two orthogonal directions. Expression (5.3.60) is the simplest generalisation of this model. Polarisation of the molecule is defined by the expression $\vec{p} = ex\vec{e}_1 + ey\vec{e}_2$, and the total polarisation is the product of the molecules density n_A and the polarisation of one molecule.

The propagation of the USP will be considered in the unidirectional wave approximation, so the wave equations for electric field corresponding to the different polarisation components of the USP can be written as follows

$$\frac{\partial E_1}{\partial z} + \frac{1}{c}\frac{\partial E_1}{\partial t} = -\frac{2\pi n_A e}{c}\frac{\partial x}{\partial t}, \quad \frac{\partial E_2}{\partial z} + \frac{1}{c}\frac{\partial E_2}{\partial t} = -\frac{2\pi n_A e}{c}\frac{\partial y}{\partial t}. \tag{5.3.61}$$

The equations of motion for anharmonic oscillator follow from the classical Newton equations

$$\frac{\partial^2 x}{\partial t^2} + \omega_1^2 x + \kappa_2 x y^2 + \kappa_4 x^3 = \frac{e}{m} E_1(z,t), \tag{5.3.62a}$$

$$\frac{\partial^2 y}{\partial t^2} + \omega_2^2 y + \kappa_2 y x^2 + \kappa_4 y^3 = \frac{e}{m} E_2(z,t). \tag{5.3.62b}$$

Now let us introduce the new independent variables $\zeta = z/ct_0, \tau = (t - z/c)/t_0$ and $q_1 = (2\pi e n_A)x$, $q_2 = (2\pi e n_A)y$, $p_j = \partial q_j / \partial \tau$, Thus, the two-component (vector) Duffing model is described by the following system of equations:

$$\frac{\partial E_j}{\partial \zeta} = -p_j, \quad \frac{\partial q_j}{\partial \tau} = p_j, \quad j = 1, 2, \tag{5.3.63a,b}$$

$$\frac{\partial p_1}{\partial \tau} = aE_1 - \nu_1^2 q_1 - b_2 q_1 q_2^2 - b_4 q_1^3, \tag{5.3.63c}$$

$$\frac{\partial p_2}{\partial \tau} = aE_2 - \nu_2^2 q_2 - b_2 q_2 q_1^2 - b_4 q_2^3. \tag{5.363d}$$

Here $a = 2\pi n_A e^2 t_0^2 / m$, $b_{2,4} = \kappa_{2,4}(2\pi e n_A t_0^2)^{-2}$, $\nu_{1,2} = \omega_{1,2} t_0$.

It is possible to find a steady state solution of this system of equations. Let us suppose the fields depend only on $\eta = \tau - \zeta / V$. Integrating equations (5.3.63a,b) and taking account of the boundary conditions, which correspond to the vanishing electric field of the USP and polarisation of molecules at infinity, we find that $E_j = Vq_j$. Substitution of this result into (5.3.63c,d) leads to the following equations

$$\frac{d^2 q_1}{d\eta^2} + (b_4 q_1^2 + b_2 q_2^2) q_1 = (aV - \nu_1^2) q_1, \tag{5.364a}$$

$$\frac{d^2 q_2}{d\eta^2} + (b_2 q_1^2 + b_4 q_2^2) q_2 = (aV - \nu_2^2) q_2. \tag{5.3.64b}$$

These equations appear in many works, so we can easily find their solutions. Let suppose that the eigen-frequencies ω_1 and ω_2 are equal. The partial solution of system (5.3.64) in this case is

$$q_1(\eta) = q_2(\eta) = \sqrt{\frac{2(aV - \nu^2)}{b_2 + b_4}} \operatorname{sech}\left[\sqrt{(aV - \nu^2)}(\eta - \eta_0)\right],$$

where η_0 is the integration constant. This solution similar to that above describes the linear polarised video pulse propagating without distortion in the considered non-linear medium.

The more interesting solution exists if the frequencies ω_1 and ω_2 are different, but factors of anharmonicity coincide $b_2 = b_4$. In this case the system of equations (5.3.64) complies with that considered in section 4.3.2, and we can simply use the solutions derived there:

$$q_1(\eta) = \frac{2\sqrt{2b_2^{-1}}\mu_1 \exp(\theta_1)\{1+\exp[2\theta_2+\mu_{12}]\}}{1+\exp(2\theta_1)+\exp(2\theta_2)+\exp(2\theta_1+2\theta_2+\mu_{12})},$$

$$q_2(\eta) = \frac{2\sqrt{2b_2^{-1}}\mu_2 \exp(\theta_2)\{1+\exp[2\theta_1+\mu_{12}]\}}{1+\exp(2\theta_1)+\exp(2\theta_2)+\exp(2\theta_1+2\theta_2+\mu_{12})},$$ (5.3.65)

where

$$\exp(\mu_{12}) = \frac{\mu_1-\mu_2}{\mu_1+\mu_2}.$$

In these expressions the following parameters $\mu_{1,2}^2 = (aV - \nu_{1,2}^2)$ and $\theta_{1,2} = \mu_{1,2}(\eta - \eta_{1,2})$ are used, where $\eta_{1,2}$ are the integration constants. Solutions (5.3.65) describe the steady-state propagation of the USP where one of these component corresponds to a unipolar spike of the electric field, sub-cycle pulse. The other component is a bipolar impulse or mono-cycle pulse.

5.3.5. OPTICAL SHOCK-WAVES

Let the wave package be located in the frequency band, where the group-velocity dispersion is absent. The electric displacement vector $\vec{D}(t,z)$ at each point of space and at each moment of time is defined by the magnitude of the electric field strength $\vec{E}(t,z)$ at the same moment of time and in the same spatial point. It appears that in this case we can solve Maxwell equations exactly. The corresponding solution in general describes the non-linear evolution of the plane electromagnetic wave [80].

Let the wave propagate in the isotropic dielectric medium along z-axis, electrical field be directed along y-axis and magnetic field is directed along x-axis. The Maxwell equations take the form

$$-\frac{\partial H}{\partial z} = \frac{1}{c}\frac{\partial D}{\partial t}, \quad -\frac{\partial E}{\partial z} = \frac{1}{c}\frac{\partial H}{\partial t}.$$ (5.3.66)

We assume that the dielectric permeability can be written as

$$\varepsilon(E) = \frac{dD}{dE}.$$

and the electric displacement vector is proportional to the electric field at $\vec{E}(t,z)\to 0$. In this limit the dielectric permeability tends to the constant value, which is equal to a linear dielectric permeability. If the magnetic field strength can be expressed in terms of electric field strength, then Maxwell equations may be rewritten in the form

$$\frac{\varepsilon}{c}\frac{\partial E}{\partial t}+\frac{dH}{dE}\frac{\partial E}{\partial z}=0\,,\qquad \frac{1}{c}\frac{dH}{dE}\frac{\partial E}{\partial t}+\frac{\partial E}{\partial z}=0\,, \tag{5.3.67}$$

This is the homogeneous system of linear equations with respect to partial derivatives of the electric field strength. The non-trivial solutions exist only when its determinant is equal to zero. As a result we have an equation

$$\varepsilon(E)-\left(\frac{dH}{dE}\right)^2=0\,,$$

or

$$H(E)=\pm\int_0^E\sqrt{\varepsilon(e)}\;de\,.$$

That means that the solution of our problem is possible only under condition $\varepsilon(E)>0$. Substitution of this expression into one of the equations (5.3.67) brings about to the homogeneous quasi-linear partial differential equation

$$\frac{\partial E}{\partial z}\pm\frac{\sqrt{\varepsilon(E)}}{c}\frac{\partial E}{\partial t}=0\,. \tag{5.3.68}$$

Hence, we obtain

$$-\frac{(\partial E/\partial t)_z}{(\partial E/\partial z)_t}=\left(\frac{\partial z}{\partial t}\right)_E=\mp\frac{c}{\sqrt{\varepsilon(E)}}\,.$$

Integrating of this equation results in

$$z\mp\frac{c}{\sqrt{\varepsilon(E)}}\,t=G(E)\,,$$

where $G(E)$ is an arbitrary function, which can be determined by the initial conditions. If we denote the reciprocal value of this function as G^{-1}, then the solution of the equation (5.3.68) can be written as

$$E = G^{-1}\left(z \mp \frac{c}{\sqrt{\varepsilon(E)}} t \right). \tag{5.3.69}$$

Sign in the argument of G^{-1} defines a direction of the wave propagation. This expression is implicitly determines a desired solution of (5.3.69). Notice that in the weak field limit, when $\varepsilon(E) = \varepsilon_0$, we obtain the plane linear wave.

In the course of wave propagation its initial profile distorts, because its different points move with different velocities. When $\varepsilon(E)$ decreases with the growing of the electric field strength, the points of the profile where E is stronger, move faster than those with weaker E. So there is a moment when the profile of the wave turns over. At this moment an electromagnetic shock wave arises.

5.4. Conclusion

Here we have considered some models of the non-linear medium in which the USP propagates. In all cases the slowly varying pulse envelope approximation has not been applied. But when the resonant (two or three level) systems were considered, some restrictions on the pulse duration have occurred. The criterion of admissibility for an employing of the resonance approximation for description of the medium is the condition that spectral half-width of the video pulse $\Delta\omega$ is less than or of the order of the frequency of a transition from the ground state to the nearest (on the energy scale) excited state, whereas the other excited states are farther away than $\hbar\Delta\omega$ from the ground state. The spectrums of the similar structure can be found among the spectra of atom of alkaline and ions of rare earth metals.

Duffing model and its vector generalisation are free from the restrictions of the resonance approximation. This advantage was noted earlier [70] when the electromagnetic "bubbles" and shock waves were analysed. It is noteworthy that there are other models of a non-linear medium, for instance: the free electrons model, anharmonic oscillators, multi-frequencies oscillator models, exciton matter models, models of the "photonic band gap" medium [81]. It would be interesting to generalise these models in the case of polarised electromagnetic radiation.

References

1. Vlasov, R.A., and Doctorov E.V.: Nonuniform optical solitons in resonant Kerr media, *Dokl. Akad. Nauk BSSR*, **26**, (1982), 322-324.
2. Maimistov, A.I., and Manykin, E.A.: Propagation of ultrashort optical pulses in resonant non-linear light guides, *Zh.Eksp.Teor.Fiz.* **85** (1983), 1177-1181 [*Sov.Phys. JETP* **58** (1983), 685-687].

3. Basharov, A.M., and Maimistov, A.I.:. Self-induced transparency in a Kerr medium, *Opt.Spektrosk.* **66** (1989), 167-173.
4. Kozlov, V.V., and Fradkin, E.E.: Distortion of self-induced transparency solitons as a result of self-phase modulation in ion-doped fibers, *Opt.Letts.* **21** (1995), 2165-2167.
5. Guzman, A., Locati, F.S., and Wabnitz, S.: Coupled-mode analysis of the self-induced transparency soliton switch, *Phys. Rev.* A, **46** (1992), 1594-1605.
6. Matulic, L., Torres-Cisneros, G.E., and Sabchez-Mondragon J.J.: Pulse Propagation in a resonant absorber with Kerr-type non-linearity, *J.Opt.Soc.Amer.* **B8** (1991) 1276-1283
7. Nakazawa, M., Yamada, E., and Kubota, H.: Coexistence of self-induced transparency soliton and non-linear Schrodinger soliton, *Phys.Rev.Lett.* **66** (1991), 2625-2628.
8. Nakazawa, M., Kimura, Y., Kurokawa, K., and K.Suzuki, K.: Self-induced-transparency solitons in an Erbium-doped fiber waveguide, *Phys.Rev.* A **45** (1992), R23-R26.
9. Moskalenco, S.A., Rotaru, A.H., and Khadzhi, P.I.: Superfluidity of Bose condensed dipole-active excitons and photons and the phenomenon of self-induced transparency, *Opt.Commun.* **23** (1977), 367-368.
19. Agranovich, V.M., and Rupasov, V.I.: Self-induced transparency in media with space dispersion, *Fiz.Tverd.Tela (Leningrad)* **18** (1976), 801-807.
20. J.Goll, J., and Haken, H.: Exciton self-induced transparency and the dispersion law of steady-state exciton-photon pulses, *Opt.Commun.* **24** (1978), 1-4.
21. Adamashvili, G.T.: Self-induced transparency of excitons in anisotropic media, *Opt.Spektrosk.* **54** (1983), 668-672.
22. Huhn, W.: Self-induced transparency of excitons in semiconductors, *Opt.Commun.* **68** (1988), 153-156.
23. De Moura, M.A., and De Oliveira, J.R.: Self-induced transparency of excitons, *Phys.Stat.Solidi* **B 158**, (1990), K9-K11.
24. Stroud, C.R., Jr., Bowden, C.M., and Allen, L.: Self-induced transparency in self-chirped media, *Opt.Commun.* **67** (1988), 387-390.
25. Bowden, C.M., Postan, A., and Inguva, R.: Invariant pulse propagation and self-phase modulation in dense media, *J.Opt.Soc.Amer.* **B8** (1991), 1081-1084.
26. Bowden, Ch.M., and Dowling, J.P.: Near-dipol-dipol effects in dense media: generalized Maxwell-Bloch equations, *Phys.Rev.* **A47** (1993), 1247-1251.
27. Maimistov, A.I., and Elyutin S.O.: Non-stationary propagation of ultra-short light pulses under exciton absorption in semiconductors, in *Non-stationary processes in semiconductors and dielectrics*, Energoatomizdat, Moscow, 1986, p.65- 70
28. Bullough, R.K., and Ahmad, F.: Exact solutions of the self-induced transparency equations, *Phys.Rev.Lett.* **27** (1971), 330-333.
29. Lee C.T.,: Self-induced transparency of an extremely short pulse, *Opt.Commun.* **9** (1973), 1-3.
30. Lee C.T.,: Four possible types of pulses for self-induced transparency, *Opt.Commun.* **10** (1974), 111-113.
31. Caudrey, P.J., and Eilbeck, J.C.: Numerical evidence for breakdown of soliton behavior in solutions of the Maxwell-Bloch equations, *Phys.Lett.* **A62** (1977), 65-66.
32. Kujawski, A.: Self-induced transparency of very short optical pulses, *Zs.Phys.* **B66**, (1987), 271-274.
33. Branis, S.V., Martin, O, and Birman, J.L.: Solitary-wave velocity selection in self-induced transparency, *Phys.Rev.Lett.* **65** (1990), 2638-2641.
34. Branis, S.V., Martin, O, and Birman, J.L.: Discrete velocities for solitary-wave solutions selected by self-induced transparency, *Phys.Rev.* **A43** (1991), 1549-1563.
35. Andreev, A.V. : Non-reduced Maxwell-Bloch equations and a chirped soliton. *Phys.Lett.* **A179** (1993), 23-26.
36. Fork, R.L., Brito Cruz, C.H., Becker, P.C., and Shank, Ch.V.: Compression of optical pulses to six femtosecond by using cubic phase compensation, *Opt.Lett.* **12** (1987), 483-485.

37. Christov, I.P., and Danailov M.B.: Pulse compression by free electrons, *Opt.Commun.* **69** (1988), 291-294.
38. Tai, K., and Tomita, A.: 1100 X optical fiber pulse compression using grating pair and soliton effect at 1,319 mkm, *Appl.Phys.Lett.* **48** (1986) 1033-1035.
39. Halbout, J.-M., and Grischkowsky D.: 12-fs ultrashort optical pulse compression at a high repetition rate, *Appl.Phys.Lett.* **45** (1984), 1281-1283.
40. Gouveia-Neto, A.S., Gomes, A.S.L., and Taylor, J.R.: Generation of 33-fsec pulses at 1,32 mkm through a high-order soliton effect in a single-mode optical fiber, *Opt.Lett.* **12** (1987), 395-397.
41. Kumar, A.: Soliton dynamics in a monomode optical fibre, *Phys.Rept.* **C 187** (1990), 63-108.
42. Ablowitz, M.J., Kaup, D.J., Newell, A.C., and Segur, H.: Non-linear evolution equations of physical significance, *Phys.Rev. Letts.* **31**, (1973), 125 -127.
43. Ablowitz, M.J., Kaup, D.J., Newell, A.C., and Segur, H.: The inverse scattering transform - Fourier analysis for non-linear problems, *Stud.Appl.Math.* **53** (1974) 249-315.
44. Zakharov, V.E., Manakov, S.V., Novikov, S.P., and Pitaevskii, L.P.: *Theory of Solitons: The Inverse Problem Method* [in Russian], Nauka, Moscow, 1980. *Theory of Solitons: The Inverse Scattering Method*, Plenum, New York, 1984.
45. Ablowitz, M.J., and Segur, H.: *Solitons and the Inverse Scattering Transform*, SIAM, Philadelphia, 1981.
46. Zakharov, V.E., and Schabat, A.B.: Zakharov, V.E., and Schabat, A.B.: The exact theory of two-dimensional self-focussing and one-dimensional self-modulating of waves in non-linear medium, *Zh.Eksp.Teor.Fiz.* **61** (1971), 118-134 [*Sov.Phys. JETP* **34** (1972), 62-69].
47. Maimistov, A.I., Basharov, A.M., Elyutin S.O., and Sklyarov Yu.M.: Present state of self-induced transparency theory, *Phys.Rept*, **C 191** (1990), 1-108.
48. Manakov, S.V.: On the theory of two-dimensional stationary self-focusing of electromagnetic waves, *Zh.Eksp.Teor.Fiz.* **65** (1973), 505-516 [*Sov.Phys. JETP* **38** (1974), 248-]
49. Basharov, A.M., and Maimistov, A.I.: Polarized solitons in three-level media, *Zh.Eksp.Teor.Fiz.* **94** (1988), 61-75 [*Sov.Phys. JETP* **67** (1988), 2426-2433]
50. Eilbeck, J.L., Gibbon, J.D., Caudrey, P.J., and Bullough, R.K.,: Solitons in non-linear optics. I. A more accurate description of the 2π-pulse in self-induced transparency, *J.Phys.* A, **6**, (1973), 1337-1347.
51. Bullough, R.K., Jack, P.M., Kitchenside, P.W., and Saudders, R.: Solitons in laser physics, *Phys.Scr.* **20** (1979), 364-381.
52. McCall, S.L., and Hahn, E.L.: Self-induced transparency, *Phys.Rev.***183** (1969), 457-485.
53. Allen, L., and Eberly, J.H.: *Optical Resonance and Two-Level Atoms*, Wiley, New York, 1975.
54. Slusher, R.E.: Self-induced transparency, *Progr.Optics*, **12** (1974), 53-100.
55. Courtens, E.: Giant Faradey rotations in self-induced transparency. *Phys.Rev.Letts.* **21** (1968), 3-5.
56. Lamb, G.L., Jr.: Analytical descriptions of ultrashort optical pulse propagation in a resonant medium, *Rev.Mod Phys.* **43** (1971), 99-124.
57. Gibbon, J.D., Caudrey, P.J., Bullough, R.K., and Eilbeck, J.L.: An N-soliton solution of a non-linear optics equation derived by a general inverse method, *Letts.Nuovo Cimento* **8**, (1973), 775-779.
58. Maimistov, A.I.: On the theory of self-induced transparency without approximation of slowly varying amplitudes and phases, *Kvantov. Elektronika (Moscow)* **10** (1983), 360-364.
59. Bullough, R.K., and Ahmad, F.: Exact solutions of the self-induced transparency equations, *Phys.Rev.Lett.* **27** (1971), 330-333.
60. Basharov, A.M., and Maimistov, A.I. :On self-induced transparency under condition of degeneration of resonant energy levels, *Zh.Eksp.Teor.Fiz.* **87** (1984), 1594-1605 [*Sov.Phys. JETP* **60** (1984), 913-]
61. Basharov, A.M., Maimistov, A.I., and Sklyarov Yu.M.: Self-induced transparency on the transition $1 \rightarrow 1$ is a exactly solvable polarization model of non-linear optics, *Opt.Spektrosk.* **63** (1987), 707-709 [*Opt.Spectrosc.(USSR)* **62** (1987), 418-].
62. Tratnik M.V., and Sipe J.E.: Bound solitary waves in a birefringent optical fiber, *Phys.Rev.* **A38** (1988), 2011-2017.

63. Radhakrishnan R., and Lakshmanan M.: Bright and dark soliton solutions to coupled Non-linear Schrodinger equations, *J.Phys.* **A28** (1995), 2683-2692.
64. Hirota R., and Satsuma J.: A variety of non-linear network equations generated from the Bдcklund transformation for the Toda lattice, *Progr.Theor.Phys., Suppl.* **59** (1976), 64-100.
65. Hirota R.: Direct Method of Finding Exact Solutions of Non-linear Evolution Equations, in R.M.Miura (ed.) *Baclund Transformations, the Inverse Scattering Method, Solitons and Their Applications* (Lect. Notes in Math. **515**), Springer-Verlag, Berlin, 1976, p.40-68
66. Kazuhiro Akimoto: Properties and applications of ultra-short electromagnetic mono- and sub-cycle waves, *J. Phys.Soc.Japan* **65** (1996), 2020-2032.
67. *Higher Transcendental Functions*, eds. H.Bateman and A.Erdelyi, McGraw-Hill, New York, 1955.
68. Sazonov S.V., Trifonov E.V.: Solutions for Maxwell-Bloch equations without using the approximation of a slowly varying envelope: Circularly polarized video pulses, *J.Phys.* **B27** (1994), L7-L12.
69. Kaplan A. E., and Shkolnikov, P. L.: Subfemtosecond pulses in the multicascade stimulated Raman scattering. *J.Opt.Soc.Amer.* **B13** (1992), 347-354.
70. Kaplan A. E., and Shkolnikov, P. L.: Electromagnetic "bubbles" and shock waves: Unipolar, nonoscillating EM-solitons, *Phys. Rev. Lett.* **75** (1995), 2316-2319.
71. Wadati, M.: The Modified Korteweg-de Vries equation, *J.Phys.Soc.Japan* **34** (1973), 1289-1296.
72. Gredeskul, S.A., Kivshar, Yu.S., and Yanovskaya, M.V.: Dark-pulse solitons in non-linear-optical fibers, *Phys.Rev.* **A41** (1990), 3994-4008.
73. Kivshar, Yu.S., and Afanasjev, V.V.: Decay of dark solitons due to the stimulated Raman effect, *Opt.Lett.* **16** (1991), 285-287.
74. Ono, H,: Algebraic solution of the modified Korteweg-de Vries equation, *J.Phys.Soc.Japan* **41** (1976), 1817-1818.
75. Ablowitz, M.J., and Satsuma, J.: Solitons and rational solitons of non-linear evolution equations, *J.Math.Phys.* **19** (1978), 2180-2186.
76. Belenov, E.M., and Nazarkin, A.V.: On some solutions of the equations in non-linear optics without approximation of slowly varying amplitudes and phases, *Piz'ma Zh.Eksp.Teor.Fiz* **51** (1990), 252-255.
77. Vuzhva, A.D.: Self-induced transparency in ionic crystals, *Fiz.Tverd.Tela (Leningrad)* **20** (1978), 272-273
78. Sazonov, S.V., and Yakupova L.S.: Non-linear video pulses in a two-level sigma-transition medium, *J.Phys.* **B27** (1994), 369-375.
79. Maimistov, A.I., and Elyutin, S.O.: Ultrashort optical pulse propagation in non-linear non-resonance medium. *J.Mod.Opt.* **39** (1992), 2201-2208.
80. Landau, L.D., and Lifshitz E.M.: *Electrodinamika sploshnikh sred*, Nauka, Moscow, 1982, (in Russian).
81. Soukoulis, C.M. (ed.), *Photonic Band Gap Materials*, Kluwer Academic Publisher, Dordrecht, 1996.

CHAPTER 6

OPTICAL SOLITONS IN FIBERS

It is well known that the rate of the information transfer by means of fibre optical communication systems (FOCS) with the mode pulse-code modulation is limited mainly by the effect of dispersion of group velocities. The influence of this effect can be suppressed efficiently and ideally can be completely excluded if one uses sufficiently powerful pulses of light. Because of the non-linear effect of self-influence such pulses in the certain conditions are transformed in solitons and their propagation in FOCS does not accompany by dispersive broadening. Today the researches in the FOCS field focus on the development of extremely high bit-rates, of the order of 100 Gbit/s.

It is important to emphasise that soliton does not exist in real communication systems in the true sense of the word. The influence of group velocities dispersion of the high orders, optical loss and some other effects break a dynamic balance between the non-linear compression of pulse and its dispersion broadening. As a result of these effects the optical pulse suffers an envelope distortion and damping. But the distance, passed by the soliton in a fibre considerably exceeds one, which weak pulse should be passed in the linear regime of propagation. The experiments made with powerful optical pulses confirm this. We can consider the soliton as a good approach for the real non-linear pulses in a fibre under certain conditions.

The suggestion to use optical solitons for the information transfer along the fibre was made in the works [1,2] and it was demonstrated [3-9]. Later on the equation describing the optical pulse propagation in a fibre with account of only the second-order group-velocities dispersion was received in works [10-13] and was investigated in [14,15]. In the soliton theory it was known as the *Nonlinear Schrödinger* (NLS) equation. Besides the classical works on complete integrability of the NLS [14-16], the Refs. [17-41] should be mentioned. In these works the modulation instability was investigated [17-27], the Painleve property of the NLS was established [29-32], and the exact periodic solutions of NLS were found [33-38]. The expression for multi-soliton solutions of the NLS has been obtained in [39,40]. The new type of the solutions named multiple-pole solutions have been found in [41]. These solutions correspond to the multiple points of a discrete spectrum of the Zakharov-Shabat problem in IST method. It is difficult to realise these solutions in practice because any small stir of the initial conditions will remove degeneration in the multiple points of a discrete spectrum. It means that only the exclusive initial optical pulses can be transformed into multiple-pole optical solitons.

The problem of solitons formation from the initial pulses carrying phase and/or amplitude modulation [42-61], or from the stochastic pulses can be considered in the framework of the NLS equation model. It is known [62] that the soliton has not any phase/amplitude modulation. Thus, we can expect that the initially modulated pulses (with regular or stochastic modulation) transform either into a non-modulated soliton, or disappear as any weak signal due to dispersion broadening. This phenomenon was referred to as the non-linear filtration of the optical pulse [44,47,49,51,52,61]. Due to complete integrability of the NLS equation we can use the IST method to determine the threshold amplitude of the modulation. The excess of this threshold makes soliton formation impossible. The non-linear filtration of the optical pulse takes place in an opposite case.

If the pulse duration is shorter than one picosecond, the higher-order dispersion of the group velocities and the dispersion of the non-linear susceptibility becomes important [63-107]. A theory, describing the propagation of such short pulses, is based on the generalisation of the non-linear Schrödinger equation [79-83]. In [63-78] the third-order group-velocities dispersion was involved into consideration. Among the analytic solutions there are only steady-state ones [78,94], and there are not any which describe the evolution of the initial pulse. Reduction of the pulse duration results in the necessity to take into account the fourth-order dispersion of the group-velocities [100-104] and even the higher orders [105-107].

The research of the pulses propagation in a spatially non-uniform non-linear medium has been provided formerly for the waves in the plasma [108]. However, such problem arises in non-linear fibre optics when the properties of a fibre vary along its axis [10,13,109-121]. This problem becomes especially urgent at present, when the long distance pulses propagation in the fibre is investigated in the framework of *Dispersion-Managed Soliton Transmission Systems* technology [122-133]. So in [122,127] the fibres with variable dispersion were proposed to reduce the effect of the frequency jitter. This effect is due to the soliton collisions and the amplification of the spontaneous emission by the fibre amplifiers [134-136].

It is known that the optical fibre guides the several modes of electromagnetic radiation [137-140]. So the evolution of the pulse is governed by the evolution of the partial amplitudes of the excited modes [140]. Even in a one-mode fibre the optical wave can propagate in two opposite directions. Thus, there are two components of the running wave. During the manufacture the optical fibre is drawn from molten glass preform. It results in the mechanical stresses, which are frozen at cooling of the glass. For this reason all optical fibres as a rule have a birefringence [141]. Two circular polarised components of the electromagnetic radiation propagate with different phase velocities. In the case of a non-linear dielectric medium these velocities depend on the intensity of an electromagnetic field. The optical pulse evolution in such birefringent fibres was investigated in [142-158]. These researches are based on the equations belonging to class of the vector NLS equations [145,146,159,160]. Soliton propagation in multi-mode (or multi-core) optical fibres is described in terms of the multi-component NLS equation [161-164]. The investigation of the optic pulses propagating in two opposite directions [165,166], a pulse propagation in the non-linear two-mode fibres [168-173], and the pulse propagation in a two-channel (twin-core) fibre [174-183] result in the necessity to discuss the two-component NLS equation. This equation belongs to the same class of equations, i.e. vector NLS equations. There is the generalisation of the theory of non-linear waves in twin-core fibres in the case of multi-channel fibres [183-195], which are named as *non-linear optical waveguide array* (NOWA).

In some special case the vector NLS equation is completely integrable [159]. In the general case we must use various approximate approach to the analysis of the optic waves evolution. When the equation of interest is not integrable due to a small term, one can develop a perturbation theory in which soliton is used as a zero-approximation. The most systematic perturbation theory is based on the IST method [196-202]. In a particular case, this approach leads to a very effective and popular adiabatic perturbation theory. This method and its development have been reviewed in detail in [202]. Another effective method to investigate the non-linear waves is the variational approach [203-216], which does not exploit the complete integrability of the equation of interest. This method was successfully used for various generalisations of the NLS equation. It is worth to note that, if one chooses a soliton solution as a trial function, then the system of equations of adiabatic perturbation theory follows from the corresponding Euler-Lagrange equation [203,204]. It is quite natural to suggest that the choice of a more general trial function can lead to the more accurate description of the non-linear waves. The restriction of the space of trial functions, where the approximate solution is found, is the drawback of this method. It should be mentioned the works [210,217], where validity of this method has been investigated.

In this chapter we shall consider some results concerning the theory of optical pulses propagation in non-linear fibres. Standard non-linear Schrödinger equation will be considered in section 6.1, where the steady-state solution of this equation will be found. We shall demonstrate that this equation can be solved by the IST method. The phenomenon of non-linear filtration of the optical soliton will be exhibited as one of the applications of this method. In section 6.2 we shall consider the theory of the optical pulses propagation in non-linear fibres when the effects of the high-order dispersion become important. Section 6.3. is devoted to the two-component optical solitons theory. The perturbation theory for optical soliton and the variation method will be presented in section 6.4. Some examples of application of these approximate methods are given there.

6.1. Picosecond solitons in the single-mode optical fibres

Let us consider the optical pulses propagation in fibre. We assume that the refraction index changes only in the transverse direction. In the longitudinal direction the medium is uniform. Such situation is a good prototype of the optical fibre used in FOCS.

It is known that a solution of Maxwell equations, describing the guided (directed) waves, is a linear superposition of the waveguide modes $\vec{\Psi}_m(\vec{\rho})$, where $\vec{\rho}=(x,y)$ is a radius-vector in the transverse section of the waveguide:

$$\vec{E}(\vec{\rho},z;\omega_0)=\sum_m A_m\vec{\Psi}_m(\vec{\rho},\omega_0)\exp[i\beta_m(\omega_0)z]. \tag{6.1.1}$$

The mode functions $\vec{\Psi}_m(\vec{\rho})$ are eigenfunctions of the boundary problem, which appears under the determination of eigenmodes of the cylindrical dielectric fibre [137-139]. The numbers β_m are the eigenvalues of this boundary problem and they are named the propagation constants. Both the propagation constants β_m and the mode functions $\vec{\Psi}_m(\vec{\rho})$ depend parametrically on the carrier wave frequency ω_0. This dependency will indicated when it is needed.

In the uniform isotropic medium the mode functions may be harmonic functions $\exp(\pm i\vec{k}\vec{r})$. The spatial non-uniformity provided by the waveguide makes these functions unsuitable as an orthonormal basis. The Fourier expansion in terms of harmonic functions must be changed for the generalised Fourier expansion in terms of the orthogonal waveguide mode functions. Thus the spatial non-uniformity turns out to be taken into account immediately. The possibility to use these eigenfunctions as a basis in the Fourier expansion is based on the orthogonality of the functions $\vec{\Psi}_m(\vec{\rho})$ that is expressed by the following formula:

$$\int_{\Sigma} \vec{\Psi}_m(\vec{\rho})\vec{\Psi}_n(\vec{\rho})d\vec{\rho} = \delta_{mn} \,. \qquad (6.1.2)$$

In (6.1.2) Σ is a surface normally crossing the waveguide. The norms of the mode functions chosen here are convenient for the further calculations. Therewith, another normalisation can be exploited too. For instance, the norm of the mode function is equal to the power flow related with this mode.

Let only one pulse propagate in the ideal linear fibre. We assume that the condition of the slowly varying pulse envelopes holds. The expression (6.1.1) is correct for the monochromatic radiation, and it can be used to obtain the necessary approximate expression. To find the Fourier components of the slowly varying pulse envelope $\mathcal{E}(\omega)$, it is simply necessary to shift the frequency $\omega_0 \to \omega' = \omega_0 + \omega$:

$$\vec{\mathcal{E}}(\vec{r};\omega) = \sum_m A_m(z\,,\omega')\vec{\Psi}_m(\vec{\rho},\omega')\exp[i\beta_m(\omega_0)z].$$

Here the propagation constant β_m is taken at the frequency ω_0 because the total dependency on ω' has been taken into account in A_m. By repeating the procedure of derivation of equations for the slowly varying pulse envelope (section 1.3.1.) and taking into account characteristics of the mode functions we can get the system of equations for A_m

$$\frac{\partial^2 A_l}{\partial z^2} + 2i\beta_l \frac{\partial A_l}{\partial z} + \Delta\beta_l^2 A_l = -\sum_m \Delta\kappa_{lm}^2 A_m \exp\{iz(\beta_m - \beta_l)\},$$

where

$$\Delta\beta_l^2 = \beta_l^2(\omega') - \beta_l^2(\omega_0),\ \Delta\kappa^2 = \kappa^2(\omega') - \kappa_0^2(\omega'),$$

and

$$\Delta\kappa_{lm}^2 = \int d\vec{\rho}\vec{\Psi}_l(\vec{\rho})\Delta\kappa^2\vec{\Psi}_m(\vec{\rho}) .$$

As far as $\omega << \omega_0$, the quantity $\Delta\beta_l^2$ can been expressed as the Taylor series about the carrier frequency:

$$\Delta\beta_l^2 = \sum_N \alpha_{lN}(\omega_0)\omega^N ,$$

where

$$\alpha_{lN}(\omega_0) = \frac{1}{N!}\frac{d^N}{d\omega^N}\left[\beta_l^2(\omega)\right]_{\omega=\omega_0} . \tag{6.1.3}$$

For the ideal linear waveguide (or fibre) $\Delta\kappa^2 = 0$.

Now, the evolution of mode envelopes is governed by the following equations

$$\frac{\partial^2 A_l}{\partial z^2} + 2i\beta_l\frac{\partial A_l}{\partial z} + \sum_N \alpha_{lN}(\omega_0)\left(i\frac{\partial}{\partial t}\right)^N A_l = 0 .$$

The physical meaning of the $\alpha_{iN}(\omega_0)$ can be easily found from their definitions. For instance,

$$\alpha_{l1}(\omega_0) = \left.\frac{d\beta_l^2}{d\omega}\right|_{\omega=\omega_0} = 2\beta_l(\omega_0)\left(\frac{d\beta_l}{d\omega}\right)_{\omega=\omega_0}$$

For the uniform medium the group velocity of light pulse is $v_g = d\omega / dk$. In the case of the optical fibre it is $v_g = (d\beta / d\omega)^{-1}$ [138-140]. Thereby,

$$\alpha_{l1}(\omega_0) = 2\beta_l(\omega_0)v_{gl}^{-1}(\omega_0) .$$

Further, we can find that the parameters $\alpha_{iN}(\omega_0)$ are related with the group velocity v_{gl} by considering definition of these parameters, i.e., (6.1.3). For instance, at N=2 we get

$$\alpha_{l2}(\omega_0) = v_{gl}^{-2}\left[1 - \beta_l\frac{dv_{gl}}{d\omega}\right]_{\omega=\omega_0} .$$

Consequently, for $N > 1$ parameter $\alpha_{iN}(\omega_0)$ describes the *group-velocities dispersion* in a fibre. With taking into account $\alpha_{iN}(\omega_0)$ with $N > 2$ we can discuss the high-order group-velocities dispersion effects. As usual, $\alpha_{i3} << \alpha_{l2}$ and all the terms with $N > 2$ are rejected in the sum over N.

The evolution of mode envelope $A = A_l$ in fibre is described by the equation

$$\frac{\partial^2 A}{\partial z^2} + 2i\beta \frac{\partial A}{\partial z} + i\alpha_1 \frac{\partial A}{\partial t} - \alpha_2 \frac{\partial^2 A}{\partial t^2} = 0 \ , \tag{6.1.4}$$

were only second-order group-velocities dispersion was taken into account. If function $A(z, t)$ is slowly varying with the increase of z, so $|\partial A / \partial z| << \beta |A|$, then the second-order derivative with respect to z in equation (6.1.4) can be omitted, and we get

$$2i\beta \frac{\partial A}{\partial z} + i\alpha_1 \frac{\partial A}{\partial t} - \alpha_2 \frac{\partial^2 A}{\partial t^2} = 0 . \tag{6.1.5}$$

Going over to the new variables $z = L\zeta$, $t = t_p \tau + z / V$ in this equation we obtain

$$i \frac{\partial A}{\partial \zeta} + s_1 \left(\frac{L}{L_D} \right) \frac{\partial^2 A}{\partial \tau^2} = 0 , \tag{6.1.6}$$

where $V = 2\beta / \alpha_1$. It is easy to verify that $V = v_{gl}$. Here $s_1 = \mathrm{sgn}(\alpha_2)$, t_p is a pulse duration at the input of fibre. $L_D = 2\beta t_p^2 |\alpha_2^{-1}|$ will be named the dispersion length. The arbitrary parameter L should be considered as spatial length, where an essential changing of a pulse in fibre occurs.

If the group-velocities dispersion may be neglected, i.e., the fibre length is much less than L_D, then all mode envelopes are expressed through initial values $A_l^{(0)}(t)$ at $z = 0$ in a simple way:

$$A_l(z,t) = A_l^{(0)}(t - z / v_g).$$

Hence the optical pulse is represented by the superposition of pulses of the modes and each lth mode propagates with its individual group velocity v_{gl}. If it is a single-mode fibre, then an optical pulse propagates without broadening. For multiple-mode fibre this pulse suffers distortion, which is due to the group velocity differences:

$$\vec{\mathcal{E}}(\vec{r};t) = \sum_m A_m^{(0)}(t - z / v_{gm}) \vec{\Psi}_m(\vec{\rho}, \omega_0) \exp[i\beta_m(\omega_0) z].$$

Thereby in a linear single-mode fibre without dispersion group-velocities for each mode, the optical pulse does not distort in the course of its propagation. However, this pulse does distort in multiple-mode fibre. This mechanism of the optical pulse distortion was named the *inter-mode dispersion of group velocities*.

Let now dispersion length be finite. The natural scale of the spatial variable z is L_D. Thus we can assign $L = L_D$, and the equation (6.1.6) can be rewritten as

$$i\frac{\partial A}{\partial \zeta} + s_1 \frac{\partial^2 A}{\partial \tau^2} = 0 .$$

This is the well-known the Schrödinger equation in quantum mechanics for the wave functions of the free moving particle. It is known that with an increase in the distance z the initial distribution $A_l^{(0)}(z=0,t)$ (profile of the incident optical pulse) broadens and its amplitude decreases.

Hence, in an ideal linear fibre the optical pulse inevitably spreads as a result of group-velocities dispersion. This is due to dispersive characteristics of material of fibre and the inter-mode dispersion of group velocities.

6.1.1. NON-LINEAR SCHRÖDINGER EQUATION

We now consider the pulse propagation in a non-linear fibre. To take the non-linear optical processes into account, we can proceed as follows. Let us divide the vector of polarisation $\vec{P}(\vec{E})$ into linear and non-linear parts:

$$\vec{P}(\vec{E}) = 4\pi\chi_L \vec{E} + \vec{P}^{NL}(\vec{E}) .$$

Wave equation for slowly varying envelopes A_m keeps its form, but now the coefficient $\Delta\kappa^2$ should be changed $\Delta\kappa^2 = 4\pi(\omega'/c)^2 \chi_{NL}(\mathbf{r},\omega')$, where $\chi_{NL} = P^{NL}/E$ in the case of scalar waves. For the vector waves we can use the definition $P_j^{NL} = (\chi_{NL})_{jk} E_k$. Hereafter, for the sake of simplicity we shall consider only the scalar case. If the dispersion of the non-linear polarizability is neglected, i.e., $\chi_{NL}(\mathbf{r},\omega') \approx \chi_{NL}(\mathbf{r},\omega_0)$, then the system of equations for slowly varying mode envelopes should be taken in the form

$$\begin{aligned} &\frac{\partial^2 A_l}{\partial z^2} + 2i\beta_l \frac{\partial A_l}{\partial z} + \sum_N \alpha_{lN}(\omega_0)\left(i\frac{\partial}{\partial t}\right)^N A_l = \\ &= -\sum_m \mu_{lm}(\omega_0) A_m \exp\left[iz(\beta_m - \beta_l)\right] \end{aligned} \qquad (6.1.7)$$

where $\mu_{lm}(\omega_0) = \mu_{lm}(\omega_0;\{A_k\}) = 4\pi(\omega_0/c)^2 \int \vec{\Psi}_l \chi_{NL} \vec{\Psi}_m d\vec{\rho}$.

Here dependency of non-linear connection coefficients of lth and mth modes on their envelopes is indicated.

In a resonance approximation the connected modes equations (6.1.7) can be simplified

$$\frac{\partial^2 A_l}{\partial z^2} + 2i\beta_l \frac{\partial A_l}{\partial z} + \sum_N \alpha_{lN}(\omega_0)\left(i\frac{\partial}{\partial t}\right)^N A_l + \mu_{ll}(\omega_0)A_l = 0 \ .$$

When we are restricted only by the effect of the second-order group-velocity dispersion, then the corresponding equations (comparing with (6.1.4)) result in

$$\frac{\partial^2 A}{\partial z^2} + 2i\beta\frac{\partial A}{\partial z} + i\alpha_1\frac{\partial A}{\partial t} - \alpha_2\frac{\partial^2 A}{\partial t^2} + \mu(\omega_0;\{A_k\})A = 0 \ , \qquad (6.1.8a)$$

where the interior mode index was omitted. Under condition of slowly varying of the envelope the term with a second spatial derivation in (6.1.8a) can be rejected. In terms of dimensionless independent variables ζ, τ: $z = L\zeta, t = t_p\tau + z/V$ equation (6.1.8a) can be written in the form of (6.1.6). It is natural to choose L as a dispersion length and to define $L_{NL} = 2\beta|\mu^{-1}|, s_2 = \mathrm{sgn}(\mu)$. Now, the evolution of the waveguide mode envelopes in a non-linear fibre can be described by the non-linear system of equations:

$$i\frac{\partial A_l}{\partial \zeta} + s_1\frac{\partial^2 A_l}{\partial \tau^2} + s_2\left(\frac{L_g}{L_{NL}}\right)A_l = 0,$$

In this non-linear equation the L_{NL} depends on the envelopes of the all waveguide modes of the fibre. Hereafter we shall consider cubic non-linear susceptibility (refer to section 1.3.2 and the equation (1.3.10.)). The evolution of the waveguide mode envelopes in this case can be governed by the equation

$$i\frac{\partial A}{\partial \zeta} + s_1\frac{\partial^2 A}{\partial \tau^2} + s_2\mu_K |A|^2 A = 0 \ , \qquad (6.1.9a)$$

where the third-order non-linear susceptibility $\chi^{(3)}$ defines the Kerr coupling constant

$$\mu_K = (4\pi\omega_0^2/c^2)\int \chi^{(3)}(\vec{\rho})\Psi^4(\vec{\rho})dxdy\left(\int \Psi^2(\vec{\rho})dxdy\right)^{-1}.$$

It is convenient to introduce the normalised pulse envelope $q(\tau,\zeta) = A(t,z)/A_0$, so the equation (6.1.9a) becomes

$$i\frac{\partial q}{\partial \zeta} + s_1 \frac{\partial^2 q}{\partial \tau^2} + s_2 \mu |q|^2 q = 0, \qquad (6.1.9b)$$

where $\mu = \mu_K A_0^2$ is a normalised Kerr coupling constant. Both equation (6.1.9a) and the equation (6.1.9b) are named the *Nonlinear Schrödinger equation.*

When the optical pulse duration is sufficiently short, being comparable with the response time of systems defined by the optical characteristics of fibre material, it is necessary to take a dispersion of non-linear susceptibility into account simultaneously with the third- (or above-) order group-velocity dispersion. As a rule, it is essential to pulses of a femtosecond duration range. Let us represent a non-linear polarizability as a sum $\chi_{NL}(\vec{r},\omega') \approx \chi_{NL}(\vec{r},\omega_0) + \chi'_{NL}(\vec{r},\omega_0)\omega$, where $\chi'_{NL}(\vec{r},\omega_0)$ is a derivative of a non-linear polarizability with respect to frequency, being taken at the carrier wave frequency. So far the effects of higher-order group-velocity dispersion have not been taken into account, when the propagation of the femtosecond pulses into a non-linear fibre is considered. Now, it is necessary to reproduce all preceding manipulations, so that the resulting equation can be obtained as follows

$$\begin{aligned} &\frac{\partial^2 A}{\partial z^2} + 2i\beta \frac{\partial A}{\partial z} + i\alpha_1 \frac{\partial A}{\partial t} + \alpha_2 \frac{\partial^2 A}{\partial t^2} - i\alpha_3 \frac{\partial^3 A}{\partial t^3} + \\ &+ \mu(\omega_0;\{A_k\})A + i\frac{\partial}{\partial t}\left(\mu'(\omega_0;\{A_k\})A\right) = 0, \end{aligned} \qquad (6.1.8b)$$

where factor $\mu'(\omega_0;\{A_k\})$ is determined by the expression

$$\mu'(\omega_0;\{A_k\}) = 4\pi(\omega_0/c)^2 \int \vec{\Psi}_l \chi'_{NL}(\vec{\rho},\omega_0)\vec{\Psi}_l d\vec{\rho}.$$

It should be remarked that there are also the alternative approaches to the reductions of the Maxwell equations to the non-linear evolution equations for the optical pulses in fibre [84-86].

6.1.2. STEADY-STATE SOLUTIONS OF THE NONLINEAR SCHRÖDINGER EQUATION

It is useful to find stationary solutions of the Nonlinear Schrödinger (NLS) equation (6.1.9), which previously were the only exact solutions of this equation. A stationary solution or steady-state solution is that which does not change the form under the translation along the characteristics, being dependent on the spatial and time variable z and t as a function of only one variable $\eta = \zeta - \vartheta\tau = (L_D^{-1} + \vartheta / Vt_p)z - (\vartheta / t_p)t$.

Let NLS equation be presented in a standard form ($s_2 = 1$ and $s_1 = s$, $s^2 = 1$)

$$i\frac{\partial q}{\partial \zeta} + s\frac{\partial^2 q}{\partial \tau^2} + \mu |q|^2 q = 0, \tag{6.1.10}$$

We shall find the solution of the equation (6.1.10) in the following form

$$q(\zeta, \tau) = \exp(i\mathrm{K}\zeta - i\Omega\tau)u(\tau + \vartheta\zeta) .$$

Substituting this expression in (6.1.10) and separating the resulting equation into the real and imaginary parts, we get two equations in real-valued variables:

$$(2s\Omega - \vartheta)\frac{du}{d\eta} = 0, \qquad \frac{d^2u}{d\eta^2} - (s\mathrm{K} + \Omega^2)u + \mu s u^3 = 0.$$

The former equation is satisfied if we choose the parameter Ω to be equal to $s\vartheta/2$. In the latter equation it is convenient to design $p = (s\mathrm{K} + \Omega^2)$ and to multiply both sides by the derivative $du / d\eta$. Integration of the resulting expression with respect to η gives rise to the equation

$$(du / d\eta)^2 = pu^2 - \kappa u^4 + C,$$

where $\kappa = s\mu/2$ and C is an integrating constant, which is defined from the boundary conditions. These conditions are defined by the behaviour of the solution as $\eta \to \pm\infty$. There are essentially two different kinds of such conditions.

Solitary wave with zero asymptotic values

This solution is associated with the following boundary condition

$$\lim_{|\eta| \to \infty} u(\eta) = 0, \quad \lim_{|\eta| \to \infty} du / d\eta = 0.$$

Taking into account these conditions it is possible to define that integrating constant as equal to zero. Thereby, it is necessary to find the solution of the equation

$$(du / d\eta)^2 = pu^2 - \kappa u^4. \tag{6.1.11}$$

It is possible to notice that negative values of parameter p are not acceptable because they give rise to solutions, which do not agree with the boundary conditions. According to these boundary conditions the solution and its derivative approach to zero simultaneously. Consequently, we need to take only positive value of p, but the sign of κ is not fixed.

Let κ >0. It is seen from (6.1.11) that the possible values of the functions u are limited by the value $(p/\kappa)^{1/2}$. If we define a new variable

$$u=(p/\kappa)^{1/2}y, \quad \xi=\eta p^{1/2},$$

that corresponds to the positive κ, then the envelope of solitary wave is defined by the equation

$$\frac{dy}{d\xi}=\pm y\sqrt{1-y^2}\,.$$

The solution of this equation is a hyperbolic secant. The real envelope is given by

$$u(\zeta-\vartheta\tau)=\sqrt{p/\kappa}\,\mathrm{sech}\left[\sqrt{p}(\zeta-\vartheta\tau-\zeta_0)\right], \tag{6.1.12a}$$

where ζ_0 is a new integrating constant that determines the initial position of a solitary wave. This constant is often supposed to be zero.

The case of $\kappa<0$ can be considered in the same way. From equation (6.1.11) it is seen that the admissible values of the functions u are unlimited. By using the same normalisation it is possible to get the equation for y:

$$\frac{dy}{d\xi}=\pm y\sqrt{1+y^2}\,.$$

The solution of this equation is

$$u(\zeta-\vartheta\tau)=\sqrt{p/\kappa}\,\mathrm{cosech}\left[\sqrt{p}(\zeta-\vartheta\tau-\zeta_0)\right]. \tag{6.1.12b}$$

This function has a singularity at $\zeta=\vartheta\tau+\zeta_0$. Therefor, from the physical point of view this solution is unacceptable as in the vicinity of this point approximation under consideration, being the basis of theories of non-linear optical waves in a Kerr medium, is invalid. In this situation the high order of non-linear susceptibility should be taken into account.

Thus, the steady-state solution of NLS with zero asymptotic values describing solitary optical wave in a Kerr medium can exist only under condition $s\mu > 0$. It is important to emphasise that only under this condition NLS has also soliton solutions.

Solutions with nonzero asymptotes. Cnoidal wave and dark solitary waves
These solutions are consistent with the boundary conditions

$$\lim_{|\eta|\to\infty} u(\eta) = u_0 , \quad \lim_{|\eta|\to\infty} du / d\eta = 0 .$$

That allows to define an integrating constant *C*. Then we can rewrite the equation for the real-valued envelope of the non-linear wave as

$$(du / d\eta)^2 = p(u^2 - u_0^2) - \kappa(u^4 - u_0^4) .$$

The right-hand side of this equation may be transformed as the product of multiplicands

$$(du / d\eta)^2 = (u^2 - u_0^2)\left[(\kappa u_0^2 - p) + \kappa u^2\right].$$

It can be seen now that the solution of this equation is expressed through the elliptic functions. Depending on the sign of the nonlinearity parameter κ, the value and sign of *p* the resulting expressions will be different.

Let us consider the case of $\kappa < 0$ and denote $\bar{p} = -p$. Then the equation for the real-valued envelope of a non-linear wave can be rewritten as

$$(dy / d\eta)^2 = (1 - y^2)\left[1 - k^2 y^2\right], \tag{6.1.13}$$

where $y = u / u_0$ and $k^2 = |\kappa| u_0^2 (\bar{p} - |\kappa| u_0^2)^{-1}$.

In the region of values of the parameters $\bar{p} > 2|\kappa| u_0^2$, a module of the elliptic integral, which results from the solution of (6.1.13), is less than unit, i.e., $k < 1$. The solution of equation (6.1.13)can be written through the Jacoby elliptic function:

$$y(\eta) = \mathrm{sn}\left[(\eta - \eta_0) , k\right] . \tag{6.1.14a}$$

where η_0 is an integrating constant.

In the region of the values of parameters $2|\kappa| u_0^2 > \bar{p} > |\kappa| u_0^2$ the module of elliptic integral is greater than unity. If we define $\tilde{y} = ky$, then equation (6.1.13) will take the form

$$(d\tilde{y} / dk\eta)^2 = (1 - \tilde{y}^2)\left[1 - k^{-2} \tilde{y}^2\right].$$

This equation has the same form as the equation (6.1.13), but with the modified argument and module. Thus, the solution of the initial equation may be written as

$$y(\eta) = k^{-1}\,\mathrm{sn}\left[k(\eta - \eta_0)\ ,\ k^{-1}\right]. \tag{6.1.14b}$$

When $\bar{p} = 2|\kappa|u_0^2$ the module of elliptic integral is equal to unity and the solution of equation (6.1.13) is

$$y(\eta) = \tanh(\eta - \eta_0)\ . \tag{6.1.14c}$$

This solution of (6.1.13) corresponds to *dark solitary wave*, which complies with dark soliton of NLS equation.

In the region of values of parameters $0 < \bar{p} < |\kappa|u_0^2$ equation (6.1.13) takes the form

$$(dy/d\eta)^2 = (1 - y^2)\left[1 + k^2 y^2\right],$$

where $1 < k^2 < \infty$ and $k^2 = |\kappa|u_0^2(|\kappa|u_0^2 - \bar{p})^{-1}$. The integral was appeared here may be interpreted as an elliptic integral of the first kind with imaginary module. Hence we can write a solution of this equation as

$$y(\eta) = \mathrm{sn}\left[(\eta - \eta_0)\ , i\,k\right].$$

Now, by using the transformation rules of Jacobian elliptic functions [243] this expression can be written through the elliptic function with real modules and arguments:

$$y(\eta) = \frac{\mathrm{sd}\left(\sqrt{1+k^2}\,y\,,\,k\sqrt{(1+k^2)^{-1}}\right)}{\sqrt{1+k^2}}\ , \tag{6.1.14d}$$

where

$$\mathrm{sd}(z,k) = \frac{\mathrm{sn}(z,k)}{\mathrm{dn}(z,k)}$$

is the elliptic function incorporated by J. Glaisher [243].

6.1.3. SIMILARITY SOLUTIONS OF THE NONLINEAR SCHRÖDINGER EQUATION

Besides steady-state solutions of NLS equation in the form of solitary and cnoidal waves, there are other solutions, which are non-stationary ones. The solutions, which can be expressed through Painleve transcendents are the particular case of those.

Let NLS equation be written as

$$i\frac{\partial q}{\partial \zeta}+s\frac{\partial^2 q}{\partial \tau^2}+\mu|q|^2 q=0 . \qquad (6.1.15)$$

Let the solution of this equation be in the following form [30-32]

$$q(\zeta,\tau)=\exp[i(a\zeta+b\tau+c\zeta\tau+r\zeta^2+g\zeta^3)]u(\tau+f\zeta+h\zeta^2) . \qquad (6.1.16)$$

Substituting (6.1.16) into (6.1.15) and separating resulting equation into real and imaginary parts we obtain the two equations

$$[(f+2h\zeta)+2s(b+c\zeta)]\frac{du}{d\eta}=0,$$

$$s\frac{d^2u}{d\eta^2}+(a+c\tau+2r\zeta+3g\zeta^2+sb^2+$$

$$+2sbc\zeta+sc^2\zeta^2)u+\mu u^3=0.$$

Here we introduce a new variable $\eta=\tau+f\zeta+h\zeta^2$. The first equation above will be satisfied if we choose the following relation between parameters

$$b=-sf/2, \quad c=-sh . \qquad (6.1.17)$$

The second equation is a non-uniform ordinary differential equation containing the variable ζ to the second power. Consequently, it is possible to choose the rest of parameters so as the power of non-uniformity will be reduced. That means that the multiplier in the second term becomes the linear function of the independent variable η. To do this we require $2(r+sbc)=cf$, $3g+sc^2=hc$. These relations with regard to (6.1.17) allow us to define the relationship between the remainder of the parameters $r=-sfh$, $gs=-2sh^2/3$ whereupon the equation for the real-valued functions $u(\eta)$ takes the form

$$s\frac{d^2u}{d\eta^2}-[(a+sf^2/4)-sh\eta]u+\mu u^3=0 . \qquad (6.1.18)$$

If we now denote $w=Au$ and $\xi=(\eta-q)/p$, where

$$A^2=\mp\frac{1}{2}\left(\frac{\mu}{s}\right)\left(\frac{1}{h}\right)^{2/3}, \quad q=\left(\frac{s}{h}\right)\left(a+s\frac{f^2}{4}\right), \quad p=-\left(\frac{1}{h}\right)^{1/3},$$

then the equation (6.1.18) can be transformed to the equation having the standard form of the *Painleve equation of second kind* (or P-II) [32, 244]:

$$\frac{d^2 w}{d\xi^2} = \xi w \pm 2w^3 . \tag{6.1.19}$$

Solution of this equation can not be expressed through any elementary functions. It is a special function, which has been named the *Painleve transcendent* of second kind.

If only the limited solutions of the NLS equation are of interest, then it is possible to omit the second term in the right-hand side of the equation (6.1.19) at $|\xi|>>1$ that leads thereby to the Airy equation. Consequently, under great values of arguments the limited solution of (6.1.19) behaves like the Airy function $\mathrm{Ai}(\xi)$, which has the following asymptotic behavior

$$\mathrm{Ai}(\xi) \cong \begin{cases} \dfrac{1}{2\sqrt{\pi}} \xi^{-1/4} \exp\left(-\dfrac{2}{3}\xi^{3/2}\right), & \xi \to +\infty \\ \dfrac{1}{\sqrt{\pi}} |\xi|^{-1/4} \sin\left(\dfrac{2}{3}|\xi|^{3/2} + \dfrac{\pi}{4}\right), & \xi \to -\infty \end{cases}$$

Equation (6.1.19) is the equation of second order and it has the second linear independent solution, which in the linear limit behaves as another Airy function labelled $\mathrm{Bi}(\xi)$. This function is a singular one, consequently it is unsatisfactory for our purpose from the physical point of view.

6.1.4. SOLITONS OF THE NONLINEAR SCHRÖDINGER EQUATION

We now turn to the Nonlinear Schrödinger equation in the following form

$$i\frac{\partial q}{\partial \zeta} + \frac{1}{2}\frac{\partial^2 q}{\partial \tau^2} + |q|^2 q = 0 . \tag{6.1.20}$$

The *U-V*-pair for (6.1.20) can be found according to the procedure described in Chapter 4. However, another algorithm will be described here. If the spectral problem is chosen in the form (3.2.1), then the zero-curvature representation can be reduced to the system of equations (4.1.45).

We assume that the functions $A(\lambda), B(\lambda)$ and $C(\lambda)$ are some polynomials of degree *N*:

$$A(\lambda) = \sum_{n=0}^{N} a_n(\zeta,\tau)\lambda^n , \quad B(\lambda) = \sum_{n=0}^{N} b_n(\zeta,\tau)\lambda^n , \quad C(\lambda) = \sum_{n=0}^{N} c_n(\zeta,\tau)\lambda^n . \tag{6.1.21}$$

Substituting (6.1.21) into (4.1.45) and equating the coefficients of various powers of λ in resulting equations, we obtain

$$\frac{\partial a_n}{\partial \tau} = qc_n - rb_n, \quad n = 0, 1, 2, \ldots, N; \tag{6.1.22a}$$

$$\frac{\partial b_n}{\partial \tau} + 2qa_n + 2ib_{n-1} = 0, \quad n = 1, 2, \ldots, N; \tag{6.1.22b}$$

$$\frac{\partial c_n}{\partial \tau} - 2ra_n - 2ic_{n-1} = 0, \quad n = 1, 2, \ldots, N; \tag{6.1.22c}$$

$$b_N = 0, \quad c_N = 0; \tag{6.1.22d}$$

$$\frac{\partial q}{\partial \zeta} = \frac{\partial b_0}{\partial \tau} + 2qa_0; \qquad \frac{\partial r}{\partial \zeta} = \frac{\partial c_0}{\partial \tau} - 2ra_0. \tag{6.1.22e}$$

It is useful to consider the simplest example when N=1. The system of equations (6.1.22) in this case takes the form

$$\frac{\partial a_0}{\partial \tau} = qc_0 - rb_0, \tag{6.1.23a}$$

$$\frac{\partial a_1}{\partial \tau} = qc_1 - rb_1, \tag{6.1.23b}$$

$$\frac{\partial b_1}{\partial \tau} + 2qa_1 + 2ib_0 = 0, \tag{6.1.23c}$$

$$\frac{\partial c_1}{\partial \tau} - 2ra_1 - 2ic_0 = 0, \tag{6.1.23d}$$

$$b_1 = 0, \quad c_1 = 0; \tag{6.1.23e}$$

$$\frac{\partial q}{\partial \zeta} = \frac{\partial b_0}{\partial \tau} + 2qa_0; \qquad \frac{\partial r}{\partial \zeta} = \frac{\partial c_0}{\partial \tau} - 2ra_0. \tag{6.1.23f}$$

From (6.1.23d) and (6.1.23b) it follows that $a_1 = a_1(\zeta)$. Then we can find from the equations (6.1.23c) and (6.1.23d) that

$$b_0(\zeta, \tau) = ia_1(\zeta)q(\zeta, \tau), \quad c_0(\zeta, \tau) = ia_1(\zeta)r(\zeta, \tau). \tag{6.1.24}$$

Hereinafter, the dependence of the functions a_n, b_n, c_n upon variables τ and ζ will be omitted sometimes. Substitution of the equations (6.1.24) into (6.1.23a) allows to define function $a_0(\zeta, \tau)$

$$a_0 = a_0(\zeta). \tag{6.1.25}$$

If the expressions (6.1.24) and (6.1.25) substitute into equation (6.1.23e), then the evolution equations for q and r can be obtained. These equations are solved by the IST method with the spectral problem (3.2.1), where matrix $\hat{U}$ is chosen in the form of (4.1.44a) and the matrix $\hat{V}$ has been found above. These equations are

$$\frac{\partial q}{\partial \zeta} = ia_1 \frac{\partial q}{\partial \tau} + 2a_0 q, \qquad \frac{\partial r}{\partial \zeta} = ia_1 \frac{\partial r}{\partial \tau} - 2a_0 r.$$

The system of these linear equations has a senior degree of derivation with respect to co-ordinate x which is equal to one. Note that the coefficient $A(\lambda) = a_0(\zeta) + \lambda a_1(\zeta)$ is a polynomial of degree one, whereas the coefficients $B(\lambda)$ and $C(\lambda)$ are polynomials of degree zero with respect to λ

$$B(\lambda) = ia_0(\zeta)q(\zeta,\tau), \quad C(\lambda) = ia_0(\zeta)r(\zeta,\tau).$$

Thus, the degree of polynomials $B(\lambda)$ and $C(\lambda)$ is one unit less than degree of $A(\lambda)$. This is a consequence of a choice of the concrete form of matrix $\hat{U}$ in equation (3.2.1) resulting in the condition (6.1.22d).

This example shows that if a non-linear evolution equation has senior degree of spatial derivation equal to N, then we can find $A(\lambda)$ as the N-degree polynomial with respect to λ, whereas $B(\lambda)$ and $C(\lambda)$ are the polynomials of degree (N-1). In the case of the NLS equation the senior spatial derivative degree is two. Consequently, we can choose

$$A(\lambda) = a_0(\zeta,\tau) + a_1(\zeta,\tau)\lambda + a_2(\zeta,\tau)\lambda^2,$$
$$B(\lambda) = b_0(\zeta,\tau) + b_1(\zeta,\tau)\lambda, \quad C(\lambda) = c_0(\zeta,\tau) + c_1(\zeta,\tau)\lambda,$$

and find the NLS equation among the resulting systems of the evolution equations.

System (6.1.22) in this case will take the form

$$\frac{\partial a_2}{\partial \tau} = 0, \tag{6.1.26a}$$

$$\frac{\partial a_1}{\partial \tau} = qc_1 - rb_1, \tag{6.1.26b}$$

$$\frac{\partial a_0}{\partial \tau} = qc_0 - rb_0, \tag{6.1.26c}$$

$$2qa_2 + 2ib_1 = 0, \quad -2ra_2 - 2ic_1 = 0, \tag{6.1.26d}$$

$$\frac{\partial b_1}{\partial \tau} + 2qa_1 + 2ib_0 = 0, \quad \frac{\partial c_1}{\partial \tau} - 2ra_1 - 2ic_0 = 0. \tag{6.1.26e}$$

Equations for q and r comply with (6.1.22e). From (6.1.26a) it follows that $a_2 = a_2(\zeta)$. The equation (6.1.26d) yields

$$b_1(\zeta,\tau) = ia_2(\zeta)q(\zeta,\tau), \quad c_1(\zeta,\tau) = ia_2(\zeta)r(\zeta,\tau) \tag{6.1.27}$$

Formulas (6.1.27) and (6.1.26b) result in the condition $a_1 = a_1(\zeta)$. With taking this condition into account the equation (6.1.26e) allows to define b_0 and c_0, i.e.,

$$b_0 = -\frac{1}{2}a_2\frac{\partial q}{\partial \tau} + ia_1 q, \qquad c_0 = \frac{1}{2}a_2\frac{\partial r}{\partial \tau} + ia_1 r.$$

Substitution of these expressions into (6.1.26c) results in

$$a_0(\zeta,\tau) = \frac{1}{2}a_2 qr + \alpha(\zeta),$$

where $\alpha(\zeta)$ is an integrating constant.

The non-linear evolution equations, which can been solved by the IST method with the considered U-V-pair, follow from (6.1.22e):

$$\frac{\partial q}{\partial \zeta} = a_2 qrq - \frac{1}{2}a_2\frac{\partial^2 q}{\partial \tau^2} + ia_1\frac{\partial q}{\partial \tau} + 2\alpha q, \tag{6.1.28a}$$

$$\frac{\partial r}{\partial \zeta} = -a_2 rqr + \frac{1}{2}a_2\frac{\partial^2 r}{\partial \tau^2} + ia_1\frac{\partial r}{\partial \tau} - 2\alpha r. \tag{6.1.28b}$$

Thus, the matrix $\hat{V}$ has the following matrix elements:

$$A(\lambda) = \alpha(\zeta) + \frac{1}{2}a_2(\zeta)qr + a_1(\zeta)\lambda + a_2(\zeta)\lambda^2,$$

$$B(\lambda) = ia_1(\zeta)q - \frac{1}{2}a_2(\zeta)\frac{\partial q}{\partial \tau} + ia_2(\zeta)q\lambda,$$

$$C(\lambda) = ia_1(\zeta)r + \frac{1}{2}a_2(\zeta)\frac{\partial r}{\partial \tau} + ia_2(\zeta)r\lambda.$$

If we put $\alpha = a_1 = 0$, $a_2 = i$ in the equations (6.1.28) and, besides

$$r = -q^*, \quad q = u \text{ or } q = iu, \tag{6.1.29}$$

then the equations (6.1.28) convert into the NLS equation in the form of (6.1.20).

It should be noted that the reduction (6.1.29) transforms the system (6.1.28) to the one non-linear equation, namely NLS one with the time varying coefficients. The soliton solutions of such an equation exhibit nontrivial behaviour. For instance, they can be accelerated or slowed down, they can have the complex phase modulation and etc. However, we shall assume that this coefficients are constants.

Now we can represent a system of linear equations of IST method in the explicit form

$$\frac{\partial\psi_1}{\partial\tau} = -i\lambda\psi_1 + q\psi_2, \quad \frac{\partial\psi_2}{\partial\tau} = i\lambda\psi_2 - q^*\psi_1\,, \tag{6.1.30}$$

$$\begin{aligned} \frac{\partial\psi_1}{\partial\zeta} &= \left(i\lambda^2 - \frac{i}{2}|q|^2\right)\psi_1 - \left(\lambda q + \frac{i}{2}\frac{\partial q}{\partial\tau}\right)\psi_2, \\ \frac{\partial\psi_2}{\partial\zeta} &= \left(\lambda q^* - \frac{i}{2}\frac{\partial q^*}{\partial\tau}\right)\psi_1 - \left(i\lambda^2 - \frac{i}{2}|q|^2\right)\psi_2. \end{aligned} \tag{6.1.31}$$

Evolution of matrix elements of the transfer matrix is defined by the expression:

$$\begin{aligned} &T_{12}(\lambda;\zeta) = T_{12}(\lambda;0)\exp\{2i\zeta^2 t\}, \\ &T_{21}(\lambda;\zeta) = T_{21}(\lambda;0)\exp\{-2i\zeta^2 t\}, \\ &T_{11}(\lambda;\zeta) = T_{11}(\lambda;0), \quad T_{22}(\lambda;\zeta) = T_{22}(\lambda;0)\,, \end{aligned} \tag{6.1.32}$$

provided that $u(\zeta,\tau)\to 0$ as $|\tau|\to\infty$.

Let us consider the particular case of initial condition for the NLS equation, which allows to find the exact solutions of the spectral problems of the IST method. It is

$$r(\zeta,\tau) = -q^*(\zeta,\tau) \text{ and } q(\zeta=0,\tau) = iA\,\mathrm{sech}(\tau)\,.$$

In this case, as it was found in section 3.4

$$a(\lambda) \equiv T_{11}(\lambda) = \frac{\Gamma(1/2 - i\lambda)\Gamma(1/2 - i\lambda)}{\Gamma(1/2 - i\lambda - A)\Gamma(1/2 - i\lambda + A)},$$

$$b(\lambda) \equiv T_{12}(\lambda) = \frac{i\sin(\pi A)}{\cosh(\pi\lambda)}\,.$$

From these expressions it follows that the purely soliton solution appears under condition $A = N$. As this takes place, the continuum spectrum of the Zakharov-Shabat inverse spectral problem is absent.

Here several examples of soliton solutions of the NLS equation are represented, which can be obtained by solving the Gelfand-Levitan-Marchenko equations for small values of "soliton numbers" N [39].

1. Let $N = 1$. The single-soliton solution has the form

$$q_{S1}(\zeta,\tau) = i\,\mathrm{sech}(\tau)\exp\{-i\zeta/2\ .$$

2. Let $N = 2$. The two-soliton solution is a more complex one

$$q_{S2}(\zeta,\tau) = 4i\exp(-i\zeta/2)\frac{\cos(3\tau) + 2\cosh(\tau)\exp(-4i\zeta)}{\cosh(4\tau) + 4\cosh(2\tau) + 3\cos(4\zeta)}\ .$$

The space-time evolution of this two-soliton optical pulse shown in Fig. 6.1.1, is characterised by the periodic variation of the amplitude and width.

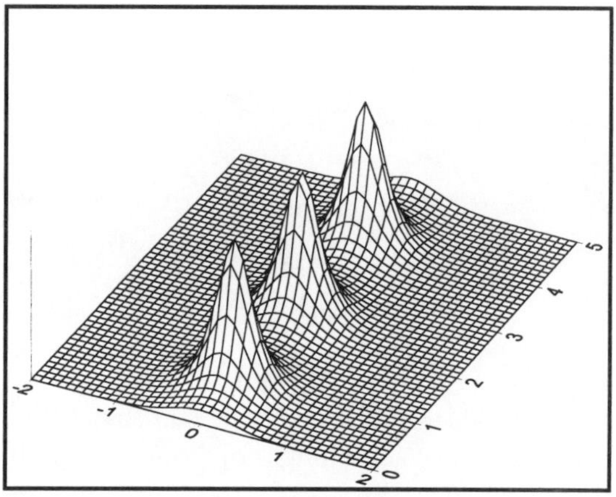

Fig. 6.1.1

The two-soliton solution of the NLS equation, describing the pulse propagation in non-linear fibre, N=2.

3. Let $N = 3$. A three-soliton solution is given by the more cumbersome expression:

$$q_{S3}(\zeta,\tau) = 3i\exp(-i\zeta/2)\frac{M(\zeta,\tau)}{D(\zeta,\tau)},$$

where

$$\begin{aligned} M(\zeta,\tau) = {} & 2\cosh(8\tau) + 32\cosh(2\tau) + \\ & + \exp(-4i\zeta)\{36\cosh(4\tau) + 16\cosh(6\tau)\} + \\ & + \exp(-12i\zeta)\{20\cosh(4\tau) + 80\cosh(2\tau)\} + \\ & + 5\exp(8i\zeta) + 45\exp(-8i\zeta) + 20\exp(-16i\zeta), \end{aligned}$$

$$D(\zeta,\tau) = \cosh(9\tau) + 9\cosh(7\tau) + 64\cosh(3\tau) + \\ + 36\cosh(\tau) + 36\cosh(5\tau)\cos(4\zeta) + \\ + 20\cosh(3\tau)\cos(12\zeta) + 90\cosh(\tau)\cos(8\zeta).$$

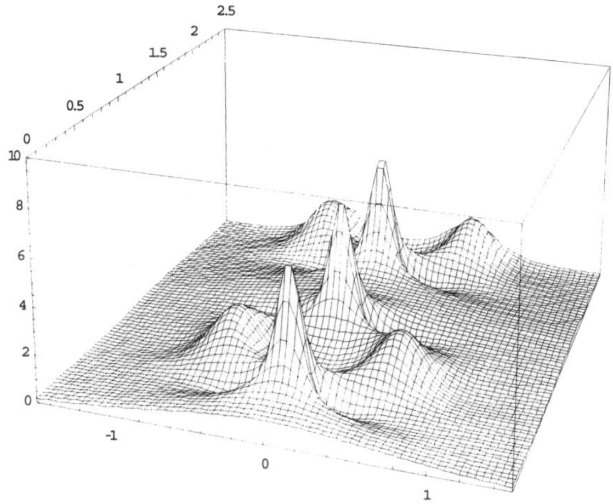

Fig. 6.1.2

The three-soliton solution of the NLS equation, describing the pulse propagation in non-linear fibre, N=3.

In the paper by Buryak, and Akhmediev [245], the two-soliton solution of the NLS equation, has been presented. It is

$$q_{S2}(\zeta,\tau) = 2\sqrt{2}\frac{M(\zeta,\tau)}{D(\zeta,\tau)}\exp\{i\eta_2^2\tau + i[(\eta_1^2 - \eta_2^2)\zeta]\},$$

where

$$M(\zeta,\tau) = (\eta_1^2 - \eta_2^2)\{\eta_2\cosh(\eta_1\zeta) + \eta_1\cosh(\eta_2\zeta)\},$$

$$D(\zeta,\tau) = (\eta_1 - \eta_2)^2\cosh[(\eta_1 + \eta_2)\zeta] + \\ + (\eta_1 + \eta_2)^2\cosh[(\eta_1 - \eta_2)\zeta] + \\ + 4\eta_1\eta_2\cos[(\eta_1^2 - \eta_2^2)\tau].$$

Both three-soliton optical pulse [39] and the two-soliton pulse by Buryak and Akhmediev are represented in Fig.6.1.2 and 6.1.3. These pictures show the general behaviour of the multi-soliton solutions of the NLS equation, i.e., the periodical variations of the pulse amplitude and width (or a pulse duration, what is more adequate to the situation under consideration here).

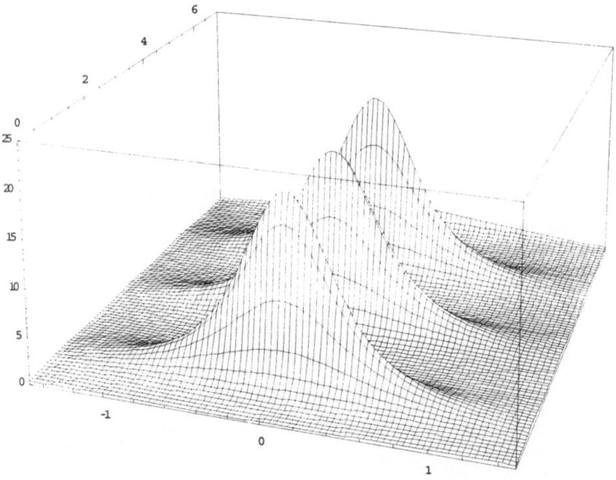

Fig. 6.1.3

The two-soliton solution of the NLS equation which has been obtained in [245]

It is important to emphasise that multi-soliton solutions of the NLS equations considered above relate to the pure imaginary eigenvalues of the Zakharov-Shabat spectral problem in the IST method. Only special initial optical pulses can result in such eigenvalues. For instance, those are the pulses with time-symmetric envelopes. The more complicated form of the pulse envelope can result in complex eigenvalues. That means that the N-soliton solution of the NLS equation can be split into N single solitons, which propagate with different velocities. As the time passes we can see the array of the single pulses propagating independently from each other.

6.1.5. NONLINEAR FILTRATION OF OPTICAL PULSES

It is known that solitons cannot be modulated: any modulation, including phase modulation (PM) disappears as a soliton optical pulse travels in a non-linear medium. Duration, velocity, or amplitude of the pulse can change, but it retains its standard form of a hyperbolic secant, which is a typical for a soliton. However, there is one further possibility of evolution of an initial pulse characterised by phase modulation: a soliton may not be formed from such a pulse, although the energy needed for this purpose is sufficient. Here we shall consider the role of phase modulation in the formation of soliton.

The propagation of the optical pulse with carrier frequency ω_0, duration t_p and an electric field strength envelope $\mathcal{E}(t,z)$ in a single-mode non-linear fibre without losses and high-order group-velocities dispersions are described by the Nonlinear Schrödinger equation (see section 6.1.1). We shall consider the standard form of this equation:

$$i\frac{\partial q}{\partial \zeta}+\frac{1}{2}\frac{\partial^2 q}{\partial \tau^2}+|q|^2 q=0. \tag{6.1.33}$$

Furthermore, here we shall discuss only the region of anomalous dispersion.

The significant properties of the NLS equation are its complete integrability [14-16] and the ability to solve the Cauchy problem for equation (6.1.33) by the IST method. In the frame of this method the problem of existence of solitons, their amplitudes, propagation velocities, and phase shift can be solved in principle by using only the initial conditions, i.e., from the given function $q_0(\tau)=q(\zeta=0,\tau)$. For this purpose we must consider an auxiliary linear system of equations

$$\frac{\partial \psi_1}{\partial \tau}=-i\lambda\psi_1+q_0(\tau)\psi_2,\quad \frac{\partial \psi_2}{\partial \tau}=i\lambda\psi_2-q_0^*(\tau)\psi_1, \tag{6.1.34}$$

and find the scattering data for equation (6.1.34).

It is possible to solve this spectral problem for the given potential $q_0(\tau)$ only as an exception. In general, one could start from a certain potential $q_0'(\tau)$ for which the solution of (6.1.34) is known and develop the perturbation theory on assumption that the difference of potentials $\Delta q_0(\tau)=q_0(\tau)-q_0'(\tau)$ is small. The eigenvalues λ_n' corresponding to $q_0'(\tau)$ are shifted in complex plane by the action of this perturbation on initial potential [16,45]. These shifts are small for sufficiently weak perturbation $\Delta q_0(\tau)$. Therefore, we can expect that PM of a pulse, which evolves to soliton (or solitons) in the absence of modulation, causes the change of parameters of a resultant soliton if the phase modulation depth is not too great. However, if PM is sufficiently deep, it may be an obstacle to the formation of soliton [44-57].

Illustrative example

Now we shall consider a simple example, when the exact solution of the spectral problem is possible. It allows us to show the influence of the PM on soliton parameters. Let

$$q_0(\tau)=iq_0\,\mathrm{sech}(2\eta\tau)\exp\{i\phi(\tau)\}.$$

If $\phi(\tau)=0$, then for $q_0\ge\eta$ such a pulse evolves into a soliton. If $\phi(\tau)$ is chosen in the form

$$\phi(\tau)=(f/2\eta)\ln\left[\cosh(2\eta\tau)\right],$$

then the solution of equations (6.1.34) can be found analytically.

After changing the variables $x = 2\eta\tau$, $\psi_1 = u_1 \exp(i\phi/2)$, $\psi_2 = u_2 \exp(-i\phi/2)$, system (6.1.34) is presented in the following form

$$\frac{\partial u_1}{\partial x} = -i[\xi + \mu \tanh(x)]u_1 + iA \operatorname{sech}(x) u_2 , \tag{6.1.35a}$$

$$\frac{\partial u_2}{\partial x} = i[\xi + \mu \tanh(x)]u_2 + iA \operatorname{sech}(x) u_1 , \tag{6.1.35b}$$

where $\mu = f/4\eta$, $\xi = \lambda/2\eta$, $A = q_0/2\eta$. Following Ref. [16], we can eliminate $u_2(x)$ from (6.1.35) and thus obtain the equation for $u_1(x)$. The substitution

$$s = [1 - \tanh(x)], \quad u_1 = s^k (1-s)^l w(s)$$

into the resultant equation yields the hypergeometric equation

$$s(1-s)\frac{d^2 w}{ds^2} + [c - (1 + a + b)s]\frac{dw}{ds} - abw = 0 ,$$

where $c = 2k + 1/2$, $a + b = 2(k + l)$. The requirement $ab = \text{const}$ fixes the exponents k and l:

$$k_\pm = \frac{1}{4}\{1 \pm [1 - 2i(\xi + \mu)]\}, \quad l_\pm = \frac{1}{4}\{1 \pm [1 + 2i(\xi - \mu)]\} . \tag{6.1.36}$$

The boundary conditions for the Jost functions determine whether values $k_\pm$ and $l_\pm$ from expressions (6.1.36) should be taken to find the function $u_1(s)$. For $k = k_-$ and $l = l_-$

$$c = \frac{1}{2} + i(\xi + \mu), \quad a_\pm = i\mu \pm (A^2 - \mu^2)^{1/2}, \quad b_\pm = i\mu \mp (A^2 - \mu^2)^{1/2} .$$

If we express $u_1(x)$ in terms of the hypergeometric functions and define $u_2(x)$ in terms of $u_1(x)$ from the system (6.1.35), we will finally be able to obtain scattering data for spectral problem related to the system (6.1.34). In our analysis we need only a diagonal element of transfer matrix $T_{11}(\xi) = a(\xi)$ (3.4.16). This element is expressed by the formula

$$a(\xi) = \frac{\Gamma\left(\frac{1}{2} + i\xi + i\mu\right)\Gamma\left(\frac{1}{2} + i\xi - i\mu\right)}{\Gamma\left(\frac{1}{2} - i\xi + \sqrt{A^2 - \mu^2}\right)\Gamma\left(\frac{1}{2} - i\xi - \sqrt{A^2 - \mu^2}\right)} .$$

Zeroes of $a(\xi)$ lying in the upper half-plane of the complex variable ξ (it is just they define the solitons amplitudes) coincide with the poles of Γ-function under condition $\mathrm{Im}\xi > 0$:

$$\xi_n = i\left(\sqrt{A^2 - \mu^2} - n - 1/2\right).$$

The index of ξ_n is an integer number ranging from zero to some maximum value N defined by the condition $\mathrm{Im}\xi_N > 0$. If there is no phase modulation, then $\xi_n(\mu = 0)$ = $i(A - n - 1/2)$, that corresponds to the known result [16, 62]. Therefore, the eigenvalues of the initial spectral problem λ_n are purely imaginary:

$$\lambda_n = i\left[\sqrt{(q_0^2 - f^2)/4} - 2\eta(n + 1/2)\right].$$

Let the amplitude of the initial pulse be $q_0 = 2\eta N$. If $\phi(\tau)$ = 0, then the N-soliton pulse forms from such initial pulse. If $\phi(\tau) \neq 0$, then

$$\lambda_n = 2i\eta\left[N\sqrt{1 - (f\,t_p / 2N)^2} - (n + 1/2)\right].$$

where $t_p = (2\eta)^{-1}$ is the duration of a single-soliton pulse.

For f = 0 there is a range of values of $\mathrm{Im}\lambda_n = \lambda_n''$, such that $\lambda_0'' > \lambda_1'' > \ldots > \lambda_{N-1}''$. With the increase of f the series of these numbers shifts along imaginary axis λ and for the certain values of $f_C^{(n)}$ the eigenvalue λ_n'' becomes zero. Therefore, as long as PM depth obeys the condition $f < f_C^{(N)} = t_p^{-1}(4N - 1)^{1/2}$, the initial pulse transforms into N-soliton pulse. If $f_C^{(N)} < f < f_C^{(N-1)} = t_p^{-1}(12N - 9)^{1/2}$, the pulse of the same energy converts into (N-1)-soliton, whereas for $f_C^{(N-1)} < f < f_C^{(N-2)} = t_p^{-1}(10N - 25)^{1/2}$, we obtain ($N$-2)-soliton pulse, and so on. If the depth of PM obeys the inequality $f > f_C^{(0)}$ = $t_p^{-1}(4N^2 - 1)^{1/2}$, a soliton is not formed, though the energy of the initial pulse is high enough. (In the considered scenario this energy is constant and it equals the energy of N-soliton pulse.)

Since all eigenvalues are purely imaginary, a multi-soliton pulse does not split into series of single-solitons. This picture of N-soliton pulse destruction by PM will be called scenario A.

In general, we can find the critical phase modulation depths $f_C^{(n)}$, (n = $N - 1$, $N - 2, \ldots$, 1, 0):

$$f_C^{(n)} = 2t_p^{-1}\sqrt{(q_0 t_p)^2 - (n-1/2)^2}\ ,$$

where N is the integral part of $(q_0 t_p - 1/2)$.

Dissociation of a soliton pulse under the influence of PM
Numerical simulation of the evolution of pulses carrying initial PM [49] confirmed scenario A, but at the same time it indicated that it was not the only one possible. The choice of another form of initial envelope and other types of PM demonstrated the examples where the initial pulse disintegrated in another way. If the modulation depth exceeds a certain threshold value, which was determined by the form of $q_0(\tau)$, the initial pulse splits into two parts each of which moves at its own velocity. The resulting pulses could develop into solitons if the depth of PM did not reach the second (higher) threshold value. Otherwise, the resulting pulses disappeared due to dispersion broadening. In terms of the spectral problem of the IST method that means that the pair of eigenvalues λ_1 and λ_2 (that corresponds to a double-soliton) leaves the imaginary axis and acquires real parts of opposite signs but with the same modulus, whereas imaginary parts remain the same. The picture of distortion of the multi-soliton pulses under the influence of PM, where the resultant solitons move with different velocities, will be called scenario *B*.

We can estimate the threshold value of the depth of PM $\tilde{f}_C^{(1)}$, the excess of which leads to the splitting of two-soliton solution of the NLS equation, by using the approximate *method of the integrals of motion* (ID, for short). This method has been used to estimate the amplitudes of soliton solutions of the Korteweg-de Vries equations and the reduced Maxwell-Bloch equation. In [248] this method was used to evaluate the amplitudes and velocities of the NLS solitons, but on the condition that the initial pulses have no phase modulation.

It is known that the NLS equation has an infinite sequence of integrals of motion $\{I_n\}$ if its solutions vanish (or it tends to the same asymptotic values) as $\tau \to \pm\infty$. The first four IDs are

$$I_1[q] = \int_{-\infty}^{\infty} |q|^2\, d\tau\ ,\quad I_2[q] = \frac{1}{2}\int_{-\infty}^{\infty}\left(q^*\frac{\partial q}{\partial \tau} - q\frac{\partial q^*}{\partial \tau}\right)d\tau\ ,\ I_3[q] = \int_{-\infty}^{\infty}\left(\left|\frac{\partial q}{\partial \tau}\right|^2 - |q|^4\right)d\tau\ ,$$

$$I_4[q] = \int_{-\infty}^{\infty}\left(q^*\frac{\partial^3 q}{\partial \tau^3} - q\frac{\partial^3 q^*}{\partial \tau^3} + 3|q|^2 q^*\frac{\partial q}{\partial \tau} - 3|q|^2 q\frac{\partial q^*}{\partial \tau}\right)d\tau\ .$$

The single-soliton solution of the NLS equation is characterised by the parameters A, V, θ_1 and θ_2:

$$q_S(\zeta,\tau) = iA\,\mathrm{sech}\left[A(\tau - V\zeta + \theta_1)\right]\exp\left\{i\left[V\tau - (V^2 - A^2)\zeta/2 + \theta_2\right]\right\}. \quad (6.1.37)$$

The ID method is based on two assumptions. If the initial condition $q_0(\tau)$ splits in the process of evolution into N solitons with different parameters $(A_j, V_j, \theta_{1j}, \theta_{2j})$, where j = 1,2,..., N, then in the limit $\zeta \to \infty$ there is no overlap between neighbouring solitons. The overlap integrals of two solitons decrease exponentially with the increasing of ζ or z. Therefore,

- the first assumption is that the total number of solitons is known,
- the second assumption is that all solitons are separate and the contribution of the non-soliton part of the solution of the NLS equation can be ignored.

Under this approximation any ID I_n for the given initial condition $q_0(\tau)$ can be written as

$$I_n[q_0] = \sum_{j=1}^{N} I_n[q_S^{(j)}], \tag{6.1.38}$$

where $I_n[q_S^{(j)}]$ are calculated for the jth soliton of the (6.1.37) type. The system of equations (6.1.38), where n = 1,2,...,2N allows us to find N pairs of A_j and V_j. The constant phase shifts θ_{1j} and θ_{2j} are not determined within the frames of the presented variant of ID method.

In [248] the ID method has been used to analyse the evolution of NLS solutions in the case of real $q_0(\tau)$. If $q_0(\tau)$ is complex, no analytic solution of equations (6.1.38) is obtained. However, in some cases it is possible to find amplitudes and velocities of solitons, even though the envelope of the initial optical pulse $q_0(\tau)$ possesses the phase modulation. We shall assume that $q_0(\tau)$ is complex function and the phase of the input pulse $\phi(\tau) = \arg q_0(\tau)$ is a symmetric function.

Let pulse energy be sufficiently high to form two solitons in the absence of PM. In this case we have $I_2 = I_4 = 0$, and the system (6.1.38) with N = 2 transforms into four equations

$$A_1 + A_2 = W_1 \ , \quad V_1 A_1 + V_2 A_2 = 0 \ , \quad (A_1^3 + A_2^3) - 3(A_1 V_1^2 + A_2 V_2^2) = 3W_3 \ ,$$
$$(A_1^3 V_1 + A_2^3 V_2) - (A_1 V_1^3 + A_2 V_2^3) = 0 \ ,$$

where we introduce the following notations: $I_1 = 2W_1$ and $I_3 = 2W_3$. Depending on the relationship between W_1 and W_3, the ID method predicts different evolution of initial pulse. Let us define

$$\Delta = \frac{W_3}{W_1} - \frac{W_1^2}{12}. \tag{6.1.39}$$

Then, if $\Delta > 0$, then the two solitons have equal velocities, but different amplitudes: $V_1 = V_2 = 0$, $A_{1,2} = (W_1/2) \pm \sqrt{\Delta}$. If $\Delta < 0$, then the amplitudes are the same but velocities are different: $A_1 = A_2 = W_1/2$, $V_1 = -V_2 = \sqrt{-\Delta}$. Therefore, we can expect dissociation (or decay) of double-soliton pulse into two single-solitons if the increase in the PM depth brings the change of the sign of Δ in (6.1.39) and makes it negative.

Let us now turn to the special case, when

$$q_0(\tau) = iq_0 \operatorname{sech}(\tau) \exp\{i\phi(\tau)\}.$$

For such input pulse we have

$$W_1 = q_0^2, \quad W_3 = \frac{1}{3} q_0^2 \left(2q_0^2 - 1\right) - q_0^2 Z[\phi(\tau)],$$

where

$$Z[\phi(\tau)] = \int_{-\infty}^{\infty} (d\phi / d\tau)^2 \operatorname{sech}^2(\tau) d\tau.$$

The parameter Δ considered as the function of Z has the form

$$\Delta(Z) = \Delta(0) - \frac{1}{2} Z = \frac{3 - (q_0^2/2 - 2)^2}{3} - \frac{1}{2} Z. \tag{6.1.40}$$

If in the absence of PM we have $\Delta(0) > 0$, then under the phase modulation $\Delta(Z)$ decreases and for a certain value of $Z[\phi(\tau)]$ we can obtain $\Delta(Z) < 0$. In formula (6.1.40) parameter q_0 should be chosen in such a way that for $\phi(\tau) = 0$ the double-soliton pulse is formed from initial condition $q_0(\tau)$, i.e., it should be $3/2 < q_0 < 5/2$. The critical value of parameter Z_C is defined by the formula

$$Z_C = \frac{2}{3} \left[3 - \left(\frac{q_0^2}{2} - 2 \right)^2 \right]. \tag{6.1.41}$$

The "linear chirp" $\phi(\tau) = f\tau^2$ is the example of the PM often discussed in literature. In this case $Z[\phi(\tau)] = (2/3)(f\pi)^2$. Formula (6.1.41) provides the critical value of PM depth f_C

$$f_C = \frac{1}{\pi} \sqrt{3 - \left(\frac{q_0^2}{2} - 2 \right)^2}. \tag{6.1.42}$$

If $q_0 = 2$, then from (6.1.42) $f_C = \sqrt{3} / \pi \approx 0.551$. The numerical simulation of the evolution of such pulse [49] gave $0.4 \le f_C \le 0.7$. When the phase modulation is described by the function

$$\phi(\tau) = f \operatorname{sech} \tau ,$$

the numerical simulation [49] yields the following estimate $1 \le f_C \le 3$. The ID method considered above in this case gives $f_C = \sqrt{7.5} \approx 2.74$. In Fig.6.1.4 the evolution of the pulse with the PM of the considered type for two different values of depth modulation is presented. Fig.6.1.4a illustrates the stability of the two-soliton pulse if the magnitude of the PM depth is less than the critical value. Moreover, if the magnitude of the PM depth

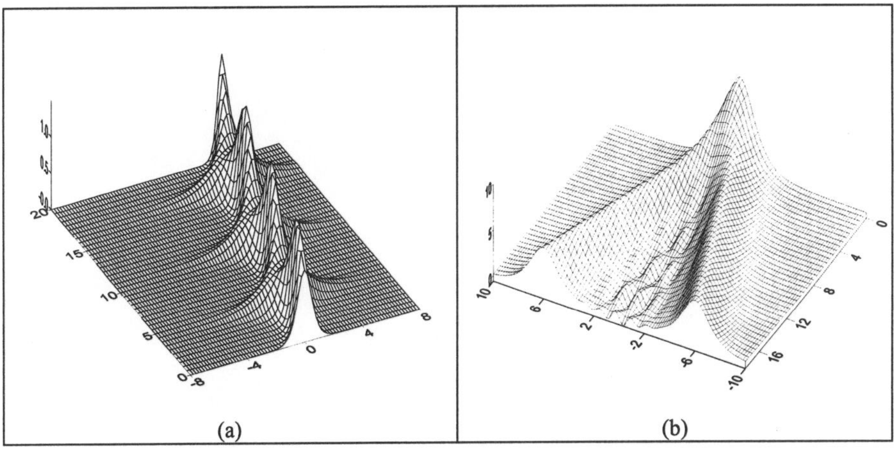

Fig.6.1.4 Evolution of two-soliton pulse with chirp in initial condition

$f > 3$ i.e., it is greater than f_C, the initial two-soliton pulse splits into two single pulses moving with different velocities (see Fig.6.1.4b).

6.1.6. CUBIC-QUINTIC NONLINEAR SCHRODINGER EQUATION

Now we consider the propagation of the optical pulse in the medium, which is characterised by the higher order nonlinearity than Kerr-type one. The simplest of these media is the quintic non-linear medium. It is a very convenient model for the study of the two- and three-dimensional solitary waves, when the cubic nonlinearity is self-focusing, while the quintic nonlinearity is self-defocusing, precluding the wave collapse. However, in this section we restrict our attention to one-dimensional case.

Let us take a look at the wave equation for slowly varying envelope of the optical pulse (6.1.8a). By omitting the space derivative of second order we can write

$$2i\beta\frac{\partial A}{\partial z}+i\alpha_1\frac{\partial A}{\partial t}-\alpha_2\frac{\partial^2 A}{\partial t^2}+\mu(\omega_0;\{A_k\})A=0\,. \qquad (6.1.43)$$

The non-linear polarisation can be represented as

$$\mathcal{P}^{NL}=\chi_{NL}\mathcal{E}=\left(\chi^{(3)}\,|\mathcal{E}|^2+\chi^{(5)}\,|\mathcal{E}|^4\right)\mathcal{E}\,.$$

Substitution of this expression into the definition of the non-linear term in (6.1.43) yields

$$\mu(\omega_0;\{A_k\})=\frac{4\pi\omega^2}{c^2}\left(\chi_{eff}^{(3)}\,|A|^2+\chi_{eff}^{(5)}\,|A|^4\right),$$

where the following effective susceptibilities were introduced

$$\chi_{eff}^{(3)}=\int\chi^{(3)}(x,y)\Psi^4(x,y)dxdy\left(\int\Psi^2(x,y)dxdy\right)^{-1},$$
$$\chi_{eff}^{(5)}=\int\chi^{(5)}(x,y)\Psi^6(x,y)dxdy\left(\int\Psi^2(x,y)dxdy\right)^{-1}.$$

Thus the wave equation (6.1.43) takes the form

$$2i\beta\frac{\partial A}{\partial z}+i\alpha_1\frac{\partial A}{\partial t}-\alpha_2\frac{\partial^2 A}{\partial t^2}+\frac{4\pi\omega^2}{c^2}\left(\chi_{eff}^{(3)}\,|A|^2+\chi_{eff}^{(5)}\,|A|^4\right)A=0\,. \qquad (6.1.44)$$

In terms of dimensionless independent variables ζ, τ: $z=L_D\zeta$, $t=t_p\tau+z/v_g$, this equation can be written as

$$i\frac{\partial q}{\partial\zeta}-s_2\frac{\partial^2 q}{\partial\tau^2}+\mu\,|q|^2\,q+\mu_5\,|q|^2\,q=0\,, \qquad (6.1.45)$$

where

$$\mu=\frac{2\pi\omega_0^2\chi_{eff}^{(3)}A_o^2}{c^2\beta}L_D,\quad \mu_5=\frac{2\pi\omega_0^2\chi_{eff}^{(5)}A_o^4}{c^2\beta}L_D\,.$$

Equation (6.1.45) was named the *cubic-quintic nonlinear Schrödinger equation* .

It is useful to find any solution of the non-linear evolution equation under consideration. Every so often such problem can be solved only at the cost of a some simplifications. So, let us assume that

$$q(\zeta,\tau)=u(\zeta,\tau)\exp[i\phi(\zeta,\tau)].$$

The two new equations in terms of the real-valued variables are derived from (6.1.45)

$$\frac{\partial u}{\partial \zeta}-2s_2\frac{\partial \phi}{\partial \tau}\frac{\partial u}{\partial \tau}-s_2 u\frac{\partial^2 \phi}{\partial \tau^2}=0, \tag{6.1.46}$$

$$s_2\left(\frac{\partial^2 u}{\partial \tau^2}-u\frac{\partial \phi}{\partial \tau}\frac{\partial \phi}{\partial \tau}\right)+u\frac{\partial \phi}{\partial \zeta}-\mu u^3-\mu_5 u^5=0. \tag{6.1.47}$$

Let $\partial\phi/\partial\tau=\Omega$ and $\partial\phi/\partial\zeta=\mathrm{K}$ be some constant parameters. Then the equation (6.1.46) is reduced to

$$\frac{\partial u}{\partial \zeta}-2s_2\Omega\frac{\partial u}{\partial \tau}=0.$$

From this equation it follows that $u(\zeta,\tau)=u(\eta)$, where $\eta=(\tau+2s_2\Omega\zeta)$. Taking this result into account the equation (6.1.47) converts to

$$d^2u/d\eta^2-p^2u-s_2\mu u^3-s_2\mu_5u^5=0, \tag{6.1.48}$$

where parameter $p^2=\Omega^2-s_2\mathrm{K}$ is introduced for the sake of simplicity. The first integral of this equation can be obtained by the regular method. With taking into account the vanishing of the u and its derivative at infinity, we obtain

$$(du/d\eta)^2-p^2u^2-\frac{1}{2}s_2\mu u^4-\frac{1}{3}s_2\mu_5u^6=0.$$

The substitution $u=w^{-1/2}$ converts this equation to

$$\left(\frac{dw}{d\eta}\right)^2=4p^2w^2+2s_2\mu w+\frac{4}{3}s_2\mu_5. \tag{6.1.49}$$

This equation can be represented in the following form

$$(dw/d\eta)^2=4p^2\left[(w+a)^2-\Delta^2\right],$$

where $a = s_2\mu/4p^2$, $\sigma = 16s_2\mu_5 p^2/3\mu^2$, and $\Delta^2 = a^2(1-\sigma)$. It should be pointed that the sign of the parameter σ is defined by the sign of non-linear susceptibility $\chi^{(5)}$ and by the sign of dispersion coefficient $s_2 = \mathrm{sgn}(\alpha_2)$, since

$$\sigma = \frac{2p^2c^2\beta}{3\pi\omega_0 L_D}\left(\frac{s_2\chi_{eff}^{(5)}}{\chi_{eff}^{(3)2}}\right).$$

Let us consider the case of a negative parameter $s_2\mu_5$, corresponding, for example, to anomalous dispersion region and self-focusing quintic nonlinearity. The solution of equation (6.1.49) is

$$w(\eta) = |a|\left\{\sqrt{1+|\sigma|}\cosh[2p(\eta-\eta_0)] - s_2\right\},$$

with η_0 as an integrating constant. The solution of the initial equations (6.1.48) can be written as

$$u^2(\eta) = \frac{1}{|a|\left\{\sqrt{1+|\sigma|}\cosh[2p(\eta-\eta_0)] - s_2\right\}}. \tag{6.1.50}$$

This solution describes the steady-state solitary wave with the amplitude depending on the sign of the dispersion coefficient. The amplitude of the wave in normal dispersion region is less than one in the anomalous dispersion region.

Now let us consider the case of the positive parameter $s_2\mu_5$ that corresponds to the normal dispersion region and self-focusing quintic nonlinearity or to the anomalous dispersion and self-defocusing quintic nonlinearity. In these cases we have to make distinction between $0 < \sigma < 1$ and $\sigma > 1$. If $0 < \sigma < 1$, then the solution of equation (6.1.49) writes as $w(\eta) = |a|\left\{\sqrt{1-\sigma}\cosh[2p(\eta-\eta_0)] - s_2\right\}$. For the anomalous dispersion, where $s_2 = -1$ we can write the solution of the initial equation as

$$u^2(\eta) = \frac{1}{|a|\left\{\sqrt{1-\sigma}\cosh[2p(\eta-\eta_0)] + 1\right\}}. \tag{6.1.51}$$

However, when $s_2 = +1$, the function $w(\eta)$ is not positive at all values of the argument. Thus, $u^2(\eta)$ describes the solitary wave with a singular core. If $\sigma > 1$, then the solution of the equation (6.1.49) can be written as

$$w(\eta) = |a|\sqrt{\sigma - 1}\sinh[2p(\eta-\eta_0)] - a.$$

The regular (or bounded, non-singular) solution of the equation (6.1.48) does not exist in this case.

6.2. Femtosecond optical solitons

When the optical pulse is shorter than one picosecond, the higher-order dispersion of the group velocities and the dispersion of the non-linear susceptibility become important. A theory describing the propagation of such short pulses is based on the equations solved primarily by numerical methods. Analytical solutions are usually represented by the steady-state solutions and some approximate expressions.

If we assume that the slowly-varying envelope approximation is held true, then the generation of the basic equations may be divided into two steps. First, we must take into account the next contributions of the dispersion of the group velocities. The equation (6.1.8a) takes the form

$$2i\beta\frac{\partial A}{\partial z}+i\alpha_1\frac{\partial A}{\partial t}-\alpha_2\frac{\partial^2 A}{\partial t^2}-i\alpha_3\frac{\partial^3 A}{\partial t^3}+\alpha_4\frac{\partial^4 A}{\partial t^4}+\mu(\{A_k\})A=0\,, \quad (6.2.1)$$

where the effect of the third- and fourth-order group-velocity dispersions are presented by the coefficients $\alpha_3(\omega_0)$ and $\alpha_4(\omega_0)$. These parameters are defined by the expression (6.1.3). Usually only the effect of the third-order group-velocity dispersions is considered under femtosecond pulse propagation investigation.

The second step is more difficult. Here we must determine the parameter $\mu(\{A_k\})$ in equation (6.2.1). In the time domain the non-linear cubic polarisation is expressed in the most general way by the convolution integral (see section 1.2.):

$$P_i^{NL}(\vec{r},t)=3\int_{-\infty}^{+\infty}dt_1\int_{-\infty}^{+\infty}dt_2\int_{-\infty}^{+\infty}dt_3\chi_{ijkl}^{(3)}(t-t_1,t-t_2,t-t_3)E_j(t_1)E_k(t_2)E_l(t_3)\,.$$

In optical fibres there are two dominant contributions to this non-linear polarisation, namely the electronic Kerr-effect and the Raman effect. The Kerr-effect is due to polarisation of the electron cloud around the individual atoms. The response time τ_R for this effect is extremely short, being of the order of femtosecond (τ_R = 2÷6 fs). The Raman contribution to the non-linear polarisation in a silica glass originated from the interaction between the electric field of the light wave and optical phonons, which are transversally oscillating molecular vibration modes in the medium [218-220]. Assuming a linearly polarised electromagnetic field in the *x*-direction, we can obtain the non-linear polarisation as the convolution

$$P_x^{NL}(t)=E_x(t)\int_{-\infty}^{+\infty}f(t-t')E_x^2(t')dt'\,, \quad (6.2.2)$$

where $f(t)$ is the non-linear response function, which indicates how fast the medium responds to an applied external electromagnetic field. It is convenient to consider some model of a non-linear medium to find this response function. Furthermore, we assume that the resonant vibration frequency of the molecules is much less than the spectral width of the ultrashort optical pulse.

6.2.1. EXTENDED NON-LINEAR SCHRÖDINGER EQUATION

We shall consider the optical pulse propagation in a weakly non-linear medium taking into account the group-velocity dispersion of second and third orders. The dispersion of non-linear susceptibility will be taken into consideration as well.

Raman response function

We shall consider the non-linear response according to Placzek model [218]. The molecules in the Raman medium are represented by classical harmonic oscillators. Polarisation of this medium can be expressed as

$$P = n_A \alpha(Q) E, \tag{6.2.3}$$

where n_A is the concentration of the molecules, $\alpha(Q)$ is the molecular polarizability, Q is the vibrational co-ordinate of a molecule, determining the magnitude of deflection from equilibrium. In the case of small vibrations the molecular polarizability is expressed by the first two terms of Taylor series, i.e.,

$$\alpha(Q) = \alpha_0 + \left(\frac{\partial \alpha}{\partial Q}\right)_0 Q = \alpha_0 + \alpha_D Q.$$

Hence, the non-linear polarisation is $P^{NL} = n_A \alpha_D Q E$. The Hamiltonian describing the photon-phonon interaction can be written as

$$\mathcal{H}_{ph-ph} = -P^{NL} E/2 = -n_A \alpha_D Q E^2/2.$$

The equation of motion for this oscillator is

$$\frac{d^2 Q}{dt^2} + \gamma \frac{dQ}{dt} + \omega_V^2 Q = \frac{\alpha_D}{2m} E^2, \tag{6.2.4}$$

where $\gamma = 2/T_R$ is a vibration line width, ω_V is the resonant vibration frequency, and m is the molecular mass. One assumes that $\gamma << \omega_V << \omega_0$.

Let us introduce the slowly varying envelopes for an oscillator co-ordinate $u(t)$ and an electric field $\mathcal{E}$

$$Q = u(t)\exp(-i\omega_V t) + u^*(t)\exp(+i\omega_V t),$$
$$E = \mathcal{E}\exp(-i\omega_0 t) + \mathcal{E}^* \exp(+i\omega_0 t).$$

Equation (6.2.4) describes the slow frequency oscillation, which is induced by the electric strength in square E^2. This imposed force contains both slow and high frequency vibrations. However, only the former of these two motions is important. Under slowly varying amplitudes approximation equation (6.2.4) takes the form

$$(\gamma - 2i\omega_V)\frac{du}{dt} - i\omega_V \gamma u = \frac{\alpha_D}{m}|\mathcal{E}|^2. \qquad (6.2.5)$$

With taking into account inequality $\omega_0 >> \omega_V$, and the condition that the spectral width of the optical pulse $\Delta\omega_p$ is greater than ω_V, we can write the slowly varying envelope of polarisation as

$$\mathcal{P}^{NL} = \mathcal{P}_R = n_A \alpha_D (u + u^*)\mathcal{E}. \qquad (6.2.6)$$

The solution of equation (6.2.5) is given by

$$u(t) = \frac{\alpha_D}{m(\gamma - 2i\omega_V)} \int_0^\infty |\mathcal{E}(t-t')|^2 \exp\{-\Gamma_R t'\} dt',$$

where

$$\Gamma_R = -\frac{2\omega_V^2 \gamma}{\gamma^2 + 4\omega_V^2} + i\frac{\omega_V \gamma^2}{\gamma^2 + 4\omega_V^2}.$$

In the sharp vibration line limit ($\gamma << \omega_V$) we have

$$\Gamma_R = -\frac{\gamma}{2} + i\omega_V \left(\frac{\gamma}{2\omega_V}\right)^2 = \Gamma_R' + i\Gamma_R''.$$

Now, we can find the Raman polarisation envelope. It is

$$\mathcal{P}_R(t) = \mathcal{E}(t) \int_{-\infty}^{+\infty} g_R(t') |\mathcal{E}(t-t')|^2 dt', \qquad (6.2.7)$$

where the non-linear Raman response function $g_R(t)$ is defined as

$$g_R(t) = \left(\frac{\alpha_D}{m}\right)\frac{2n_A\alpha_D}{\gamma^2 + 4\omega_V^2}\left(\gamma\cos\Gamma''_R t - 2\omega_V\sin\Gamma''_R t\right)\exp(-\Gamma'_R t),$$

or, with taking into account inequality $\gamma << \omega_V$,

$$g_R(t) = \frac{n_A\alpha_D^2}{m\omega_V^2 T_R}\exp\left(-\frac{t}{T_R}\right) = \alpha_R\frac{1}{T_R}\exp\left(-\frac{t}{T_R}\right).$$

Modified NLS equation

Let us now consider a wave equation for slowly varying envelope of the ultrashort pulse from section 6.1.1.

$$2i\beta_l\frac{\partial A_l}{\partial z} + \sum_N \alpha_{lN}(\omega_0)\left(i\frac{\partial}{\partial t}\right)^N A_l = -\sum_m \mu_{lm}(\omega')A_m\exp\left[iz(\beta_m - \beta_l)\right].$$

In this system of equations the term with the second-order spatial deviation is omitted. Here the non-linear coupling terms contain the frequency variable, i.e.,

$$\mu_{lm}(\omega') = \mu_{lm}(\omega' = \omega_0 + \omega; \{A_k\}) = 4\pi(\omega'/c)^2\int\Psi_l\chi_{NL}\Psi_m dxdy.$$

Since the relevant frequency domain is defined by the inequality $\omega << \omega_0$, we can write

$$\mu_{lm}(\omega')A_m \approx \frac{4\pi}{c^2}\left(\omega_0^2 + 2i\omega_0\frac{\partial}{\partial t}\right)\left\{\int\Psi_l\chi_{NL}\Psi_m dxdy\, A_m\right\}. \qquad (6.2.8)$$

The non-linear polarisation was divided into different parts, i.e., the Kerr and the Raman ones. Thus we can represent this polarisation as

$$\mathcal{P}^{NL} = \mathcal{P}_K + \mathcal{P}_R = \chi_K\mathcal{E} + \chi_R\mathcal{E},$$

where

$$\chi_K = \chi^{(3)}(\omega_0, \omega_0, -\omega_0, \omega_0)|\mathcal{E}(t)|^2,$$

$$\chi_R = \int_{-\infty}^{+\infty} g_R(t')|\mathcal{E}(t-t')|^2\,dt'.$$

Substitution of these susceptibilities into (6.2.8) with exploiting the resonance approximation yields

$$\sum_m \mu_{lm}(\omega')A_m \exp[iz(\beta_m - \beta_l)] \approx$$

$$\approx \frac{4\pi\omega_0^2}{c^2}\left(\chi_{K,eff} \mid A_l \mid^2 A_l + 2i\omega_0^{-1}\chi_{K,eff} \frac{\partial}{\partial t}(\mid A_l \mid^2 A_l)\right) +$$

$$+\frac{4\pi\omega_0^2}{c^2} A_l \int_0^\infty g_{R,eff}(t') \mid A(t-t') \mid^2 dt' .$$

where we have introduced the effective parameters

$$\chi_{K,eff} = \int \chi_K(x,y)\Psi^4(x,y)dxdy\left(\int \Psi^2(x,y)dxdy\right)^{-1},$$

$$g_{R,eff} = \int g_R(x,y)\Psi^4(x,y)dxdy\left(\int \Psi^2(x,y)dxdy\right)^{-1}.$$

Taking into account the fast Raman response in relation to pulse envelope varying, we can reduce the Raman contribution to the non-linear polarisation:

$$\int_{-\infty}^{+\infty} g_{R,eff}(t') \mid A(t-t') \mid^2 dt' \approx \alpha_{R,eff} \mid A(t) \mid^2 - \alpha_{R,eff} T_R \frac{\partial}{\partial t}\left(\mid A(t) \mid^2\right), \qquad (6.2.9)$$

where

$$\alpha_{R,eff} = \int \alpha_R(x,y)\Psi^4(x,y)dxdy\left(\int \Psi^2(x,y)dxdy\right)^{-1}.$$

The wave equation for slowly varying envelope of the ultrashort pulse (6.2.1) in this limit becomes

$$2i\beta\frac{\partial A}{\partial z} + i\alpha_1\frac{\partial A}{\partial t} - \alpha_2\frac{\partial^2 A}{\partial t^2} - i\alpha_3\frac{\partial^3 A}{\partial t^3} + \alpha_4\frac{\partial^4 A}{\partial t^4} =$$

$$= -\frac{4\pi\omega_0^2}{c^2}\chi_{K,eff}\left[\left(1 + \frac{\alpha_{R,eff}}{\chi_{K,eff}}\right) \mid A \mid^2 A + \frac{2i}{\omega_0}\frac{\partial}{\partial t}\left(\mid A \mid^2 A\right) - \frac{\alpha_{R,eff}}{\chi_{K,eff}} T_R A \frac{\partial}{\partial t}\left(\mid A \mid^2\right)\right].$$

It is convenient to introduce the normalised variables

$$\zeta = z / L_D, \quad \tau = (t - z / v_g)t_{p0}^{-1}, \quad q(\zeta,\tau) = A(z,t) / A_0,$$

where $v_g = 2\beta / |\alpha_1|$ is the group velocity, t_{p0} is the initial pulse duration. If $\alpha_2 \neq 0$, then the length L_D can be defined as

$$L_D = 2\beta t_{p0}^2 / |\alpha_2|.$$

This is the dispersion length. The third- and fourth-order dispersion lengths are defined by similar expressions

$$L_3 = 2\beta t_{p0}^3 / |\alpha_3|, \quad L_4 = 2\beta t_{p0}^4 / |\alpha_4| .$$

In terms of the normalised variables the equation for the slowly varying envelope of the ultrashort pulse takes the form

$$i\frac{\partial q}{\partial \zeta} - s_2\frac{\partial^2 q}{\partial \tau^2} - is_3\eta_3\frac{\partial^3 q}{\partial \tau^3} + s_4\eta_4\frac{\partial^4 q}{\partial \tau^4} + \\ + \mu\left[(1+\rho)|q|^2 q + i\chi\frac{\partial}{\partial \tau}\left(|q|^2 q\right) - \gamma q\frac{\partial}{\partial \tau}\left(|q|^2\right)\right] = 0, \tag{6.2.10}$$

where $s_j = \mathrm{sgn}(\alpha_j)$, $j = 2,3,4$, $\eta_{3,4} = L_D / L_{3,4}$, and

$$\rho = \frac{\alpha_{R,eff}}{\chi_{K,eff}}, \quad \mu = \frac{2\pi\omega_0^2\chi_{K,eff}A_o^2}{c^2\beta}L_D, \quad \chi = \frac{2}{\omega_0 t_{p0}}, \quad \gamma = \rho\frac{T_R}{t_{p0}}.$$

Equation (6.2.10) is a modified Nonlinear Schrödinger equation describing the ultrashort pulse propagation in a single-mode fibre It takes account of the high-order dispersion effects, the self-steeping or optical shock formation, and the self-induced frequency shift. The optical shock formation is represented by term, which is proportional to parameter χ. The Raman term with parameter γ generates the self-induced carrier frequency shift. From the definition of the parameters χ and γ we can see that the self-steeping effect is important when the pulse duration appreciably exceeds the optical period $T_0 = 2\pi / \omega_0$. The self-induced carrier frequency shift becomes important when the pulse duration approaches the Raman response time T_R. Its magnitude lies in the interval from 6 femtosecond to 3 femtosecond.

In Ref. [93] some estimates for the parameters of non-linear terms in equation (6.2.10) were presented. To show their importance we indicate that

$$\eta_3 \approx 3\cdot 10^{-3}, \quad \mu\chi \approx 6\cdot 10^{-3}, \quad \mu\gamma \approx 1{,}4\cdot 10^{-2}, \quad \text{at } t_{p0} = 500\,fs .$$

The ratio $\rho \le 1$, hence in the equation (6.2.10) the Kerr susceptibility can be omitted or re-normalised by including parameter ρ into its magnitude.

Some variants of modified NLS equation

Frequently the high-order dispersion of group velocities are neglected. In this case the equation (6.2.10) is reduced to

$$i\frac{\partial q}{\partial \zeta} - s_2\frac{\partial^2 q}{\partial \tau^2} + \mu\left[|q|^2 q + i\chi\frac{\partial}{\partial \tau}\left(|q|^2 q\right) - \gamma q\frac{\partial}{\partial \tau}\left(|q|^2\right)\right] = 0, \qquad (6.2.11)$$

In the case when the pulse duration is longer than one picosecond $\chi = \gamma = 0$ we obtain the NLS equations (6.1.10), which is a good model in this time domain.

If we assume that $\gamma = 0$, then the equation (6.2.10) becomes

$$i\frac{\partial q}{\partial \zeta} - s_2\frac{\partial^2 q}{\partial \tau^2} + \mu|q|^2 q + i\tilde{\chi}\frac{\partial}{\partial \tau}\left(|q|^2 q\right) = 0, \qquad (6.2.12)$$

where $\tilde{\chi} = \mu\chi$. Sometimes this non-linear equation is titled "*Modified Nonlinear Schrödinger equation*" (for example, [94]). It should be remarked that this equation can be transformed to DNLS equation by the change of the variables. Let $s_2 = -1$, that is, we consider the anomalous dispersion region. If one defines

$$\tilde{q}(\zeta, s) = q(\zeta, \tau)\exp[-ia\tau - ib\zeta], \qquad s = \tau - (2\mu/\tilde{\chi})\zeta,$$

where $a = \mu/\tilde{\chi}$, $b = -(\mu/\tilde{\chi})^2$, then we get an equation

$$i\frac{\partial \tilde{q}}{\partial t} + \frac{\partial^2 \tilde{q}}{\partial s^2} + i\tilde{\chi}\frac{\partial(|\tilde{q}|^2\tilde{q})}{\partial s} = 0 \qquad (6.2.13)$$

Thus, the modified NLS equation is equivalent to the DNLS equation. Equation (6.2.13) was used to investigate the pulse distortion, i.e., self-phase modulation and self-steepening in non-linear optical fibre [79-83].

In number of works [63-78] the effect of third-order group-velocity dispersion was considered. These investigations were based on the following equation

$$i\frac{\partial q}{\partial \zeta} - s_2\frac{\partial^2 q}{\partial \tau^2} + \mu|q|^2 q - is_3\eta_3\frac{\partial^3 q}{\partial \tau^3} + i\tilde{\chi}\frac{\partial}{\partial \tau}\left(|q|^2 q\right) = 0. \qquad (6.2.14)$$

It was found that the third-order dispersion and optic shock effect lead to forming the highly asymmetric power spectra of a propagating pulse. Its band-width increases with increasing pulse intensity. In addition, the peak of spectrum shifts towards the short wavelength end.

The third-order dispersion effects are important for fibres, which has a zero-dispersion crossing at $1.27 \div 1.3\ \mu\text{m}$. However, there are dispersion-flattened fibres with an extremum of the second-order dispersion at the operating wavelength facilitating the propagation of femtosecond optical pulses, since the effect of the third-order dispersion is minimised in this case. A correct description of ultrashort pulses in such fibres must include the combined effects of the second-order and fourth-order dispersions in combination with the derivative nonlinearity. The first of these models is based on the following equation

$$i\frac{\partial q}{\partial \zeta} - s_2\frac{\partial^2 q}{\partial \tau^2} + \mu|q|^2 q - is_4\eta_4\frac{\partial^4 q}{\partial \tau^4} = 0, \tag{6.2.15}$$

which was considered in [102]. The more correct model considered in [103] is based on the generalised equation

$$i\frac{\partial q}{\partial \zeta} - s_2\frac{\partial^2 q}{\partial \tau^2} + \mu|q|^2 q - is_4\eta_4\frac{\partial^4 q}{\partial \tau^4} - i\tilde{\chi}\frac{\partial}{\partial \tau}\left(|q|^2 q\right) = 0, \tag{6.2.16}$$

under condition $s_2 = s_4 = -1$.

6.2.2. STEADY-STATE SOLUTION OF DERIVATIVE NONLINEAR SCHRЦDINGER EQUATION

Let us consider the Derivative Non-linear Schrцdinger equation in a standard form:

$$i\frac{\partial q}{\partial \zeta} + \frac{\partial^2 q}{\partial \tau^2} + i\tilde{\chi}\frac{\partial(|q|^2 q)}{\partial \tau} = 0. \tag{6.2.17}$$

This equation describes the ultrashort optical pulse propagation in a non-linear fibre provided the boundary condition $q(\zeta,\tau) \to 0$ as $|\tau| \to \infty$ holds. Until the IST-method was developed the exact solution of this equations may be found only in some specific cases. For example, it is possible to find the stationary solution describing propagation of a pulse with an invariant envelope, i.e., the steady-state pulse. This case may be considered as didactic example of the solution of non-linear equation.

We assume the solution $q(\zeta,\tau)$ has the form

$$q(\zeta,\tau) = a(y)\exp[i\phi]. \tag{6.2.18}$$

Where $y = \tau - \vartheta\zeta$, $a(y)$ and $\phi(\zeta, \tau)$ are real functions. After substitution (6.2.18) into (6.2.17) the DNLS equation is decomposed into real and imaginary parts, i.e., into the pair of non-linear equations

$$\frac{\partial a}{\partial \zeta} + 2\frac{\partial \phi}{\partial \tau}\frac{\partial a}{\partial \tau} + a\frac{\partial^2 \phi}{\partial \tau^2} + \tilde{\chi}\frac{\partial a^3}{\partial \tau} = 0, \tag{6.2.19}$$

$$\frac{\partial^2 a}{\partial \tau^2} - a\frac{\partial \phi}{\partial \zeta} - a\frac{\partial \phi}{\partial \tau}\frac{\partial \phi}{\partial \tau} - \tilde{\chi}a^3\frac{\partial \phi}{\partial \tau} = 0, \tag{6.2.20}$$

We assume that the function $F(y) = \partial\phi/\partial\tau$ depends only on $y = \tau - \vartheta\zeta$. Taking into account

$$\frac{\partial a}{\partial \zeta} = -\vartheta\frac{da}{dy}, \quad \frac{\partial a}{\partial \tau} = \frac{da}{dy},$$

we can consider the equation (6.2.19) as one for $F(y)$

$$-\vartheta\frac{da}{dy} + 2F\frac{da}{dy} + a\frac{dF}{dy} + \tilde{\chi}\frac{d}{dy}(a^3) = 0$$

or

$$\frac{dF}{dy} = -2\left(\frac{1}{a}\frac{da}{dy}\right)F + \vartheta\left(\frac{1}{a}\frac{da}{dy}\right) - \frac{3\tilde{\chi}}{2}\frac{d}{dy}(a^2). \tag{6.2.21}$$

Let $F_0(y)$ be a solution of the uniform equation

$$\frac{dF_0}{dy} = -2\left(\frac{1}{a}\frac{da}{dy}\right)F_0.$$

After integration of this equation we obtain $F_0(y) = Ca^{-2}(y)$ According to the standard method of solution of the ordinary differential equations, we can represent the solution of non-uniform equation in the form:

$$F(y) = C(y)a^{-2}(y). \tag{6.2.22}$$

Substitution of the expression (6.2.22) into (6.2.21) leads to

$$\frac{dC}{dy} = \vartheta a\frac{da}{dy} - 3\tilde{\chi}a^3\frac{da}{dy}.$$

Integration of this equation gives rise to $C(y)$ and the function $F(y)$ becomes

$$F(y) = C_0 a^{-2}(y) + \vartheta/2 - 3\tilde{\chi}a^2/4 .$$

Using the boundary condition $a(y) \to 0$ as $|\tau| \to \infty$ we get integrating constant C_0. So, the function $F(y)$ is

$$F(y) = \vartheta/2 - 3\tilde{\chi}a^2/4 \tag{6.2.23}$$

From this expression it is possible to conclude that

$$\phi(\zeta,\tau) = \phi_0(\zeta) - (3\tilde{\chi}/4)\int^{\tau} a^2(y)dy + \vartheta\tau/2 .$$

This expression yields

$$\frac{\partial\phi}{\partial\zeta} = \frac{\partial\phi_0}{\partial\tau} - \frac{3}{4}\tilde{\chi}\int^{\tau}\frac{\partial}{\partial\zeta}a^2 dy .$$

As $d(a^2)/d\zeta = -\vartheta d(a^2)/dy$, this expression can be rewritten as

$$\frac{\partial\phi}{\partial\zeta} = \frac{\partial\phi_0}{\partial\zeta} + \frac{3\tilde{\chi}\vartheta}{4}\int^{\tau}\frac{\partial}{\partial y}\left(a^2\right)dy = \frac{\partial\phi_0}{\partial\zeta} + \frac{3\tilde{\chi}\vartheta}{4}a^2(y) . \tag{6.2.24}$$

Let $\partial\phi_0/\partial\zeta = \tilde{\omega}$ be constant (It is the correction of instant frequency of a carrier wave.). Let us denote $p = (3/4)\tilde{\chi}\vartheta$, so that one has

$$\partial\phi/\partial\zeta = \tilde{\omega} + pa^2 .$$

Now from (6.2.20) it is possible to determine the function $a(y)$. Substitution of the expressions (6.2.23) and (6.2.24) into (6.2.20) leads to

$$a'' - aR(a^2) = 0, \tag{6.2.25a}$$

where

$$R(a^2) = (\tilde{\omega} + \vartheta^2/4) + (\tilde{\chi}\vartheta/2)a^2 - (3\tilde{\chi}^2/16)a^4 . \tag{6.2.25b}$$

Equation (6.2.25) may be integrated after multiplication of all its terms by da/dy with taking into account the boundary condition for $a(y)$

$$(a')^2 = \int R(a^2)\,da^2 .$$

Now, taking (6.2.25b) into account we can find

$$\frac{da}{dy} = \sqrt{b_1 a^2 + b_2 a^4 - b_3 a^6}\,, \tag{6.2.26}$$

where $b_1 = (\tilde{\omega} + \vartheta^2/4)$, $b_2 = \tilde{\chi}\vartheta/4$, and $b_3 = \tilde{\chi}^2/16$. Substitution of the expression $w(y) = a^{-2}(y)$ into equation (6.2.26) yields

$$\frac{dw}{dy} = -2\sqrt{b_1\left[\left(w + b_2/2b_1\right)^2 - \Delta^2\right]}\,, \tag{6.2.27}$$

where $\Delta^2 = (4b_1 b_3 + b_2^2)/4b_1^2$. The integration of this equation gives

$$w(y) + b_2/2b_1 = \Delta\cosh\left[2\sqrt{b_1}\,(y - y_0)\right].$$

Finally, the expression for real-valued pulse envelope can be written as

$$\begin{aligned} a^2(y) &= \frac{2b_1}{\sqrt{4b_1 b_3 + b_2^2}\cosh\left[2\sqrt{b_1}\,(y - y_0)\right] - b_2} = \\ &= \frac{4(\tilde{\omega} + \vartheta^2/4)}{\tilde{\chi}\sqrt{\tilde{\omega} + \vartheta^2/2}\cosh\left[2\sqrt{\tilde{\omega} + \vartheta^2/4}(y - y_0)\right] - (\tilde{\chi}\vartheta/2)}\,. \end{aligned} \tag{6.2.28}$$

It is necessary to note that $\tilde{\omega}$ is an arbitrary constant. If we choose $\tilde{\omega} = -\vartheta^2/4$, then the coefficient b_1 in (6.2.26) will be equal to zero and instead of (6.2.27) we shall have

$$\frac{dw}{dy} = -2\sqrt{b_2 w - b_3}\;.$$

For $u = b_2 w$ that leads to some differential equation, which has a solution

$$\sqrt{u - b_3} = -b_2(y - y_0)\,.$$

So, the expression for a square of the real pulse envelope is

$$a^2(y) = \frac{b_2}{b_3 + b_2^2(y - y_0)^2} .$$

If we use the definitions of the parameters b_2 and b_3, then this formula takes the form

$$\tilde{\chi}^2 a^2(y) = \frac{4\vartheta}{1 + 4\vartheta^2 (y - y_0)^2} . \tag{6.2.29}$$

This is the solution corresponding to the familiar "*algebraic soliton*".

Both expression (6.2.28) and the formula (6.2.29) for solutions of the DNLS equation can be obtained by using the IST method with the special spectral problem. This method was developed in [221-224] to solve the DNLS equation.

6.2.3. SOLITONS OF THE DERIVATIVE NONLINEAR SCHRODINGER EQUATION

The DNLS equation is the new example of the whole class of the integrable equations. These equations are not embedded into AKNS hierarchy, i.e., DNLS equation needs different spectral problem of the IST method. The appropriate spectral problem was found by D.J. Kaup and A.C. Newell [221]. The development of the IST based on the Kaup-Newell spectral problem was made in [222-224].

Zero-curvature representation of DNLS equation

According to [221] let us consider the following system of the linear equations

$$\frac{\partial \psi_1}{\partial \tau} = -i\lambda^2 \psi_1 + q(\tau)\lambda\psi_2 , \tag{6.2.30a}$$

$$\frac{\partial \psi_2}{\partial \tau} = i\lambda^2 \psi_2 + r(\tau)\lambda\psi_1 , \tag{6.2.30.b}$$

and the second system of the linear equations

$$\frac{\partial \psi_1}{\partial \zeta} = A\psi_1 + B\psi_2 , \tag{6.2.31a}$$

$$\frac{\partial \psi_2}{\partial \zeta} = C\psi_2 - A\psi_1 . \tag{6.2.31b}$$

The integrability conditions of these systems result in

$$\frac{\partial A}{\partial \tau} = \lambda(qC - rB) \ , \tag{6.2.32a}$$

$$\frac{\partial B}{\partial \tau} - 2i\lambda^2 B - 2\lambda qA = \lambda \frac{\partial q}{\partial \zeta} \ , \tag{6.2.32b}$$

$$\frac{\partial C}{\partial \tau} + 2i\lambda^2 C + 2\lambda rA = \lambda \frac{\partial r}{\partial \zeta} \ . \tag{6.2.32c}$$

It is convenient to rewrite the DNLS equation as two equations

$$\frac{\partial q}{\partial \zeta} = i\frac{\partial^2 q}{\partial \tau^2} - \varepsilon\tilde{\chi}\frac{\partial (qrq)}{\partial \tau} \quad \text{and} \quad \frac{\partial r}{\partial \zeta} = -i\frac{\partial^2 r}{\partial \tau^2} - \varepsilon\tilde{\chi}\frac{\partial (rqr)}{\partial \tau} , \tag{6.2.33}$$

where $r = \varepsilon q^*$, $\varepsilon = \pm 1$. Substitution of these equations into the equations (6.2.32b) and (6.2.32c) leads to

$$\frac{\partial B}{\partial \tau} - 2i\lambda^2 B - 2\lambda qA = \lambda\left(i\frac{\partial^2 q}{\partial \tau^2} - \varepsilon\tilde{\chi}\frac{\partial (qrq)}{\partial \tau}\right),$$

$$\frac{\partial C}{\partial \tau} + 2i\lambda^2 C + 2\lambda rA = -\lambda\left(i\frac{\partial^2 r}{\partial \tau^2} + \varepsilon\tilde{\chi}\frac{\partial (rqr)}{\partial \tau}\right). \tag{6.2.34}$$

These equations contain the partial derivatives of q , qrq, rqr and r with respect to τ in the right-hand part. Consequently, we can propose the following Ansatz:

$$B = b_0 q + b_1 \frac{\partial q}{\partial \tau} + b_2(qrq), \qquad C = c_0 r + c_1 \frac{\partial r}{\partial \tau} + c_2(rqr)$$

where coefficients $b_{0,1,2}$ and $c_{0,1,2}$ are the unknown functions of λ. Substitution of these expansions in the equations (6.2.34) and comparison of the factors at equal degrees of q , qrq, rqr and r yields

$$b_0 = -2\lambda^3 , \quad b_1 = i\lambda , \quad b_2 = -\varepsilon\tilde{\chi}\lambda ,$$
$$c_0 = -2\lambda^3 , \quad c_1 = -i\lambda , \quad c_2 = -\varepsilon\tilde{\chi}\lambda ,$$

and the relation

$$A = 2i\lambda^4 + i\varepsilon\tilde{\chi}(rq)\lambda^2 .$$

Equation (6.2.32a) is also satisfied. Thus, we get the expressions for matrix elements of V-matrix of the IST method. They are

$$A = 2i\lambda^4 + i\varepsilon\widetilde{\chi}(rq)\lambda^2, \tag{6.2.35a}$$

$$B = -2\lambda^3 q + i\lambda\frac{\partial q}{\partial\tau} - \lambda\varepsilon\widetilde{\chi}(qrq), \tag{6.2.35b}$$

$$C = -2\lambda^3 r - i\lambda\frac{\partial r}{\partial\tau} - \lambda\varepsilon\widetilde{\chi}(rqr), \tag{6.2.35c}$$

where $r = \varepsilon q^*$, $\varepsilon = \pm 1$. The linear problems (6.2.30) and (6.2.31) provide the zero-curvature representation of the DNLS equation under conditions of (6.2.35).

Inverse scattering transformation

In the usual manner, we introduce the Jost functions by

$$\Phi(\tau,\lambda) \to \begin{pmatrix} 1 \\ 0 \end{pmatrix} \exp(-i\lambda^2\tau), \quad \overline{\Phi}(\tau,\lambda) \to \begin{pmatrix} 0 \\ -1 \end{pmatrix} \exp(+i\lambda^2\tau), \quad \text{as } \tau \to -\infty \tag{6.2.36a}$$

$$\Psi(\tau,\lambda) \to \begin{pmatrix} 0 \\ 1 \end{pmatrix} \exp(+i\lambda^2\tau), \quad \overline{\Psi}(\tau,\lambda) \to \begin{pmatrix} 1 \\ 0 \end{pmatrix} \exp(-i\lambda^2\tau), \quad \text{as } \tau \to +\infty \tag{6.2.36b}$$

and the transition (or scattering) matrix coefficients by

$$\Phi = a(\lambda)\overline{\Psi} + b(\lambda)\Psi,$$
$$\overline{\Phi} = -\bar{a}(\lambda)\Psi + \bar{b}(\lambda)\overline{\Psi}$$

where $a(\lambda)\bar{a}(\lambda) + b(\lambda)\bar{b}(\lambda) = 1$. We note that, for $r = \pm q^*$

$$\overline{\Phi}(\lambda) = \begin{pmatrix} 0 & \pm 1 \\ 1 & 0 \end{pmatrix} \Phi^*(\lambda^*), \quad \overline{\Psi}(\lambda) = \begin{pmatrix} 0 & 1 \\ \pm 1 & 0 \end{pmatrix} \Psi^*(\lambda^*).$$

From whence it follows that $\bar{a}(\lambda) = a^*(\lambda^*), \quad \bar{b}(\lambda) = \mp b^*(\lambda^*)$.

It is convenient to designate

$$\mu^+ = (1/2)\int_\tau^\infty rq\,d\tau', \quad \mu^- = (1/2)\int_{-\infty}^\tau rq\,d\tau', \quad \mu = \mu^+ + \mu^-$$

Now we can formulate the theorem of analytical properties of the Jost functions and matrix elements of the transition matrix [221, 222].

If for $\tilde{Q} = q\exp(-i\mu^-)$

$$\int_{-\infty}^{\infty}\left(|\tilde{Q}| + |\tilde{Q}|^2 + |\partial\tilde{Q}/\partial\tau|\right)d\tau < \infty,$$

then $\Phi\exp(i\lambda^2 x)$, $\Psi\exp(-i\lambda^2 x)$, *and* $a(\lambda)$ *are analytic functions of* λ *for* λ *in the upper half of* $\varsigma = \lambda^2$ *plane. As* $|\lambda| \to \infty$, $\operatorname{Im}\varsigma > 0$ *we have*

$$\Phi(\lambda)\exp(i\lambda^2\tau) \to \begin{pmatrix}1\\0\end{pmatrix}\exp(i\mu^-) + \ldots,$$

$$\Psi(\lambda)\exp(-i\lambda^2\tau) \to \begin{pmatrix}0\\1\end{pmatrix}\exp(i\mu^+) + \ldots,$$

$$a(\lambda) \to \exp(i\mu) + \ldots$$

The zeros of $a(\lambda)$ in the upper half of the $\varsigma = \lambda^2$ plane are the bound state eigenvalue. Let us designate these zeros by λ_j $(j = 1,2,\ldots,2N)$. The matrix element $a(\lambda)$ is an even function of λ. Thus, we will adopt the convention that λ_{2j} lies in the first quadrant of the λ plane, and λ_{2j+1} lies in the third quadrant, where $\lambda_{2j+1} = -\lambda_{2j}$ and $j = 1, 2, \ldots, N$. At the zero of $a(\lambda)$, we have

$$\Phi(\lambda_j) = b_j\Psi(\lambda_j), \quad \Phi(\lambda_j) = b_j\Psi(\lambda_j)$$

There are the integral representation of the Jost functions (i.e., triangular representation)

$$\Psi(\tau;\lambda) = \begin{pmatrix}0\\1\end{pmatrix}\exp[i(\lambda^2\tau + \mu^+) + \int_{\tau}^{\infty}\begin{pmatrix}\lambda K_1(\tau,s)\exp[-i\mu^+(\tau)]\\ K_2(\tau,s)\exp[i\mu^+(\tau)]\end{pmatrix}\exp(i\lambda^2 s)ds.$$

The kernel of the triangular representation $\hat{K}(\tau,s)$ satisfies the following relations

$$\lim_{s\to\infty} K_m(\tau,s) = 0, \quad m = 1,2; \tag{6.2.37a}$$

$$K_1(\tau,\tau) = (-1/2)q(\tau)\exp[2i\mu^+(\tau)], \tag{6.2.37b}$$

$$\left(\frac{\partial}{\partial\tau} - \frac{\partial}{\partial s}\right)K_1(\tau,s) = q(\tau)L(\tau,s)\exp[2i\mu^+(\tau)], \text{ (6.2.37b)} \tag{6.2.37c}$$

$$\left(\frac{\partial}{\partial\tau} + \frac{\partial}{\partial s}\right)L(\tau,s) = -\frac{i}{2}K_1(\tau,s)\exp[-i\mu^+(\tau)]\frac{\partial}{\partial\tau}\{r(\tau)\exp[-i\mu^+(\tau)]\}, \tag{6.2.37d}$$

where

$$L(\tau,s)=K_2(\tau,s)-(i/2)r(\tau)K_1(\tau,s)\exp[-2i\mu^+(\tau)]$$

Due to the characteristics in (6.2.37c,d), given $q(\tau)$, there exists a solution of (6.2.37c,d) which satisfies (6.2.37a) and (6.2.37b). Thus, the triangular representation of the Jost function $\Psi(\lambda;x)$ is valid. From (6.2.37b) we can obtain the solution of the non-linear equation by solving the Gelfand-Levitan equations for $y>\tau$

$$K_2^*(\tau,y)-i\int_x^\infty K_1(\tau,s)F'(s+y)ds=0\,, \tag{6.2.38a}$$

$$\pm K_1(\tau,y)+F^*(\tau+y)+\int_x^\infty K_2^*(\tau,s)F^*(s+y)ds=0\,, \tag{6.2.38b}$$

for $r=\pm q^*$, where

$$F(z)=\frac{1}{2\pi}\int_C\frac{b(\lambda)}{a(\lambda)}\exp(i\lambda^2 z)\,dz\;,\qquad F'(z)=\frac{dF}{dz}$$

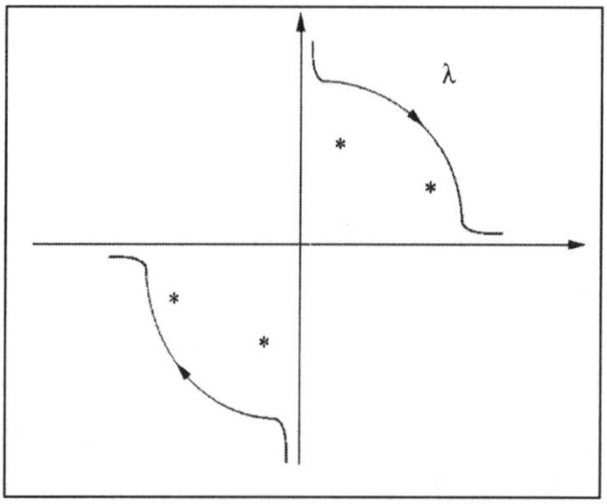

Fig.6.2.1.

The contour C in the complex λ plane. The asterisks indicate where the zeros of $a(\lambda)$ lie.

The contour C is shown in Fig. 6.2.1 . Define for real $\varsigma=\lambda^2$ we introduce

$$\rho(\varsigma)=\frac{b(\varsigma)}{\varsigma a(\varsigma)}\,,$$

and for the bound states

$$C_k=iB_k\,/\,a_k'$$

where

$$a'_k = \frac{da}{d\lambda}\bigg|_{\lambda=\lambda_k} \quad \text{and} \quad B_k = b_{2k}/\lambda_{2k} = b_{2k+1}/\lambda_{2k+1}.$$

Then we can find that

$$F(z) = \sum_{m=1}^{N} C_m \exp(i\varsigma_m z) + \frac{1}{2\pi}\int_{-\infty}^{+\infty} \rho(\xi)\exp(i\xi z)d\xi, \tag{6.2.39}$$

where $\varsigma_m = \lambda_{2m}^2 = \lambda_{2m+1}^2$. From (6.2.39) we can see that the scattering data for this spectral problem are

$$\mathbf{S}(\zeta) = \{\rho(\xi), \xi \in \text{Re}; \{C_m, \varsigma_m\}, m = 1,2,\ldots,N\}$$

Thus, for given $\mathbf{S}(\zeta)$, we can construct function $F(z)$ as in (6.2.39), then one can solve the equations (6.2.38) for $K_1(\tau, y)$, and by using (6.2.37b) one can obtain $q(\zeta,\tau)$.

Soliton solutions of the DNLS equation
To obtain soliton solutions it is necessary to set $\rho(\varsigma) = 0$. Let $N = 1$. It is convenient to designate the single eigenvalue of the spectral problem $\varsigma_1 = \lambda_1^2$ by

$$\varsigma_1 = i\Delta^2 \exp\left[\pm i\left(\frac{\pi}{2} - \gamma\right)\right] = \xi + i\eta,$$

where $\eta = \Delta^2 \sin\gamma$, $\xi = \Delta^2 \cos\gamma$, and $\gamma \in (0,\pi)$. The coefficient C_1 in expression (6.2.39) can be written as

$$C_1 = 2\eta\Delta^{-1}\exp(2i\sigma_0 + 2\eta\tau_0).$$

Then the solution of the Gelfand-Levitan equation leads to

$$q(\tau)\exp[2i\mu^+(\tau)] = \pm 4\Delta\sin\gamma \frac{\exp(2\theta - 2i\sigma)}{\exp(4\theta) + \exp(\mp i\gamma)}, \tag{6.2.40}$$

where $\theta = \eta(\tau - \tau_0)$, $\sigma = \xi\tau + \sigma_0$. From the definition of the function $\mu^+(\tau)$ and (6.2.40), we have

$$\exp(i\mu^+) = \frac{\exp(4\theta) + \exp(\pm i\gamma)}{\exp(4\theta) + \exp(\mp i\gamma)}.$$

Thus, the single-soliton solution of the DNLS equation is

$$q(\tau) = \pm 4\Delta \sin\gamma \frac{\exp(2\theta - 2i\sigma)\exp[-2i\mu^+(\tau)]}{\exp(4\theta) + \exp(\mp i\gamma)} . \tag{6.2.41}$$

The time dependence of the scattering data follows from the system of linear equations (6.2.31) and definitions (6.2.35)

$$\frac{\partial \rho}{\partial \zeta} = -4\xi^2 \rho , \quad \frac{\partial \varsigma_m}{\partial \zeta} = 0 , \quad \frac{\partial C_m}{\partial \zeta} = -4\varsigma_m^2 C_m , \quad \text{for} \quad m = 1, 2, \ldots, N .$$

If one defines

$$C_m = 2\eta_m \Delta^{-1} \exp(2i\sigma_{0,m} + 2\eta\tau_{0,m}) ,$$

then we have

$$\frac{\partial \sigma_{0,m}}{\partial \zeta} = 2\Delta^4 \cos 2\gamma , \quad \frac{\partial \tau_{0,m}}{\partial \zeta} = \pm 4\Delta^2 \cos\gamma .$$

Despite the solitons discussed above, there are the algebraic or rational solitons. These solution arise whenever $\gamma \to \pi$. In this limit we get

$$\exp(i\mu^+) = \frac{4\Delta^2(\tau - \tau_0) \pm i}{4\Delta^2(\tau - \tau_0) \mp i} ,$$

and

$$q(\tau) = \pm \frac{4\Delta \exp\left[-2i(\pm\Delta^2\tau + \sigma_0)\right] \exp[-3i\mu^+(\tau)/2]}{\left[1 + 16\Delta^4(\tau - \tau_0)^2\right]^{1/2}} . \tag{6.2.42}$$

In the usual manner, we may obtain an infinite number of the conserved quantities, i.e., integrals associated with DNLS equation. The first three integrals are

$$I_1 = -\int_{-\infty}^{+\infty} q(\tau) r(\tau) d\tau , \quad I_2 = \int_{-\infty}^{+\infty} \left[q^2(\tau) r^2(\tau) - i\left(q(\tau)\frac{\partial r(\tau)}{\partial \tau} - r(\tau)\frac{\partial q(\tau)}{\partial \tau} \right) \right] d\tau ,$$
$$I_3 = -\int_{-\infty}^{+\infty} \left[\frac{\partial r(\tau)}{\partial \tau}\frac{\partial q(\tau)}{\partial \tau} + 3i\left(r^2(\tau)\frac{\partial q^2(\tau)}{\partial \tau} - q^2(\tau)\frac{\partial r^2(\tau)}{\partial \tau} \right) + q^3(\tau) r^3(\tau) \right] d\tau .$$

If one defines the conserved densities $q\Gamma_n[q]$ so that

$$\frac{\partial}{\partial \zeta} \int_{-\infty}^{+\infty} q(\tau)\Gamma_n[q] d\tau = 0 ,$$

then the integrals can be expressed in terms of these conserved densities. The functionals $\Gamma_m[q]$ can be obtained from the recursion relation

$$2i\Gamma_{n+2} = \frac{\partial}{\partial\tau}\Gamma_n + q(\tau)\sum_{j=1}^{n}\Gamma_j\Gamma_{n+1-j}\,, \quad n \geq 1\,, \qquad \Gamma_0 = 0\,, \Gamma_1 = (i/2)r(\tau)\,, \Gamma_2 = 0\,.$$

In [225,226] the different kinds of the solutions of the DNLS equation were found. These solutions correspond to a non-vanishing boundary condition, $|q(\tau)|^2 \to \pm q_0^2$ as $|\tau| \to \infty$. The IST method for the non-vanishing case is more difficult than the vanishing case. However one-soliton solution was obtained with the closed form.

Let DNLS equation have the form as in [225]:

$$i\frac{\partial q}{\partial\zeta} + \frac{\partial^2 q}{\partial\tau^2} - im\frac{\partial(|q|^2 q)}{\partial\tau} = 0\,.$$

At first we describe the case of $m = 1$. Denote $\varsigma_1 = (\lambda_1^2 - q_0^2)^{1/2}$, where λ_1 is the single eigenvalue of the spectral problem. If λ_1 is real as $0 < \xi_1 < q_0$ (in general case $\lambda_1 = \xi_1 + i\eta_1$), the quantity ς_1 becomes pure imaginary, $\varsigma_1 = i\vartheta$, where $\vartheta^2 = q_0^2 - \xi_1^2$. In this case one can express the soliton solution of the DNLS equations as

$$|q(\tau)|^2 = q_0^2 + 4\vartheta^2\frac{2(\xi_1/q_0)\,\mathrm{sgn}\,\mathrm{Re}(\xi_1 C_1)}{\cosh(2\xi_1\vartheta\tau - \phi_1) - 2(\xi_1/q_0)\,\mathrm{sgn}\,\mathrm{Re}(\xi_1 C_1)}\,, \tag{6.2.43}$$

where

$$\exp\phi_1 = 2q_0^2\,|\mathrm{Re}(\xi_1 C_1)|/(2\xi_1\vartheta)^2\,.$$

If $\mathrm{Re}(\xi_1 C_1)$ is positive (negative), then the expression (6.2.43) represents the bright (dark) modulation, i.e., the positive or negative solitary wave on pedestal

If $\mathrm{sgn}\,\mathrm{Re}(\xi_1 C_1) = +1$, we can rewrite expression (6.2.43) as

$$|q(\tau)|^2 = q_0^2 + 4\vartheta^2\frac{(\xi_1/q_0)}{\cosh(2\xi_1\vartheta\tau - \phi_1) - (\xi_1/q_0)}\,.$$

For sufficiently small ϑ using the approximation $\cosh z \approx 1 + z^2/2$ and $\xi_1/q_0 \approx 1 - (1/2)(\vartheta/q_0)^2$ we obtain

$$|q(\tau)|^2 = q_0^2 + \frac{8q_0^2}{1 + (2q_0^4\tau - \phi_1)^2}\,. \tag{6.2.44}$$

This is the algebraic soliton on pedestal. If $\operatorname{sgn}\operatorname{Re}(\xi_1 C_1) = -1$, we can obtain the algebraic soliton as a local depression of the pedestal depth.

Now we describe the case of $m = -1$. Assuming λ_1 to be pure imaginary, $0 < \eta_1 < q_0$, the authors of [225] obtain

$$|q(\tau)|^2 = q_0^2 + 4\rho^2 \frac{2(\eta_1 / q_0)\operatorname{sgn}\operatorname{Re}(\xi_1 C_1)}{\cosh(2\eta_1\rho\tau - \psi_1) - 2(\eta_1 / q_0)\operatorname{sgn}\operatorname{Re}(\xi_1 C_1)}, \tag{6.2.45}$$

where $\rho^2 = q_0^2 - \eta_1^2$ and $\exp\psi_1 = 2q_0^2 \,|\operatorname{Re}(\xi_1 C_1)|/(2\eta_1\rho)^2$. For sufficiently small ρ this expression results in

$$|q(\tau)|^2 = q_0^2 + \frac{8q_0^2}{1 + 4q_0^4\tau^2}. \tag{6.2.46}$$

Here $\operatorname{sgn}\operatorname{Re}(\xi_1 C_1) = +1$. This is the algebraic soliton on pedestal as well. In [225] the multi-soliton solution of the DNLS equation was also represented under the non-vanishing boundary conditions.

6.2.4. ZERO OF SECOND-ORDER DISPERSION

The second-order group-velocity dispersion in optical fibres has two main contributions. There are the material and waveguide dispersions. These contributions can be modified in the fibre drawing process. For instance, the waveguide dispersion can be modified by the relevant changing of the index profile of the fibre. The zero-dispersion wavelength can be shifted to higher wavelength domain in the so called dispersion-shifted fibres by this way. It is also possible to manufacture fibres in which material dispersion nearly compensates waveguide one over the wide spectral region. When the second-order group-velocity dispersion is nearly zero, the higher-order dispersion and the dispersion of the non-linear susceptibilities become essential.

Generalised complex mKdV equation

In order to discuss the pulse propagation under consideration of the zero second-order group-velocity dispersion, we can exploit the equation (6.2.14):

$$i\frac{\partial q}{\partial \zeta} - s_2 \frac{\partial^2 q}{\partial \tau^2} + \mu |q|^2 q - i s_3 \eta_3 \frac{\partial^3 q}{\partial \tau^3} + i\widetilde{\chi}\frac{\partial}{\partial \tau}\left(|q|^2 q\right) = 0 .$$

Substitution of the variables, $\xi = \tau - \vartheta_2 \zeta$ and $q(\zeta, \tau) = u(\zeta, \xi)\exp(-i\mu\xi / \widetilde{\chi} + i\vartheta_1 \zeta)$

where

$$\vartheta_1 = \left(\frac{\mu}{\tilde{\chi}}\right)^2 \left(s_2 - s_3 \frac{\mu}{\tilde{\chi}}\right), \quad \vartheta_2 = \frac{\mu}{\tilde{\chi}}\left(3s_3 \frac{\mu}{\tilde{\chi}} - 2s_2\right),$$

makes it possible to replace equation (6.2.14) with the equation for $u(\zeta,\xi)$:

$$i\frac{\partial u}{\partial \zeta} - s_2' \frac{\partial^2 u}{\partial \xi^2} - is_3\eta_3 \frac{\partial^3 u}{\partial \xi^3} + i\tilde{\chi}\frac{\partial}{\partial \xi}\left(|u|^2 u\right) = 0. \qquad (6.2.47)$$

In the above equation the second-order group-velocity dispersion is allowed for by a new parameter $s_2' = (s_2 - 3s_3\mu/\tilde{\chi})$. If the carrier frequency is selected so as $s_2' = 0$, the influence of this dispersion can be eliminated. Thus, the propagation of an optical pulse with the carrier frequency in the region of zero second-order dispersion of the group velocities can be described by the equation

$$\frac{\partial u}{\partial \zeta} - s_3\eta_3 \frac{\partial^3 u}{\partial \xi^3} + \tilde{\chi}\frac{\partial}{\partial \xi}\left(|u|^2 u\right) = 0. \qquad (6.2.48)$$

This equation can be regarded as a complex generation of the famous modified Korteweg- de Vries equation [62]. An analytical solution of the Cauchy problem for equation (6.2.48) would make it possible to describe in detail the evolution of the optical pulse propagating in the non-linear fibre. However, this solution is unknown, and we can consider only some partial solutions.

Let the envelope of optical pulse be a real-valued function. Then from the equation (6.2.48) the modified Korteweg-de Vries (mKdV) equation follows

$$\frac{\partial u}{\partial \zeta} + 3\tilde{\chi}u^2 \frac{\partial u}{\partial \xi} - s_3\eta_3 \frac{\partial^3 u}{\partial \xi^3} = 0. \qquad (6.2.49)$$

This equation is an example of completely integrable equations and, hence, it can be solved by the IST method. In section 5.33. we have considered the solitons of mKdV equation. As it was mentioned this equation has the breather solution that represents the connected pair of one-soliton pulses.

Integrable models

It is interesting to obtain the evolution equations describing the pulse propagation on condition that second-order dispersion is absent, and which can give rise to solitons. For this purpose it is convenient to use the IST method. Let us consider the system of equations (6.1.22) in assuming that $N = 3$. By solving the resulting system of equations in accordance with the procedure presented in section 6.1.4, we can obtain the following expressions

$$\frac{\partial q}{\partial \zeta} = \frac{i}{4} a_3 \left(6rq \frac{\partial q}{\partial \tau} - \frac{\partial^3 q}{\partial \tau^3} \right) + \frac{1}{2} a_2 \left(2qrq - \frac{\partial^2 q}{\partial \tau^2} \right) + ia_1 \frac{\partial q}{\partial \tau} + 2\alpha q \,, \quad (6.2.50a)$$

$$\frac{\partial r}{\partial \zeta} = \frac{i}{4} a_3 \left(6rq \frac{\partial r}{\partial \tau} - \frac{\partial^3 r}{\partial \tau^3} \right) - \frac{1}{2} a_2 \left(2rqr - \frac{\partial^2 r}{\partial \tau^2} \right) + ia_1 \frac{\partial r}{\partial \tau} + 2\alpha r \,. \quad (6.2.50b)$$

Now, the matrix $\hat{V}$ has the following matrix elements:

$$A(\lambda) = \alpha(\zeta) + \frac{1}{2} a_2 qr - \frac{i}{4} a_3 \left(q \frac{\partial r}{\partial \tau} - r \frac{\partial q}{\partial \tau} \right) + \frac{1}{2} (a_3 qr + a_1) \lambda + a_2 \lambda^2 + a_3 \lambda^3 \,,$$

$$B(\lambda) = ia_1 q + \frac{i}{2} a_3 qrq - \frac{1}{2} a_2 \frac{\partial q}{\partial \tau} - \frac{i}{4} a_3 \frac{\partial^2 q}{\partial \tau^2} + \left(ia_2 q - \frac{1}{2} a_3 \frac{\partial q}{\partial \tau} \right) \lambda + ia_3 q \lambda^2 \,,$$

$$C(\lambda) = ia_1 r + \frac{i}{2} a_3 rqr + \frac{1}{2} a_2 \frac{\partial r}{\partial \tau} - \frac{i}{4} a_3 \frac{\partial^2 r}{\partial \tau^2} + \left(ia_2 r + \frac{1}{2} a_3 \frac{\partial r}{\partial \tau} \right) \lambda + ia_3 r \lambda^2 \,.$$

The reduction $r = \mp q^*$ yields

$$\frac{\partial q}{\partial \zeta} - \frac{i}{4} a_3 \left(\frac{\partial^3 q}{\partial \tau^3} \pm 6 |q|^2 \frac{\partial q}{\partial \tau} \right) - \frac{1}{2} a_2 \left(\frac{\partial^2 q}{\partial \tau^2} \pm 2 |q|^2 q \right) - ia_1 \frac{\partial q}{\partial \tau} - 2\alpha q = 0 \,.$$

If $a_3 = a_1 = \alpha = 0$ and $a_2 = i$ this equation converts to the NLS equation (6.1.20). The evolution equation we are interested in can be obtained by taking that $a_2 = 0$. For $a_1 = \alpha = 0$ and $a_3 = 4i$ or $a_3 = 4$ we get the complex versions of the modified Korteweg-de Vries equations

$$\frac{\partial q}{\partial \zeta} + \left(\frac{\partial^3 q}{\partial \tau^3} \pm 6 |q|^2 \frac{\partial q}{\partial \tau} \right) = 0 \,, \quad (6.2.51a)$$

or

$$i \frac{\partial q}{\partial \zeta} + \left(\frac{\partial^3 q}{\partial \tau^3} \pm 6 |q|^2 \frac{\partial q}{\partial \tau} \right) = 0 \,, \quad (6.2.51b)$$

respectively. If we suppose $a_1 = \alpha = 0$, $a_3 = 4i$ and $a_2 = i$, then the Hirota equation

[250] follows

$$i\frac{\partial q}{\partial \zeta}+i\left(\frac{\partial^3 q}{\partial \tau^3}\pm 6|q|^2\frac{\partial q}{\partial \tau}\right)+\frac{1}{2}\left(\frac{\partial^2 q}{\partial \tau^2}\pm 2|q|^2 q\right)=0. \qquad (6.2.52)$$

From this equation follows that the effect of the second-order group velocity dispersion can be compensated by the cubic (i.e., Kerr-type) nonlinearity. Whereas the only special-purpose non-linear correction of the group velocity suppresses the influence of the third-order dispersion.

The more adequate description of the pulse propagation in the femtosecond region was proposed in [84,86,93]. Unfortunately, higher-order NLS equation is not a completely integrable one and has not soliton solutions in the strict sense.

6.2.5 SOLITONS OF THE HIGHER-ORDER NLS EQUATION

In order to describe the non-linear phenomena associated with femtosecond optical solitons in non-linear fibres, the higher-order Nonlinear Schrödinger (HNLS) equation has been proposed [84,86] (see also [93]). Let us consider the following generalisation of the equation (6.2.14)

$$i\frac{\partial q}{\partial \zeta}-s_2\frac{\partial^2 q}{\partial \tau^2}+\mu|q|^2 q+i\left(s_3\eta_3\frac{\partial^3 q}{\partial \tau^3}+\mu_2|q|^2\frac{\partial q}{\partial \tau}+\mu_3 q\frac{\partial |q|^2}{\partial \tau}\right)=0, \qquad (6.2.53)$$

The parameter η_3 corresponds to the third-order group-velocity dispersion, parameters μ_2 and μ_3 represent the two inertial contributions to the non-linear polarisation, i.e., Raman self-scattering and self-steeping formation. If $\eta_3 = 0$, $\mu_2 = \mu_3 = 1$, then the DNLS comes out from this equation. If $\eta_3 = 1$, $\mu_2 = \pm 6$, $\mu_3 = 0$, then the equation (6.2.53) is reduced to the Hirota equations [250]. In general, equation (6.2.53) may not be completely integrable, i.e., may not be solved by the IST method. However, in [97] it was shown that this equation can be solvable by means of IST under some alternative of the parameters. Hereafter, we shall consider HNLS equation following Ref. [97].

Reduction to the Sasa-Satsuma equation
It is convenient to make the following variable transformation

$$q(\tau,\zeta)=u(\tau,\zeta)\exp\left(-ia\tau+ib\zeta\right). \qquad (6.2.54a)$$

The substitution of this expression into equation (6.2.53) yields

$$i\frac{\partial u}{\partial \zeta}+\left(-2ias_2+3ia^2s_3\eta_3\right)\frac{\partial u}{\partial \tau}+\left(-s_2+3as_3\eta_3\right)\frac{\partial^2 u}{\partial \tau^2}+\left(-b+s_2a^2-s_3\eta_3a^3\right)u+$$
$$+(\mu-a\mu_2)\,|\,u\,|^2\,u+i\left(\mu_2\,|\,u\,|^2\,\frac{\partial u}{\partial \tau}+\mu_3 u\frac{\partial\,|\,u\,|^2}{\partial \tau}-s_3\eta_3\frac{\partial^3 u}{\partial \tau^3}\right)=0\,.$$

The parameters a and b must be chosen to vanish the terms with $\partial^2 u/\partial\tau^2$, $|\,u\,|^2\,u$, and u. Thus, we obtain that

$$a=s_2(3s_3\eta_3)^{-1},\quad a=\mu/\mu_2,\quad \text{and}\quad b=a^2(s_2-s_3\eta_3 a)\,.$$

These conditions can be satisfied only if $\mu_2=3s_2s_3\eta_3\mu$. Then $b=2s_2(27\eta_3^2)^{-1}$. The next step is the introduction of the new independent variable

$$\xi=\tau+\zeta/3s_3\eta_3\,. \tag{6.2.54b}$$

The resulting equation takes the form

$$i\frac{\partial u}{\partial \zeta}-s_3\eta_3\frac{\partial^3 q}{\partial \xi^3}+i\left(\mu_2\,|\,u\,|^2\,\frac{\partial u}{\partial \xi}+\mu_3 u\frac{\partial\,|\,u\,|^2}{\partial \xi}\right)=0\,. \tag{6.2.55}$$

If we suppose that $s_3\eta_3=-\varepsilon$, $\mu_2=6\varepsilon$, and $\mu_3=3\varepsilon$, where $\varepsilon=\pm 1$, then the HNLS equation (6.2.55) reduces to the Sasa-Satsuma equation [97]

$$\frac{\partial u}{\partial \zeta}+\varepsilon\left(\frac{\partial^3 u}{\partial \xi^3}+6\,|\,u\,|^2\,\frac{\partial u}{\partial \xi}+3u\frac{\partial\,|\,u\,|^2}{\partial \xi}\right)=0\,. \tag{6.2.56}$$

Zero-curvature representation of the Sasa-Satsuma equation
Now let us consider the Manakov spectral problem (4.2.13c). Zero-curvature condition results in the equations (4.2.15), i.e.,

$$\frac{\partial A}{\partial \tau}=\vec{Q}\cdot\vec{C}-\vec{R}\cdot\vec{B},\quad \frac{\partial \hat{D}}{\partial\ \tau}=\vec{R}\otimes\vec{B}-\vec{C}\otimes\vec{Q}\,,$$
$$\frac{\partial \vec{B}}{\partial \tau}+2i\lambda\vec{B}+A\vec{Q}-\vec{Q}\cdot\hat{D}=\frac{\partial \vec{Q}}{\partial \zeta},\quad \frac{\partial \vec{C}}{\partial \tau}-2i\lambda\vec{C}-A\vec{R}+\hat{D}\cdot\vec{R}=\frac{\partial \vec{R}}{\partial \zeta}\,. \tag{6.2.57}$$

We assume that the functions $A(\lambda)$, $\vec{B}(\lambda)$, $\vec{C}(\lambda)$ and $\hat{D}(\lambda)$ are the polynomials of degree 3

$$A(\lambda)=\sum_{n=0}^{3} a_n(\zeta,\tau)\lambda^n, \quad \hat{D}(\lambda)=\sum_{n=0}^{3}\hat{D}_n(\zeta,\tau)\lambda^n,$$
$$\vec{B}(\lambda)=\sum_{n=0}^{3}\vec{b}_n(\zeta,\tau)\lambda^n, \quad \vec{C}(\lambda)=\sum_{n=0}^{3}\vec{c}_n(\zeta,\tau)\lambda^n .$$

Substituting these expansions into equations (6.2.57) and equating the coefficients of the different powers of λ in the resulting equations, we obtain the net results

$$a_3(\zeta,\tau)=a_3(\zeta), \quad \hat{D}_3(\zeta,\tau)=d_3(\zeta)\hat{\mathrm{I}},$$
$$\vec{b}_2(\zeta,\tau)=(1/2)\rho_3(\zeta)\vec{Q}, \quad \vec{c}_2(\zeta,\tau)=(1/2)\rho_3(\zeta)\vec{R},$$
$$a_2(\zeta,\tau)=a_2(\zeta), \quad \hat{D}_2(\zeta,\tau)=d_2(\zeta)\hat{\mathrm{I}},$$
$$\vec{b}_1(\zeta,\tau)=\frac{1}{2}\rho_2(\zeta)\vec{Q}-\frac{1}{2}\rho_3(\zeta)\frac{\partial\vec{Q}}{\partial\tau}, \quad \vec{c}_1(\zeta,\tau)=\frac{1}{2}\rho_2(\zeta)\vec{R}-\frac{1}{2}\rho_3(\zeta)\frac{\partial\vec{R}}{\partial\tau},$$
$$a_1(\zeta,\tau)=a_1(\zeta)+(1/2)\rho_3(\zeta)(\vec{Q}\cdot\vec{R}), \quad \hat{D}_1(\zeta,\tau)=\hat{d}_1(\zeta)-(1/2)\rho_3(\zeta)(\vec{Q}\otimes\vec{R}),$$
$$a_0(\zeta,\tau)=a_0(\zeta)+(1/4)\rho_2(\zeta)(\vec{Q}\cdot\vec{R})+(1/4)\rho_3(\zeta)\left(\vec{Q}\frac{\partial\vec{R}}{\partial\tau}-\vec{R}\frac{\partial\vec{Q}}{\partial\tau}\right),$$
$$\hat{D}_0(\zeta,\tau)=d_0(\zeta)\hat{\mathrm{I}}-\frac{1}{4}\rho_2(\zeta)(\vec{R}\otimes\vec{Q})-\frac{1}{4}\rho_3(\zeta)\left(\frac{\partial\vec{R}}{\partial\tau}\otimes\vec{Q}-\vec{R}\otimes\frac{\partial\vec{Q}}{\partial\tau}\right),$$
$$\vec{b}_0(\zeta,\tau)=\frac{1}{2}\left(\rho_1(\zeta)\vec{Q}-\frac{1}{2}\rho_2(\zeta)\frac{\partial\vec{Q}}{\partial\tau}+\frac{1}{2}\rho_3(\zeta)\frac{\partial^2\vec{Q}}{\partial\tau^2}-\rho_3(\zeta)\vec{Q}(\vec{R}\cdot\vec{Q})\right),$$
$$\vec{c}_0(\zeta,\tau)=\frac{1}{2}\left(\rho_1(\zeta)\vec{R}+\frac{1}{2}\rho_2(\zeta)\frac{\partial\vec{R}}{\partial\tau}+\frac{1}{2}\rho_3(\zeta)\frac{\partial^2\vec{R}}{\partial\tau^2}-\rho_3(\zeta)\vec{R}(\vec{Q}\cdot\vec{R})\right).$$

Here the following designations are used:

$$\rho_1(\zeta)=a_1(\zeta)-d_1(\zeta), \quad \rho_2(\zeta)=a_2(\zeta)-d_2(\zeta), \text{ and } \rho_3(\zeta)=a_3(\zeta)-d_3(\zeta)$$

In terms of these coefficient functions the evolution equations can be represented as

$$\frac{\partial\vec{Q}}{\partial\zeta}=\frac{\partial\vec{b}_0}{\partial\tau}+a_0(\zeta,\tau)\vec{Q}-\vec{Q}\cdot\hat{D}_0(\zeta,\tau),$$
$$\frac{\partial\vec{Q}}{\partial\zeta}=\frac{\partial\vec{b}_0}{\partial\tau}+a_0(\zeta,\tau)\vec{Q}-\vec{Q}\cdot\hat{D}_0(\zeta,\tau).$$

By using the explicit expressions for the coefficient functions we can deduce the desired system from these equations

$$\frac{\partial \vec{Q}}{\partial \zeta} = -\frac{1}{2}\rho_0(\zeta)\vec{Q} + \frac{1}{2}\rho_1(\zeta)\frac{\partial \vec{Q}}{\partial \tau} - \frac{\rho_2(\zeta)}{4}\left(\frac{\partial^2 \vec{Q}}{\partial \tau^2} - 2(\vec{Q}\cdot\vec{R})\vec{Q}\right) + \\ + \frac{\rho_3(\zeta)}{4}\left(\frac{\partial^3 \vec{Q}}{\partial \tau^3} - 3(\vec{Q}\cdot\vec{R})\frac{\partial \vec{Q}}{\partial \tau} - 3(\vec{R}\cdot\frac{\partial \vec{Q}}{\partial \tau})\vec{Q}\right), \tag{6.2.58}$$

$$\frac{\partial \vec{R}}{\partial \zeta} = +\frac{1}{2}\rho_0(\zeta)\vec{R} + \frac{1}{2}\rho_1(\zeta)\frac{\partial \vec{R}}{\partial \tau} + \frac{\rho_2(\zeta)}{4}\left(\frac{\partial^2 \vec{R}}{\partial \tau^2} - 2(\vec{Q}\cdot\vec{R})\vec{R}\right) + \\ + \frac{\rho_3(\zeta)}{4}\left(\frac{\partial^3 \vec{R}}{\partial \tau^3} - 3(\vec{Q}\cdot\vec{R})\frac{\partial \vec{R}}{\partial \tau} - 3(\vec{Q}\cdot\frac{\partial \vec{R}}{\partial \tau})\vec{R}\right), \tag{6.2.59}$$

where $\rho_0(\zeta) = a_0(\zeta) - d_0(\zeta)$. Equations (6.2.58) and (6.2.59) provide the base for the description of the non-linear wave propagation in an inhomogeneous media. However, we shall consider only the homogeneous medium.

Let $\rho_0 = \rho_1 = \rho_2 = 0$ and $\rho_3 = 4\varepsilon$, where $\varepsilon = \pm 1$. For real $\vec{Q}$ the reduction $\vec{R} = \varepsilon\vec{Q}$, yields

$$\frac{\partial \vec{Q}}{\partial \zeta} = \varepsilon\frac{\partial^3 \vec{Q}}{\partial \tau^3} - 3(\vec{Q}\cdot\vec{Q})\frac{\partial \vec{Q}}{\partial \tau} - 3(\vec{Q}\cdot\frac{\partial \vec{Q}}{\partial \tau})\vec{Q}. \tag{6.2.60}$$

It is the vector modified Korteweg-de Vries equation. If we designate $\psi = Q_1 + iQ_2$, then the complex generalisation of the modified Korteweg-de Vries equation can be obtained

$$\frac{\partial \psi}{\partial \zeta} = \varepsilon\frac{\partial^3 \psi}{\partial \tau^3} - \frac{3}{2}\left(\frac{\partial(|\psi|^2\psi)}{\partial \tau} + |\psi|^2\frac{\partial \psi}{\partial \tau}\right). \tag{6.2.61}$$

The reduction $\vec{R} = \varepsilon\vec{Q}^*$ yields the following equation

$$\frac{\partial \vec{Q}}{\partial \zeta} = \varepsilon\frac{\partial^3 \vec{Q}}{\partial \tau^3} - 3(\vec{Q}\cdot\vec{Q}^*)\frac{\partial \vec{Q}}{\partial \tau} - 3(\vec{Q}^*\cdot\frac{\partial \vec{Q}}{\partial \tau})\vec{Q}. \tag{6.2.62}$$

Let us consider the two additional reductions. The assumption $Q_1 = \varepsilon Q_2 = u$ reduces the equation (6.2.62) to

$$\frac{\partial u}{\partial \zeta} = \varepsilon \frac{\partial^3 u}{\partial \tau^3} - 12 |u|^2 \frac{\partial u}{\partial \tau}. \tag{6.2.62}$$

This equation is equivalent to (6.2.51). If we assume that $Q_1 = u$ and $Q_2 = \varepsilon u^*$, then the equation (6.2.62) can be converted to

$$\frac{\partial u}{\partial \zeta} = \varepsilon \frac{\partial^3 u}{\partial \tau^3} - 6 |u|^2 \frac{\partial u}{\partial \tau} - 3u \frac{\partial |u|^2}{\partial \tau}. \tag{6.2.63}$$

The other reduction $\vec{R} = -\vec{Q}^*$ yields the equation

$$\frac{\partial \vec{Q}}{\partial \zeta} = \varepsilon \left(\frac{\partial^3 \vec{Q}}{\partial \tau^3} + 3(\vec{Q} \cdot \vec{Q}^*) \frac{\partial \vec{Q}}{\partial \tau} + 3(\vec{Q}^* \cdot \frac{\partial \vec{Q}}{\partial \tau}) \vec{Q} \right). \tag{6.2.64}$$

Setting $Q_1 = u$ and $Q_2 = \varepsilon u^*$, we can convert equation (6.2.64) into

$$\frac{\partial u}{\partial \zeta} = \varepsilon \left(\frac{\partial^3 u}{\partial \tau^3} + 6 |u|^2 \frac{\partial u}{\partial \tau} + 3u \frac{\partial |u|^2}{\partial \tau} \right). \tag{6.2.65}$$

Thus, we obtain the zero-curvature representation of the Sasa-Satsuma equation. The U-matrix takes the form

$$\hat{U}(\lambda) = \begin{pmatrix} -i\lambda & -u^* & -\varepsilon u \\ u & i\lambda & 0 \\ \varepsilon u^* & 0 & i\lambda \end{pmatrix}, \tag{6.2.66}$$

and the V-matrix is expressed in terms of the functions $A(\lambda)$, $\vec{B}(\lambda)$, $\vec{C}(\lambda)$ and $\hat{D}(\lambda)$ have been found above. By using the properties of the Manakov scattering problem one can obtain that

$$K_2^{(1)}(x,y) = -\varepsilon K_3^{(1)*}(x,y), \quad \text{and} \quad F_{12}(x) = -\varepsilon F_{13}^*(x). \tag{6.2.67}$$

The solution of the evolution equation (6.2.65) is expressed in terms of the $K_2^{(1)}(x,y)$:

$$u(\zeta,\tau) = -2K_2^{(1)}(\tau,\tau). \tag{6.2.68}$$

The function $K_2^{(1)}(x,y)$ is the solution of the following equation

$$-K_2^{(1)}(x,y) = F_{12}(x+y) + \int_x^\infty K_2^{(1)}(x,z)\mathcal{R}(x;z,y)dz\,, \tag{6.2.69}$$

where

$$\mathcal{R}(x;x',y) = \sum_{a=2,3} \int_x^\infty F_{1a}^*(x'+z)F_{1a}(z+y)dz.$$

This equation follows from the system of Gelfand-Levitan-Marchenko integral equations (see section 4.2.5). It should be remarked that the kernel of the integral equation (6.2.69) with taking account the formulae (6.2.67) can be written as

$$\mathcal{R}(x;x',y) = 2\int_x^\infty \mathrm{Re}\, F_{12}^*(x'+z)F_{12}(z+y)\,dz.$$

Evolution of the scattering data is defined by the matrix $\hat{V}(\lambda)$ of the *U-V*-pair. The relevant equation is

$$\frac{\partial T_{12}}{\partial \zeta} = (V_{11}^{(-)} - V_{22}^{(-)})\, T_{12},$$

where matrix elements of the $\hat{V}(\lambda)$ should be determined with account for $\rho_3 = 4\varepsilon$ and $\rho_0 = \rho_1 = \rho_2 = 0$. The necessary manipulations give rise to $(V_{11}^{(-)} - V_{22}^{(-)}) = 4\varepsilon\lambda^3$. Thus, one can write the general expression for the kernel of the system of the Gelfand-Levitan-Marchenko integral equations as

$$F_{12}(y) = -i\sum_{n=1}^{N} \frac{C_{12}^{(n)}(0)}{T_{11}'(\lambda_n)} \exp(i\lambda_n y + 4\varepsilon\lambda_n^3\zeta) + \int_{-\infty}^{\infty} \frac{T_{12}(\xi,0)\exp\{i\xi y + 4\varepsilon\xi^3\zeta\}}{2\pi T_{11}(\xi)} d\xi\;.$$

Solitons of the Sasa-Satsuma equation

Soliton solutions of equation (6.2.65) can be obtained by the regular procedure of the IST method. If we suppose that the solution of this equation is a real-valued function, then the equation (6.2.65) can be reduced to mKdV equation. Hence there are solitons and breathers among the solutions of the Sasa-Satsuma equation. We consider here only multi-soliton solution which has been proposed in [97].

Let us assume that matrix element of the transfer matrix is represented as

$$T_{11}(\lambda) = \prod_{j=1}^{N} \frac{(\lambda - \lambda_j^*)(\lambda + \lambda_j)}{(\lambda - \lambda_j)(\lambda + \lambda_j^*)},$$

and $T_{12}(\lambda) = 0$ at the real axis. The roots $\lambda_n = (-\xi_n + i\eta_n)/2$ (where $n = 1, 2, \ldots, N$) forms the discrete spectrum of the spectral problem. Assuming $K_2^{(1)}(x, y)$ to be

$$K_2^{(1)}(x, y) = \sum_{j=1}^{N} \left[L_j(x) \exp\{-i\lambda_j^* y\} + M_j(x) \exp\{i\lambda_j y\} \right],$$

we obtain a system of linear algebraic equations for L_j and M_j from Gelfand-Levitan-Marchenko equations. Then, solution of this algebraic system result in

$$u(\zeta, \tau) = -\frac{2}{\det \hat{D}} \sum_{j=1}^{N} \det \hat{D}^{(j)} \exp[-i\lambda_j^* \zeta].$$

In the above expression, $\hat{D}$ is the $2N \times 2N$ matrix, whose the matrix elements are given by

$$\text{for } 1 \le m \le N \qquad D_{km} = \sum_{j=1}^{N} \frac{p_m p_{j+N} \exp\{-i(\lambda_k^* + 2\lambda_{j+N}^* + \lambda_m^*)\}}{(\lambda_k^* + \lambda_{j+N}^*)(\lambda_m^* + \lambda_{j+N}^*)} - \delta_{km},$$

$$\text{for } N+1 \le m \le 2N \quad D_{km} = \sum_{j=1}^{N} \frac{p_m p_j \exp\{-i(\lambda_k^* + 2\lambda_j^* + \lambda_m^*)\}}{(\lambda_k^* + \lambda_j^*)(\lambda_m^* + \lambda_j^*)} - \delta_{km},$$

where $\lambda_{j+N} = -\lambda_j^*$, $p_j = -iT_{12}^{(j)}(\zeta)/T_{11}'(\lambda_j^*)$, $p_{j+N} = p_j^*$. Matrix $\hat{D}^{(j)}$ is obtained by replacing the j-th row of with

$$(p_1 \exp[-i\lambda_1^* x], p_2 \exp[-i\lambda_2^* x], \ldots, p_N \exp[-i\lambda_N^* x], 0, \ldots, 0).$$

The special case of the two-soliton solution of the equation (6.2.65) has been discussed in [97]. In the specific case this solution has a two maxima. The distance between these maxima becomes large under some conditions. It is noteworthy that this type of solitons is non-usual for soliton solutions of the NLS equation and its generalisation such as the DNLS equation.

6.3. Vector solitons

In the general case the *vector soliton* is the soliton solution of the non-linear system of the evolution equations, which can be presented as the one-dimension array. For example, the vector soliton of the NLS equation is the solution of the following vector equation

$$i\frac{\partial\vec{q}}{\partial\zeta}+\frac{1}{2}\frac{\partial^2\vec{q}}{\partial\tau^2}+(\vec{q}\cdot\vec{q}^*)\vec{q}=0,$$

where $\vec{q}=\{q_1,q_2,\dots q_M\}$. Hereafter, this equation will be referred to as the *v-NLS equation*. Other examples were discussed in section 4.2. The vector index of this soliton can be resulted from the different physical origin. We shall consider some situations connected with the non-linear wave in the Kerr medium and in optical birefringent fibres.

6.3.1. POLARISED WAVES IN KERR MEDIUM

In this section we shall follow the paper by A.D.Bordman and G.S.Cooper [143,144], where the polarised waves propagation in Kerr medium is presented in detail. Let us suppose that the optical wave have two field components E_x and E_y. The total vector of an electric field strength is written as

$$\vec{E}=E_x\vec{e}_x+E_y\vec{e}_y,$$

where $\vec{e}_x$ and $\vec{e}_y$ are the orthonormal vectors in x and y directions, and the wave propagates along z axis. These components may be written in terms of quasi-harmonic waves taking into account the slowly varying complex envelopes:

$$E_x=\mathcal{E}_x(z,t)\Psi(x,y)\exp[i(\beta_x z-\omega_0 t)],\quad E_y=\mathcal{E}_y(z,t)\Psi(x,y)\exp[i(\beta_y z-\omega_0 t)],$$

where ω_0 is the carrier frequency, β_x and β_y are the propagation constants. Radial distribution of the electric field in fibre is defined by the mode function $\Psi(x,y)$. For an optical fibre both propagation constants will be close to the average value β, so that $\beta_x=\beta+\Delta\beta$, and $\beta_y=\beta-\Delta\beta$. The complex envelopes can be also expressed in terms of real-valued amplitudes and phases, as

$$\mathcal{E}_x=R_x\exp[i\varphi_x],\quad\text{and}\quad\mathcal{E}_y=R_y\exp[i\varphi_y],$$

where the phases can be expressed, in terms of deviations from a mean value $\tilde{\varphi}$, as

$$\varphi_x = \tilde{\varphi} + \phi \quad \text{and} \quad \varphi_y = \tilde{\varphi} - \phi .$$

So, the electric field components finally become

$$E_x = A_x(z,t)\Psi(x,y)\exp[i\beta z - i\omega_0 t], \quad \text{and} \quad E_y = A_y(z,t)\Psi(x,y)\exp[i\beta z - i\omega_0 t],$$

where

$$A_x = R_x \exp[i(\tilde{\varphi} + \phi + \Delta\beta z)], \quad \text{and} \quad A_y = R_y \exp[i(\tilde{\varphi} - \phi - \Delta\beta z)].$$

These complex amplitudes modify the complex envelopes and show that the phase modification is attributable to the following effects:

- the intrinsic birefringence (i.e., the birefringence that would be present in the linear limit) is represented by the $\pm\Delta\beta z$;
- the self-phase modulation caused by the no-linearity effects is represented by the $\pm\phi(t,z)$.

In a weakly non-linear and weakly birefringent medium such as the typical glass of an optical fibre, it is reasonable to assume an instantaneous nonlinearity. The validity of this assumption will depend on the pulse rise time. It will break down if the optical pulse becomes narrow. This type of effect is ignored. We shall consider the Kerr type non-linear medium, assuming that (i) the dielectric medium is isotropic, (ii) third-harmonic generation can be neglected and (iii) the second-order non-linear susceptibility is equal to zero. So, the slowly varying envelope of the cubic polarisation $\vec{\mathcal{P}}^{NL}(z,t)$ is

$$\vec{\mathcal{P}}^{NL} = 2\chi^{(3)}_{1122}(\mathcal{E}\cdot\mathcal{E})\mathcal{E} + \chi^{(3)}_{1221}(\mathcal{E}\cdot\mathcal{E})\mathcal{E}^*,$$

where spatial and temporal dispersion are assumed to be absent. This expression leads to

$$\mathcal{P}_x^{NL} = \left[(a+b)|A_x|^2 + \{a + b\exp[-4i(\phi + \Delta\beta z)]\}|A_y|^2\right]A_x, \tag{6.3.1}$$
$$\mathcal{P}_y^{NL} = \left[(a+b)|A_y|^2 + \{a + b\exp[+4i(\phi + \Delta\beta z)]\}|A_x|^2\right]A_y, \tag{6.3.2}$$

where we introduce the designation $2\chi^{(3)}_{1122} = a$ and $\chi^{(3)}_{1221} = b$.

The non-linear polarisations (6.3.1) and (6.3.2) can be used to obtain the coupled evolution equations for the polarised optical pulse in a fibre. We must repeat the derivation of the connected wave equations as in section 5.11. The interior index l in equation (6.1.8) now is the polarisation component index, x or y. The evolution of amplitudes A_x and A_y with account of the second-order group-velocity dispersion, is determined by the system of coupled wave equation

$$i\frac{\partial A_x}{\partial z} + iv_x^{-1}\frac{\partial A_x}{\partial t} - \sigma_x\frac{\partial^2 A_x}{\partial t^2} + \Delta\beta A_x + \\ + m_x\left(a_{11}\,|\,A_x\,|^2 + a_{12}\,|\,A_y\,|^2\right)A_x = 0, \qquad (6.3.3a)$$

$$i\frac{\partial A_y}{\partial z} + iv_y^{-1}\frac{\partial A_y}{\partial t} - \sigma_y\frac{\partial^2 A_y}{\partial t^2} - \Delta\beta A_y + \\ + m_y\left(a_{21}\,|\,A_x\,|^2 + a_{22}\,|\,A_y\,|^2\right)A_y = 0, \qquad (6.3.3b)$$

where dumping is omitted and the coefficients are introduced as

$$v_{x,y}^{-1} = d\beta_{x,y}/d\omega, \qquad \sigma_{x,y} = (1/2)d^2\beta_{x,y}/d\omega^2, \qquad m_{x,y} = \omega_0^2/2c^2\beta_{x,y}.$$

The effect of the self-interaction E_x (E_y) is represented by the coefficients a_{11} (a_{22}). The effect of E_x on E_y (E_y on E_x) is represented by a_{12} (a_{21}). The effective non-linear interaction parameter χ_{eff} can therefore be taken out as a factor, which leaves only a material susceptibility tensor element ratio, i.e., the ratio $\chi_{1221}^{(3)}/\chi_{1122}^{(3)}$. Thus, we have

$$a_{11} = a_{22} = \chi_{eff},$$

$$a_{12} = \frac{a + b\exp[-4i(\phi + \Delta\beta z)]}{a+b}\chi_{eff}, \qquad a_{21} = \frac{a + b\exp[+4i(\phi + \Delta\beta z)]}{a+b}\chi_{eff},$$

where an effective non-linear interaction parameter χ_{eff} is defined as in section 6.1.1:

$$\chi_{eff} = \int \chi_{1122}^{(3)}(\boldsymbol{\rho})\,|\,\Psi(\boldsymbol{\rho})\,|^4\,d\boldsymbol{\rho} \Big/ \int |\,\Psi(\boldsymbol{\rho})\,|^2\,d\boldsymbol{\rho}.$$

If the optical fibre is made from silica, then third-order susceptibility is predominantly due to non-linear electronic response, and $a = 2b$, so that

$$a_{12} = \chi_{eff}\left(\frac{2}{3} + \frac{1}{3}\exp[-4i(\phi + \Delta\beta z)]\right), \qquad a_{21} = \chi_{eff}\left(\frac{2}{3} + \frac{1}{3}\exp[+4i(\phi + \Delta\beta z)]\right).$$

Now, we consider the same simplifications used in a number of works [145-158].

Linearly polarised pulse

Let us consider the simple case of a pulse propagation where

$$v_x = v_y = v, \quad \sigma_x = \sigma_y = \sigma, \quad m_x = m_y = m.$$

It is convenient to introduce the new variables as follows $z = \zeta L_D$, $\tau = (t - z/v)t_{p0}^{-1}$, and $A_x = A_0 q_1$, $A_y = A_0 q_2$, where t_{p0} is the initial pulse duration, L_D is dispersion length. In terms of these variables equations (6.3.3) can be written as

$$i\frac{\partial q_1}{\partial \zeta} - s\frac{\partial^2 q_1}{\partial \tau^2} + \delta q_1 + mA_0^2 L_D\left(a_{11}|q_1|^2 + a_{12}|q_2|^2\right)q_1 = 0, \qquad (6.3.4a)$$

$$i\frac{\partial q_2}{\partial \zeta} - s\frac{\partial^2 q_2}{\partial \tau^2} - \delta q_2 + mA_0^2 L_D\left(a_{21}|q_1|^2 + a_{22}|q_2|^2\right)q_2 = 0, \qquad (6.3.4b)$$

where $s = \operatorname{sgn}\sigma$, $\delta = \Delta\beta L_D$, and the dispersion length was defined as $L_D = t_{p0}^2/|\sigma|$.

If we use the relations

$$\exp[4i(\phi + \Delta\beta z)] = \frac{E_x E_y^*}{E_x^* E_y} = \frac{q_1 q_2^*}{q_1^* q_2}, \quad \exp[-4i(\phi + \Delta\beta z)] = \frac{E_y E_x^*}{E_y^* E_x} = \frac{q_2 q_1^*}{q_2^* q_1},$$

then the non-linear terms in equations (6.3.4) can be transformed to the following expressions

$$\left(a_{11}|q_1|^2 + a_{12}|q_2|^2\right)q_1 = \chi_{eff}\left(|q_1|^2 q_1 + \frac{2}{3}|q_2|^2 q_1 + \frac{1}{3}q_1^* q_2^2\right),$$

$$\left(a_{21}|q_1|^2 + a_{22}|q_2|^2\right)q_2 = \chi_{eff}\left(|q_2|^2 q_2 + \frac{2}{3}|q_1|^2 q_1 + \frac{1}{3}q_2^* q_1^2\right).$$

By taking into account these expressions we can write the system of equations describing the linear polarised pulse propagation in a birefringent fibre:

$$i\frac{\partial q_1}{\partial \zeta} - s\frac{\partial^2 q_1}{\partial \tau^2} + \delta q_1 + \mu\left(|q_1|^2 q_1 + \frac{2}{3}|q_2|^2 q_1 + \frac{1}{3}q_1^* q_2^2\right) = 0, \qquad (6.3.5a)$$

$$i\frac{\partial q_2}{\partial \zeta} - s\frac{\partial^2 q_2}{\partial \tau^2} - \delta q_2 + \mu\left(|q_2|^2 q_2 + \frac{2}{3}|q_1|^2 q_2 + \frac{1}{3}q_2^* q_1^2\right) = 0, \qquad (6.3.5b)$$

where $\mu = \chi_{eff} A_0^2 L_D \omega_0^2 /(2c^2\beta)$ is a non-linear coupling coefficient. This is the famous example of vector NLS equation of the non-linear fibre optics.

Circularly polarised pulse

The circular nature of birefringence can be seen by expressing the evolution equations in terms of circularly polarised propagating fields. Casting the equations in this form also facilitates the inclusion of intrinsic circular birefringence into the equations. It is known [145,146], the right and left circularly polarised fields are

$$E^{(+)} = E_x + iE_y, \quad E^{(-)} = E_x - iE_y,$$

respectively, and their corresponding complex envelopes may be written as

$$A^{(+)} = A_x + iA_y, \quad A^{(-)} = A_x - iA_y.$$

Hence,

$$q^{(+)} = q_1 + iq_2, \quad q^{(-)} = q_1 - iq_2.$$

It should be emphasised that

$$q_1^2 + q_2^2 = q^{(+)}q^{(-)}, \text{ and } |q_1|^2 + |q_2|^2 = \frac{1}{2}\left(|q^{(+)}|^2 + |q^{(-)}|^2\right).$$

The system of equations (6.3.5) with taking into account these formulae transforms to the following one

$$i\frac{\partial q^{(+)}}{\partial \zeta} - s\frac{\partial^2 q^{(+)}}{\partial \tau^2} + \delta q^{(-)} + \frac{1}{3}\mu\left(|q^{(+)}|^2 + 2|q^{(-)}|^2\right)q^{(+)} = 0, \tag{6.3.6a}$$

$$i\frac{\partial q^{(-)}}{\partial \zeta} - s\frac{\partial^2 q^{(-)}}{\partial \tau^2} + \delta q^{(+)} + \frac{1}{3}\mu\left(2|q^{(+)}|^2 + |q^{(-)}|^2\right)q^{(-)} = 0. \tag{6.3.6b}$$

It is another form of the v-NLS equation, describing the same process of the pulse propagation in a birefringent fibre.

In dispersionless and linear dielectric limit these equations lead to the known result

$$q^{(+)}(\zeta) = -i\sin(\delta\zeta), \quad q^{(-)}(\zeta) = \cos(\delta\zeta).$$

This well known result demonstrates the intrinsic birefringence effect.

Both system of equations (6.3.4) and system (6.3.5) were obtained under assumption that the group velocities of different polarised components of an optical pulse are equal. However, if $v_x \neq v_y$, then these systems can be modified. For example, equations (6.3.4) become

$$i\frac{\partial q_1}{\partial \zeta} + i\eta\frac{\partial q_1}{\partial \tau} - s\frac{\partial^2 q_1}{\partial \tau^2} + \delta q_1 + \mu\left(|q_1|^2 q_1 + \frac{2}{3}|q_2|^2 q_1 + \frac{1}{3}q_1^* q_2^2\right) = 0, \quad (6.3.7a)$$

$$i\frac{\partial q_2}{\partial \zeta} - i\eta\frac{\partial q_2}{\partial \tau} - s\frac{\partial^2 q_2}{\partial \tau^2} - \delta q_2 + \mu\left(|q_2|^2 q_2 + \frac{2}{3}|q_1|^2 q_2 + \frac{1}{3}q_2^* q_1^2\right) = 0, \quad (6.3.7b)$$

where parameter η and variable τ are:

$$\eta = \frac{L_D}{2t_{p0}}\left(\frac{1}{v_1} - \frac{1}{v_2}\right), \quad \tau = \frac{1}{t_{p0}}\left(t - z\frac{v_1 + v_2}{2v_1 v_2}\right).$$

The similar modifications must be made in the v-NLS equation (6.3.5).

$$i\frac{\partial q^{(+)}}{\partial \zeta} + i\eta\frac{\partial q^{(-)}}{\partial \tau} - s\frac{\partial^2 q^{(+)}}{\partial \tau^2} + \delta q^{(-)} + \frac{1}{3}\mu\left(|q^{(+)}|^2 + 2|q^{(-)}|^2\right)q^{(+)} = 0, \quad (6.3.8a)$$

$$i\frac{\partial q^{(-)}}{\partial \zeta} + i\eta\frac{\partial q^{(+)}}{\partial \tau} - s\frac{\partial^2 q^{(-)}}{\partial \tau^2} + \delta q^{(+)} + \frac{1}{3}\mu\left(2|q^{(+)}|^2 + |q^{(-)}|^2\right)q^{(-)} = 0. \quad (6.3.8b)$$

It is important to note that difference of the group velocities results in additional pulse distortion. It is due to spatial divergence of the orthogonal polarised components of this optical pulse.

Frequently the following dispersion constant [139,219,220] is used

$$D(\lambda_0) = \frac{\pi c^2}{\lambda_0}\sigma .$$

The dispersion length may be expressed as $L_D = \pi c^2 t_{p0}^2 (\lambda_0 |D(\lambda_0)|)^{-1}$. The typical value of the dispersion constant $D(\lambda_0)$ at $\lambda_0 = 1{,}5\ \mu\text{m}$ is 15 ps/nm/km. With the assumption that $(v_1^{-1} - v_2^{-1}) \approx \lambda_0 \Delta\beta / 2\pi c$, we can obtain

$$\eta \approx \frac{ct_{p0}\Delta\beta}{4|D(\lambda_0)|}, \quad \text{and} \quad \delta = \frac{\pi c^2 t_{p0}^2 \Delta\beta}{\lambda_0 |D(\lambda_0)|}.$$

In [149] these parameters were estimated for the typical characteristics of optical fibres. From these expressions we get $\eta/\delta \approx \lambda / 4(ct_{p0})$. Considering the picosecond pulses the terms related with the group velocities difference can be neglected. However, the group-velocity mismatch effect may be important in the case of femtosecond solitary wave propagation in optical fibers.

6.3.2. MULT I-COMPONENT NONLINEAR WAVES

We shall consider some examples of the physical problems where the simplest vector solitons appear. As usual, they are the two-component waves, but sometimes we can obtain more complete ones.

Non-linear waves in periodic medium

We are discussing now the optical pulse propagation in the medium where the refractive index is altering along one direction. This situation is pertinent to a number of topics in guided wave optics and fibre optics [137-139,227-230], including grating couplers, distributed feedback structures and gap medium [231-233]. As an example, one can consider the fibre with the refractive index periodically varying along the fibre-axis. Let dielectric penetrability be defined as

$$\varepsilon(\vec{r},\omega) = \varepsilon_0(x,y;\omega) + 2\Delta\varepsilon(x,y)\cos(Qz),$$

where $Q = 2\pi/\Lambda$, Λ - is the period of this alteration. The $\varepsilon_0(x,y;\omega)$ is consistent with the ideal fibre, so that $\Delta\varepsilon$ can be interpreted as the perturbation term in the wave equation for slowly varying envelope guided modes [228] (see section 6.1.1)

$$2i\beta_l \frac{\partial A_l(\omega)}{\partial z} + \Delta\beta_l^2 A_l(\omega) = -\sum_m \Delta\kappa_{lm}^2 A_m(\omega)\exp\{iz(\beta_m - \beta_l)\}, \tag{6.3.9}$$

where

$$\Delta\kappa_{lm}^2 = \int dxdy \vec{\Psi}_l(x,y)\Delta\kappa^2 \vec{\Psi}_m(x,y),$$

and

$$\Delta\kappa^2 = \Delta\varepsilon(\exp[iQz] + \exp[-Qz]) + 4\pi(\omega_0/c)^2 \chi_{NL}(\omega_0)$$

is the measure of the deflection of the fibre from the ideal linear fibre. Substitution of this expression into equation (6.3.9) results in

$$\begin{aligned} 2i\beta_l \frac{\partial A_l}{\partial z} + \Delta\beta_l^2 A_l &= -\sum_m \mu_{lm}(\omega_0) A_m \exp\{iz(\beta_m - \beta_l)\} - \\ &- \sum_m K_{lm} A_m \left[\exp\{iz(\beta_m - \beta_l + Q)\} + \exp\{iz(\beta_m - \beta_l - Q)\}\right], \end{aligned} \tag{6.3.10}$$

where

$$K_{lm} = K_{ml} = (\omega_0/c)^2 \int \vec{\Psi}_l(x,y)\Delta\varepsilon(x,y)\vec{\Psi}_m(x,y)dxdy$$

is the coupling coefficient of two modes. For any of the two counter-propagating modes with β_m and β_l, we have $\beta_m = -\beta_l$. Let us denote the propagation constants of incident (propagating in right direction) mode and reflected (propagating in opposite direction) mode as β_1 and β_2. From equations (6.3.10) it follows

$$2i\beta_1 \frac{\partial A_1}{\partial z} + \Delta\beta_1^2 A_1 = -\sum_m \mu_{1m}(\omega_0) A_m \exp\{iz(\beta_m - \beta_1)\} -$$
$$-\sum_m K_{1m} A_m \left[\exp\{iz(\beta_m - \beta_1 + Q)\} + \exp\{iz(\beta_m - \beta_1 - Q)\}\right],$$

and

$$2i\beta_2 \frac{\partial A_2}{\partial z} + \Delta\beta_2^2 A_1 = -\sum_m \mu_{2m}(\omega_0) A_m \exp\{iz(\beta_m - \beta_2)\} -$$
$$-\sum_m K_{2m} A_m \left[\exp\{iz(\beta_m - \beta_2 + Q)\} + \exp\{iz(\beta_m - \beta_2 - Q)\}\right].$$

Let the propagation constants β_1 and β_2 satisfy the Bragg condition

$$\beta_1 - \beta_2 = \frac{2\pi}{\Lambda} + \Delta\beta_B. \tag{6.3.11}$$

This condition allows us to distinguish the terms in the right-hand side of the equations above into the resonant and non-resonant ones.

$$2i\beta_1 \frac{\partial A_1}{\partial z} + \Delta\beta_1^2 A_1 + \mu_{11}(\omega_0) A_1 = -K_{12} A_2 \exp\{iz\Delta\beta_B\} + \ldots,$$
$$-2i\beta_1 \frac{\partial A_2}{\partial z} + \Delta\beta_2^2 A_2 + \mu_{22}(\omega_0) A_2 = -K_{21} A_1 \exp\{-iz\Delta\beta_B\} + \ldots,$$

where dots denote the non-resonant terms. Hereafter all non-resonant effects are neglected. So, taking into account only the second-order group-velocity dispersion and Kerr nonlinearity, these equations can be reduced to

$$2i\beta_1 \frac{\partial A_1}{\partial z} + i\alpha_1 \frac{\partial A_1}{\partial t} - \alpha_2 \frac{\partial^2 A_1}{\partial t^2} + \frac{4\pi\omega_0^2}{c^2} \chi_{K,eff} \left(|A_1|^2 + |A_2|^2\right) A_1 + \tag{6.3.12a}$$
$$+ K_{12} A_2 \exp\{iz\Delta\beta_B\} = 0,$$

$$-2i\beta_1 \frac{\partial A_2}{\partial z} + i\alpha_1 \frac{\partial A_2}{\partial t} - \alpha_2 \frac{\partial^2 A_2}{\partial t^2} + \frac{4\pi\omega_0^2}{c^2} \chi_{K,eff} \left(|A_1|^2 + |A_2|^2\right) A_2 + \tag{6.3.12b}$$
$$+ K_{21} A_1 \exp\{-iz\Delta\beta_B\} = 0,$$

where we introduce the effective Kerr susceptibility as in section 6.2.1:

$$\chi_{K,eff} = \int \chi_K(x,y)\Psi_1^4(x,y)dxdy\left(\int \Psi_1^2(x,y)dxdy\right)^{-1}.$$

Here we take into account that the mode functions of the incident and reflected waves are identical. As in section 6.3.1. it is convenient to introduce the normalised variables $\zeta = z/L_D$, $\tau = t/t_{p0}$, and

$$q_1(\zeta,\tau) = A_1(z,t)A_0^{-1}\exp[-i\Delta\beta_B z/2], \quad q_2(\zeta,\tau) = A_2(z,t)A_0^{-1}\exp[+i\Delta\beta_B z/2],$$

where $v_g = 2\beta/|\alpha_1|$ is the group velocity, t_{p0} is the initial pulse duration. If $\alpha_2 \neq 0$, then the dispersion length can be defined, as usual

$$L_D = 2\beta t_{p0}^2/|\alpha_2|.$$

In terms of these variables the system of equations (6.3.12) can be rewritten as

$$i\frac{\partial q_1}{\partial \zeta} + i\eta_1\frac{\partial q_1}{\partial \tau} - s\frac{\partial^2 q_1}{\partial \tau^2} - \delta q_1 + Kq_2 + \mu\left(|q_1|^2 + |q_2|^2\right)q_1 = 0, \quad (6.3.13a)$$

$$i\frac{\partial q_2}{\partial \zeta} - i\eta_1\frac{\partial q_2}{\partial \tau} + s\frac{\partial^2 q_2}{\partial \tau^2} + \delta q_2 - Kq_1 - \mu\left(|q_1|^2 + |q_2|^2\right)q_2 = 0, \quad (6.3.13b)$$

where $s = \mathrm{sgn}\,\sigma_2$, $\delta = \Delta\beta_B L_D/2$, $\eta_1 = L_D/v_g t_{p0}$, $K = K_{12}L_D/2\beta_1$ is the normalised coupling constant and $\mu = 2\pi\omega_0^2\chi_{K,eff}A_0^2/c^2\beta_1$ is the normalised nonlinearity constant.

These equations describe pulse propagation in the periodic non-linear medium [166,227,228]. They often say that it is the non-linear distributed feedback structure. The solitary waves in this case are called the *gap solitons*. However, this system of equations in not completely integrable and the term "soliton" is a sheer synonym of a "solitary wave".

Waves in a tunnel-coupled non-linear waveguides

In integrated optics the exchange of power between guided modes of parallel wave guides is known as a directional coupling [227,228]. The twin-core fibre is a device which can be considered as the fibre analogue of the directional coupler. Theory of this device is based on the system of two non-linear equation, which we may obtain here.

Consider the case of a two wave guided channel in a non-linear fibre as it is shown in Fig.6.3.1. Refractive index distributions for the two channels in the absence of coupling are given by $n_L(x,y)$ and $n_R(x,y)$. The dielectric penetration distribution of the two-channel fibre is $\varepsilon(x,y)$. The mode function for a particular m-th mode of channel L for a particular l-th mode of channel R are denoted by $\Psi_m^{(L)}(x,y)$ and $\Psi_l^{(R)}(x,y)$. Propagation constants will be denoted by $\beta_m^{(L)}$ and $\beta_l^{(R)}$. The orthogonality condition for the mode functions

$$\int dxdy \vec{\Psi}_l^{(J)}(x,y)\vec{\Psi}_m^{(I)}(x,y) = \delta_{lm} \int dxdy \,|\, \vec{\Psi}_l^{(J)}(x,y)\,|^2$$

holds only at J=I. The electric field of the optical wave propagating in this two-channel waveguide structure in the positive z direction is approximated as following

$$E(x,y,z;t) = \sum_m A_m^{(L)}(z,t)\Psi_m^{(L)}(x,y)\exp[-i\omega_0 t + i\beta_m^{(L)} z] + \\ + \sum_l A_l^{(R)}(z,t)\Psi_l^{(R)}(x,y)\exp[-i\omega_0 t + i\beta_l^{(R)} z] \quad . \tag{6.3.14}$$

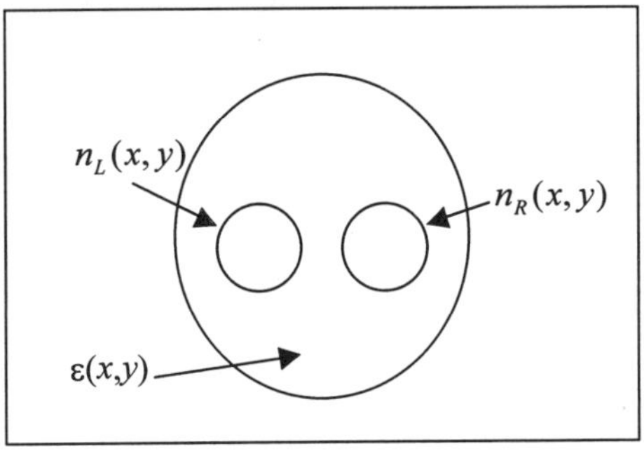

Fig. 6.3.1.

The cross-section of two channel fibre.

By using the approach developed in sections 1.2.2 and 6.1.1. we can obtain the system of equations, which is the generalisation of the equations (6.3.9)

$$2i\beta_m^{(L)} \frac{\partial A_m^{(L)}(\omega)}{\partial z} + \Delta\beta_m^{2(L)} A_m^{(L)}(\omega) = \\ = -\sum_{J=L,R} \sum_k \Delta\kappa_{mk}^{(JL)} A_k^{(J)}(\omega) \exp\{iz(\beta_k^{(J)} - \beta_m^{(L)})\}, \tag{6.3.15a}$$

$$2i\beta_l^{(R)} \frac{\partial A_l^{(R)}(\omega)}{\partial z} + \Delta\beta_l^{2(R)} A_l^{(R)}(\omega) = \\ = -\sum_{J=L,R} \sum_k \Delta\kappa_{lk}^{(JR)} A_k^{(J)}(\omega) \exp\{iz(\beta_k^{(J)} - \beta_l^{(R)})\}, \tag{6.3.15b}$$

where

$$\Delta\kappa_{lm}^{(JI)} = \int dxdy \bar{\Psi}_l^{(J)}(x,y)\Delta\kappa^{(JI)}\bar{\Psi}_m^{(I)}(x,y),$$

and

$$\Delta\kappa^{(JI)} = 4\pi(\omega/c)^2\chi_{NL}(x,y) + (\omega/c)^2\Delta\varepsilon^{(JI)},$$

$$\varepsilon^{(LL)} = \varepsilon^{(RL)} = \varepsilon(x,y) - n_R^2(x,y), \quad \varepsilon^{(RR)} = \varepsilon^{(LR)} = \varepsilon(x,y) - n_L^2(x,y).$$

The linear parts of the terms $\Delta\kappa_{mm}^{(LL)}$ and $\Delta\kappa_{ll}^{(RR)}$ represent the small corrections to the propagation constants. Let these coefficients be Δ_L and Δ_R. The non-linear parts of the terms $\Delta\kappa_{mm}^{(LL)}$ and $\Delta\kappa_{ll}^{(RR)}$ are due to Kerr type nonlinearity. It leads to self-modulation and cross-modulation effects.

Let us assume that there are two modes, for which the difference

$$\Delta\beta_{12} = \beta_m^{(L)} - \beta_l^{(R)}$$

is small enough, whereas the other propagation constants differences are further away than $\Delta\beta_{12}$. This situation can be realised, for example, in the case of two identical channels. In taking this resonance condition into account we can omit the non-resonant terms in summation over a mode index k in (6.3.15). Furthermore, we shall consider only the second-order group-velocity dispersion effects. The resulting system of the equations becomes

$$\begin{aligned} i\frac{\partial A_m^{(L)}}{\partial z} + i\frac{1}{v_m^{(L)}}\frac{\partial A_m^{(L)}}{\partial t} - \sigma_m^{(L)}\frac{\partial^2 A_m^{(L)}}{\partial t^2} + K_{12}A_k^{(R)}\exp\{iz\Delta\beta_{12}\} + \\ + K_L A_m^{(L)} + \frac{2\pi\omega_0^2}{c^2\beta_m^{(L)}}\chi_L\left(|A_m^{(L)}|^2 + \vartheta|A_l^{(R)}|^2\right)A_m^{(L)} = 0, \end{aligned} \tag{6.3.16a}$$

$$\begin{aligned} i\frac{\partial A_l^{(R)}}{\partial z} + \frac{i}{v_l^{(R)}}\frac{\partial A_l^{(R)}}{\partial t} - \sigma_l^{(R)}\frac{\partial^2 A_l^{(R)}}{\partial t^2} + K_{12}A_m^{(L)}\exp\{-iz\Delta\beta_{12}\} + \\ + K_R A_l^{(R)} + \frac{2\pi\omega_0^2}{c^2\beta_l^{(R)}}\chi_R\left(|A_l^{(R)}|^2 + \vartheta|A_m^{(L)}|^2\right)A_l^{(R)} = 0, \end{aligned} \tag{6.3.16b}$$

where $v_m^{(L)} = 2\beta_m^{(L)}/|\alpha_{m1}^{(L)}|$ and $v_l^{(R)} = 2\beta_l^{(R)}/|\alpha_{l1}^{(R)}|$ are the group velocities of the waves in channels, $1/\sigma_m^{(L)} = 2\beta_m^{(L)}/|\alpha_{m2}^{(L)}|$ and $1/\sigma_l^{(R)} = 2\beta_l^{(R)}/|\alpha_{l2}^{(R)}|$ represent the second-order dispersion in each channel. Linear and non-linear corrections to the propagation constants are determined by parameters $K_L = \Delta_L/2\beta_m^{(L)}$, $K_R = \Delta_R/2\beta_l^{(R)}$,

and

$$\chi_J = \int \chi_K(x,y)\Psi_m^{(J)4}(x,y)dxdy\left(\int \Psi_m^{(J)2}(x,y)dxdy\right)^{-1}, \text{ at } J = L \text{ or } R.$$

Here again we introduce the normalised variables $\zeta = z/L_D$, $\tau = (t - z/v_g)t_{p0}^{-1}$, and

$$q_1(\zeta,\tau) = A_m^{(L)}(z,t)A_0^{-1}\exp[-i\Delta\beta_{12}z/2],$$
$$q_2(\zeta,\tau) = A_l^{(R)}(z,t)A_0^{-1}\exp[+i\Delta\beta_{12}z/2],$$

where $v_g = 2v_m^{(L)}v_l^{(R)}(v_m^{(L)} + v_l^{(R)})$ is the average group velocity, t_{p0} is the initial pulse duration. L_D is some specific length. In terms of these variables the system of equations (6.3.16) can be rewritten as

$$i\frac{\partial q_1}{\partial \zeta} + i\eta\frac{\partial q_1}{\partial \tau} - \sigma_1\frac{\partial^2 q_1}{\partial \tau^2} + \delta_{12}q_2 + \delta_1 q_1 + \mu_1\left(|q_1|^2 + \vartheta|q_2|^2\right)q_1 = 0, \quad (6.3.17a)$$

$$i\frac{\partial q_2}{\partial \zeta} - i\eta\frac{\partial q_2}{\partial \tau} - \sigma_2\frac{\partial^2 q_2}{\partial \tau^2} + \delta_{12}q_1 + \delta_2 q_2 + \mu_2\left(|q_2|^2 + \vartheta|q_1|^2\right)q_2 = 0, \quad (6.3.17b)$$

where

$$\eta = \frac{L_D}{2t_{p0}}\left(\frac{1}{v_m^{(L)}} - \frac{1}{v_l^{(R)}}\right), \quad \sigma_1 = \frac{\sigma_m^{(L)}L_D}{2t_{p0}^2\beta_m^{(L)}}, \quad \sigma_2 = \frac{\sigma_l^{(R)}L_D}{2t_{p0}^2\beta_l^{(R)}},$$
$$\delta_{12} = \delta_{21} = K_{12}L_D, \quad \delta_1 = (\Delta\beta_{12} - \Delta_L/\beta_m^{(L)})L_D/2, \quad \delta_2 = (\Delta\beta_{12} - \Delta_R/\beta_l^{(R)})L_D/2,$$
$$\mu_1 = 2\pi\omega_0^2 L_D\chi_L A_0^2/c^2\beta_m^{(L)}, \quad \mu_2 = 2\pi\omega_0^2 L_D\chi_R A_0^2/c^2\beta_l^{(R)}.$$

The specific length L_D can be expressed in terms of the particular dispersion length L_1 and L_2

$$\frac{2}{L_D} = \frac{1}{L_1} + \frac{1}{L_2} = \frac{|\sigma_{m2}^{(L)}|}{2t_{p0}^2\beta_m^{(L)}} + \frac{|\sigma_{l2}^{(R)}|}{2t_{p0}^2\beta_l^{(R)}}.$$

The parameter ϑ represents the non-linear inter-channel interaction. This interaction leads to a phase modulation of the wave in one channel that is affected by the wave in another channel. This effect was named the *cross-phase modulation*. From the definition of the parameter ϑ it follows that it is determined by the overlapping integral of the square of mode functions. The strong localisation of the mode functions results in a small magnitude of ϑ, i.e., $\vartheta \ll 1$. As usual the parameter ϑ is neglected [174-183].

If we can consider the two channels as the identical ones, then the system of equations (6.3.17) can be simplified:

$$i\frac{\partial q_1}{\partial \zeta} - s\frac{\partial^2 q_1}{\partial \tau^2} + \delta_{12} q_2 + \delta q_1 + \mu\left(|q_1|^2 + \vartheta |q_2|^2\right)q_1 = 0\,, \qquad (6.3.18a)$$

$$i\frac{\partial q_2}{\partial \zeta} - s\frac{\partial^2 q_2}{\partial \tau^2} + \delta_{12} q_1 + \delta q_2 + \mu\left(|q_2|^2 + \vartheta |q_1|^2\right)q_2 = 0\,, \qquad (6.3.18b)$$

where $s = \operatorname{sgn}\sigma$.

These equations describe the optical picosecond pulse propagation in the twin-core non-linear fibre with account of the Kerr-type nonlinearity and the second-order group-velocity dispersion. The generation of these equations in the case of femtosecond pulse duration can be obtained by insertions

$$\mu(|q_1|^2 + \vartheta |q_2|^2)q_1 \quad \to \quad \mu\left[(|q_1|^2 + \vartheta |q_2|^2)q_1\right] +$$
$$+\mu\left[\frac{2i}{\omega_0 t_{p0}}\frac{\partial}{\partial \tau}\{(|q_1|^2 + \vartheta |q_2|^2)q_1\} - \frac{T_R}{t_{p0}} q_1 \frac{\partial}{\partial \tau}(|q_1|^2 + \vartheta |q_2|^2)\right]$$

and

$$\mu(|q_2|^2 + \vartheta |q_1|^2)q_2 \quad \to \quad \mu\left[(|q_2|^2 + \vartheta |q_1|^2)q_2\right] +$$
$$+\mu\left[\frac{2i}{\omega_0 t_{p0}}\frac{\partial}{\partial \tau}\{(|q_2|^2 + \vartheta |q_1|^2)q_2\} - \frac{T_R}{t_{p0}} q_2 \frac{\partial}{\partial \tau}(|q_2|^2 + \vartheta |q_1|^2)\right].$$

Thus, the pulse propagation in a twin-core non-linear fibre represents the example of the vector NLS equation, which is similar to the equations in the theory of pulse propagation in the non-linear birefringent fibre.

Waves in non-linear optical waveguide arrays

Consider the generalisation of the double-channel non-linear coupler by assuming that the number of channels is infinite (big enough) and that this waveguide structure is comprised of identical, regular spaced channels. Let these channels be lossless. Furthermore, let us assume that these channels are made from the Kerr-type non-linear material. This structure is named the non-linear optical waveguide array (NOWA) [193-195]. It was pointed out [184-187] that NOWA may be envisaged as a potential all-optical device such as, e.g., a switch, or a logical element. The existence and stability of solitary wave solutions localised both temporarily and across the array have been established.

By using the formalism of coupled wave theory [228] and by taking into account only the nearest-neighbour coupling, we can obtain the basic equations of NOWA.

According to the case of a double-channel non-linear coupler, the mode function for a particular m-th mode of channel J will be denoted by $\Psi_m^{(J)}(x,y)$, where $J = 0, \pm 1, \pm 2, \ldots$. Propagation constants are denoted as $\beta_m^{(J)}$. The electric field of the optical wave propagating in NOWA in the positive z direction is approximated by

$$E(x,y,z;t) = \sum_{J=-\infty}^{\infty} \sum_m A_m^{(J)}(z,t)\Psi_m^{(J)}(x,y)\exp[-i\omega_0 t + i\beta_m^{(J)} z] \qquad (6.3.19)$$

By using the approach developed in sections 1.2.2 and 5.1.1. and above, we can obtain the system of equations, which is the generalisation of equations (6.3.15)

$$\begin{aligned} 2i\beta_m^{(J)} \frac{\partial A_m^{(J)}(\omega)}{\partial z} + \Delta\beta_m^{2(J)} A_m^{(J)}(\omega) = \\ = -\sum_{M=-\infty}^{\infty} \sum_k \kappa_{mk}^{(MJ)} A_k^{(M)}(\omega) \exp\{iz(\beta_k^{(M)} - \beta_m^{(J)})\} \end{aligned} \qquad (6.3.20)$$

for $J = 0, \pm 1, \pm 2, \ldots$. The coupling constants are defined as in the case of a double-channel non-linear coupler. Taking into account only the nearest-neighbour coupling, we can reduce the right part of this system:

$$\begin{aligned} 2i\beta_m^{(J)} \frac{\partial A_m^{(J)}(\omega)}{\partial z} + \Delta\beta_m^{2(J)} A_m^{(J)}(\omega) = \\ = -\sum_k \Delta\kappa_{mk}^{(J\,J)2} A_k^{(J)}(\omega) \exp\{iz(\beta_k^{(J)} - \beta_m^{(J)})\} - \\ - \sum_k \Delta\kappa_{mk}^{(J+1\,J)2} A_k^{(J+1)}(\omega) \exp\{iz(\beta_k^{(J+1)} - \beta_m^{(J)})\} - \\ - \sum_k \Delta\kappa_{mk}^{(J-1\,J)2} A_k^{(J-1)}(\omega) \exp\{iz(\beta_k^{(J-1)} - \beta_m^{(J)})\}. \end{aligned}$$

In order to simplify these equations we need to choose the pairs of modes with close propagation constants and to omit all the others. Let these relevant propagation constants be $\beta_m^{(J)}$, $\beta_m^{(J-1)}$ and $\beta_m^{(J+1)}$. Furthermore, assume that

$$\beta_m^{(J)} - \beta_m^{(J+1)} = \beta_m^{(J-1)} - \beta_m^{(J)} = \Delta\beta_{12}.$$

In taking this resonance condition into account we can omit the non-resonant, fast varying terms in summation over a mode index k in (6.3.20). The resulting system of the equations becomes

$$i\frac{\partial A_m^{(J)}}{\partial z}+i\frac{1}{v_m^{(J)}}\frac{\partial A_m^{(J)}}{\partial t}-\sigma_m^{(J)}\frac{\partial^2 A_m^{(J)}}{\partial t^2}+\frac{2\pi\omega_0^2}{\beta_m^{(J)}c^2}\chi_{K,eff}\,|A_m^{(J)}|^2 A_m^{(J)}+$$
$$+K_{12}\left(A_k^{(J+1)}\exp\{iz\Delta\beta_{12}\}+A_k^{(J-1)}\exp\{iz\Delta\beta_{12}\}\right)+K_{11}A_m^{(J)}+ \qquad (6.3.21)$$
$$+\frac{2\pi\omega_0^2}{\beta_m^{(J)}c^2}\left(\chi_{K,eff}^{(J+!)}\,|A_l^{(J+1)}|^2+\chi_{K,eff}^{(J-!)}\,|A_l^{(J-1)}|^2\right)A_m^{(J)}=0,$$

where the effective non-linear susceptibilities and the coupling constant are defined by

$$\chi_{K,eff}=\int\chi_K(x,y)\Psi_m^{(J)4}(x,y)dxdy\left(\int\Psi_m^{(J)2}(x,y)dxdy\right)^{-1},$$
$$\chi_{K,eff}^{(J\pm1)}=\int\chi_K(x,y)\Psi_m^{(J\pm1)4}(x,y)dxdy\left(\int\Psi_m^{(J)2}(x,y)dxdy\right)^{-1},$$

$$K_{11}=\Delta\kappa_{mm}^{(JJ)}/2\beta_m^{(J)},\quad\text{and}\quad K_{12}=\Delta\kappa_{mm}^{(J\pm1J)}/2\beta_m^{(J)}.$$

Here we shall consider only the second-order group-velocity dispersion effects.

In the case of the identical channels of the NOWA, the difference of the propagation constants is equal to zero. The group velocities and dispersion coefficients in (6.3.21) are equal for all channels. Thus the system of equations (6.3.21) becomes

$$i\frac{\partial A_m^{(J)}}{\partial z}+i\frac{1}{v_g}\frac{\partial A_m^{(J)}}{\partial t}-\sigma\frac{\partial^2 A_m^{(J)}}{\partial t^2}+\frac{2\pi\omega_0^2}{c^2\beta_m^{(J)}}\chi_{K,eff}\,|A_m^{(J)}|^2 A_m^{(J)}+K_{11}A_m^{(J)}+$$
$$+K_{12}\left(A_k^{(J+1)}+A_k^{(J-1)}\right)+\frac{2\pi\omega_0^2}{c^2\beta_m^{(J)}}\left(\chi_{K,eff}^{(J+!)}\,|A_l^{(J+1)}|^2+\chi_{K,eff}^{(J-!)}\,|A_l^{(J-1)}|^2\right)A_m^{(J)}=0,$$

where $v_g=v_m^{(J)}$ and $\sigma=\sigma_m^{(J)}$. One can again introduce the standard normalised variables $\tau=(t-z/v_g)t_{p0}^{-1}$, $\zeta=z/L_D$ and $q_J(\zeta,\tau)=A_m^{(J)}(z,t)A_0^{-1}$, where t_{p0} is the initial pulse duration, and the dispersion length can be defined as $L_D=2\beta t_{p0}^2/|\alpha_2|$ provided $\alpha_2\neq0$. In terms of these variables the system of equations (6.3.21) can be rewritten as

$$i\frac{\partial q_J}{\partial\zeta}-s\frac{\partial^2 q_J}{\partial\tau^2}+K_{11}q_J+K_{12}\left(q_{J+1}+q_{J-1}\right)+ \qquad (6.3.22)$$
$$+\mu_{11}\,|q_J|^2 q_J+\mu_{12}\left(|q_{J+1}|^2+|q_{J-!}|^2\right)q_J=0,$$

where

$$\mu_{11} = \frac{2\pi\omega_0^2}{c^2\beta_m^{(J)}}\chi_{K,eff} A_0^2, \quad \text{and} \quad \mu_{12} = \frac{2\pi\omega_0^2}{c^2\beta_m^{(J)}}\chi_{K,eff}^{(J\pm1)} A_0^2 .$$

Localisation of the mode function $\Psi_m^{(J)}$ in J-th channel leads to the inequality $\mu_{12} << \mu_{11}$. Hence, the equations (6.3.22) become

$$i\frac{\partial q_J}{\partial \zeta} - s\frac{\partial^2 q_J}{\partial \tau^2} + K_{11}q_J + K_{12}\left(q_{J+1} + q_{J-1}\right) + \mu_{11} \,|\, q_J \,|^2 \, q_J = 0 . \qquad (6.3.23)$$

Those are the basic systems of equations for describing the pulse propagation in non-linear optical waveguide array [184-195].

If the electric field of an optical wave varies slowly over a number of channels, then it is possible to study equation (6.3.23) in a long wavelength approximation or in a continuum approximation. We can replace the difference-differential equation (6.3.23) with a partial differential one by letting $\xi = J\Delta x$ and using

$$q_{J\pm1}(\zeta,\tau) \to q(\zeta,\tau,\xi \pm \Delta x) \approx q(\zeta,\tau,\xi) \pm \frac{\partial q(\zeta,\tau,\xi)}{\partial \xi}\Delta x - \frac{1}{2}\frac{\partial^2 q(\zeta,\tau,\xi)}{\partial \xi^2}\Delta x^2 ,$$

where Δx is the distance between the nearest-neighbour channels. Thus equation (6.3.23) can be reduced into two-dimensional NLS equation. This equation was used to describe self-focusing of laser beam. Thus, the system of equations (6.3.23) may be considered as the model of discrete self-focusing [184,187,190]. It was pointed out in [193] that the NOWA represents an ideal laboratory for the study of the dynamical behaviour of discrete systems.

Waves in non-linear bimodal fibres

It is known that the optical fibres are the multi-mode waveguides. In this case description of non-linear pulse propagation becomes more involved since modal dispersion, associated with the different group velocities of the various modes, comes into play. Formally, the case of pulse propagation in birefringent fibre is the instance of the pulse propagation in the bimodal fibre. Hence, we can use the system of equations (6.3.3) as basic one to describe this process. However, there are some differences. The system of equations governing the evolution of the two slowly varying envelopes of propagating modes A_1 and A_2 is

$$i\frac{\partial A_1}{\partial z}+\frac{i}{v_1}\frac{\partial A_1}{\partial t}-\sigma_1\frac{\partial^2 A_1}{\partial t^2}+\Delta\beta A_1+ \\ +\mu_1\left(a_{11}|A_1|^2+a_{12}|A_2|^2\right)A_1=0, \tag{6.3.24a}$$

$$i\frac{\partial A_2}{\partial z}+\frac{i}{v_2}\frac{\partial A_2}{\partial t}-\sigma_2\frac{\partial^2 A_2}{\partial t^2}-\Delta\beta A_2+ \\ +\mu_2\left(a_{21}|A_1|^2+a_{22}|A_2|^2\right)A_2=0, \tag{6.3.24b}$$

where $\Delta\beta$ is the difference of propagating constants, σ_1 and σ_2 are the dispersion constants: $\sigma_{1,2}=\alpha_{21,22}/2\beta_{1,2}$, μ is constant of nonlinearity and a_{ml} are the mode overlapping integrals

$$\mu_{1,2}=\frac{2\pi\omega_0^2}{c^2\beta_{1,2}}\chi_K, \qquad a_{ml}=\frac{\int\Psi_m^2(x,y)\Psi_l^2(x,y)dxdy}{\int\Psi_m^2(x,y)dxdy}.$$

The same system of equations has been proposed for description of the pulse propagation in photorefractive materials [234-236].

Another example of the bimodal pulse propagation in fibres is the *stimulated Raman scattering*. This process has been considered in section 1.3.3. and section 5.2.1. If the frequency of molecular vibrations is much greater than the width of the optical pulse spectrum, then we can distinguish the pump and the Stokes solitary waves. It allows us to formulate the system of equations describing the stimulated Raman scattering in optical fibre as well as the equations (6.3.24). One mode is a pump wave, the second mode corresponds to Stokes wave. Non-linear coupling between these waves is determined by the third-order non-linear susceptibilities (see section 1.3.3.).

Optical solitons in quadratic medium

It is known that the second-order optical susceptibilities are larger than the third-order susceptibilities in the materials without the centre of inversion. The non-linear phenomena in such medium are more available to investigate non-linear wave propagation by experiment. One of the popular instances is the second harmonic generation [237-242]. We shall consider the system of equations describing this phenomenon with taking into account the second-order group-velocity dispersion.

As usual, the electric field of the optical wave propagating in a quadratic non-linear fibre in the positive z direction is approximated by

$$E(x,y,z;t)=\sum_m A_{1m}(z,t)\Psi_m^{(1)}(x,y)\exp[-i\omega_0 t+i\beta_m^{(1)}z]+ \\ +\sum_l A_{2l}(z,t)\Psi_l^{(2)}(x,y)\exp[-i2\omega_0 t+i\beta_l^{(2)}z] \quad , \tag{6.3.25}$$

where the m-th mode function for a pump wave and the l-th mode of second harmonic wave are denoted by $\Psi_m^{(1)}(x,y)$ and $\Psi_l^{(2)}(x,y)$, respectively, and $\beta_m^{(1)}$ and $\beta_l^{(2)}$ are the propagation constants. Substitution of this expression into wave equations results in the analogues of the equations (6.1.7) or (6.3.9):

$$2i\beta_m^{(1)}\frac{\partial A_{1m}(\omega)}{\partial z}+\Delta\beta_{1m}^2 A_{1m}(\omega)=-\sum_k P_{mk}^{(NL)}(\omega)\exp\{-iz\beta_m^{(1)}\}, \tag{6.3.26a}$$

$$2i\beta_l^{(2)}\frac{\partial A_{2l}(\omega)}{\partial z}+\Delta\beta_{2l}^2 A_{2l}(\omega)=-\sum_k P_{lk}^{(NL)}(2\omega)\exp\{-iz\beta_l^{(2)}\}, \tag{6.3.26b}$$

where Fourier components of the non-linear polarisation $P_{mk}^{(NL)}(\omega)$ and $P_{lk}^{(NL)}(\omega)$ can be determined according to equations (1.3.14) from section 1.3.3.

$$P_{mk}^{(NL)}(\omega)=\frac{8\pi\omega_0^2}{c^2}\chi_{mk}^{(2)}(\omega_0)A_{1k}^*(\omega)A_{2k}(\omega)\exp[iz(\beta_k^{(2)}-\beta_k^{(1)})],$$

$$P_{lk}^{(NL)}(2\omega)=\frac{16\pi\omega_0^2}{c^2}\chi_{lk}^{(2)}(2\omega_0)A_{1l}(\omega)A_{1k}(\omega)\exp[i2z\beta_k^{(1)}],$$

where effective susceptibilities are defined as

$$\chi_{mk}^{(2)}(\omega_0)=\frac{\int\chi^{(2)}(2\omega_0,-\omega_0)\Psi_m^{(1)}(x,y)\Psi_k^{(1)}(x,y)\Psi_k^{(2)}(x,y)dxdy}{\int\Psi_m^{(1)}(x,y)\Psi_m^{(1)}(x,y)dxdy},$$

and

$$\chi_{lk}^{(2)}(2\omega_0)=\frac{\int\chi^{(2)}(\omega_0,\omega_0)\Psi_k^{(1)}(x,y)\Psi_k^{(1)}(x,y)\Psi_l^{(2)}(x,y)dxdy}{\int\Psi_k^{(2)}(x,y)\Psi_k^{(2)}(x,y)dxdy}.$$

Interaction of the waves leads to the essential effect if the phase-matching condition occurs. In this case the phase-matching condition is $\Delta\beta=2\beta_m^{(1)}-\beta_l^{(2)}\approx 0$, that means the phase velocities of pump and harmonic waves are equal. The phase-matching condition allows us to use the resonance approximation to reduce the equations (6.3.26) to

$$2i\beta_m^{(1)}\frac{\partial A_{1m}(\omega)}{\partial z}+\Delta\beta_{1m}^2A_{1m}(\omega)=-\frac{8\pi\omega_0^2}{c^2}\chi_{ml}^{(2)}(\omega_0)A_{1m}^*(\omega)A_{2l}(\omega)\exp[iz(\beta_l^{(2)}-2\beta_k^{(1)})],$$
$$2i\beta_l^{(2)}\frac{\partial A_{2l}(\omega)}{\partial z}+\Delta\beta_{2l}^2A_{2l}(\omega)=-\frac{16\pi\omega_0^2}{c^2}\chi_{lm}^{(2)}(2\omega_0)A_{1m}(\omega)A_{1m}(\omega)\exp[iz(2\beta_m^{(1)}-\beta_l^{(2)})].$$

By taking into account only the second-order group-velocity dispersion we can obtain the following system of equations

$$i\frac{\partial A_{1m}}{\partial z}+\frac{i}{v_1}\frac{\partial A_{1m}}{\partial t}-\frac{\alpha_{2m}}{2\beta_m^{(1)}}\frac{\partial^2A_{1m}}{\partial t^2}=-\frac{4\pi\omega_0^2}{\beta_m^{(1)}c^2}\chi_{ml}^{(2)}(\omega_0)A_{1m}^*A_{2l}\exp[-iz\Delta\beta],\quad (6.3.27a)$$
$$i\frac{\partial A_{2l}}{\partial z}+\frac{i}{v_2}\frac{\partial A_{2l}}{\partial t}-\frac{\alpha_{2l}}{2\beta_l^{(2)}}\frac{\partial^2A_{2l}}{\partial t^2}=-\frac{8\pi\omega_0^2}{\beta_l^{(2)}c^2}\chi_{ml}^{(2)}(2\omega_0)A_{1m}^2\exp[+iz\Delta\beta],\quad (6.3.27b)$$

Again introduce the normalised variables $\zeta=z/L_D$, $\tau=(t-z/v_g)t_{p0}^{-1}$, and

$$q_1(\zeta,\tau)=A_{1m}(z,t)A_0^{-1}\exp[+i\Delta\beta z/2],$$
$$q_2(\zeta,\tau)=A_{2l}(z,t)A_0^{-1}\exp[-i\Delta\beta z/2],$$

where $v_g=2v_1v_2(v_1+v_2)$ is the average group velocity, t_{p0} is the initial pulse duration. L_D is some specific length. In terms of these variables the system of equations (6.3.27) can be represented as

$$i\frac{\partial q_1}{\partial\zeta}+i\eta\frac{\partial q_1}{\partial\tau}-\sigma_1\frac{\partial^2q_1}{\partial\tau^2}+\delta q_1+\mu_1q_2q_1^{\circ}=0,\quad (6.3.28a)$$
$$i\frac{\partial q_2}{\partial\zeta}-i\eta\frac{\partial q_2}{\partial\tau}-\sigma_2\frac{\partial^2q_2}{\partial\tau^2}-\delta q_2+\mu_2q_1^2=0,\quad (6.3.28b)$$

where $\delta=\Delta\beta L_D$ and

$$\eta=\frac{L_D}{2t_{p0}}\left(\frac{1}{v_1}-\frac{1}{v_2}\right),\quad \sigma_1=\frac{\alpha_{2m}L_D}{2t_{p0}^2\beta_m^{(1)}},\quad \sigma_2=\frac{\alpha_{2l}L_D}{2t_{p0}^2\beta_l^{(2)}},$$
$$\mu_1=\frac{4\pi\omega_0^2L_D}{\beta_m^{(1)}c^2}\chi_{ml}^{(2)}(\omega_0)A_0,\quad \mu_2=\frac{8\pi\omega_0^2L_D}{\beta_l^{(2)}c^2}\chi_{ml}^{(2)}(2\omega_0)A_0.$$

Thus, the equations (6.3.27), or (6.3.28), describe the second harmonic generation in the field of picosecond optical pulse propagating in the quadratic non-linear fibre with account of the second-order group-velocity dispersion. These equations can be simplified if the phase-matching condition is held and the group velocities of pump and harmonic waves are equal.

6.3.3. STATIONARY VECTOR WAVES

Next consider the same example for an exact solution of the vector NLS equation. These solutions describe the stationary pulse propagation in several waveguide structures. For instance, these are a birefringent fibres, two-mode fibres, twin-core fibres, and a fibres with distributed feedback. There are no exact analytical solutions of the two-component NLS equation, with the exception of steady state solutions and the soliton solutions of the special kind of that equation [159]. Now it is convenient to consider these particular examples.

Cnoidal waves

Take the system of the equations describing the linear polarised pulse propagation in a birefringent fibre (6.3.5)

$$i\frac{\partial q_1}{\partial \zeta} - s_1\frac{\partial^2 q_1}{\partial \tau^2} + \delta q_1 + \mu\left(|q_1|^2 q_1 + \frac{2}{3}|q_2|^2 q_1 + \frac{1}{3}q_1^* q_2^2\right) = 0,$$

$$i\frac{\partial q_2}{\partial \zeta} - s_2\frac{\partial^2 q_2}{\partial \tau^2} - \delta q_2 + \mu\left(|q_2|^2 q_2 + \frac{2}{3}|q_1|^2 q_2 + \frac{1}{3}q_2^* q_1^2\right) = 0,$$

where s_1 and s_2 are the signs of the dispersion coefficients. According to [169] we shall find the solution of this system in the following form

$$q_1(\zeta,\tau) = a_1(\tau)\exp\{i\Omega\zeta\}, \quad q_2(\zeta,\tau) = a_2(\tau)\exp\{i\Omega\zeta\}, \tag{6.3.29}$$

where $a_1(\tau)$ and $a_2(\zeta)$ are real functions, Ω is an arbitrary real constant. Substitution of (6.3.29) into equations (6.3.5) yields the equations for $a_1(\tau)$ and $a_2(\zeta)$:

$$\frac{d^2 a_1}{d\tau^2} + \mu\left(a_1^2 + a_2^2\right)a_1 = (\Omega - \delta)a_1, \tag{6.3.30a}$$

$$\frac{d^2 a_2}{d\tau^2} + \mu\left(a_1^2 + a_2^2\right)a_2 = (\Omega + \delta)a_2\ . \tag{6.3.30b}$$

Here we select the case $s_1 = s_2 = -1$ that corresponds to the anomalous dispersion region. Let us assume that the solution of these equations may be represented as

$$a_1(\tau) = B\,\mathrm{sn}(p\tau)\,\mathrm{dn}(p\tau), \qquad (6.3.31a)$$
$$a_1(\tau) = B\,\mathrm{cn}(p\tau)\,\mathrm{dn}(p\tau), \qquad (6.3.31b)$$

where $\mathrm{sn}(p\tau)$, $\mathrm{cn}(p\tau)$, and $\mathrm{dn}(p\tau)$ are the Jacobian elliptic functions [243]. It is suitable to use the formulae for these functions

$$\mathrm{sn}^2(z) + \mathrm{cn}^2(z) = 1, \qquad \mathrm{dn}^2(z) + k^2\,\mathrm{sn}^2(z) = 1,$$
$$\frac{d}{dz}\mathrm{sn}(z) = \mathrm{cn}(z)\,\mathrm{dn}(z), \quad \frac{d}{dz}\mathrm{cn}(z) = -\mathrm{sn}(z)\,\mathrm{dn}(z), \quad \frac{d}{dz}\mathrm{dn}(z) = -k^2\,\mathrm{cn}(z)\,\mathrm{dn}(z),$$
$$\frac{d^2}{dz^2}\mathrm{sn}(z) = -\mathrm{sn}(z)\left(\mathrm{dn}^2(z) + k^2\,\mathrm{cn}^2(z)\right),$$
$$\frac{d^2}{dz^2}\mathrm{cn}(z) = -\mathrm{cn}(z)\left(\mathrm{dn}^2(z) - k^2\,\mathrm{sn}^2(z)\right),$$
$$\frac{d^2}{dz^2}\mathrm{dn}(z) = -k^2\,\mathrm{dn}(z)\left(\mathrm{cn}^2(z) - \mathrm{sn}^2(z)\right),$$

where k is the modulus of the elliptic functions.

If we substitute the expression (6.3.31) into equations (6.3.30) with taking into account these formulae, we shall obtain some expression. By comparing the factors at equal degrees of $\mathrm{sn}(p\tau)$, one can find the relations between unknown parameters p and B

$$6p^2 = \mu B^2, \quad p^2(1+k^2) - \mu B^2 = -(\Omega - \delta), \quad p^2(1+4k^2) - \mu B^2 = -(\Omega + \delta).$$

From these relation it follows

$$k^2 = \frac{10\delta}{(3\Omega + 5\delta)}, \quad p^2 = \frac{(3\Omega + 5\delta)}{15}, \quad \mu B^2 = \frac{2}{5}(3\Omega + 5\delta).$$

Thus, we get the periodic solution of the (6.3.31) that represents the cnoidal wave of polarisation radiation in a non-linear fibre. The spatial period of this wave is expressed in term of a complete elliptic integral, being equal to $\zeta_0 = 4\mathbf{K}(k)/p$ [243].

Variation of the parameter Ω reduces this solution to the partial ones. When modulus $k^2 = 1$, the elliptic functions take the forms

$$\mathrm{sn}(z) = \tanh(z), \quad \mathrm{cn}(z) = \mathrm{dn}(z) = \mathrm{sech}(z).$$

In this case (at $\Omega = 5\delta/3$) the cnoidal waves are reduced to solitary waves

$$a_1(\zeta) = B\tanh(p\zeta)\,\mathrm{sech}(p\zeta), \tag{6.3.32a}$$
$$a_2(\zeta) = B\,\mathrm{sech}^2(p\zeta) \tag{6.3.32b}$$

These waves are shown in Fig.6.3.2 , where $\delta = 1{,}5$.

In opposite case, when modulus $k = 0$ elliptic functions become

$$\mathrm{sn}(z) = \sin(z), \quad \mathrm{cn}(z) = \cos(z), \quad \mathrm{dn}(z) = 1.$$

In this case (at $\delta = 0$) the cnoidal waves are reduced to usually harmonic waves.

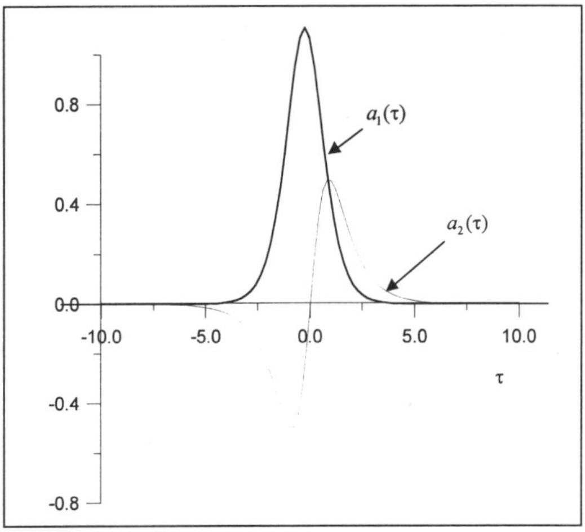

Fig. 6.3.2

Steady-state profiles of the orthogonal components of the linear polarised pulse

Solitary waves

The trivial solution of the equations (6.3.5) can be obtained if one supposes that one of the components q_1 or q_2 equals zero. In this case the system of equations (6.3.5) reduces to one the NLS equation. Thus the solitary wave is the soliton solution of the NLS equation. However, it is easy to show that this solution is not stable. Non-linear coupling induces the polarisation vector rotation. In terms of equations (6.3.6) this result is obvious. More interesting solution can be obtain from (6.3.30) following to [246]. On the other hand in section 4.3.2, under consideration of the ultrashort optical pulse propagation in the resonant medium the equations (6.3.30) have appeared. Thus, we can use all results from that section. So, the solitary polarised wave are represented by the expressions

$$a_1(\zeta) = \frac{2\sqrt{2/\mu}\exp(\theta_1)\{1+\exp[2\theta_2+\Delta_{12}]\}}{1+\exp(2\theta_1)+\exp(2\theta_2)+\exp(2\theta_1+2\theta_2+\Delta_{12})}, \qquad (6.3.33a)$$

$$a_2(\zeta) = \frac{2\sqrt{2/\mu}\exp(\theta_2)\{1-\exp[2\theta_1+\Delta_{12}]\}}{1+\exp(2\theta_1)+\exp(2\theta_2)+\exp(2\theta_1+2\theta_2+\Delta_{12})}, \qquad (6.3.33b)$$

where $\theta_1 = \Omega - \delta(\zeta - \zeta_{01})$, $\theta_2 = \Omega + \delta(\zeta - \zeta_{02})$, and

$$\exp(\Delta_{12}) = -\frac{\delta}{\Omega + \sqrt{\Omega^2 - \delta^2}}.$$

To illustrate asynchronous behaviour of the polarisation component of the steady state solitary wave, we represent the profiles of this wave at $\Omega = 1.5$ and $\delta = 1$ in Fig. 6.3.3.

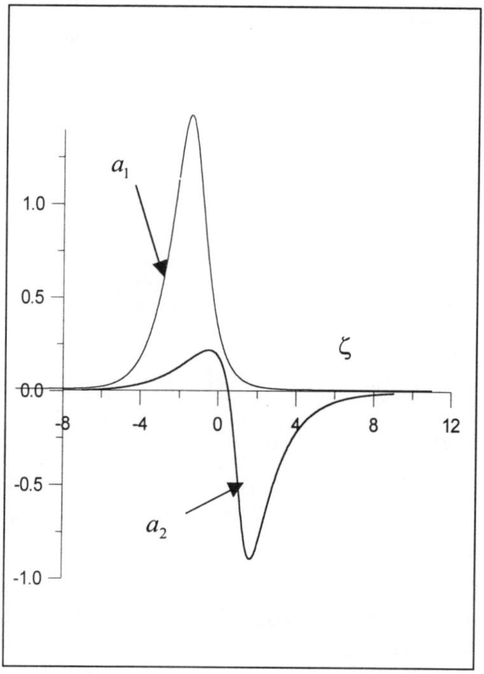

Fig. 6.3.3

Components of the steady state optical pulse corresponding to (6.3.33)

Let us consider two pulses of the coherent radiation co-propagating in a fibre in the anomalous and normal dispersion regime, respectively. This process can be described by the system of equations (6.3.24), where the slowly varying amplitudes of the modes correspond to the envelopes of these pulses. Let $v_1 = v_2$ that means one can ignore the "walk-off" effects. Furthermore, it implies that the two pulses overlap, and we can equate the mode overlapping integrals. After normalisation the system of equations (6.3.24) becomes

$$i\frac{\partial q_1}{\partial \zeta} - s_1 \frac{\partial^2 q_1}{\partial \tau^2} + \delta q_1 + \mu\left(|q_1|^2 + |q_2|^2\right)q_1 = 0, \tag{6.3.34a}$$

$$i\frac{\partial q_2}{\partial \zeta} - s_2 \frac{\partial^2 q_2}{\partial \tau^2} - \delta q_2 + \mu\left(|q_1|^2 + |q_2|^2\right)q_2 = 0, \tag{6.3.34a}$$

where $\delta = \Delta\beta L_D$ is the normalised phase mismatch, other parameters were defined in section 6.3.2. Let us suppose that q_1 and q_2 are taken in the form of (6.3.29). Then the system (6.3.34) can be reduced to

$$\frac{d^2 a_1}{d\tau^2} - s_1\mu\left(a_1^2 + a_2^2\right)a_1 = -s_1(\Omega - \delta)a_1, \tag{6.3.35a}$$

$$\frac{d^2 a_2}{d\tau^2} - s_2\mu\left(a_1^2 + a_2^2\right)a_2 = s_2(\Omega + \delta)a_2 . \tag{6.3.35b}$$

Let us suppose that the real amplitudes $a_1(\tau)$ and $a_2(\tau)$ are the following trial functions

$$a_1(\zeta) = B_1 \operatorname{sech}(p\zeta), \quad a_2(\zeta) = B_2 \tanh(p\zeta).$$

Substituting of these expression into (6.3.35) and comparing the factors at the equal degrees of $\operatorname{sech}(p\tau)$, we can find the relations between parameters p, B_1 and B_2:

$$\mu B_1^2 = (\Omega - \delta) - s_1 p^2, \quad \mu B_2^2 = (\Omega - \delta) + s_1 p^2, \quad \mu B_2^2 = (\Omega + \delta).$$

These relations yield $\mu B_2^2 = (\Omega + \delta)$, $\mu B_1^2 = (\Omega - 3\delta)$, $p^2 = 2s_1\delta$. Thus, we obtain the partial solution of the equations (6.3.34)

$$a_1(\zeta) = (\Omega - 3\delta)^{1/2}\mu^{-1/2}\operatorname{sech}(\sqrt{2s_1\delta}\zeta), \tag{6.3.36a}$$

$$a_2(\zeta) = (\Omega + \delta)^{1/2}\mu^{-1/2}\tanh(\sqrt{2s_1\delta}\zeta) . \tag{6.3.36b}$$

This solution of the system of equations (6.3.34) was obtained in the [165,167]. It shows that in the region of anomalous dispersion ($s_1 = -1$ and $s_2 = -1$) for the first wave and at $\delta < 0$ the vector solitary wave consists of the bright ($a_1(\tau)$) and dark ($a_2(\tau)$) components. In this case the bright wave dominates. Parameter Ω must be greater than $|\delta|$. In the regime of normal dispersion ($s_1 = +1$ and $s_2 = +1$) expression (6.3.36) requires that $\delta > 0$, and $\Omega > 3\delta$. In this case the dark component dominates in the vector solitary steady state wave.

If one considers the mixed situation where one of the components of the solitary wave corresponds to the region of anomalous dispersion but another one corresponds to normal dispersion ($s_1 = -1$ and $s_2 = +1$), then we can try to obtain the following solution

$$a_1(\zeta) = -B_1 \operatorname{sech}(p\zeta), \quad a_2(\zeta) = B_2 \tanh(p\zeta).$$

Substitution of these expressions into (6.3.35) results in

$$\mu B_2^2 = (\Omega + \delta), \quad \mu B_1^2 = (\Omega - 3\delta), \quad p^2 = -2\delta.$$

Thus, we find that the expressions

$$a_1(\zeta) = s_1(\Omega - 3\delta)^{1/2} \mu^{-1/2} \operatorname{sech}(\sqrt{2s_1\delta}\zeta), \tag{6.3.36c}$$

$$a_2(\zeta) = (\Omega + \delta)^{1/2} \mu^{-1/2} \tanh(\sqrt{2s_1\delta}\zeta), \tag{6.3.36d}$$

are the solution of the equations (6.3.35) at $s_2 = +1$ and arbitrary sign of dispersion for the first component of the optical pulse under consideration.

6.3.4. SOLITONS OF THE VECTOR NONLINEAR SCHRÖDINGER EQUATION

We have only one example where the v-NLS equation is the completely integrable one. This equation is embedded into AKNS hierarchy that allows to exploit the IST method to find the soliton solution. The suitable spectral problem was found by Manakov [159]. The IST method based on this spectral problem was discussed in sections 4.2.4-4.2.6, hence we can use all the results from these paragraphs.

Zero-curvature representation of v-NLS equation

Let us consider the two systems of the linear equations (4.2.13)

$$\partial\Psi / \partial\tau = \hat{U}(\lambda)\Psi \quad \text{and} \quad \partial\Psi / \partial\zeta = \hat{V}(\lambda)\Psi, \tag{6.3.37}$$

where $\Psi = \text{colon}(\psi_1, \psi_2, \dots, \psi_N)$, $\hat{U}$ and $\hat{V}$ are $(N\times N)$ matrices, depending on spectral parameter λ and the dependent variables of the non-linear equation under consideration. As in section 4.2.4 we shall consider the U-V pair in the following form

$$\hat{U}(\lambda) = \begin{pmatrix} -i\lambda & \vec{q} \\ \vec{r} & i\lambda \mathrm{I} \end{pmatrix}, \quad \hat{V}(\lambda) = \begin{pmatrix} A & \vec{B} \\ \vec{C} & \hat{D} \end{pmatrix}, \tag{6.3.38}$$

where $\vec{r} = \varepsilon \vec{q}^{\,+}$, $\vec{q} = (q_1, q_2, \dots, q_N)$, $\varepsilon = \pm 1$. The integrability conditions of these systems result in

$$\frac{\partial A}{\partial \tau} = \vec{q}\cdot\vec{C} - \vec{r}\cdot\vec{B}, \qquad \frac{\partial \hat{D}}{\partial \tau} = \vec{r}\otimes\vec{B} - \vec{C}\otimes\vec{q}, \tag{6.3.39a,b}$$

$$\frac{\partial \vec{B}}{\partial \tau} + 2i\lambda\vec{B} + A\vec{q} - \vec{q}\cdot\hat{D} = \frac{\partial \vec{q}}{\partial \zeta} \tag{6.3.39b}$$

$$\frac{\partial \vec{C}}{\partial \tau} - 2i\lambda\vec{C} - A\vec{r} + \hat{D}\cdot\vec{r} = \frac{\partial \vec{r}}{\partial \zeta} \tag{6.3.39c}$$

We assume that the coefficients of V-matrix are some polynomials of degrees 2 and 1 :

$$A = a_0(\tau,\zeta) + a_1(\tau,\zeta)\lambda + a_2(\tau,\zeta)\lambda^2, \quad \hat{D} = \hat{d}_0(\tau,\zeta) + \hat{d}_1(\tau,\zeta)\lambda + \hat{d}_2(\tau,\zeta)\lambda^2,$$
$$B = b_0(\tau,\zeta) + b_1(\tau,\zeta)\lambda, \quad C = c_0(\tau,\zeta) + c_1(\tau,\zeta)\lambda .$$

Substituting these expansions into (6.3.39) and equating the coefficients of various powers of λ in the resulting equations, we obtain the system of equations for these coefficients

$$\frac{\partial a_2}{\partial \tau} = 0 , \quad \frac{\partial \hat{d}_2}{\partial \tau} = 0 \tag{6.3.40a}$$

$$\frac{\partial a_1}{\partial \tau} = \vec{q}\cdot\vec{c}_1 - \vec{r}\cdot\vec{b}_1 , \quad \frac{\partial \hat{d}_1}{\partial \tau} = \vec{r}\otimes\vec{b}_1 - \vec{c}_1\otimes\vec{q} , \tag{6.3.40b}$$

$$\frac{\partial a_0}{\partial \tau} = \vec{q}\cdot\vec{c}_0 - \vec{r}\cdot\vec{b}_0 , \quad \frac{\partial \hat{d}_0}{\partial \tau} = \vec{r}\otimes\vec{b}_0 - \vec{c}_0\otimes\vec{q} , \tag{6.3.40c}$$

$$\vec{q}a_2 - \vec{q}\cdot\hat{d}_2 + 2i\vec{b}_1 = 0 , \ -\vec{r}a_2 + \hat{d}_2\cdot\vec{r} - 2i\vec{c}_1 = 0 \tag{6.3.40d,e}$$

$$\frac{\partial \vec{b}_1}{\partial \tau} + \vec{q}a_1 - \vec{q}\cdot\hat{d}_1 + 2ib_0 = 0 , \quad \frac{\partial \vec{c}_1}{\partial \tau} - \vec{r}a_1 + \hat{d}_1\cdot\vec{r} - 2ic_0 = 0 , \tag{6.3.40f}$$

$$\frac{\partial \vec{b}_0}{\partial \tau} + \vec{q}a_0 - \vec{q}\cdot\hat{d}_0 = \frac{\partial \vec{q}}{\partial \zeta} , \quad \frac{\partial \vec{c}_0}{\partial \tau} - \vec{r}a_0 + \hat{d}_0\cdot\vec{r} = \frac{\partial \vec{r}}{\partial \zeta} . \tag{6.3.40g}$$

From equation (6.3.40a) it follows that $a_2 = \alpha_2(\zeta)$ and $\hat{d}_2 = \hat{d}_2(\zeta)$. We can choose $\hat{d}_2 = \delta_2(\zeta)\hat{I}$, where $\hat{I}$ is a unit matrix. The equations (6.3.40d) and (6.3.40e) lead to

$$\vec{b}_1 = \frac{1}{2i}(\delta_2 - \alpha_2)\vec{q}, \qquad \vec{c}_1 = \frac{1}{2i}(\delta_2 - \alpha_2)\vec{r}. \tag{6.3.41}$$

The substitution of these expressions in equations (6.3.40b) and (6.3.40f) result in

$$a_1 = \alpha_1(\zeta) \text{ and } \hat{d}_1 = \hat{d}_1(\zeta),$$

$$\vec{b}_0 = -\frac{1}{2i}\left[\frac{1}{2i}(\delta_2 - \alpha_2)\frac{\partial \vec{q}}{\partial \zeta} + (\delta_1 - \alpha_1)\vec{q}\right], \quad \vec{c}_0 = +\frac{1}{2i}\left[\frac{1}{2i}(\delta_2 - \alpha_2)\frac{\partial \vec{r}}{\partial \zeta} + (\delta_1 - \alpha_1)\vec{r}\right].$$

Here we assume that $\hat{d}_1 = \delta_1(\zeta)\hat{I}$. Integrating the equations (6.3.40c) with account for these expressions we get

$$a_0(\tau,\zeta) = \alpha_0(\zeta) - (\delta_1 - \alpha_1)(\vec{q}\cdot\vec{r})/4, \quad \hat{d}_0(\tau,\zeta) = \delta_0(\zeta)\hat{I} + (\delta_1 - \alpha_1)(\vec{q}\otimes\vec{r})/4.$$

Now, we can write the non-linear evolution equations, which are solved by the IST method with the chosen *U-V*-pair. They are

$$\frac{\partial \vec{q}}{\partial \zeta} = +\frac{1}{4}(\delta_2 - \alpha_2)\left[\frac{\partial^2 \vec{q}}{\partial \tau^2} - 2(\vec{q}\cdot\vec{r})\vec{q}\right] + \frac{1}{2i}(\delta_1 - \alpha_1)\frac{\partial \vec{q}}{\partial \tau} - (\delta_0 - \alpha_0)\vec{q},$$
$$\frac{\partial \vec{r}}{\partial \zeta} = -\frac{1}{4}(\delta_2 - \alpha_2)\left[\frac{\partial^2 \vec{r}}{\partial \tau^2} - 2(\vec{q}\cdot\vec{r})\vec{r}\right] + \frac{1}{2i}(\delta_1 - \alpha_1)\frac{\partial \vec{r}}{\partial \tau} + (\delta_0 - \alpha_0)\vec{r}$$

Let us assume that $\alpha_0 = \alpha_1 = \delta_0 = \delta_1$ and $\alpha_2 = -\delta_2 = i$. The reduction $\vec{r} = \varepsilon\vec{q}^*$ leads to the *v*-NLS equation in the following form

$$i\frac{\partial \vec{q}}{\partial \zeta} = \frac{1}{2}\frac{\partial^2 \vec{q}}{\partial \tau^2} - \varepsilon(\vec{q}\cdot\vec{q}^*)\vec{q}.$$

In the case of $N=2$ this equation was solved in [159] by the IST method. As it have been stated in section 3.2.6 the evolution of the scattering data is governed by the equations

$$\frac{\partial}{\partial \zeta}S_{kn} = (V_{nn} - V_{kk})S_{kn},$$

if $\hat{V}^{(+)} = \hat{V}^{(-)} = \text{diag}(V_{11}, V_{22}, V_{33})$. In our case $V_{22} = V_{33} = -V_{11} = -i\lambda^2$. Let $\lambda_1 = \xi_1 + i\eta_1$ be the single point of a discrete spectrum in the Manakov spectral problem, and the reflection coefficients are zero. By using the results from section 4.2.6 and the presented solution of the Gelfand-Levitan-Marchenko equations, we can write the one-soliton solution of the v-NLS equation

$$q_k(\tau,\zeta) = 2\eta_1 l_k \,\text{sech}[2\eta_1(\tau - \tau_0 + 2\xi_1\zeta)]\exp[i\phi(\tau,\zeta)], \qquad (6.3.41)$$

where $\phi(\tau,\zeta) = -2\xi_1\tau + 4(\eta_1^2 - \xi_1^2)\zeta + \phi_0$. The constants τ_0, ϕ_0 and unit (polarisation) vector $\vec{l}$ are determined by the initial pulse profile.

The main properties of the solitons of v-NLS equations in this case are the same as those of the vector solitons in the case of self-induced transparency for polarised USP (section 4.4). For example, the scattering of two solitons leads to rotation of their polarisation vectors according to expression (4.2.47).

Zero-curvature representation of v-DNLS equation

The generalisation of the DNLS equation to the case of N-component field gives rise to the following equation

$$i\frac{\partial \vec{q}}{\partial \zeta} = \frac{\partial^2 \vec{q}}{\partial \tau^2} + i\varepsilon\frac{\partial}{\partial \tau}(\vec{q}\cdot\vec{q}^*)\vec{q} .$$

This equation is named as vector Derivative Non-linear Schrödinger (v-DNLS) equation. To obtain a zero-curvature representation of this equation we must consider the generation of the Kaup-Newell spectral problem. It is the pair of linear equations (6.3.37) with U-V-matrices

$$\hat{U}(\lambda) = \begin{pmatrix} -i\lambda^2 & \lambda\vec{q} \\ \lambda\vec{r} & i\lambda^2 \text{I} \end{pmatrix}, \quad \hat{V}(\lambda) = \begin{pmatrix} A & \vec{B} \\ \vec{C} & \hat{D} \end{pmatrix}. \qquad (6.3.42)$$

The condition of zero-curvature of the connection defined by these matrices, gives the following system of equations

$$\frac{\partial A}{\partial \tau} = \lambda(\vec{q}\cdot\vec{C} - \vec{r}\cdot\vec{B}), \qquad \frac{\partial \hat{D}}{\partial \tau} = \lambda(\vec{r}\otimes\vec{B} - \vec{C}\otimes\vec{q}), \qquad 6.3.43a,b)$$

$$\frac{\partial \vec{B}}{\partial \tau} + 2i\lambda^2\vec{B} + \lambda A\vec{q} - \lambda\vec{q}\cdot\hat{D} = \lambda\frac{\partial \vec{q}}{\partial \zeta} , \qquad (6.3.43c)$$

$$\frac{\partial \vec{C}}{\partial \tau} - 2i\lambda^2\vec{C} - \lambda A\vec{q} + \lambda\hat{D}\cdot\vec{r} = \lambda\frac{\partial \vec{r}}{\partial \zeta} . \qquad (6.3.43d)$$

In section 6.2.3 the explicit expressions of the matrix elements of V-matrix were obtained as the polynomials of λ. Hence, let us assume that the coefficients of V-matrix are the same polynomials:

$$A = a_0(\tau,\zeta) + a_2(\tau,\zeta)\lambda^2 + a_4(\tau,\zeta)\lambda^4, \quad \hat{D} = \hat{d}_0(\tau,\zeta) + \hat{d}_2(\tau,\zeta)\lambda^2 + \hat{d}_4(\tau,\zeta)\lambda^4,$$
$$\vec{B} = \vec{b}_1(\tau,\zeta)\lambda + \vec{b}_3(\tau,\zeta)\lambda^3, \quad \text{and} \quad \vec{C} = \vec{c}_1(\tau,\zeta)\lambda + \vec{c}_3(\tau,\zeta)\lambda^3.$$

As above, substituting these expansions into (6.3.43) and equating the coefficients of various powers of λ in the resulting equations, we obtain the system of equations for these coefficients:

$$\partial a_4 / \partial\tau = \vec{q}\cdot\vec{c}_3 - \vec{r}\cdot\vec{b}_3 \ , \quad \partial\hat{d}_4 / \partial\tau = \vec{r}\otimes\vec{b}_3 - \vec{c}_3\otimes\vec{q} \ , \tag{6.3.44a}$$
$$\partial a_2 / \partial\tau = \vec{q}\cdot\vec{c}_1 - \vec{r}\cdot\vec{b}_1 \ , \quad \partial\hat{d}_2 / \partial\tau = \vec{r}\otimes\vec{b}_1 - \vec{c}_1\otimes\vec{q} \ , \tag{6.3.44b}$$
$$\partial a_0 / \partial\tau = 0 \ , \quad \partial\hat{d}_0 / \partial\tau = 0 \ , \tag{6.3.44c}$$
$$\vec{q}a_4 - \vec{q}\cdot\hat{d}_4 + 2i\vec{b}_3 = 0 \ , \ -\vec{r}a_4 + \hat{d}_4\cdot\vec{r} - 2i\vec{c}_3 = 0 \tag{6.3.44d}$$
$$\partial\vec{b}_3 / \partial\tau + \vec{q}a_2 - \vec{q}\cdot\hat{d}_2 + 2ib_1 = 0 \ , \quad \partial\vec{c}_3 / \partial\tau - \vec{r}a_2 + \hat{d}_2\cdot\vec{r} - 2ic_1 = 0 . \tag{6.3.44e}$$
$$\partial\vec{b}_1 / \partial\tau + \vec{q}a_0 - \vec{q}\cdot\hat{d}_0 = \partial\vec{q} / \partial\zeta , \quad \partial\vec{c}_1 / \partial\tau - \vec{r}a_0 + \hat{d}_0\cdot\vec{r} = \partial\vec{r} / \partial\zeta , \tag{6.3.44f}$$

which one can consider as the equations under consideration. From (6.3.d) the functions $\vec{b}_3(\tau,\zeta)$ and $\vec{c}_3(\tau,\zeta)$ can be expressed and substituted into equations (6.3.44a). If we assume that the matrix $\hat{D}$ is a symmetric one, than equations (6.3.44a) are integrable. Thus, one can obtain $a_4(\tau,\zeta) = \alpha_4(\zeta)$, $\hat{d}_4(\tau,\zeta) = \delta_4(\tau)\hat{I}$, $\vec{b}_3(\tau,\zeta) = (\gamma_4 / 2i)\vec{q}$, $\vec{c}_3(\tau,\zeta) = (\gamma_4 / 2i)\vec{r}$, where $\gamma_4 = \delta_4(\zeta) - \alpha_4(\zeta)$ and $\delta_4(\zeta)$, $\alpha_4(\zeta)$ are the arbitrary functions of ζ. From the expressions (6.3.44e) and (6.3.44b) we can get the equations for a_2 and $\hat{d}_2$

$$\partial a_2 / \partial\tau = -(\gamma_4 / 4)\partial(\vec{q}\cdot\vec{r}) / \partial\tau , \qquad \partial\hat{d}_2 / \partial\tau = (\gamma_4 / 4)\partial(\vec{q}\otimes\vec{r}) / \partial\tau .$$

Thus,

$$a_2(\tau,\zeta) = -(\gamma_4 / 4)(\vec{q}\cdot\vec{r}) + \alpha_2(\zeta),$$
$$\hat{d}_2(\tau,\zeta) = (\gamma_4 / 4)(\vec{q}\otimes\vec{r}) + \delta_2(\zeta)\hat{I},$$

where $\alpha_4(\zeta)$ and $\delta_4(\zeta)$ are the new arbitrary functions of ζ. From (6.3.44c) it follows $a_0(\tau,\zeta) = \alpha_0(\zeta)$, $\hat{d}_0(\tau,\zeta) = \delta_0(\zeta)\hat{I}$. By using the expressions for all coefficients of polynomials A, $\vec{B}$, $\vec{C}$ and $\hat{D}$ we obtain integrable equations

$$\frac{\partial \vec{q}}{\partial \zeta} = \gamma_0 \vec{q} - i\frac{\gamma_4}{4}\frac{\partial}{\partial \tau}\left(i\frac{\partial \vec{q}}{\partial \tau} + (\vec{q}\cdot\vec{r})\vec{q} \right) - i\frac{\gamma_2}{2}\frac{\partial \vec{q}}{\partial \tau},$$

$$\frac{\partial \vec{r}}{\partial \zeta} = -\gamma_0 \vec{r} - i\frac{\gamma_4}{4}\frac{\partial}{\partial \tau}\left(-i\frac{\partial \vec{r}}{\partial \tau} + (\vec{q}\cdot\vec{r})\vec{r} \right) - i\frac{\gamma_2}{2}\frac{\partial \vec{r}}{\partial \tau},$$

where $\gamma_2 = \delta_2(\zeta) - \alpha_2(\zeta)$ and $\gamma_0 = \delta_0(\zeta) - \alpha_0(\zeta)$. If one takes into account the reduction $\vec{r} = \varepsilon \vec{q}^*$ and relations $\gamma_n = -\gamma_n^*$, $n = 0, 2, 4$, then we obtain a single non-linear equation

$$\frac{\partial \vec{q}}{\partial \zeta} + i\frac{\gamma_2}{2}\frac{\partial \vec{q}}{\partial \tau} + i\frac{\gamma_4}{4}\frac{\partial}{\partial \tau}\left(i\frac{\partial \vec{q}}{\partial \tau} + \varepsilon(\vec{q}\cdot\vec{q}^*)\vec{q} \right) - \gamma_0 \vec{q} = 0. \qquad (6.3.45)$$

Let $\delta_4(\zeta) = -\alpha_4(\zeta) = 2i$ and $\delta_0(\zeta) = \delta_2(\zeta) = \alpha_4(\zeta) = \alpha_2(\zeta) = 0$. From equation (6.3.45) the standard form of the ν-DNLS equation follows

$$i\frac{\partial \vec{q}}{\partial \zeta} + \frac{\partial}{\partial \tau}\left(\frac{\partial \vec{q}}{\partial \tau} - i\varepsilon(\vec{q}\cdot\vec{q}^*)\vec{q} \right) = 0. \qquad (6.3.46)$$

In this case V-matrix will be defined by coefficients

$$A = 2i\lambda^4 - i\varepsilon(\vec{q}\cdot\vec{q}^*)\lambda^2, \qquad \hat{D} = -2i\lambda^4\hat{1} + i\varepsilon(\vec{q}\otimes\vec{q}^*)\lambda^2,$$

$$\vec{B} = 2\vec{q}\lambda^3 + \left(i\frac{\partial \vec{q}}{\partial \tau} + \varepsilon(\vec{q}\cdot\vec{q}^*)\vec{q} \right)\lambda, \qquad \vec{C} = 2\vec{q}^*\lambda^3 - \left(i\frac{\partial \vec{q}^*}{\partial \tau} - \varepsilon(\vec{q}\cdot\vec{q}^*)\vec{q}^* \right)\lambda.$$

It should be remarked that the analogous result was obtained by Morris and Dodd [247]. They take the following matrix elements of V-matrix

$$A = -9i\lambda^4 - i\varepsilon(\vec{q}\cdot\vec{q}^*)\lambda^2, \qquad \hat{D} = 2i\varepsilon(\vec{q}\otimes\vec{q}^*)\lambda^2,$$

$$\vec{B} = 3\vec{q}\lambda^3 + \left(i\frac{\partial \vec{q}}{\partial \tau} + \varepsilon\frac{2}{3}(\vec{q}\cdot\vec{q}^*)\vec{q} \right)\lambda, \qquad \vec{C} = 3\varepsilon\vec{q}^*\lambda^3 - \left(i\varepsilon\frac{\partial \vec{q}^*}{\partial \tau} - \frac{2}{3}(\vec{q}\cdot\vec{q}^*)\vec{q} \right)\lambda.$$

In this case the ν-DNLS equation becomes

$$i\frac{\partial \vec{q}}{\partial \zeta} + \frac{\partial}{\partial \tau}\left(\frac{\partial \vec{q}}{\partial \tau} - i\varepsilon\frac{2}{3}(\vec{q}\cdot\vec{q}^*)\vec{q} \right) = 0.$$

It should be kept in mind that the reduction $\vec{r} = \varepsilon\vec{q}^*$ was used here.

6.4. Variational approach to soliton phenomena and perturbation method

The IST method allows to solve the Cauchy problem for the NLS equation, however practically this way is very difficult and it seldom leads to the final cause. The spectral problem of the IST method can been solved exactly for very a small number of potentials, which are the initial conditions for the Cauchy problem. It goes without saying the approximate approaches are attractive ones to investigate the non-linear wave evolution. Amongst the approximate analytical methods of study of the differential equations the variational methods are known. It is for instance the Ritz method. The variational method was used broadly in the works by Whitham. Its results are presented in detail in the book [249]. Using this method for certain rather broad class of initial conditions, Anderson et al. [203-205] obtain a number of useful results describing the behaviour of the optical pulse, which is governed by the non-linear Schrödinger equation. In the following works [206-214] this approach was prolonged on the non-integrable versions of generalised NLS. It is useful, however, to begin from the most simplest case.

6.4.1. FORMATION OF SCALAR NLS-SOLITON

So, let the complex pulse envelope $q(\zeta,\tau)$ be governed by the non-linear Schrödinger equation ($s^2=1$):

$$i\frac{\partial q}{\partial \zeta}+s\frac{\partial^2 q}{\partial \tau^2}+\mu|q|^2 q=0.$$

This equation is the Euler-Lagrange equation for a variational problem with action functional

$$\mathbf{S}[q]=\int_0^\infty d\zeta\int_{-\infty}^\infty d\tau\mathcal{L}[q]=\int_0^\infty d\zeta\mathbf{L}[q],$$

where Lagrangian density $\mathcal{L}[q]$ is given by the expression

$$\mathcal{L}[q]=\frac{i}{2}\left(q\frac{\partial q^*}{\partial \zeta}-q^*\frac{\partial q}{\partial \zeta}\right)+s\left|\frac{\partial q}{\partial \zeta}\right|^2-\frac{\mu}{2}|q|^4.$$

Lagrangian density integrated on “time” $\mathbf{L}[q]$ is a Lagrangian function.

In the Ritz optimisation procedure the variation of the action functional is made to vanish within a set of suitable chosen functions. These functions are considered as a trial functions of the considered variational problem. Here we are interested only in solitary waves vanishing at infinity. This problem can be named as the problem of “bright” soliton, unlike "dark" soliton, for which asymptotic is not zero at infinity. It is supposed that trial functions may be represented by the following ones

$$\bar{q}(\zeta,\tau) = a(\zeta)\,\mathrm{sech}[Y(\zeta,\tau)]\exp\{i\Phi(\zeta,\tau)\}, \qquad (6.4.1)$$

where

$$Y(\zeta,\tau) = [\tau - \tau_C(\zeta)]/\tau_p(\zeta),$$

$$\Phi(\zeta,\tau) = \phi(\zeta) + C(\zeta)[\tau - \tau_C(\zeta)] + b(\zeta)[\tau - \tau_C(\zeta)]^2,$$

where the real normalised amplitude $a(\zeta)$, the normalised pulse duration $\tau_p(\zeta)$ = $t_p(z)/t_{p0}$, the centre of mass of a pulse $\tau_C(\zeta)$, the normalised frequency shift $C(\zeta)$, the frequency chirp $b(\zeta)$ and phase $\phi(\zeta)$ are allowed to vary with the distance of propagation. Substituting this trial function into the Lagrangian density and integrating with respect to τ, we obtain Lagrangian function restricted on the class of these trial functions:

$$\mathbf{L}[q] = 2\tau_p a^2\left[\frac{d\phi}{d\zeta} - C\frac{d\tau_C}{d\zeta} + \frac{\pi^2\tau_p^2}{12}\frac{db}{d\zeta}\right] + 2s\tau_p a^2\left[\frac{1}{3\tau_p^2} + C^2 + \frac{\pi^2\tau_p^2}{3}b^2\right] - \frac{2\mu}{3}\tau_p a^4.$$

The action $\mathbf{S}[q]$ is restricted on the chosen class of trial functions also. This reduced action is a functional of the small number of the variables: $a(\zeta)$, $\tau_p(\zeta)$, $\tau_C(\zeta)$, $C(\zeta)$, $b(\zeta)$ and $\phi(\zeta)$. Calculation of all necessary variational derivatives reduces the Euler-Lagrange equations to a system of the ordinary differential equations for the parameters of the trial functions. After a number of simplifying transformations, the system of these equations becomes

$$\frac{d(\tau_p a^2)}{d\zeta} = 0\,, \quad \frac{dC}{d\zeta} = 0\,, \qquad (6.4.2)$$

$$\frac{d\tau_C}{d\zeta} = 2sC, \qquad (6.4.3)$$

$$\frac{d\tau_p}{d\zeta} = 4sb\tau_p, \qquad (6.4.4)$$

$$\frac{db}{d\zeta} = \frac{4s}{\pi^2}\left(\tau_p^{-4} - \pi^2 b^2\right) - \frac{2\mu a^2}{\pi^2\tau_p^2}. \qquad (6.4.5)$$

Except for these equations, there is the equation for the phase shift $\phi(\zeta)$

$$\frac{d\phi}{d\zeta} - sC^2 + \frac{2s}{3\tau_p^2} - \frac{5\mu a^2}{6} = 0, \qquad (6.4.6)$$

which we can integrate after the determination $a(\zeta)$, $\tau_p(\zeta)$, $\tau_C(\zeta)$, $C(\zeta)$ and $b(\zeta)$.

The system of equations (6.4.2)-(6.4.5) allows to consider an evolution of an initial pulse envelope, which can be described by such a trial function. Foremost, it shows a limit of possibilities of this method. If we take other symmetrical in relation to the own argument function instead of the secant in the expression (6.4.1), then the resulting system of equations will differ from (6.4.2)-(6.4.5) by only numerical values of coefficients into right parts of these equations. Expanding the class of trial functions, it is possible to perfect this approach. But the qualitative results we may try to get within the framework of the available approach.

As an example of using this variational approach, we shall consider the optical soliton formation from the initial pulse under assumption that it is sufficiently smooth. We also assume that the initial pulse can be approximated by the formula (6.4.1) and it has no initial phase modulation. From equations (6.4.2) and (6.4.3) it follows that phase and group velocities of the pulse are not changed. Besides, we obtain the integral of motion, which will be identified as "energy". In fact, for the NLS equation the canonical Hamiltonian of the this equation is energy. However, we shall use this term for the sake of convenience only.

$$\tau_p a^2 = W = \text{const}. \qquad (6.4.7)$$

Introduction of the new variables:

$$X = \tau_p(\zeta), \quad \Psi = \pi b(\zeta), \quad \xi = s(\pi/4)\zeta,$$

allows us to rewrite equations (6.4.4) and (6.4.5) in the form

$$\frac{dX}{d\xi} = X\Psi, \quad \frac{d\Psi}{d\xi} = \left(X^{-4} - \Psi^2\right) - \gamma X^{-3}, \qquad (6.4.8)$$

where $\gamma = s\mu W/2$.

This system of equations has a point of rest on the phase plane (X, Ψ) with coordinates $(\gamma^{-1}, 0)$. As far as the pulse duration is a positive value, the negative values γ are inadmissible. Consequently the stable optical pulse will not exist in the region of the normal second-order group-velocity dispersion for self-focusing media. Opposite, in the region of anomalous dispersion such pulse exists. The linear analysis of stability shows that this point of rest is the "centre". There is no asymptotic stability in this case. That is due to the considered class of trial functions, which do not contain dispersive waves that carry away part of the optical radiation energy. In the chosen approximation the pulse duration changes periodically in increasing of the distance. But actually these oscillations fade, and stationary pulse appears. That corresponds to soliton solution of NLS equation. Integral curves of the system (6.4.8) are defined by equation

$$\frac{d\Psi}{dX} = \frac{X^{-4} - \gamma X^{-3} - \Psi^2}{X\Psi}.$$

The solution of this equation can be written as

$$\Psi^2(X) = C_0 X^{-2} + 2\gamma X^{-3} - X^{-4}.$$

As the left part of this expression is positive, these curves do not exist for all values of parameter γ and integrating constant C_0.

The further discussion of the solutions of system (6.4.8) is convenient to do after excluding the Ψ. It results in one equation

$$\frac{d^2X}{d\xi^2} = \frac{1}{X^3} - \frac{\gamma}{X^2},$$

which may be interpreted as an Newton equation for a point-particle on X axis. Integrating this equation leads to the conservation law of the "total energy of this particle"

$$(dX / d\xi)^2 + U(X) = U_0 , \tag{6.4.9}$$

where $U_0 = 1 - 2\gamma + \Psi^2(0)$. Here the first term in the left part presents "kinetic energy", the second summand corresponds to "potential energy" and

$$U(X) = X^{-2} - 2\gamma X^{-1}.$$

The graph $U(X)$ at $\gamma = 8$ is presented in Fig. 6.4.1.

This potential also appears in the Kepler problem, where the second term corresponds to the attraction between two material points, so we can use the familiar results [251] for this problems. Integration (6.4.9) gives rise to

$$\xi = \int \frac{X dX}{\sqrt{U_0 X^2 + 2\gamma X - 1}}. \tag{6.4.10}$$

If $U_0 \geq 0$, the moving of material point is infinite, that means that optical pulse duration is unlimitedly increases, and its amplitude decreases. Finite motion takes place under condition $U_0 < 0$. In this case the variations of pulse duration are limited. Hence there is a constrain of the positive parameter γ. This constrain means that the energy of pulse must provide $2\gamma > [1 + \pi^2 b^2(0)]$.

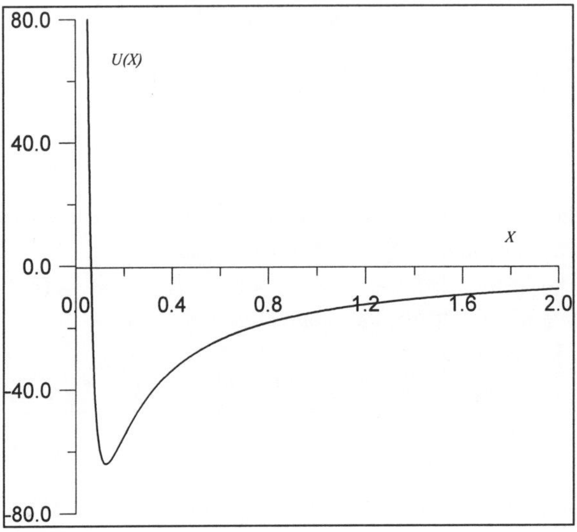

Fig.6.4.1

Potential energy of point particle on X axis

Let $U_0 < 0$. We can introduce the new variables φ as $X = X_0(1 + \hat{e}\sin\varphi)$, where $X_0 = \gamma |U_0^{-1}|$ and $\hat{e} = \sqrt{1 - |U_0|\gamma^{-2}}$ is excentrisitet of the orbit of material point in the Kepler problem. Integration of the equation (6.4.9) leads to the parametric form of the solution

$$\xi - \xi_0 = X_0 |U_0|^{-1/2} [\varphi - \hat{e}\cos\varphi], \tag{6.4.11a}$$

$$X = X_0(1 + \hat{e}\sin\varphi). \tag{6.4.11b}$$

Here ξ_0 is a new integrating constant defined from the initial conditions. For instance, we can take that $X = 1$ at $\xi = 0$. The variation of the pulse duration with the distance increasing is shown in Fig.6.4.2, where: (a) $\gamma = 5\,(\hat{e} = 4/5)$ and (b) $\gamma = 2\,(\hat{e} = 0.5)$. In these graphs the co-ordinate ς is defined by the expression $\varsigma = |U_0|^{-1/2} X_0^{-1}\xi$.

For the case of infinite motion, i.e., $U_0 \geq 0$, the similar way allows to find solution of (6.4.9) in the form

$$\xi - \xi_0 = X_0 U_0^{-1/2} \left[\hat{e} \sinh \varphi - \varphi \right], \tag{6.4.12a}$$

$$X = X_0 (\hat{e} \cosh \varphi - 1), \tag{6.4.12b}$$

where now $\hat{e} = \sqrt{1 + U_0 \gamma^{-2}}$ and $X_0 = \gamma U_0^{-1}$. The case of the defocusing non-linear medium and the anomalous group velocities dispersion (or medium focusing and dispersion normal) the parameter γ is negative.

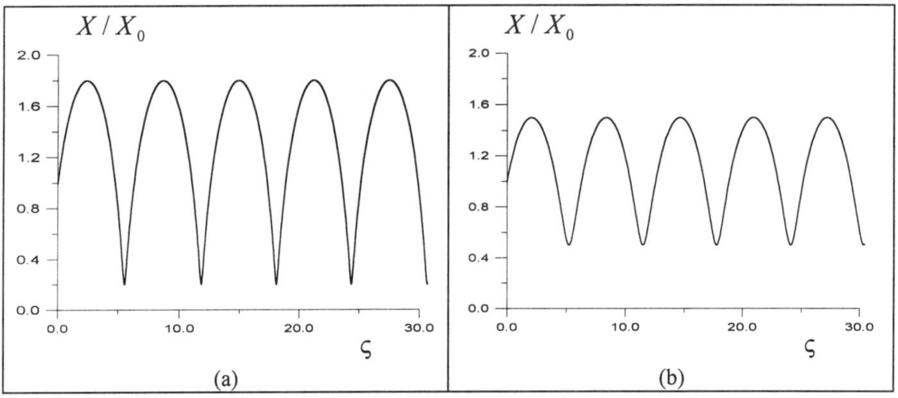

Fig. 6.4.2 The pulse duration vs normalised distance

It corresponds to a Kepler problem with repulsing. In this case the solution of (6.4.9)

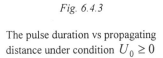

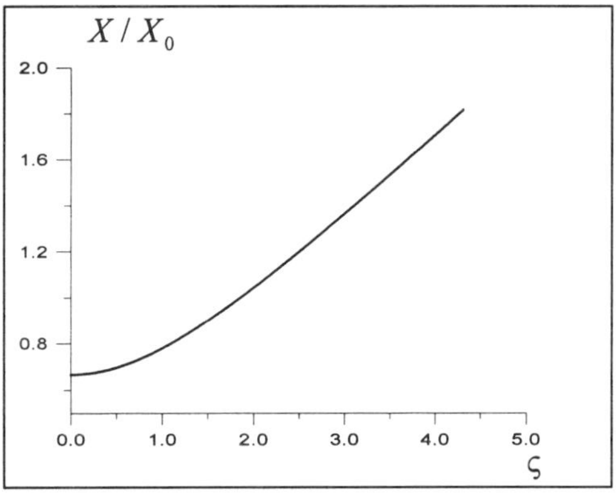

Fig. 6.4.3

The pulse duration vs propagating distance under condition $U_0 \geq 0$

can be obtained in parametric form

$$\xi - \xi_0 = X_0 U_0^{-1/2} \left[\hat{e} \sinh \varphi + \varphi \right], \tag{6.4.13a}$$

$$X = X_0 (\hat{e} \cosh \varphi + 1). \tag{6.4.13b}$$

Like the linear limit corresponding to $U_0 \geq 0$ and the pulse propagation in the defocusing non-linear medium, the pulse duration increases with propagating distance. This behaviour is illustrated by the plot in Fig.6.4.3, where $\gamma = 0.25\ (\hat{e} = 3)$ was exploited.

Thus, we obtain the threshold power (or energy) of the optical pulse to form the soliton P_{th} (E_{th}) from the condition $\mu W_{th} = 1 + \pi^2 b^2(0)$. It is worth noting that the initial phase modulation leads to increasing these values.

6.4.2. SCALAR PERTURBED NONLINEAR SCHRÖDINGER EQUATION

Let the complex pulse envelope $q(\zeta, \tau)$ be defined by the non-linear Schrödinger equation with the perturbation

$$i \frac{\partial q}{\partial \zeta} + s \frac{\partial^2 q}{\partial \tau^2} + \mu |q|^2 q = R[q]. \tag{6.4.14}$$

Here $R[q]$ is an arbitrary function of $q(\zeta, \tau)$ and its time derivatives. Under condition of $R[q] = 0$ the equation (6.4.14) has soliton solutions. If $R[q]$ is a weak perturbation, then the solution of the equation (6.4.14) in the form of a solitary wave can be considered as one, which is close to soliton.

This case can be considered within the framework of the variational approach, as this has been made in works [204-212, 252]. The derived here equations for the parameters of solitary wave contain the equations of the adiabatic perturbation theory for soliton of the NLS equation as a particular case. Thereby the variation approach may be considered as the simplest generalisation of the soliton adiabatic perturbation theory [196,198-202]. But just the same system of equations can be obtained in another way, which was named by the integrals of motion method. The variation method was used in the case of unperturbed NLS equation for the determination of parameters of the soliton formed from different initial conditions [205,206].

The system of equations of adiabatic perturbation theory is derived from the IST. However, it is known [253] that a simpler way to obtain it is to apply the equations for the integrals of motion, which are not constants in the presence of perturbation terms. Along this approach, it is necessary to substitute a soliton solution of a completely integrable equation into the exact integro-differential equations. It leads to approximative equations describing the soliton parameters. Another similar method, the method of moments, has been used by authors of work [254-256] at the analysis of self-focusing phenomena.

If it becomes necessary to study the behaviour of a solitary wave in the system which is approximately integrable (or for which nothing is known about integrability), then one can attempt to use the method of integrals of motion in combination with a moment method [248,257]. The method may be named a "*roughening procedure*". The transition from the evolution equation describing the system with the infinite degrees of freedom to a system of ordinary differential equations for finite number of variable is reminiscent of the transition from a microscopic description of a system to a macroscopic one. The analogue of macroscopic variables are the parameters of solitary wave, which is essence of meaning of the term "*roughening*".

Integrals of motion for the NLS equation

When $R[q] = 0$, the NLS equation has an infinite number of the integrals of motion (under suitable boundary conditions) [14-17,62]. These integrals can be obtained by using the standard method represented in section 3.2.5 (or in [17]). However, it is convenient to use the expression (3.2.43a) to find the recursion relation for integrals. The spectral problem of the IST method in the case of NLS equation is the linear system of equations (6.1.30). If we introduce the complex function $\Gamma = \psi_2 / \psi_1$, where one assumes $\psi_1 \neq 0$, then (6.1.30) results in

$$\frac{\partial \Gamma}{\partial \tau} = 2i\lambda\Gamma - q\Gamma^2 - q^* .$$

This equation is similar to the vector equation (4.2.29a). Hence, new results can be obtained by substituting (q, q^*, Γ) into (4.2.29) and (4.2.42) instead of $(Q_a, -R_a, \Gamma_a)$. So, we obtain the pseudo-potential $\Gamma(q, q^*; \lambda)$ as the power series of $(2i\lambda)^{-1}$:

$$\Gamma(q, q^*; \lambda) = \sum_{n=1}^{\infty} (2i\lambda)^{-n} \Gamma^{(n)}[q, q^*].$$

The coefficients of this series $\Gamma^{(n)}[q, q^*]$ is determined from the recursion relation

$$\Gamma^{(n+1)} = \frac{\partial \Gamma^{(n)}}{\partial \tau} + \sum_{m=1}^{n-1} \Gamma^{(m)} \Gamma^{(n-m)} q, \quad n \geq 3$$

$$\Gamma^{(1)} = q^*, \quad \Gamma^{(2)} = \frac{\partial q^*}{\partial \tau}.$$

As we have $\lim B = 0$, $\lim A = A^{\pm}$ as $\tau \to \pm\infty$ and $A^+ = A^-$, then the quantities

$$I_n = \int_{-\infty}^{+\infty} (\Gamma^{(n)} q) d\tau, \quad n = 1,2,3,...$$

form an infinite set of the integrals of motion. So, we have the first three integrals

$$I_1 = \int_{-\infty}^{+\infty} q q^* d\tau, \quad I_2 = \frac{1}{2} \int_{-\infty}^{+\infty} \left(q \frac{\partial q^*}{\partial \tau} - q^* \frac{\partial q}{\partial \tau} \right) d\tau,$$

$$I_3 = \int_{-\infty}^{+\infty} \left(q \frac{\partial^2 q^*}{\partial \tau^2} + (q q^*)^2 \right) d\tau .$$

Only integrals I_1 and I_2 hold interest. The integral I_3 is the Hamiltonian of the system describing by the non-linear Schrödinger equation.

Roughening procedure
For the perturbed NLS equation these integrals are not conserved. One can verify directly with the help of (6.4.14) that the following equations hold

$$\frac{d I_1}{d\zeta} = i \int_{-\infty}^{\infty} \left(q R^* - q^* R \right) d\tau , \tag{6.4.15a}$$

$$\frac{d I_2}{d\zeta} = -i \int_{-\infty}^{\infty} \left(\frac{\partial q}{\partial \tau} R^* + \frac{\partial q^*}{\partial \tau} R \right) d\tau \tag{6.4.15b}$$

The integrands in expressions of I_1 and I_2 can be understood as a certain distribution. The moments of these distribution are of interest since they are related to parameters of the solitary wave. These important moments are

$$D_1 = \int_{-\infty}^{\infty} \tau |q|^2 d\tau \qquad D_2 = \int_{-\infty}^{\infty} (\tau - \tau_C)^2 |q|^2 d\tau$$

$$M = \int_{-\infty}^{\infty} (\tau - \tau_C) \left(q^* \frac{\partial q}{\partial \tau} - q \frac{\partial q^*}{\partial \tau} \right) d\tau$$

where $\tau_C = D_1/I_1$ – the centre of mass of the "distribution", $|q(\zeta,\tau)|^2$ – is the number density of particles in the theories of solitons of the NLS equation. Using (6.4.14), we find equations, which specify the changes in these integrals as a function of ζ:

$$\frac{dD_1}{d\zeta} = 2isI_2 - i\int_{-\infty}^{\infty}\tau\left(q^*R - qR^*\right)d\tau \tag{6.4.16a}$$

$$\frac{dD_2}{d\zeta} = -2isM - i\int_{-\infty}^{\infty}(\tau-\tau_C)^2\left(q^*R - qR^*\right)d\tau \tag{6.4.16b}$$

$$\begin{aligned}\frac{dM}{d\zeta} &= 2I_2\frac{d\tau_C}{d\zeta} + i\int_{-\infty}^{\infty}\left(q^*R + qR^*\right)d\tau + \\ &+ i\int_{-\infty}^{\infty}\left(4s\left|\frac{\partial q}{\partial \tau}\right|^2 - \mu|q|^4\right)d\tau + 2i\int_{-\infty}^{\infty}(\tau-\tau_C)\left(R^*\frac{\partial q}{\partial \tau} - R\frac{\partial q^*}{\partial \tau}\right)d\tau.\end{aligned} \tag{6.4.16c}$$

We need also to supplement these equations with the following identity [249]

$$\begin{aligned}\int_{-\infty}^{\infty}\left(q\frac{\partial q^*}{\partial \zeta} - q^*\frac{\partial q}{\partial \zeta}\right)d\tau &= 2i\int_{-\infty}^{\infty}\left(s\left|\frac{\partial q}{\partial \tau}\right|^2 - \mu|q|^4\right)d\tau \\ &+ i\int_{-\infty}^{\infty}\left(qR^* + q^*R\right)d\tau\end{aligned} \tag{6.4.17}$$

which allows us to determine the law governing the phase of the pulse envelope.

Starting from the system of exact relations (6.4.15), (6.4.16) and (6.4.17), we can construct the systems of equations for parameters of a solitary wave with the various degrees of accuracy. If we assume that the solitary wave retains the shape of a soliton of the NLS equation as it evolves, then we can obtain the equations of adiabatic perturbations theory for solitons [196,198-202]. The only distinction from these equations which we need to modify integrand in the equation for the phase of soliton.

To get a more accurate description of the evolution of the solitary wave, we should choose a more general trial function $\tilde{q}(\zeta,\tau)$. As in section 6.4.1. let us choose the trial function as follows

$$\tilde{q}(\zeta,\tau) = a(\zeta)\,\mathrm{sech}[Y(\zeta,\tau)]\exp\{i\Phi(\zeta,\tau)\}, \tag{6.4.18}$$

where

$$Y(\zeta,\tau) = [\tau - \tau_C(\zeta)]/\tau_p(\zeta),$$

$$\Phi(\zeta,\tau) = \phi(\zeta) + C(\zeta)[\tau - \tau_C(\zeta)] + b(\zeta)[\tau - \tau_C(\zeta)]^2,$$

then we can obtain the system of equation for parameters $a(\zeta)$, $\tau_p(\zeta)$, $\tau_C(\zeta)$, $C(\zeta)$, $b(\zeta)$ and $\phi(\zeta)$. These equations are generalised ones of adiabatic perturbations theory for solitons.

Using (6.4.18), we can evaluate the integrals in (6.4.15) and (6.4.16):

$$I_1 = 2\tau_p a^2, \quad I_2 = -2i\tau_p a^2 C, \quad D_1 = 2\tau_C\tau_p a^2,$$
$$D_2 = (\pi^2/6)(\tau_p a^2)\tau_p^2, \quad M_1 = i(2\pi^2/3)(\tau_p a^2)\tau_p^2 b.$$

From (6.4.15a) and the expression for I_1 we find the equation

$$\frac{d(\tau_p a^2)}{d\zeta} = \tau_p a \int_{-\infty}^{\infty} \mathrm{sech}(y)\,\mathrm{Im}(\rho)dy, \tag{6.4.19a}$$

where $\rho = R[q]\exp(-i\Phi)$. Now considering I_2, from (6.4.15b) and (6.4.19a) we find the equation

$$\frac{dC}{d\zeta} = -(\tau_p a)^{-1}\int_{-\infty}^{\infty} \tanh(y)\,\mathrm{Re}(\rho)dy + 2b\tau_p a^{-1}\int_{-\infty}^{\infty} y\,\mathrm{sech}(y)\,\mathrm{Im}(\rho)dy, \tag{6.4.19b}$$

Using (6.4.16a) and (6.4.19a) we find an equation for $\tau_C(\zeta)$:

$$\frac{d\tau_C}{d\zeta} = 2sC + \tau_p a^{-1}\int_{-\infty}^{\infty} y\,\mathrm{sech}(y)\,\mathrm{Im}(\rho)dy. \tag{6.4.19c}$$

From (6.4.16b) and (6.4.19a) we can obtain the equation for $\tau_p(\zeta)$:

$$\frac{d\tau_p}{d\zeta} = 4sb\tau_p - \tau_p(2a)^{-1}\int_{-\infty}^{\infty}\left(1 - \frac{12y^2}{\pi^2}\right)\mathrm{sech}(y)\,\mathrm{Im}(\rho)dy. \tag{6.4.19d}$$

And from (6.4.16c), using everything derived so far, we get the equation for $b(\zeta)$:

$$\frac{db}{d\zeta} = \frac{4s}{\pi^2}\left(\tau_p^{-4} - \pi^2 b^2\right) - \frac{2\mu a^2}{(\pi\tau_p)^2} -$$
$$- 6(\pi^2 a\tau_p^2)^{-1}\int_{-\infty}^{\infty}[y\tanh(y) - 1/2]\,\mathrm{sech}(y)\,\mathrm{Re}(\rho)dy. \tag{6.4.19e}$$

The equation for the phase $\phi(\zeta)$ follows from identity (6.4.17):

$$\frac{d\phi}{d\zeta} = C\frac{d\tau_C}{d\zeta} - \frac{\pi^2\tau_p^2}{12}\frac{db}{d\zeta} - sC^2 + \frac{2\mu a^2}{3} - \frac{s}{3}\left(\tau_p^{-2} + \pi^2 b^2 \tau_p^2\right) - \\ - (2a)^{-1}\int_{-\infty}^{\infty}\text{sech}(y)\,\text{Re}(\rho)dy, \tag{6.4.19g}$$

The system of equations (6.4.19) make it possible to describe evolution of an optical pulse (and not only solitons) for the NLS equation with perturbation, (6.4.14), in the generalised adiabatic approximation adopted here, (6.4.18). As a particular case this system of equations contains the equations of the known adiabatic perturbations theory for solitons, which results from (6.4.19) if we discard (6.4.19d) and (6.4.19e) and if we impose the conditions $b = 0$ and $\tau_p a = 1$ in the remaining equations.

6.4.3. VECTOR PERTURBED NONLINEAR SCHRÖDINGER EQUATION

The similar approach to study the evolution of the solitary non-linear waves can be developed also for the vector NLS equation with perturbations

$$i\frac{\partial q_1}{\partial \zeta} + s_1\frac{\partial^2 q_1}{\partial \tau^2} + \mu\left(|q_1|^2 + r|q_2|^2\right)q_1 = R_1[q_1, q_2], \\ i\frac{\partial q_2}{\partial \zeta} + s_2\frac{\partial^2 q_2}{\partial \tau^2} + \mu\left(|q_2|^2 + r|q_1|^2\right)q_2 = R_2[q_1, q_2]. \tag{6.4.20}$$

Here we shall consider the variational approach by Anderson and Lisak [205,258] as an alternative to the methods of the integrals of motion and of the moments. We shall generalise a system of equations (6.4.19) on the vector NLS equation (in particular case of two-component one). It is important to note that the system of equations (6.4.20) can be obtained as Euler-Lagrange equations in the variational problem with Lagrangian density $\mathcal{L} = \mathcal{L}_0 + \mathcal{L}_1$, where

$$\mathcal{L}_0 = \frac{i}{2}\sum_{n=1,2}\left[q_n\frac{\partial q_n^*}{\partial \zeta} - q_n^*\frac{\partial q_n}{\partial \zeta}\right] + \sum_{n=1,2} s_n\left|\frac{\partial q_n}{\partial \tau}\right|^2 - \\ - \frac{\mu}{2}\left(|q_1|^4 + |q_2|^4 + 2r|q_1|^2|q_2|^2\right), \tag{6.4.21a}$$

$$\mathcal{L}_1 = \sum_{n=1,2}\left(q_n R_n^* + q_n^* R_n\right) \tag{6.4.21b}$$

In calculating the variational derivatives $\delta\mathcal{L}/\delta q_n$ and $\delta\mathcal{L}/\delta q_n^*$ we have to assume that R_1 and R_2 are constant [205]. The action functional $\mathbf{S}[\vec{q}]$ is defined by

$$\mathbf{S}[\vec{q}] = \int_0^\infty d\zeta \int_{-\infty}^\infty d\tau \mathcal{L}[\vec{q}] = \int_0^\infty d\zeta \mathbf{L}[\vec{q}],$$

and $\mathbf{S}[\vec{q}]$ can be calculated for the trial functions $\bar{q}_n(\zeta,\tau)$, where

$$\bar{q}_n(\zeta,\tau) = a_n(\zeta)\,\mathrm{sech}[Y(\zeta,\tau)]\exp\{i\Phi_n(\zeta,\tau)\},$$

$$Y(\zeta,\tau) = [\tau - \tau_C(\zeta)]/\tau_p(\zeta),$$
$$\Phi_a(\zeta,\tau) = \phi_a(\zeta) + C(\zeta)[\tau - \tau_C(\zeta)] + b(\zeta)[\tau - \tau_C(\zeta)]^2. \tag{6.4.22}$$

This trial function corresponds to a two-component solitary wave, which is close to vector soliton of completely integrable version of the vector NLS equation, rather than interaction of two such solitary waves described by the two-solution solution of this equation.

The calculation of $\delta\mathbf{S}[\bar{q}_1,\bar{q}_2]$ now reduces to the determination of variational derivatives of the Lagrangian function $\mathbf{L}[\bar{q}_1,\bar{q}_2] = \mathbf{L}_0[\bar{q}_1,\bar{q}_2] + \mathbf{L}_1[\bar{q}_1,\bar{q}_2]$ with respect to the parameters $a_n(\zeta)$, $\tau_p(\zeta)$, $\tau_C(\zeta)$, $C(\zeta)$, $b(\zeta)$ and $\phi_n(\zeta)$, which depend on ζ. If we use (6.4.22) and the definition of $\mathbf{L}[\bar{q}_1,\bar{q}_2]$, we find that

$$\mathbf{L}_0[\bar{q}_1,\bar{q}_2] = 2t_p\left(\sum_n a_n^2\right)\left[\frac{\pi^2\tau_p^2}{12}\frac{\partial b}{\partial\zeta} - C\frac{\partial\tau_C}{\partial\zeta}\right] + 2\tau_p\sum_n a_n^2\frac{\partial\phi_n}{\partial\zeta} + \frac{2\mu\tau_p}{3}\sum_n \bar{a}_n^2 a_n^2 +$$
$$+ 2\tau_p\left(\sum_n s_n a_a^2\right)\left[\frac{1}{3\tau_p^2} + C^2 + \frac{\pi^2}{3}b^2\tau_p^2\right],$$
$$\mathbf{L}_1[\bar{q}_1,\bar{q}_2] = \sum_n a_n \int_{-\infty}^\infty \left[R_n\exp(-i\Phi_n) + R_n^*\exp(i\Phi_n)\right]\mathrm{sech}(y)dy$$

where $\bar{a}_1^2 = a_1^2 + ra_2^2$, $\bar{a}_2^2 = a_2^2 + ra_1^2$.

Calculation of all necessary variational derivatives reduces the Euler-Lagrange equations to the system of equations for parameters of the trial function (6.4.22). After a number of simplifying transformations, this system becomes

$$\frac{d(\tau_p a_n^2)}{d\zeta} = \tau_p a_n \int_{-\infty}^\infty \mathrm{sech}(y)\,\mathrm{Im}(\rho_n)dy, \tag{6.4.23a}$$

$$W\frac{dC}{d\zeta} = -\sum_n a_n \int_{-\infty}^{\infty} \tanh(y)\,\mathrm{sech}(y)\,\mathrm{Re}(\rho_n)dy + \\ + 2b\tau_p^2 \sum_n a_n \int_{-\infty}^{\infty} y\,\mathrm{sech}(y)\,\mathrm{Im}(\rho_n)dy, \tag{6.4.23b}$$

$$W\frac{d\tau_C}{d\zeta} = 2\tau_p \sum_n s_n a_n^2 C + \tau_p^2 \sum_n a_n \int_{-\infty}^{\infty} y\,\mathrm{sech}(y)\,\mathrm{Im}(\rho_n)dy, \tag{6.4.23c}$$

$$W\frac{d\tau_p}{d\zeta} = 4\sum_n s_n a_n^2 b\tau_p^2 - \frac{\tau_p^2}{2}\sum_n a_n^2 \int_{-\infty}^{\infty}\left(1-\frac{12y^2}{\pi^2}\right)\mathrm{sech}(y)\,\mathrm{Im}(\rho_n)dy, \tag{6.4.23d}$$

$$W\frac{db}{d\zeta} = \frac{4}{\pi^2}\left(\sum_n s_n \tau_p a_n^2\right)\left(\tau_p^{-4} - \pi^2 b^2\right) - \frac{2\mu}{\pi^2\tau_p}\sum_n \bar{a}_n^2 a_n^2 - \\ - 6(\pi^2\tau_p)^{-1}\sum_n a_n \int_{-\infty}^{\infty}[y\tanh(y) - 1/2]\,\mathrm{sech}(y)\,\mathrm{Re}(\rho_n)dy, \tag{6.4.23e}$$

$$\frac{d\phi_n}{d\zeta} = C\frac{d\tau_C}{d\zeta} - \frac{\pi^2\tau_p^2}{12}\frac{db}{d\zeta} - \sigma_n C^2 + \frac{2\mu\bar{a}_n^2}{3} - \frac{s_n}{3}\left(\tau_p^{-2} + \pi^2 b^2 \tau_p^2\right) - \\ - (2\bar{a}_n^2)^{-1}\int_{-\infty}^{\infty}\mathrm{sech}(y)\,\mathrm{Re}(\rho_n)dy. \tag{6.4.23g}$$

The following designations are used in the above equations:

$$\rho_n = R_n[\vec{q}]\exp(-i\Phi_n) \quad \text{and} \quad W = \tau_p \sum_{n=1,2} a_n^2 .$$

This system of equations includes the Anderson's equations [205] for optical soliton, (if we assume that $R_a[\vec{q}] = 0$ and $q_2 = 0$) and the equations of the adiabatic perturbations theory for solitons [202], if we assume that $q_2 = 0$, $b(\zeta) = 0$ and $\tau_p a = 1$ for all values of ζ. In general, this system of equations yields rough description of the evolution of an optical polarised pulse or a two-mode scalar solitary wave, which is similar to simulton.

6.4.4. COLLAPSE OF OPTICAL SOLITONS

We shall now consider the examples of application of an perturbation theory discussed above to analyse the solitary wave evolution describing the propagation of the optical pulse in a strongly non-linear fibre [259-262]. It was found [261] that an optical pulse can be self-compressed due to high-order nonlinearity. Hereafter we use designation for the sign of second-order group-velocity dispersion $s = -\mathrm{sgn}\,\alpha_2$.

It should be noted that when the optical pulse begins to compress, so that its duration approaches zero, a theory based on the NLS equation becomes inadequate. In this case it is necessary to take into account higher-order derivatives of the pulse envelope and the same is true of non-linear effects. However, if we are interested in the time dependence of the pulse parameters close to the threshold determining collapse, then it is permissible to restrict the treatment to this model.

Collapse of the scalar picosecond optical solitons

It is useful to consider the collapse of optical solitons as an example, in which an analysis of the evolution of a solitary wave on the basis of the system of equations (6.4.19) leads to a result which differs from the adiabatic soliton perturbation theory result.

We assume that the optical pulse propagates in non-linear waveguide with strongly nonlinearity and the following equations can be used

$$i\frac{\partial q}{\partial \zeta}+s\frac{\partial^2 q}{\partial \tau^2}+\mu|q|^2 q+\mu_5|q|^4 q=0, \qquad (6.4.24)$$

where $\mu_5=(2\pi\omega_0^2 L_D/c^2\beta)\chi_{eff}^{(5)}A_0^4$, and the effective fifth-order non-linear susceptibility $\chi_{eff}^{(5)}$ is defined by the expression

$$\chi_{eff}^{(5)}=\int\chi^{(5)}(x,y)\Psi^6(x,y)dxdy\left(\int\Psi^2(x,y)dxdy\right)^{-1}.$$

This is the cubic-quintic Nonlinear Schrödinger equation that has been considered in section 6.1.6. We now take a look at this equation as the NLS with perturbation term. In this case the perturbation term is $R[q]=-\mu_5|q|^4 q$.

Since we have $\mathrm{Im}\rho=0$ and $\mathrm{Re}\rho=-\mu_5 a^5\,\mathrm{sech}^5 y$, we find a conservation law from (6.4.19a)

$$\tau_p a^2=W=\mathrm{const}.$$

If we define the energy of the optical pulse as

$$G=\frac{1}{4\pi}\int_{-\infty}^{+\infty}|\mathcal{E}|^2\,dt,$$

then the one-soliton pulse energy is

$$G_{sol}=\frac{t_{p0}A_0^2 S_{eff}}{2\pi},$$

where $S_{eff} = \int \Psi^2(x,y)dxdy$ is effectively cross-section of the fibre. In terms of these parameters W is either the energy ration G/G_{sol}, or the power ration P/P_{sol}, where P_{sol} is the power of the fundamental optical soliton (i.e., one-soliton solution of the NLS equation).

The other equations from (6.4.19) become

$$\frac{dC}{d\zeta} = 0, \qquad \frac{d\tau_C}{d\zeta} = 2sC, \tag{6.4.25a)}$$

$$\frac{d\tau_p}{d\zeta} = 4sb\tau_p, \tag{6.4.25b)}$$

$$\frac{db}{d\zeta} = \frac{4s}{\pi^2}\left(\tau_p^{-4} - \pi^2 b^2\right) - \frac{2\mu W}{\pi^2 \tau_p^3} - \frac{2\tilde{\mu}_5 W^2}{\pi^2 \tau_p^4}, \tag{6.4.25c)}$$

with $\tilde{\mu}_5 = (16\mu/15)$. Making the substituting $X = \tau_p(\zeta)$, $\xi = s(\pi/4)\zeta$, and eliminating the variable $b(\zeta)$ from (6.4.25c), we find an equation for X

$$\frac{d^2X}{d\xi^2} = \frac{(1-\delta)}{X^3} - \frac{\gamma}{X^2}, \tag{6.4.26}$$

where $\delta = s\tilde{\mu}_5 W/2$ and $\gamma = s\mu W/2$. Equation (6.4.26) leads to

$$\left(\frac{dX}{d\xi}\right)^2 + \frac{(1-\delta)}{X^2} - \frac{2\gamma}{X} = U_0, \tag{6.4.27}$$

where $U_0 = 1 - \delta - 2\gamma + [\pi b(0)]^2$.

Equation (6.4.26) describes the motion of a material point in one-dimensional space in a potential $U(X)$ in accordance with Newton's law, where

$$U(X) = \frac{(1-\delta)}{X^2} - \frac{2\gamma}{X}.$$

More precisely, that is the Kepler problem [251]. With $\delta = 0$ equations (6.4.26) and (6.4.27) correspond to the case studied in [205,206] and discussed above in section 6.4.1. If the initial parameters of a solitary wave are such that we have $U_0>0$, the pulse duration will increase (this is an analogue of an open orbit in the Kepler problem). If $U_0<0$, the normalised pulse duration periodically changes with increasing ξ. Here we have an analogue of closed orbits in the Kepler problem. If we introduce $X_0 = \gamma/|U_0|$

and $p_0 = [1-(1-\delta)|U_0|\gamma^{-2}]^{1/2}$, then the solution of equation (6.4.26) corresponding to the finite motion of the point particle can be represented parametrically as

$$X = X_0(1+p_0\sin\varphi), \quad \xi = \xi_0 + |U_0|^{-1/2} X_0(\varphi - p_0\cos\varphi). \qquad (6.4.28)$$

These expressions coincide with those found previously (see equations (6.4.11)) where parameter p_0 is the renormalised excenrtisitet. The implicit dependence of the pulse duration on space co-ordinate ξ can be obtained from (6.4.28). It is

$$(|U_0|)^{1/2}\xi = X_0\left\{\arcsin\left(\frac{X-X_0}{X_0 p_0}\right) - \arcsin\left(\frac{1-X_0}{X_0 p_0}\right) + \right.$$
$$\left. + p_0\left[1-\left(\frac{1-X_0}{X_0 p_0}\right)^2\right]^{1/2} - p_0\left[1-\left(\frac{X-X_0}{X_0 p_0}\right)^2\right]^{1/2}\right\}. \qquad (6.4.29)$$

Under condition $\delta > 1$ (or $p_0 > 1$), however, there is a case of "falling on a centre", i.e., a collapse of the optical pulse. The critical value $\delta_c = 1$ is reached when the energy W is equal to the threshold value

$$W_{col} = \sqrt{2s/\tilde{\mu}_5}\,. \qquad (6.4.28)$$

We are assuming s>0 and μ_5>0 here.

Fig. 6.4.4

The self-compression (collapse) of the optical pulse ($p_0 = 1.02$)

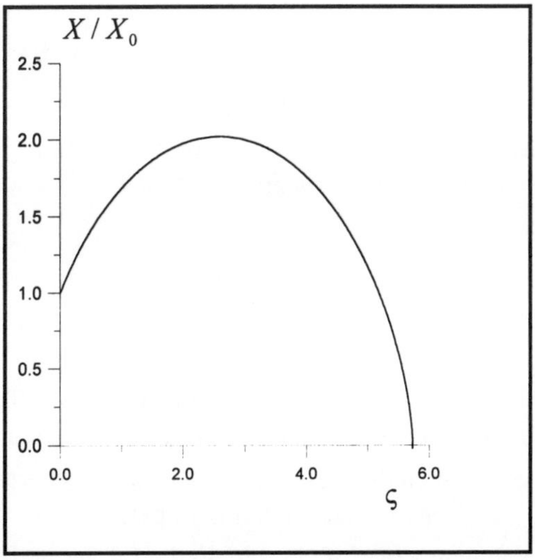

We would like to point out that pulse duration drops to zero in a finite length of the fibre. The expression defining this length $z_{col} = 4L_D\xi_{col}/\pi$ can be found from the expression (6.4.29) by assuming that $X = 0$:

$$\xi_{col} = \frac{X_0}{(|U_0|)^{1/2}}\left\{\arcsin\left(\frac{1-X_0}{X_0 p_0}\right) - \arcsin\left(\frac{1}{p_0}\right) + \right. \\ \left. + p_0\left[1-\left(\frac{1-X_0}{X_0 p_0}\right)^2\right]^{1/2} - p_0\left[1-\left(\frac{1}{p_0}\right)^2\right]^{1/2}\right\}. \tag{6.4.30}$$

Such a self-compression (collapse) of an optical pulse due to a high-order nonlinearity as in (6.4.24) was pointed out by Azimov *et al.* [261]. Note that this result does not follow in the adiabatic soliton perturbation theory. In ref. [263] it has been observed that the dissipation of the electromagnetic wave can been obstacle to such a process.

Collapse of the scalar femtosecond optical solitons

The ultrashort (i.e., femtosecond) pulse propagation in non-linear fibre with account of dispersion of non-linear susceptibility is described by the modified NLS equation [83-93]. Let us consider the model equation discussed in section 6.2.1:

$$i\frac{\partial q}{\partial \zeta} + s\frac{\partial^2 q}{\partial \tau^2} + \mu |q|^2 q + i\mu_2\frac{\partial}{\partial \tau}\left(|q|^2 q\right) + i\mu_3 q\frac{\partial}{\partial \tau}\left(|q|^2\right) = 0. \tag{6.4.31}$$

We have now

$$\operatorname{Im}\rho = (3\mu_2 + 2\mu_3)\tau_p^{-1}a^3 \tanh y \operatorname{sech}^3 y,$$
$$\operatorname{Re}\rho = \mu_2 a^3 (C + 2b\tau_p y)\operatorname{sech}^3 y.$$

The conservation law $\tau_p a^2 = W = \text{const}$ follows from the equation (6.3.19a). The other equations can be rewritten with account of this result as

$$\frac{dC}{d\zeta} = \frac{4}{3}(\mu_2 - \mu_3)Wb\tau_p^{-1}, \tag{6.4.32a}$$

$$\frac{d\tau_p}{d\zeta} = 4sb\tau_p, \tag{6.4.32b}$$

$$\frac{db}{d\zeta} = \frac{4s}{\pi^2}\left(\tau_p^{-4} - \pi^2 b^2\right) - \frac{2\mu W}{\pi^2 \tau_p^3} + \frac{2\mu_2 WC}{\pi^2 \tau_p^3}, \tag{6.4.32c}$$

$$\frac{d\tau_C}{d\zeta} = 2sC + \left(\mu_2 W - \frac{2}{3}\mu_3 W\right)\tau_p^{-1} \tag{6.4.32d}$$

From (6.4.32a) and (6.4.32b) we find

$$\frac{dC}{d\zeta} = -\frac{1}{3s}[(\mu_2 - \mu_3)W]\frac{d}{d\zeta}\tau_p^{-1} .$$

Hence

$$C + \frac{\tilde{\mu}W}{3s}\tau_p^{-1} = C_0 = \text{const} , \tag{6.4.33}$$

where $\tilde{\mu} = \mu_2 - \mu_3$. The integrating constant C_0 is found from the initial values of C and τ_p. Eliminating $b(\zeta)$ from (6.4.32b) and (6.4.32c) and using (6.4.33), we find the equation for the pulse duration

$$\frac{d^2\tau_p}{d\zeta^2} = \left[\left(\frac{4}{\pi}\right)^2 - \frac{2}{3}\left(\frac{2\tilde{\mu}W}{\pi}\right)^2\right]\frac{1}{\tau_p^3} - \frac{8sW(\mu - \tilde{\mu}C_0)}{\pi^2}\frac{1}{\tau_p^2} . \tag{6.3.34}$$

From this equation and the discussion above it follows that the optical pulse can collapse. The threshold energy W_{col} is now given by

$$W_{col} = \sqrt{\frac{3}{2}}\left(\frac{2s}{|\tilde{\mu}|}\right) . \tag{6.3.35}$$

Here we assume $s = -\operatorname{sgn}\alpha_2 = 1$.

Since the pulse duration is a periodic function of co-ordinate (when the initial conditions correspond to closed orbits in Kepler problem), we find from (6.3.18) that the centre of mass of the optical pulse “shakes”, i.e., there is periodic change in $\tau_C(\zeta)$ with increasing normalised co-ordinate ζ. From the expression (6.4.33) we can see that there is the shift of the carrier wave frequency.

Collapse of the vector optical solitons

Now we consider the propagation of a two-frequency optical pulse in a non-linear fibre. Two-frequency situation arises in the stimulated Raman scattering process. Both this process and the pulse propagation in non-linear bimodal fibre can be described by the system of equations (6.3.24). If the nonlinearity of fibre medium characterised by the third- and fifth-order susceptibilities, optical pulse may be collapsed under certain conditions [259-261].

We shall assume that the main contribution to the pulse broadening comes from the second-order dispersion of the group velocities and we shall ignore the effect of difference between the group velocities. The optical shock formation is take into account despite quintic nonlinearity. Thus, the system of equations under consideration becomes

$$i\frac{\partial q_1}{\partial \zeta}+s_1\frac{\partial^2 q_1}{\partial \tau^2}+\mu\left(|q_1|^2+r|q_2|^2\right)q_1+ \\ +i\tilde{\chi}\frac{\partial}{\partial \tau}\left[\left(|q_1|^2+r|q_2|^2\right)q_1\right]+\mu_5\left(|q_1|^4+\tilde{r}|q_2|^4\right)q_1=0, \tag{6.4.36a}$$

$$i\frac{\partial q_2}{\partial \zeta}+s_2\frac{\partial^2 q_2}{\partial \tau^2}+\mu\left(|q_2|^2+r|q_1|^2\right)q_2+ \\ +i\tilde{\chi}\frac{\partial}{\partial \tau}\left[\left(|q_2|^2+r|q_1|^2\right)q_2\right]+\mu_5\left(|q_2|^4+\tilde{r}|q_1|^4\right)q_2=0, \tag{6.4.36b}$$

where the overlap integrals of the mode functions for different values of the frequencies are accounted for in this system by the parameters r and $\tilde{r}$. Other parameters in system (6.4.21) were defined above.

If we introduce a matrix $\hat{\mathrm{P}}$ such that

$$\hat{\mathrm{P}}=\begin{pmatrix}0 & 1\\ 1 & 0\end{pmatrix},$$

we can then write

$$\mathrm{Re}\,\rho_n=-\mu_5 a_n\left(a_n^4+\tilde{r}\sum_k \mathrm{P}_{nk}a_k^4\right)\mathrm{sech}^5 y+ \\ +\tilde{\chi}_n a_n\left(a_n^2+r\sum_k \mathrm{P}_{nk}a_k^2\right)\left(C+2b\tau_p y\right)\mathrm{sech}^3 y, \tag{6.4.37a}$$

and

$$\mathrm{Im}\,\rho_n=3\tilde{\chi}_n a_n\left(a_n^2+r\sum_k \mathrm{P}_{nk}a_k^2\right)\tau_p^{-1}\tanh y\,\mathrm{sech}\, y, \tag{6.4.37b}$$

Substitution of these expressions into equations (6.4.23) gives the following system of equations:

$$\frac{d}{d\zeta}(\tau_p a_n^2) = 0 \tag{6.4.38a}$$

$$W\frac{dC}{d\zeta} = \frac{4}{3} b\tau_p \sum_n \tilde{\chi}_n a_n^2 \left(a_n^2 + r\sum_k \mathrm{P}_{nk} a_k^2 \right), \tag{6.4.38b}$$

$$W\frac{d\tau_p}{d\zeta} = 4b\tau_p \sum_n s_n W_n \,, \tag{6.4.38c}$$

$$W\frac{d\tau_C}{d\zeta} = 2\sum_n s_n W_n + \tau_p \sum_n \tilde{\chi}_n a_n^2 \left(a_n^2 + r\sum_k \mathrm{P}_{nk} a_k^2 \right), \tag{6.4.38d}$$

$$W\frac{db}{d\zeta} = \frac{4}{\pi^2}\left(\tau_p^{-4} - \pi^2 b^2\right)\sum_n s_n a_n^2 - \frac{2\mu}{\pi^2\tau_p^3}\left(a_1^4 + a_2^4 + 2ra_1^2 a_2^2\right) -$$
$$- \frac{2}{\pi^2\tau_p}\left(\frac{16}{15}\chi_5\right)\sum_n a_n^2\left(a_n^4 + \tilde{r}\sum_k \mathrm{P}_{nk} a_k^4 \right) - \tag{6.4.38e}$$
$$- \frac{2C}{\pi^2\tau_p}\sum_n \tilde{\chi}_n a_n^2 \left(a_n^2 + r\sum_k \mathrm{P}_{nk} a_k^2 \right).$$

The notation $W_n = \tau_p a_n^2$ is used here and it follows from equation (6.4.38a) that W_n are constant quantities governed by the initial conditions at $\zeta = 0$: $W = W_1 + W_2$. The equation for the phases $\phi_n(\zeta)$ is not used because these phases do not occur in system (6.4.38) that describes the parameters of a solitary wave under consideration.

It is convenient to introduce a new dispersion constant $\bar{s}$ and the effective interaction constants $\bar{\chi}, \bar{\mu}, \bar{\mu}_5$:

$$\bar{s} = (s_1 W_1 + s_2 W_2)W^{-1}, \quad \bar{\mu} = \mu(W_1^2 + W_2^2 + 2rW_1W_2)W^{-2},$$
$$\bar{\chi} = \left[\tilde{\chi}_1(W_1^2 + rW_1W_2) + \tilde{\chi}_2(W_2^2 + rW_1W_2)\right]W^{-2},$$
$$\bar{\mu}_5 = (16\mu_5/15)[W_1(W_1^2 + \tilde{r}W_2^2) + W_2(W_2^2 + \tilde{r}W_1^2)]W^{-3}.$$

The system of equations (6.4.38) can now be rewritten in a more convenient form:

$$\frac{dC}{d\zeta} = \frac{4\bar{\chi}}{3} Wb\tau_p^{-1}\,, \tag{6.4.39a}$$

$$\frac{d\tau_p}{d\zeta} = 4\bar{s}b\tau_p\,, \tag{6.4.39b}$$

$$\frac{db}{d\zeta}=\frac{4\bar{s}}{\pi^2}\left(\tau_p^{-4}-\pi^2 b^2\right)-\frac{2(\bar{\mu}-\bar{\chi}C)W}{\pi^2\tau_p^3}-\frac{2\bar{\mu}_5 W^2}{\pi^2\tau_p^4} \tag{6.4.39c}$$

$$\frac{d\tau_C}{d\zeta}=2\bar{s}C+\bar{\chi}W\tau_p^{-1} \tag{6.4.39d}$$

From (6.4.39a) and (6.4.39b) we obtain again

$$C+\frac{\bar{\chi}W}{3\bar{s}}\tau_p^{-1}=C_0=\text{const}, \tag{6.4.40}$$

The integration constant C_0 is governed by the values of $C(\zeta)$ and $\tau_p(\zeta)$ at $\zeta=0$. Eliminating $b(\zeta)$ from equations (6.4.39b) and (6.4.39c) and allowing for equation (6.4.40) gives the equation alike (6.3.34) for $\tau_p(\zeta)$:

$$\frac{d^2\tau_p}{d\zeta^2}=\left[\left(\frac{4\bar{s}}{\pi}\right)^2-\frac{8}{3}\left(\frac{\bar{\chi}W}{\pi}\right)^2-\frac{8\bar{s}\bar{\mu}_5 W^2}{\pi^2}\right]\frac{1}{\tau_p^3}-\frac{8\bar{s}W(\bar{\mu}-\bar{\chi}C_0)}{\pi^2}\frac{1}{\tau_p^2} \tag{6.4.41}$$

So, we conclude that the conditions

$$1-\frac{2}{3}\left(\frac{\bar{\chi}W}{2\bar{s}}\right)^2-\frac{\bar{\mu}_5 W^2}{2\bar{s}}-\frac{W(\bar{\mu}-\bar{\chi}C_0)}{2\bar{s}}+[\pi b(0)]^2=0$$

can be used to determine the normalised pulse energy W_{sol} the excess of which can lead to the optical soliton formation. Strictly speaking it is not a soliton, since the initial system of equations is not completely integrable, but a solitary wave which does not spread out. If $W > W_{sol}$, the solution of the equation (6.4.41) describes a periodic variation of the pulse duration near some value t_s that corresponds to the point of rest of the system (6.4.39).

Let us define the parameter ρ as

$$\rho=1-\frac{2}{3}\left(\frac{\bar{\chi}W}{2\bar{s}}\right)^2-\frac{\bar{\mu}_5 W^2}{2\bar{s}}.$$

This parameter depends on the normalised pulse energy W and may be negative. If $\rho<0$, equation (6.4.26) describes collapse, i.e., the pulse duration tends to zero. Thus, critical value of the parameter $\rho_{col} = 0$ is obtained when the energy op the initial pulse is varied from W_{sol} to W_{col}, where

$$W_{sol}^2 = 6\left(\frac{\bar{s}}{\bar{\chi}}\right)\left(1+\frac{3\bar{s}\bar{\mu}_5}{\bar{\chi}^2}\right)^{-1}. \tag{6.4.42}$$

Thus, if $W \leq W_{col}$, the duration of the two-frequency pulse propagating in a strongly non-linear medium varies periodically. For $W > W_{col}$ the value $t_p(z)$ (or $\tau_p(\zeta)$) drops to zero in a finite time, i.e. in a finite length of the waveguide.

It should be pointed out that expression (6.2.42) contains the effective parameters of a non-linear dispersive medium and, since $W_{col} \propto |\bar{s}|$, we can select the frequencies of the carrier wave so as to obtain the dispersions for each wave with opposite signs, and by selecting partial pulse energy W_1 and W_2, we can try to reduce $|\bar{s}|$ as much as possible. Consequently, the collapse threshold of a two-frequency pulse can be considerably less than that of a single-frequency pulse.

6.4.5. PROPAGATION OPTICAL PULSE IN NONLINEAR BIREFRINGENT FIBRE

The propagation of short optical pulses in a non-linear birefringent single-mode fibre is one example of the phenomenon described by the vector NLS equation, which is not completely integrable [142-149]. Polarisation dynamics of light pulses in this case have been investigated in the framework of the soliton perturbation theory [155] and the Hamiltonian formalism [156]. There are some specific solutions [145,152,158-160] and numerical results [151,153,154]. The variational method discussed above allows to observe the evolution of the polarisation states analytically. So, we consider this problem for the sake of illustration applicability of this method according to [157].

The system of equations governing the polarisation components of the electric field was represented in section 6.3.1. That is

$$i\frac{\partial q^{(+)}}{\partial \zeta} - s\frac{\partial^2 q^{(+)}}{\partial \tau^2} + \delta q^{(-)} + \frac{1}{3}\mu\left(|q^{(+)}|^2 + 2|q^{(-)}|^2\right)q^{(+)} = 0, \tag{6.4.43a}$$

$$i\frac{\partial q^{(-)}}{\partial \zeta} - s\frac{\partial^2 q^{(-)}}{\partial \tau^2} + \delta q^{(+)} + \frac{1}{3}\mu\left(2|q^{(+)}|^2 + |q^{(-)}|^2\right)q^{(-)} = 0. \tag{6.4.43b}$$

It should be remind that the $q^{(+)} = q_1 + iq_2$, and $q^{(-)} = q_1 - iq_2$ are the right and left circularly polarised normalised envelopes of electric field of the optical pulse.

The system of these equations is expressible in the form of the Euler-Lagrange equations, when the Lagrangian density is chosen in the following form

$$\mathcal{L} = \frac{i}{2}\left(q^{(+)}\frac{\partial q^{(+)*}}{\partial\zeta} - q^{(+)*}\frac{\partial q^{(+)}}{\partial\zeta}\right) + \frac{i}{2}\left(q^{(-)}\frac{\partial q^{(-)*}}{\partial\zeta} - q^{(-)*}\frac{\partial q^{(-)}}{\partial\zeta}\right) - s\left|\frac{\partial q^{(+)}}{\partial\tau}\right|^2 - s\left|\frac{\partial q^{(-)}}{\partial\tau}\right|^2$$

$$-\frac{\mu}{6}\left(|q^{(+)}|^4 + |q^{(-)}|^4 + 2|q^{(+)}|^2|q^{(-)}|^2\right) - \delta\left(q^{(+)*}q^{(-)} + q^{(-)*}q^{(+)}\right) .$$

Thus, we can use the results of section 6.4.3, i.e., the equations (6.4.23) to write the relevant system of equations. By assuming the $\tau_C(\zeta) = C(\zeta) = 0$, we get

$$\frac{d}{d\zeta}(\tau_p a_1^2) = 2\delta\tau_p a_1 a_2 \sin\Phi , \tag{6.4.44a}$$

$$\frac{d}{d\zeta}(\tau_p a_2^2) = -2\delta\tau_p a_1 a_2 \sin\Phi , \tag{6.4.44b}$$

$$\frac{d}{d\zeta}\left[\tau_p^3(a_1^2 + a_2^2)\right] = 4s\tau_p^3 b(a_1^2 + a_2^2) \tag{6.4.44c}$$

$$\frac{\pi^2\tau_p^2}{b}\frac{db}{d\zeta} = \frac{s}{3\tau_p^2} - \frac{s\pi^2\tau_p^2 b^2}{3} - \mu\frac{a_1^4 + a_2^4 + 2ra_1^2a_2^2}{18(a_1^2 + a_2^2)} , \tag{6.4.44d}$$

$$\frac{d}{d\zeta}\Phi = \frac{2\mu(1-r)(a_1^2 - a_2^2)}{18} + \delta\frac{(a_1^2 - a_2^2)}{a_1 a_2}\cos\Phi , \tag{6.4.44e}$$

where $\Phi = \phi_1 - \phi_2$, $r = 2$. Functions ϕ_1 and ϕ_2 can be determined from the following equations

$$\frac{d\phi_1}{d\zeta} = -\frac{1}{2}\left(\frac{s}{3\tau_p^2} + \frac{s\pi^2\tau_p^2 b^2}{3} + \frac{\pi^2\tau_p^2}{6}\frac{db}{d\zeta}\right) - \frac{2\mu(a_1^2 + ra_2^2)}{18} - \delta\frac{a_2}{a_1}\cos\Phi ,$$

$$\frac{d\phi_2}{d\zeta} = -\frac{1}{2}\left(\frac{s}{3\tau_p^2} + \frac{s\pi^2\tau_p^2 b^2}{3} + \frac{\pi^2\tau_p^2}{6}\frac{db}{d\zeta}\right) - \frac{2\mu(a_2^2 + ra_1^2)}{18} - \delta\frac{a_1}{a_2}\cos\Phi .$$

Equations (6.4.44a) and (6.4.44b) provide the first integral

$$\tau_p(\zeta)(a_1^2 + a_2^2) = W^2 = \text{const} \tag{6.4.45}$$

From (6.4.44c) with the help of this integral one may obtain

$$\frac{d}{d\zeta}\tau_p = 2s\tau_p b \tag{6.4.46}$$

By introducing new variables

$$f_{1,2}(\zeta) = \tau_p^{1/2}(\zeta)a_{1,2}(\zeta)/W ,$$

and taking integral (6.4.45) into account the equations (6.4.44a) – (6.4.44e) can be written in a more convenient form:

$$\frac{df_1}{d\zeta} = \delta f_2 \sin\Phi , \tag{6.4.47a}$$

$$\frac{df_2}{d\zeta} = -\delta f_1 \sin\Phi , \tag{6.4.47b}$$

$$\frac{d\Phi}{d\zeta} = \delta\frac{f_2^2 - f_1^2}{f_1 f_2}\cos\Phi + \frac{2\mu(1-r)W^2}{9\tau_p}(f_2^2 - f_1^2), \tag{6.4.47c}$$

$$\frac{d\tau_p}{d\zeta} = 2s\tau_p b , \tag{6.4.47d}$$

$$\frac{db}{d\zeta} = \frac{2s}{\pi^2}\left(\frac{1}{\tau_p^4} - \pi^2 b^2\right) - 2\mu W^2 \frac{f_1^4 + f_2^4 + 2rf_1^2 f_2^2}{3\pi^2\tau_p^3} . \tag{6.4.47e}$$

It should be noticed that in the linear limit, when $\mu W^2 \to 0$, equations (6.4.47a)–(6.4.47c) and equations (6.4.47d) and (6.4.47e) constitute systems of equations independent of each other. That means that the variation of the optical pulse duration and the depth of phase modulation are not associated with the evolution of the polarisation state of this pulse. The polarisation state is characterised by two component vector $\mathbf{f} = (f_1 , f_2)$. The scale distance at which noticeable pulse broadening takes place is determined from system of equations (6.4.47d) and (6.4.47e). This is the normalised dispersion length $l_D = \pi\tau_{p0}^2/2$, where τ_{p0} is the initial pulse duration. The normalised distance at which the polarisation vector $\mathbf{f}$ turns through 2π is defined from the system of equations (6.4.47a)–(6.4.47c): $l_B = 2\pi/\delta$. The l_B is named the (normalised) beat length.

If the power of the initial pulse is sufficiently high, i.e., the normalised Kerr length l_K is of the order of l_B or l_D, then the non-linear effects should not be ignored and system of equations (6.4.47) should be considered fully. The exact analytical solution of this system is unlikely to be found. However, if $l_D >> l_B$ or $l_D << l_B$ the approximate solutions can be proposed.

Let the dispersion length be small with regards to the beat length, i.e., $l_D << l_B$. At the distances scaled by l_D the right hand parts of equations (6.4.47a) and (6.4.47b) are close to zero. Therefore the polarisation state can be considered as not altered. At the same time optical pulse duration and phase modulation may evolve significantly resulting, in particular, in a soliton. As soon as the trial solution (6.4.22) is of soliton-like form there is no use to discuss multi-soliton solutions for the reason that the class of trial functions (6.4.22) does not contain any multi-soliton solutions of the initial problem. The expression enclosed in parentheses in the second term in (6.4.47e) can be replaced by $1-2(1-r)f_1^2 f_2^2$ with the constraint of the equality $f_1^2+f_2^2=1$. As $f_{1,2}<1$ the product $f_1^2 f_2^2$ is small compared with unity and it can be omitted or replaced by the product of 'average' values, i.e., by accepting $f_1^2=f_2^2=1/2$.

For the first variant of the estimation system, equations (6.4.47d) and (6.4.47e) can be converted into the ones not containing f_1 and f_2 :

$$\frac{d\tau_p}{d\zeta}=2s\tau_p b,\quad \frac{db}{d\zeta}=2s(\tau_p^{-4}-\pi^2 b^2)/\pi^2-2\kappa W^2/\pi^2\tau_p^3, \tag{6.4.48}$$

In terms of the variable $X(\zeta)=\tau_p(\zeta)/\tau_{p0}$ the equations (6.4.48) yield

$$\frac{d^2X}{d\zeta^2}=\frac{\gamma}{X^3}-\frac{\nu}{2X^2}, \tag{6.4.49}$$

where $\gamma=4/\pi^2=l_D^{-2}$ and $\nu=(8s\mu W^2/\pi^2)$. Equation (6.4.49) has the integral of motion

$$\left(\frac{dX}{d\zeta}\right)^2+\frac{\gamma}{X^2}-\frac{\nu}{X}=G_0, \tag{6.4.50}$$

where $G_0=(4b_0+\gamma-\nu)$, $b_0=b(0)$. The equations (6.4.49) and (6.4.50) has been discussed in section 6.4.4, where the collapse of optical pulses was considered. Hence, we can now use all necessary results of this section. The case of $\nu<0$ corresponds to an unlimited broadening of the initial pulse over the range of normal group-velocity dispersion. In this frequency band the parameter ν is negative and a soliton does not exist. We can see this result consistent with that is obtained in the framework of variational approach. If $\nu>0$, which relates to the range of anomalies dispersion, variation variable X is limited. That means the optical pulse keeps a finite duration by oscillating around steady magnitude $\tau_s=1/\mu W^2$. It is seen from equations (6.4.48) that this steady pulse is not phase modulated. The oscillation of the pulse duration is not consistent with the numerical results because the trial function (6.4.22) does not account radiation part of the non-soliton solution of the initial equations (6.4.43).

As long as the scale of polarisation state evolution l_B exceeds l_D significantly, then by assigning $\tau_p(\zeta)$ in (6.4. a)-(6.4. c) to be equal to τ_s one can obtain a complete system of equations determining f_1, f_2 and Φ. Let us introduce the notations $f_2(\zeta) = f(\zeta)$ and $f_1(\zeta) = [1 - f^2(\zeta)]^{1/2}$, then the equations (6.4.47b) and (6.4.47c) can be written as

$$\frac{df}{d\zeta'} = -(1 - f^2)^{1/2} \sin\Phi, \tag{6.4.51a}$$

$$\frac{d\Phi}{d\zeta'} = -\frac{2f - 1}{f(1 - f^2)^{1/2}} \cos\Phi + m(1 - 2f^2), \tag{6.4.51b}$$

where $\zeta' = \delta\zeta = 2\pi\zeta / l_B$, $m = 2\mu(1 - r)W^2 / 3\tau_s\delta$. Boundary conditions are chosen in the form

$$f(\zeta = 0) = 0, \quad \Phi(\zeta = 0) = \pi / 2. \tag{6.4.52}$$

From (6.4.51) it follows

$$\frac{d\cos\Phi}{df} = \frac{2f - 1}{f(1 - f^2)} \cos\Phi + m\frac{(1 - 2f^2)}{(1 - f^2)^{1/2}}.$$

Solution of the above equation is

$$\cos\Phi = (m/2)f(1 - f^2)^{1/2}. \tag{6.4.53}$$

Let us introduce the new variable ψ by the formula $f(\zeta') = \sin(\psi / 2)$. Then equation (6.4.51a) and equality (6.4.53) allow to formulate the equation for ψ:

$$\frac{d\psi}{d\zeta'} = -2\left[1 - (m/4)^2 \sin^2\psi\right]^{1/2}.$$

Hence it follows the expression for ζ' through the elliptic integral of the first kind [243]

$$-2\zeta' = \mathrm{F}(\psi, |m| / 4). \tag{6.4.54}$$

Equality (6.4.54) yields

$$\cos\psi(\zeta') = \mathrm{cn}(2\zeta', |m|/4) \quad \text{and} \quad \sin\psi(\zeta') = \mathrm{sn}(2\zeta', |m|/4),$$

so that the functions f_1 and f_2 can be expressed through the Jacobian elliptic functions by the formulas:

$$f_1^2(\zeta) = \frac{1}{2}[1 + \mathrm{cn}(2\delta\zeta, |m|/4)], \quad f_2^2(\zeta) = \frac{1}{2}[1 - \mathrm{cn}(2\delta\zeta, |m|/4)]. \tag{6.4.55}$$

As far as the elliptic Jacobian cosine is a periodical function with a period $4\mathbf{K}(|m|/4)$ ($\mathbf{K}(x)$ is the complete elliptic integral of the first kind [243]), then as it follows from (6.4.55) each of the components of polarisation state vector $\mathbf{f}$ oscillates in space with the period Z_0, where

$$Z_0 = 2\mathbf{K}(|m|/4). \tag{6.4.56}$$

When the pulse normalised power W grows, parameter $|m|$ and Z_0 increase as well. When $|m| = |m_c| = 4$, period Z_0 becomes infinite. In this case the functions f_1 and f_2 are defined by the formulas:

$$f_1^2(\zeta) = \frac{1}{2}[1 + \mathrm{sech}(2\delta\zeta)], \quad f_2^2(\zeta) = \frac{1}{2}[1 - \mathrm{sech}(2\delta\zeta)]. \tag{6.4.57}$$

If $|m| > 4$, then either the solution of equations (6.4.51) should be searched over again, or by using the properties of Jacobian functions [243] one can try to find the expressions for f_1 and f_2 starting from (6.4.55). It results in

$$f_1^2(\zeta) = \frac{1}{2}[1 + \mathrm{dn}(|m|\delta\zeta/2, 4/|m|)], \tag{6.4.58a}$$

$$f_2^2(\zeta) = \frac{1}{2}[1 - \mathrm{dn}(|m|\delta\zeta/2, 4/|m|)]. \tag{6.4.58b}$$

In this case the oscillation period for these functions is given by the formula

$$Z_0 = 4\mathbf{K}(4/|m|)/m\delta. \tag{6.4.59}$$

which means that with increasing W^2, period Z_0 approaches zero. As it follows from (6.4.58) at $|m|>4$ the state of polarisation is being gradually frozen, i.e., maximum value of $f_2(\zeta)=f(\zeta)$ decreases:

$$\max f_2^2=\frac{1}{2}\left[1-\sqrt{1-\frac{16}{m^2}}\right].$$

It is worth noting that the polarisation behaviour discussed above was observed in numerical simulation [151,153,154].

Let the dispersion length be much greater than the beat length: $l_D >> l_B$. In this case the pulse broadening at the distance scaled by l_B is negligible and polarisation state evolution is governed by equations (6.4.47a)-(6.4.47c) at $\tau=\tau_{p0}$. Their dependence on W^2 functions f_1 and f_2 is defined by expressions (6.4.55), (6.4.57) and (6.4.58) with parameter $m=2\mu(1-r)W^2/3\tau_{p0}\delta$. However, the actual beat length herein is Z_0 and with the growth of W^2 condition $l_D >> Z_0$ can be broken. The optical pulse broadening through group-velocity dispersion causes the relaxation of a non-linear effect contribution (which is proportional to W^2) to the dynamics of $f(\zeta)$ that is formally displayed in the decrease of parameter $|m|$ and presumably to a cessation of "non-linear beat length" Z_0 growth.

It is noteworthy that if one changes the boundary condition (6.4.52) by taking

$$f(\zeta=0)=1/\sqrt{2}, \quad \Phi(\zeta=0)=0,$$

then equations (6.4.51) will describe the soliton trapping effect [160], which may be treatment in the same way.

It is convenient to present the dynamics of the polarisation state of the optical pulse propagating in a birefringent fibre by the phase plane of equations (6.4.51). In the general case these integral curves are defined by the equations

$$f\sqrt{1-f^2}\cos\Phi-(m/2)f^2(1-f^2)=C,$$

where C is any integration constant. These constants enumerate the integral curves. Fig. 6.4.5 shows pictures of the phase plane under alteration of $|m|$ round the critical value. The functions $f(\zeta)$ and $\cos\Phi(\zeta)$ are chosen as co-ordinates in phase plane because they are just what define the first integrals of the system (6.4.51). The integral curve passing through the point $(f=0, \cos\Phi=0)$ is discussed in detail above. In Fig. 6.4.5 the cases (a) and (b) correspond to the critical values of the parameter $|m|$, namely

$|m|=1$ and $|m|=3$, respectively. Figures for the case of critical value $|m|=4$ and for over critical one $|m|=6$ are shown in Fig.6.4.5 (*c*) and (*d*), respectively. We would like to draw attention to the fact that at $|m|>4$ a new stable stationary points with the co-ordinates

$$f=\sqrt{0.5[1\pm(1-4/m^2)^{1/2}]} \text{ and } \cos\Phi=1$$

appeared instead of a stable point $(f=1/\sqrt{2},\cos\Phi=1)$ which vanished. However, the point of rest with $f=1/\sqrt{2}$ and $\cos\Phi=-1$ is retained.

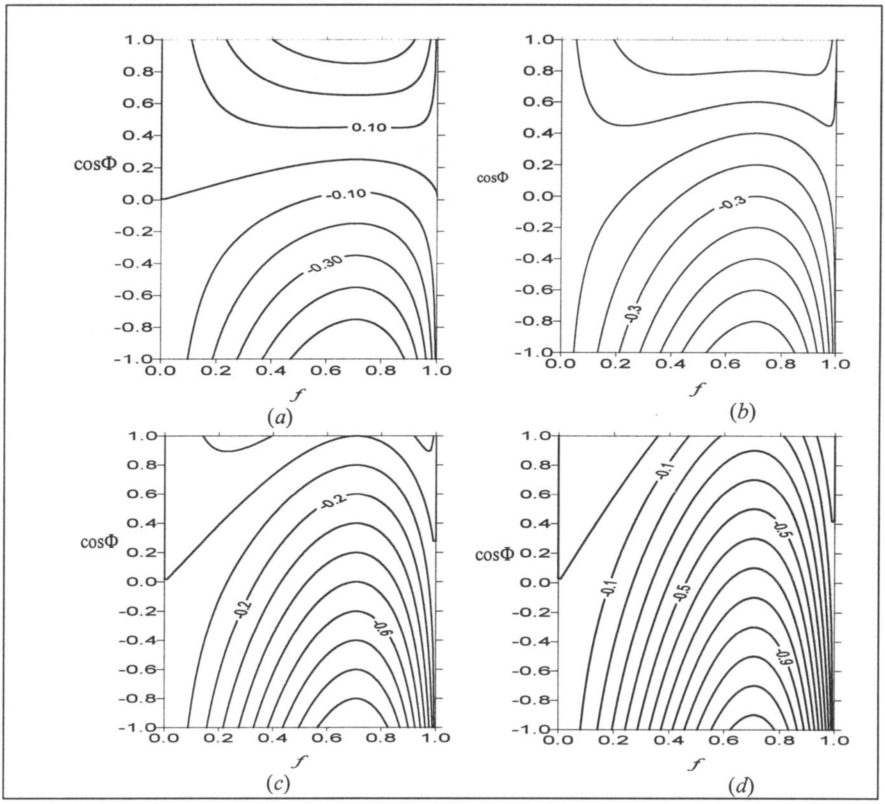

Fig. 6.4.5 Phase plane of system (6.4.51).

6.5 Conclusion

We have considered the propagation of the optical pulse in the medium, which is characterised by the weak nonlinearity, as usual it has been the cubic nonlinearity. The simplest model describing relevant phenomena is based on the Nonlinear Schrödinger equation. This equation is the completely integrable one, so the soliton solutions of this equation is the excellent mathematical tool to investigate the optical pulses in the nonlinear fibres. The higher order nonlinearity, more than Kerr-type one, and the high-order group-velocity dispersion represent the next step in the development of the fibre optics theory. Here there are a few analytical methods for investigation of the non-linear pulse propagation. The variational approach is the one of them. We have discussed the base of this method and some of its generalisation in the frame of the one-dimensional models. Recently the variational approach has been extended into the field of the two- and three-dimensional wave structures, i.e., optical beams and "optical bullets" [264]. We would like to point out that these optical structures exist in bulk non-linear media, however, they can be realised as surface waves too.

Recently, great attention has been attracted to the use of the so-called dispersion management (DM) in long-haul fibre communication systems [122-136]. The term of the *dispersion management* denotes variation of the dispersion characteristic of the optical fibre along the axis. As usual, the periodic variation of the dispersion constant is provided by the connection in line fibre pieces with the normal and anomalous dispersion of the group-velocities. The propagation of a short optical pulse in such fibre system is described by the generalised (non-uniform) cubic NLS equations

$$i\frac{\partial q}{\partial \zeta} + s(\zeta)\frac{\partial^2 q}{\partial \tau^2} + \mu |q|^2 = i\gamma(\zeta)q\,, \tag{6.5.1}$$

where $\gamma(\zeta)$ accounts for contributes from the homogeneously distributed losses and the periodically located amplifiers. The function $s(\zeta)$ periodically varies around its average value. A typical example of this function is sketched in Fig. 6.5.1.

Fig. 6.5.1.

Sketch of a non-symmetric element for dispersion compensation.

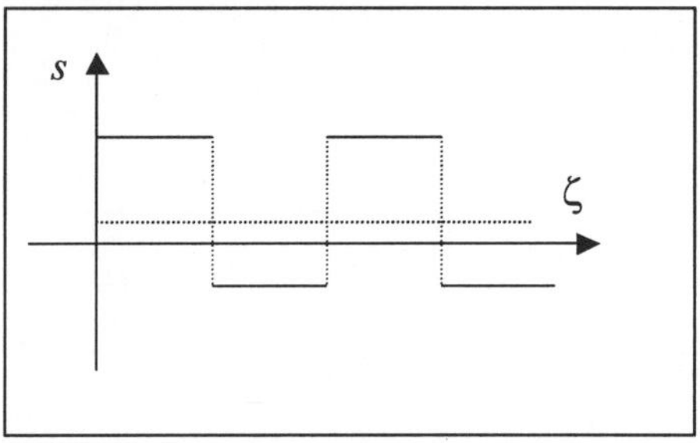

Under propagation of the optical pulse in this fibre the pulse duration periodically increases and decreases due to dispersion broadening and self-induced chirp. On the average the pulse duration increases more slowly than in the case of the homogeneous fibreline. Soliton transmission using dispersion compensating fibres has attracted great attention. The DM soliton propagation has been investigated by both the numerical simulation and experimentally. It should be pointed out that the non-uniform NLS equation (6.5.1) is not completely integrable, thus there is no soliton solution of this equation. Only approximate method to the solution of this equation can be exploited. On the basis of the variational approach, which was considered in the previous section, the effective method has been proposed [129-133, 265,266]. So, in [266] the analytical formula for the soliton power enhancement has been obtained. The analytical expression is in good agreement with numerical results, in the limit when residual dispersion and nonlinearity only slightly affected the pulse dynamics over one compensation period. The found results also allow to describe the shape of the DM soliton.

In finishing this chapter we should like to advise the following surveys and books [84, 219,220, 267,268], which represent the wide field of the solitons and solitary waves application in fibre optics.

Reference

1. Hasegawa, A., and Tappert, F.: Transmission of stationary non-linear optical pulses in dispersive dielectric fibres. I. Anomalous dispersion, *Appl.Phys.Lett.* **23** (1973), 142-144.
2. Hasegawa, A., and Tappert, F.: Transmission of stationary non-linear optical pulses in dispersive dielectric fibres. II. Normal dispersion. *Appl.Phys.Lett.* **23** (1973), 171-172.
3. Chu, P.L., Whitbread, T.: Application of solitons to communication system. *Electron.Lett.* **14** (1978), 531-532.
4. Bloom D.M., Mollenauer L.F., Lin Ch., Taylor N., and Del Gaudio A.M.: Direct demonstration of distortionless picosecond-pulse propagation in kilometre-length optical fibres, *Opt.Letts.* **4** (1979), 297-299.
5. Mollenauer, L.F., Stolen, R.H., and Gordon, J.P.: Experimental observation of picosecond pulse narrowing and solitons in optical fibres. *Phys.Rev.Letts.* **45** (1980), 1095-1098.
6. Blow, K.J., and Doran, N.J.: High bit rate communication systems using non-linear effects. *Opt.Commun.* **42** (1982), 403-406.
7. Doran, N.J., and Blow, K.J.: Solitons in optical communications, *IEEE J. Quant. Electron.* **QE-19** (1983), 1883-1888.
8. Stolen, R.H., Mollenauer, L.F., and Tomlinson, W.J.: Observation of pulse restoration at the soliton period in optical fibres, *Opt.Lett.* **8** (1983), 186-188.
9. Mollenauer, L.F., Stolen, R.H., and Islam, M.N.: Experimental demonstration of soliton propagation in long fibres: Loss compensated by Raman gain, *Opt.Lett.* **10** (1985), 229-231.
10. Jain M., and Tzoar N.: Propagation of non-linear optical pulses in inhomogeneous media, *J.Appl.Phys.* **49** (1978), 4649-4654.
11. Jain M., and Tzoar N.: Non-linear pulse propagation in optical fibres, *Opt. Lett.* **3** (1978), 202-204.
12. Bendow B., Gianino P.D., Tzoar N.J., and Jain M.: Theory of non-linear pulse propagation in optical waveguides, *J.Opt.Soc.Amer.* **70** (1980), 539-546.

13. Bendow B., and Gianino P.D.: Theory of non-linear pulse propagation in inhomogeneous waveguides, *Opt. Lett.* 4 (1979), 164-166.
14. Zakharov, V.E., and Schabat, A.B.: The exact theory of two-dimensional self-focussing and one-dimensional self-modulating of waves in nonlinear medium, *Zh.Eksp.Teor.Fiz.* **61** (1971), 118-134 [*Sov.Phys. JETP* **34** (1972), 62-69].
15. Zakharov, V.E., and Manakov, S.V.: On complete integrability of the non-linear Schrödinger equation, *Teor.Mat.Fiz.* **19** (1974), 332-343 (in Russia).
16. Satsuma, J., and Yajima, N.: Initial value problems of one-dimensional self-modulation of non-linear waves in dispersive media, *Progr. Theor. Phys. Suppl.* №55 (1974), 284-306
17. Satsuma, J.: Higher conservation laws for the Nonlinear Schrödinger Equation through Baclund transformation, *Progr.Theor.Phys.* **53** (1975), 585-586.
18. Hasegawa, A.: Generation of a train of soliton pulses by induced modulational instability in optical fibers, *Opt.Lett.* **9** (1984), 288-290.
19. Anderson, D., and Lisak, M.: Modulation instability of coherent optical-fiber transmission signals, *Opt.Lett.* **9** (1984), 468-470.
20. Tai, K., Hasegawa, A., and Tomita, A.: Observation of modulational instability in optical fibers, *Phys.Rev.Lett.* **56** (1986), 135-138.
21. Shukla, P.K., and Rasmussen, J.J.: Modulational instability of short pulses in long optical fibers, *Opt.Lett.* **11** (1986), 171-173.
22. Potasek, M.J.: Modulation instability in an extended Nonlinear Schrödinger equation, *Opt.Lett.* **12** (1987), 921-923.
23. Kothari, N.C.: Theory of modulational instability in optical fibers, *Opt.Commun.* **62** (1987), 247-249.
24. Hasegawa, A., and Tai, K.: Effects of modulational instability on coherent transmission systems, *Opt.Lett.* **14** (1989), 512-513.
25. Agrawal, G.P., Baldeck, P.L., and Alfano, R.R.: Modulation instability induced by cross-phase modulation, *Phys.Rev.* **A39** (1989), 3406-3413.
26. Sudo, S., Itoh, H., Okamoto. K., and Kubodera, K.: Generation of 5 THz repetition optical pulses by modulation instability in optical fibers, *App.Phys.Lett.* **54** (1989), 993-994.
27. Greer, E.J., Patrick, D.M., Wigley, P.G.J., and Taylor, J.R.: Generation of 2 THz repetition rate pulse trains through induced modulational instability, *Electron.Lett.* **25** (1989), 1246-1248.
28. Gouveia-Neto A.S., Faldon M.E., Sombra, A.S.B., Wigley, P.G.J., and Taylor, J.R.: Subpicosecond-pulse generation through cross-phase-modulation-induced modulational instability in optical fibers, *Opt.Lett.* **13** (1988), 901-903.
29. Boiti, M., and Pempinelli, F.: Non-linear Schrödinger equation, Backlund transformations and Painleve transcendents, *Nuovo Cimento* **B59** (1980), 40-58.
30. Giannini, J.A., and Joseph, R.I.: The role of the second Painleve transcendent in non-linear optics, *Phys.Letts.* **A141** (1989), 417-419.
31. Can, M.: On the relation between Non-linear Schrödinger Equation and Painleve-IV equation, *Nuovo Cimento* **B106** (1991), 205-207.
32. Ablowitz, M.J.: Remarks on nonlinear evolution equations and ordinary differential equations of Painleve type, *Physica* **D3** (1981), 129-141.
33. Tracy, E.R., Chen, H.H., and Lee, Y.C.: Study of quasi-periodic solutions of the Nonlinear Schrödinger equation and the nonlinear modulation instability, *Phys.Rev.Lett.* **53** (1984), 218-221.
34. Tracy, E.R., Chen, H.H., Non-linear self-modulation: An exactly solvable model, *Phys.Rev..* **A 37** (1988), 815-839.
35. Mertsching, J.: Quasiperiodic solutions of the Nonlinear Schrödinger equation, *Fortschr.Phys.* 35 (1987), 519-536.
36. Akhmediev, N.N., and Kornev, V.I.: Propagation of periodic pulse train in optical fiber, *Izv. VUZov, radiofiz.* **30** (1987) 1249-1254 (in Russia).
37. Kamchatnov, A.M.: On propagation of ultra-short periodic pulses in non-linear fibers, *Zh.Eksp.Teor.Fiz.* **97** (1990), 144-153 (in Russia).

38. Kamchatnov, A.M.: On improving the effectiveness of periodic solutions of the NLS and DNLS equations, *J.Phys.* **A 23** (1990), 2945-2960.
39. Schrader, D.: Explicit calculation of N-soliton solutions of the Non-linear Schrödinger equation. *IEEE J.Quant.Electron.* **QE-31** (1995), 2221-2225.
40. Liu, Sh.-liang, and Wang, Wen-zheng: Exact N-solitons of the extended Non-linear Schrödinger equation, *Phys.Rev.* **E 49** (1994), 5726-5730.
41. Olmedilla, E.: Multiple pole solutions of the Nonlinear Schrödinger equation, *Physica* **D 25** (1987), 330-346.
42. Fattakhov, A.M., and Chirkin, A.S.: Noise influence on the light pulse propagation in optical fibres, *Kvantov.Electron.* **10** (1983), 1989-1996 (in Russia).
43. Fattakhov, A.M., and Chirkin, A.S.: Non-linear propagation of the phase-modulated light pulses, *Kvantov.Electron.* **11** (1984), 2349-2355 (in Russia).
44. Elgin, J.N., Kaup, D.J.: Inverse scattering theory with stochastic initial potentials, *Opt.Commun.* **43** (1982), 233-236.
45. Elgin, J.N.: Inverse scattering theory with stochastic initial potentials, *Phys.Lett.* **A 110** (1985), 441-443.
46. Blow, K.J., and Wood, D.: The evolution of solitons from non-transform limited pulses, *Opt.Commun.* **58** (1986), 348-354.
47. Maimistov, A.I., Manykin, E.A., and Sklyarov, Yu.M.: Noise influence on the optical soliton formation in light-guides, *Kvantov.Electron.* **13** (1986), 2243-2248 (in Russia).
48. Vysloukh, V.A., and Fattakhov, A.M.: On non-linear compensation of dispersive broadening of noise pulses, *Izv. VUZov, radiofiz.* **29** (1986) 545-550 (in Russia).
49. Maimistov, A.I., and Sklyarov, Yu.M.: Influence of regular phase modulation on the optical solitons formation, *Kvantov.Electron.* **14** (1987), 796-803 (in Russia).
50. Belov, M.H.: Influence of the initial phase modulation on the optical soliton formation, *Kvantov.Electron.* **14** (1987), 1627-1629 (in Russia).
51. Konotop, V.V., and Vekslerchik, V.E.: Dynamics of femtosecond optical pulses. Solitons with initial phase modulation, *Phys.Lett.* **A 131** (1988), 357-360.
52. Konotop, V.V.: On influence of stochastic phase modulation on soliton formation and propagation, *Kvantov.Electron.* **16** (1989), 1032-1037 (in Russia).
53. Konotop, V.V.: On influence of initial modulation on the information transmission by solitons in FOCS, *Kvantov.Electron.* **16** (1989), 1911-1914 (in Russia).
54. Bass, F.G., Kivshar, Yu.S., Konotop, V.V., and Puzenko, S.A.: Propagation of non-linear incoherent pulses in single-mode optical fibres, *Opt.Commun.* **68** (1988), 385-390.
55. Bass, F.G., Kivshar, Yu.S., Konotop, V.V., and Pritula, G.M.: On stochastic dynamics of solitons in inhomogeneous optical fibres, *Opt.Commun.* **70** (1989), 309-314.
56. Gordon, J.P., and Mollenauer, L.F.: Phase noise in photonic communications systems using linear amplifiers, *Opt. Lett.* **15** (1990), 1351-1353.
57. Konotop, V.V.: Non-linear Schrödinger equation with random initial conditions at small correlation radius, *Phys.Lett.* **A 145** (1990), 50-54.
58. Elgin, J.N.: Stochastic perturbations of optical solitons, *Opt.Lett.* **18** (1993), 10-12.
59. Gouveia-Neto, A.S., Gomes, A.S.L., and Taylor, J.R.: Experimental study of the evolution of chirped soliton pulses in single mode optical fibres, *Opt.Commun.* **64** (1987), 383-386.
60. Gouveia-Neto, and Taylor, J.R.: Soliton evolution from noise bursts, *Electron. Lett.* **25** (1989), 736-737.
61. Dianov, E.M., Korobkin, D.B., and Strelsov, A.M.: Noise discrimination at quasi-soliton pulse propagation in single-mode light-guide, *Pizma v Zh.Eksp.Teor.Fiz.* **49** (1989), 141-143 (in Russia).
62. Ablowitz, M.J., and Segur, H.: *Solitons and the Inverse Scattering Transform*, SIAM, Philadelphia, 1981.
63. Hermansson, B., and Yevick, D.: Numerical investigation of soliton interaction, *Electron. Lett.* **19** (1983), 570-571.

64. Vysloukh, V.A.: Pulse propagation in optical fibres in the neighborhood of the dispersion minimum: influence of the high order dispersion and nonlinearity, *Kvantov.Electron.* **10** (1983), 1688-1690 (in Russia).
65. Chu, P.L., and Desem, C.: Effect of third-order dispersion of optical fiber on soliton interaction, *Electron. Lett.* **21** (1985), 228-229.
67. Potasek, M.J., Agarwal, G., and Pinault, S.C.: Analytic and numerical study of pulse broadening in nonlinear dispersive optical fibers. *J.Opt.Soc.Amer.* **B3** (1986), 205-211.
68. Agarwal, G., and Potasek, M.J.: Nonlinear pulse distortion in single-mode optical fibers at the zero-dispersion wavelength, *Phys.Rev.* **A 33** (1986), 1765-1776.
69. Wai, P.K.A., Menyuk, C.R., Lee, Y.C., and Chen, H.H.: Nonlinear pulse propagation in the neighborhood of the zero-dispersion wavelength of monomode optical fibers, *Opt.Lett.* **11** (1986), 464-466.
70. Boyer, G.R., and Carlotti, X.F.: Nonlinear propagation in a single-mode optical fiber in case of small group velocity dispersion, *Opt.Commun.* **60** (1986), 18-22.
71. Wai, P.K.A., Menyuk, C.R., Chen, H.H., and Lee, Y.C.: Soliton at the zero-group-dispersion wavelength of a single-mode fiber, *Opt.Lett.* **12** (1987), 628-630.
72. Vysloukh, V.A., and Sukhotskova, N.A.: Effect of third-order dispersion on the train picosecond pulses generation in fibre due to self-modulation instability, *Kvantov. Electron.* **14** (1987), 2371-2374 (in Russia).
73. Aleshkevich, V.A., Vysloukh, V.A., Kozhoridze, G.D., Matveev, A.N., and Terzieva, S.B.: Nonlinear propagation of the partial coherent pulse in fiber and influence of high order dispersion, *Kvantov.Electron.* **15** (1988), 325-332 (in Russia).
73. Zhao, W., and Bourkoff, E.: Femtosecond pulse propagation in optical fibers: Higher order effects, *IEEE J.Quant.Electron.* **QE-24** (1988), 365-372.
74. Wen, Senfar, and Chi, Sien: Approximate solution of optical soliton in lossless fibers with third-order dispersion, *Opt. and Quant. Electron.* **21** (1989), 335-341.
75. Uzunov, I.M., Mitev, V.M., and Kovachev, L.M.: Propagation of one-soliton pulses successions in monomode optical fibers, *Opt.Commun.* **70** (1989), 389-392.
76. Uzunov, I.M., and Mitev, V.M.: Succession of one-soliton pulses in the zero-dispersion region of silica optical fibers: the role of the phase difference between the adjacent pulses, *Opt. and Quant. Electron.* **21** (1989), 507-510.
77. Wai, P.K.A., Chen, H.H., and Lee, Y.C.: Radiation by "solitons" at the zero group-dispersion wavelength of single-mode optical fibers, *Phys.Rev.* **A 41** (1990), 426-439.
78. Mezentsev, V.K., and Turitsyn, S.K.: New class of solitons in fiber light-guide in the zero-dispersion region, *Kvantov.Electron.* **18** (1991), 610-612 (in Russia).
79. Anderson, D., Lisak, M.: Non-linear asymmetric pulse distortion in long optical fibres. *Opt.Lett.* **7** (1982), 394-396.
80. Anderson, D., Lisak, M.: Non-linear asymmetric self-phase modulation and self-steeping of pulses in long optical waveguides, *Phys.Rev.* **A 27** (1983), 1393-1398.
81. Ohkuma, K., Ichikawa, Y.H., and Abe, Y.: Soliton propagation along optical fibers, *Opt.Lett.* **12** (1987), 516-518.
82. Tzoar N., and Jain M.,: Self-phase modulation in long-geometry optical waveguides, *Phys.Rev.* **A 23** (1981), 1266-1270.
83. Christodoulides, D.N., and Joseph, R.I.: Femtosecond solitary waves in optical fibers, *Electron. Lett.* **20** (1984), 659-660.
84. Kodama, Y.: Optical solitons in a monomode fiber, *J.Stat.Phys.* **39** (1985), 597-614.
85. Kodama, Y., and Nozaki, K.: Soliton interaction in optical fibers, *Opt.Lett.* **12** (1987), 1038-1040.
86. Kodama, Y., A.Hasegawa, A.: Nonlinear pulse propagation in a monomode dielectric guide, *IEEE J.Quant.Electron.* **QE-23** (1987), 510-524.
87. Golovchenko, E.A., Dianov, E.M., Pilipetskii, A.H., Prokhorov, A.M., and Serkin, B.N.: Self-interaction and limit of compression of the femtosecond optical wave pockets in non-linear dispersive medium, *Pisma v Zh.Eksp.Teor.Fiz.* **45** (1987), 73-76.

88. Serkin, V.H.: Coloured solitons of envelope in fibre , *Pisma v Zh.Tekhn.Phys.* **13** (1987), 772-775.
89. Hodel, W., and Weber, H.P.: Decay of femtosecond higher-order solitons in an optical fibre induced by Raman self-pumping, *Opt.Lett.* **12** (1987), 924-926.
90. Bourkoff, E., Zhao, W., Joseph, R.I., and Christodoulides, D.N.: Intensity-dependent spectra of pulses propagating in optical fibres, *Opt.Commun.* **62** (1987), 284-288.
91. Bourkoff, E., Zhao, W., Joseph, R.I., and Christodoulides, D.N.: Evolution of femtosecond pulses in single-mode fibres having higher-order nonlinearity and dispersion, *Opt.Lett.* **12** (1987), 272-274.
92. Chi, S., and Wen, S.: Raman cross talk of soliton collision in a lossless fibre, *Opt.Lett.* **14** (1989), 1216-1218.
93. Potasek, M.J.: Novel femtosecond solitons in optical fibres, photonic switching, and computing, *J.Appl.Phys.* **65** (1989), 941-953.
94. Grudinin, A.B., and Fursa, T.N., Exactly solution of the equation of pulse propagation in FOCS in femtosecond region, *Opt. spectrosk.* **68** (1990), 210-213.
95. Agrawal, G.P.: Effect of intrapulse stimulated Raman scattering on soliton - effect pulse compression in optical fibers, *Opt.Lett.* **15** (1990), 224-226.
96. Potasek, M.J., and Tabor, M.: Exact solutions for an extended Non-linear Schrödinger equation, *Phys.Lett.* **A 154** (1991), 449-452.
97. Sasa, N., Satsuma, J.: New-type of soliton solutions for a higher-order Non-linear Schrödinger equation, *J.Phys.Soc.Japan* **60** (1991), 409-417.
98. Mihalache, D., Panoiu, N.-C., Moldlveanu, F., and Baboiu, D.-M.: The Riemann problem method for solving a perturbed Non-Linear Schrödinger equation describing pulse propagation in optical fibers, *J.Phys.* **A 27** (1994), 6177-6189.
99. Maimistov, A.I.: Propagation of an ultrashort light pulse in the region of zero second-order dispersion of group velocities, *Quantum Electronics* **24** (1994), 687-691.
100. Christodoulides, D.N., and Joseph, R.I.: Femtosecond solitary waves in optical fibers – beyond the slowly varying envelope approximation, *Appl.Phys.Lett.* **47**, (1985), 76-78.
101. Cavalcanti, S.B., Cressoni, J.C., da Cruz, H.R., and Gouveia-Neto, A.S.: Modulation instability in the region of minimum group-velocity dispersion of single-mode optical fibers via an extended Nonlinear Schrödinger equation, *Phys.Rev.* **A 43** (1991), 6162-6165.
102. Karlsson, M., and Hook, A.: Soliton-like pulses governed by fourth order dispersion in optical fibers, *Opt. Commun.* **104** (1994), 303-307
103 Hook, A., and Karlsson, M.: Ultrashort solitons at the minimum-dispersion wavelength effects of fourth-order dispersion, *Opt. Lett.* **18** (1993), 1388-1390.
104. Hook, A., and Karlsson, M.: Soliton instabilities and pulse compression in minimum dispersion fibers, *IEEE J.Quant.Electron.* **QE-30** (1994), 1831-1841.
105. Kolchanov, I.G.: Inelastic interaction of optical solitons, *Opt.Commun.* **79** (1990), 353-360.
106. Mamyshev, P.V., and Chernikov, S.V.: Ultrashort-pulse propagation in optical fibers, *Opt.Lett.* **15** (1990), 1076-1078.
107. Mamyshev, P.V., Chernikov, S.V., E.M.Dianov, and Prokhorov A.M.: Generation of a high-repetition-rate train of practically non-interacting solitons by using the induced modulational instability and Raman self-scattering effects, *Opt.Lett.* **15** (1990), 1365-1367.
108. Chen, H-H., and Liu, Ch-Sh. : Solitons in nonuniform media, *Phys.Rev.Lett.* **37** (1976), 693-696.
109. Balakrishnan, R.: Inverse spectral transform analysis of a Nonlinear Schrödinger equation with x-dependent coefficients. *Physica* **D16** (1985), 405-413.
110. Zheng, Yu-kun, and Chan, W.L.: Backlund transformation for the non-isospectral and variable-coefficient Non-linear Schrödinger equation, *J.Phys.* **A 22** (1989), 441-449.
111. Tajima, K.: Compensation of soliton broadening in non-linear optical fibers with loss, *Opt.Lett.* **12** (1987), 54-56.
112. Qiushi Ren, Hsiung Hsu, Soliton in dispersion-compensated optical fibres with loss, *IEEE J.Quant.Electron.* **QE-24** (1988), 2059-2062.
113. Kuehl, H.H.: Solitons on an axially nonuniform optical fibre, *J.Opt.Soc.Amer.* **B 5** (1988), 709-713.

114. Bordon, E.E., and Anderson, W.L.: Dispersion-adapted monomode fibre for propagation of non-linear pulses, *J.Lightwave Technol.* **7** (1989), 353-357.
115. Kivshar, Yu.S., and Konotop, V.V.: Solitons in fibre light-guides with slowly varying parameters, *Kvantov.Electron.* **16** (1989), 868-871 (in Russia).
116. Chernikov, S.V., and Mamyshev, P.V.: Femtosecond soliton propagation in fibres with slowly decreasing dispersion, *J.Opt.Soc.Amer.* **B 8** (1991), 1633-1641.
117. Malomed, B.A.: Modulational instability in a non-linear optical fibre induced by a spatial inhomogeneity, *Phys.Scr.* **47** (1993), 311-314.
118. Malomed, B.A.: Formation of a soliton in an inhomogeneous non-linear wave guide, *Phys.Scr.* **47** (1993), 797-799.
119. Malomed, B.A.: Ideal amplification of an ultrashort soliton in a dispersion-decreasing fibre, *Opt.Lett.* **19** (1997), 341-343.
120. Papaioannou, E., Frantzeskakis, D.J., and Hizanidis, K.: An analytical treatment of the effect of axial inhomogeneity on femtosecond solitary waves near the zero dispersion point, *IEEE J.Quant.Electron.* **QE-32** (1996), 145-154.
121. Abdullaev, F.Kh., Caputo, J.G., and Flytzanis, N.: Envelope soliton propagation in media with temporally modulated dispersion, *Phys.Rev.* **E 50** (1994), 1552-1558.
122. Hasegawa, A., Kumar, Sh., and Kodama, Y.: Reduction of collision-induced time jitters in dispersion-managed soliton transmission systems, *Opt.Lett.* **21** (1996), 39-41.
123. Yu, T., Golovchenko, E.A., Pilipetskii, A.N., and Menyuk, C.R.: Dispersion-managed soliton interactions in optical fibres. *Opt.Lett.* **22** (1997), 793-795.
124. Grigoryan, V.S., Yu, T, Golovchenko, E.A., Menyuk, C.R., and Pilipetskii, A.N.: Dispersion-managed soliton dynamics *Opt.Lett.* **22** (1997), 1609-1611.
125 Golovchenko, E.A., Pilipetskii, A.N., and Menyuk, C.R.: Periodic dispersion management in soliton wavelength-division multiplexing transmission with sliding filters. *Opt.Lett.* **22** (1997), 1156-1158.
126 Yang, T.S., and Kath, W.L.: Analysis of enhanced-power solitons in dispersion-managed optical fibres. *Opt.Lett.* **22** (1997), 985-987.
127. Malomed, B.A.: Suppression of soliton jitter and interactions by means of dispersion management, *Opt.Commun.* **147** (1998), 157-162.
128. Wald, M., Uzunov, I.M., Lederer, F. and Wabnitz, S.: Optimisation of soliton transmissions in dispersion-managed fibre links. *Opt.Commun.* **145** (1998), 48-52.
129. Shapiro, E.G. and Turitsyn, S.K.: Enhanced power breathing soliton in communication systems with dispersion management, *Phys.Rev.* **E56** (1997), 4951R-4955R.
130. Shapiro, E.G. and Turitsyn, S.K.: Theory of guiding-centre breathing soliton propagation in optical communication systems with strong dispersion management. *Opt.Lett.* **22** (1997), 1544-1546.
131. Turitsyn, S.K., and Shapiro, E.G.: Dispersion-managed solitons in optical amplifier transmission systems with zero average dispersion. *Opt.Lett.* **23** (1998), 682-684.
132. Yang, T.S., Kath, W.L., and Turitsyn, S.K.: Optimal dispersion maps for wavelength-division- multiplexed soliton transmission. *Opt.Lett.* **23** (1998), 597-599.
133. Devaney, J.F.L., Forysiak, W., Niculae, A.M., and Doran, N.J.: Soliton collisions in dispersion-managed wavelength-division-multiplexed systems. *Opt.Lett.* **22** (1997), 1695-1697.
134. Gordon, J.P.: Theory of the soliton self-frequency shift, *Opt.Lett.* **11** (1986), 665-667.
135. Gordon, J.P., and Haus, H.A.: Random walk of coherently amplified solitons in optical fiber transmission, *Opt.Lett.* **11** (1986), 665-667.
136. Mitschke, F.M., and Mollenauer, L.F.: Discovery of the soliton self-frequency shift, *Opt.Lett.* **11** (1986), 659-661.
137. Adams, M.J.: *An introduction to optical waveguides*, Wiley, New York, 1981.
138. Marcuse, D.: *Light Transmission Optics*, van Nostrand Reinhold, New York, 1982
139. Snyder, A.W., and Love, J.D.: *Optical waveguide theory*, Chapman and Hall, London, 1983.
140. Kapron, F.P., Keck, D.B.: Pulse transmission through a dielectric optical waveguide, *App.Opt.* **10** (1971), 1519-1523.

141. Kapron, F.P., Borrelli, N.F., and Keck, D.B.: Birefringence in dielectric optical waveguides, *IEEE J.Quant.Electron.* **QE-8** (1972), 222-225.
142. Crosignani, B., and Di Porto, P.: Intensity-induced rotation of the polarisation ellipse in low-birefringence, single-mode optical fibres, *Opt. Acta* **32** (1985), 1251-1258.
143. Boardman, A.D., and Cooper, G.S.: Power-dependent polarisation of optical pulses, *J.Opt.Soc.Amer.* **B5** (1988), 403-418.
144. Boardman, A.D., and Cooper, G.S.: Evolution of circularly polarised pulses in non-linear optical fibres, *J. Mod. Opt.* **35** (1988), 407-417.
145. Christodoulides, D.N., and Joseph, R.I.: Vector solitons in birefringent non-linear dispersive media, *Opt.Lett.* **13** (1988), 53-55.
146. Blow, K.J., Doran, N.J., and Wood, D.: Polarisation instabilities for solitons in birefringent fibres, *Opt.Lett.* **12** (1987), 202-204.
147. Menyuk, C.R.: Non-linear pulse propagation in birefringent optical fibres, *IEEE J.Quant.Electron.* **QE-23** (1987), 174-176.
148. Menyuk, C.R.: Stability of solitons in birefringent optical fibres. I. Equal propagation amplitudes, *Opt.Lett.* **12** (1987), 614-616.
149. Menyuk, C.R.: Stability of solitons in birefringent optical fibres. II. Arbitrary amplitudes, *J.Opt.Soc.Amer.* **B5** (1988), 392-402.
150. Menyuk, C.R.: Pulse propagation in an elliptically birefringent Kerr medium, *IEEE J.Quant.Electron.* **QE-25** (1989), 2674-2682.
151. Trillo, S., Wabnitz, S., Banyai, W.C., Finlayson, N., Seaton, C.T., Stegeman, G.I., and Stolen, R.H.: Observation of ultrafast non-linear polarisation switching induced by polarisation instability in a birefringent fibre rocking filter, *IEEE J.Quant.Electron.* **QE-25** (1989), 104-112.
152. Pare', C., and Florjan'czyk, M.: Approximate model of soliton dynamics in all-optical couplers, *Phys.Rev.* **A 41** (1990), 6287-6295.
153. Wabnitz, S., Wright, E.M, and Stegeman, G.I.: Polarisation instabilities of dark and bright coupled solitary waves in birefringent optical fibres, *Phys.Rev.* **A 41** (1990), 6415-6424.
154. Wabnitz, S., Trillo, S., Wright, E.M, and Stegeman, G.I.: Wavelength-dependent soliton self-routing in birefringent fibre filters, *J.Opt.Soc.Amer.* **B8** (1991), 602-613.
155. Malomed, B.A.: Polarisation dynamics and interactions of solitons in a birefringent optical fibre, *Phys.Rev.* **A 43** (1991), 410-423.
156. Muraki, D.J., Kath, W.L.: Hamiltonian dynamics of solitons in optical fibres, *Physica* **D 48** (1991), 53-64.
157. Maimistov, A.I., and Elyutin, S.O.: Propagation of short light pulses in non-linear birefringent fibre. Variational approach, *J. Mod. Opt.* **39** (1992), 2193-2200.
158. Kaup, D.J., Malomed, B.A., and Tasgal, R.S.: Internal dynamics of a vector soliton in non-linear optical fibre, *Phys.Rev.* **E 48** (1993), 3049-3053.
159. Manakov, S.V.: On the theory of two-dimensional stationary self-focusing of electromagnetic waves, *Zh.Eksp.Teor.Fiz.* **65** (1973), 505-516 [*Sov.Phys. JETP* **38** (1974), 248]
160. Kaup, D.J., and Malomed, B.A.: Soliton trapping and daughter waves in the Manakov model, *Phys.Rev.* **A 48** (1993), 599-604.
161. Hasegawa, A.: Self-confinement of multimode optical pulse in a glass fibre, *Opt.Lett.* **5** (1980), 416-417.
162. Crosignani, B., Papas, C.H., and Di Porto, P.: Coupled-mode theory approach to non-linear pulse propagation in optical fibres, *Opt.Lett.* **6** (1981), 61-63.
163. Crosignani, B., and Di Porto, P.: Soliton propagation in multimode optical fibres, *Opt.Lett.* **6** (1981), 329-330.
164. Crosignani, B., Cutolo, A., and Di Porto, P.: Coupled-mode theory of non-linear propagation in multimode and single-mode fibres: Envelope soliton and self-confinement, *J.Opt.Soc.Amer.* **72** (1982), 1136-1141.

165. Trillo, S., Wabnitz, S., Wright, E.M, and Stegeman, G.I.: Optical solitary waves induced by cross-phase modulation, *Opt.Lett.* **13** (1988), 871-873.
166. Crosignani, B., and Yariv, A.: Non-linear interaction of copropagating and counterpropagating waves in straight and highly twisted single-mode fibres, *J.Opt.Soc.Amer.* **B 5** (1988), 507-510.
167. Christodoulides, D.N.: Black and white vector solitons in weakly birefringent optical fibers, *Phys.Lett.* **A 132** (1988), 451-452.
168. Wright, E.M, Stegeman, G.I., and Wabnitz, S.: Solitary-wave decay and symmetry-breaking instabilities in two-mode fibres, *Phys.Rev.* **A 40** (1989), 4455-4466.
169. Florjanczyk, M., and Tremblay, R.: Periodic and solitary waves in bimodal optical fibers, *Phys.Lett.* **A 141** (1989), 34-36.
170. Ueda, T., and Kath, W.L.: Dynamics of coupled solitons in non-linear optical fibres, *Phys.Rev.* **A 42** (1990), 563-571.
171. Belanger, P.A., and Pare, C.: Soliton switching and energy coupling in two-mode fibers. Analytical results, *Phys.Rev.* **A 41** (1990), 5254-5256.
172. de Sterke, C.M., and Sipe, J.E.: Coupled modes and the non-linear Schrödinger equation, *Phys.Rev.* **A 42** (1990), 550-555.
173. Millot, G., Pitois, S., Dinda, P.T., and Haelterman, M.: Observation of modulational instability induced by velocity-matched cross-phase modulation in a normally dispersive bimodal fibre, *Opt.Lett.* **22** (1997), 1686-1688.
174. Andruschko, L.M., Karplyuk, K.S., and Ostrovskii, S.B.: On soliton propagation in coupled optical fibres, *Radiotechn.& Electron.* **32** (1987), 427-429 (in Russia).
175. Heatley, D.R., Wright, and E.M, Stegeman, G.I.: Soliton coupler, *Appl.Phys.Lett.* **53** (1988), 172-174.
176. Trillo, S., Wabnitz, S., Wright, E.M, and Stegeman, G.I.: Soliton switching in fibre non-linear directional couplers, *Opt.Lett.* **13** (1988), 672-674.
177. Friberg, S.R., Weiner, A.M., Silberberg, Y., Sfez, B.G., and Smith, P.S.: Femtosecond switching in a dual-core-fibre non-linear coupler, *Opt.Lett.* **13** (1988), 904-906.
178. Kivshar, Y.S., and Malomed, B.A.: Interaction of solitons in tunnel-coupled optical fibres, *Opt.Lett.* **14** (1989), 1365-1367.
179. Trillo, S., Wabnitz, S., and Stegeman, G.I.: Non-linear propagation and self-switching of ultrashort optical pulses in fibre non-linear directional couplers: The normal dispersion regime, *IEEE J.Quant.Electron.* **QE-25** (1989), 1907-1916.
180. Chen,Y.: Solution to full coupled wave equations of non-linear coupled systems, *IEEE J.Quant.Electron.* **QE-25** (1989), 2149-2154.
181. Nayar, B.K., Finlayson, N., Doran, N.J., Davey, S.T., Williams, D.L., and Arkwright, J.W.: All-optical switching in a 200-m twin-core fibre non-linear Mach-Zehnder interferometer, *Opt.Lett.* **16** (1991), 408-410.
182. Maimistov, A.I.: On light pulse propagation in tunnel-coupled optical waveguides, *Kvantov.Electron.* **18** (1991), 758-761 (in Russia).
183. Chu, P.L., and Wu, B.: Optical switching in twin-core erbium-doped fibres, *Opt.Lett.* **17** (1992), 255-257.
184. Christodoulides, D.N., and Joseph, R.I.: Discrete self-focusing in non-linear arrays of coupled waveguides, *Opt.Lett.* **13** (1988), 794-796.
185. Abdullaev, F.Kh., Abrarov, R.M., and Darmanyan, S.A.: Dynamics of solitons in coupled optical fibres, *Opt.Lett.* **14** (1989), 131-133.
186. Abrarov, R.M., Christiansen, P.L., Darmanyan, S.A., Scott, A.C., and Soerensen, M.P.: Soliton propagation in three coupled Nonlinear Schrödinger equations, *Phys.Lett.* **A 171** (1992), 298-302.
187. Kivshar, Yu.S.: Self-localisation in arrays of defocusing waveguides, *Opt.Lett.* **18** (1993), 1147-1149.
188. Abdullaev, F.Kh., Abrarov, R.M., Goncharov, V.I., and Darmanyan, S.A.: Interaction of solitons in non-linear directional couplers, *Zh.Eksp.Teor.Fiz.* **64** (1994), 101-109.

189. Jones, D.J., and Molter, L.A.: Generalised switching properties of three-guide circular fibre arrays. Using coupled-mode analysis, *IEEE J.Quant.Electron.* **QE-30** (1994), 119-125
190. Aceves, A.B., Luther, G.G., DeAngelis, C., Rubenchik, A.M., and Turitsyn, S.K.: Energy localisation in non-linear fibre arrays: Collapse-effect compressor, *Phys.Rev. Lett.* **79** (1995), 73-76.
191. Buryak, A.V., Akhmediev, N.N.: Stationary pulse propagation in *N*-core non-linear fibre arrays, *IEEE J.Quant.Electron.* **QE-31** (1995), 682-688.
192. Aceves A. B., and Santagiustina, M.: Bistable and tristable soliton switching in collinear arrays of linearly coupled waveguides, *Phys.Rev.* **E 56** (1997), 1113-1124.
193. Darmanyan, S., Relke, I., Lederer, F.: Instability of continuous waves and rotating solitons in waveguide arrays, *Phys.Rev.* **E 55** (1997), 7662-7668.
194. Darmanyan, S., Kobyakov, A., Schmidt, E., and Lederer, F.: Strongly localised vectorial modes in non-linear waveguide arrays. *Phys.Rev.* **E 57** (1998), 3520-3531.
195. Darmanyan, S., Kobyakov, A., and Lederer, F.: Stability of strongly localised excitation in discrete media with cubic nonlinearity. *Zh.Eksp.Teor.Fiz.* **113** (1998), 1253-1261.
196. Karpman, V.I., and Maslov, E.M.: Perturbation theory for solitons, *Zh.Eksp.Teor.Fiz.* **73** (1977), 537-539.
197. Wyller, J., and Mjolhus, E.: A Perturbation theory for Alfven solitons, *Physica* **D 13** (1984), 234-246.
198. Kodama, Y., and Ablowitz, M.J.: Perturbations of solitons and solitary waves, *Stud.Appl.Math.* **64** (1981), 225-245.
199. Herman, R.L.: A direct approach to study soliton perturbations, *J.Phys.* **A 23** (1990), 2327-2362.
200. Kaup, D.J.: Perturbation theory for solitons in optical fibers, *Phys.Rev.* **A 42** (1990), 5689-5694.
201. Lakoba, T.I., and Kaup, D.J.: Perturbation theory for the Manakov soliton and its applications to pulse propagation in randomly birefringent fibers, *Phys.Rev.* **E 56** (1997), 6147-6165.
202. Kivshar, Yu.S., and Malomed, B.A.: Dynamics of solitons in nearly integrable systems, *Rev.Mod.Phys.* **61** (1989), 763-915.
203. Anderson, D., Bondeson, A., and Lisak, M.: Variational approach to strongly non-linear Schrödinger equations, *Phys.Fluids* **22** (1979), 788-789.
204. Bondeson, A., Lisak, M., and Anderson, D.: Soliton perturbations: A variational principle for the soliton parameters, *Phys.Scr.* **20** (1979), 479-485.
205. Anderson, D.: Variational approach to non-linear pulse propagation in optical fibres. *Phys.Rev.* **A 27** (1983), 3135-3145.
206. Anderson, D.: Nonlinear pulse propagation in optical fibres: A variational approach, *IEE Proc. Pt.J.* **132** (1985), 122-125.
207. Anderson, D., and Lisak, M.: Bandwidth limits due to incoherent soliton interaction in optical-fibre communication systems, *Phys.Rev.* **A 32** (1985), 2270-2274.
208. Anderson, D., and Lisak, M.: Asymptotic linear dispersion of pulses in the presence of fiber nonlinearity and loss, *Opt.Lett.* **10** (1985), 390-392.
209. Anderson, D., Lisak, M., and Reichel, T.: Asymptotic propagation properties of pulses in a soliton-based optical-fiber communication system, *J.Opt.Soc.Amer.* **B 5** (1988), 207-210.
210. Anderson, D., Lisak, M., and Reichel, T.: Approximate analytical approach to nonlinear pulse propagation in optical fibers: A comparison, *Phys.Rev.* **A 38** (1988), 1618-1620.
211. Desaix, M., Anderson, D., and Lisak, M.: Solitons emerging from pulses launched in the vicinity of the zero-dispersion point in a single-mode optical fibre, *Opt.Lett.* **15** (1990), 18-20.
212. Desaix, M., Anderson, D., and Lisak, M.: Variational approach to collapse of optical pulses, *J.Opt.Soc.Amer.* **B 8** (1991), 2082-2086.
213. Karlsson, M., Anderson, D., Desaix, M., and Lisak, M.: Dynamic effects of Kerr nonlinearity and spatial diffraction on self-phase modulation of optical pulses, *Opt.Lett.* **16** (1992), 1373-1375.
214. Karlsson, M., Anderson, D., and Desaix, M.: Dynamics of self-focusing and self-phase modulation in a parabolic index optical fibres, *Opt.Lett.* **17** (1992), 22-24.
215. Karlsson, M.: Optical beams in saturable self-focusing media, *Phys.Rev.* **A 46** (1992), 2726-2734.

216. Afanasjev, V.V., Malomed, B.A., Chu, P.L., and Islam, M.K.: Generalised variational approximations for the optical soliton, *Opt.Commun.* **147** (1998), 317-322
217. Lakoba, T.I., and Kaup, D.J.: Variational method: How it can generate false instabilities, Institute for Nonlinear Studies, Clarkson University, Potsdam, #263, (1995)
218. Pantell, R.H., and Puthoff, H.E.: *Fundamentals of quantum electronics*, John Wiley & Sons, New York, London, Sidney, Toronto, 1969.
219. Agrawal, G.P.: *Nonlinear Fiber Optics*, Academic Press, Inc., Boston, San Diego, New York, London, Sydney, Tokyo, Toronto, 1989.
220. Hasegawe, A.: *Optical Solitons in Fibers*. Springer-Verlag, Berlin, 1990.
221. Kaup, D.J., and Newell, A.C.: An exact solution for a Derivative Nonlinear Schrödinger equation, *J.Math.Phys.* **19** (1978), 798-801.
222. Nakamura, A., and Chen, H.-H.: Multi-soliton solutions of a Derivative Nonlinear Schrödinger equation, *J.Phys.Soc.Japan* **49** (1980), 813-816.
223. Kawata, T., and Sakai, J.-I.: Linear problems associated with the Derivative Nonlinear Schrödinger equation, *J.Phys.Soc.Japan* 49 (1980), 2407-2414.
224. Iino, K.-h., Ichikawa, Y.-H., and Wadati, M.: Interrelation of alternative sets of Lax-pairs for a generalised Nonlinear Schrödinger equation, *J.Phys.Soc.Japan* **51** (1982), 3724-3728.
225. Kawata, T., and Inoue, H.: Exact solutions of the Derivative Nonlinear Schrödinger equation under the non-vanishing conditions, *J.Phys.Soc.Japan* **44** (1978), 1968-1976.
226. Kawata, T., Kobayashi, N., and Inoue, H.: Soliton solutions of the Derivative Nonlinear Schrödinger equation, *J.Phys.Soc.Japan* **46** (1979), 1008-1015.
227. Yariv, A., and Yeh, P.: *Optical waves in crystals*. John Wiley & Sons, New York, Chichester, Brisbane, Toronto, Singapore, 1984.
228. Yariv, A.: Coupled mode theory for guided wave optics, *IEEE J.Quant.Electron,* **9** (1973), 919-933.
229. Wabnitz, S.: Forward mode coupling in periodic nonlinear-optical fibers: modal dispersion cancellation and resonance solitons, *Opt.Lett.* **14** (1989), 1071-1073.
230. Christodoulides, D.N., and Joseph, R.I.: Slow Bragg solitons in nonlinear periodic structures, *Phys.Rev.Lett.* **62** (1989), 1746-1749.
231. de Sterke, C.M., and Sipe, J.E.: Self-localized light: launching of low-velocity solitons in corrugated nonlinear waveguides, *Opt.Lett.* **14** (1989), 871-873.
232. Neset A., and Sajeev, J.: Optical solitary waves in two- and three-dimensional nonlinear photonic band gap structures, *Phys.Rev.* **E57** (1998), 2287-2320.
233. Conti, C., Trillo, S., and Assanto,G.: Optical gap solitons via second-harmonic generation: Exact solitary solutions, *Phys.Rev.* **E57** (1998),1251R-1255R.
234. Segev, M., Valley, G.C., Singh, S.R., Carvalho, M.I., and Christodoulidis, D.N.: Vector photorefractive spatial solitons, *Opt. Lett.* **20** (1995), 1764 -1766.
235. Christodoulides, D.N., Singh, S.R., Carvalho, M.I., and Segev, M.: Incoherently coupled soliton pairs in biased photorefractive crystals, *Appl.Phys.Lett.* **68** (1996), 1763-1765.
236. Kutuzov, V,. Petnikova, V.M., Shuvalov, V.V., and Vysloukh, V.A.: Cross-modulation coupling of incoherent soliton modes in photorefractive crystals, *Phys.Rev.* **E 57** (1998), 6056-6066.
237. Buryak, A.V., and Kivshar, Yu.S.: Dark solitons in dispersive quadratic media, *Opt. Lett.* **20** (1995), 834-836.
238. Pelinovsky, D.E., Buryak, A.V., and Kivshar, Yu.S.: Instability of solitons governed by quadratic nonlinearities, *Phys.Rev.Lett.* **75** (1995), 591-595.
239. Baek, Y., Schiek, R., Stegeman, G.I., Baumann, I, and Sohler, W.: Interactions between one-dimensional quadratic solitons, *Opt. Lett.* **22** (1997), 1550-1552.
240. Derossi, A., Trillo, S., Buryak, A.V., and Kivshar, Yu.S.: Snake instability of one-dimensional parametric spatial solitons, *Opt. Lett.* **22** (1997), 868-870.
241. Etrich, C., Peschel, U., Lederer, F., and Malomed, B.A.: Stability of temporal chirped solitary waves in quadratically nonlinear media, *Phys.Rev.* **E 55** (1997), 6155-6162.

242. Costantini, B., DeAngelis, C., Barthelemy, A., Bourliaguet, B., and Kermene, V.: Collisions between type II two-dimensional quadratic solitons, *Opt. Lett.* **23** (1998), 424-426.
243. Bateman, H., and Erdelyi, A. (eds): *Higher Transcendental Functions*, v.3. Mc Graw-Hill Book Company, Inc., New York, Toronto, London, 1955.
244. Ince, E.L.: *Ordinary Differential equations*, Dover, New York, 1956.
245. Buryak, A.V., and Akhmediev, N.N.: Internal friction between Solitons in near-integrable systems. *Phys.Rev.***E 50** (1994), 3126-3133.
246. Tratnik M.V., and Sipe J.E.: Bound solitary waves in a birefringent optical fibre, *Phys.Rev.* **A38** (1988), 2011-2017.
247. Morris, H.C., and Dodd, R.K.: The two component derivation nonlinear Schrödinger equation, *Phys.Scr.* **20** (1979), 505-508.
248. Watanabe, Sh., Miyakawa, M., and Yajima, N.: Method of conservation laws for solving non-linear Schrödinger equation, *J. Phys. Soc. Japan* **46** (1979), 1653-1659.
249. Whitham, G.B.: *Linear and nonlinear waves*, Wiley-Interscience Publication, John Wiley & Sons, New York, London, Sydney, Toronto, 1974.
250. Hirota, R.: Exact envelope-soliton solutions of non-linear wave equation, *J.Math.Phys.* **14** (1973), 805-809
251. Landau, L.D., and Lifshitz, E.M.: *Mechanics*, Nauka, Moscow, 1983.
252. Anderson, D., Bondeson, A., and Lisak, M.: A variational approach to perturbed soliton equations, *Phys.Lett.* **A 67** (1978), 331-334.
253. Karpman, V.I., and Maslov, E.M.: Inverse problem method for the perturbed non-linear Schrödinger equation, *Phys. Lett.* **A 61** (1977), 355-357.
254. Vlasov, S.N., Petrishchev, V.A., and Talanov, V.I.: Averaging description of wave beams in linear and non-linear media (method of moments), *Izv.Vyssh.Uchebn. Zaved., Radiofiz.* **14** (1971), 1353-1363.
255. Zakharov, V.I.: Collapse of Langmuir waves, *Sov.Phys. JETP* **35** (1972), 908-914.
256. Lam, J.F., Lippman, B., and Tappert, F.: Moment theory of self-trapped laser beams with non-linear saturation, *Opt. Commun.* **15** (1975), 419-421.
257. Maimistov, A.I.: Evolution of solitary waves which are approximately solitons of a non-linear Schrödinger equation, *JETP* **77** (1993), 727-731.
258. Anderson, D., and Lisak, M.: Variational approach to incoherent two-soliton interaction, *Phys.Scr.* **33** (1986), 193-196.
259. Azimov, B.S., Sagatov, A.P., and Sukhorukov, A.P.: Formation and propagation of steady-state laser pulses in a media with third- and fifth-order nonlinearities, *Kvantov.Electron.* **18** (1991), 104-106 (in Russia).
260. De Angelis, C.: Self-trapped propagation in the nonlinear cubic-quintic Schrödinger equation: A variational approach, *IEEE J.Quant.Electron.* **QE-30** (1994), 818-821.
261. Azimov, B.S., and Sagatov, A.P.: On effect of the explosive compression of laser pulses, *Vestnik MSU, ser.3,* **32** (1991), 96-97 (in Russia).
262. Turitsyn, S.K.: Wave collapse and optical pulse compression, *Phys.Rev.* **A 47** (1993), R27-R29.
263. Astrakharchik, A.E., and Maimistov, A.I.: Effect of dissipation on the collapse of a solitary wave in a non-linear weakly dispersive medium, *JETP* **81** (1995), 275-279.
264. Silberberg Y.: Collapse of optical pulses, *Opt.Lett.* **15** (1990), 1282-1284.
265. Turitsyn, S.K.: Breathing self-similar dynamics and oscillatory tails of the chirped dispersion-managed soliton, *Phys.Rev.* **E58** (1998), 1256R-1259R.
266. Turitsyn, S.K., Gabitov, I., Laedke,E.W., Mezentsev, V.K., Musher, S.L., Shapiro, E.G., Schafer, T., and Spatschek, K.H.: Variational approach to optical pulse propagation in dispersion compensated transmission systems, *Opt.Commun.* **151** (1998), 117-135.
267. Hasegawa, A., and Kodama, Y.: Signal transmission by optical solitons in monomode fibre, *Proc.IEEE* **69** (1981), 1145-1152.
268. Cotter, D.: Fibre nonlinearities in optical communications, *Opt.and Quant.Electron.* **19** (1987), 1-17.

CHAPTER 7

PARAMETRIC INTERACTION OF OPTICAL WAVES

One of the broadest classes of phenomena in non-linear optics is the transformation of frequency of an electromagnetic radiation propagating in the non-linear medium. Harmonics generation of the fundamental wave (pump), sum-frequency and difference-frequency mixing are classified among these phenomena [1]. Under sufficiently high intensity of a pump the polarisation of a medium is not a linear function of the electric field strength of the wave. If the frequencies of an electromagnetic field are not in resonance with atomic transition frequencies, one can use a standard perturbation theory to reveal this dependency. So we can expand polarisation $\vec{P}$ in a power series of electrical field strength. This expansion is non-local due to the spatial and time dispersion of a non-linear medium. In the case of quasi-monochromatic fields, the polarisation is presented as a sum of its Fourier components. For the order n of the perturbations theory the correction term to the Fourier-component of the polarisation can be written as [2]:

$$\vec{\mathcal{P}}(\omega) = \sum_{\omega_1+\omega_2+\ldots+\omega_n=\omega} \hat{\chi}^{(n)}(\omega_1,\omega_2,\ldots,\omega_n)\vec{\mathcal{E}}(\omega_1)\otimes\vec{\mathcal{E}}(\omega_2)\otimes\ldots\otimes\vec{\mathcal{E}}(\omega_n).$$

where sign $\otimes$ marks a tensor product. Tensors of the *n*th rank $\hat{\chi}^{(n)}$, named as non-linear susceptibility, describe different processes of the electromagnetic waves interaction. The non-linear effects, described by the *n*th rank tensors of non-linear susceptibility, are often interpreted as the interaction of the $(n+1)$ waves.

Among all similar interactions the simplest are the three waves interactions. One can say that non-linear optics began from the study of these particular processes, which reduce to the sum-frequency mixing and difference-frequency mixing. These phenomena will be observed in section 7.1. In the undepleted pump approximation the known results of the parametric amplification and parametric generation [3,4] will be derived. When the group and phase velocities are equal, the equations of the 3-wave interaction can be analytically solved [1,5]. But this exact solution describes a stationary regime of interaction and propagation of three collinear waves. When frequencies of two waves from this triad are equal, then the third wave has twice as large frequency. In this case they say about second harmonic generation (SHG) [1-3]. This is the well-studied process [6-11], in which all peculiarities of the 3-wave interaction exhibit to a large scale. The resonance Raman scattering [12-15] under condition of weak variation of the energy levels population of a medium and the scattering of optical waves by an acoustical wave [16-26] can be considered from one position as the specific realisations of the 3-wave interaction. In both cases the systems of equations, describing these processes, can be transformed into one universal system [27,28]. We will consider these phenomena as an example of the 3-wave interaction being referred to as Raman scattering process. Following [12-15, 27,28] one can demonstrate that under certain conditions the reduced Maxwell-Bloch equations appear here, which describe the propagation of the ultrashort pulse in a resonance medium (see Chapter 4). Due to this property the Raman scattering can be analysed in terms of IST method [12,14,15,18].

Section 7.2 is devoted to the consideration of 3-wave interaction under condition of phase matching in terms of completely integrable systems [19-26, 29-35]. A vast amount of useful and important results were obtained in this direction of investigation [26, 36,37]. It is remarkable that the system of equations, describing the parametric interaction of two waves, can be solved by IST method for a 3D case [38-42], whereas the most of soliton equations are one-dimensional ones. The main results concerning the parametric interaction of the 3D solitary wave packets are presented in the surveys [43,44]. It should be noted, that the non-collinear parametric interaction which corresponds to the case where two waves are similar oblique waves propagating at equal and opposite angles to one direction can be integrated analytically [45] without the use of IST method. The equations for 3-wave interaction permit both the infinite number of conservation laws [30,41] and the Bäcklund transformation [34,39]. They can be presented as the Hamiltonian equations by employing the r-matrix [46]. It was shown in [47-49] that these equations pass the Painleve test and that there is a class of self-similar solutions, expressed in terms of Painleve transcendents P-V and P-VI. All that is the characteristic properties of the completely integrable systems of equations [50]. So the integrability of the quantum problem of three wave mixing [51-56] is not surprised.

It should be emphasised, that the 3-wave interaction gives an example of the dispersionless propagation of the non-linear waves. They often say that soliton is the result of the compensation of dispersion broadening and non-linear compression of the wave packet. The 3-wave interaction just demonstrates "the narrowness" of this statement. Due to the absence of the phase and group velocities dispersion the solitons in this process do not detach from the non-soliton part of the solution (which is often named radiation). It seems very difficult to study the process analytically, so one has to confine the investigation to some particular solutions. In this connection it is worth to refer to publications [45,57]. In [57] the particular solution of the non-stationary SHG was obtained by the bilinearization method. This method was developed by Hirota in [58,59] and has been successfully applied to the wide class of non-linear evolutionary equations [60]. The non-collinear SHG provides an example of a specific case of the 3-wave interaction. In [45] an exact solution of this problem was found without employing the IST method. The solution obtained explicitly illustrates the non-separability of the soliton and non-soliton parts of the solution of these three wave interaction equations.

The natural generalisation of a theory of 3-wave interaction is an account of the group velocities dispersion of the interacting solitary waves [61-64]. Attention was paid to a 3-wave interaction of four [46,47, 65-67] and more waves [68-71]. More complex processes, where parametric interaction is accompanied by resonance transitions in a multi-level medium, was discussed in [72-74]. In [75,76] a three-wave mixing was explored in detail in an inhomogeneous medium presented by a thin layer at the interface of two linear dielectric media.

The four-wave interaction should be considered as the next one in the hierarchy of complicated non-linear effects (see for example [65, 68,69,71]. It will be noted in section 7.3 that all processes of this sort can be reduced to the sum-frequency mixing (when the sum of the frequencies of three waves is equal to the frequency of the output fourth wave), and the hyperparametric scattering (when the sum of the frequencies of two waves is equal to the sum of the frequencies of another pair of waves). The systems of equations describing these processes can be unified in such a way that one may obtain valid results for all other processes having considered only one process. We omit a simple case of the given pump approximation and turn immediately to a stationary wave mixing, which can be considered rigorously under condition of phase matching. A specific case of the 4-wave interaction is the third harmonic generation, which will be observed in this section.

7.1. Three-wave parametric interaction

Let the non-linear characteristics of a medium be described by non-linear susceptibility of the second order $\hat{\chi}^{(2)}$. They call it quadratic non-linear media. Let the waves with the carrier frequencies ω_1 and ω_2 propagate along z axis. As the polarisation is the non-linear (quadratic) function of the electrical field strengths, the waves with carrier frequencies $\omega = \omega_1 \pm \omega_2$, $\omega = 2\omega_1$ and $\omega = 2\omega_2$ appear in such medium. These waves, in their turn, can cause the generation of new waves with the frequencies $\omega = 2\omega_1 \pm \omega_2$, $\omega = \omega_1 \pm 2\omega_2$, and etc. But in a dispersive medium all these processes are not equally efficient. There is a condition of phase matching, which selects a certain type of interaction of three waves, leaving all other unaffected. Sometimes such phase matching takes place for the waves propagating in the same direction. In this case they say about collinear parametric interaction. In this case the distance where the interaction of waves occurs can be made sufficiently long and, consequently, the effective frequency transformation will take place. On the contrary, when the phase matching is achievable only for the waves propagating in different directions, their interaction occurs only in the field of overlapping of the wave beams. Non-collinear parametric interaction is worth to draw special attention as the number of interesting results concerning the integrability of three-wave mixing equations have been found.

7.1.1. THE UNIFICATION OF EVOLUTION EQUATIONS

Let us consider the situation when only the collinear propagating wave with sum-frequency or difference frequency is generated. In the slowly varying envelopes and phases approximation, the system of equations describing the interaction of the three waves has the following form

$$\left(\frac{\partial}{\partial z}+\frac{1}{v_1}\frac{\partial}{\partial t}\right)\mathcal{E}_1 = i\frac{4\pi\omega_1\chi^{(2)}(\omega_3,-\omega_2)}{cn(\omega_1)}\mathcal{E}_2^*\mathcal{E}_3\exp(+i\Delta k\, z)\,, \tag{7.1.1a}$$

$$\left(\frac{\partial}{\partial z}+\frac{1}{v_2}\frac{\partial}{\partial t}\right)\mathcal{E}_2 = i\frac{4\pi\omega_2\chi^{(2)}(\omega_3,-\omega_1)}{cn(\omega_2)}\mathcal{E}_1^*\mathcal{E}_3\exp(+i\Delta k\, z)\,, \tag{7.1.1b}$$

$$\left(\frac{\partial}{\partial z}+\frac{1}{v_3}\frac{\partial}{\partial t}\right)\mathcal{E}_3 = i\frac{4\pi\omega_3\chi^{(2)}(\omega_1,\omega_2)}{cn(\omega_3)}\mathcal{E}_1\mathcal{E}_2\exp(-i\Delta k\, z)\,, \tag{7.1.1c}$$

where $\Delta k = k_3-(k_1+k_2)$, for a sum-frequency mixing process $\omega_3=\omega_1+\omega_2$, and

$$\left(\frac{\partial}{\partial z}+\frac{1}{v_1}\frac{\partial}{\partial t}\right)\mathcal{E}_1 = i\frac{4\pi\omega_1\chi^{(2)}(\omega_3,\omega_2)}{cn(\omega_1)}\mathcal{E}_2\mathcal{E}_3\exp(+i\Delta k\, z)\,, \tag{7.1.2a}$$

$$\left(\frac{\partial}{\partial z}+\frac{1}{v_2}\frac{\partial}{\partial t}\right)\mathcal{E}_2 = i\frac{4\pi\omega_2\chi^{(2)}(\omega_1,-\omega_3)}{cn(\omega_2)}\mathcal{E}_3^*\mathcal{E}_1\exp(-i\Delta k\, z)\,, \tag{7.1.2b}$$

$$\left(\frac{\partial}{\partial z}+\frac{1}{v_3}\frac{\partial}{\partial t}\right)\mathcal{E}_3 = i\frac{4\pi\omega_3\chi^{(2)}(\omega_1,-\omega_2)}{cn(\omega_3)}\mathcal{E}_2^*\mathcal{E}_1\exp(-i\Delta k\, z)\,, \tag{7.1.2c}$$

where $\Delta k = k_3-(k_1-k_2)$, for difference-frequency mixing $\omega_3=\omega_1-\omega_2$. In these equations v_k, $k=1,2,3$ are the group velocities of the corresponding wave. Here the effects of group-velocity dispersion are neglected.

If we choose new variables $\mathcal{E}_1=\sqrt{\gamma_1}Q_1$, $\mathcal{E}_2=\sqrt{\gamma_2}Q_2$, $\mathcal{E}_3=-\sqrt{\gamma_3}Q_3^*$, for the case of *sum-frequency mixing*, and $\mathcal{E}_1=\sqrt{\gamma_1}Q_3$, $\mathcal{E}_2=\sqrt{\gamma_{21}}Q_2^*$, $\mathcal{E}_3=\sqrt{\gamma_3}Q_3^*$, for the case of *difference-frequency mixing* (the indices at the velocities of the first and the third waves should be changed in this case), then one can write both systems of equations in unified form. This form is used in the theory of the three-wave interaction by Kaup [3,4] and also in the book [5].

$$\left(\frac{\partial}{\partial z}+\frac{1}{v_1}\frac{\partial}{\partial t}\right)Q_1 = i\tilde{\beta}Q_2^*Q_3^*\exp(+i\Delta k\, z)\,, \tag{7.1.3a}$$

$$\left(\frac{\partial}{\partial z}+\frac{1}{v_2}\frac{\partial}{\partial t}\right)Q_2 = i\tilde{\beta}Q_3^*Q_1^*\exp(+i\Delta k\, z)\,, \tag{7.1.3b}$$

$$\left(\frac{\partial}{\partial z}+\frac{1}{v_3}\frac{\partial}{\partial t}\right)Q_3 = -i\tilde{\beta}Q_1^*Q_2^*\exp(+i\Delta k\, z)\,, \tag{7.1.3c}$$

where $\widetilde{\beta} = \sqrt{\gamma_1\gamma_2\gamma_3}$. Coefficients γ_n serve as constant factors in right-hand parts of equations (7.1.1) and (7.1.2). In the transparent dielectrics the non-linear susceptibility included in these factors are identical. So the expression for γ_n can be written as

$$\gamma_n = \frac{4\pi\omega_n \chi^{(2)}(\omega_1,\omega_2)}{cn(\omega_n)}, \quad n = 1,2,3.$$

It should be noted, that this system of equations describes the case of decay instability in the theories of three-wave interaction [3,6].

Parametric three-wave interaction, considered here, refers to the collinear propagation of the interacting waves. If to assume that the directions of all three waves are different, then spatial derivatives in the equations (7.1.3) are to be changed for the directional derivatives in accordance with the wave vectors of each wave. We can arbitrarily assign directions of only two waves, then the direction of the third wave will be fixed by the wave synchronism condition. In other directions the generated wave is very weak because of a phase mismatch of all the three waves.

7.1.2. PARAMETRIC INTERACTION UNDER APPROXIMATION OF UNDEPLETED PUMP

Let us consider a parametric sum-frequency mixing and difference-frequency mixing provided that the difference of group velocities of interacting waves is unessential (it is a condition of group synchronism, $v_1 = v_2 = v_1$) and that the amplitude of the pump does not change. Both conditions are satisfied for sufficiently fine crystals. The second condition implies small efficiency of frequency transformation that usually happens if the phase matching condition does not hold. Under this approximation (undepleted pump approximation) the system of equations describing the parametric frequencies transformation can be linearised and solved.

It is the best of all to consider the unified system (7.1.3) separately for the processes of sum-frequency mixing and difference-frequency mixing. Indeed, assuming the amplitude of the pump to be constant from equations (7.1.3) for parametric generations of a wave with sum frequency we can get that

$$\frac{\partial}{\partial\xi} Q_2 = i\widetilde{\beta} Q_{10}^* Q_3^* \exp(+i\Delta k\,\xi), \tag{7.1.4a}$$

$$\frac{\partial}{\partial\xi} Q_3 = -i\widetilde{\beta} Q_{10}^* Q_2^* \exp(+i\Delta k\,\xi), \tag{7.1.4b}$$

and for the generation of the wave with difference frequency

$$\frac{\partial}{\partial \xi} Q_1 = i\widetilde{\beta} Q_{30}{}^* Q_2{}^* \exp(+i\Delta k\,\xi)\,, \tag{7.1.5a}$$

$$\frac{\partial}{\partial \xi} Q_2 = i\widetilde{\beta} Q_{30}{}^* Q_1{}^* \exp(+i\Delta k\,\xi)\,, \tag{7.1.5b}$$

Here $\xi = z$, $\tau = t - v^{-1} z$ are new variables.

In equation (7.1.4) the amplitude of the pump wave Q_{10} is a constant value. By denoting $\widetilde{\beta} Q_{10}^* = \mu$, the system of equations (7.1.4) becomes linear non-uniform equations. Substitution

$$Q_2 = A \exp[i\Delta k \xi / 2]\,, \quad Q_3 = B^* \exp[i\Delta k \xi / 2]$$

allows us to rewrite these linearised equations in the following form:

$$\frac{\partial A}{\partial \xi} + i\left(\frac{\Delta k}{2}\right) A - i\mu B = 0\,, \qquad \frac{\partial B}{\partial \xi} - i\left(\frac{\Delta k}{2}\right) B - i\mu^* A = 0\,.$$

This is a set of equations with constant coefficients which can be solved by standard procedure. The solution has the following form

$$A(\xi) = A_0 \cos\Omega\xi + i\frac{\mu B_0 - (\Delta k / 2) A_0}{\Omega} \sin\Omega\xi\,, \tag{7.1.6a}$$

$$B(\xi) = B_0 \cos\Omega\xi + i\frac{\mu^* A_0 + (\Delta k / 2) B_0}{\Omega} \sin\Omega\xi\,, \tag{7.1.6b}$$

where A_0 and B_0 are initial values of the normalised amplitudes of idler and signal wave. From expressions (7.1.6) we can see that these waves vary in space with the space-frequency $\Omega = \left(|\mu|^2 + (\Delta k / 2)^2\right)^{1/2}$.

Parametric difference-frequency mixing process can be considered similarly. The pump wave amplitude is Q_{30} in this case. The system of equations (7.1.5) in terms of variables

$$Q_1 = A \exp[i\Delta k \xi / 2]\,, \quad Q_2 = B^* \exp[i\Delta k \xi / 2]$$

and $\widetilde{\beta} Q_{30}^* = \mu$ takes the form of linear equations

$$\frac{\partial A}{\partial \xi} + i\left(\frac{\Delta k}{2}\right) A - i\mu B = 0\,, \qquad \frac{\partial B}{\partial \xi} - i\left(\frac{\Delta k}{2}\right) B + i\mu^* A = 0\,.$$

Solution of these equations can be presented as following

$$A(\xi) = A_0 \cosh \Omega\xi + i\frac{\mu B_0 - (\Delta k/2)A_0}{\Omega}\sinh \Omega\xi, \qquad (7.1.7a)$$

$$B(\xi) = B_0 \cosh \Omega\xi - i\frac{\mu^* A_0 - (\Delta k/2)B_0}{\Omega}\sinh \Omega\xi, \qquad (7.1.7b)$$

where now $\Omega = \left(|\mu|^2 - (\Delta k/2)^2\right)^{1/2}$, and A_0 and B_0 are initial values of normalised amplitudes of the idler wave and signal wave as it was previously.

Zero value of the idler wave amplitude $B_0 = 0$ often serves as initial amplitudes. In this case expressions (7.1.7) transform in those often quoted in different books (for example [1-4]).

7.1.3. STEADY-STATE PARAMETRIC PROCESS

If the spread of non-linear medium is not large enough to notice a spatial separation of the interacting waves, then the difference in group velocities can be neglected in equations (7.1.3). The system of equations (7.1.3) then writes

$$\frac{\partial}{\partial \xi}Q_1 = i\tilde{\beta}Q_2^*Q_3^*, \quad \frac{\partial}{\partial \xi}Q_2 = i\tilde{\beta}Q_1^*Q_3^*, \quad \frac{\partial}{\partial \xi}Q_3 = -i\tilde{\beta}Q_2^*Q_1^*, \qquad (7.1.8)$$

where $\xi = z$, $\tau = t - v^{-1}z$, and $v = v_1 = v_2 = v_3$. This system can be solved exactly [1]. Let us define the real variables according to the formulae:

$$Q_k = u_k \exp(i\varphi_k), \quad \Phi = \varphi_1 + \varphi_2 + \varphi_3 .$$

It follows from equations (7.1.8) that

$$\frac{\partial u_1}{\partial \xi} = \tilde{\beta}u_2u_3 \sin\Phi, \quad \frac{\partial u_2}{\partial \xi} = \tilde{\beta}u_1u_3 \sin\Phi, \quad \frac{\partial u_3}{\partial \xi} = -\tilde{\beta}u_2u_1 \sin\Phi, \qquad (7.1.9a)$$

$$\frac{\partial \Phi}{\partial \xi} = \tilde{\beta}\left(\frac{u_2u_3}{u_1} + \frac{u_1u_3}{u_2} - \frac{u_2u_1}{u_3}\right)\sin\Phi. \qquad (7.1.9b)$$

By using the amplitude equation (7.1.9a) we can write the phase equation as

$$\frac{\partial \Phi}{\partial \xi} = \tilde{\beta}\frac{\cos\Phi}{\sin\Phi}\frac{\partial}{\partial \xi}\ln(u_1u_2u_3).$$

After being integrated this equation yields

$$u_1 u_2 u_3 \cos\Phi = \text{const}\,. \tag{7.1.10}$$

Besides this integral of motion three more integrals result from equations (7.1.9a)

$$u_1^2 + u_3^2 = m_1^2, \qquad u_2^2 + u_3^2 = m_2^2, \qquad u_1^2 + u_2^2 = m_3^2, \tag{7.1.11}$$

where the following expression comes from

$$u_1^2 + u_2^2 + 2u_3^2 = \text{const}\,.$$

This formula is known as Manley-Rowe relation.

The constants in all these expressions are defined by the initial or boundary conditions. Let one of the fields take zero value at least in one point of the characteristic. Then the constant of integrating in (7.1.10) is chosen to be zero, and, consequently, everywhere we take $\Phi = \pi/2$.

Substitution of variables u_1 and u_2 from (4) into the equation (7.1.9a) gives an equation:

$$\frac{du_3}{d\xi} = -\tilde{\beta}\sqrt{(m_1^2 - u_3^2)(m_2^2 - u_3^2)}\,.$$

Let $m_2 > m_1$ for the sake of certainty. The solution of this equation is expressed in terms of Jacobian elliptic functions:

$$u_3(\xi) = m_1\,\mathrm{sn}\left(\tilde{\beta} m_2 \xi,\ k\right), \tag{7.1.12a}$$

where $k = m_1/m_2$ is a module of these functions. The rest of solutions of (7.1.2) are expressed by formulae

$$u_1(\xi) = m_1\,\mathrm{cn}\left(\tilde{\beta} m_2 \xi,\ k\right), \tag{7.1.12b}$$

$$u_2(\xi) = m_2\,\mathrm{dn}\left(\tilde{\beta} m_2 \xi,\ k\right). \tag{7.1.12c}$$

In the case of $m_2 = m_1$ these solutions degenerate to the following

$$u_3(\xi) = m_1\,\tanh\left(\tilde{\beta} m_1 \xi\right), \tag{7.1.13a}$$

$$u_1(\xi) = m_1\,\mathrm{sech}\left(\tilde{\beta} m_1 \xi\right), \tag{7.1.13b}$$

$$u_2(\xi) = m_1\,\mathrm{sech}\left(\tilde{\beta} m_1 \xi\right). \tag{7.1.13c}$$

Expressions (7.1.1.3) describe the propagation of two solitary waves in the form of hyperbolic secant pulse, and the dark solitary wave associated with them.

7.1.4. SECOND HARMONIC GENERATION

There is a specific case of parametric process of sum-frequencies mixing. For a long time optical harmonics generation was the basic phenomenon to produce coherent radiation in the frequency band where there were no laser sources. Though nowadays the development of laser technology has made this application less actual, it is necessary to remind, however, that coherent radiation of the record short wavelength (for example, 38 nm) was achieved by means of third-harmonic generation under two-photon resonance conditions with four-harmonic of Nd^{+3}-glass laser radiation pumping [77]. Harmonic generation remains so far as an actual mean of non-linear spectroscopy, which allows to get data on non-linear susceptibilities of different media.

The theory of parametric processes can be developed on the base of reduced Maxwell equations (under non-resonance conditions) or Maxwell-Bloch equations (under resonance conditions). Stationary processes of harmonic generation present the most complete part of the theory. In non-stationary regime the analytical results can be obtained under the approximation of an undepleted pump. It turned out that the IST method can be applied in some exclusive cases too, and then one manages to find an explicit solutions of the reduced Maxwell equation for the interacting waves. Some of these solutions are solitons.

Phase matching

We will consider a stationary process of the second-harmonic generation (SHG). This effect occurs both in the field of continuous wave and under the pulsed pump when the length of a non-linear crystal is much less then the dimension of a space occupied by pulse or when the propagation velocities of pump and harmonic pulses are equal. In this case reduced Maxwell equations (7.1.1) are as follows

$$\frac{d\mathcal{E}_1}{dz} = i\frac{4\pi\omega^2}{c^2 k_1}\chi^{(2)}(2\omega,-\omega)\mathcal{E}_1^*\mathcal{E}_2\exp(-i\Delta k z), \tag{7.1.14}$$

$$\frac{d\mathcal{E}_2}{dz} = i\frac{8\pi\omega^2}{c^2 k_2}\chi^{(2)}(\omega,\omega)\mathcal{E}_1^2\exp(i\Delta k z), \tag{7.1.15}$$

where $\Delta k = 2k_1 - k_2$ and we define that

$$k_1 = \omega c^{-1}\sqrt{\varepsilon(\omega)} = \omega c^{-1} n(\omega), \quad k_2 = 2\omega c^{-1}\sqrt{\varepsilon(2\omega)} = 2\omega c^{-1} n(2\omega).$$

If $\Delta k = 0$, then they say that SHG takes place under conditions of phase synchronism. It is named the phase-matching process. The physical meaning of this condition can be understood if to re-write Δk in the form:

$$\Delta k = 2\frac{\omega}{c}\left[n(\omega) - n(2\omega)\right],$$

where $n(\omega)$ and $n(2\omega)$ are linear refractive indices of the medium at the frequencies of pump (ω) and harmonic (2ω) waves. Hence

$$\Delta k = 2\omega\left[v_{ph}^{-1}(\omega) - v_{ph}^{-1}(2\omega)\right],$$

where $v_{ph}(\omega)$ and $v_{ph}(2\omega)$ are the phase velocities of the corresponding waves. So the phase-matching condition means that the phase velocities of interacting waves are equal. If $\Delta k \neq 0$, interacting waves are not synchronised in phases. In a distance $L_c = |c\Delta k^{-1}|$, known as the coherence length, the phase mismatch becomes equal to π,. It is important to emphasise that this condition has no relations to SHG. This is a general result for interaction of waves of any nature. Wave interaction is the most effective if the phases difference of the interacting waves is constant. For instance, for the Nth harmonic generation the phase-matching condition yields

$$\Delta k = Nk(\omega) - k(N\omega) = N\omega\left[v_{ph}^{-1}(\omega) - v_{ph}^{-1}(N\omega)\right].$$

In a more general case, if the parametric interaction of the waves with frequencies ω_1 and ω_2 gives rise to the wave with frequency $\omega_3 = \omega_1 \pm \omega_2$, then the phase-matching condition is represented by the expression:

$$\Delta k = \frac{\omega_1 n(\omega_1) \pm \omega_2 n(\omega_2) - \omega_3 n(\omega_3)}{c}.$$

Let the phase-matching condition not hold. In a dispersive medium refractive indices are different at different frequencies, if only special measures are not taken. The efficiency of SHG will be insufficient, as far as the transformation into harmonic changes for reverse process within a distance of coherency. Let us suppose that the amplitude of the pump wave does not alter, which means the undepleted pump approximation. Then we have one equation for the second harmonic amplitude instead of two (7.1.14) and (7.1.15):

$$\frac{d\mathcal{E}_2}{dz} = i\frac{8\pi\omega^2}{c^2 k_2}\chi^{(2)}(\omega,\omega)\mathcal{E}_{10}^2 \exp(i\Delta k z), \tag{7.1.16}$$

where $\mathcal{E}_{10} = \mathcal{E}_1(z=0)$. The solution of this equation can be obtained by integrating (7.1.16) with respect to z from 0 to L, where L is a length of non-linear medium (for instance, non-linear crystal):

$$\mathcal{E}_2(L) = -\frac{8\pi\chi^{(2)}(\omega,\omega)}{cn(2\omega)}\mathcal{E}_{10}^2 \exp(i\Delta kL/2)\frac{\sin(\Delta kL/2)}{\Delta k}.$$

Second harmonic intensity, averaged over the period of quick oscillations, is

$$I_2(L) = \frac{4(4\pi)^3\omega^2|\chi^{(2)}(\omega,\omega)|^2}{c^3n^2(\omega)n(2\omega)} I_{10}^2 \frac{\sin^2(\Delta kL/2)}{\Delta k^2}. \tag{7.1.17}$$

One can see that the intensity of second harmonic varies periodically in a non-linear medium with the period $L_c = |c\Delta k^{-1}|$. Under condition $L << L_c$ it is seen that harmonic intensity I_2 changes as L^2, and it does not depend on Δk. Square-law dependency of second harmonic intensity on pump intensity I_{10} is characteristic for the process of SHG, and, as it will be seen further, this dependency does not follow from the undepleted pump approximation. This dependency is violated when SHG occurs in focused optical beams. But for the case of plane waves under current consideration such dependency means that two photons of pump wave give birth to one photon of harmonica.

Let the uniaxial crystal be used as a non-linear medium. Denote the angle between the direction of waves propagation and the optical axis of crystal as θ. The refractive indices $n(\omega)$ and $n(2\omega)$ depend on this angle by the formula

$$\frac{\sin^2\theta}{n_{\parallel}^2} + \frac{\cos^2\theta}{n_{\perp}^2} = \frac{1}{n_e^2(\theta)}. \tag{7.1.18a}$$

for an extraordinary wave and do not depend on θ for an ordinary wave. If $n_e < n_0$, they say that this is a negatively-birefringent uniaxial crystal. The crystal with $n_e > n_0$ is named a positively-birefringent uniaxial one. Let us consider the negative-birefringent uniaxial crystal (for instance, in the crystal of tellurium and KDP) and assume that $n_0(\omega) > n_e(2\omega)$. Then there is an angle θ_m, which indicates the direction of synchronism, so that

$$n_0(\omega) = n_e(2\omega, \theta_m). \tag{7.1.18b}$$

Here SHG corresponds to the scheme $o + o \to e$, i.e., two quanta of ordinary pump wave give one quantum of the extraordinary wave of harmonica. This is an illustration of *type* I *phase matching*. The phase-matching angle is determined from (7.1.18) by the expression:

$$\sin\theta_m = \sqrt{\frac{n_0^{-2}(\omega) - n_0^{-2}(2\omega)}{n_e^{-2}(2\omega) - n_0^{-2}(2\omega)}}\,.$$

In the positively-birefringent uniaxial crystal the phase-matching angle is given by the similar formula:

$$\sin\theta_m = \sqrt{\frac{n_0^{-2}(\omega) - n_0^{-2}(2\omega)}{n_0^{-2}(\omega) - n_e^{-2}(\omega)}}\,.$$

It is interesting that phase-matching condition may be satisfied in the scheme $e + o \rightarrow e$ too, when two photons of pump ordinary and extraordinary waves give one extraordinary second-harmonic photon. This is the *type* II *phase matching*.

Thereby, in anisotropic crystals the phase-matching condition can be satisfied and highly efficient SHG can be achieved. There is the more complicated phase matching condition in the case of the biaxial crystals [3].

System of equations (7.1.14) and (7.1.15) under condition $\Delta k = 0$ can be solved exactly. Let us present complex functions $\mathcal{E}_1$, $\mathcal{E}_2$ as

$$\mathcal{E}_1 = \rho_1 \exp(i\varphi_1),\ \mathcal{E}_2 = \rho_2 \exp(i\varphi_2)\,.$$

Real amplitudes and phase difference $\Phi = 2\varphi_1 - \varphi_2$ are determined by the solutions of equations which follow from (7.1.14) and (7.1.15) :

$$\frac{d\rho_1}{dz} = -a\rho_1\rho_2 \sin\Phi\,, \tag{7.1.19}$$

$$\frac{d\rho_2}{dz} = a\rho_1{}^2 \sin\Phi\,, \tag{7.1.20}$$

$$\frac{d\Phi}{dz} = -a\left(2\rho_2 - \rho_1{}^2 / \rho_2\right)\cos\Phi\,, \tag{7.1.21}$$

where $a = 4\pi\omega c^{-1} n^{-1}(\omega)\chi^{(2)}(\omega,\omega)$. Equations (7.1.19) and (7.1.20) possess an integral of motion

$$\rho_1^2(z) + \rho_2^2(z) = \rho_0^2 = \text{const}\,. \tag{7.1.22}$$

Here ρ_0 is a value of real amplitude of pump at the boundary of non-linear medium. For the sake of certainty we will consider that harmonic wave is absent there. By excluding variable ρ_1 by means of (7.1.22) from (7.1.21) we can find the second integral of motion

$$\rho_2\left(\rho_0^2-\rho_2^2\right)\cos\Phi = C = \text{const.} \tag{7.1.23}$$

When the undepleted pump approximation is valid, then at $z \to 0$, the phase difference $\Phi \to \pi/2$ as $\rho_2 \to 0$. Hence the integrating constant in (7.1.23) can be found exactly:

$$\rho_2\left(\rho_0^2-\rho_2^2\right)\cos\Phi = 0.$$

With taking (7.1.22), (7.1.23) into account, one can find from equation (7.1.20) the equation for ρ_2, which solution under initial condition $\rho_2 = 0$ is :

$$\rho_2(z) = \rho_0 \tanh[a\rho_0 z]. \tag{7.1.24}$$

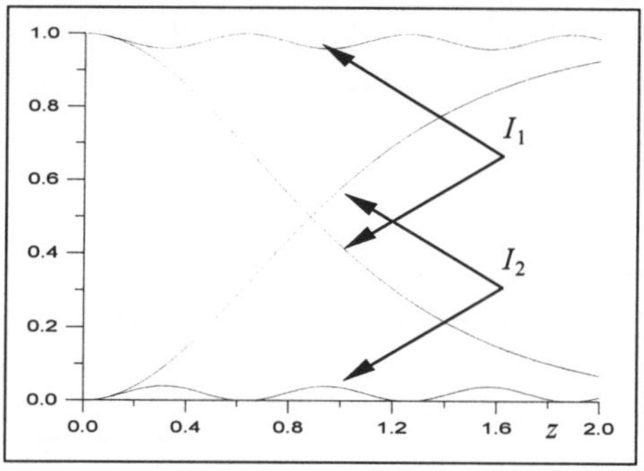

Fig. 7.1.1.

Normalised plots of second harmonic intensity and fundamental intensity vs length of medium *z*

The dependencies of the normalised intensity of the second harmonic $I_2(z)=\rho_2^2(z)/\rho_0^2$ and the normalised intensity of the fundamental wave $I_1(z)=\rho_1^2(z)/\rho_0^2$ versus length of the non-linear medium z for perfect phase matching $\Delta k = 0$ differ noticeably from these dependencies for a finite phase mismatch (Fig.7.1.1). This expression shows that the second harmonic intensity is proportional to a square of pump intensity, as it has been already found in the frame of the undepleted pump approximation. The distance z_c, which is equal to

$$z_c \approx 1{,}47(a\rho_0)^{-1},$$

gives a length of a non-linear crystal, on which second harmonica amplitude reaches 0,9 part of the initial value of pump amplitude. It should be noted that if Δk is not zero, systems (7.1.14) and (7.1.15) yet have an exact solution, but it is rather cumbersome to analyse it here.

Solution of SHG equations by Hirota's method

Now we can go over to the analysis of SHG under the phase-matching condition, when pump depletion should not be ignored. System of equations (7.1.14) and (7.1.15) in terms of normalised complex envelopes of pump and harmonic pulses has the form

$$\frac{\partial q_1}{\partial \tau} = i q_1^* q_2 , \qquad \frac{\partial q_2}{\partial \varsigma} = i q_1^2 , \tag{7.1.25}$$

where $\mathcal{E}_1(t,z) = \mathcal{E}_{01} q_1(t,z)$, $\mathcal{E}_2(t,z) = \mathcal{E}_{01} q_2(t,z)$ are the envelopes of pulses. Directional derivatives are

$$\left(\frac{\partial}{\partial z} + \frac{1}{v_1}\frac{\partial}{\partial t}\right) = K\frac{\partial}{\partial \tau}, \qquad \left(\frac{\partial}{\partial z} + \frac{1}{v_2}\frac{\partial}{\partial t}\right) = K\frac{\partial}{\partial \varsigma},$$

where K is a coupling coefficient:

$$K = \frac{4\pi\omega^2}{k_1 c^2}\chi^{(2)}(\omega,\omega)\mathcal{E}_{10}.$$

In ref. [57] the solution of the SHG equations was derived by Hirota bilinear transform method. Let q_1 be in the form

$$q_1 = \frac{G}{F}, \quad \operatorname{Im} F = 0. \tag{7.1.26a}$$

From the first equation of (7.1.25) we obtain

$$i q_2 = \frac{1}{F G^*}\left(F\frac{\partial G}{\partial \tau} - G\frac{\partial F}{\partial \tau}\right).$$

In terms of Hirota's D-operators [58-60] the above expression can be written in the form

$$q_2 = \frac{-i}{F G^*} D_\tau (G \cdot F). \tag{7.1.26b}$$

Substitution of this expression into the second equation of the system (7.1.25) leads to

$$D_\zeta\left(D_\tau(G\cdot F)\cdot G^* F\right) = -(GG^*)^2. \tag{7.1.27}$$

Complex conjugation of the equation (7.1.27) gives rise to

$$D_\zeta\left(GF\cdot D_\tau\left(F\cdot G^*\right)\cdot\right) = -(GG^*)^2.$$

The summation of the two above expressions yields

$$D_\zeta\left(D_\tau(G\cdot F)\cdot G^* F\right) + D_\zeta\left(GF\cdot D_\tau\left(G^*\cdot F\right)\right) = -2(GG^*)^2. \tag{7.1.28}$$

A useful property of Hirota's operators [60]

$$\begin{aligned} D_x\Big[D_y(a\cdot d)\cdot cb + ad\cdot D(c\cdot b)\Big] &= \\ &= D_x D_y(a\cdot b)\,cd - abD_x D_y(c\cdot d) + \\ &+ D_x(a\cdot b)D_y(c\cdot d) - D_y(a\cdot b)D_x(c\cdot d)\,. \end{aligned}$$

and the equality $D_x(a\cdot a) = 0$ can be exploited to transform expression (7.1.28) into the following

$$D_\zeta D_\tau\left(G\cdot G^*\right)F^2 - GG^* D_\zeta D_\tau(F\cdot F) = -2(GG^*)^2.$$

This expression can be rewritten in the form of two bilinear equations with respect to functions F and G:

$$D_\zeta D_\tau\left(G\cdot G^*\right) = 0, \tag{7.1.29a}$$

$$D_\zeta D_\tau(F\cdot F) = 2(GG^*)^2. \tag{7.1.29b}$$

According to the general approach to bilinear equations [58-60], we expand F and G in terms of the powers of a parameter ε,

$$\begin{aligned} F &= 1 + \varepsilon f_1 + \varepsilon^2 f_2 + \varepsilon^3 f_3 + \ldots, \\ G &= \varepsilon g_1 + \varepsilon^2 g_2 + \varepsilon^3 g_3 + \ldots. \end{aligned}$$

Substitution of this expansion into (7.1.29) and equating the terms with the same powers of ε yields in the first order of ε

$$D_\zeta D_\tau (f_1 \cdot 1) = 0, \tag{7.1.30a}$$

in the second order of ε

$$D_\zeta D_\tau (g_1 \cdot g_1^*) = 0, \tag{7.1.30b}$$

$$D_\zeta D_\tau (f_2 \cdot 1 + 1 \cdot f_2) = 2g_1 g_1^* - D_\zeta D_\tau (f_1 \cdot f_1), \tag{7.1.30c}$$

in the third order of ε

$$D_\zeta D_\tau (g_1 \cdot g_2^* + g_2 \cdot g_1^*) = 0 \tag{7.1.30d}$$

$$D_\zeta D_\tau (f_3 \cdot 1 + 1 \cdot f_3) = 2g_1 g_2^* + 2g_2 g_1^* - D_\zeta D_\tau (f_1 \cdot f_2 + f_2 \cdot f_1), \tag{7.1.30e}$$

and so on. Equation (7.1.30a) in a conventional form writes as

$$\frac{\partial^2 f_1}{\partial \tau \partial \zeta} = 0.$$

The solution of this equation is $f_1(\zeta,\tau) = h(\tau) + \tilde{h}(\zeta)$, where $h(\tau)$ and $\tilde{h}(\zeta)$ are arbitrary functions. In order to solve the equation (7.1.30b) we denote $g_1 = R\exp(i\varphi)$ and rewrite the equation (7.1.30b) in terms of real variables

$$R\frac{\partial^2 R}{\partial \tau \partial \zeta} - \frac{\partial R}{\partial \tau}\frac{\partial R}{\partial \zeta} = 2R^2 \frac{\partial \varphi}{\partial \tau}\frac{\partial \varphi}{\partial \zeta}.$$

If function φ is chosen as $\varphi = \varphi_1(\tau)$ (or as $\varphi = \varphi_2(\zeta)$), then the solution of this equation is $R(\tau,\zeta) = R_1(\tau)R_2(\zeta)$, where $R_{1,2}$ are arbitrary functions. Then

$$g_2 = R_1(\tau)R_2(\zeta)\exp[i\varphi_1(\tau)].$$

Equation (7.1.30c) allows to find $f_2(\zeta,\tau)$. We can rewrite it, taking into account the properties of D-operators and the representation of $f_1(\zeta,\tau)$,

$$\frac{\partial^2 f_2}{\partial \tau \partial \zeta} = R_1^2(\tau)R_2^2(\zeta) + \frac{\partial h}{\partial \tau}\frac{\partial \tilde{h}}{\partial \zeta}.$$

Let the functions $R_1(\tau)$, $R_2(\zeta)$ be chosen as

$$R_1^2(\tau)=\sigma\frac{\partial h}{\partial\tau},\quad R_2^2(\zeta)=\sigma\frac{\partial\tilde{h}}{\partial\zeta},$$

where σ is some constant. Hence, $f_2(\zeta,\tau)$ obeys the equation

$$\frac{\partial^2 f_2}{\partial\tau\partial\zeta}=(1+\sigma^2)\frac{\partial h}{\partial\tau}\frac{\partial\tilde{h}}{\partial\zeta},$$

which results in $f_2(\zeta,\tau)=(1+\sigma^2)h(\tau)\tilde{h}(\zeta)$.

One can find that $D_\zeta D_\tau\left(f_1\cdot f_2\right)=0$. If we assume that $g_2=f_3=0$, then the power series in terms of ε for F and G are the finite sums. Thus, one obtains

$$F(\zeta,\tau)=1+h(\tau)+\tilde{h}(\zeta)+(1+\sigma^2)h(\tau)\tilde{h}(\zeta),$$
$$G(\zeta,\tau)=\sigma\left(\frac{\partial h}{\partial\tau}\frac{\partial\tilde{h}}{\partial\zeta}\right)^{1/2}\exp[i\varphi(\tau)]. \tag{7.1.31}$$

The solution of SHG equations is given by the formulae (7.1.26)

It is interesting that the SHG equations lead to the conservation law

$$\frac{\partial|q_1|^2}{\partial\tau}+\frac{\partial|q_2|^2}{\partial\zeta}=0. \tag{7.1.32}$$

That means that some "potential function" $\mathcal{W}$ can be introduced, such that

$$|q_1|^2=-\frac{\partial\mathcal{W}}{\partial\zeta},\quad |q_2|^2=\frac{\partial\mathcal{W}}{\partial\tau}. \tag{7.1.33}$$

By the direct evaluating of the partial derivatives we can find

$$|q_1|^2=\frac{|G|^2}{F^2}=\frac{\partial^2\ln F}{\partial\tau\partial\zeta}, \tag{7.1.34a}$$

since the "potential function" is

$$\mathcal{W}=-\frac{\partial\ln F}{\partial\tau}.$$

We can also write an expression for $|q_2|$ which is similar to (7.1.34a):

$$|q_2|^2 = -\frac{\partial^2 \ln F}{\partial \tau^2} . \tag{7.1.34b}$$

Let us consider the particular forms of the functions $h(\tau)$ and $\widetilde{h}(\zeta)$

$$h(\tau) = \exp[\vartheta_1], \quad \widetilde{h}(\zeta) = \exp[\vartheta_2], \quad \vartheta_1 = \Omega_1 \tau, \quad \vartheta_2 = \Omega_2 \zeta .$$

From (7.1.34) one can obtain the following expressions

$$|q_1|^2 = \frac{\Omega_1 \Omega_2 \sigma^2 \exp(\vartheta_1 + \vartheta_2)}{\left(1 + \exp(\vartheta_1) + \exp(\vartheta_2) + (1+\sigma^2)\exp(\vartheta_1 + \vartheta_2)\right)^2} , \tag{7.1.35a}$$

$$|q_2|^2 = \frac{\Omega_1^2 \exp(\vartheta_1)\left(1 + \exp(\vartheta_2)\right)\left(1 + (1+\sigma^2)\exp(\vartheta_2)\right)}{\left(1 + \exp(\vartheta_1) + \exp(\vartheta_2) + (1+\sigma^2)\exp(\vartheta_1 + \vartheta_2)\right)^2} . \tag{7.1.35b}$$

We would like to remind at this point that the characteristic coordinates τ and ζ are defined as

$$\tau = \frac{Kv_1}{v_1 - v_2}\left(z - v_2 t\right) \quad \text{and} \quad \zeta = \frac{Kv_2}{v_2 - v_1}\left(z - v_1 t\right).$$

Generation of second harmonica as inverse problem

The system of SHG equations has a zero-curvature representation [50,57], that allows to find its solutions by the IST method. It is useful to demonstrate here how this representation can be found.

Let equations of IST method be in the form

$$\frac{\partial \psi}{\partial \tau} = U(\lambda)\psi , \quad \frac{\partial \psi}{\partial \varsigma} = V(\lambda)\psi ,$$

so the system of equations (7.1.25) is to coincide with matrix equation

$$\frac{\partial U}{\partial \varsigma} = \frac{\partial V}{\partial \tau} + \left[V(\lambda), U(\lambda)\right]. \tag{7.1.36}$$

Evidently, matrix U must depend only on q_2, Matrix V depends only on q_1. Let these matrices be in the form of Zakharov-Shabat spectral problem:

$$U(\lambda)=\begin{pmatrix}-i\lambda & Q\\ R & i\lambda\end{pmatrix},\quad V(\lambda)=\begin{pmatrix}A & B\\ C & -A\end{pmatrix}.$$

Let us assume that $Q=\alpha q_2$, $R=\beta q_1^*$. Matrix equation (7.1.36) under these assumptions and equation for q_2 may be written as

$$\frac{\partial A}{\partial\tau}=\alpha q_2 C-\beta q_2^* B, \tag{7.1.37a}$$

$$\frac{\partial B}{\partial\tau}+2i\lambda B=-i\alpha q_1^2-2\alpha q_2 A, \tag{7.1.37b}$$

$$\frac{\partial C}{\partial\tau}-2i\lambda C=i\beta q_1^{*2}+2\beta q_2^* A. \tag{7.1.37c}$$

If in addition we expect that $B=b(\lambda)q_1^2$, $C=c(\lambda)q_1^{*2}$, then from the two last equations of this system one can find that

$$b(\lambda)=-\alpha(2\lambda)^{-1},\quad c(\lambda)=-\beta(2\lambda)^{-1},\quad A=-i(2\lambda)^{-1}|q_1|^2. \tag{7.1.38}$$

To determine the remained parameters α and β one should use equation (7.1.37a) and (7.1.38). The integrating in (7.1.37a) can be carried out if to engage the equation for q_1.

$$\frac{\partial A}{\partial\tau}=\alpha q_2\left(-\frac{\beta}{2\lambda}\right)q_1^{*2}-\beta q_2^*\left(-\frac{\alpha}{2\lambda}\right)q_1^2=$$

$$=\left(-\frac{\alpha\beta}{2\lambda}\right)\left(q_1^* q_2 q_1^*-q_2^* q_1 q_1\right)=\left(-\frac{i\alpha\beta}{2\lambda}\right)\left(\frac{\partial q_1}{\partial\tau}q_1^*+\frac{\partial q_1^*}{\partial\tau}q_1\right)=\left(-\frac{i\alpha\beta}{2\lambda}\right)\frac{\partial|q_1|^2}{\partial\tau}.$$

By comparing this result with the preceding one, we can conclude that $\alpha\beta=1$ and integrating constant is zero. Here we have two possibilities to choose the values of the parameters α and β: either $\alpha=\beta=1$ or $\alpha=-\beta=i$. In any case we have a condition (reduction) for the spectral problem: $R=Q^*$. So, the U-V matrix of zero-curvature representation has the form [57]:

$$U(\lambda)=\begin{pmatrix}-i\lambda & q_2\\ q_2^* & i\lambda\end{pmatrix},\quad V(\lambda)=\left(-\frac{i}{2\lambda}\right)\begin{pmatrix}|q_1|^2 & -iq_1^2\\ -iq_1^{*2} & -|q_1|^2\end{pmatrix}, \tag{7.1.39a}$$

or

$$U(\lambda)=\begin{pmatrix}-i\lambda & iq_2\\ -iq_2^* & i\lambda\end{pmatrix},\quad V(\lambda)=\left(-\frac{i}{2\lambda}\right)\begin{pmatrix}|q_1|^2 & q_1^2\\ q_1^{*2} & -|q_1|^2\end{pmatrix}. \tag{7.1.39b}$$

Here it is useful to note that in both cases the spectral problem with U-matrix deals with Hermitian operator. Therefore there are no complex values in its discrete spectrum and, consequently, equations (7.1.25) have no soliton solutions vanishing at infinity. It does not mean that there are no stationary (steady-state) solutions, or there are no Bäcklund transformations, or infinite set of conservation laws.

Let solutions of the equations (7.1.25) depend on the single variable $\eta=K(z-Vt)$ only. Here V is a common group velocity of the coupled pair of interacting waves. In this case equations (7.1.25) reduce to the system of the ordinary differential equations, which differ from (7.1.14) and (7.1.15) only by the notations. Turning back to real amplitudes and phase difference, we obtain the system of equations (7.1.19)-(7.1.21). So the stationary solutions of (7.1.25) can be written as

$$q_1(\eta)=\operatorname{sech}\left(\eta\sqrt{\frac{v_1v_2}{(v_1-V)(v_2-V)}}\right),$$

$$q_2(\eta)=-i\sqrt{\frac{v_2(v_1-V)}{v_1(v_2-V)}}\tanh\left(\eta\sqrt{\frac{v_1v_2}{(v_1-V)(v_2-V)}}\right).$$

In so far as the phase difference does not change, the field of the pump can be chosen to be real, and the field of harmonic has the $\pi/2$ phase shift. It is worth to note that the pump envelope does not vanish at infinity that reminds a dark soliton. Velocity V should satisfy inequality $V<\min(v_1,v_2)$.

The form of conservation laws for non-linear evolution equations solved by means of the IST method, is determined by a sort of spectral problem. As Zakharov-Shabat's spectral problem is concerned in this case, we have the same conservation laws. If the boundary conditions are such that the V-matrix at plus infinity is equal to V-matrix at minus infinity, then there is an infinite series of integrals of motion. The expressions for conserving densities are assigned by standard formulae, which are obtained for the whole class of AKNS of evolution equations under the corresponding reduction.

Let us consider U-V-matrices as the functions of complex variable λ. They have the same singular points as U-V-matrices of zero-curvature representations of the Maxwell-Bloch equations in SIT theory. As a result we have a chiral equivalence of SHG equations (7.1.25) and equations of motion in the principal chiral fields theory.

Backlund transformation for (7.1.25) can be found by the same procedure as for Maxwell-Bloch equations or Sin-Gordon equation. Let us transfer from linear equations of IST method to Riccati equations

$$\frac{\partial\Gamma}{\partial\tau} = 2i\lambda\Gamma - q_2\Gamma^2 + q_2^*, \qquad \frac{\partial\Gamma}{\partial\varsigma} = -2A\Gamma - B\Gamma^2 + C$$

where $\Gamma = \psi_2 / \psi_1$. Changing over to new variables Γ' and λ' so that the form of Riccati equations does not alter, we have to change variables q_2, A, B, and C in some special way. Let $\Gamma' = \Gamma^{-1}$. Then it can be shown that the forms-invariance of Riccati equations is ensured by additional transformation q_2 according to the formula

$$q_2' = q_2^* + \frac{2i(\lambda - \lambda')\Gamma}{1+|\Gamma|^2}.$$

This expression is named the Bäcklund transformations of q_2. Sometimes they call it the Darboux transformations.

U-V matrices obtained above correspond to a zero-curvature representation of the SHG equations (7.1.25). If we return to the initial co-ordinate frame, which is connected with the non-linear crystal, these equations can be written as:

$$\frac{\partial q_1}{\partial Z} + \frac{1}{v_1}\frac{\partial q_1}{\partial T} = iq_1^* q_2, \qquad \frac{\partial q_2}{\partial Z} + \frac{1}{v_2}\frac{\partial q_2}{\partial T} = iq_1^2,$$

where independent variable Z and T are normalised as $Z = Kz, T = Kt$, where K is the coupling coefficient. In equations of IST method we can go over to this new independent variable, thus

$$\left(\frac{\partial}{\partial Z} + \frac{1}{v_1}\frac{\partial}{\partial T}\right)\psi = U(\lambda)\psi, \quad \left(\frac{\partial}{\partial Z} + \frac{1}{v_2}\frac{\partial}{\partial T}\right)\psi = V(\lambda)\psi,$$

and to rewrite them in the following form

$$\frac{\partial\psi}{\partial T} = \tilde{U}(\lambda)\psi, \tag{7.1.40a}$$

$$\frac{\partial\psi}{\partial Z} = \tilde{V}(\lambda)\psi, \tag{7.1.40b}$$

where

$$\widetilde{U}(\lambda)=\left(\frac{v_1v_2}{v_2-v_1}\right)\begin{pmatrix}-i\lambda+\frac{i}{2\lambda}|q_1|^2 & q_2+\frac{1}{2\lambda}q_1^2\\ q_2^*+\frac{1}{2\lambda}q_1^{*2} & i\lambda-\frac{i}{2\lambda}|q_1|^2\end{pmatrix},$$

$$\widetilde{V}(\lambda)=\left(\frac{1}{v_2-v_1}\right)\begin{pmatrix}i\lambda v_1-\frac{iv_2}{2\lambda}|q_1|^2 & -v_1q_2-\frac{v_2}{2\lambda}q_1^2\\ -v_1-\frac{v_2}{2\lambda}q_1^{*2} & -i\lambda v_1+\frac{iv_2}{2\lambda}|q_1|^2\end{pmatrix} \quad (7.1.41)$$

Now it is possible to consider equation (7.1.40a) as a spectral problem of IST method. Evolution of spectral data is found from equation (7.1.40b). As usual at $z = 0$ the envelope of pump wave is defined under the assumption that the field of harmonic is absent.

7.1.5. RAMAN SCATTERING PROCESS

Consider the optical waves propagating in an infinite medium under consideration the scattering of these waves by optical phonons. This process is titled as Raman scattering in contrary to Mandelschtam-Brillouin scattering, where acoustic phonons take part. As usually the phonon frequency ω_V is much less than the carrier frequencies of optical waves. The interaction of the optical and vibration modes results in shift of the frequencies of the former. Thus, besides an incident wave with the carrier frequency ω_0 there are two new waves. They is the Stokes wave, with the frequency $\omega_S=\omega_0-\omega_V$, and anti-Stokes wave, $\omega_{AS}=\omega_0+\omega_V$. As a rule, the intensity of the anti-Stokes wave is less than Stokes one. Hence hereafter, we consider only incident and Stokes waves. Furthermore, we shall consider the stimulated Raman scattering.

The Raman scattering can be described in the frame of the classical Placzek model (see section 6.2.1 in more details) or the matrix density technique (see sections 1.1.1 and 1.3.4). We will consider both approaches.

Classical oscillator model

Let us denote the spectral half-width of the ultrashort optical pulse as $\Delta\omega_p$, and introduce the slowly varying envelopes for the oscillator co-ordinate $u(z,t)$ and electric fields $\mathcal{E}_0$ and $\mathcal{E}_S$:

$$Q=u(z,t)\exp(-i\omega_{V0}t+ik_{V0}z)+u^*(z,t)\exp(+i\omega_{V0}t-ik_{V0}z),$$

$$E=\mathcal{E}_0\exp(-i\omega_0t+ik_0z)+\mathcal{E}_0^*\exp(+i\omega_0t-ik_0z)+$$
$$+\mathcal{E}_S\exp(-i\omega_St+ik_Sz)+\mathcal{E}_S^*\exp(+i\omega_St-ik_Sz).$$

Here the wave vector k_{V0} is connected with ω_{V0} by a dispersion relation for optical phonons. The frequency ω_{V0} is the centre of the vibration line. Other frequencies ω_V form this inhomogeneous broadened line. We assume that the Stokes frequency is defined as $\omega_S = \omega_0 - \omega_{V0}$ Starting from the equation (6.2.3) and taking into account the assumption $\Delta\omega_p << \omega_V$, we can write the analogy of the equation (6.25)

$$\frac{\partial u}{\partial t} - i(\omega_V - \omega_S)u = i\frac{\alpha_D}{2\omega_V m}\mathcal{E}_0^*\mathcal{E}_S \exp[i\Delta kz], \qquad (7.1.42a)$$

where $\Delta k = k_S - k_0 - k_{V0}$ is a wave mismatch. As a suitable approximation one can take $\Delta k \approx k_S - k_0$. Now, inequality $\Delta\omega_p << \omega_V$ allows to write non-linear slowly varying polarisation (6.2.6) as a sum of well distinguished terms $\mathcal{P}_0 = n_A\alpha_D u^*\mathcal{E}_S$ and $\mathcal{P}_S = n_A\alpha_D u\mathcal{E}_0$.

The slowly varying envelopes of the incident and Stokes pulses are governed by the system of the reduced Maxwell equations (see section 1.3.4)

$$2ik_0\left(\frac{\partial\mathcal{E}_0}{\partial z} + \frac{1}{v_0}\frac{\partial\mathcal{E}_0}{\partial t}\right) = -\frac{4\pi\omega_0^2}{c^2}\mathcal{P}_0 \exp[i\Delta kz], \qquad (7.1.42b)$$

$$2ik_S\left(\frac{\partial\mathcal{E}_S}{\partial z} + \frac{1}{v_S}\frac{\partial\mathcal{E}_S}{\partial t}\right) = -\frac{4\pi\omega_S^2}{c^2}\mathcal{P}_S \exp[-i\Delta kz]. \qquad (7.1.42c)$$

So, the system of equations (7.1.42) is the base of the USP Raman scattering theory.

Two-levels system model

In some case the resonant Raman medium can be considered as an ensemble of two-level atoms. Then we used the results of Chapter 1 (section 13.4). For the sake of simplicity we ignore the dynamical Stark shift effect and inhomogeneously broadening of the resonant line. Thus the Maxwell-Bloch equations (1.3.17) and (1.3.21) result in

$$i\left(\frac{\partial\mathcal{E}_0}{\partial z} + \frac{1}{v_0}\frac{\partial\mathcal{E}_0}{\partial t}\right) = -\frac{2\pi\omega_0 n_A}{c}\Pi_{ca}^*(-\omega_S)R_{21}^*\mathcal{E}_S \exp[i\Delta kz], \qquad (7.1.43a)$$

$$i\left(\frac{\partial\mathcal{E}_S}{\partial z} + \frac{1}{v_S}\frac{\partial\mathcal{E}_S}{\partial t}\right) = -\frac{2\pi\omega_S n_A}{c}\Pi_{ca}^*(-\omega_0)R_{21}\mathcal{E}_0 \exp[-i\Delta kz], \qquad (7.1.43b)$$

$$\frac{\partial R_{21}}{\partial t} = i(\omega_S - \omega_0 - \omega_{V0})R_{21} + i\hbar^{-1}\Pi_{ca}(\omega_S)\mathcal{E}_S\mathcal{E}_0^* N_{21} \exp[i\Delta kz], \qquad (7.1.43c)$$

$$\frac{\partial N_{21}}{\partial t} = -2i\hbar^{-1}\Pi_{ca}(\omega_S)\mathcal{E}_S\mathcal{E}_0^* R_{21}^* \exp[i\Delta kz] - c.c.. \qquad (7.1.43d)$$

Assuming that population difference N_{21} is approximately constant we can omit the equation (7.1.43d). That means the optical pulse is weak enough and the energy levels populations variation can be ignored. In this case the system of equations follows from (7.1.43)

$$i\left(\frac{\partial \mathcal{E}_0}{\partial z}+\frac{1}{v_0}\frac{\partial \mathcal{E}_0}{\partial t}\right)=-\frac{2\pi\omega_0 n_A}{c}\Pi^*_{ca}(-\omega_S)R^*_{21}\mathcal{E}_S\exp[i\Delta kz], \qquad (7.1.44a)$$

$$i\left(\frac{\partial \mathcal{E}_S}{\partial z}+\frac{1}{v_S}\frac{\partial \mathcal{E}_S}{\partial t}\right)=-\frac{2\pi\omega_S n_A}{c}\Pi^*_{ca}(-\omega_0)R_{21}\mathcal{E}_0\exp[-i\Delta kz], \qquad (7.1.44b)$$

$$\frac{\partial R_{21}}{\partial t}-i\Delta\omega R_{21}=i\hbar^{-1}\Pi_{ca}(\omega_S)\mathcal{E}_S\mathcal{E}^*_0\exp[i\Delta kz], \qquad (7.1.44c)$$

where $\Delta\omega=(\omega_S-\omega_0-\omega_{V0})$ is the frequency detuning. Both these equations and the system of equations (7.1.42) describe the *stimulated Raman scattering* (SRS) under condition of the weak excitation of the resonant Raman medium.

Reduction of the Raman scattering equation

In order to demonstrate the applicability of the IST method in analysis of SRS we must deduce the systems of equations (7.1.42) and (7.1.43) to Maxwell-Bloch one. Following [12-15] we assume $v_0=v_S$ and define the new variable: $\tau=(t-z/v_0)t^{-1}_{p0}$, $\zeta=z/L_C$. We introduce additionally:

- for an oscillator model

$$u=\frac{l_{ph}Q_3}{\sqrt{\omega_0\omega_S}},\quad \mathcal{E}_0=\frac{A_0Q_2}{\sqrt{\omega_S}},\quad \mathcal{E}_S=\frac{A_0Q_1}{\sqrt{\omega_0}},$$

$$l_{ph}=\frac{\alpha_D A_0^2 i_{p0}}{2m\omega_V},\quad L_C=\left(\frac{\pi n_A\alpha_D^2A_0^2t_{p0}}{cm\omega_V}\right)^{-1},$$

- and for a two-level system model

$$R_{21}=\frac{t_{p0}\Pi_{ca}(\omega_S)A_0^2}{\hbar\sqrt{\omega_S\omega_0\Pi^*_{ca}(\omega_S)\Pi^*_{ca}(\omega_0)}}Q_3,\quad \mathcal{E}_0=\frac{A_0Q_2}{\sqrt{\omega_S\Pi^*_{ca}(\omega_S)}},\quad \mathcal{E}_S=\frac{A_0Q_1}{\sqrt{\omega_0\Pi^*_{ca}(\omega_0)}},$$

$$L_C=\left(\frac{2\pi n_A\Pi_{ca}(\omega_S)A_0^2t_{p0}}{c\hbar}\right)^{-1}.$$

The unified system of equations takes the form

$$\frac{\partial Q_1}{\partial \zeta} = iQ_2Q_3 \exp(-i\Delta\zeta)\,, \quad \frac{\partial Q_2}{\partial \zeta} = iQ_1Q_3^* \exp(+i\Delta\zeta)\,, \quad \frac{\partial Q_3}{\partial \tau} = i\delta Q_3 + iQ_2^*Q_1 \exp(+i\Delta\zeta)\,,$$

where $\delta = \Delta\omega t_{p0}$ is the normalised frequency detuning and $\Delta = \Delta k L_C$ is the normalised wave mismatch. In terms of new variables

$$q_1 = Q_1 \exp(i\Delta\zeta)\,, \quad q_2 = Q_2 \exp(-i\Delta\zeta)\,, \quad q_3 = Q_3 \exp(i\Delta\zeta)\,,$$

this system becomes

$$\frac{\partial q_1}{\partial \zeta} = i\Delta q_1 + iq_2q_3\,, \tag{7.1.45a}$$

$$\frac{\partial q_2}{\partial \zeta} = -i\Delta q_2 + iq_3^*q_1\,, \tag{7.1.45b}$$

$$\frac{\partial q_3}{\partial \tau} = i\delta q_3 + iq_1q_2^*\,. \tag{7.1.45c}$$

Now we can say that the first two equations of (7.1.45) are the components of Schrödinger equation for a two-level atom and the last equation of (7.1.45) is the reduced Maxwell equation for the pulse envelope of the pseudo-electromagnetic field q_3. It is known that the Schrödinger equation for a two-level atom may be converted into Bloch equations if the squares of the wave functions are defined

$$w_{12} = q_1q_2^*\,, \quad w_{11} = |q_1|^2 - |q_2|^2\,.$$

In terms of these variables system (7.1.45) can be rewritten as

$$\frac{\partial q_3}{\partial \tau} = i\delta q_3 + iw_{12}\,. \tag{7.1.46a}$$

$$\frac{\partial w_{12}}{\partial \zeta} = 2i\Delta w_{12} - iq_3w_{11}\,, \tag{7.1.46b}$$

$$\frac{\partial w_{11}}{\partial \zeta} = 2i(q_3w_{12}^* - q_3^*w_{12})\,, \tag{7.1.46c}$$

Let the linear equations of the IST method take the following form

$$\frac{\partial\psi}{\partial\zeta}=\hat{U}(\lambda)\psi,\quad \frac{\partial\psi}{\partial\tau}=\hat{V}(\lambda)\psi\,. \tag{7.1.47}$$

By using the results of section 3.1.3. we can write the *U-V*-pare for zero-curvature representation of the system of equations (7.1.46):

$$\hat{U}(\lambda)=\begin{pmatrix}-i\lambda & q_3\\ -q_3^* & i\lambda\end{pmatrix},$$

$$\hat{V}(\lambda)=\begin{pmatrix}\frac{i\delta}{2}+\frac{i(|q_1|^2-|q_2|^2)}{4(\Delta k+\lambda)} & \frac{q_1q_2^*}{2(\Delta k+\lambda)}\\ \frac{-q_1^*q_2}{2(\Delta k+\lambda)} & -\frac{i\delta}{2}-\frac{i(|q_1|^2-|q_2|^2)}{4(\Delta k+\lambda)}\end{pmatrix}. \tag{7.1.48}$$

It is worth noting that this *U-V*-pair is not suitable to solve initial problem for system of equations (7.1.45) describing the SRS by the IST method with the Zakharov-Schabat spectral problem. Here the *V*-matrix must be considered to determine the spectral date. However, we can obtain the several partial results related with the SRS. Furthermore, this zero-curvature representation was used to solve the problem, which related with the creation of solitons in the SRS [18].

7.2. Three-wave interaction and soliton formation

7.2.1. ZERO-CURVATURE REPRESENTATION OF THE 3-WAVE INTERACTION EQUATIONS

Three-waves interaction in a quadratic non-linear medium under condition of the phase matching leads to the system of equations, which can be solved (or at least can be analysed) by IST method. The simplest generalisation of the zero-curvature representation, which appears here, is a simple change of Zakharov-Shabat-AKNS equations with the matrices of 2×2 dimension to the equations with the matrices of greater dimension.

Let us consider the $N\times N$ - matrix generalisation of the equations of IST method

$$\frac{\partial\psi}{\partial t}=\hat{U}\psi=i\hat{\Lambda}\psi+\hat{P}\psi\,, \tag{7.2.1}$$

$$\frac{\partial\psi}{\partial z}=\hat{V}\psi\,. \tag{7.2.2}$$

Here ψ is N component matrix-column, $\hat{\Lambda} = \mathrm{diag}(\Lambda_1, \Lambda_2, \ldots, \Lambda_N)$ is a constant matrix, $\Lambda_j \neq \Lambda_k$ at any j and k, $\hat{P}(t,z)$ is the matrix potentials of IST spectral problem and $\hat{V}(t,z)$ is the matrix depending on the spectral parameter λ and $\hat{P}(t,z)$.

The integrability condition for the pair of equations (7.2.1) and (7.2.2) results in matrix equation

$$\frac{\partial \hat{V}}{\partial t} = \frac{\partial \hat{P}}{\partial z} + i\lambda\left[\hat{\Lambda}, \hat{V}\right] + \left[\hat{P}, \hat{V}\right], \tag{7.2.3}$$

which presents the zero-curvature representation of the equation (or set of equations) solved by IST method discussed above.

Taking obvious dependency of $\hat{V}(t,z)$ on λ into account, we may substitute such matrix into equation (7.2.3). Then by equating the coefficients of various powers of λ in resultant equation to zero, we obtain the desired system of equations.

The simplest case is when $\hat{V}(t,z)$ is a linear function of spectral parameter λ. Let

$$\hat{V}(t,z;\lambda) = \lambda \hat{V}^{(1)}(t,z) + \hat{V}^{(0)}(t,z).$$

Then from equation (7.2.3) it follows

$$\lambda \frac{\partial \hat{V}^{(1)}}{\partial t} + \frac{\partial \hat{V}^{(0)}}{\partial t} = \frac{\partial \hat{P}}{\partial z} + i\lambda\left[\hat{\Lambda}, \hat{V}^{(0)}\right] + i\lambda^2\left[\hat{\Lambda}, \hat{V}^{(1)}\right] + \left[\hat{P}, \hat{V}^{(0)}\right] + \lambda\left[\hat{P}, \hat{V}^{(1)}\right].$$

Equalisation of the similar terms in this expression gives the matrix equations

$$\left[\hat{\Lambda}, \hat{V}^{(1)}\right] = 0, \tag{7.2.4a}$$

$$i\left[\hat{\Lambda}, \hat{V}^{(0)}\right] + \left[\hat{P}, \hat{V}^{(1)}\right] = \frac{\partial \hat{V}^{(1)}}{\partial t}, \tag{7.2.4b}$$

$$\frac{\partial \hat{P}}{\partial z} + \left[\hat{P}, \hat{V}^{(0)}\right] = \frac{\partial \hat{V}^{(0)}}{\partial t}. \tag{7.2.4c}$$

It follows from (7.2.4a) that $\hat{V}^{(1)}(t,z)$ is an arbitrary diagonal matrix. Besides, let $\hat{V}^{(1)}(t,z)$ be not dependent on space co-ordinates and time. Let us denote

$$\hat{V}^{(1)} = \mathrm{diag}(a_1, a_2, \ \ldots, a_N).$$

In this case equation (7.2.4b) brings about the relation for matrix elements of $\hat{V}^{(0)}(t,z)$ and $\hat{P}(t,z)$:

$$\hat{V}^{(0)}{}_{kj} = \frac{a_k - a_j}{i(\Lambda_k - \Lambda_j)} P_{kj} = a_{kj} P_{kj} . \tag{7.2.5}$$

Here we assume that $\Lambda_j \neq \Lambda_k$. But if such indices j' and k' for which $\Lambda_{j'} = \Lambda_{k'}$ are existed, then by choosing $a_{j'} = a_{k'}$ one can satisfy the appropriate matrix component of equation (7.2.4b).

Substitution (7.2.5) into (7.2.4c) results in the system of equations:

$$\frac{\partial P_{kj}}{\partial z} - a_{kj} \frac{\partial P_{kj}}{\partial t} = \sum_{m=1}^{N} \left(a_{km} - a_{jm} \right) P_{km} P_{mj} . \tag{7.2.6}$$

We will consider now the unified system of equations describing the parametric three-waves interaction (7.1.3) under conditions of phase matching

$$\left(\frac{\partial}{\partial z} + \frac{1}{v_1} \frac{\partial}{\partial t} \right) Q_1 = i\tilde{\beta} Q_2^* Q_3^* , \tag{7.2.7a}$$

$$\left(\frac{\partial}{\partial z} + \frac{1}{v_2} \frac{\partial}{\partial t} \right) Q_2 = i\tilde{\beta} Q_3^* Q_1^* , \tag{7.2.7b}$$

$$\left(\frac{\partial}{\partial z} + \frac{1}{v_3} \frac{\partial}{\partial t} \right) Q_3 = -i\tilde{\beta} Q_1^* Q_2^* , \tag{7.2.7c}$$

For the sake of certainty let us choose the following relation between velocities

$$v_1 < v_2 < v_3 .$$

The comparison of equations (7.2.6) with (7.2.7) gives rise to

$$P_{12} = \frac{-i\tilde{\beta} Q_3}{\sqrt{b_{13} b_{23}}}, \quad P_{21} = \frac{i\tilde{\beta} Q_3^*}{\sqrt{b_{13} b_{23}}}, \quad P_{13} = \frac{i\tilde{\beta} Q_2^*}{\sqrt{b_{23} b_{12}}},$$

$$P_{31} = \frac{-i\tilde{\beta} Q_2}{\sqrt{b_{23} b_{12}}}, \quad P_{23} = \frac{i\tilde{\beta} Q_1}{\sqrt{b_{12} b_{13}}}, \quad P_{32} = \frac{i\tilde{\beta} Q_1^*}{\sqrt{b_{12} b_{13}}},$$

where $a_{12} = -v_3^{-1}$, $a_{13} = -v_2^{-1}$, $a_{23} = -v_1^{-1}$ and $b_{kj} = \left(v_k^{-1} - v_j^{-1} \right)$.

Evolution of the scattering data is defined by the second linear equation of the IST method (7.2.2) and matrix $\hat{V}(t,z)$, which is defined by their matrix elements:

$$V_{kj} = \lambda a_k \delta_{kj} + a_{kj} P_{kj} \ . \qquad (7.2.8)$$

If, following [31,50], we take $\Lambda_k = v_k^{-1}$, then one gets

$$a_k = \frac{iv_k}{v_1 v_2 v_2}, \quad a_{kj} = -\frac{v_k v_j}{v_1 v_2 v_2} \ .$$

Now it is useful to consider the spectral problem which related with the linear equation (7.1.1) under vanishing boundary conditions.

7.2.2. INVERSE PROBLEM BY ZAKHAROV-MANAKOV-KAUP

The solution of the inverse problem for (7.2.1) has been discussed in detail in [30,31]. Here we find expedient to bring only the minimum information, which is required for solution of the equations (7.2.7). Data on properties of operator into (7.2.1) may be found in special articles or book [50].

As usual, Jost functions are introduced as solutions of the equations (7.2.1), which have the following asymptotics :

$$\Phi^{(1)}(t,\lambda) \to \begin{pmatrix}1\\0\\0\end{pmatrix} \exp(i\lambda\Lambda_1 t) \ , \ \Phi^{(2)}(t,\lambda) \to \begin{pmatrix}0\\1\\0\end{pmatrix} \exp(i\lambda\Lambda_2 t) \ , \ \Phi^{(1)}(t,\lambda) \to \begin{pmatrix}0\\0\\1\end{pmatrix} \exp(i\lambda\Lambda_3 t)$$

as $t \to -\infty$, and

$$\Psi^{(1)}(t,\lambda) \to \begin{pmatrix}1\\0\\0\end{pmatrix} \exp(i\lambda\Lambda_1 t) \ , \ \Psi^{(2)}(t,\lambda) \to \begin{pmatrix}0\\1\\0\end{pmatrix} \exp(i\lambda\Lambda_2 t) \ , \ \Psi^{(1)}(t,\lambda) \to \begin{pmatrix}0\\0\\1\end{pmatrix} \exp(i\lambda\Lambda_3 t)$$

as $t \to +\infty$.

Transfer matrix $\hat{T}(\lambda)$ is defined by means of these Jost functions:

$$\Phi^{(k)}(t,\lambda) = \sum_{n=1}^{3} T_{kn}(\lambda) \Psi^{(n)}(t,\lambda) \ , \qquad (7.2.9a)$$

and the inverse matrix $\hat{R}(\lambda)$ is defined as

$$\Psi^{(k)}(t,\lambda) = \sum_{n=1}^{3} R_{kn}(\lambda)\Phi^{(n)}(t,\lambda). \qquad (7.2.9b)$$

If the potentials in the spectral problem (7.2.1) are absolutely integrable, then $T_{11}(\lambda)$ and $R_{33}(\lambda)$ are analytical functions of spectral parameter λ under condition $\mathrm{Im}\,\lambda < 0$, but $T_{33}(\lambda)$ and $R_{11}(\lambda)$ are analytical functions of λ under condition $\mathrm{Im}\,\lambda > 0$.

The Jost functions $\Psi^{(1)}$ and $\Psi^{(3)}$ allow a triangular representation:

$$\Psi^{(1)} \exp(-i\lambda\Lambda_1 t) = \begin{pmatrix}1\\0\\0\end{pmatrix} + \int_t^{\infty} K^{(1)}(t,s)\exp\{i\lambda\gamma_{12}(s-t)\}ds\ , \qquad (7.2.10a)$$

$$\Psi^{(3)} \exp(-i\lambda\Lambda_3 t) = \begin{pmatrix}0\\0\\1\end{pmatrix} + \int_t^{\infty} K^{(3)}(t,s)\exp\{-i\lambda\gamma_{23}(s-t)\}ds\ , \qquad (7.2.10b)$$

where $\gamma_{nk} = \Lambda_n - \Lambda_k$, $K^{(1)}$ and $K^{(3)}$ are matrices-columns, not containing variable λ. These functions define potentials $\hat{P}(t,z)$ of the spectral problem under consideration.

The system of linear integral equations for $K^{(1)}$ and $K^{(3)}$, obtained in [31], is often named by Gelfand-Levitan-Marchenko. This system generalises the corresponding equations of Zakharov-Shabat-AKNS spectral problem. Under condition $y > x$ these equations have the form

$$K^{(1)}(x,y) + \begin{pmatrix}0\\1\\0\end{pmatrix}F_1(y) + \begin{pmatrix}1\\0\\0\end{pmatrix}F_3(x,y) + \begin{pmatrix}0\\0\\1\end{pmatrix}F_5(x,y) + $$
$$+ \int_x^{\infty}\left[K^{(1)}(x,s)F_3(s,y) + K^{(3)}(x,s)F_5(s,y)\right]ds = 0 \qquad (7.2.11a)$$

$$K^{(3)}(x,y) + \begin{pmatrix}0\\1\\0\end{pmatrix}F_2(y) + \begin{pmatrix}1\\0\\0\end{pmatrix}F_6(x,y) + \begin{pmatrix}0\\0\\1\end{pmatrix}F_4(x,y) + $$
$$+ \int_x^{\infty}\left[K^{(1)}(x,s)F_6(s,y) + K^{(3)}(x,s)F_4(s,y)\right]ds = 0\,. \qquad (7.2.11b)$$

where the kernels of these equations F_n are expressed in terms of scattering data, the functions $\rho_n(\lambda)$, and are assigned by the following formulae:

$$F_1(x)=\gamma_{12}\oint_{\bar{C}}\frac{d\lambda}{2\pi}\rho_1(\lambda)\exp\{-i\lambda\gamma_{12}x\},\quad F_2(x)=\gamma_{23}\oint_{C}\frac{d\lambda}{2\pi}\rho_2(\lambda)\exp\{i\lambda\gamma_{23}x\}$$

$$F_3(x,y)=\gamma_{12}\oint_{C}\frac{d\lambda}{2\pi i}\rho_5(\lambda)\exp\{i\lambda\gamma_{12}x\}\oint_{\bar{C}}\frac{d\lambda'}{2\pi}\frac{\rho_1(\lambda')\exp\{-i\lambda'\gamma_{12}y\}}{\lambda'-\lambda},$$

$$F_4(x,y)=-\gamma_{23}\oint_{\bar{C}}\frac{d\lambda}{2\pi i}\rho_6(\lambda)\exp\{-i\lambda\gamma_{23}x\}\oint_{C}\frac{d\lambda'}{2\pi}\frac{\rho_2(\lambda')\exp\{i\lambda'\gamma_{23}y\}}{\lambda'-\lambda},$$

$$F_5(x,y)=\gamma_{12}\oint_{\bar{C}}\frac{d\lambda}{2\pi}\rho_3(\lambda)\exp\{-i\lambda(\gamma_{12}y+\gamma_{23}x)\}-$$

$$-\gamma_{12}\oint_{\bar{C}}\frac{d\lambda}{2\pi i}\rho_6(\lambda)\exp\{-i\lambda\gamma_{23}x\}\oint_{\bar{C}}\frac{d\lambda'}{2\pi}\frac{\rho_1(\lambda')\exp\{-i\lambda'\gamma_{12}y\}}{\lambda'-\lambda+i\varepsilon},$$

$$F_6(x,y)=\gamma_{23}\oint_{C}\frac{d\lambda}{2\pi}\rho_4(\lambda)\exp\{i\lambda(\gamma_{12}y+\gamma_{23}x)\}-$$

$$+\gamma_{23}\oint_{C}\frac{d\lambda}{2\pi i}\rho_5(\lambda)\exp\{i\lambda\gamma_{12}x\}\oint_{C}\frac{d\lambda'}{2\pi}\frac{\rho_2(\lambda')\exp\{i\lambda'\gamma_{23}y\}}{\lambda'-\lambda+i\varepsilon}.$$

In the integrals above contour C passes on the complex plane λ from $-\infty+i\varepsilon$ to $+\infty+i\varepsilon$, turning round from above the zeroes of functions $T_{33}(\lambda)$ and $R_{11}(\lambda)$. Contour $\overline{C}$ passes from $-\infty-i\varepsilon$ to $+\infty-i\varepsilon$ turning round from below all zeroes of $T_{11}(\lambda)$ and $R_{33}(\lambda)$.

Fig.7.2.1

Contours C and $\overline{C}$ in complex plane of spectral parameter

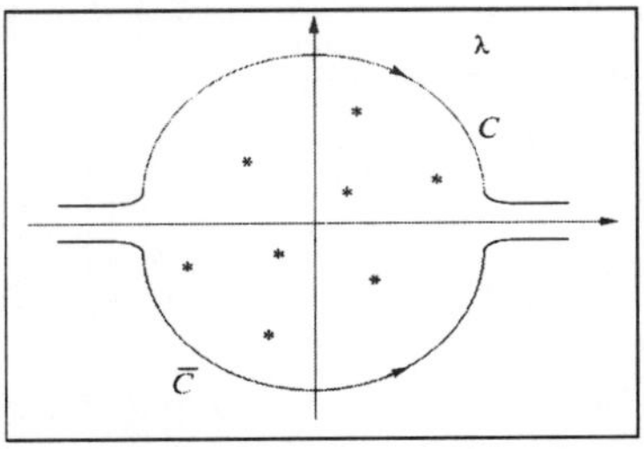

Solutions of Gelfand-Levitan-Marchenko equations allow to determine the potentials of

spectral problem according to the following expressions

$$P_{21}(t) = -K_2^{(1)}(t,t) \ , \qquad P_{31}(t) = -\frac{\gamma_{13}}{\gamma_{12}} K_3^{(1)}(t,t) \, , \tag{7.2.12a}$$

$$P_{13}(t) = -\frac{\gamma_{13}}{\gamma_{23}} K_1^{(3)}(t,t) \ , \quad P_{23}(t) = -K_2^{(3)}(t,t) \, , \tag{7.2.12b}$$

$$P_{12}(t) = -\gamma_{12}\left\{ E_1(t) - \int_t^\infty \left[K_1^{(3)}(t,s)E_2(s) - K_1^{(1)}(t,s)E_1(s) \right] ds \right\} \, , \tag{7.2.12c}$$

$$P_{32}(t) = -\gamma_{23}\left\{ E_2(t) + \int_t^\infty \left[K_3^{(3)}(t,s)E_2(s) - K_3^{(1)}(t,s)E_1(s) \right] ds \right\} , \tag{7.2.12d}$$

where

$$E_1(x) = \oint_C \frac{d\lambda}{2\pi} \rho_5(\lambda) \exp\{i\lambda\gamma_{12}x\} \ , \quad E_2(x) = \oint_{\bar{C}} \frac{d\lambda}{2\pi} \rho_6(\lambda) \exp\{-i\lambda\gamma_{12}x\} \, .$$

Expressions for F_n and E_n contain functions $\rho_n(\lambda)$, which are defined through the matrix elements of transfer matrices $\hat{T}(\lambda)$ and $\hat{R}(\lambda)$:

$$\rho_1(\lambda) = \frac{T_{12}(\lambda)}{T_{11}(\lambda)}, \quad \rho_2(\lambda) = \frac{T_{32}(\lambda)}{T_{33}(\lambda)}, \quad \rho_3(\lambda) = \frac{T_{13}(\lambda)}{T_{11}(\lambda)},$$
$$\rho_4(\lambda) = \frac{T_{31}(\lambda)}{T_{33}(\lambda)}, \quad \rho_5(\lambda) = \frac{R_{21}(\lambda)}{R_{11}(\lambda)}, \quad \rho_6(\lambda) = \frac{R_{23}(\lambda)}{R_{33}(\lambda)}. \tag{7.2.13}$$

These functions, which can be called the scattering coefficients, together with the set of zeroes of the functions $T_{33}(\lambda)$ and $R_{11}(\lambda)$, $T_{11}(\lambda)$ and $R_{33}(\lambda)$ (this set forms a discrete spectrum of operator in (7.2.1)) are identified as scattering data for a spectral problem (7.2.1).

Evolution of the transfer matrix $\hat{T}(\lambda, z)$ has the form of [50] (also see section 3.5)

$$\frac{d}{dz}\hat{T} = \hat{A}^{(+)}\hat{T} - \hat{T}\hat{A}^{(-)} .$$

If we accept the condition that at $t \to \pm\infty$ all interacting waves vanish, then

$$\hat{V}^{(+)} = \hat{V}^{(-)} = \text{diag}(\lambda a_1, \lambda a_2, \lambda a_3) \ .$$

Thereby

$$\frac{d}{dz}\hat{T} = \left[\hat{V}^{(-)}, \hat{T}\right].$$

By a similar way we can get

$$\frac{d}{dz}\hat{R} = \left[\hat{V}^{(-)}, \hat{R}\right].$$

These equations can be solved for matrix elements of transfer matrices, so the resultant expressions are as follows:

$$T_{nm}(\lambda, z) = T_{nm}(\lambda, 0)\exp\{\lambda z(a_n - a_m)\},$$
$$R_{nm}(\lambda, z) = R_{nm}(\lambda, 0)\exp\{\lambda z(a_n - a_m)\}.$$

In addition the symmetry relations should be mentioned, which take place here

$$R_{nm}(\lambda) = \varepsilon_n \varepsilon_m T^*_{mn}(\lambda^*) ,$$

and

$$\rho_5(\lambda) = -\rho_1^*(\lambda^*), \quad \rho_6(\lambda) = \rho_2^*(\lambda^*), \\ \rho_4(\lambda) - \rho_3^*(\lambda^*) = \rho_1^*(\lambda^*)\rho_2(\lambda), \tag{7.2.14}$$

where $\varepsilon_1 = 1, \quad \varepsilon_2 = \varepsilon_3 = -1$. These relations in their turn lead to other relations, which provide the simplification of calculations:

$$F_3^*(x,y) = F_3(x,y), \quad F_4^*(x,y) = F_4(y,x), \\ F_5^*(x,y) = -\frac{\gamma_{12}}{\gamma_{23}} F_6(y,x), \\ E_1(x) = \frac{1}{\gamma_{12}} F_1^*(x), \quad E_2(x) = \frac{1}{\gamma_{23}} F_2^*(x). \tag{7.2.15}$$

7.2.3. SOLITONS OF 3-WAVE INTERACTION

It is known that the points of a discreet spectrum, in which $T_{11}(\lambda)$ and $T_{33}(\lambda)$ vanish, determine soliton solutions of the equations of parametric three wave interaction (7.1.3). Let us consider the case of simple zeroes of $T_{11}(\lambda)$ and $T_{33}(\lambda)$:

$$T_{11}(\bar{\lambda}_1) = 0, \quad \mathrm{Im}\,\bar{\lambda}_1 < 0; \quad T_{33}(\lambda_3) = 0, \quad \mathrm{Im}\,\lambda_3 > 0 \ :$$

Let us introduce four real parameters

$$\gamma_{12}\bar{\lambda}_1 = \xi_1 - i\eta_1, \quad \gamma_{23}\lambda_3 = \xi_3 + i\eta_3,$$

and let

$$\operatorname{Res}\rho_2(\lambda_3) = \frac{T_{32}(\lambda_3)}{T'_{33}(\lambda_3)} = A \text{ and } \operatorname{Res}\rho_1(\bar{\lambda}_1) = \frac{T_{12}(\bar{\lambda}_1)}{T'_{11}(\bar{\lambda}_1)} = \bar{A}.$$

All other residues are fixed by the symmetry properties of the transfer matrix, so

$$\operatorname{Res}\rho_4(\lambda_3) = \frac{T_{31}(\lambda_3)}{T'_{33}(\lambda_3)}, \qquad \operatorname{Res}\rho_3(\bar{\lambda}_1) = \frac{T_{13}(\bar{\lambda}_1)}{T'_{11}(\bar{\lambda}_1)},$$

$$\operatorname{Res}\rho_6(\lambda_3^*) = A^*, \qquad \operatorname{Res}\rho_5(\bar{\lambda}_1^*) = -\bar{A}^*.$$

Then from the definition of the F_n it follows

$$F_1(x) = i\gamma_{12}\bar{A}\exp\{-i\gamma_{12}\lambda_1 x\}, \quad F_2(x) = -i\gamma_{23}A\exp\{i\gamma_{23}\lambda_3 x\},$$

$$F_3(x,y) = -\gamma_{12}\frac{\bar{A}^*\bar{A}}{2\eta_1}\exp\left\{i\gamma_{12}\left(\bar{\lambda}_1^* x - \bar{\lambda}_1 y\right)\right\},$$

$$F_4(x,y) = \gamma_{23}\frac{A^*A}{2\eta_3}\exp\left\{-i\gamma_{23}\left(\lambda_3^* x - \lambda_3 y\right)\right\},$$

$$F_5(x,y) = -i\gamma_{12}\frac{\bar{A}^*A}{\lambda_1^* - \lambda_3^*}\exp\left\{-i\gamma_{23}\lambda_3^* x - i\gamma_{12}\bar{\lambda}_1 y\right\},$$

$$F_6(x,y) = -i\gamma_{23}\frac{\bar{A}^*A}{\lambda_3 - \bar{\lambda}_1^*}\exp\left\{i\gamma_{12}\bar{\lambda}_1^* x + i\gamma_{23}\lambda_3 y\right\}.$$

The equations of evolution of scattering data result in

$$A = A_0\exp\left\{-i\gamma_{23}\lambda_3 v_1^{-1} z\right\}, \quad \bar{A} = \bar{A}_0\exp\left\{i\gamma_{13}\bar{\lambda}_1 v_3^{-1} z\right\}.$$

It is convenient to introduce

$$A_0 = 2\eta_3\gamma_{23}^{-1}\exp\left\{\eta_3 t_1 - i\xi_3\bar{t}_1\right\}, \quad \bar{A}_0 = 2\eta_1\gamma_{12}^{-1}\exp\left\{\eta_1 t_3 + i\xi_1\bar{t}_3\right\}.$$

Gelfand-Levitan-Marchenko equations under these conditions transform into a set of ordinary linear equations, whose solutions can be found by a conventional technique, thus providing $K^{(1)}$ and $K^{(3)}$. Then, using expressions (7.2.12), we can find:

$$P_{23} = \frac{2\eta_3 \exp\{i\xi_3\theta_1)}{D(t,z)} \left\{ \exp(\eta_1\theta_3) - \left(\frac{\bar{\lambda}_1 - \lambda_3}{\bar{\lambda}_1^* - \lambda_3} \right) \exp(-\eta_1\theta_3) \right\},$$

$$P_{31} = \frac{4\eta_1\eta_3\gamma_{13}}{\gamma_{12}\gamma_{23} D(t,z)} \exp\left\{ -i\xi_1\theta_3 - i\xi_3\theta_1 \right\},$$

$$P_{12} = \frac{2\eta_1 \exp\{i\xi_1\theta_3)}{D(t,z)} \left\{ \exp(\eta_3\theta_1) + \left(\frac{\bar{\lambda}_1^* - \lambda_3^*}{\bar{\lambda}_1^* - \lambda_3} \right) \exp(-\eta_3\theta_1) \right\},$$

$$P_{32} = -P_{23}^*, \quad P_{13} = P_{31}^*, \quad P_{21} = P_{12}^*.$$

Where we denoted $\theta_1 = t - z / v_1 - \bar{t}_1$, $\theta_3 = t - z / v_3 - \bar{t}_3$ and

$$D(t,z) = \left[\exp(\eta_1\theta_3) - \exp(-\eta_1\theta_3) \right] \left[\exp(\eta_3\theta_1) - \exp(-\eta_3\theta_1) \right] + \\ + (-1) \frac{(\bar{\lambda}_1 - \bar{\lambda}_1^*)(\lambda_3 - \lambda_3^*)}{(\bar{\lambda}_1 - \lambda_3^*)(\bar{\lambda}_1^* - \lambda_3)} \exp(-\eta_1\theta_3 - \eta_3\theta_1)$$

As $z \to -\infty$ we have a solution in the form of two pulses: pulse Q_1 moving along the characteristic $t - z / v_1$ and pulse Q_3 moving along the characteristic $t - z / v_3$, whereas $Q_2 = 0$. At the moment when the pulses Q_1 and Q_3 collide, the optical pulse with envelope Q_2 appears. Then Q_2 disappears, but pulses Q_1 and Q_3 diverge, moving along their characteristic. The envelopes of the Q_1 and Q_3 do not alter in this process.

7.2.4. NON-COLLINEAR SECOND HARMONIC GENERATION

We shall consider a process of the SHG, when pump waves and harmonic propagate in arbitrary directions. In this case the evolution of slowly varying optical pulse envelopes are described by the equations, which are similar to (7.1.1), but where the spatial derivative must be replaced by the directional derivative with respect to a corresponding wave and the argument of the exponent functions in the right-hand side of these equations change to $\pm i(\vec{k}_1 + \vec{k}_2 - \vec{k}_3) \cdot \vec{r}$. Consequently, the phase-matching condition will now look as

$$(\vec{k}_1 + \vec{k}_2 - \vec{k}_3) = 0 \ . \tag{7.2.16}$$

Let the harmonic wave propagate along z-axis, and the wave vector of a pump wave lies in the xz- plane, so that $\vec{k}_1 = k_1(\eta_x, \eta_z)$ and $\vec{k}_2 = k_1(-\eta_x, \eta_z)$. The projection of the vector equality (7.2.16) on the z-axis yields an angle θ_m, under which the phase-matching condition is valid:

$$n(2\omega) = n(\omega)\cos\theta_m .$$

Obviously, it is possible to reach a phase-matching for a non-collinear SHG only under condition $n(2\omega) < n(\omega)$.

The system of equations, describing the SHG under the phase-matching reads

$$\left(\eta_z \frac{\partial}{\partial z} + \eta_x \frac{\partial}{\partial x} + \frac{1}{v_1}\frac{\partial}{\partial t}\right)\mathcal{E}_1^{(+)} = i\frac{4\pi\omega\chi^{(2)}(2\omega,-\omega)}{cn(\omega)}\mathcal{E}_1^{(-)*}\mathcal{E}_2, \qquad (7.2.17a)$$

$$\left(\eta_z \frac{\partial}{\partial z} - \eta_x \frac{\partial}{\partial x} + \frac{1}{v_1}\frac{\partial}{\partial t}\right)\mathcal{E}_1^{(-)} = i\frac{4\pi\omega\chi^{(2)}(2\omega,-\omega)}{cn(\omega)}\mathcal{E}_1^{(+)*}\mathcal{E}_2 , \qquad (7.2.17b)$$

$$\left(\frac{\partial}{\partial z} + \frac{1}{v_2}\frac{\partial}{\partial t}\right)\mathcal{E}_2 = i\frac{8\pi\omega\chi^{(2)}(\omega,\omega)}{cn(2\omega)}\mathcal{E}_1^{(+)}\mathcal{E}_1^{(-)} \qquad (7.2.17c)$$

where the pump wave is presented as

$$E(t,\vec{r}) = \mathcal{E}_1^{(+)}(t,x,z)\exp[-i\omega t + k_1(\eta_z z + \eta_x x)] + \\ + \mathcal{E}_1^{(-)}(t,x,z)\exp[-i\omega t + k_1(\eta_z z - \eta_x x)]$$

and the second harmonic wave is represented as

$$E_2(t,\vec{r}) = \mathcal{E}_2(t,x,z)\exp[-i2\omega t + k_2 z].$$

Define the normalised envelopes of interaction waves in the form

$$\mathcal{E}_1^{(+)} = \mathcal{E}_{10} q_1 , \quad \mathcal{E}_2 = \mathcal{E}_{10} q_2 , \quad \mathcal{E}_1^{(-)} = \mathcal{E}_{10} q_3 ,$$

and let again the coupling constant be

$$K = \frac{4\pi\omega\chi^{(2)}(\omega,\omega)}{cn(\omega)}\mathcal{E}_{10} .$$

By introducing the characteristic co-ordinates in the form

$$\zeta = \frac{2v_2 K\, n(\omega)}{n(2\omega)(v_2 - \eta_z v_1)}\left(z - \eta_z v_1 t\right),$$

$$\tau = \frac{v_1 K}{(v_2 - \eta_z v_1)} \left(\eta_x v_1 t - \eta_x \frac{v_1}{v_2} z + x - \eta_z \frac{v_1}{v_2} x \right),$$

$$\xi = \frac{v_1 K}{(v_2 - \eta_z v_1)} \left(\eta_x v_1 t - \eta_x \frac{v_1}{v_2} z - x + \eta_z \frac{v_1}{v_2} x \right),$$

these equations reduce to

$$\frac{\partial q_1}{\partial \tau} = i q_2 q_3^*, \quad \frac{\partial q_2}{\partial \zeta} = i q_1 q_3, \quad \frac{\partial q_3}{\partial \xi} = i q_2 q_1^*. \tag{7.2.18}$$

Now we shall consider these equations referring to [45]. It is interesting that the solutions, obtained in [45], are identical to those found by the IST method. The present solutions will be obtained without the IST technique, but by the elementary method, constituting the simplest class of the closed-form solutions.

From (7.2.18) it follows

$$\frac{\partial |q_1|^2}{\partial \tau} = -\frac{\partial |q_2|^2}{\partial \zeta} = \frac{\partial |q_3|^2}{\partial \xi}. \tag{7.2.19}$$

Let the normalised complex amplitudes be expressed in terms of real-valued amplitudes and phases as $q_n = a_n \exp(i\varphi_n)$, $n = 1,2,3$. Then the equations (7.2.18) can be converted into the system of real equations

$$\frac{\partial a_1}{\partial \tau} = -a_2 a_3 \sin\Phi, \quad \frac{\partial a_2}{\partial \zeta} = a_1 a_3 \sin\Phi, \quad \frac{\partial a_3}{\partial \xi} = -a_2 a_1 \sin\Phi, \tag{7.2.20}$$

and

$$a_1 \frac{\partial \varphi_1}{\partial \tau} = a_2 a_3 \cos\Phi, \quad a_2 \frac{\partial \varphi_2}{\partial \zeta} = a_1 a_3 \cos\Phi, \quad a_3 \frac{\partial \varphi_3}{\partial \xi} = a_2 a_1 \cos\Phi,$$

where $\Phi = \varphi_2 - \varphi_1 - \varphi_3$. The solutions to be obtained are such that either all phases φ_n remain constant, or the phase difference Φ is equal to $\pi/2$ at all space-time coordinates. These are the same condition as those for SHG under phase-matching condition (see Sec. 6.1.4). Equations (7.2.19) lead to the conclusion that there is a potential $\mathcal{W}$ to satisfy the following relations

$$a_1^2 = -\frac{\partial^2 \mathcal{W}}{\partial \xi \partial \zeta}, \quad a_2^2 = \frac{\partial^2 \mathcal{W}}{\partial \xi \partial \tau}, \quad a_3^2 = -\frac{\partial^2 \mathcal{W}}{\partial \tau \partial \zeta}. \tag{7.2.21}$$

If to insert expressions (7.2.21) into any of the equation from (7.2.20), we obtain the resulting equation with respect to potential $\mathcal{W}$ [45]:

$$\frac{\partial^3 \mathcal{W}}{\partial\xi\partial\zeta\,\partial\tau} = 2\left(\frac{\partial^2 \mathcal{W}}{\partial\xi\partial\zeta}\cdot\frac{\partial^2 \mathcal{W}}{\partial\zeta\partial\tau}\cdot\frac{\partial^2 \mathcal{W}}{\partial\tau\partial\xi}\right)^{1/2} \qquad (7.2.22)$$

The task to find the solutions of SHG equations has been now reduced to the search for potential function $\mathcal{W}(\tau,\zeta,\xi)$, which satisfies the equation (7.2.22). An appropriate function is found [45] to be

$$\mathcal{W}(\tau,\zeta,\xi) = -\ln\left[f(\tau) + h(\zeta) + \tilde{h}(\xi)\right]. \qquad (7.2.23)$$

Potential function (7.2.23) satisfies (7.2.22) for all differentiable functions $f(\tau), h(\zeta)$ and $\tilde{h}(\xi)$. The corresponding amplitudes follow from (7.2.21) and (7.2.23)

$$a_1^2 = -\frac{1}{\left(f(\tau)+h(\zeta)+\tilde{h}(\xi)\right)^2}\frac{\partial\tilde{h}}{\partial\xi}\frac{\partial h}{\partial\zeta}, \qquad (7.2.24a)$$

$$a_2^2 = \frac{1}{\left(f(\tau)+h(\zeta)+\tilde{h}(\xi)\right)^2}\frac{\partial\tilde{h}}{\partial\xi}\frac{\partial f}{\partial\tau}, \qquad (7.2.24b)$$

$$a_3^2 = -\frac{1}{\left(f(\tau)+h(\zeta)+\tilde{h}(\xi)\right)^2}\frac{\partial f}{\partial\tau}\frac{\partial h}{\partial\zeta}. \qquad (7.2.24c)$$

The permissible choice of $f(\tau), h(\zeta)$ and $\tilde{h}(\xi)$ is restricted to differentiable functions, so that the sum $f(\tau)+h(\zeta)+\tilde{h}(\xi)$ is non-zero and the right hand sides of (7.2.24) are all positive. Accordingly, $(\partial\tilde{h}/\partial\xi)(\partial f/\partial\tau)$ must be positive, but $(\partial\tilde{h}/\partial\xi)(\partial h/\partial\zeta)$ and $(\partial h/\partial\zeta)(\partial f/\partial\tau)$ are negative. Some particular solutions were discussed in [45].

7.3. Four-wave parametric interaction

Let us consider the case when $\hat{\chi}^{(2)} = 0$ and non-linear properties of the medium are described by non-linear susceptibility of the next order, i.e. by $\hat{\chi}^{(3)}$. Such medium is referred to as cubic-non-linear. Let the waves with carrier frequencies ω_1, ω_2 and ω_3 propagate along z-direction. Due to the non-linear dependency of polarisation on electric field strengths of these waves, the new waves appear with the different carrier frequencies.

We select two cases among other possible ones

- sum-frequency mixing: $\omega_1 + \omega_2 + \omega_3 = \omega_4$,
- hyper-parametric scattering: $\omega_1 + \omega_2 = \omega_3 + \omega_4$.

The first case is a simple generalisation of parametric sum-frequency mixing under three-wave interaction. Degeneration ($\omega_1 = \omega_2 = \omega_3 = \omega$, $\omega_4 = 3\omega$) gives third-harmonic generation. The second case has no analogies in quadratic non-linear media.

7.3.1. THE UNIFICATION OF EVOLUTION EQUATIONS

Consider the interaction of collinear propagating waves. Under slowly-varying envelopes and phases approximation the system of equations for four waves interaction has the following form:

$$\left(\frac{\partial}{\partial z} + \frac{1}{v_1}\frac{\partial}{\partial t}\right)\mathcal{E}_1 = i\frac{12\pi\omega_1\chi^{(3)}(\omega_4, -\omega_2, -\omega_3)}{cn(\omega_1)}\mathcal{E}_2^*\mathcal{E}_3^*\mathcal{E}_4 \exp(+i\Delta k z), \quad (7.3.1a)$$

$$\left(\frac{\partial}{\partial z} + \frac{1}{v_2}\frac{\partial}{\partial t}\right)\mathcal{E}_2 = i\frac{12\pi\omega_2\chi^{(3)}(\omega_4, -\omega_3, -\omega_1)}{cn(\omega_2)}\mathcal{E}_1^*\mathcal{E}_3^*\mathcal{E}_4 \exp(+i\Delta k z), \quad (7.3.1b)$$

$$\left(\frac{\partial}{\partial z} + \frac{1}{v_3}\frac{\partial}{\partial t}\right)\mathcal{E}_3 = i\frac{12\pi\omega_3\chi^{(3)}(\omega_4, -\omega_1, -\omega_2)}{cn(\omega_3)}\mathcal{E}_1^*\mathcal{E}_2^*\mathcal{E}_4 \exp(+i\Delta k z), \quad (7.3.1c)$$

$$\left(\frac{\partial}{\partial z} + \frac{1}{v_4}\frac{\partial}{\partial t}\right)\mathcal{E}_4 = i\frac{12\pi\omega_4\chi^{(3)}(\omega_1, \omega_2, \omega_3)}{cn(\omega_4)}\mathcal{E}_1\mathcal{E}_2\mathcal{E}_3 \exp(-i\Delta k z), \quad (7.3.1d)$$

where $\Delta k = k_4 - (k_1 + k_2 + k_3)$, for sum-frequency mixing $\omega_4 = \omega_1 + \omega_2 + \omega_3$ and

$$\left(\frac{\partial}{\partial z} + \frac{1}{v_1}\frac{\partial}{\partial t}\right)\mathcal{E}_1 = i\frac{12\pi\omega_1\chi^{(3)}(\omega_4, -\omega_2, \omega_3)}{cn(\omega_1)}\mathcal{E}_2^*\mathcal{E}_3\mathcal{E}_4 \exp(+i\Delta k z), \quad (7.3.2a)$$

$$\left(\frac{\partial}{\partial z} + \frac{1}{v_2}\frac{\partial}{\partial t}\right)\mathcal{E}_2 = i\frac{12\pi\omega_2\chi^{(3)}(\omega_4, \omega_3, -\omega_1)}{cn(\omega_2)}\mathcal{E}_1^*\mathcal{E}_3\mathcal{E}_4 \exp(+i\Delta k z), \quad (7.3.2b)$$

$$\left(\frac{\partial}{\partial z} + \frac{1}{v_3}\frac{\partial}{\partial t}\right)\mathcal{E}_3 = i\frac{12\pi\omega_3\chi^{(3)}(-\omega_4, \omega_1, \omega_2)}{cn(\omega_3)}\mathcal{E}_1\mathcal{E}_2\mathcal{E}_4^* \exp(+i\Delta k z), \quad (7.3.3c)$$

$$\left(\frac{\partial}{\partial z} + \frac{1}{v_4}\frac{\partial}{\partial t}\right)\mathcal{E}_4 = i\frac{12\pi\omega_4\chi^{(3)}(\omega_1, \omega_2, -\omega_3)}{cn(\omega_4)}\mathcal{E}_1\mathcal{E}_2\mathcal{E}_3^* \exp(-i\Delta k z), \quad (7.3.4d)$$

where $\Delta k = k_4 + k_1 - (k_2 + k_3)$, for hyper-parametric scattering: $\omega_1 + \omega_2 = \omega_3 + \omega_4$.

Here we do not take into account the effects of self-action, which are described by the same type of susceptibilities $\hat{\chi}^{(3)}$, that gives rise to the following contribution into polarisation :

$$\mathcal{P}(\omega_n) = 12\pi \sum_{m=1}^{4} \chi^{(3)}(\omega_n, \omega_m, -\omega_m) |\mathcal{E}(\omega_m)|^2 \, \mathcal{E}(\omega_n) .$$

If as in the case of quadratic non-linear media, we introduce here the constants

$$\gamma_i = \frac{12\pi\omega_i \chi^{(3)}(\omega_1, \omega_2, \omega_3)}{cn(\omega_i)}, \quad i = 1,2,3,4$$

and $\tilde{\beta} = \sqrt{\gamma_1\gamma_2\gamma_3\gamma_4}$, then the systems of equations (7.3.1) and (7.3.2) can be written in the unified form

$$\left(\frac{\partial}{\partial z} + \frac{1}{v_1}\frac{\partial}{\partial t}\right) Q_1 = i\tilde{\beta} Q_2^* Q_3^* Q_4^* \exp(+i\Delta k\, z) , \qquad (7.3.3a)$$

$$\left(\frac{\partial}{\partial z} + \frac{1}{v_2}\frac{\partial}{\partial t}\right) Q_2 = i\tilde{\beta} Q_3^* Q_4^* Q_1^* \exp(+i\Delta k\, z) , \qquad (7.3.3b)$$

$$\left(\frac{\partial}{\partial z} + \frac{1}{v_3}\frac{\partial}{\partial t}\right) Q_3 = i\tilde{\beta} Q_4^* Q_1^* Q_2^* \exp(+i\Delta k\, z) , \qquad (7.3.3c)$$

$$\left(\frac{\partial}{\partial z} + \frac{1}{v_4}\frac{\partial}{\partial t}\right) Q_4 = -i\tilde{\beta} Q_1^* Q_2^* Q_3^* \exp(+i\Delta k\, z) , \qquad (7.3.3d)$$

where for sum-frequency mixing one has

$$\mathcal{E}_1 = \sqrt{\gamma_1} Q_1 , \quad \mathcal{E}_2 = \sqrt{\gamma_2} Q_2 , \quad \mathcal{E}_3 = \sqrt{\gamma_3} Q_3 , \quad \mathcal{E}_4 = -\sqrt{\gamma_4} Q_4^* ,$$

and for hyper-parametric scattering :

$$\mathcal{E}_1 = \sqrt{\gamma_1} Q_1 , \quad \mathcal{E}_2 = \sqrt{\gamma_2} Q_2 , \quad \mathcal{E}_3 = \sqrt{\gamma_3} Q_3^* , \quad \mathcal{E}_4 = -\sqrt{\gamma_4} Q_4^* .$$

7.3.2. STEADY-STATE PARAMETRIC PROCESS

If the phase-matching condition holds and it is possible to neglect the difference of propagation velocities of interacting waves (or the non-linear medium is not sufficiently long to notice spatial separation of interacting pulses), the equations, describing four-waves parametric interaction (7.3.3) can be simplified so that

$$\frac{\partial}{\partial \xi} Q_1 = i\tilde{\beta} Q_2^* Q_3^* Q_4^*, \qquad \frac{\partial}{\partial \xi} Q_2 = i\tilde{\beta} Q_3^* Q_4^* Q_1^*, \tag{7.3.4a,b}$$

$$\frac{\partial}{\partial \xi} Q_3 = i\tilde{\beta} Q_4^* Q_1^* Q_2^*, \qquad \frac{\partial}{\partial \xi} Q_4 = -i\tilde{\beta} Q_1^* Q_2^* Q_3^*, \tag{7.3.4c,d}$$

where $\xi = z$, $\tau = t - v^{-1} z$. If to transfer real variables, introduced by the formulae:

$$Q_k = a_k \exp(i\varphi_k), \quad \Phi = \varphi_1 + \varphi_2 + \varphi_3 + \varphi_4,$$

then (7.3.4) yields

$$\frac{\partial a_1}{\partial \xi} = \tilde{\beta} a_2 a_3 a_4 \sin\Phi, \qquad \frac{\partial a_2}{\partial \xi} = \tilde{\beta} a_1 a_3 a_4 \sin\Phi, \tag{7.3.5a,b}$$

$$\frac{\partial a_3}{\partial \xi} = \tilde{\beta} a_4 a_1 a_2 \sin\Phi, \qquad \frac{\partial a_4}{\partial \xi} = -\tilde{\beta} a_1 a_2 a_3 \sin\Phi, \tag{7.3.5c,d}$$

$$\frac{\partial \Phi}{\partial \xi} = \tilde{\beta}\left(\frac{a_2 a_3 a_4}{a_1} + \frac{a_1 a_3 a_4}{a_2} + \frac{a_1 a_2 a_4}{a_3} - \frac{a_1 a_2 a_3}{a_4} \right) \sin\Phi . \tag{7.3.6}$$

By using the amplitude equation (7.3.5) we can write the equation for the phase

$$\frac{\partial \Phi}{\partial \xi} = \tilde{\beta} \frac{\cos\Phi}{\sin\Phi} \frac{\partial}{\partial \xi} \ln(a_1 a_2 a_3 a_4).$$

This equation allows us to obtain the first integral of motion

$$a_1 a_2 a_3 a_4 \cos\Phi = \text{const}. \tag{7.3.7}$$

Besides this integral, other integrals of motion follow from equations (7.3.)

$$a_1^2 + a_4^2 = m_1^2, \quad a_2^2 + a_4^2 = m_2^2, \quad a_3^2 + a_4^2 = m_3^2, \tag{7.3.8}$$

$$a_1^2 + a_2^2 + a_3^2 + 3a_4^2 = \text{const}. \tag{7.3.9}$$

These integrals are the variety of the known Manley-Rowe relations.

If to be interested in solutions, which vanish in some spatial points, then the constant of integrating in the expression (7.3.7) can be fixed as zero. That means $\Phi = \pi / 2$ everywhere. Accounting (7.3.8), equation for $a_4 = a(\xi)$ writes:

$$\frac{da}{d\xi} = -\tilde{\beta}\sqrt{(m_1^2 - a^2)(m_2^2 - a^2)(m_3^2 - a^2)} . \tag{7.3.10}$$

For the sake of certainty we shall choose the condition

$$m_1^2 > m_2^2 > m_3^2 .$$

This condition can always be achieved by renumbering the interacting fields. Exclusion is the degeneration of parametric process, when either two or all three of these integrals are equal. For example it realises in third-harmonic generation (THG).

Let the degeneration be absent. The change of variable $y = a^{-2}$ brings equation (7.3.10) to the following one

$$\frac{dy}{d\xi} = 2\tilde{\tilde{\beta}}\sqrt{(y - y_1)(y - y_2)(y - y_3)} , \tag{7.3.11}$$

where $\tilde{\tilde{\beta}} = \tilde{\beta}(m_1 m_2 m_3)$, $y_n = m_n^{-2}, n = 1,2,3$. Equation (7.3.11) belongs to the class of equations with elliptic functions serving as solutions. One more variable change

$$y \to x = \sqrt{\frac{y - y_3}{y - y_2}} ,$$

transfers (7.3.11) to a canonical form

$$\frac{dx}{d\xi} = \tilde{\tilde{\beta}}\sqrt{y_3 - y_1}\sqrt{(1 - x^2)(1 - m^2 x^2)} ,$$

with

$$m^2 = \frac{y_2 - y_1}{y_3 - y_1} = \frac{m_3^2 (m_1^2 - m_2^2)}{m_2^2 (m_1^2 - m_3^2)} .$$

This parameter is a modulus of the elliptic integral of the first kind (or Jacobian elliptic functions). Finally we can write the solution of this equation with ξ_0 denoting the integrating constant:

$$x(\xi) = \mathrm{sn}\left(\tilde{\tilde{\beta}}\sqrt{y_3 - y_1}(\xi - \xi_0), m\right) .$$

Thus, in terms of initial variables we can write the solutions of the system (7.3.5):

$$a_1^2(\xi) = \left(m_2^2 - m_3^2\right)\frac{m_1^2 - m_3^2 \operatorname{cn}^2(X,m)}{m_2^2 - m_3^2 \operatorname{sn}^2(X,m)}, \tag{7.3.12a}$$

$$a_2^2(\xi) = \frac{m_2^2\left(m_2^2 - m_3^2\right)}{m_2^2 - m_3^2 \operatorname{sn}^2(X,m)}, \tag{7.3.12b}$$

$$a_3^2(\xi) = \frac{m_3^2\left(m_2^2 - m_3^2\right)\operatorname{sn}^2(X,m)}{m_2^2 - m_3^2 \operatorname{sn}^2(X,m)}, \tag{7.3.12c}$$

$$a_4^2(\xi) = \frac{m_2^2 m_3^2 \operatorname{cn}^2(X,m)}{m_2^2 - m_3^2 \operatorname{sn}^2(X,m)}, \tag{7.3.12d}$$

where $X = \widetilde{\beta} m_3 \sqrt{m_1^2 - m_3^2}\,(\xi - \xi_0)$.

Let us turn now to the case when degeneration is involved. To be more certain, let $m_2^2 = m_1^2 > m_3^2$. In this case equation (7.3.11) provides

$$\frac{dy}{d\xi} = 2\tilde{\tilde{\beta}}(y - y_1)\sqrt{(y - y_3)}\,.$$

The variable changing $y \to z = y - y_1$ tends to equation

$$\frac{dz}{d\xi} = 2\tilde{\tilde{\beta}} z \sqrt{(z - a)}\,,$$

where $a = y_3 - y_1 > 0$. The solution is in the form $z = a\left(1 + \tan^2 X\right)$, $X = \tilde{\tilde{\beta}}\sqrt{a}(\varsigma - \varsigma_0)$. In terms of initial variables, the solutions of system (7.3.5) under degeneration of the parametric process can be written as:

$$a_1^2(\xi) = a_2^2(\xi) = \frac{m_1^2\left(m_1^2 - m_3^2\right)\left(1 + \tan^2 X\right)}{m_1^2 + (m_1^2 - m_3^2)\tan^2 X}, \tag{7.3.13a}$$

$$a_3^2(\xi) = \frac{m_3^2\left(m_1^2 - m_3^2\right)\tan^2 X}{m_1^2 + (m_1^2 - m_3^2)\tan^2 X}, \tag{7.3.13b}$$

$$a_4^2(\xi) = \frac{m_1^2 m_3^2}{m_1^2 + (m_1^2 - m_3^2)\tan^2 X}, \tag{7.3.13c}$$

where $X = \tilde{\beta} m_1 \sqrt{m_1^2 - m_3^2}(\xi - \xi_0)$.

Assume that $m_1^2 > m_2^2 = m_3^2$. Then equation (7.3.11) will take the form

$$\frac{dy}{d\xi} = 2\tilde{\tilde{\beta}}(y - y_2)\sqrt{(y - y_1)}\,.$$

A new variable changing $y \to z = y - y_2$ leads to equation

$$\frac{dz}{d\xi} = 2\tilde{\tilde{\beta}} z \sqrt{(z + a)}\,,$$

where $a = y_2 - y_1 > 0$. Integrating this equation yields

$$z = a\left(1 + \tanh^2 X\right), \quad X = \tilde{\tilde{\beta}}\sqrt{a}(\varsigma - \varsigma_0)\,.$$

Thus, the solutions of equations (7.3.5) can be written as

$$a_1^2(\xi) = \frac{m_1^2\left(m_1^2 - m_2^2\right)\left(2 + \tanh^2 X\right)}{m_1^2 + (m_1^2 - m_2^2)(1 + \tanh^2 X)}, \tag{7.3.14a}$$

$$a_2^2(\xi) = a_3^2(\xi) = \frac{m_2^2\left(m_1^2 - m_2^2\right)(1 + \tanh^2 X)}{m_1^2 + (m_1^2 - m_2^2)(1 + \tanh^2 X)}, \tag{7.3.14b}$$

$$a_4^2(\xi) = \frac{m_1^2 m_2^2}{m_1^2 + (m_1^2 - m_2^2)(1 + \tanh^2 X)}, \tag{7.3.14c}$$

with $X = \tilde{\beta} m_1 \sqrt{m_1^2 - m_2^2}(\xi - \xi_0)$.

Another case corresponds to the initial conditions, which can be chosen to satisfy $m_1^2 = m_2^2 = m_3^2$. Equation (7.3.11) then writes

$$\frac{dy}{d\xi} = 2\tilde{\tilde{\beta}}(y - y_1)^{3/2}\,.$$

The solution of this equation is $y(\xi) = y_1 + X^{-2}$, where $X = \tilde{\tilde{\beta}}(\xi - \xi_0)$.

Solutions of initial equations (7.3.5) are expressed by the following formulae:

$$a_1^2(\xi) = a_2^2(\xi) = a_3^2(\xi) = \frac{m_1^2 m_1^2}{m_1^2 + X^2}, \quad a_4^2(\xi) = \frac{m_1^2 X^2}{m_1^2 + X^2}\,. \tag{7.3.15}$$

These solutions describes the propagation of the "bright" solitary wave on the background of the "dark " solitary wave.

The solution of equation (7.3.11) can be alternatively obtained by changing the variable

$$y \to x = y - \frac{1}{3}(y_1 + y_2 + y_3)\,,$$

In terms of this variable the equation (7.3.11) transforms to a canonical form of the equation for $\wp$ - function by Weierstrass:

$$\frac{dx}{d\xi} = 2\tilde{\tilde{\beta}}\sqrt{(x-e_1)(x-e_2)(x-e_3)}\,,$$

where $3e_1 = 2y_1 - y_2 - y_3$, $3e_2 = 2y_2 - y_1 - y_3$, $3e_3 = 2y_3 - y_2 - y_1$. The solution of this equation is expressed in terms of $\wp$ -functions: $x(X) = \wp(X; g_2, g_3)$, where

$$X = \tilde{\tilde{\beta}}(\xi - \xi_0)\,,\ g_2 = -4(e_1e_2 + e_2e_3 + e_3e_1)\,,\ g_3 = 4e_1e_2e_3\,.$$

First of all one can write the solution of equation (7.3.11) as

$$a_4^2(\xi) = \frac{3}{(y_1 + y_2 + y_3) + 3\wp(X; g_2, g_3)}\,. \tag{7.3.16a}$$

Weierstrass function is less popular than Jacobian functions and this is the reason to change for these elliptic functions by the formulae

$$\wp(X; g_2, g_3) = e_3 + \frac{(e_1 - e_2)}{\mathrm{sn}^2(u,k)}\,,$$

where

$$u = \tilde{\tilde{\beta}}\sqrt{e_1 - e_3}(\xi - \xi_0),\quad k^2 = \frac{e_2 - e_3}{e_1 - e_3}\,.$$

Expression for $a_4 = a(\xi)$ is written as

$$a_4^2(\xi) = \frac{\mathrm{sn}^2(u,k)}{(y_1 - y_2) + y_3\,\mathrm{sn}^2(u,k)}\,. \tag{7.3.16b}$$

Let $y_1 > y_2 > y_3$ (or $m_1^2 < m_2^2 < m_3^2$). (If it is necessary to consider another order of these parameters, we can simply renumber them). Then modulus of Jacobian elliptic functions is

$$k^2 = \frac{y_2 - y_3}{y_1 - y_3} = \frac{m_3^2 - m_2^2}{m_3^2 - m_1^2}.$$

In terms of initial variables the solutions of system (7.3.5) with the account of (7.3.8) can be written as

$$a_1^2(\xi) = \frac{m_1^2(m_2^2 - m_1^2)[m_3^2 - m_2^2 \operatorname{sn}^2(u,k)]}{m_3^2(m_2^2 - m_1^2) + m_1^2 m_2^2 \operatorname{sn}^2(u,k)}, \tag{7.3.17a}$$

$$a_2^2(\xi) = \frac{m_2^2[m_3^2(m_2^2 - m_1^2) - m_2^2(m_2^2 - m_3^2)\operatorname{sn}^2(u,k)]}{m_3^2(m_2^2 - m_1^2) + m_1^2 m_2^2 \operatorname{sn}^2(u,k)}, \tag{7.3.17b}$$

$$a_3^2(\xi) = \frac{m_3^4(m_2^2 - m_1^2)}{m_3^2(m_2^2 - m_1^2) + m_1^2 m_2^2 \operatorname{sn}^2(u,k)}, \tag{7.3.17c}$$

$$a_4^2(\xi) = \frac{m_1^2 m_2^2 m_3^2 \operatorname{sn}^2(u,k)}{m_3^2(m_2^2 - m_1^2) + m_1^2 m_2^2 \operatorname{sn}^2(u,k)}, \tag{7.3.17d}$$

where $u = \tilde{\beta} m_1 \sqrt{m_3^2 - m_1^2}(\xi - \xi_0)$. The degeneration of four-wave interaction can be considered as it was done above. There are three such situations here: (a) $m_1 = m_2 < m_3$, (b) $m_1 < m_2 = m_3$, and (c) $m_1 = m_2 = m_3$.

7.3.3. THIRD HARMONIC GENERATION

The third harmonic generation presents an example of a degenerated four-wave parametric process (i.e., $\omega_1 = \omega_2 = \omega_3 = \omega, \omega_4 = 3\omega$) This is the well known example of optical waves interaction, which has attracted not less attention than SHG. This is the reason to observe the process of the third harmonic generation in more details. Unlike the interaction of waves in a quadratic medium, under third harmonic generations self-action processes occur along with the sum-frequency mixing. These are, for example, self-phase modulations and cross-phase modulations.

Equations, describing process of third harmonic generations (TGH), write

$$\left(\frac{\partial}{\partial z}+\frac{1}{v_3}\frac{\partial}{\partial t}\right)\mathcal{E}_3 = i\frac{3\pi\omega_1}{cn(\omega_3)}\left\{\chi^{(3)}(\omega,\omega,\omega)\mathcal{E}_1^3\exp(-i\Delta k\, z)+\right.$$
$$\left.+2\chi^{(3)}(\omega,-\omega,3\omega)|\mathcal{E}_1|^2\,\mathcal{E}_3+\chi^{(3)}(3\omega,-3\omega,3\omega)|\mathcal{E}_3|^2\,\mathcal{E}_3\right\}$$

Here $\Delta k = k_3 - 3k_1$ is a phase mismatch. Denote the following normalised variables:

$$q_{1,3} = \mathcal{E}_{1,3}(z,t)/\mathcal{E}_0\,,\quad \varsigma = zL_C^{-1} = z\left(\frac{3\pi\omega_1\chi^{(3)}(\omega,\omega,\omega)}{cn(\omega_3)}\right)\mathcal{E}_0^2,$$

$$\mu_1 = \frac{n(\omega_3)\chi^{(3)}(\omega,\omega,-\omega)}{n(\omega_1)\chi^{(3)}(\omega,\omega,\omega)},\quad \mu_3 = \frac{\chi^{(3)}(3\omega,3\omega,-3\omega)}{\chi^{(3)}(\omega,\omega,\omega)},\quad r_3 = \frac{\chi^{(3)}(3\omega,\omega,-\omega)}{\chi^{(3)}(\omega,\omega,\omega)},$$

$$r_1 = \frac{n(\omega_3)\chi^{(3)}(\omega,3\omega,-3\omega)}{n(\omega_1)\chi^{(3)}(\omega,\omega,\omega)},\quad \gamma_1 = \frac{n(\omega_3)\chi^{(3)}(3\omega,-\omega,-\omega)}{n(\omega_1)\chi^{(3)}(\omega,\omega,\omega)}.$$

By means of these parameters the normalised system of equations describing TGH is as follows

$$L_C^{-1}\left(\frac{\partial}{\partial z}+\frac{1}{v_1}\frac{\partial}{\partial t}\right)q_1 = i\gamma_1 q_1^{*2}q_3\exp(+i\Delta k\, z)+i\mu_1|q_1|^2\,q_1+2ir_1|q_3|^2\,q_1\,, \quad (7.3.18a)$$

$$L_C^{-1}\left(\frac{\partial}{\partial z}+\frac{1}{v_3}\frac{\partial}{\partial t}\right)q_3 = iq_1^3\exp(-i\Delta k\, z)+i\mu_3|q_3|^2\,q_3+2ir_3|q_1|^2\,q_3\,, \quad (7.3.18b)$$

If the condition of phase matching is not satisfied and the transformation of a pump wave into the harmonic one is very weak, we can approximately consider TGH by neglecting both the pump wave amplitude depleting and the influence of harmonic wave on the altering of refraction index of the non-linear medium. Like in the preceding paragraph we shall consider the TGH in the stationary regime. Under such approximation of the given field we have a set of normalised equations, describing TGH:

$$\frac{\partial q_1}{\partial\varsigma} = i\mu_1|q_1|^2\,q_1\,, \quad (7.3.19a)$$

$$\frac{\partial q_3}{\partial\varsigma} = iq_1^3\exp(-i\Delta\varsigma)+ir_3|q_1|^2\,q_3\,. \quad (7.3.19b)$$

Here we assume that group velocities of the interaction waves are the same. If now to turn to real amplitudes and phases of interacting waves, we obtain

$$\frac{\partial a_1}{\partial \varsigma} = 0, \quad \frac{\partial \varphi_1}{\partial \varsigma} = \mu_1 a_1^2, \tag{7.3.20a}$$

$$\frac{\partial a_3}{\partial \varsigma} = a_1^3 \sin(\Delta + \varphi_3 - 3\varphi_1), \quad \frac{\partial \varphi_3}{\partial \varsigma} = \frac{a_1^3}{a_3}\cos(\Delta + \varphi_3 - 3\varphi_1) + r_3 a_1^2 . \tag{7.3.20b}$$

It follows from (7.3.20) that the amplitude of a pump wave does not change as it is expected under given approximation. But due to self-phase modulations the plane pump wave front exhibits the modulation instability, which develops under the excess of a certain threshold of power. Let this threshold be not reaching in our analysis of TGH and the pump wave is stable.

Let $a_1(\varsigma) = a_1(0) = 1$ and $a_3(\varsigma) = a, \Phi = -\Delta + 3\varphi_1 - \varphi_3$. System of equation then follows from (7.3.20b):

$$\frac{\partial a}{\partial \varsigma} = \sin(\Phi), \quad \frac{\partial \Phi}{\partial \varsigma} = -\Delta + \frac{1}{a}\cos(\Phi) + (r_3 - 3\mu_1) . \tag{7.3.21}$$

It is seen from the second equation of this system, that the phase synchronism condition, i.e. phase matching condition, is to be modified to take into account the non-linear self- and cross-phase modulations. Let the value $\Delta - (r_3 - 3\mu_1) = \tilde{\Delta}$ measure the phase mismatch of interacting waves. Alike the SHG, for these equations, the first integral of motion takes place: $a\cos(\Phi) + (\tilde{\Delta}/2)a^2 = \text{const}$. Assume there is no third harmonic wave on the border of a non-linear medium, then one can find that the constant is zero. By excluding phase difference Φ from the amplitude equation (7.3.21) by means of the integral of motion, we now are able to integrate the resultant equation and to get an expression for the harmonic amplitude:

$$a_3(\varsigma) = a(\varsigma) = \frac{2}{\tilde{\Delta}}\sin\left(\frac{\tilde{\Delta}\varsigma}{2}\right) .$$

Hence, after returning to initial variables, one can obtain the following expression for the intensity of the third harmonic wave:

$$I_3 = \frac{(24\pi^2)^2 \omega^2 |\chi^{(3)}(\omega,\omega,\omega)|^2}{c^4 n(\omega_3) n^2(\omega_1)} I_{10}^3 \frac{\sin^2(\Delta\tilde{k} z/2)}{(\Delta\tilde{k}/2)^2} . \tag{7.3.22}$$

The length of coherence is defined in the same way as it was done for SHG. But now this parameter depends on pump intensity. Indeed, if to analyse the expression for $\widetilde{\Delta}$, one can be convinced that this parameter is proportional to the difference of phase velocities of interacting waves. We should take into account that refractive indices include non-linear corrections proportional to the second power of the amplitude of a pump wave as it also takes place in the high-frequency Kerr-effect.

Usually this undepleted pump approximation satisfactorily describes the process of TGH when the phase velocity mismatch is sufficiently large. Due to the infinitesimal of the non-linear corrections of refraction indices, their effect will be essential only under conditions of phase matching. So it is possible to change $\Delta\widetilde{k}$ for Δk here.

Everything regarding the phase matching condition for TGH stays unaffected as in the case of SHG. The phase matching angle is expressed by the same formulae as previously but the change $n(\omega_2) \rightarrow n(\omega_3)$ only has to be done. Besides that, according to symmetry relations by Kleiman, we have $\gamma_1 = 1$, $r_1 = r_2 = r$.

When the phase matching condition is satisfied, it is suitable to use the real variables

$$q_1 = a_1 \exp\{i\varphi_1\}, \quad q_3 = a_3 \exp\{i\varphi_3\} .$$

Then (7.3.19) yields

$$\frac{\partial a_1}{\partial \varsigma} = -a_1^2 a_3 \sin\Phi \ , \tag{7.3.23a}$$

$$\frac{\partial a_3}{\partial \varsigma} = a_1^3 \sin\Phi \ , \tag{7.3.23b}$$

$$\frac{\partial \Phi}{\partial \varsigma} = \left(\frac{a_1^3}{a_3} - 3a_1 a_3\right)\cos\Phi - \beta a_1^2 - \alpha a_3^2 \tag{7.3.23c}$$

where $\Phi = 3\varphi_1 - \varphi_3$ and $\alpha = 6r - \mu_3, \beta = 3\mu_1 - 2r$.

Equations (7.3.23a) and (7.3.23b) result in the conservation law

$$a_1^2 + a_3^2 = 1. \tag{7.3.24}$$

Then from (7.3.23b) and (7.3.23c) with the account of the expression (7.3.24) we can find the second integral of motion:

$$a_3(1 - a_3^2)^{3/2}\cos\Phi - \frac{1}{2}a_3^2\left[\beta + \frac{1}{2}(\alpha - \beta)a_3^2\right] = C_1 = \text{const} . \tag{7.3.25}$$

It can be seen from (7.3.25) that in TGH process the phase synchronism (i.e., phase matching) does not hold due to the changing of refraction indices for the indices at the pump and harmonic frequencies. It occurs because of the high-frequency Kerr effect. This effect involved into consideration by the cubic non-linear susceptibilities $\chi^{(3)}(\omega,\omega,-\omega)$, $\chi^{(3)}(3\omega,3\omega,-3\omega)$, and $\chi^{(3)}(\omega,3\omega,-3\omega)=\chi^{(3)}(3\omega,\omega,-\omega)$, which, in their turn, ascertain parameters α and β in (7.3.24) and (7.3.25). However, if this effect of the phase matching changing is neglected, then instead of (7.3.25) we have

$$a_3(1-a_3^2)^{3/2}\cos\Phi = C_1 .$$

If at the entrance of non-linear medium the wave of third harmonic is absent, then the constant $C_1 = 0$. Consequently, wherever the harmonic amplitude be non-zero, phase difference $\Phi = \pi/2$. Taking into account relation (7.3.24), the equation (7.3.23b) can be written as follows

$$\frac{\partial a_3}{\partial \varsigma} = (1-a_3^2)^{3/2} .$$

Solution of this equation is

$$a_3(\varsigma) = \frac{\varsigma}{\sqrt{1+\varsigma^2}} .$$

Thereby, the amplitude of the third harmonic varies with the length of the non-linear medium according to the expression

$$\mathcal{E}_3(z) = \frac{\mathcal{E}_0(zL_C^{-1})}{\sqrt{1+(zL_C^{-1})^2}} . \tag{7.3.26}$$

For the intensity of the third harmonic $I_3(z)$ we obtain $I_3(z) \propto I_{10}^3 z^2$ under condition that $z \ll L_C$. It is necessary to recall that for intensity of the second harmonic $I_2(z)$ we found that $I_2(z) \propto I_{10}^2 z^2$. Thus, we may assume that the intensity of *n*th-order harmonic varies as the square of non-linear medium length and one is proportional to pump intensity to the *n*th power.

From (7.3.26) it is possible to find the length L^* of the 90% transformations, introduced by the condition

$$I_3(L^*) = 0{,}9 I_{10} ,$$

whence it follows that

$$L^* = 3L_C = \frac{cn(\omega)}{\pi\omega\chi^{(3)}(\omega,\omega,\omega)\mathcal{E}_0^2} .$$

7.4. Conclusion

In this chapter we have considered several models describing the interaction of the wave with different carrier frequencies. The sum-frequency and difference-frequency mixing are classical examples of this interaction. In a number of cases parametric wave interaction can be described by the system of equations which admits the zero-curvature representation. We would like to point out that three-wave interaction provides a unique instance in nonlinear optics in which the relevant system of equations can be solved by IST method for a 3D case [38-42].

It is well known that non-linear effects can be produced in thin-film optical waveguides [77,78] and optical fibres. Theory of the SHG in optical waveguides based on the connected modes formalism (see section 6.1.1, 8.4, and 8.5) leads to a system of equations which is similar to (7.1.3) or (7.1.14) and (7.1.15). Some progress in this field has been motivated by the reports of SHG and generation of the sum-frequency radiation in fibres [79,80]. What is surprising thing is that the second harmonic generation occurs there. In the fibres made of glass the second-order non-linear susceptibility is zero by symmetry. The suggestion has been made that the SHG could be due to the nonlinearity at the core-cladding interface or to a non-linear polarisation proportional to $\vec{E}\nabla\vec{E}$ terms. Optical mixing in the fibres due to these effects has been analysed in [81]. Let us consider the electric quadrupolar contribution into the non-linear polarisation [6, 81]

$$P_k(2\omega) = (\delta - \beta - 2\gamma)E_i\nabla_i E_k + \beta E_k \nabla_i E_i + 2\gamma E_i \nabla_k E_i ,$$

where we follow the convention that double indices indicate summation, and

$$\nabla_1 = \frac{\partial}{\partial x}, \quad \nabla_2 = \frac{\partial}{\partial y}, \quad \nabla_3 = \frac{\partial}{\partial z} .$$

The coefficients $(\delta - \beta - 2\gamma)$, β, and 2γ are effective quadrupole polarisabilities. In the low-frequency limit $\delta = 0$ and $\beta = -2\gamma$. These low-frequency relations are quite general and apply to optical range of frequencies. Substitution of these expressions into wave equations for fundamental and harmonic waves leads to the system of equations for interacting waves in the slowly varying envelope approximation. Here we have left aside this problem as the simple exercise for readers.

Recently the parametric processes attract the attention of the researchers in context of the two-dimensional solitary waves (*quadratic space solitons*) in a medium with the quadratic nonlinearity [82-93]. Here the diffraction broadening is suppressed by the parametric interaction between fundamental and harmonic waves. Furthermore, it is well known that cubic or Kerr-type self-focusing nonlinearity gives rise to collapse in two- and three-dimensional cases. But collapse does not take place in the medium with the quadratic nonlinearity. Thus, the parametric interaction permits of the stable two-dimension and three-dimensional solitons

The theory of the quadratic space solitons is based on the following system of equations

$$\frac{\partial \mathcal{E}_1}{\partial z} + \frac{i}{2k_1}\Delta_\perp \mathcal{E}_1 = i\frac{4\pi\omega^2}{c^2 k_1}\chi^{(2)}(2\omega,-\omega)\mathcal{E}_1^* \mathcal{E}_2 \exp(-i\Delta k z)\,,$$

$$\frac{\partial \mathcal{E}_2}{\partial z} + \frac{i}{2k_2}\Delta_\perp \mathcal{E}_2 = i\frac{8\pi\omega^2}{c^2 k_2}\chi^{(2)}(\omega,\omega)\mathcal{E}_1^2 \exp(i\Delta kz)\,.$$

It is the two- and three-dimensional generation of the SHG equations (7.1.14) and (7.1.15). The term with transverse Laplasian $\Delta_\perp$ takes account of the diffraction effect. This system of equations is not likely to be integrable by the IST method, and hence, in actual truth, it has not soliton solutions. However, the investigation of these equations by the numerical simulation illustrates the nearly steady-state regime of the wave pockets propagation. That is the reason why such wave pockets have been named as quadratic spatial solitons.

References

1. Shen, Y.R.: *The principles of non-linear optics*, John Wiley & Sons, New York, Chicester, Brisbane, Toronto, Singapore, 1984.
2. Bertein, F.: *Bases de l'electronique quatnique*, Editions Eyrolles, Pasis, 1969.
3. Bloembergen, N.: *Non-linear optics*, Benjamin, New York, 1965.
4. Yariv, A., and Yeh, P.: *Optical waves in crystals*. John Wiley & Sons, New York, Chichester, Brisbane, Toronto, Singapore, 1984
5. Armstrong, J.A., Bloembergen, N., Ducing, J., and Pershan, P.S.: *Phys. Rev.* **127** (1962), 1918
6. Yariv, A., and Louisell, W.H.: Theory of the optical parametric oscillator, *IEEE J.Quant.Elect.* **QE-2** (1966), 418
7. Harris, S.E.: Tuneable optical parametric oscillators, *Proc.IEEE* **57** (1969), 2096-2113.
8. Bjorkholm, J.E.: Optical SHG using a focused laser beam, *Phys.Rev.***142** (1966), 126-136.
9. Boyd, G.D., and Kleiman, D.A.: Parametric interaction of focused Gaussian light beams, *J.Appl.Phys.* **39** (1968), 3597-
10. Fisher, R., and Kulevskii L.A.: Optical parametric generators of light (Review), *Kvantov. Electron. (Moscow)* **4** (1977), 245-289.
11. Butcher, P.N., and Cotter, D.: *The Elements of Non-linear Optics*" University Press, Cambridge, 1990.
12. Chu, F.Y.F , and Scott, A.C.: Inverse scattering transform for the wave-wave scattering, *Phys.Rev.* **A12** (1975), 2060-2064.
13. Chu, F.Y.F.: Bäcklund transformation for the wave-wave scattering equations, *Phys.Rev.* **A12** (1975), 2065-2067.
14. Chiu, S.C.: On the self-induced transparency effect of the three-wave resonance process, *J.Math.Phys.* **19** (1978), 168-176.
15. Steudel, H.: Solitons in stimulated Raman scattering, *Ann.Phys.(DDR)* **34** (1977), 188-202.
16. Gursey, Y.: Soliton solutions in stimulated Brillouin scattering, *Phys.Rev.* **B24** (1981), 6147-6150.
17. Gursey, Y.: Soliton solutions in stimulated Brillouin scattering. II., *Phys.Rev.* **B26** (1982), 7015-7018.

18. Kaup, D.J.: Creation of a soliton out of dissipation, *Physica* **D19** (1986), 123-134.
19. Enns, R.H., and Rangnekar, S.S.: An application of ISTM to asymptotic laser pulse scattering, *Phys.Lett.* **A81** (1981), 313-314.
20. Enns, R.H., and Rangnekar, S.S.: Application of the inverse scattering transform method to stimulated Brillouin backscattering in a generator set-up: The Zakharov-Manakov solution, *Canad.J.Phys.* **59** (1981), 1817-1828.
21. Enns, R.H.: Zakharov-Manakov solution of the SBBS amplifier problem for non-rectangular envelopes, *Phys.Lett.* **A88** (1982), 222-224.
22. Enns, R.H.: Zakharov-Manakov solution of the stimulated Brillouin backscattering generator problem for nonrectangular envelopes, *Canad.J.Phys.* **60** (1982), 1404-1413.
23. Enns, R.H.: Zakharov-Manakov solution of the stimulated Brillouin backscattering generator problem for "spliced potentials", *Canad.J.Phys.* **60** (1982), 1620-1629.
24. Enns, R.H.: Application of the inverse scattering transform method to SBBS in an inhomogeneous medium: Zakharov-Manakov solution, *Canad.J.Phys.* **61** (1983), 604-611.
25. Enns, R.H., and Rangnekar, S.S.: Zakharov-Manakov solution of the 3-wave explosive interaction problem, *Canad.J.Phys.* **61** (1983), 1386-1400.
26. Enns, R.H., and Rangnekar, S.S.: Inverse scattering and the three-wave interaction in non-linear optics: A review, *IEEE J.Quant.Electron.* **QE-22** (1986), 1204-1214.
27. Sudhanshu S. Jha : Envelope -soliton propagation for three interacting coherent excitations in a dispersive medium, *Pramana* **11** (1978), 313-322.
28. Ray, D. Shankar : 2π-"pulse" for three-wave mixing, *Phys.Lett.* **A102** (1984), 99-101.
29. Zakharov, V.E., and Manakov S.V.: On the resonant interaction of wave packets in non-linear media, *Pis'ma v Zh.Eksp.Teor.Fiz.* **18** (1973), 413-416.
30. Zakharov, V.E., and Manakov S.V.: On the theory of resonant interaction of wave packets in non-linear media, *Zh.Eksp.Teor.Fiz.* **69** (1975), 1654-1673.
31. Kaup, D.J.: The three-wave interaction - A nondispersive phenomenon, *Stud. Appl. Math.* **55** (1976), 9-44.
32. Rieman, A.H., Bers, A., and Kaup, D.J.: Non-linear interactions of three wave packets in an inhomogeneous medium. *Phys.Rev.Letts.* **39** (1977), 245-248.
33. Rieman, A.H., and Kaup, D.J.: Multi-shock solutions of random phase three-wave interactions, *Phys.Fluid* **24** (1981), 228-232.
34. Case, K.M., and Chiu, S.C.: Backlund transformation for the resonant three-wave process. *Phys.Fluids* **20** (1977), 746-749.
35. Enns, R.H., Guenther, D.B., and Rangnekar, S.S.: An application of the inverse scattering method to the 3-wave interaction in non-linear optics, *Canad.J.Phys.* **58** (1980), 1468-1476.
36. Kaup, D.J., Rieman, A.H., and Bers, A.: Space-time evolution of non-linear three-wave interactions. I. Interaction in homogeneous medium, *Rev.Mod.Phys.* **51** (1979), 275-310.
37. Rieman, A.H.: Space-time evolution of non-linear three-wave interactions. II. Interaction in a inhomogeneous medium, *Rev.Mod.Phys.* **51** (1979), 311-330.
38. Kaup, D.J.: The soliton of the general initial value problem for the full three dimensional three-wave resonant interaction, *Physica* **D3** (1981), 374-395.
39. Kaup, D.J.: The lump solitons and the Bäcklund transformation for the three-dimensional three-wave resonant interaction, *J.Math.Phys.* **22** (1981), 1176-1181.
40. Kanashov, A.A., and Rubenchik, A.M.: On diffraction and dispersion effect on three wave interaction, *Physica* **D4** (1981), 122-134.
41. Shulman, E.I.: On existence of numerable series of the integral of motion for system of three-dimensional resonant interacting wave packets, *Teor.Mat.Fiz.* **44** (1980), 224-228.
42. Bakurov, V.G., The method of inverse problem for three-dimensional theory of the three-wave resonant interaction, *Teor.Mat.Fiz.* **76** (1988), 18-30.
43. Kaup, D.J.: Determining the final profiles from initial profiles for the full three-dimensional three-wave resonant interaction, *Lect. Notes Phys.* **130** (1980), 247-254

44. Kaup, D.J.: A Method for solving the separable initial-value problem of the full three-dimensional three-wave interaction, *Stud.Appl.Math.* **62** (1980), 75-83.
45. Craik, A.D.D.: Evolution in space and time of resonant wave triads. II. A class of exact solutions, *Proc.Roy.Soc.(London)*, **A363** (1978), 256-269.
46. Jurco, B.: Integrable generalizations of non-linear multiple three-wave interaction models, *Phys.Lett.* **A138** (1989), 497-501.
47. Menyuk, C.R., Chen, H.H., and Lee, Y.C.: Restricted multiple three-wave interactions: Painleve analysis, *Phys.Lett.* **A27** (1983), 1597-1611.
48. Gromak, V.I., and Czegelnik V.V.: System of 3-wave resonant interaction and equations of P-type, *Teor.Mat.Fiz.* **78** (1989), 22-34.
49. Kitaev, A.V.: On similarity reductions of the three-wave resonant system to the Painleve' equations, *J.Phys.* **A23** (1990), 3453-3553.
50. Ablowitz, M.J., and Segur, H.: *Solitons and the Inverse Scattering Transform*, SIAM Phil., 1981.
51. Ohkuma, K., and Wadati, M.: Quantum three wave interaction models, *J.Phys.Soc. Japan* **53** (1984), 2899-2907.
52. Wadati, M, and Ohkuma, K.: Bethe states for the quantum three wave interaction equation, *J.Phys.Soc. Japan* **53** (1984), 1229-1237.
53. Ohkuma, K.: Thermodynamics of the quantum three wave interaction model, *J.Phys.Soc.Japan* **54** (1985), 2817-2828.
54. Jurco, B.: On quantum integrable models related to non-linear quantum optics. An algebraic Bethe ansatz approach, *J.Math.Phys.* **30** (1989), 1739-1743.
55. Kulish, P.P.: Quantum non-linear wave interaction system, *Physica* **D18** (1986), 360-364.
56. Jurco, B.: Quantum integrable multiple three-wave interaction models. *Phys.Lett.* **A143** (1990), 47-51.
57. Hirota, R.: An exact solution to "Simple Harmonic Generation" *J.Phys.Soc.Japan* **46** (1979), 1927-1928.
58. Hirota, R.: Exact solution of the sine-Gordon equation for multiple collisions of solitons, *J.Phys.Soc.Japan* **33** (1972), 1459-1463.
59. Hirota, R.: Exact solution of the modified Korteweg- de Vries equation for multiple collisions of solitons, *J.Phys.Soc.Japan* **33** (1972), 1456-1458.
60. Hirota R.: Direct method of finding exact solutions of non-linear evolution equations, in R.M.Miura (ed.) *Baclund Transformations, the Inverse Scattering Method, Solitons and Their Applications* (Lect. Notes in Math. **515**), Springer-Verlag, Berlin, 1976, p.40-68
61. Karamzin, Yu. N., and Filipchuk, N.S.: On existence parametric connected waveguides and solitons under tree-frequency interaction of waves, *Zh. Prikl. Matemat. Tekhn.Fiz.(in Russia)* (1977), №1, 47-52
62. Karamzin, Yu. N., Sukhorukov A.P., and Filipchuk, N.S.: On existence of tree-frequency soliton solutions at second order approximation of dispersion theory, *Izv.VUZov, radiofiz. (in Russia)*, **21** (1978), 456-458.
63. Karamzin, Yu. N., Sukhorukov A.P., and Filipchuk, N.S.: On the new class of connected solitons in dispersive medium with quadratic non-linearity, *Vestn. MGU, ser.fiz. and astronom. (in Russia)*, **19** (1978), 91-98.
64. Azimov, B.S., Sukhorukov A.P., and Trukhov D.V.: Parametrical multifrequensy solitons: creation, collisions and decay, *Izv. AN SSSR, ser.fiz., (in Russia)*, **51** (1987), 229-233.
65. Romeiras, F.J.: Integrability of double three-wave interaction, *Phys.Lett.* **A93** (1983), 227-229.
66. McKinstrie, C.J., and Luther, G.G.: Solitary-wave solitons of the generalized three-wave and four-wave equations, *Phys.Lett.* **A127** (1988), 14-18.
67. Menyuk, C.R., Chen, H.H., and Lee, Y.C.: Restricted multiple three-wave interactions: Integrable cases of this system and other related systems, *J.Math.Phys.* **24** (1983), 1073-1079.
68. Cheng, Y.: Symmetries and hierarchies of equations for the (2+1)-dimensional *N*-wave interaction, *Physica* **D34** (1989), 277-288.

69. Calogero, F.: Universality and integrability of the non-linear evolution PDE's describing *N*-wave interactions, *J.Math.Phys.* **30** (1989), 28-40.
70. Verheest, F.: Proof of integrability for five-wave interactions in a case with unequal coupling constants, *J.Phys.* **A21** (1988), L545-L549.
71. McKinstrie, C.J., and Luther, G.G.: Solitary-wave solitons of the generalized three-wave and four-wave equations, *Phys.Lett.* **A127** (1988), 14-18.
72. Zabolotski, A.A.: Dynamics of periodical wave in model with quadratic and cubic non-linearity, *Zh.Eksp.Teor.Fiz.* **107** (1995), 1100-1121.
73. Zabolotski, A.A.: Dense configuration of solitons in resonant four-wave mixing, *Phys.Rev.* **A50** (1994), 3384-3393.
74. Zabolotski, A.A.: Coherent four-wave mixing of the light pulses, *Zh.Eksp.Teor.Fiz.* **97** (1990), 127-135.
75. Kaup, D.J., and Malomed, B.A.: The Resonant three-wave interaction in an inhomogeneous medium, *Phys.Lett.* **A 169** (1992), 335-340.
76. Kaup, D.J., and Malomed, B.A.: Three-wave resonant interaction in a thin layer. *Phys.Lett.* **A183** (1993), 283-288.
77. Conwell, E.M.: Theory of second-harmonic generation in optical waveguides, *IEEE J.Quant.Electron.* **QE-9** (1973), 867-879.
78. Uesugi, N., and Kimura, T.: Efficient second-harmonic generation in three dimensional $LiNbO_3$ optical waveguide, *Appl.Phys.Lett.* **29** (1976), 572-574.
79. Sasaki, Y., and Ohmori, Y.: Phase-matched sum-frequency light generation in optical fibres, *Appl.Phys.Lett.* **39** (1981), 466-468.
80. Ohmori, Y., and Sasaki, Y.: Two-wave sum-frequency light generation in optical fibres, *IEEE J.Quant. Electron.* **QE-18** (1982), 758-762.
81. Terhune, R.W., and Weinberger, D.A.: Second-harmonic generation in fibres, *J.Opt.Soc.Amer.* **B4** (1987), 661-673.
82. Karpierz, M.A.: Coupled solitons in waveguides with second- and third- order nonlinearities, *Opt. Letts.* **20** (1995), 1677-1679.
83. Torruellas, W.E., Zuo Wang, Torner, L., and Stegeman, G.I.: Observation of mutual trapping and dragging of two-dimensional spatial solitary waves in a quadratic medium, *Opt. Letts.* **20** (1995), 1949-1951.
84. Torner, L., Torruellas, W.E., Stegeman, G.I., and Menyuk, C.R.: Beam steering by $\chi^{(2)}$ trapping, *Opt. Letts.* **20** (1995), 1952-1954.
85. Torner, L., Mihalache, D., Mazilu, D., and Akhmediev, N.N.: Stability of spatial solitary waves in quadratic media, *Opt. Letts.* **20** (1995), 2183-2185.
86. Trillo, S., Haelterman, M. , and Sheppard, A.: Stable topological spatial solitons in optical parametric oscillators, *Opt. Letts.* **22** (1997), 970-972.
87. Canva, M.T.G., Fuerst, R.A., Baboiu, S., Stegeman, G.I., and Assanto, G.: Quadratic spatial soliton generation by seeded downconversion of a strong harmonic pump beam, *Opt. Letts.* **22** (1997), 1683-1685.
89. Steblina, V.V., Kivshar, Yu.S., and Buryak, A.V.: Scattering and spiralling of solitons in a bulk quadratic medium, *Opt.Lett.* **23** (1998), 156-158.
90. Alexander, T.J., Buryak, A.V., and Kivshar, Y.S.: Stabilisation of dark and vortex parametric spatial solitons, *Opt. Letts.* **23** (1998), 670-672.
91. Malomed, B.A., Drummond, P., He, H., Berntson, A., Anderson, D., and Lisak, M.: Spatiotemporal solitons in multidimensional optical media with a quadratic nonlinearity, *Phys.Rev.* **E56** (1997), 4725-4736.
92. Conti, Cl., Trillo, S., and Assanto, G.: Optical gap solitons via second-harmonic generation: Exact solitary solutions, *Phys.Rev.* **E57** (1998), 1251R-1255R.
93. Sammut, R.A., Buryak, A.V., and Kivshar, Y.S.: Modification of solitary waves by third-harmonic generation, *Opt. Letts.* **22** (1997), 1385-1387.

CHAPTER 8

NON-LINEAR WAVEGUIDE STRUCTURES

It is known [1,2] that an electromagnetic wave can propagate along the planar interface of two linear dielectric media. The strengths of both electric and magnetic field vanish as $|x| \to \infty$. This is the reason why these waves were called *surface waves*. It can be shown that such a surface wave exists only if a magnetic field strength vector is parallel to the interface plane, i.e. it is a TM-wave (sometimes it is called a *p*-polarised wave). If the dielectric permeability of one of the media is a square-law function of electrical field strength, then both propagation of TM wave and the TE wave are possible [3-9]. Such waves were named either non-linear surface waves or non-linear surface polaritons.

At the early stage of investigation of non-linear surface waves (NLSW), the dispersion relations were established for TM-waves [5-7,10-14] and TE-waves [8,15]. It was found that the excitation of the TE-waves demands exceeding of some threshold power. Therefore, this wave is essentially non-linear one. And on the contrary, there are no threshold conditions for excitation of the TM-wave. But all exact solutions of Maxwell equations for these types of NLSW's were obtained for rather specific cases of the anisotropy of non-linear susceptibility.

Analogous to NLSW, the non-linear surface plasmons were investigated in [16]. These waves run along the interface of a dielectric and metal. TE-plasmons are featured by a threshold excitation energy. The propagation of plasmons is possible only for the limited magnitudes of the propagation constant [17,18]. In [20] the propagation of surface non-linear plasmons of TM-type along the interface of the Kerr dielectric and metal film were studied. The thickness of a film periodically altered along the interface. An explicit expression was found for the frequencies of surface plasmons.

Further the investigations of NLSW are transferred to the non-linear media of more general nature than Kerr media. In [21] for example the non-linear susceptibility of the fifth order is taken into account. The dispersion relation is found for NLSW propagating along the interface of the linear and the quadratic non-linear dielectric [22]. The non-linear polaritons are studied in a thin ferromagnetic [23] and paramagnetic [24] film. In [25] the NLSW of TM-type are discussed with the account of light interaction with excitons and exciton-biexciton conversion. In these cases the exact dispersion relations were derived. In [26] the description of NLSW is provided when the refraction index of the linear substrate is decreasing exponentially deep into the medium to some constant magnitude. The dispersion relation and the flux of power along the interface were obtained. On the contrary, the case of NLSW of the TM-type when the dielectric susceptibility of a non-linear medium changes exponentially was investigated numerically [27]. A propagation of NLSW along the semi-infinite super-lattice provides an example of non-linear optical phenomena in the low-dimensional systems, which are popular now. For this case in [28,29] the dispersion relations and the transverse distributions of the electric field of TE-wave have been found.

It is known that plane waves are unstable in bulk Kerr media. In excess of a certain power threshold they convert into 3D (or temporal) solitary waves. Similar phenomenon takes place in the case of NLSW. Both numerical [30] and analytical [31] investigation of this process was executed in terms of the non-linear Schrödinger equation solution, where the co-ordinate along the interface plays the role of time. It was found that the threshold of the modulation instability is different for different branches of non-linear surface polaritons. The similar studies of the diffusive Kerr media [32] demonstrate that the criterion for stability is identical to the case arising in a non-diffusive limit.

The NLSW may propagate along the interface of two [33,34] and more [35] non-linear media as well.

It should be noted that the NLSW can be excited due to parametric interaction of four waves. Three of them are incident waves onto the surface of interface. The theory of this type of excitation is presented in the works [34,36-38]. In [38] the emergence of non-propagating stable kinks is predicted under condition of four wave mixing. This solution of the corresponded system of equation describes a domain wall between two homogeneous states of electromagnetic wave, whose phases of oscillations differ by π.

In section 8.1 we will consider exact solutions of Maxwell equations for the simplest cases of the interfaces of homogeneous linear and Kerr dielectrics. The main results will be the dispersion relations and the transverse profiles of the NLSW of TE- and TM-types.

The simplest example of a non-linear planar waveguide is a thin film on a non-linear (Kerr) substrate [39-42]. In this case the Maxwell equation for TE-mode can be solved exactly and dispersion relations for all guided waves can be found. The NLSW can propagate along the interface of a linear film and non-linear substrate. It is known that in order to excite this wave a power threshold should be exceeded [39]. The eigenmodes of a planar waveguide keep but their propagation constants depend now on the power transferred by the wave. In [44,43] the numerical methods are described to explore the guided waves (NLGW) in the waveguides, where the refraction index profiles for all three layers are arbitrary and the non-linear absorption of radiation is taken into account.

The linear waveguide surrounded with non-linear media presents more general case of non-linear waveguide structure [45-54]. As in this case there are two interfaces of linear and non-linear media, then two NLSWs can exist, each coupled with one of the interface [46]. Then there is a guided wave localised mainly in non-linear media. The minimum of the power flow is in a linear layer. This wave can be represented by a coupled state of two NLSWs mentioned above. Other NLGWs are modified modes of a linear waveguide.

The study of non-stationary regime of the guided TE-waves in a waveguide in Kerr-media environment was carried out in [54].

From the theoretical point of view a structure formed by a non-linear film in a linear environment is a more complex example [55-65]. There is an interesting scheme of classification of all TE-waves in such waveguides based on the conic classification. The existing analytical expressions of the dispersion relations can be analysed only numerically [60-64]. The numerical analysis of TE- and TM- waves in a stripe waveguide was done in [65].

A new class of solutions for a non-linear waveguide in linear surrounding was found in [66]. For the stationary TE waves the electric field profile has a minimum of a field strength in the centre of the non-linear film. The stability analysis was carried out numerically and it was found that these waves were not stable.

The non-linear guided structures with more than three layers were considered in [67-69]. The dispersion relations were found and conditions determined, under which the propagation of NLGW was characterised by the bistable behaviour.

As for the non-linear surface waves, the role of the higher order non-linear susceptibility was investigated [70]. The non-linear plasmons in a Kerr waveguide (with one of the surfaces coated by a metal film) were observed in [71]. In [72-80] the non-Kerr-like guided media were studied. In the case of saturating non-linear media it was found that some modes existing in a pure Kerr-waveguide could disappear with the increasing of power flow [73-75]. A theoretical amplification of the TE-wave was investigated when a substrate contained homogeneously distributed two-level atoms pumped by an external source [80].

As with the NLSW, the modulation instability can occur [81-85]. As a rule the theory was based on the non-linear Schrödinger equation. The results obtained qualitatively agree with the ones known for wave propagation in the two-dimensional Kerr media.

An interesting result was obtained in [86]. It has been found that non-stationary regime can arise when the guided wave propagates in a linear waveguide on a non-linear substrate. (or in a non-linear environment). For such a structure the radiation comes out of a waveguide to a substrate (or to an environment) in the form of the narrow beams - *spatial solitons*. If one employs a waveguide with saturating non-linearity, then such soliton emission keeps, but soliton becomes an ordinary solitary wave [87-89]. The results of the investigation of solitary waves emission are represented in the review [90].

An essentially non-linear dependence of the NLGW propagation constant on the power flow results in the instability of modes under certain conditions. It leads to the hysteresis loops in the dependence of the output power versus input power [91]. The hysteresis indicates the existence of two stable states of the wave propagation regimes This phenomenon sometimes called the bistability was investigated in [91-93].

A sufficiently good survey of the results concerning NLSW and NLGW can be found in [94-96].

In section 8.2 the NLSW and NLGW will be considered in a planar waveguide formed by a linear dielectric film on a non-linear substrate with the non-linear coating. The non-linearity of Kerr type is taken for the sake of simplicity. In this case Maxwell equation can be solved exactly and all important characteristics of TE- and TM- waves can also be found in such guided structure. Analogously, in 8.3 section a non-linear waveguide surrounded by linear dielectric media is discussed. Both dispersion relations and electric field distributions in NLGW and power flux for TE- and TM-waves will be obtained there.

A more complex waveguide structure is the *directional coupler*. This structure is formed by two parallel waveguides separated by such a short distance that radiation can penetrate from one waveguide to another due to violated total internal reflection. Sometimes they say that such two waveguides are tunnel-connected [98,104]. Waves in both waveguides, or channels spread in one direction. The distance at which radiation from one channel completely transfers to another is called a *coupling length*. If waveguides are made of non-linear dielectrics or they are immersed in non-linear medium, then such structure is called the *non-linear directional coupler* (NLDC). The principal characteristic of NLDC – the dependence of coupling length on an input power – was determined in the first works [105-107]. As a result, by varying the radiation launched to a NLDC with the fixed length, it is possible to switch the output radiation between the output ports of a coupler. The analytical solution of the NLDC model with Kerr non-linearity was found and studied in [108-113] as well. The experiments [114-116] demonstrated all-optical switching in such NLDCs.

In further investigation attention was drawn to the role of different factors in switching. The non-linearity of the higher order than Kerr non-linearity was considered in [117]. The effect of small longitudinal inhomogeneities on the optical response of NLDC was observed [118,119]. The parameters of the directional coupler for non-Kerr non-linearity were discussed in [120-126]. The effect of saturation of non-linearity and losses was examined in [120-123]. The influence of the diffusion non-linearity on switching characteristics of NLDC were studied in [124,125] numerically. It was found that switching could still occur even when a diffusion length equated to the distance between two channels. Under this condition the switching threshold increases and the output versus input power rate of change aggravates. It is demonstrated in [126] that the two-photon absorption can noticeably attenuate the effectiveness of switching in NLDC.

The main application of NLDC is concerned with all-optical switching devices. It determines the principal direction of investigation in this field. The experimental observation of the bistability [127] dates from the early investigations. It was shown in the following works [128-134] that NLDC permits creation of the high contrast switch.

The generalisation of a two-channel directional coupler is a waveguide structure with a great number of channels. The first step in this direction was the generalisation of an N-channel directional coupler ($N >> 1$) for the case of Kerr non-linearity [135]. In a continuous limit (when $N \to \infty$) the system of non-linear equations of the coupled waves theory becomes the non-linear Schrödinger equation that describes one-dimensional spatial solitons or two-dimensional self-focusing. The exploring of multi-channel NLDC [135-137] showed that in such waveguide structure a "discrete" self-focusing could be implemented. In [138] the numerical simulation for $N<6$ demonstrated an operation of this arrangement as an all-optical switching. In [139] a considerable decrease in switching energy was achieved for the case $N=3$. For the greater number of channels in a similar structure (called now Non-linear Waveguide

Array), the problem of modulation instability of a transversal distribution of radiation over channels arises [135,140-142].

A very interesting application of non-linear directional coupler was found in [143]. It was shown that NLDC allows to carry out the generation of squeezed light with a low level of pumping and a short length of interaction. So this non-linear integrated optics device can be useful in the quantum optics investigations.

In section 8.4 we will consider a simple model of NLDC based on a two-channel waveguide, supposing that the non-linear properties of the channels are described by the third order non-linearity as it was done in [105]. The non-linear equations of the coupled waves permit the exact solutions in this case. This allows to observe rigorously the main regimes of NLDC operation. The dependence of the coupling length on the input power and on the tunnel-connected waveguides parameters will be obtained.

Another widespread elementary integrated optics device is the distributed feedback structure (DBF-structure) [144-146]. The feedback is implemented by periodic changes of either a refraction index or a waveguide thickness. When the difference between the propagation constants of two guided waves (waveguide modes) is equal to the wave number of some Fourier-component of the periodical perturbations of the waveguide, an effective inter-mode energy exchange takes place. This process can be considered as a three-waves parametric interaction, while one of them is a non-propagating wave. If the modes of a waveguide differ only by the directions of propagation, then the DBF-structure displays the properties of the distributed reflector. The resonance condition here resembles Bragg condition for a reflection grating under condition that the diffraction angle is π. As soon as the condition of Bragg resonance holds not for all modes, a DBF-structure operates as a selective reflector or distributed filter. The mentioned properties explain the wide spreading of this structure in integrated optics.

It is natural that if the DBF-structure is made of non-linear dielectric, then only non-linear waves can propagate there. Under the action of radiation the waveguide refractive index will change and, consequently, the Bragg condition will be broken. The simplest example of non-linear DBF-structure (which will be called now as NLDBF-structure) was considered in [147]. A periodical structure was formed by alternating linear and non-linear dielectric layers in the direction of wave propagation. The NLDBF-structure as an integrated optics element was observed in [148,149], where a transmission-reflection switching was investigated. A prospect to use a dielectric with a saturating non-linearity was also considered there.

The development of the NLDBF-structures is presented in [150-152]. In these works the interaction of an external Gaussian beam with the waveguide modes was studied. Both the two-photon absorption and its role in the saturation of non-linearity was discussed in [151]. The dependence of both transmission and reflection on the wave detuning was derived in [153].

As the most of non-linear waveguides, the optical bistability phenomenon takes place for the NLDBF-structures as well. The NLDBF transmission and reflection coefficients are the multivalued functions of the input power [154,155]. An experimental study of the bistable behaviour of the NLDBF-structures was carried out in

[156,157]. The optical bistability in NLDBF-structure on the base of the coupled waves theory was investigated numerically in [158].

Propagation of ultra-short optical pulses through NLDBF-structure gives an example of a new type of solitary waves – a *gap soliton* [159,160]. These solitons propagate in the periodic medium when their power spectrum is located in the forbidden gap which arose due to the periodical alterations of waveguide properties. The frequency of the carrier wave may be close to the frequency of Bragg resonance, but, unlike the case of linear waves, reflection does not occur. The equations describing propagation of the gap soliton in NLDBF-structure are similar to the equations of a massive Thirring model [161-163], but unlike them, these equations can not be integrated by the inverse scattering method. At least we do not know the results concerning the complete integrability of the equations for a gap soliton. Some specific solutions are presented in [164], where the analogy with the Thirring model is essentially exploited. The perturbation theory for a gap soliton, developed in [165], is also based on this analogy.

In section 8.5 we will consider the equations describing dynamics of the guided waves amplitudes in the non-linear DBF-structure. The exact solutions of the non-linear system of equations under rigorous Bragg resonance condition will be found in the case of continuous radiation. We will also observe the system of equations of the evolution of the short optical pulse envelope the in NLDBF-structure. These equations represent the modified massive Thirring model. The steady state solutions of this system of equations will be obtained. These solutions are often called the gap solitons or Bragg solitons.

8.1. Non-linear surface waves

This section is concerned with the discussion of some phenomena arising when an electromagnetic wave propagates along a planar interface of two dielectric media or in a planar dielectric waveguide. Let the direction of the wave propagation be the *z*-axes of the Cartesian frame. The normal to the surface is fixed in the *x*-direction while the *y*-axis lies in the interface plane. There are two types of the surface waves: (a) transverse electric waves (TE wave) with $\vec{E}=(0,E_y,0)$ and $\vec{H}=(H_x,0,H_z)$, and (b) transverse magnetic waves (TM wave) with $\vec{E}=(E_x,0,E_z)$ and $\vec{H}=(0,H_y,0)$ .If the dielectric permeability of one of the media is a square-law function of electrical field strength

$$\hat{\varepsilon}(\omega,\vec{E})=\hat{\varepsilon}(\omega)+\hat{\varepsilon}_{nl}(\omega):\vec{E}\vec{E} \ , \qquad (8.1.1)$$

then both the TM waves and the TE waves, i.e., non-linear surface polaritons propagation are possible.

8.1.1. NON-LINEAR SURFACE WAVES OF TE-TYPE

The electrical field vector of the TE-waves has only one nonzero component [97,98], for instance E_y. Thus, the wave equation looks like following:

$$\frac{\partial^2 E_y}{\partial z^2}+\frac{\partial^2 E_y}{\partial x^2}+k_0^2\varepsilon\left(\omega,|\vec{E}|^2\right)E_y=0\,, \qquad (8.1.2)$$

where $k_0 = \omega/c$. The magnetic field strength vector components for this sort of waves are connected with E_y by the relations: $H_x = ik_0^{-1}\partial E_y/\partial z$, $H_y = 0$, $H_z = -ik_0^{-1}\partial E_y/\partial x$. On the interface plane at $x = 0$ E_y, H_x and H_z are continuous. The boundary conditions

$$\lim_{|x|\to\infty} E_y(\omega,x,z)=0\,,\quad \lim_{|x|\to\infty}\left[\frac{\partial E_y}{\partial x}\right]=0\,. \qquad (8.1.3)$$

select only those solutions of equation (8.1.2) which correspond to the surface waves.

Due to the translation symmetry along z-axis the y-component of the electric field $E_y(\omega, x, z)$ can be chosen as

$$E_y(\omega,x,z)=\Phi(x)\exp\{i\beta(\omega)z\}\,, \qquad (8.1.4)$$

where $\beta(\omega)$ is a propagation constant [98]. Equation for the transverse profile of the electrical field of NLSW $\Phi(x)$ follows from the equations (8.1.2) and (8.1.4) :

$$\Phi,_{xx}+[k_0^2\varepsilon(\omega,x)-\beta^2]\Phi=0\,, \qquad (8.1.5)$$

where $\Phi(x)$ and $\Phi,_x(x)$ are continuous at $x = 0$ and approach to zero when $x\to\pm\infty$. In this chapter the derivative $d\Phi(x)/dx$ is symbolised as $\Phi,_x(x)$. Dielectric permeability $\varepsilon(\omega,x) = \varepsilon_{yy}(\omega,x;|E|^2)$ is given by the condition:

$$\varepsilon(\omega,x)=\begin{cases}\varepsilon_1 & x<0\\ \varepsilon_2+\varepsilon_{nl}\Phi^2, & x>0\end{cases}$$

Equation (8.1.5) may be rewritten as a pair of equations with the constant coefficients:

$$\Phi,_{xx}+(k_0^2\varepsilon_1-\beta^2)\Phi=0\,, \qquad x<0 \qquad (8.1.6a)$$

$$\Phi,_{xx}+(k_0^2\varepsilon_2-\beta^2)\Phi+k_0^2\varepsilon_{nl}\,\Phi^3=0\,, \quad x>0. \qquad (8.1.6b)$$

Solution of the equation (8.1.6a), satisfying the boundary condition at $x \to -\infty$, is

$$\Phi^{(1)}(x) = A_1 \exp(qx)\,, \tag{8.1.7}$$

where $q^2 = (\beta^2 - k_0^2\varepsilon_1) > 0$ and we choose q >0 for the sake of certainty.

If we multiply the left part of equation (8.1.6b) by $\Phi,_x(x)$, then the resulting expression will become the total derivative of some function, and after integrating, the first integral of equation (8.1.6b) will be expressed in the following form

$$\left(\Phi,_x\right)^2 + (k_0^2\varepsilon_2 - \beta^2)\Phi^2 + (k_0^2\varepsilon_{nl}/2)\Phi^4 = I\,. \tag{8.1.8}$$

Where $I = 0$ is in accordance with the boundary conditions. Expression (8.1.8) allows to make certain qualitative conclusions about the properties of the solutions of (8.1.6b).

If $\varepsilon_{nl} > 0$ (this is the case of a self-focusing non-linear medium), $(d\Phi/dx)^2$ can be positive when the condition $p^2 = (\beta^2 - k_0^2\varepsilon_2) > 0$ only holds. Then the function $\Phi(x)$ is limited and it reaches a zero value along with its derivatives and the boundary conditions under $x \to +\infty$ are satisfied.

If $\varepsilon_{nl} < 0$ (self-defocusing medium), $\Phi(x)$ does not reach zero under condition $(k_0^2\varepsilon_2 - \beta^2) > 0$ and fails to satisfy the boundary conditions. If $(k_0^2\varepsilon_2 - \beta^2) < 0$, $\Phi(x)$ and $\Phi,_x(x)$ can reach zero simultaneously, but the function itself is not limited.

For a self-focusing medium equation (8.1.8) reduces to the equation:

$$(\Phi,_x)^2 = p^2\Phi^2\left(1 - \alpha\Phi^2\right)\,, \tag{8.1.9}$$

where $\alpha = k^2_0\varepsilon_{nl}/2p^2$, and its solution gives

$$\Phi^{(2)}(x) = \alpha^{-1/2}\,\mathrm{sech}\left[p(x - x_2)\right], \tag{8.1.10}$$

Here x_2 is the location of the maximum of the electrical field strength. It appears here as a constant of integrating for equation (8.1.9).

For a self-defocusing medium equation (8.1.8) may be presented in the form:

$$(\Phi,_x)^2 = p^2\Phi^2\left(1 + |\alpha|\Phi^2\right)\,, \tag{8.1.11}$$

Solution of this equation is given by the expression

$$\Phi^{(2)}(x) = \pm(|\alpha|)^{-1/2}\,\mathrm{cosech}\left[p(x - x_2)\right], \tag{8.1.12}$$

This solution has a singularity at $x = x_2$. The strength of electrical field grows unlimitedly near this point. If $x_2 < 0$, this singularity is fictitious, but the derivative of $\Phi(x)$ has a break-up at $x = 0$ leading to the discontinuity of the magnetic field strength, when crossing the interface of the two dielectric media. Consequently, the condition $x_2 > 0$ only holds. In the half-space filled with the non-linear dielectric matter, an electrical field strength grows infinitely. The approximation the present theory of non-linear guided waves is based on, is no longer valid as it is limited by an assumption of a weak non-linearity (i.e. $\varepsilon_{yy}(\omega;|E|^2)$ in the form of (8.1.1)). Besides that, it would become necessary to take the losses into account too. These losses can be of a non-linear nature (two-photon absorption etc.). Thereby, within the framework of the model accepted here, the non-linear surface waves of the TE-type spreading along a planar interface of linear and non-linear self-focusing media are only possible. Hereinafter, only this case will be the subject of consideration.

The condition of continuity at the interface $x = 0$ requires $\Phi^{(1)}(0) = \Phi^{(2)}(0)$, $\Phi^{(1)}{}_{,x}(0) = \Phi^{(2)}{}_{,x}(0)$ and yields a result for:

$$A_1 = \alpha^{-1/2}\,\mathrm{sech}(px_2)\,, \tag{8.1.13a}$$

$$q = p\tanh(px_2)\,. \tag{8.1.13b}$$

Formula (8.1.13a) can be considered as the determination of the x_2 by an electric field strength in the NLSW at $x = 0$. Equality (8.1.13b) is a dispersion relation for this TE - NLSW. As far as it is chosen $q > 0$, it follows from (8.1.13b) that $x_2 > 0$. That means the maximum of the electric field strength in NLSW is located in the non-linear dielectric medium (Fig.8.1.1).

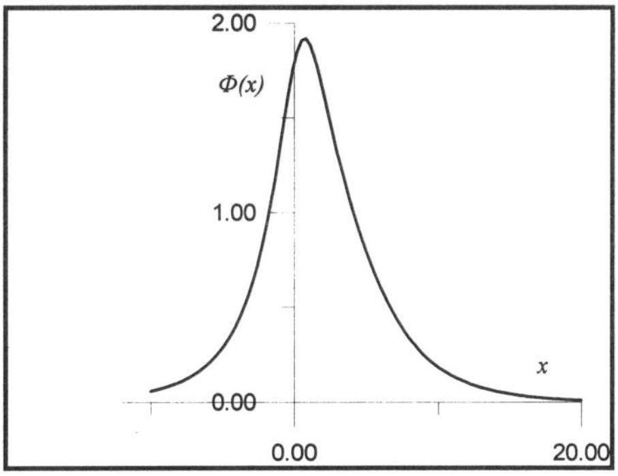

Fig.8.1.1.

Profile of the electric field of TE-NLSW

It is possible to define an electric field strength in the NLSW on the interface from

(8.1.13), which yields:

$$A^2 = \frac{p^2 - q^2}{\alpha p^2} = 2\frac{\Delta\varepsilon}{\varepsilon_{nl}}$$

where $\Delta\varepsilon = \varepsilon_1 - \varepsilon_2$. It comes out that the NLSW of the TE-type can propagate under condition $\Delta\varepsilon > 0$. The power flow is defined by the expression

$$P = \int_{-\infty}^{+\infty} S_z dx ,$$

where S_Z is a z-component of Poynting vector averaged over the fast time oscillations. For the TE-wave the power flow reads as

$$P = \frac{c^2\beta}{8\pi\omega} \int_{-\infty}^{+\infty} |\Phi(x)|^2 dx$$

Using the expressions (8.1.7) and (8.1.11) for $\Phi(x)$ and relations (8.1.13), one can obtain the result in the form:

$$P/P_0 = A^2/2q + [1 + \tanh(px_2)]/\alpha p = \beta(p+q)^2/2p = \beta[p + q + k_0^2\Delta\varepsilon/2q].$$

where $P_0 = c(4\pi k_0 \varepsilon_{nl})^{-1}$. It is useful to analyse the dependence of P/P_0 on propagation constant β keeping in mind the definition of p and q in terms of β.

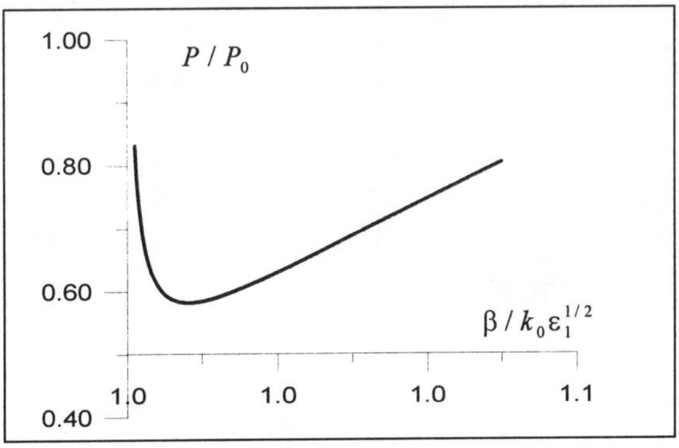

Fig. 8.1.2. Variation of the normalised power flow with propagation constant for with $\Delta\varepsilon/\varepsilon_1 = 0{,}05$.

First of all it is worth to note that the inequality $\beta^2 > k_0^2\varepsilon_1 > k_0^2\varepsilon_2$ always holds. When $\beta^2 \to k_0^2\varepsilon_1$, the ratio P/P_0 increases as $(\beta^2 - k_0^2\varepsilon_1)^{-1/2}$. When β approaches infinity, the P/P_0 grows as β^2. Consequently, the ratio P/P_0 as a function of β has at least one minimum. In order to find the value of β, where P/P_0 is extreme, it is necessary to solve the equation $d(P/P_0)/d\beta=0$. Having done this, one may be convinced that for a certain value of β_c P/P_0 takes minimum denoted as P/P_0. Thereby, in order to excite the NLSW, the flow of power must exceed the critical value $P_c = P(\beta_c)$. Then, under condition $P > P_c$, two NLSWs propagate along the interface carrying equal power, but having different propagating constants (Fig.8.1.2).

Following [4] we shall consider how the NLSW parameters depend on the maximum of the electric field $A_m = \Phi(x = x_2)$. Let us define a dimensionless parameter D by the relation $A_m^2 = (1+D)A^2$. It appears that all features of the NLSW can be expressed in terms of D. For instance, $p^2 = (1+D)Q^2$, $q^2 = DQ^2$, $\beta^2 = (\Delta\varepsilon^{-1}\varepsilon_{nl} + D)Q^2$, where $Q^2 = \Delta\varepsilon k_0^2$. As it follows from these relations and also from (8.1.13):

$$x_2 = \frac{1}{Q\sqrt{1+D}}\,\mathrm{artanh}\left(\frac{\sqrt{D}}{\sqrt{1+D}}\right) = \frac{1}{2Q\sqrt{1+D}}\ln\left(\frac{\sqrt{1+D}+\sqrt{D}}{\sqrt{1+D}-\sqrt{D}}\right),$$

$$P/P_0 = Q^2\sqrt{\varepsilon_1\Delta\varepsilon^{-1} + D}\left(\sqrt{D} + \sqrt{1+D} + 1/2\sqrt{D}\right).$$

We can regard these expressions as a parametric definition of the function x_2 versus P. This dependence is shown on Fig.8.1.3, where $\Delta\varepsilon = 0{,}06$ and $\varepsilon_1 = 2{,}25$.

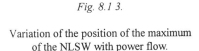

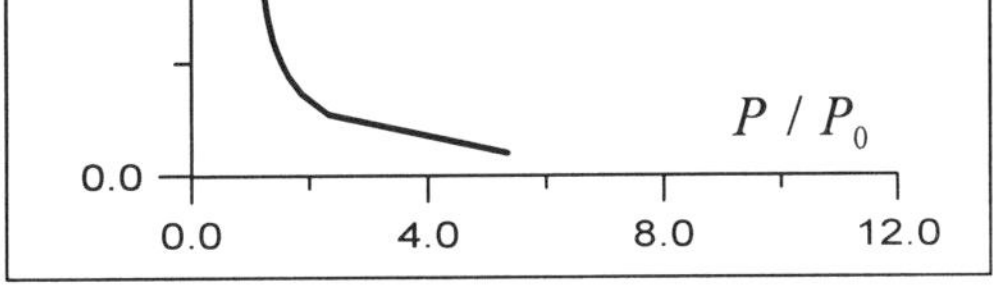

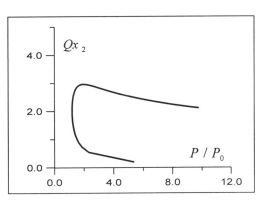

Fig. 8.1 3.

Variation of the position of the maximum of the NLSW with power flow.

It is noteworthy that the position of the maximum of the NLSW electric field strength changes non-monotonically with the increase of D, x_2 reaches its maximum value $\max(x_2) = 0{,}663/Q$ at $D = 2{,}277$. Furthermore, the dependence x_2 on power flow is ambiguous.

In conclusion we note that there is no linear partner for the considered NLSW. When the non-linear permeability correction ε_{nl} approaches zero, the NLSW disappears simply.

8.1.2. NON-LINEAR SURFACE WAVES OF TM-TYPE

The TM-waves have nonzero tangent E_z and normal E_x components of the electric field strength vector and only one component of the magnetic field vector – H_y is nonzero [98]. Let only the diagonal components of the dielectric permeability tensor be nonzero:

$$\hat{\varepsilon} = \mathrm{diag}\left(\varepsilon_{xx}, \varepsilon_{yy}, \varepsilon_{zz}\right), \tag{8.1.14a}$$

and let these components not depend on the electric field strength of the light wave when $x < 0$. But when $x > 0$, the diagonal elements are defined as follows:

$$\begin{aligned} &\varepsilon_{xx}(\omega, |\vec{E}|^2) = \varepsilon_{\perp}(\omega), \\ &\varepsilon_{yy}(\omega, |\vec{E}|^2) = \varepsilon_{zz}(\omega, |\vec{E}|^2) = \tilde{\varepsilon}_{||} = \varepsilon_{||}(\omega) + \varepsilon_{nl} |\vec{E}|^2 . \end{aligned} \tag{8.1.14b}$$

The NLSW propagates along z-axis. Due to the translation symmetry to this axis vectors of the electric and magnetic fields can be written as

$$\vec{H}(x,z,\omega) = \vec{h}(x,\omega)\exp[i\beta(\omega)z], \quad \vec{E}(x,z,\omega) = \vec{e}(x,\omega)\exp[i\beta(\omega)z].$$

Maxwell equations for TM-waves reduce to the equations for the y-independent components of the electric and magnetic fields

$$\beta h_y = k_0\varepsilon_{\perp} e_x, \; h_{y,x} = -ik_0\tilde{\varepsilon}_{||} e_z, \quad ik_0 h_y = i\beta e_x - e_{z,x} . \tag{8.1.15}$$

Hence it follows

$$h_y(x,\omega) = \frac{ik_0\varepsilon_{\perp}}{k_0^2\varepsilon_{\perp} - \beta^2}\left(\frac{de_z}{dx}\right). \tag{8.1.16}$$

If we denote $e_z(x,\omega) = \Phi(x)$, then the equations (8.1.15) and (8.1.16) result in the wave equation for $\Phi(x)$

$$\Phi,_{xx} + \varepsilon_{\perp}^{-1}[k_0^2\varepsilon_{\perp} - \beta^2]\tilde{\varepsilon}_{\|}\Phi = 0\,, \tag{8.1.17}$$

The boundary conditions for $\Phi(x)$ are similar to those ones for the TE-waves, but the continuity condition of the tangent components of the vectors $\vec{h}$ and $\vec{e}$ at $x = 0$ on the interface plane looks as

$$[\Phi] = 0\,,\quad \left[\varepsilon_{\perp}(\beta^2 - \varepsilon_{\perp}k_0^2)^{-1}\Phi_{,xx}\right] = 0 \tag{8.1.18}$$

where the square brackets $[f]$ mean the magnitude of a jump of the functions $f(x)$ at the point $x = 0$.

The strong restriction of this models is a choice of anisotropy of the non-linear dielectric permeability in the form of (8.1.14b). However, in a more general case there is no exact analytical solution of equation (8.1.15).

In the case of only one interface, which is being considered here, equation (8.1.17) decomposes to the pair of equations

$$\Phi,_{xx} + (\varepsilon_{\|}^{(1)} / \varepsilon_{\perp}^{(1)})\left(k_0^2\varepsilon_{\perp}^{(1)} - \beta^2\right)\Phi = 0\,, \tag{8.1.19a}$$

$$\Phi,_{xx} + (\varepsilon_{\|}^{(2)} / \varepsilon_{\perp}^{(2)})\left(k_0^2\varepsilon_{\perp}^{(2)} - \beta^2\right)\Phi + (\varepsilon_{nl} / \varepsilon_{\perp}^{(2)})\left(k_0^2\varepsilon_{\perp}^{(2)} - \beta^2\right)\Phi^3 = 0\,, \tag{8.1.19b}$$

Solution of the equation (8.1.19a), satisfying the boundary condition as $x \to -\infty$, is

$$\Phi^{(1)}(x) = A_1 \exp(qx)\,, \tag{8.1.20}$$

where $q^2 = (\varepsilon_{\|}^{(1)} / \varepsilon_{\perp}^{(1)})\left(\beta^2 - k_0^2\varepsilon_{\perp}^{(1)}\right) > 0$, and it is chosen $q > 0$ (i.e. plus at the square root from q^2).

Condition (8.1.19b) then gives the result similar to the case of TE-waves:

$$(\Phi,_x)^2 = p^2\Phi^2\left(1 + \alpha\Phi^2\right)\,, \tag{8.1.21}$$

where notations $p^2 = (\varepsilon_{\|}^{(2)} / \varepsilon_{\perp}^{(2)})\left(\beta^2 - k_0^2\varepsilon_{\perp}^{(2)}\right)$, $\alpha = \varepsilon_{nl} / 2\varepsilon_{\|}^{(2)}$ are introduced. As the left part of the expressions (8.1.21) is positive, then $p^2 > 0$. But as with for TE-waves, here it necessary to distinguish between self-focusing ($\varepsilon_{nl} > 0$) and self-defocusing ($\varepsilon_{nl} < 0$) media.

When $\varepsilon_{nl} > 0$ the solution of equation (8.1.21) is

$$\Phi^{(2)}(x) = \pm(|\alpha|)^{-1/2}\,\mathrm{cosech}\left[p(x - x_2)\right], \tag{8.1.22}$$

The continuity conditions (8.1.18) result into the relations

$$A_1 = \pm(|\alpha|)^{-1/2}\,\mathrm{cosech}\big[px_2\big], \qquad (8.1.23a)$$

$$\frac{\varepsilon_{\parallel}^{(1)}}{q} = \frac{\varepsilon_{\parallel}^{(2)}}{p\tanh(px_2)}. \qquad (8.1.23b)$$

As far as $q > 0$, it follows from (8.1.23) that $x_2 > 0$ when $\varepsilon_{\parallel}^{(1)} > 0$, $\varepsilon_{\parallel}^{(2)} > 0$. That means that the electric field has a singularity at the point x_2 located in the non-linear medium. If either $\varepsilon_{\parallel}^{(1)}$ or $\varepsilon_{\parallel}^{(2)}$ is negative, then $x_2 < 0$ and there is no any singularity. A transverse profile of the electric field strength of NLSW has the form shown in the Fig.8.1.4.

If $\varepsilon_{nl} < 0$ solution of the equation (8.1.21) is

$$\Phi^{(2)}(x) = (|\alpha|)^{-1/2}\,\mathrm{sech}\big[p(x - x_2)\big], \qquad (8.1.24)$$

and the continuity condition (8.1.18) leads to the expressions

$$A_1 = (|\alpha|)^{-1/2}\,\mathrm{sech}\big[px_2\big], \qquad (8.1.25a)$$

$$\frac{\varepsilon_{\parallel}^{(1)}}{q} = \frac{\varepsilon_{\parallel}^{(2)}}{p\coth(px_2)}. \qquad (8.1.25b)$$

Formula (8.1.25a) gives a relationship between the parameter x_2 (that is an integrating constant of (8.1.21) or a co-ordinate of maximum of the electric field strength of the TM NLSW) and the amplitude of NLSW on the interface plane $x = 0$. Formula (8.1.25b) is a dispersion law for this wave. If both $\varepsilon_{\parallel}^{(1)}$ and $\varepsilon_{\parallel}^{(2)}$ are positive, then $x_2 > 0$ (refer to Fig.8.1.1). If either $\varepsilon_{\parallel}^{(1)}$ or $\varepsilon_{\parallel}^{(2)}$ is negative, then $x_2 < 0$ means that the electric field is "ejected" from a non-linear dielectric (refer to. Fig.8.1.4).

Similar to the case of the TE-waves, one can find here a power flow associated with the TM-type NLSW, and one can define its dependence on the propagating constant β. As far as these waves are possible both with $\varepsilon_{nl} > 0$ and with $\varepsilon_{nl} < 0$, the analysis will be more tedious but still there are no principle difficulties involved. The reader may be recommended to do this procedure on his own.

The qualitative difference should be noted between the two types of NLSW, the TM-type and the TE-type. As it has already been discussed, the NLSW of the TM-type under condition $\varepsilon_{nl} > 0$ can exist only when the dielectric permeability's $\varepsilon_{\parallel}^{(1)}$ and $\varepsilon_{\parallel}^{(2)}$ have the opposite signs. In the limit $\varepsilon_{nl} \to 0+$ the NLSW changes over to the linear surface

Fig. 8.1.4.

Profile of the electric field of TM-NLSW

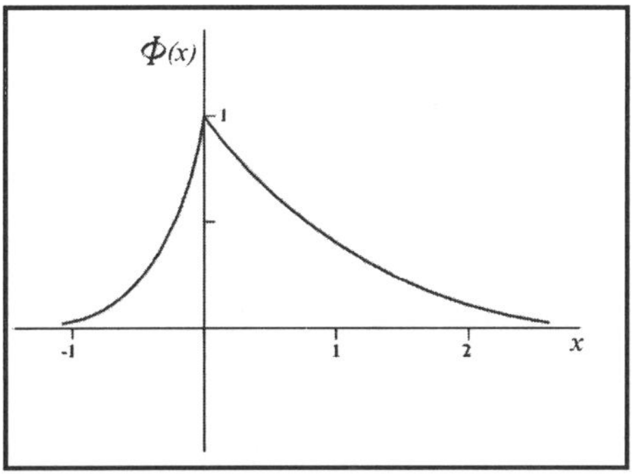

wave. That means that it has a linear partner. If $\varepsilon_{nl} < 0$, then where $\varepsilon_{\parallel}^{(1)}$ and $\varepsilon_{\parallel}^{(2)}$ are both positive, the NLSW has no linear partner, but when the signs of $\varepsilon_{\parallel}^{(1)}$ and $\varepsilon_{\parallel}^{(2)}$ are opposite, a linear partner of the NLSW does exist.

8.1.3. ANOTHER CASE OF NON-LINEAR SURFACE WAVES

An analysis of the surface waves running along the interface between two non-linear dielectric is the natural generalisation of the NLSW theory. This case may be easily analysed on the basis of the results established above. As an example let us consider the NLSW of the TE-type. In both cases $x < 0$ and $x > 0$, the equation determining the transverse profile of an electric field of NLSW has the form of (8.1.6b). If one chooses $\varepsilon_{nl}^{(1)}$ and $\varepsilon_{nl}^{(2)}$ to be positive, then the solution of the equation (8.1.6b) writes

$$\Phi(x) = \begin{cases} \Phi^{(1)}(x) = \alpha_1^{-1/2}\operatorname{sech}[q(x - x_1)] & , \quad x < 0 \\ \Phi^{(2)}(x) = \alpha_2^{-1/2}\operatorname{sech}[p(x - x_2)] & , \quad x > 0 \end{cases}$$

Here p, q, α_1, α_2 are determined in the same way as in section 8.1.1. Index 1(or 2) refers to negative (or positive) x. Continuity condition for $\Phi(x)$ and $d\Phi(x)/dx$ at the interface $x = 0$ results into two relations [8]:

$$\alpha_1^{-1/2}\operatorname{sech}(qx_1) = \alpha_2^{-1/2}\operatorname{sech}(px_2) \tag{8.1.26a}$$

$$q\tanh(qx_1) = p\tanh(px_2) \tag{8.1.26b}$$

The electric field strength $A = |E(\omega, x = 0)|$ on the interface is defined by the expression (8.1.26a). Having excluded the constants of integrating x_1 and x_2, it is possible to show from (8.1.26) that $A^2 = 2(\varepsilon_1 - \varepsilon_2)/(\varepsilon_{nl}^{(2)} - \varepsilon_{nl}^{(1)})$. Hence it follows that NLSW exists in two situations: (a) when $\varepsilon_1 > \varepsilon_2$, $\varepsilon_{nl}^{(2)} > \varepsilon_{nl}^{(1)}$ and (b) when $\varepsilon_1 < \varepsilon_2$, $\varepsilon_{nl}^{(2)} < \varepsilon_{nl}^{(1)}$. From (8.1.26) x_1 can be expressed in terms of x_2. The dispersion relation will depend on x_2 parametrically. However, it is more suitable to take the power flow as a parameter. Similarly to the case considered in section 8.1.1 one manages to derive an expression for P [3]. For $\varepsilon_{nl}^{(1)}$ and $\varepsilon_{nl}^{(2)} > 0$ we have

$$P / P_0 = N\left[\sqrt{N^2 - \varepsilon_2} + \gamma\sqrt{N^2 - \varepsilon_1} \mp (\gamma - 1)\left(N^2 - \frac{\gamma\varepsilon_1 - \varepsilon_2}{\gamma - 1}\right)^{1/2}\right], \quad (8.1.27)$$

where $N = \beta / k_0$ is an efficient refraction factor, $\gamma = \varepsilon_{nl}^{(2)} / \varepsilon_{nl}^{(1)}$, $P_0 = c(4\pi k_0 \varepsilon_{nl}^{(2)})^{-1}$. Sign plus corresponds to the NLSW with the maximum at $x < 0$, but minus - to the NLSW located in the region $x > 0$.

An analysis of (8.1.27) shows that P/P_0 as a function of N has the minimum at some N_k. Thereby, NLSW exists only under $P > P_{k1} = P(N_k)$. Herewith the propagation of two NLSWs is possible.

The position of the maximum of electrical field in NLSW x_m depends on P and on the second critical value of power flow P_{k2}. So, if $P_{k1} < P < P_{k2}$, then $x_m < 0$. But under the condition $P_{k2} < P$ the value of x_m can be either positive or negative [3];

$$P_{k2} = P_0\left(\gamma + \sqrt{\gamma}\right)\left[(\varepsilon_1 - \varepsilon_2)(\gamma\varepsilon_1 - \varepsilon_2)\right]^{1/2}(\gamma - 1)^{-1}.$$

The case when $\varepsilon_{nl}^{(1)}$ and $\varepsilon_{nl}^{(2)} < 0$ leads to the results which are alike. Indices 1 and 2 in all formulae should be only interchanged. The NLSW under consideration has no analogue in the linear media and this is what distinguishes it from the NLSW of the TM-type. An extended study of the NLSW along the interface of two non-linear media was given in [33]. In the same work the NLSW along the "non-linear dielectric - metal" interface was explored too.

8.2. Linear waveguide in a non-linear environment

Let us consider a guide structure consisting of a thin dielectric film of thickness h characterised by the tensor of dielectric permeability $\hat{\varepsilon}_f$, which does not depend on the electrical field of a light wave. Let the dielectric permeability tensor of a substrate on which this film is placed be given in the form:

$$\hat{\varepsilon}_S(\omega, \vec{E}) = \hat{\varepsilon}_1 + \hat{\varepsilon}_{nl}^{(1)} : \vec{E}\vec{E}$$

and an infinite (or very thick) cladding layer has a dielectric permeability tensor

$$\hat{\varepsilon}_C(\omega, \vec{E}) = \hat{\varepsilon}_3 + \hat{\varepsilon}_{nl}^{(3)} : \vec{E}\vec{E}$$

As far as there are two interfaces (at x =0 and x= h), two types of guided waves are possible. These are the waves in the film and the bounded surface waves. The first type of waves will be named non-linear guided modes (NLGM), and the second one by a non-linear surface waves (NLSW). Both of them can be either TE- or TM-waves. Their consideration can be made independently.

8.2.1. NONLINEAR SURFACE WAVES OF TE-TYPE (DISPERSION RELATIONS)

We shall consider an isotropic, or cubic symmetry medium. For the TE waves the equation for $\Phi(x)$ describing the transverse electric field profile of the wave in a three-layer structure can be decomposed into a system of simple equations:

$$\Phi,_{xx} + (k_0^2\varepsilon_1 - \beta^2)\Phi + k_0^2\varepsilon_{nl}^{(1)}\Phi^3 = 0, \qquad x < 0 \quad (8.2.1a)$$

$$\Phi,_{xx} + (k_0^2\varepsilon_2 - \beta^2)\Phi = 0, \qquad 0 < x < h \quad (8.2.1b)$$

$$\Phi,_{xx} + (k_0^2\varepsilon_3 - \beta^2)\Phi + k_0^2\varepsilon_{nl}^{(3)}\Phi^3 = 0, \qquad h < x \quad (8.2.1c)$$

In the region $0<x<h$ solutions of the equation (8.2.1b) defining the NLSW yields

$$\beta^2 - k_0^2\varepsilon_2 = \kappa^2 > 0, \qquad (8.2.2)$$

then

$$\Phi^{(2)}(x) = A_2\exp(-\kappa x) + B_2\exp(\kappa x). \qquad (8.2.3a)$$

Under conditions $x < 0$ and $x > h$ the analysis of non-linear equations (8.2.1a) and (8.2.1c) can be carried out in the same way as it was done for non-linear surface polaritons. It follows from (8.2.1a) and (8.2.1c) that

$$(\Phi,_x)^2 + (k_0^2\varepsilon - \beta^2)\Phi^2 + (k_0^2\varepsilon_{nl}/2)\Phi^4 = 0, \qquad (8.2.4)$$

where indices at ε and ε_{nl} are omitted. In a self-focusing medium (ε_{nl} >0) equation (8.2.4) gives a solution which satisfies the boundary conditions as $x \to \pm\infty$ if only

$(\beta^2 - k_0^2\varepsilon) > 0$. In a self-defocusing medium (ε_{nl}<0) under condition $(\beta^2 - k_0^2\varepsilon) > 0$ and boundary conditions (8.1.3) the singular solution of (8.2.4) is possible, but there are no such solutions under condition $(\beta^2 - k_0^2\varepsilon) < 0$. Consequently, the guided waves do not exist, but the existence of the radiation waves or the waves in the substrate is possible.

If ε_{nl} >0, solution of the equation (8.2.4) has the form (i =1,3)

$$\Phi^{(i)}(x) = \alpha_i^{-1/2} \operatorname{sech}\left[p_i(x - x_i)\right], \tag{8.2.5a}$$

and for $\varepsilon_{nl} < 0$

$$\Phi^{(i)}(x) = |\alpha_i|^{-1/2} \operatorname{cosech}\left[p_i(x - x_i)\right], \tag{8.2.5b}$$

where

$$p_1^2 = p^2 = \beta^2 - k_0^2\varepsilon_1, \quad p_3^2 = q^2 = \beta^2 - k_0^2\varepsilon_3$$
$$\alpha_i = k_0^2\, \varepsilon_{nl}^{(i)} / 2p_i^2 .$$

and x_i are the integrating constants which are equal to co-ordinates of the electric field strength maximum positions. Depending on the signs of ε_{nl}, four cases are possible.

$\varepsilon_{nl}^{(1)} > 0$ *and* $\varepsilon_{nl}^{(3)} > 0$.

Then we have:

$$\Phi^{(1)}(x) = \alpha_1^{-1/2} \operatorname{sech}\left[p(x - x_1)\right], \quad x<0$$

$$\Phi^{(3)}(x) = \alpha_3^{-1/2} \operatorname{sech}\left[q(x - x_3)\right] \quad x>h \tag{8.2.6}$$

The conditions of continuity of the electrical and magnetic fields strengths at the interfaces result it the equations:

$$\begin{aligned} A_1 &= A_2 + B_2, \qquad \tilde{p}A_1 = -\kappa(A_2 - B_2), \\ A_3 &= A_2 \exp(-\kappa h) + B_2 \exp(\kappa h), \\ \tilde{q}A_3 &= \kappa[A_2 \exp(-\kappa h) - B_2 \exp(\kappa h)], \end{aligned} \tag{8.2.7a}$$

where $\tilde{p} = p\tanh(px_1)$, $\tilde{q} = q\tanh[q(h - x_1)]$, $A_1 = \Phi^{(1)}(x=0)$, $A_3 = \Phi^{(3)}(x=h)$.

From equation (8.2.6) we obtain

$$A_1 = \alpha_1^{-1/2}\,\text{sech}(px_1)\,,$$
$$A_3 = \alpha_3^{-1/2}\,\text{sech}[q(h-x_3)]\,. \qquad (8.2.8a)$$

The uniform system of linear equations in A_1, A_2, A_3, B_2 has a nonzero solution if its determinant is equal to zero. This condition leads to the dispersion relation for NLSW:

$$\exp(2\kappa h)\left(1+\tilde{p}/\kappa\right)\left(1+\tilde{q}/\kappa\right) = \left(1-\tilde{p}/\kappa\right)\left(1-\tilde{q}/\kappa\right) \qquad (8.2.9)$$

If one arranges parameters ϕ_p and ϕ_q in the form

$$\tanh(\phi_p/2) = \tilde{p}/\kappa\,, \quad \tanh(\phi_q/2) = \tilde{q}/\kappa\,,$$

then the relation (8.2.9) reads

$$2\kappa h + \phi_p + \phi_q = 0 \qquad (8.2.10)$$

Hence it follows that either ($\phi_p < 0$ or $\phi_q < 0$) or (ϕ_p and $\phi_q < 0$). As far as the signs of p, q and κ are chosen positive then, either $x_1 < 0$, $h > x_3$ or $x_1 > 0$, $h < x_3$, or $x_1 < 0$, $h < x_3$.

The coefficients A_1 and A_3 are coupled by a relationship

$$A_3 = A_1\left[\cosh(\kappa h) - (\tilde{p}/\kappa)\sinh(\kappa h)\right], \qquad (8.2.11)$$

that is a relationship between x_1 and x_3. In this way the dispersion law (8.2.9) contains x_1 as a parameter. On the other hand, expression (8.2.8) determines A_1 in terms of x_1, so the electric field strength of NLSW at $x = 0$ can be chosen as a parameter instead of x_1. Thereby (8.2.9) serves as an implicit expression for the dependency of the propagation constant β on the frequency ω: $\beta = \beta(\omega\,; A_1^2)$.

Sometimes the dispersion relation (8.2.9) is presented in the form [96]:

$$\tanh(\kappa h) = \frac{\kappa(\tilde{p}+\tilde{q})}{\kappa^2 + \tilde{p}\tilde{q}} \qquad (8.2.12)$$

$\varepsilon_{nl}^{(1)} < 0$ *and* $\varepsilon_{nl}^{(3)} > 0$.

In this case

$$\Phi^{(1)}(x) = |\alpha_1|^{-1/2}\,\text{cosech}\left[p(x-x_1)\right], \quad \Phi^{(3)}(x) = \alpha_3^{-1/2}\,\text{sech}\left[q(x-x_3)\right].$$

Continuity conditions for the fields on the boundary planes $x = 0$ and $x = h$ give rise to the system of the equations, which is similar to the (8.2.7), but where

$$\tilde{p} = p\coth(-px_1), \quad \tilde{q} = q\tanh[q(h - x_3)]$$

and

$$A_1 = |\alpha_1|^{-1/2} \operatorname{cosech}(-px_1), \quad A_3 = \alpha_3^{-1/2} \operatorname{sech}[q(h - x_3)]. \tag{8.2.8b}$$

The dispersion relation in this case has the form of either (8.2.9) or (8.2.10), or (8.2.12).

$\varepsilon_{nl}^{(1)} < 0$ *and* $\varepsilon_{nl}^{(3)} < 0$.
Here function $\Phi(x)$ looks as

$$\Phi^{(1)}(x) = |\alpha_1|^{-1/2} \operatorname{cosech}[p(x - x_1)], \qquad x < 0,$$
$$\Phi^{(3)}(x) = |\alpha_3|^{-1/2} \operatorname{cosech}[q(x - x_3)], \qquad x > h \tag{8.2.6c}$$

and if one defines

$$\begin{gathered} \tilde{p} = p\coth(-px_1), \quad \tilde{q} = q\coth[q(h - x_3)] \\ A_1 = |\alpha_1|^{-1/2} \operatorname{cosech}(-px_1), \\ A_3 = |\alpha_3|^{-1/2} \operatorname{cosech}[q(h - x_3)], \end{gathered} \tag{8.2.8c}$$

then the expression (8.2.9) becomes the dispersion relation for this case.

$\varepsilon_{nl}^{(1)} > 0$ *and* $\varepsilon_{nl}^{(3)} < 0$.
In this case function $\Phi(x)$ can be written as

$$\Phi^{(1)}(x) = \alpha_1^{-1/2} \operatorname{sech}[p(x - x_1)], \qquad x < 0$$
$$\Phi^{(3)}(x) = |\alpha_3|^{-1/2} \operatorname{cosech}[q(x - x_3)], \qquad x > h \tag{8.2.6d}$$

Dispersion relation complies with (8.2.9), but now

$$\tilde{p} = p\tanh(px_1), \quad \tilde{q} = q\coth[q(h - x_3)].$$

The amplitude of the electrical field of the NLSW at the interface $h = 0$ A_1 is given by the equality

$$A_1 = \alpha_1^{-1/2} \operatorname{sech}(px_1),$$

It corresponds to the electrical field amplitude A_3 at the second interface $x = h$ by the relation (8.2.11).

8.2.2. NONLINEAR GUIDED WAVES OF TE-TYPE (DISPERSION RELATIONS)

Analysis of the NLGW of the TE-type is based on the equations (8.2.1), but now one should assign

$$k_0^2 \varepsilon_2 - \beta^2 = \kappa^2 > 0. \tag{8.2.13}$$

in the equation (8.2.1b).

Then the following function serves as a solution of these equations for $0 < x < h$

$$\Phi^{(2)}(x) = A_2 \exp(i\kappa x) + B_2 \exp(-i\kappa x).$$

For $x>h$ and $x<0$ $\Phi(x)$ takes correspondingly the form of $\Phi^{(3)}(x)$ and $\Phi^{(1)}(x)$ from the preceding section.

The continuity conditions for $\Phi(x)$ and $\Phi_{,x}(x)$ at the interfaces $x = 0$ and $x = h$ yields

$$\begin{aligned} &A_1 = A_2 + B_2\,, \qquad \tilde{p}A_1 = i\kappa(A_2 - B_2)\,, \\ &A_3 = A_2 \exp(i\kappa h) + B_2 \exp(-i\kappa h)\,, \\ &-\tilde{q}A_3 = i\kappa[A_2 \exp(i\kappa h) - B_2 \exp(-i\kappa h)]\,, \end{aligned} \tag{8.2.7b}$$

hence a dispersion relation for NLGW follows as

$$\exp(2i\kappa h)\left(\frac{1 - i\tilde{p}/\kappa}{1 + i\tilde{p}/\kappa}\right)\left(\frac{1 - i\tilde{q}/\kappa}{1 + i\tilde{q}/\kappa}\right) = 1, \tag{8.2.14}$$

or

$$\tan(\kappa h) = \frac{\kappa(\tilde{p} + \tilde{q})}{\kappa^2 - \tilde{p}\tilde{q}}. \tag{8.2.15}$$

If we define the parameters ϕ_p and ϕ_q by the formulae

$$\tan(\phi_p/2) = \tilde{p}/\kappa\,, \quad \tan(\phi_q/2) = \tilde{q}/\kappa\,, \tag{8.2.16}$$

then equation (8.2.14) can be written as

$$2\kappa h = \phi_p + \phi_q + 2\pi m\,, \quad m = 0,1,2,\ldots . \tag{8.2.17}$$

This expression has a simple physical meaning: the full phase shift of the zigzag wave [98] $2\pi m$ is a sum of $2\kappa h$ arising from the linear medium of the dielectric film, and the phase shifts $-\phi_p$ and $-\phi_q$, which occur in the total internal reflection at the linear-non-linear interfaces (*non-linear Goos-Hanchen effect*).

The quantities $\tilde{p}$, $\tilde{q}$, $\grave{A}_1$ and $\grave{A}_3$ in the expressions (8.2.7b) - (8.2.17) are determined by formulae (8.2.8) according to the choice of the signs of $\varepsilon_{nl}^{(1)}$ and $\varepsilon_{nl}^{(3)}$.

It is known that in linear integrated optics there are several unification parameters in terms of which the dispersion relation for TE-modes can be written in a universal form. We can use these parameters in our expressions too. Let us set the effective refraction factor $N = \beta / k_0$ and unification parameters as

$$V = k_0 h \sqrt{\varepsilon_2 - \varepsilon_1}\,, \qquad b = \frac{N^2 - \varepsilon_1}{\varepsilon_2 - \varepsilon_1}\,, \qquad a = \frac{\varepsilon_1 - \varepsilon_3}{\varepsilon_2 - \varepsilon_1}.$$

Then the dispersion relation for linear guided TE -wave can be written as

$$V\sqrt{1-b} = \arctan\left[\sqrt{\frac{b}{1-b}}\right] + \arctan\left[\sqrt{\frac{b+a}{1-b}}\right] + \pi m\ .$$

The universal expression for non-linear dispersion relation holds

$$V\sqrt{1-b} = \arctan\left[\mu_1\sqrt{\frac{b}{1-b}}\right] + \arctan\left[\mu_3\sqrt{\frac{b+a}{1-b}}\right] + \pi m\,, \qquad (8.2.18)$$

where

$$\mu_1 = \tilde{p} / p = \pm(1 - \alpha_1 A_1^2)^{1/2}\,, \qquad \mu_3 = \tilde{q} / q = \pm(1 - \alpha_3 A_3^2)^{1/2}$$

under any choice of signs of $\varepsilon_{nl}^{(1)}$ and $\varepsilon_{nl}^{(3)}$. It would be useful to compare below the two expressions of the dispersion relations for TE-waves.

8.2.3. NONLINEAR SURFACE WAVES OF TM-TYPE (DISPERSION RELATIONS)

In the case of TM-waves the equation for functions $\Phi(x)$ decomposes into three equations with the permanent coefficients

$$\Phi,_{xx} + p_1^2\Phi + \alpha_1 p_1^2 \Phi^3 = 0\,, \qquad x < 0 \qquad (8.2.19a)$$

$$\Phi,_{xx} + p_2^2\Phi = 0\,, \qquad 0 < x < h \qquad (8.2.19b)$$

$$\Phi,_{xx} + p_3^2\Phi + \alpha_3 p_3^2 \Phi^3 = 0\,, \qquad h < x \qquad (8.2.19c)$$

where

$$\alpha_i = \varepsilon_{nl}^{(i)} / \varepsilon_{\perp}^{(i)}, \quad p_i^2 = (\varepsilon_{\parallel}^{(i)} / \varepsilon_{\perp}^{(i)})(k_0^2 \varepsilon_{\perp}^{(i)} - \beta^2), \quad i = 1, 3.$$

In the case of non-linear surface waves it is necessary to assign $p_2^2 < 0$. Depending on the signs of $\varepsilon_{nl}^{(1)}$ and $\varepsilon_{nl}^{(3)}$, the different types of TM-waves are possible as it takes place in the case of TE-waves.

$\varepsilon_{nl}^{(1)} > 0$ *and* $\varepsilon_{nl}^{(3)} > 0$.
It is suitable to re-assign the coefficients into the system of the equations (8.2.19) as

$$p_1^2 = -p^2, \quad p_2^2 = -\kappa^2, \quad p_3^2 = -q^2.$$

It follows from (8.2.19b) that under condition $0 < x < h$

$$\Phi^{(2)}(x) = A_2 \exp(-\kappa x) + B_2 \exp(\kappa x). \tag{8.2.20a}$$

When $x < 0$ and $x > h$, the expression for $\Phi^{(1)}(x)$ and $\Phi^{(3)}(x)$, which is analogous with (8.2.5b), follows correspondingly from (8.2.19a,c)

$$\Phi^{(1)}(x) = \pm(2/\alpha_1)^{1/2} \operatorname{cosech}\left[p(x - x_1)\right], \tag{8.2.20b}$$
$$\Phi^{(3)}(x) = \pm(2/\alpha_3)^{1/2} \operatorname{cosech}\left[q(x - x_3)\right]. \tag{8.2.20c}$$

The way these solutions were found is the same as was used to solve the equations (8.1.18b) and (8.1.20). Continuity conditions (8.1.18) at the interfaces $x = 0$ and $x = h$ give a system of equations

$$A_1 = A_2 + B_2, \qquad (\varepsilon_{\parallel}^{(1)} / p) A_1 \coth(-p x_1) = (\varepsilon_{\parallel}^{(2)} / \kappa)(A_2 - B_2),$$
$$A_3 = A_2 \exp(-\kappa h) + B_2 \exp(\kappa h),$$
$$(\varepsilon_{\parallel}^{(3)} / q) A_3 \coth[q(h - x_3)] = (\varepsilon_{\parallel}^{(2)} / \kappa)[A_2 \exp(-\kappa h) - B_2 \exp(\kappa h)],$$

where A_1 and A_3 are determined by the expressions

$$A_1 = \pm(2/\alpha_1)^{1/2} \operatorname{cosech}(-p x_1), \quad A_3 = \pm(2/\alpha_3)^{1/2} \operatorname{cosech}[q(h - x_3)].$$

If we define the parameters

$$\tilde{p} = p \tanh(-p x_1), \tilde{q} = q \tanh[q(h - x_3)], \quad \mu_{ij} = \varepsilon_{\parallel}^{(i)} / \varepsilon_{\parallel}^{(j)}, \tag{8.2.21}$$

then the condition, under which the system of equations for A_1, A_2, A_3 and B_2 is solvable, yields

$$\exp(-2\kappa h)\left(\frac{1+\mu_{12}\kappa/\tilde{p}}{1-\mu_{12}\kappa/\tilde{p}}\right)\left(\frac{1-\mu_{32}\kappa/\tilde{q}}{1+\mu_{32}\kappa/\tilde{q}}\right)=1\,. \tag{8.2.22a}$$

This is a dispersion relation for NLSW of the TM-type for the case under consideration.

In terms of parameters ϕ_p and ϕ_q introduced as

$$\tanh(\phi_p/2)=-\mu_{12}\kappa/\tilde{p}\,,\quad \tanh(\phi_q/2)=\mu_{32}\kappa/\tilde{q}\,,$$

the dispersion relation will be written in the form

$$2\kappa h+\phi_p+\phi_q=0\,. \tag{8.2.22b}$$

As it was in the case of TE-waves, here either ϕ_p or ϕ_q or both ϕ_p and ϕ_q are to be negative. Hence it follows that the positions of the electrical field strength maximum x_1 and x_2, can be located on different sides of the linear dielectric film, i.e. either $x_1 < h$ or $x_3 > 0$, or $x_1 < h$ and $x_3 > 0$.

The amplitudes of the electric fields in the points of the planes $x = 0$ and $x = h$ are combined by the relation

$$A_3 = A_1\left[\cosh(\kappa h)-\mu_{32}\frac{\kappa}{\tilde{p}}\sinh(\kappa h)\right],$$

that provides a relationship between x_1 and x_3. Another form of the dispersion relation (8.2.22) comes out from (8.2.22b):

$$\tanh(\kappa h)=-\frac{\mu_{12}\kappa\tilde{p}-\mu_{32}\kappa\tilde{q}}{\tilde{p}\tilde{q}+\mu_{12}\mu_{32}\kappa^2}\,. \tag{8.2.22c}$$

$\varepsilon_{nl}^{(1)} < 0$ *and* $\varepsilon_{nl}^{(3)} < 0$.

In this case solution of the equations (8.2.19a) and (8.2.19c) can be found in a similar way as it was done for (8.1.19b):

$$\Phi^{(1)}(x)=(2/|\alpha_1|)^{1/2}\,\mathrm{sech}\left[p(x-x_1)\right], \tag{8.2.20d}$$

$$\Phi^{(3)}(x)=(2/|\alpha_3|)^{1/2}\,\mathrm{sech}\left[q(x-x_3)\right]. \tag{8.2.20e}$$

Continuity conditions (8.1.18) present a uniform system of linear equations to determine A_1, A_2, A_3 and B_2

$$A_1 = A_2 + B_2\,, \qquad A_3 = A_2 \exp(-\kappa h) + B_2 \exp(\kappa h),$$
$$-\mu_{12}\kappa\tilde{p}^{-1} A_1 = A_2 - B_2\,, \tag{8.2.23}$$
$$\mu_{32}\kappa\tilde{q}^{-1} A_3 = A_2 \exp(-\kappa h) - B_2 \exp(\kappa h)\,,$$

where now

$$\tilde{p} = p\coth(px_1)\,, \qquad \tilde{q} = q\coth[q(h-x_3)]\,,$$
$$A_1 = (2/|\alpha_1|)^{1/2}\,\mathrm{sech}(px_1)\,, \quad A_3 = (2/|\alpha_3|)^{1/2}\,\mathrm{sech}[q(h-x_3)]\,.$$

Dispersion relation for TM-waves in this case has the form

$$\exp(-2\kappa h)\left(\frac{1-\mu_{12}\kappa/\tilde{p}}{1+\mu_{12}\kappa/\tilde{p}}\right)\left(\frac{1-\mu_{32}\kappa/\tilde{q}}{1+\mu_{32}\kappa/\tilde{q}}\right) = 1 \tag{8.2.24a}$$

or

$$2\kappa h + \phi_p + \phi_q = 0\,, \tag{8.2.24b}$$

where ϕ_p and ϕ_q are fixed by the relations:

$$\tanh(\phi_p/2) = \mu_{12}\kappa/\tilde{p}\,, \quad \tanh(\phi_q/2) = \mu_{32}\kappa/\tilde{q}\,.$$

The above expressions and (8.2.24b) give the dispersion relation in the form:

$$\tanh(\kappa h) = -\frac{\mu_{12}\kappa\tilde{p} + \mu_{32}\kappa\tilde{q}}{\tilde{p}\tilde{q} + \mu_{12}\mu_{32}\kappa^2}\quad . \tag{8.2.24c}$$

$\varepsilon_{nl}^{(1)} < 0$ and $\varepsilon_{nl}^{(3)} > 0$.

If these inequalities hold the dispersion relation for TM-type, NLSW formally coincides with (8.2.24), but $\tilde{p}$, $\tilde{q}$, A_1 and A_3 are now defined by the expressions

$$\tilde{p} = p\coth(px_1)\,, \qquad \tilde{q} = q\tanh[q(h-x_3)]\,,$$
$$A_1 = (2/|\alpha_1|)^{1/2}\,\mathrm{sech}(px_1)\,, \quad A_3 = \pm(2/\alpha_3)^{1/2}\,\mathrm{cosech}[q(h-x_3)]\,.$$

$\varepsilon_{nl}^{(1)} > 0$ *and* $\varepsilon_{nl}^{(3)} > 0$.

This case can be offered to a reader as a simple exercise.

8.2.4. NONLINEAR GUIDED WAVES OF TM-TYPE (DISPERSION RELATIONS)

The derivation of the dispersion relations for the TM-type NLGW is based on the analysis of the solutions of the equations (8.2.19), but now for $p_2^2 = \kappa^2 > 0$. Parameters p^2 and q^2 are determined as above. Under condition $0 < x < h$ the solution of equations (8.2.19) can be written as

$$\Phi^{(2)}(x) = A_2 \exp(i\kappa x) + B_2 \exp(-i\kappa x) . \qquad (8.2.25)$$

Under conditions $x < 0$ and $x > h$ functions $\Phi^{(1)}(x)$ and $\Phi^{(3)}(x)$ are determined by the formulae similar to those of (8.2.20b) and (8.2.20e).

Continuity condition (8.1.18) then gives the equations of the (8.2.23) type in which κ should be changed for $i\kappa$. Dispersion relation for NLGW can be obtained from the resulting uniform system of the linear equations

$$2\kappa h = \phi_p + \phi_q + 2\pi m, \quad m = 0,1,2,\ldots. \qquad (8.2.26a)$$

where ϕ_p and ϕ_q are determined by the formulae

$$\tan(\phi_p / 2) = \mu_{12}\kappa / \tilde{p}, \qquad \tan(\phi_q / 2) = \mu_{32}\kappa / \tilde{q},$$

Another form of this dispersion relation is

$$\tan(\kappa h) = \frac{\mu_{32}\kappa\tilde{p} + \mu_{12}\kappa\tilde{q}}{\tilde{p}\tilde{q} - \mu_{12}\mu_{32}\kappa^2} . \qquad (8.2.26b)$$

The interpretation of the equation (8.2.26a) is exactly the same as it was in the case of TE-waves.

Quantities $\tilde{p}$, $\tilde{q}$, A_1 and A_3 should be chosen with taking signs of $\varepsilon_{nl}^{(1)}$ and $\varepsilon_{nl}^{(3)}$ into account. Analysis of all choices of the non-linear medium (either self-focusing or self-defocusing) should be made exactly in the same way as it was done in section 8.2.3 for TM-type NLSW. Herewith all the results obtained there remain unaltered.

8.2.5. ANALYSIS OF DISPERSION RELATIONS

The situation with eigenmodes of the linear planar waveguide imbedded to the non-linear medium is characterised by a broad range of possibilities. These waveguides can be either symmetric ($\varepsilon_S(\omega, \vec{E}) = \varepsilon_C(\omega, \vec{E})$) or asymmetric ($\varepsilon_S(\omega, \vec{E}) \neq \varepsilon_C(\omega, \vec{E})$). The sorts of waves can be different: TE- or TM-type. Apart from guided waves (i.e. NLGW), which have analogues in the linear limit ($\varepsilon_{nl}^{(i)} \to 0$), the NLSW are presented here which can have no linear partners. Both non-linear media can be either self-focusing or self-defocusing, or one of them is self-focusing but the other is self-defocusing.

The TE-waves for a waveguide in non-linear environment is a subject of this section. Besides this, one limiting case will be considered, when substrate is a non-linear medium. This limiting case can be easily obtained from the general analysis provided $\varepsilon_{nl}^{(3)} \to 0$. There is the experimental study in this case but for $\varepsilon_{nl}^{(1)} \to 0$, i.e. the linear waveguide with non-linear cladding. It should be noted that qualitatively there are no differences between the waveguide on the non-linear substrate and waveguide with non-linear cladding, as far as one considers a substrate as a cladding and vice versa.

The results obtained in sections 8.2.1 and 8.2.2 allow writing different expressions in the unified form with no dependence on the signs of non-linear susceptibilities. Besides that the substitution $\kappa \to i\kappa$ makes it possible to derive expressions for NLSW case from the corresponding expressions for non-linear modes of a film under the additional requirement that mode index *m* should be zero.

The electric field distribution in NLGW is given by

$$\Phi(x) = \begin{cases} A_1\left[\cosh(px) - (\tilde{p}/p)\sinh(px)\right]^{-1}, & x < 0, \\ A_1\left[\cos(\kappa x) + (\tilde{p}/p)\sin(\kappa x)\right], & 0 \le x \le h, \\ A_1 \dfrac{\cos(\kappa h) + (\tilde{p}/p)\sin(\kappa h)}{\cosh[q(x-h)] + (\tilde{q}/q)\sinh[q(x-h)]}, & h < x, \end{cases} \quad . (8.2.27)$$

where A_1 is an electric field strength in NLGW at the interface $x = 0$. The mode index *m*, which appears from dispersion relation (8.2.17), is omitted.

It is suitable to define parameters $\tilde{p}, \tilde{q}$ by analogy with *p* and *q*. Thus, introducing the *effective non-linear dielectric permeability* $\varepsilon_{i,nl}$ $(i = 1,3)$ [46-49, 99], we have

$$\tilde{p}^2 = \beta^2 - k_0^2\varepsilon_{1,nl}, \qquad \tilde{q}^2 = \beta^2 - k_0^2\varepsilon_{3,nl}. \qquad (8.2.28)$$

The introduced parameters $\varepsilon_{i,nl}$ can be expressed in terms of usual non-linear permeability:

$$\varepsilon_{i,nl} = \varepsilon_i + \varepsilon_{nl}^{(i)} A_i^2 / 2.$$

As far as the electric field strength of NLGW at the interface $x = h$ A_3 is connected with A_1 by the continuity condition, there is also a relationship between the elements of effective non-linear dielectric permeability $\varepsilon_{i,nl}$:

$$\left[\frac{\varepsilon_{3,nl}-(\varepsilon_2+\varepsilon_3)/2}{a\sqrt{\varepsilon_{nl}^{(3)}/\varepsilon_{nl}^{(1)}}}\right]^2-\left[\frac{\varepsilon_{1,nl}-(\varepsilon_2+\varepsilon_1)/2}{a}\right]^2=1, \qquad (8.2.29)$$

where parameter a is determined by expressions (8.2.30) depending on the magnitude of $\xi_{13}=(\varepsilon_{nl}^{(1)}/\varepsilon_{nl}^{(3)})$:

$$a=(1/2)\left[(\varepsilon_2-\varepsilon_3)^2\xi_{13}-(\varepsilon_2-\varepsilon_1)^2\right]^{1/2}, \qquad (8.2.30a)$$

when $0<\xi_{13}\le(\varepsilon_2-\varepsilon_1)^2/(\varepsilon_2-\varepsilon_3)^2$ and

$$a=(i/2)\left[(\varepsilon_2-\varepsilon_1)^2-(\varepsilon_2-\varepsilon_3)^2\xi_{13}\right]^{1/2}, \qquad (8.2.30b)$$

when $\xi_{13}>(\varepsilon_2-\varepsilon_1)^2/(\varepsilon_2-\varepsilon_3)^2$.
It is worth to note that expression (8.2.29) presents an equation of conic section which was taken as a basis for classification of the TE-type NLGW [100]. On Fig.8.2.1 the qualitative picture of the electrical field distributions in the NLGW is presented in compliance with four types of TE-waves.

Fig. 8.2.1.

Profiles of the electric field of NLSW

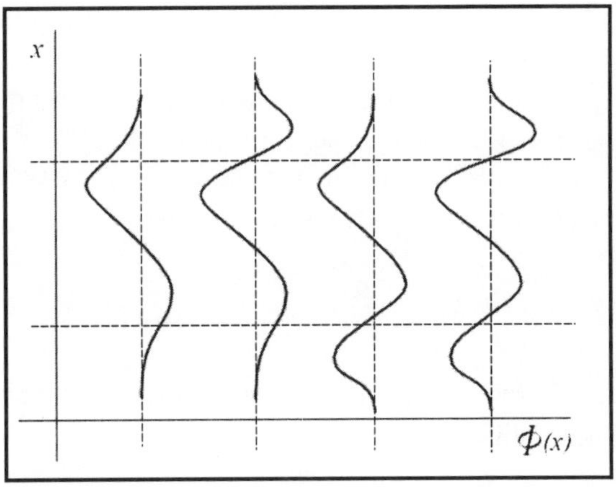

A condition of the existence of NLGW follows from the positively of κ^2, p^2, q^2 and it can be written as

$$\varepsilon_1,\varepsilon_3 \le \varepsilon_{1,nl},\ \varepsilon_{3,nl} \le \varepsilon_2 \qquad (8.2.31)$$

or

$$0 \le \varepsilon_{i,nl} A_i^2 \le 2(\varepsilon_2 - \varepsilon_i)\ .$$

For power flow P the following expression was found [48,49]:

$$P = \frac{\beta |A_1|^2}{4k_0}\left[\tilde{h}_{eff}\left(1+\frac{\tilde{p}^2}{\kappa^2}\right)+\frac{\tilde{p}(p-\tilde{p})}{\kappa^2(p+\tilde{p})}+\frac{\tilde{q}(q-\tilde{q})(\varepsilon_2-\varepsilon_{1,nl})}{\kappa^2(q+\tilde{q})(\varepsilon_2-\varepsilon_{3,nl})}\right], \qquad (8.2.32)$$

where *non-linear effective thickness* $\tilde{h}_{eff}$ was determined by the expression

$$\tilde{h}_{eff} = h + 2(p+\tilde{p})^{-1} + 2(q+\tilde{q})^{-1}. \qquad (8.2.33)$$

In the limits $\varepsilon_{nl}^{(3)} \to 0$ and $\varepsilon_{nl}^{(1)} \to 0$ one can get the known expressions for P and $\tilde{h}_{eff}$ for the linear planar waveguide. It is easy to verify that if to set $\varepsilon_{i,nl} = \varepsilon_i, \tilde{p} = p, \tilde{q} = q$ in (8.2.32) and (8.2.33), then the well known expressions for power flow and effective thickness come out.

Further consideration of the dispersion relations can be made near the boundaries of the range for the constant β by the inequality

$$\max\{k_0^2\varepsilon_{1,nl}, k_0^2\varepsilon_{3,nl}\} \le \beta^2 \le k_0^2\varepsilon_2\,.$$

Let us consider the behaviour of β as the function of parameter associated with the thickness of waveguide $V = k_0 h\sqrt{\varepsilon_2 - \varepsilon_1}$ on the basis of the equation (8.2.18). In the vicinity of the minimum of the allowed value for β there are some points named as the transition points. Let for instance $\varepsilon_{1,nl} > \varepsilon_{3,nl}$. Points x_1, and x_3 corresponding to the maximum of the electric field strength of NLGW, are located outside a waveguide ($x_1 > 0$ and $x_3 > h$). When β approaches $\beta_{\min} = k_0(\varepsilon_{1,nl})^{1/2}$, condition $\tilde{p} \to 0$ holds. That means that $x_1 \to 0$. The further changing of V causes either a cut-off of the mode or an increase of β. The value of x_1 can become neither positive nor negative herewith. The magnitude of V, under which β reaches $\beta_{\min}$, was named the *transition points* [48,49]. It is easy to find this value $V_m^{(T)}$ from (8.2.18) by assigning there $\tilde{p} = 0$

$$V_m^{(T)} = \frac{1}{\sqrt{1-\tilde{b}}}\left\{\arctan\left[\frac{\tilde{q}}{q}\sqrt{\frac{\tilde{b}+a}{1-\tilde{b}}}\right]+\pi m\right\},$$

where $\tilde{b} = (\varepsilon_{1,nl} - \varepsilon_1)/(\varepsilon_2 - \varepsilon_1)$, and m is the mode index (mode number).

In a more general case, the analysis of dispersion relations of non-linear waves is carried out numerically. Thus, to illustrate the NLGW properties it is simpler to discuss non-linear phenomena in such waveguide on the base of numerical results.

For the fixed power flow the plot of the propagation constant β vs. normalised waveguide thickness V is shown in Fig. 8.2.2. The curves marked as 1 and 2 correspond to the values of the parameters $(\varepsilon_2-\varepsilon_3)/(\varepsilon_2-\varepsilon_1)$ 0.7 and 1.0, respectively. The bold points on the curves indicate the transition points. Note that for one and the same value of V two non-linear waves can propagate simultaneously in a waveguide, carrying the same power, having an alike mode index m, but being distinguished with the propagation constant. If V changes slowly near the transition point $V_m^{(T)}$, then the fluent switching between these NLGWs will occur.

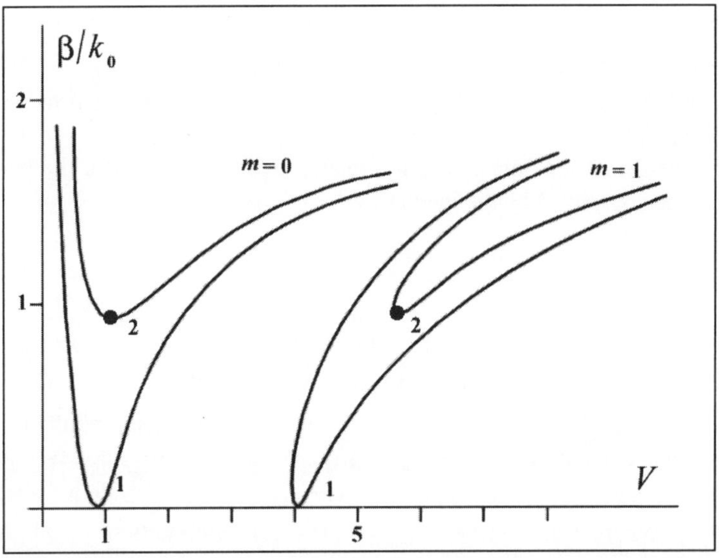

Fig 8.2.2 Propagation constant as function of normalised waveguide thickness for x_1.<0

Now let us discuss the limiting case when either substrate or cladding of the waveguide is a linear dielectric, that means that either $\varepsilon_{nl}^{(3)} \to 0$ or $\varepsilon_{nl}^{(1)} \to 0$. For this case all expressions for the characteristics of any interest can be obtained under corresponding limiting transition. For instance, let the medium (substrate) be a linear dielectric at $x<0$. Then $\widetilde{p}=p$ and the dispersion relation for NLGW obtained from

(8.2.15) has the form:

$$\tan(\kappa h) = \frac{\kappa(p + \tilde{q})}{\kappa^2 - \tilde{p}\tilde{q}} \quad .$$

From (8.2.32) it is possible to find a power flow transferred by this NLGW. It is interesting here to consider the propagation constant as a function of power flow P. Numerical solution of the dispersion relation made in [39] showed that the function β vs P had a qualitative different nature depending on the ratio h/λ (or V). Thus at $\lambda = 0{,}694$ μ and $n_1 = 1{,}52$, $n_3 = 1{,}55$ and $\varepsilon_{nl} = 10^{-9}$ cm^{3}erg^{-1} the value of β is a monotonous function

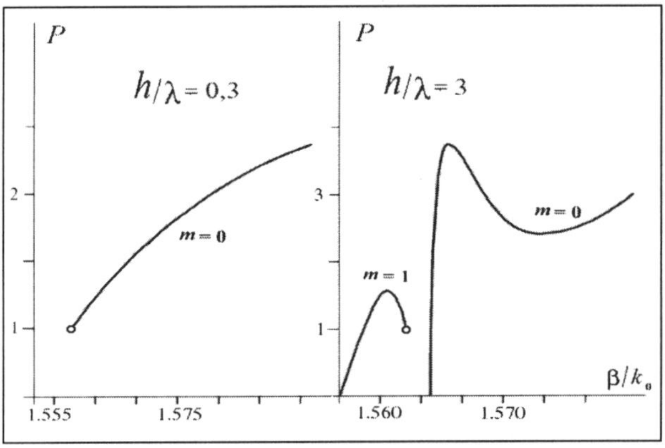

Fig.8.2.3

Propagation constant as function of the power flow P for different mode indices m

versus P for the NLGW with $m = 0$ and $h/\lambda = 0{,}3$ (Fig 8.2.3a), whereas at h/λ=3 this dependence is ambiguous (Fig 8.2.3b). This ambiguity is seen particularly well for m=1.

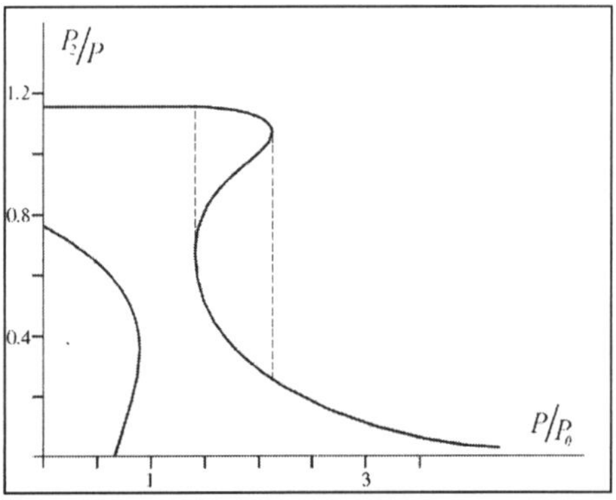

Fig.8.2.4

Ration of the power flow in waveguide to total power P_2/P in relation to the total power P for TE-mode ($h/\lambda = 3.0$)

In these plots the curves $\beta(P)$ start from certain point indicated on Fig.8.2.3 with the light circle. That means the corresponding NLGWs appear only if P exceeds a certain critical value $P_{cr}^{(m)}$, and that they can be switched off for P smaller than $P_{cr}^{(m)}$. If to denote the power flow inside the waveguide as P_2, then for sufficiently greater values of h/λ the dependence of P_2/P versus P has an ambiguous behaviour (Fig.8.2.4). Two values of P_2/P correspond to one and the same magnitude of P in some power flow interval. In ref. [39] this bistable behaviour was given the following interpretation: the state with the smaller value of $\beta(P)$ corresponds to the NLGW which power flow is comprised inside the waveguide ($0 < x < h$), but alternatively the state with the large value of $\beta(P)$ corresponds to the NLGW with the power flow located mainly in the non-linear cladding.

Bistable behaviour of the NLGW was predicted in a general case of the linear waveguide in non-linear environment [50,96,101] too. But due to more complex systematisation of the NLGWs in this case the interpretation of this phenomenon is not easy. There are different sorts of non-linear modes and there is possible switching between different modes.

The investigations of the NLGWs and the TM-type NLSWs [58] have shown that bistable behaviour of the non-linear waves takes place in this case as well (Fig.8.2.5).

Fig.8.2.5

The propagation constant as function of relative power flow in non-linear waveguide for TM-mode ($h/\lambda = 1.4$)

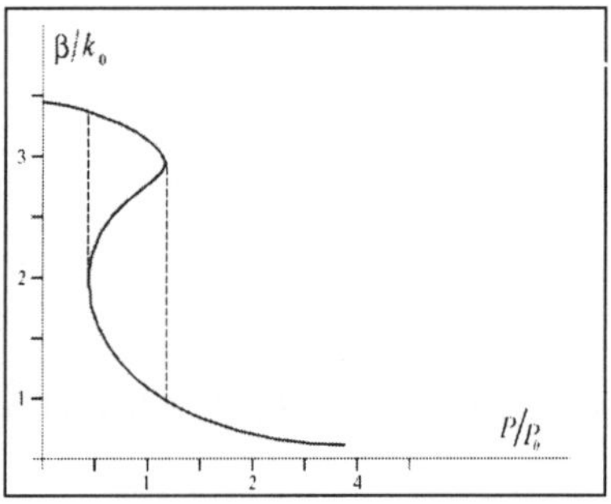

By finishing this section, attention should be drawn to the fact that non-linear modes with different indices m do not form the basis of orthogonal set of functions. This is the result of nonlinearity of the eigenvalue problem, and this makes an essential difference from linear modes of planar waveguides.

8.3. Non-linear waveguides in a linear environment

The planar waveguide that we are considering here consists of a linear dielectric, characterised by isotropic or anisotropic dielectric permeability in each of the regions $x < 0$ (substrate) and $x > h$ (cladding), and a non-linear dielectric medium in the region

$0<x<h$. Thus a dielectric permeability tensor will be assigned by a function of a transverse co-ordinate

$$\hat{\varepsilon}(x) = \begin{cases} \hat{\varepsilon}_1 & x<0 \\ \hat{\varepsilon}_2 + \hat{\varepsilon}_{nl}(\vec{E}), & 0<x<h \\ \hat{\varepsilon}_3, & x>h \end{cases}$$

The experience of the preceding chapters allows to assume that in the given case besides NLGWs the NLSWs will exist. Moreover the latter are the coupled waves propagating along an interfaces of non-linear and linear media.

8.3.1. NONLINEAR WAVES OF TE-TYPE

For these waves the nonzero component of the vector of the electric field strength obeys equation (8.1.2), whereas the transverse profile of an electric field of the TE-waves obeys equation (8.1.5). Taking into account the form of a total dielectric permeability tensor, equation (8.1.5) can be rewritten as three equations with constant coefficients:

$$\begin{aligned} &\Phi,_{xx} + (k_0^2\varepsilon_1 - \beta^2)\Phi = 0, && x<0 \\ &\Phi,_{xx} + (k_0^2\varepsilon_2 - \beta^2)\Phi + k_0^2\varepsilon_{nl}\Phi^3 = 0, && 0<x<h \\ &\Phi,_{xx} + (k_0^2\varepsilon_3 - \beta^2)\Phi = 0, && h<x \end{aligned} \tag{8.3.1}$$

The boundary conditions as $x \to \pm\infty$ give rise to

$$p^2 = \beta^2 - k_0^2\varepsilon_1 > 0, \qquad q^2 = \beta^2 - k_0^2\varepsilon_3 > 0$$

Then in the domain $x<0$ and $x>h$ function $\Phi(x)$ can be written as

$$\begin{aligned} &\Phi^{(1)}(x) = A_1 \exp(px), && x<0 \\ &\Phi^{(3)}(x) = A_3 \exp(-qx), && x>h \end{aligned} \tag{8.3.2}$$

As in section 8.2, in the non-linear layer in the field $0<x<h$ equation for $\Phi(x)$ has an integral of motion

$$(\Phi,_x)^2 + \kappa^2\Phi^2 + \kappa^2\alpha^2\Phi^4 = C, \tag{8.3.3}$$

where C is a integrating constant, $\kappa^2 = k_0^2\varepsilon_2 - \beta^2$, $\alpha^2 = k_0^2\varepsilon_{nl}/2\kappa^2$.

By exhausting different signs of parameters κ^2, α^2 and C one can observe different configurations of the non-linear modes. But it is better to choose another way. Having solved equation (8.3.3) under the certain choice of the signs of all parameters, all other cases can be found by the extension of this solution on the region where these parameters adopt other values.

Let $\kappa^2 > 0$ and $\varepsilon_{nl} < 0$. Herewith we re-assign α^2 as $\alpha^2 = k_0^2 |\varepsilon_{nl}|/2\kappa^2 > 0$. Let function $y(x) = \alpha^{1/2}\Phi(x)$, which obeys the equation obtained from (8.3.3)

$$(y_{,x})^2 = \alpha\kappa^2(\gamma + y^4 - y^2/\alpha), \tag{8.3.4}$$

where $\gamma = C/\kappa^2$ is a transformed integrating constant. Expression in brackets in (8.3.4) should be complemented to a complete square and rewritten in the form of two factors:

$$\gamma + y^4 - y^2/\alpha = (y^2 - y_1^2)(y^2 - y_2^2),$$

where

$$y_1^2 = (1/2\alpha) + \sqrt{(1/2\alpha)^2 - \gamma},$$
$$y_2^2 = (1/2\alpha) - \sqrt{(1/2\alpha)^2 - \gamma}.$$

It is expected here that $0 < \gamma < 1/4\alpha^2$. Then inequalities $y_1^2 > y_2^2 > 0$ hold. Equation (8.3.4) becomes

$$(u_{,x})^2 = (\alpha\kappa^2 y_1^2)(1 - u^2)(1 - k^2u^2), \tag{8.3.5}$$

where $u(x) = y(x)/y_2$ and $k^2 = y_2^2/y_1^2 \le 1$. Equation (8.3.5) is well known in the theories of elliptic functions [102]. It is useful to note that it is possible to forget about C/κ^2 and to consider k as a new constant of integration because k^2 is related to $\gamma = C/\kappa^2$ by the formula

$$k^2 = \left[1 - (1 - 4\alpha^2\gamma)^{1/2}\right]\left[1 + (1 - 4\alpha^2\gamma)^{1/2}\right]^{-1}.$$

This constant can be determined by continuity conditions at the interfaces $x = 0$ and $x = h$. Besides this,

$$\alpha y_1^2 = (1 + k^2)^{-1}, \qquad \alpha y_2^2 = k^2(1 + k^2)^{-1},$$

and now function $\Phi^{(2)}(x)$ can be written as a single-parameter solution of equation (8.3.5):

$$\Phi^{(2)}(x) = \frac{k}{\alpha\sqrt{1 + k^2}} u(x; k). \tag{8.3.6}$$

Now, we can integrate equation (8.3.5) resulting in:

$$\pm\frac{\kappa(x-x_0)}{\sqrt{1+k^2}}=\int_0^u\frac{d\xi}{\sqrt{(1-\xi^2)(1-k^2\xi^2)}}, \tag{8.3.7}$$

where x_0 is determined from the equation $u(x_0)=0$, but it is simpler to consider x_0 as the second integrating constant. Function

$$F=F(\varphi,k)=\int_0^{\sin\varphi}\frac{d\xi}{\sqrt{(1-\xi^2)(1-k^2\xi^2)}}$$

is an incomplete elliptic integral of first genus. Jacobian elliptic functions are defined by means of inversion of this integral:

$$\sin(\varphi)=\mathrm{sn}(F,k),\quad \cos(\varphi)=\mathrm{cn}(F,k). \tag{8.3.8}$$

The function $\varphi(F,k)$ is named as the amplitude of F, $\varphi(F,k)=\mathrm{am}(F,k)$, k is the modulus. Except functions (8.3.8) there is one more elliptic function – "delta of amplitude" – $\mathrm{dn}(F,k)$

$$\mathrm{dn}(F,k)=\left[1-k^2\sin^2\mathrm{am}(F,k)\right]^{1/2}.$$

Thereby, it follows from (8.3.7) and (8.3.8) that

$$u(x,k)=\mathrm{sn}\left[\frac{\kappa(x-x_0)}{\sqrt{1+k^2}};k\right]$$

and in view of (8.3.6) the transverse profile of electric field strength of the NLGW in the $0<x<h$ domain takes the form:

$$\Phi^{(2)}(x)=\frac{\pm k}{\alpha\sqrt{1+k^2}}\mathrm{sn}\left[\frac{\kappa(x-x_0)}{\sqrt{1+k^2}};k\right]. \tag{8.3.9}$$

Continuity conditions $\Phi(x)$ and $\Phi,_x(x)$ at $x=0$ provide

$$pA_1=\pm\frac{\kappa k\ \mathrm{cn}(l_0;k)\mathrm{dn}(l_0;k)}{\alpha(1+k^2)},\quad A_1=\pm\frac{k\mathrm{sn}(l_0;k)}{\alpha\sqrt{1+k^2}}. \tag{8.3.10}$$

The following property was used to derive the above expressions [102]

$$\frac{d\,\mathrm{sn}(x,k)}{dx} = \mathrm{cn}(x,k)\,\mathrm{dn}(x,k)$$

and also parameter $l_0 = -kx_0(1+k^2)^{-1/2}$ was introduced. Continuity conditions for $\Phi(x)$ and $\Phi_{,x}(x)$ at $x = h$ give the second pair of expressions

$$\begin{aligned} -qA_3e^{-qh} &= \pm\frac{k\,\mathrm{cn}(l_0+l,k)\,\mathrm{dn}(l_0+l,k)}{\alpha(1+k^2)}, \\ A_3e^{-qh} &= \pm\frac{k\,\mathrm{sn}(l_0+l,k)}{\alpha\sqrt{1+k^2}}. \end{aligned} \tag{8.3.11}$$

where $l = kh(1+k^2)^{-1/2}$. Having excluded A_3 from (8.3.10), the equation can be found to obtain l_0 for given k:

$$\frac{p}{\kappa}\sqrt{1+k^2} = \frac{\mathrm{cn}(l_0,k)\mathrm{dn}(l_0,k)}{\mathrm{sn}(l_0,k)} \tag{8.3.12a}$$

Having excluded $A_3\exp(-qh)$ from (8.3.11), we can find one more equation

$$\frac{q}{\kappa}\sqrt{1+k^2} = -\frac{\mathrm{cn}(l_0+l,k)\mathrm{dn}(l_0+l,k)}{\mathrm{sn}(l_0+l,k)} \tag{8.3.12b}$$

Now, there are two integrating constants, $\gamma = C/\kappa^2$ and x_0, which are simply substituted for k and l_0, and two equations (8.3.12). Expression (8.3.12a) links l_0 with k. Expression (8.3.12b) is the equation to determine k in terms of β, ω and other characteristics of the waveguide. If we express β by means of this equation in terms of all remained parameters including k, then the dispersion relation for the considered type of NLGW will be obtained.

For application purposes it is suitable to have power flow P as a parameter in dispersion relations. It is even more useful to separate P in parts: $P = P_F + P_A$, where

$$P_F = \frac{c\beta}{8\pi k_0}\int_0^h |\Phi(x)|^2\,dx \quad ,$$

$$P_A = \frac{c\beta}{8\pi k_0}\left[\int_{-\infty}^{0} |\Phi(x)|^2\,dx + \int_h^{+\infty} |\Phi(x)|^2\,dx\right].$$

These quantities P_F and P_A are deduced in [55,56]. Their dependence on β, h and the different mode indices were derived by the numerical method. By taking into account the obtained results, the analytical expressions for power flows can be written as

$$P_F = \frac{2\kappa\ \beta}{k_0^3\,|\varepsilon_{nl}|\ \ 1+k^2}\{l + E[\mathrm{am}(l_0,k)] - E[\mathrm{am}(l+l_0,k)]\},$$
$$P_A = \frac{\kappa^2\beta\ k^2}{k_0^3\,|\varepsilon_{nl}|\,(1+k^2)}\{q^{-1}\,\mathrm{sn}^2(l+l_0,k) + p^{-1}\,\mathrm{sn}^2(l_0,k)\}, \qquad (8.3.13)$$

where $E(z,k)$ is an elliptic integral of second genus [102]. On Fig.8.3.1 dependencies β on P are shown for different indices of modes under the choice of the following values

Fig.8.3.1

Propagation constant of the TE-wave versus normalised power flow in NLGW ($P_0 = 1$ GW/cm^2)

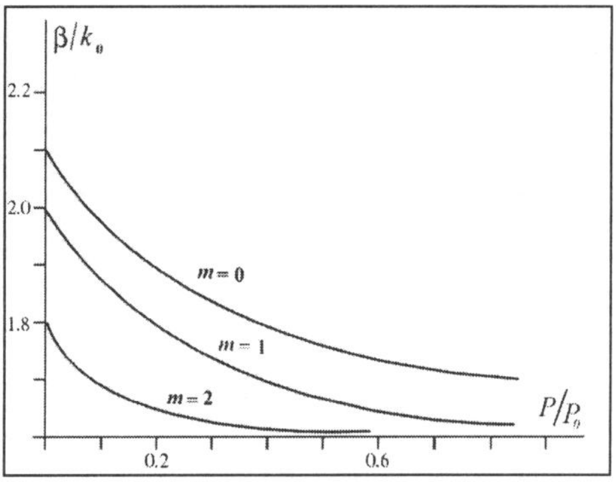

of the waveguide parameters: $\varepsilon_1 = 2{,}56$, $\varepsilon_3 = 1$, $\varepsilon_2 = 4{,}84$, $|\varepsilon_{nl}| = 10^{-11}$ erg^{-1}cm^3 and $h/\lambda = 0{,}97$ [55,56]. In this work function β vs P for the main mode (m=0) was found numerically under different values of h/λ. The results found there show that with the increase of power flow P an electromagnetic field is pushed out from a waveguide (β decreases).

This effect is confirmed by the dependencies of P_F vs P derived there (Fig. 8.3.2). Thereby, this non-linear waveguide operates as the power flow limiter.

Let us consider the case of $\kappa^2 > 0$, $\varepsilon_{nl} > 0$, (self-focusing waveguide) and let us assign $\alpha^2 = k_0^2|\varepsilon_{nl}|/2\kappa^2$. Repeating all variable transformations as it was done in the preceding case for $u(x)$, we receive the equation

$$\left(u_{,x}\right)^2 = \left(\alpha\kappa^2 y_1^2\right)(1-u^2)(1+k^2u^2), \qquad (8.3.14)$$

where $y_1^2 = -(1/2\alpha) + \sqrt{(1/2\alpha)^2 + \gamma}$, $y_2^2 = (1/2\alpha) + \sqrt{(1/2\alpha)^2 + \gamma}$,

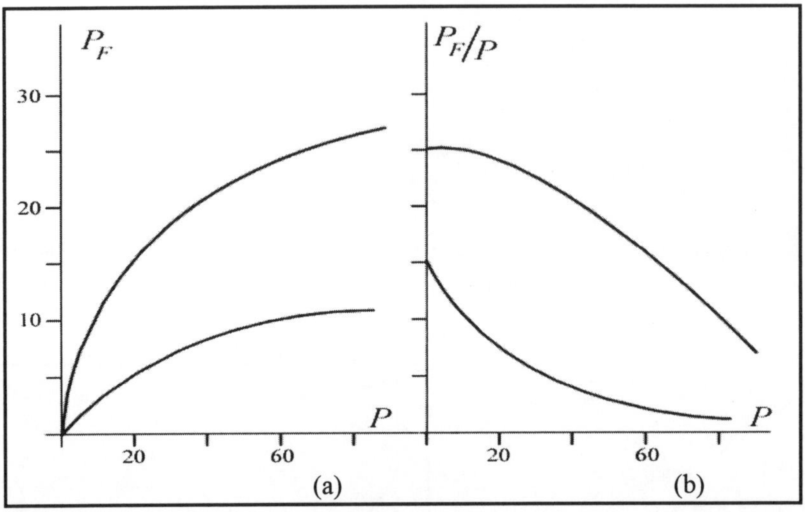

Fig. 8.3.2 Power flow P_F (a) and relative power flow P_F/P (b) versus total power in NLGW for different mode indices m

and $k^2 = y_1^2 / y_2^2 \le 1$. Unlike (8.3.5) in this equation in the last multiplicand the sign before $k^2 u^2$ is opposite. So instead of a standard expression for the elliptic integral of the first genus we have something different here:

$$\pm \frac{\kappa (x - x_0)}{\sqrt{1 + k^2}} = \int_0^u \frac{d\xi}{\sqrt{(1 - \xi^2)(1 + k^2 \xi^2)}}. \tag{8.3.15}$$

This integral is reduced to $F(\varphi, k)$ by making the substitution of variable

$$\varsigma = \xi \frac{\sqrt{1 + k^2}}{\sqrt{1 + k^2 \xi^2}}. \tag{8.3.16}$$

It can also be seen that if to make a substitution $k \to ik'$ in the integrand (k' is a real value), then the elliptic integral $F(\varphi, ik')$ will appear. So there is a recipe to evaluate the solution for $\Phi(x)$ from the known solution (8.3.6):

$$\Phi^{(2)}(x) = \frac{ik}{\alpha\sqrt{1+k^2}} u(x;ik)\,.$$

The rest of the cases when $\kappa^2 <0$, $\varepsilon_{nl} >0$ and $\kappa^2 <0$, $\varepsilon_{nl} <0$ can be considered separately, or by setting rules to build up the desirable solution from (8.3.6). Let it be an exercise to be done by the reader.

In conclusion it is useful to draw attention to the results of work [55,56], where the case $\kappa^2 >0$ and $\varepsilon_{nl}>0$ is considered. That corresponds to the NLGW in a waveguide with a self-focusing layer. The profile of the field at $0<x<h$ has the form

$$\Phi^{(2)}(x) = \frac{k}{\alpha\sqrt{1-2k^2}} \operatorname{cn}\left[\frac{\kappa(x-x_0)}{\sqrt{1-2k^2}}; k\right].$$

Dispersion relation for the NLGW derives from the continuity conditions at the interfaces $x = 0$ and $x = h$ and can be expressed by the couple of equations:

$$-\frac{p}{\kappa}\sqrt{1-2k^2} = \frac{\operatorname{sn}(l_0,k)\operatorname{dn}(l_0,k)}{\operatorname{cn}(l_0,k)},$$

$$\frac{q}{\kappa}\sqrt{1-2k^2} = \frac{\operatorname{sn}(l_0+l,k)\operatorname{dn}(l_0+l,k)}{\operatorname{cn}(l_0+l,k)},$$

where $l_0 = -kx_0(1-2k^2)^{-1/2}$, $l = kh(1-2k^2)^{-1/2}$. For the power flow represented in the form $P = P_F + P_A$ it was found that

$$P_F = \frac{2\kappa\beta}{k_0^3\,|\varepsilon_{nl}|\sqrt{1-2k^2}}\left\{E[\operatorname{am}(l_0+l,k)] - E[\operatorname{am}(l_0,k)] - (1-k^2)l\right\},$$

$$P_A = \frac{\kappa^2\beta^2k^2}{k_0^4\,|\varepsilon_{nl}|(1-2k^2)}\left\{q^{-1}\operatorname{cn}^2(l+l_0,k) + p^{-1}\operatorname{cn}^2(l_0,k)\right\}.$$

The numerical study of propagation constant β and power flow in a non-linear layer P_F as functions of P revealed more interesting properties than those of in a self-defocusing medium (Fig.8.3.3). It appears that the highest modes have some threshold value $P_K^{(m)}$ of the power flow, below which they do not excite. If $P > P_K^{(m)}$, then the mth mode ($m > 2$ in the calculations in [55,56]) has two values of propagation constant. The greater value of β corresponds to the NLGW, which is located inside a non-linear waveguide layer $0< x< h$. The smaller value of β corresponds to the NLGW, which is pushed out from a non-linear layer. It was noticed in [55,56] that these phenomena could be applied to design the power limiter.

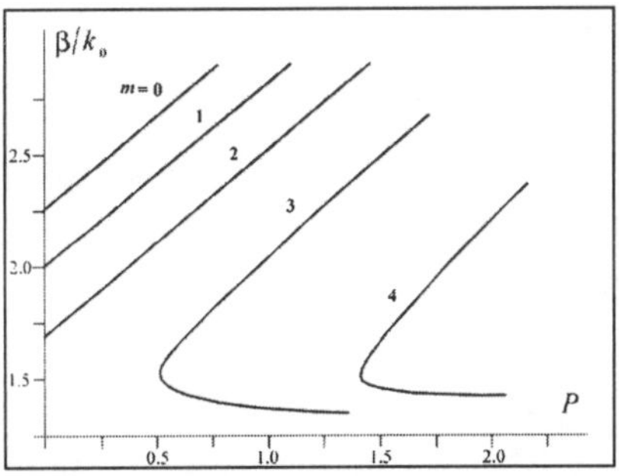

Fig.8.3.3a

Propagation constant as function of total power flow P in NLGW for different mode indices m
P measured in W/cm^2

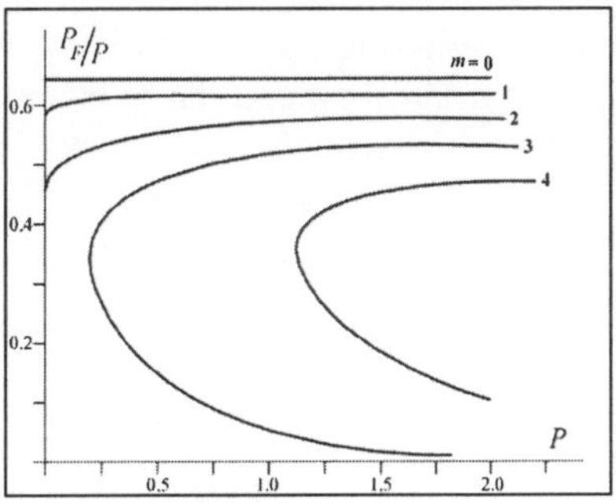

Fig.8.3.3b

The ratio of power flow in non-linear layer to total power as function of total power flow P in NLGW for different mode indices m.
P measured in W/cm^2

The investigation of the dependency of β on the thickness of a waveguide (or more precisely on the parameter h/λ) for $\varepsilon_{nl}>0$ has shown that there is an interval of h/λ where at one and the same value of a power flow P two NLGWs with different β can propagate. When the magnitude of h/λ leaves the boundaries of this interval, either both of these modes vanish disappear or one of the mode is cut off.

8.3.2. NONLINEAR WAVES OF TM-TYPE

This sort of the NLGW is explored least of all. As it takes place in general for the non-linear waves of this type, the analytical study can be only carried out when a non-linear medium has an anisotropy of non-linear susceptibility in the form as it was given in the section 8.1 formula (8.1.13b).

Three equations for $\Phi(x)$, describing the transverse distribution of longitudinal components of the vector of electric field strength of NLGW in three different intervals of x, come from (8.1.17):

$$\Phi,_{xx} + (\varepsilon_{\parallel}^{(1)} / \varepsilon_{\perp}^{(1)})(k_0^2 \varepsilon_{\perp}^{(1)} - \beta^2)\Phi = 0, \qquad x < 0$$

$$\Phi,_{xx} + (\varepsilon_{\parallel}^{(2)} / \varepsilon_{\perp}^{(2)})(k_0^2 \varepsilon_{\perp}^{(2)} - \beta^2)\Phi + k_0^2 (\varepsilon_{nl} / \varepsilon_{\perp}^{(2)})(k_0^2 \varepsilon_{\perp}^{(2)} - \beta^2)\Phi^3 = 0, \quad 0 < x < h$$

$$\Phi,_{xx} + (\varepsilon_{\parallel}^{(3)} / \varepsilon_{\perp}^{(3)})(k_0^2 \varepsilon_{\perp}^{(3)} - \beta^2)\Phi = 0, \qquad h < x$$

The solutions for $\Phi^{(1)}(x)$ and $\Phi^{(3)}(x)$ in the corresponded intervals $x < 0$ and $x > h$, are chosen in compliance with the boundary conditions in infinity: both $\Phi(x)$ and $\Phi,_x(x)$ approach zero when $|x| \to \infty$. Consequently,

$$\Phi^{(1)}(x) = A_1 \exp(px), \qquad \Phi^{(3)}(x) = A_3 \exp(-qx),$$

where

$$p^2 = (\varepsilon_{\parallel}^{(1)} / \varepsilon_{\perp}^{(1)})(\beta^2 - k_0^2 \varepsilon_{\perp}^{(1)}) > 0, \qquad q^2 = (\varepsilon_{\parallel}^{(3)} / \varepsilon_{\perp}^{(3)})(\beta^2 - k_0^2 \varepsilon_{\perp}^{(3)}) > 0,$$

for definiteness both p and q are assumed to be positive. If to determine parameters κ and α as

$$\kappa^2 = (\varepsilon_{\parallel}^{(2)} / \varepsilon_{\perp}^{(2)})(k_0^2 \varepsilon_{\perp}^{(2)} - \beta^2) > 0, \quad \alpha^2 = \varepsilon_{nl} / 2\varepsilon_{\parallel}^{(2)},$$

then the equation for $\Phi^{(2)}(x)$, i.e., equation for $\Phi(x)$ in the interval $0<x<h$ takes the following form

$$\Phi,_{xx} + \kappa^2 \Phi + 2\kappa^2 \alpha^2 \Phi^3 = 0.$$

Whence (after multiplying it on $d\Phi / dx$ and integrating) the following equation writes

$$(\Phi,_x)^2 + \kappa^2 \Phi^2 + \kappa^2 \alpha^2 \Phi^4 = C, \tag{8.3.17}$$

where C is an integrating constant. Equation (8.3.17) exactly complies with equation (8.3.3) considered in the preceding section. Thereby, we can at once turn to derivation of the dispersion relations.

As an illustration consider the case when $\kappa^2 > 0$ and $\varepsilon_{nl} < 0$. All other choices of the signs of κ^2 and ε_{nl} can be considered as exercise. This is done in [103] in detail and there is no need to consider them herein. The chosen example is interesting by comparing the found dispersion relations with (8.3.12). Function $\Phi^{(2)}(x)$ will be assigned by the formula (8.3.9). The continuity conditions (8.1.17) at the interfaces $x = 0$ and $x = h$ result in the following relations:

$$\frac{\varepsilon_{\parallel}^{(1)} A_1}{p} = -\frac{\varepsilon_{\parallel}^{(2)} k\,\mathrm{cn}(l_0,k)\mathrm{dn}(l_0,k)}{\kappa|\alpha|(1+k^2)}, \quad A_1 = -\frac{k\,\mathrm{sn}(l_0,k)}{|\alpha|\sqrt{1+k^2}}.$$

$$\frac{\varepsilon_{\parallel}^{(3)} A_3 e^{-qh}}{q} = \frac{\varepsilon_{\parallel}^{(2)} k\,\mathrm{cn}(l_0+l,k)\mathrm{dn}(l_0+l,k)}{\kappa|\alpha|(1+k^2)}, \quad A_3 e^{-qh} = \frac{k\,\mathrm{sn}(l_0+l,k)}{|\alpha|\sqrt{1+k^2}}.$$

where l_0 and l are determined in terms of k, x_0, h formally as it was done in (8.3.10) and (8.3.11). Having eliminated A_1 and $A_3\exp(-qh)$ from above, we obtain two equations:

$$\frac{\varepsilon_{\parallel}^{(1)}\kappa}{\varepsilon_{\parallel}^{(2)} p}\sqrt{1+k^2} = \frac{\mathrm{cn}(l_0,k)dn(l_0,k)}{\mathrm{sn}(l_0,k)}, \tag{8.3.18a}$$

$$\frac{\varepsilon_{\parallel}^{(3)}\kappa}{\varepsilon_{\parallel}^{(2)} q}\sqrt{1+k^2} = \frac{\mathrm{cn}(l_0+l,k)\mathrm{dn}(l_0+l,k)}{\mathrm{sn}(l_0+l,k)}. \tag{8.3.18b}$$

If to define l_0 as a function of p, q, κ from (8.3.18a) and then substitute this quantity in equation (8.3.18b), the latter will transform to the dispersion relation. It determines β in terms of k_0: $\beta = \beta(k_0, k)$, where elliptic function module k serves as a parameter.

It is interesting that under condition $k \to 0$ equation (8.3.18) leads to the dispersion law for linear waves of the TM type. In this limit parameter l_0 coincides with half of the value of a phase shift which appears under total internal reflection at the interface of two linear dielectric layers.

For the NLGW considered here it is convenient to divide the power flow P transferred along a waveguide into two parts similar to what was done in the preceding section: the power flow transferred by NLGW in a non-linear layer P_F and that outside a layer P_A: $P = P_F + P_A$. As with the case of TE-waves, the total power flow is defined through z-component of Poynting vector S_z by averaging over the fast spatial and temporal oscillations. For the TM-waves S_z has the form

$$S_z = \frac{c\beta}{8\pi k_0}\varepsilon_{\perp}^{-1}(x)|H_y(x)|^2$$

where, taking into account (8.1.15), it follows that

$$P = \frac{c\beta k_0}{8\pi} \int_{-\infty}^{+\infty} \frac{\varepsilon_\perp(x)}{[k_0^2\varepsilon_\perp(x) - \beta^2]} \left(\frac{d\Phi}{dx}\right)^2 dx$$

By means of the results obtained in this section we can now get the expressions for P_F and P_A [103]:

$$P_F = \frac{c\beta}{8\pi k_0} \left(\frac{\varepsilon_\perp^{(2)}}{|\varepsilon_{nl}|}\right) \left(\frac{\varepsilon_\parallel^{(2)}}{\varepsilon_\perp^{(2)}}\right)^{1/2} \frac{\varepsilon_\parallel^{(2)} k^2}{\kappa_1^2 (1+k^2)^{3/2}} [\tilde{I}(l+l_0) - \tilde{I}(l_0)],$$
$$P_A = \frac{c\beta\kappa\varepsilon_\parallel^{(2)} k^2}{8\pi k_0 (1+k^2)^{3/2}} \left[\frac{\varepsilon_1}{p^3} \operatorname{sn}^2(l_0, k) + \frac{\varepsilon_3}{q^3} \operatorname{sn}^2(l+l_0, k)\right], \qquad (8.3.19)$$

where for the sake of simplicity it is assumed that the media where the film is immersed are isotropic. The following notations are introduced:

$$\kappa_1^2 = (k_0^2 \varepsilon_\perp^{(2)} - \beta^2),$$

$$\tilde{I}(w) \equiv \int_0^w \operatorname{cn}^2(u,k)\operatorname{dn}^2(u,k)du = w - (1+k^2)\tilde{I}_2(w) + k^2 \tilde{I}_4(w),$$

$$\tilde{I}_m(w) = \int_0^w \operatorname{sn}^m(u,k)du.$$

The rest of parameters were determined in the preceding section. Integrals $\tilde{I}_m(w)$ are expressed in terms of elliptic functions, namely

$$\tilde{I}_2(w) = [w - E(\operatorname{am} w, k)]k^{-2},$$

$$\tilde{I}_4(w) = \frac{2(1+k^2)}{3k^4}[w - E(\operatorname{am} w, k)] + \frac{1}{3k^2}[\operatorname{sn}(w,k)\operatorname{cn}(w,k)\operatorname{dn}(w,k) - w]$$

It should be noted that in the limit $h \to 0$ equations (8.3.18) provide the known result:

$$\frac{\varepsilon_\parallel^{(1)}}{p} + \frac{\varepsilon_\parallel^{(3)}}{q} = 0 \qquad (8.3.20)$$

that is a dispersion relation for the linear surface wave of the TM-type [1].

8.4. Waves in non-linear directional couplers

A two-channel waveguide is the simplest example of the waveguide structures, which is usually used in integrated optics. It is named as a *directional coupler*. Let us assume that the each channel is the single mode waveguide. For the electric field amplitudes of these modes we have the following system of equations

$$\frac{dA_1}{dz} = iKA_2 + ir|A_1|^2 A_1 , \quad \frac{dA_2}{dz} = iKA_1 + ir|A_2|^2 A_2 . \qquad (8.4.1)$$

We assume herein that both channels are similar and that there is no phase mismatch ($\Delta\beta = \beta_1 - \beta_2$=0). Parameter K is the coupling coefficient describing the inter-channel mode overlap [104] (see section 6.3.2 also). Parameter r is used to introduce the non-linear phase self-modulation of the modes. We assume the cross-modulation to be negligible.

Let the mode amplitude in the first channel at $z = 0$ be equal to A_{10}, but there is no radiation in the second channel at all. If to transfer to real variables in (8.4.1)

$$A_1 = A_{10}u_1 \exp(i\phi_1) , \qquad A_2 = A_{10}u_2 \exp(i\phi_2) ,$$

then system (8.4.1) can be rewritten as

$$\frac{du_1}{d\xi} = u_2 \sin\Phi , \qquad (8.4.2a)$$

$$\frac{du_2}{d\xi} = -u_1 \sin\Phi , \qquad (8.4.2b)$$

$$\frac{d\Phi}{d\xi} = \tilde{\beta}(u_1^2 - u_2^2) + (u_1u_2^{-1} - u_2u_1^{-1})\cos\Phi , \qquad (8.4.2c)$$

where $\xi = Kz$, $\Phi = \phi_1 - \phi_2$, $\tilde{\beta} = r|A_{10}|^2 K^{-1}$. System of equations (8.4.2a) and (8.4.2b) possesses an integral of motion $u_1^2 + u_2^2 = 1$. Thus, we can convert system (8.4.2) to the following two equations

$$\frac{du_2}{d\xi} = -\sqrt{1-u_2^2}\,\sin\Phi , \qquad (8.4.3a)$$

$$\frac{d\Phi}{d\xi} = \tilde{\beta}(1-2u_2^2) + \frac{2u_2^2 - 1}{u_2\sqrt{1-u_2^2}}\cos\Phi , \qquad (8.4.3b)$$

For the function $f(u) = \cos\Phi(u)$ (index at variable u is omitted for brevity) system (8.4.2) results in a single equation

$$\frac{df}{du} = -\frac{1-2u^2}{u(1-u^2)} f + \tilde{\beta}\frac{1-2u}{\sqrt{1-u^2}} . \tag{8.4.4}$$

If we consider short distances, so that the amplitude in the excited first channel is constant, then the equation for complex amplitude in the second channel A_2 can be solved approximately, i.e., the nondepleting field approximation holds. The second equation in system (8.4.1) can be integrated. The solution is $A_2(z) = iKA_{10}z$, i.e., the phase difference as $z \to 0$ goes to $\pi/2$. Consequently, in general case the initial conditions for equation (8.4.4) should be taken as $\Phi(0) = \pi/2$ or $f(0) = 0$. Equation (8.4.4) can be solved by a standard method of constant variation. This solution for the arbitrary initial conditions is given by expression

$$f(u) = \frac{C_0}{u\sqrt{1-u^2}} + \frac{\tilde{\beta}}{2} u\sqrt{1-u^2} .$$

Taking into account the chosen initial conditions we obtain

$$f(u) = \frac{\tilde{\beta}}{2} u\sqrt{1-u^2} . \tag{8.4.5}$$

From (8.4.3a) and (8.4.5) equation for the mode amplitude in the second channel of coupler can be found:

$$\frac{du}{d\xi} = -\left[(1-u^2)\left\{1-\tilde{\mu}^2 u^2 (1-u^2)\right\}\right]^{1/2} , \tag{8.4.6}$$

where $\tilde{\mu} = \tilde{\beta}/2$. The change of variable $u = \sin(\varphi/2)$, allows to solve this equation. It is

$$-2\xi = \int_0^{\varphi} \frac{dx}{\sqrt{1-(\tilde{\mu}^2/4)\sin^2 x}} \equiv F(\varphi, \tilde{\mu}/2) ,$$

where the right hand side of these expressions is an elliptic integral of the first kind [102]. The definition of the Jacobian elliptical function $\mathrm{cn}(2\xi, \tilde{\mu}/2) = \cos\varphi(\xi)$ shows the way to find $\varphi(\xi)$. After some algebra we can write the expression for intensity of radiation in the second channel of a non-linear directional coupler

$$I_2(\xi) = I_0 u_2^2(\xi) = (I_0/2)\left[1 - \mathrm{cn}(2\xi, \tilde{\mu}/2)\right] . \tag{8.4.7}$$

Thereby, the intensity of radiation in the second channel is expressed in terms of periodic function (it is the Jacobian elliptic cosine) only if its modulus is not equal to unit. This picture is almost indistinguishable qualitatively from the linear directional coupler. But still there is an important difference. Here the length of coupling L_C (i.e. the distance on which radiation from one channel flows over to another channel) depends on initial intensity of radiation

$$L_C(A_{10}^2) = 2L_{CL}\mathbf{K}(\tilde{\mu}/2)/\pi, \tag{8.4.8}$$

where L_{CL} is a coupling length related to the equivalent linear directional coupler, $\mathbf{K}(*)$ is a complete elliptic integral of the first kind [102]. By taking the properties of $\mathbf{K}(*)$ into account, we can find from expression (8.4.8) that

$$L_C \approx L_{CL}[1+\tilde{\mu}^2/16],$$

when $\tilde{\mu} << 2$.

When the intensity of the input radiation in the first channel grows, one can reach the critical value $A_{10}^{(c)}$ when parameter $\tilde{\mu}$ is equal to two. Then the intensity of radiation in the second channel is expressed by the formula

$$I_2(\xi) = (I_0/2)[1-\mathrm{sech}(2\xi)]\ . \tag{8.4.9}$$

As the amplitude A_{10} increases further, this monotone dependence will change for oscillating of the intensity:

$$I_2(\xi) = (I_0/2)[1-\mathrm{dn}(\tilde{\mu}\xi, 2/\tilde{\mu})]\ , \tag{8.4.10}$$

where $\mathrm{dn}(\tilde{x},\tilde{k})$ is a Jacobian function named "delta of amplitude". It is related to Jacobian elliptic sine as $\mathrm{dn}^2(\tilde{x},\tilde{k}) = 1-\tilde{k}^2\,\mathrm{sn}^2(\tilde{x},\tilde{k})$ [102]. This is a periodic function since the intensity of radiation in the second coupler channel oscillates with increasing of $\xi = Kz$. The coupling length in this case is given by the following expression

$$L_C(A_{10}^2) = (2L_{CL}/\pi\tilde{\mu})\mathbf{K}(2/\tilde{\mu})\ . \tag{8.4.11}$$

Maximum value of I_2 now is less than I_0 and it is

$$\max I_2 = (I_0/2)\left[1-\sqrt{1-4\tilde{\mu}^{-2}}\right].$$

For critical value of amplitude in the first channel we can obtain an expression

$$|A_{10}^{(c)}|^2 = 4K/\pi r\,.$$

The formula for the maximum value of I_2 shows that in weakly bounded channels of the non-linear directional coupler the critical value of amplitude $A_{10}^{(c)}$ is less than one in the strong bounded coupler channel. Transition from the mode with the deep amplitude modulation (8.4.7) to the mode (8.4.10.), where the total switching does not occur can be qualified as locking a directional coupler.

8.5. Waves in non-linear distributed feedback structures

Among the integrated optics devices the periodic waveguides or grating waveguides and another periodic structure attract attention due to many useful applications [104, 144]. The distributed-feedback laser is one of them. Counter-propagation is a property, which distinguishes waves in such media from other wave-guiding structures. The periodic structure is able to reflect the incident wave, serving as distributed mirror. The incident and reflected waves are coupled by Bragg scattering. We shall consider below the wave propagation in the periodic non-linear medium, which for the sake of generality will be named the distributed feedback structures (DFB-structures).

The general form of the system of equation for waves propagating in the opposite directions has the form (see section 6.3.2)

$$\left(\frac{\partial A^{(+)}}{\partial z} + \frac{1}{v}\frac{\partial A^{(+)}}{\partial t}\right) + i\delta k\, A^{(+)} + iK A^{(-)} = 0\,, \qquad (8.5.1a)$$

$$\left(-\frac{\partial A^{(-)}}{\partial z} + \frac{1}{v}\frac{\partial A^{(-)}}{\partial t}\right) + i\delta k\, A^{(-)} + iK A^{(+)} = 0\,, \qquad (8.5.1b)$$

where $\delta k = (\beta - k_B)$ is a wave mismatch, $k_B = \pi/\Lambda$ is a wave number of periodic (distributed feedback) structure, Λ is a spatial period of DFB-structures, K is a coupling constant [144-146]. When $\delta k = 0$ one says about Bragg resonance. Under this condition the reflective characteristic of DFB-structure has a maximum.

The nonlinearity of waveguide medium causes parameter δk to be dependent on the electric fields strengths $E^{(\pm)} = A^{(\pm)} \exp[-i\omega t \pm i\beta z]$. The nonlinearity appears because of the dependency of Λ on the values of $A^{(\pm)}$ (it refers to the effect of photoelasticity), heat expansion due to absorption of radiation (thermoplastic effect), a high-frequency Kerr effect, which affects the propagating constant and so on. The coupling constant can depend on the electric field strength of interacting waves too, but we will not consider this case herein. So,

$$\left(\frac{\partial A^{(+)}}{\partial z}+\frac{1}{v}\frac{\partial A^{(+)}}{\partial t}\right)+i\delta k_0 A^{(+)}+i\delta k_{nl}\left(|A^{(+)}|^2+|A^{(-)}|^2\right)A^{(+)}+iKA^{(-)}=0, \quad (8.5.2a)$$

$$\left(-\frac{\partial A^{(-)}}{\partial z}+\frac{1}{v}\frac{\partial A^{(-)}}{\partial t}\right)+i\delta k_0 A^{(-)}+i\delta k_{nl}\left(|A^{(+)}|^2+|A^{(-)}|^2\right)A^{(-)}+iKA^{(+)}=0. \quad (8.5.2b)$$

In these equations the group-velocity dispersion (particularly the second order dispersion) is not taken into account. This approximation is correct when dispersion length is larger than the extent of DFB-structures. For example, in the range of picosecond durations of optical pulses the dispersion length is of the order of hundred meters in glass fibre.

System (8.5.2) presents an example of integrable system, which is close to Thirring models with the nonzero mass. Among its solutions there are solitons occurring in non-linear optics and solid state physic under the name of *gap solitons*.

8.5.1. NONLINEAR OPTICAL FILTER

If the pulse duration is sufficiently long or if the length of DFB-structures is sufficiently short, we can ignore time derivatives. Then the system of equations (8.5.1) will take the form

$$\begin{aligned}\frac{dA^{(+)}}{dz}&=-iKA^{(-)}-i\delta k_0 A^{(+)}-i\delta k_{nl}\left(|A^{(+)}|^2+|A^{(-)}|^2\right)A^{(+)},\\ \frac{dA^{(-)}}{dz}&=iKA^{(+)}+i\delta k_0 A^{(-)}+i\delta k_{nl}\left(|A^{(-)}|^2+|A^{(+)}|^2\right)A^{(-)}.\end{aligned} \quad (8.5.3)$$

If the reflection from DFB-structure is considered, then we should accept boundary conditions in the form

$$A^{(+)}(z=0)=A_0, \quad A^{(-)}(L)=0.$$

That means the absence of the reflected wave on the right boundary of DFB-structures (i.e., at $z=L$) and the given field of incident wave on the left boundary (i.e., at z = 0).

As with the case of non-linear directional coupler, it is useful here to transfer to dimensionless real variables:

$$A^{(+)}=A_0 u_1\exp(i\phi_1), \qquad A^{(-)}=A_0 u_2\exp(i\phi_2),$$

so the system of equations (8.5.3) yields

$$\frac{du_1}{d\xi} = u_2 \sin\Phi \,, \qquad (8.5.4a)$$

$$\frac{du_2}{d\xi} = u_1 \sin\Phi \,, \qquad (8.5.4b)$$

$$\frac{d\Phi}{d\xi} = \Delta + \widetilde{\beta}(u_1^2 + u_2^2) + (u_1 u_2^{-1} + u_2 u_1^{-1})\cos\Phi \,, \qquad (8.5.4c)$$

where $\xi = Kz$, $\Phi = \phi_2 - \phi_1$, $\widetilde{\beta} = \delta k_{nl} |A_0|^2 K^{-1}$, $\Delta = 2\delta k_0 K^{-1}$. The boundary conditions are in the form

$$u_1(\xi = 0) = 1\,, \quad u_2(\xi = KL = l) = 0\,, \quad \Phi(\xi = l) = \pi/2\,. \qquad (8.5.5)$$

From (8.5.4a) and (8.5.4b) the first integral of motion is obtained:

$$u_1^2(\xi) - u_2^2(\xi) = r^2 = \text{const}\,. \qquad (8.5.6)$$

It can be found from boundary conditions that this constant is a transparency factor of DFB-structures [104,144,145].

By excluding u_2 from equations (8.5.4) by means of (8.5.6) we can get a system of equations (index at u_1 is omitted):

$$\frac{du}{d\xi} = \sqrt{u^2 - r^2}\,\sin\Phi\,, \qquad (8.5.7a)$$

$$\frac{d\Phi}{d\xi} = \Delta + \widetilde{\beta}(2u^2 - r^2) + \frac{2u^2 - r^2}{u\sqrt{u^2 - r^2}}\cos\Phi\,. \qquad (8.5.7b)$$

Let us define a new variable $f(u) = \cos\Phi(u)$. For this function we can formulate an equation from (8.5.7):

$$\frac{df}{du} = -\frac{2u^2 - r^2}{u(u^2 - r^2)} f - \frac{2\widetilde{\beta}u^2 + (\Delta - \widetilde{\beta}r^2)}{\sqrt{u^2 - r^2}}\,. \qquad (8.5.8)$$

General solution of this equation has the form

$$f(u) = \frac{C_0 - \Delta u^2}{2u\sqrt{u^2 - r^2}} - \frac{\widetilde{\beta}}{2} u\sqrt{u^2 - r^2}\,. \qquad (8.5.9)$$

An integrating constant C_0 is found from the boundary condition at $\xi = l$ (i.e., on the output of DFB-structures): $C_0 = \Delta r^2$. Therefore, we have a second integral of system (8.5.4) consistent with the boundary condition (8.5.5):

$$\cos\Phi = -(\Delta + \tilde{\beta}\, u^2)(2u)^{-1}\sqrt{u^2 - r^2}. \tag{8.5.10}$$

Equation (8.5.7a) and expression (8.5.10) lead to the equation characterising the spatial dependency of normalised amplitude u_1 of the wave propagating in the positive direction. If we introduce $w = u_2$ and $\rho = r^2$, this equation can be written as:

$$(dw / d\xi)^2 = (w - \rho)\left[4w - (w - \rho)(\Delta + \tilde{\beta}\, w)^2\right]. \tag{8.5.11}$$

The right-hand side of this equation is the fourth degree polynomial, so this equation may be reduced to the equation for elliptic functions.

Under condition of Bragg resonance the solution of equation (8.5.11) can be easily obtained. Let $\Delta = 0$ in (8.5.11) and let denote $y^2 = w(w - \rho)$. Equation (8.5.11) will be now expressed in the form:

$$(dy / d\xi)^2 = 4(y^2 + a^2)(1 - b^2 y^2). \tag{8.5.12}$$

with the boundary condition $y(0) = 1 - 2a$, $y(l) = 0$. New parameters are used here:

$$a = \rho / 2 = u^2(l) / 2, \quad b = \tilde{\beta} / 2 = \delta k_{nl} A_0^2 K^{-1}.$$

It follows from (8.5.12) that

$$2(\xi - l) = \int_0^{y(\xi)} \frac{dx}{\sqrt{(a^2 + x^2)(1 - b^2 x^2)}}. \tag{8.5.13}$$

Changing of variable $x = a \sinh\psi$ converts integral in (8.5.13) into

$$2(\xi - l) = \int_0^{\psi(\xi)} \frac{d\psi}{\sqrt{(1 - a^2 b^2 \sinh^2\psi)}}. \tag{8.5.14}$$

where $y(\xi) = a\sinh\psi(\xi)$. Let $(ab)^2 = -\kappa^2$ and $\varphi = i\psi$, then expression (8.5.14) takes the form of an elliptic integral of the first genus [102]

$$2i(\xi - l) = \int_0^{-i\psi(\xi)} \frac{d\varphi}{\sqrt{(1 - \kappa^2 \sin^2\varphi)}}. \tag{8.5.15}$$

Hence a solution of (8.5.12) expressed in terms of Jacobian elliptic function of purely imaginary argument can be obtained

$$y(\xi) = i\,a\,\mathrm{sn}\left[2i(l-\xi)\,,\,i\kappa\right].$$

If we use the formulae for transformations of elliptic functions [102], then the normalised amplitude u_1 of the wave spreading in positive direction may be presented in the real form

$$u_1^2(\xi) = u_1^2(l)\frac{1+\mathrm{dn}\left[2v_1^{-1}(\xi-l),v_1\right]}{2\,\mathrm{dn}\left[2v_1^{-1}(\xi-l),v_1\right]}, \tag{8.5.16}$$

where $v_1 = (1+a^2b^2)^{-1/2}$. From this expression, by assigning $\xi = 0$, we can get a transcendental equation to determine the transparency factor, which is $r^2 = u_1^2(l)$:

$$1 = u^2(l)\frac{1+\mathrm{dn}\left[2v_1^{-1}l,v_1\right]}{2\,\mathrm{dn}\left[2v_1^{-1}l,v_1\right]}. \tag{8.5.17}$$

By using the determination of all parameters in this expression, it is convenient to re-write this equation to derive explicitly an unknown value. This provides an equation with respect to v_1:

$$\frac{bv_1}{\sqrt{1-v_1^2}} = \frac{1+\mathrm{dn}\left[2v_1^{-1}l,v_1\right]}{2\,\mathrm{dn}\left[2v_1^{-1}l,v_1\right]}, \tag{8.5.18}$$

The root of this equation $v_{1\#}$ specifies the transparency factor $r^2 = u_1^2(l)$ according to the formula

$$r^4 = u_1^4(l) = 4(1-v_{1\#}^2)(bv_{1\#})^{-2}.$$

Parameter b describes non-linear properties of DFB-structures. It determines the power of incident radiation at the boundary of DFB-structure. The non-linear DFB-structure becomes transparent at some values of this parameter. To obtain these values, we equate $u_1^2(l)$ to unity in (8.5.16). Then from resulting expression one can find an equation to obtain the desired parameter b:

$$\mathrm{dn}\left[2lv_1^{-1},v_1\right] = 1\;, \tag{8.5.19}$$

whence $l = mv_1^{(m)}\mathbf{K}(v_1^{(m)})$, $\mathbf{K}(*)$ is a complete elliptic integral of the first genus and $m = 0, 1, 2,...$ are integer numbers. Parameter $v_1^{(m)}$ is taken at $a = 1/2$, thus b is associated with $v_1^{(m)}$ by the formula $v_1^{(m)} = 1/[1+(b/2)^2]^{1/2}$. As far as the complete elliptic integral is a monotonously increasing function of its argument, solution of equation (8.5.19) exists. Moreover, these solutions form a countable set. Hence there is a countable set of values of incident power when DFB-structures become transparent.

It is useful to note that if the parameter of nonlinearity b tends to zero, i.e. the medium of DFB-structures becomes linear, then from (8.5.14) it follows

$$y(\xi) = = a \sinh[2(\xi - l)],$$

and expression (8.5.16) becomes

$$u(\xi) = \frac{\cosh(\xi - l)}{\cosh l},$$

which complies with the known result of the linear theory of waves in DFB-structure [104, 144-146].

8.5.2. STEADY-STATE SOLUTION OF THE EQUATIONS OF THE TWO-DIMENSIONAL MODIFIED MASSIVE THIRRING MODEL

We shall consider a system of equations of MMT-model. These equations describe the optical waves of periodic dielectric medium, for example the DFB-structure. If we use the normalised variables $\xi = Kz$, $\tau = Kvt$ and $\vartheta = \delta k_{nl} / K$, then under condition of Bragg resonance these equations take the form:

$$\left(\frac{\partial A^{(+)}}{\partial \xi} + \frac{\partial A^{(+)}}{\partial \tau}\right) + i\vartheta\left(|A^{(+)}|^2 + r|A^{(-)}|^2\right)A^{(+)} + iA^{(-)} = 0, \qquad (8.5.20a)$$

$$\left(\frac{\partial A^{(-)}}{\partial \xi} - \frac{\partial A^{(-)}}{\partial \tau}\right) - i\vartheta\left(|A^{(-)}|^2 + r|A^{(+)}|^2\right)A^{(-)} - iA^{(+)} = 0. \qquad (8.5.20b)$$

Here we introduce factor r for the sake of generality: at $r = 1$ system (8.5.20) describes optical pulse propagation in the non-linear DFB-structure, at $r \to \infty$, $\vartheta \to 0$, under condition $r\vartheta = g = \text{const}$ this system corresponds to equations of Massive Thirring Model [161-163]. In terms of real dependent variables

$$A^{(+)} = A_0 a_1 \exp(i\phi_1), \qquad A^{(-)} = A_0 a_2 \exp(i\phi_2),$$

system (8.5.20) converts to

$$\frac{\partial a_1}{\partial \xi}+\frac{\partial a_1}{\partial \tau}=a_2 \sin\Phi , \qquad \frac{\partial a_2}{\partial \xi}-\frac{\partial a_2}{\partial \tau}=a_1 \sin\Phi ,$$

$$\frac{\partial \phi_1}{\partial \xi}+\frac{\partial \phi_1}{\partial \tau}=-a_2 a_1^{-1}\cos\Phi-\mu\left(a_1^2+ra_2^2\right), \tag{8.5.21}$$

$$\frac{\partial \phi_2}{\partial \xi}-\frac{\partial \phi_2}{\partial \tau}=a_1 a_2^{-1}\cos\Phi+\mu\left(a_2^2+ra_1^2\right),$$

where $\Phi=\phi_2-\phi_1$ and $\mu=\delta k_{nl}A_0^2/K$.

Let us consider stationary waves propagation, when all dependent variables are the functions of only one variable $\eta=(1-V^2)^{-1/2}(\xi+V\tau)$. This assumption gives rise to the following system of equations

$$\frac{du_1}{d\eta}=u_2 \sin\Phi , \qquad \frac{du_2}{d\eta}=u_1 \sin\Phi ,$$

$$\frac{d\phi_1}{d\eta}=-u_2 u_1^{-1}\cos\Phi-\mu\sqrt{1-V^2}\left(\frac{1-V}{1+V}u_1^2+ru_2^2\right), \tag{8.5.22}$$

$$\frac{d\phi_2}{d\eta}=u_1 u_2^{-1}\cos\Phi+\mu\sqrt{1-V^2}\left(\frac{1+V}{1-V}u_2^2+ru_1^2\right),$$

where $u_1=a_1/\sqrt{1-V}$, $u_2=a_2/\sqrt{1+V}$. An integral of motion follows from equations for amplitudes

$$u_1^2(\eta)-u_2^2(\eta)=u_0^2=\text{const}. \tag{8.5.23}$$

The value of this integral is defined by the boundary conditions. If we seek for the solution in the form of solitary waves with zero asymptotic, this integral vanishes and thus we have $u_1(\eta)=\varepsilon u_2(\eta)$, $\varepsilon=\pm 1$. Nonzero value of the parameter u_0 corresponds to solutions in the form of waves on pedestal. Moreover, the case can be realised when one of the waves is "dark", but another wave is "bright".

Let us consider u_0 = 0. From (8.5.22) two equations both for $u(\eta)=u_1(\eta)$ and $\Phi(\eta)$ can be obtained:

$$\frac{du}{d\eta}=\varepsilon u \sin\Phi , \tag{8.5.24a}$$

$$\frac{d\Phi}{d\eta}=2\varepsilon\cos\Phi+2\tilde{p}u^2 , \tag{8.5.24b}$$

where $\widetilde{p} = 2\mu\sqrt{1-V^2}\left(\frac{1+V^2}{1-V^2}+r\right)$. This system provides the second integral of motion:

$$u^2 \cos\Phi + (\varepsilon\widetilde{p}/2)u^4 = \text{const}$$

For the chosen vanishing boundary values the constant is zero. Therefore,

$$\cos\Phi = -(\varepsilon/2)\widetilde{p}u^2 \tag{8.5.25}$$

and $u(\eta)$ is defined by the equation:

$$\frac{du}{d\eta} = \varepsilon u\sqrt{1-(\widetilde{p}/2)^2 u^4}\ . \tag{8.5.26}$$

This equation has the following solution

$$u^2(\eta) = (2/\widetilde{p})\cosh^{-1}[2(\eta-\eta_0)]\ . \tag{8.5.27}$$

Here η_0 is an integrating constant of the equation (8.5.26). Substitution (8.5.25) and (8.5.27) into phase equations in (8.5.22) allows integrating the system. So we have

$$\phi_1(\eta) = \phi_{10} + \mu_1 \arctan\{\exp[2(\eta-\eta_0]\}\ , \tag{8.5.28a}$$

$$\phi_2(\eta) = \phi_{20} - \mu_2 \arctan\{\exp[2(\eta-\eta_0]\}\ , \tag{8.5.28b}$$

where ϕ_{10} and ϕ_{20} are the integrating constants of the equations for phases. The parameters μ_1 and μ_2 are defined as

$$\mu_1 = 1-2\left(\frac{1-V}{1+V}+r\right)\left(\frac{1+V^2}{1-V^2}+r\right)^{-1},\quad \mu_2 = 1-2\left(\frac{1+V}{1-V}+r\right)\left(\frac{1+V^2}{1-V^2}+r\right)^{-1}.$$

These steady-state solutions present a stationary solitary two-component wave propagating in one of possible directions in a non-linear periodic medium. The velocity of this wave is less than the group velocity of a linear wave. Moreover, this steady-state solitary wave does not exist in linear medium. This is one more example of an optical gap soliton propagating in the extended DFB-structure.

8.5.3. NON-RELATIVISTIC LIMIT OF THE MODIFIED MASSIVE THIRRING MODEL

In relativistic theory of the field the free fermions are governed by the famous Dirac equation. If the kinetic energy of fermion is far smaller than the rest energy, then the Dirac equation can be reduced to the Schrödinger equation. One can say that the Schrödinger equation is the non-relativistic limit of the Dirac one. The Thirring model is the relativistic generalisation of the free fermions model to interacting fermions. Hence it is natural that we expected that in a non-relativistic limit the Thirring model leads to some kind of the Nonlinear Schrödinger equation. The unitary Foldy-Wouthuysen transformation is the convenient tool of this reduction [166]. However, we will use the simplified approach.

Let us consider the equations (8.5.2) and introduce the following designations:

$$\hat{p}_0 = \frac{i}{v}\frac{\partial}{\partial t}, \quad \hat{p}_1 = -i\frac{\partial}{\partial z}$$

for the 2-momentum operator components. The system of equations (8.5.2) can be written as

$$\begin{aligned}(\hat{p}_0 - \hat{p}_1)A^{(-)} &= KA^{(+)} - 2g\left(|A^{(-)}|^2 + r|A^{(+)}|^2\right)A^{(-)},\\ (p_0 + p_1)A^{(+)} &= KA^{(-)} - 2g\left(|A^{(+)}|^2 + r|A^{(-)}|^2\right)A^{(+)},\end{aligned} \tag{8.5.29}$$

where $2g = \delta k_{nl}$. It is convenient to write these equations in a standard Dirac spinor form, which is simpler to make a transition to a non-relativistic limit of those equations. Introduction of the new variables

$$\phi_1 = 2^{-1/2}(A^{(-)} + A^{(+)}), \quad \phi_2 = 2^{-1/2}(A^{(-)} - A^{(+)})$$

converts the system of equations (8.5.29) into a standard spinor form

$$\begin{aligned}\hat{p}_0\phi_1 - \hat{p}_1\phi_2 &= +K\phi_1 - \sqrt{2}g\left[|A^{(+)}|^2(A^{(+)} + rA^{(-)}) + |A^{(-)}|^2(A^{(-)} + rA^{(+)})\right],\\ \hat{p}_0\phi_2 - \hat{p}_1\phi_1 &= -K\phi_2 + \sqrt{2}g\left[|A^{(+)}|^2(A^{(+)} - rA^{(-)}) - |A^{(-)}|^2(A^{(-)} - rA^{(+)})\right].\end{aligned}$$

Mass parameter in case of a non-linear DFB-structure is represented by the coupling constant *K*. Hence, non-relativistic limit of MMT-model corresponds to the limit of the strong coupling. Let us introduce the variables

$$\tilde{\psi} = \phi_1 \exp(iKvt), \quad \tilde{\varphi} = \phi_2 \exp(iKvt).$$

In term of these variables the standard spinor form of the system of equations (8.5.29) is

$$\hat{p}_0\Psi - \hat{p}_1\Phi = -g\left[(1+r)\left(|\Psi|^2 + |\Phi|^2\right)\Psi + (1-r)\left(\Psi^*\Phi + \Psi\Phi^*\right)\Phi\right],$$
$$\hat{p}_0\Phi - \hat{p}_1\Psi = -2K\Phi + g\left[(1+r)\left(|\Psi|^2 + |\Phi|^2\right)\Phi + (1-r)\left(\Psi^*\Phi + \Psi\Phi^*\right)\Psi\right].$$

Assuming that $|\Phi| << |\Psi|$ and taking into account the terms only of the order $1/K$, we obtain

$$\hat{p}_0\Psi \approx \hat{p}_1\Phi - (1+r)g|\Psi|^2\Psi, \qquad \hat{p}_1\Psi \approx 2K\Phi. \tag{8.5.30}$$

Thereby, these systems of equations result in the Nonlinear Schrödinger equation:

$$\hat{p}_0\Psi = \frac{1}{2K}\hat{p}_1^2\Psi - (1+r)g|\Psi|^2\Psi.$$

In terms of initial independent co-ordinates this equation takes the form

$$i\frac{\partial\Psi}{\partial t} + \frac{v}{2K}\frac{\partial^2\Psi}{\partial z^2} + (1+r)gv|\Psi|^2\Psi = 0. \tag{8.5.31}$$

The solutions of the NLS equation (8.5.31) can be found by the familiar methods and the envelopes of the counter-propagating waves $A^{(+)}$ and $A^{(-)}$ are expressed in terms of the solutions of the NLS equation:

$$A^{(\pm)}(z,t) = \sqrt{2}\left(\Psi \mp \frac{i}{2K}\frac{\partial\Psi}{\partial z}\right)\exp(-iKvt). \tag{8.5.32}$$

Let us consider the same solution of the NLS equation, for example one-soliton solution. Assume this solution has the form

$$\Psi(z,t) = \psi_0 \operatorname{sech}(z/z_0)\exp[ivt/2Kz_0^2],$$

where z_0 is the space width of soliton. Amplitude ψ_0 is connected with parameters of the non-linear DFB-structure as

$$\psi_0^2 = 2[Kz_0^2(1+r)\delta k_{nl}]^{-1}.$$

Substitution of this expression into (8.5.32) leads to the following formula

$$A^{(\pm)} = 2\psi_0 \operatorname{sech}\left(\frac{z}{z_0}\right)\left[1 \pm \frac{i}{2Kz_0}\tanh\left(\frac{z}{z_0}\right)\right]\exp\left\{-i\left(K - \frac{1}{2Kz_0^2}\right)vt\right\}. \quad (8.5.33)$$

These solutions describe the standing solitary wave inside the non-linear DFB-structure. The real envelopes of these waves are equal, but the phase fronts are different.

Let us consider another solution of the NLS equation, i.e., the running one-soliton pulse. This solution has a form

$$\tilde{\psi}(z,t) = \psi_0 \operatorname{sech}\left(\frac{z - Vt}{z_0}\right)\exp[i\varphi_s(z,t)],$$

where

$$\varphi_s(z,t) = \left(\frac{V}{v}\right)Kz + \left[\left(\frac{1}{Kz_0}\right)^2 - \left(\frac{V}{v}\right)^2\right]\frac{vK}{2}t.$$

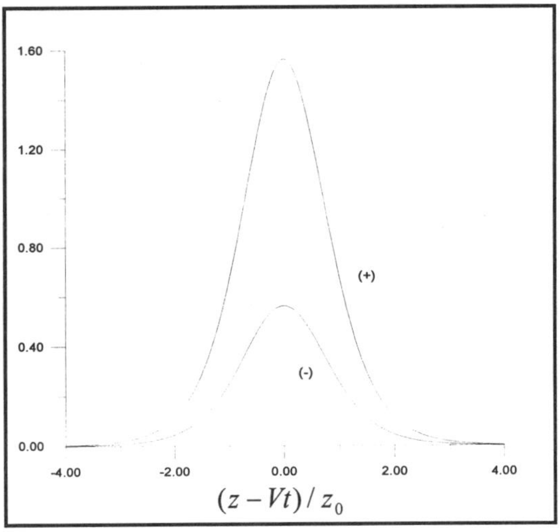

Fig. 8.5.1.

Square of the modulus of envelopes: (+) for $A^{(+)}$ and (-) for $A^{(-)}$.

In this case the envelopes of the counter-propagating waves are

$$A^{(\pm)} = 2\psi_0 \operatorname{sech}\left(\frac{z - Vt}{z_0}\right)\left[\left(1 \pm \frac{V}{2v}\right) \pm \frac{i}{2Kz_0}\tanh\left(\frac{z - Vt}{z_0}\right)\right]\exp\{-i\varphi_s(z,t) - iKvt\}.$$

In Fig. 8.5.1 we can see these envelopes of the pulses propagating inside the non-linear DFB-structure at the $V/v = 0{,}5$ and $Kz_0 = 2$.

8.6. Conclusion

In this chapter we consider the simplest examples of the non-linear surface wave and guided wave in planar waveguides. These examples have been chosen for their convenience to investigation by the analytical methods in all detail. The case of the non-Kerr-like planar waveguide is more difficult to investigate only by analytical methods. The computer simulation is the main approach in this field.

It is worth noting that the two-dimensional optical beams in thin film non-linear waveguides can be described in the framework of Nonlinear Schrödinger equation (see section 6.1). This equation has exact solutions, which we can interpret as spatial optical solitons. It is one more important example where the completely integrable model takes the physical meaning. We can say the same about the non-collinear second harmonic generation (section 7.2.4) or parametric three-wave interaction in a planar waveguide with the second-order nonlinearity. The SHG process has been analysed theoretically by evaluating the total electromagnetic fields that satisfy the polarisation driven wave equations and the boundary conditions at both interfaces [168].

Yet, another field of modern investigation of the waveguide structures is the non-linear waveguide arrays - NOWA (Sect.6.3.2). If the group-velocity dispersion can be omitted, then the system of equations describing this structure

$$\frac{dA_n}{dz} = iK(A_{n+1} + A_{n-1}) + ir\,|A_n|^2\,A_n \tag{8.6.1}$$

is the generation of equations (8.4.1). The total electric field is represented as

$$E(x,y,z) = \sum_{n=-\infty}^{\infty} A_n(z)\Psi^{(n)}(x,y)\exp[-i\omega_0 + i\beta^{(n)}z],$$

where we assumed that all channels of NOWA are identical. The details are described in section 6.3.2. Equations (8.6.1) are the particular case of the system of equations describing the localised states in NOWA [169,170]. Strongly localised solitons in this structure were investigated in [171] by both analytical and numerical methods. It has been found that there are the stable narrow soliton-like discrete configurations of the amplitudes A_n. For instance, they are the "odd solitons", where

$$A_n = A(\ldots,0,\alpha_2,\alpha_1,1,\pm\alpha_1,\pm\alpha_2,0,\ldots)\exp\{i(rA^2 + 2K^2/rA^2)z\}, \tag{8.6.2}$$

for symmetric distribution with $\alpha_1 = (rA^2/K)^{-1}$, $\alpha_2 = (rA^2/K)^{-2}$, or

$$A_n = A(\ldots,0,\alpha_2,\alpha_1,0,\pm\alpha_1,\pm\alpha_2,0,\ldots)\exp\{i(rA^2 + K^2/rA^2)z\}, \qquad (8.6.3)$$

for anti-symmetric distribution with $\alpha_1 = 1$, $\alpha_2 = (rA^2/K)^{-1}$. The "even solitons", corresponding to distributions

$$A_n = A(\ldots,0,\alpha,1,\pm 1,\pm\alpha,0,\ldots)\exp\{i(rA^2 \pm K + K^3/r^2A^4)z\}, \qquad (8.6.4)$$

where

$$\alpha = (rA^2/K)^{-1} \pm (rA^2/K)^{-2},$$

are always unstable, and are converted into "odd solitons".

The system of equations (8.6.1) is not completely integrable and there is no soliton in a rigorous means. However, there is an integrable system (viz., Ablowitz-Ladik system)

$$\frac{dA_n}{dz} = iK(A_{n+1} + A_{n-1}) + ir|A_n|^2(A_{n+1} + A_{n-1}). \qquad (8.6.5)$$

This system can be considered as zero-approximation for (8.6.1) [170], so the solutions of the equations (8.6.1) can be obtained in the framework of the perturbation theory.

References

1. Landau, L.D., and Lifshitz E.M.: *Electrodinamika sploshnikh sred*, Nauka, Moscow, 1982, p.425 (in Russian).
2. Shen, Y.R.: *The principles of nonlinear optics*, John Wiley & Sons, New York, Chicester, Brisbane, Toronto, Singapore, 1984.
3. Litvak, A.G., and Mironov, V.A.: Surface waves at the interface of nonlinear media, *Izv. Vyssh.Uchebn.Zaved. Radiofiz.* **11** (1968), 1911-1912.
4. Tomlinson, W.J.: Surface wave at a nonlinear interface, *Opt.Lett.* **5** (1980),323-325.
5. Agranovich, V.M., and Chernyak, V.Ya.: Perturbation theory for weakly nonlinear p-polarized surface polaritons, *Solid Stat.Commun.* **44** (1982), 1309-1311.
6. Agranovich, V.M., and Chernyak, V.Ya.: Dispersion of nonlinear surface polaritons at resonance with oscillations in transition layer, *Phys.Lett.* **A 88** (1982), 423-426.
7. Fedyanin, V.K., and Mihalache, D.: P-polarized nonlinear surface polaritons in layered structures, *Z.Phys.* **B 47** (1982), 167-173.
8. Lomtev, A.I. : New class of S-polarized nonlinear surface waves, *Opt. Spektrosk.* **55** (1983), 1079-1081.
9. Yu, M.Y.: Surface polaritons in nonlinear media, Phys.Rev. **A28** (1983, 1855-1856.
10. Mihalache, D., Stegeman, G.I., Seaton, C.T., Wring, E.M., Zanoni, R., Boardman, A.D., and Twardowski, T.: Exact dispersion relations for transverse magnetic polarized guided waves at a nonlinear interface, *Opt.Lett.* **12** (1987), 187-189
11. Bordman, A.D., Maradudin, A.A., Stegeman, G.I., Twardowski, T., and Wright, E.M.: Exact theory of nonlinear p-polarizad optical waves, *Phys.Rev.* **A35** (1987), 1159-1164.

12. Kushwaha, S.M.: Exact theory of nonlinear surface polaritons: TM case, *Japan J.Appl. Phys.* **29** (1990), Pt.2, L1826-1828.
13. Khadzhi, P.I., and Kiseleva, E.S.: On the theory of p-polarized surface weves, *Zh.Techn.Fiz.* **58** (1988), 1063-1070.
14. Khadzhi, P.I., Frolov L.V., and Kiseleva, E.S.: On the theory of surface TM weves *Pis'ma Zh.Tech.Fiz.* **14** (1988), 1355-1359.
15. Shen, T.P., Maradudin, A.A., and Stegeman G.I.: Low-power, single-interface guided waves mediated by high-power nonlinear guided waves: TE case, *J. Opt. Soc. Amer.* **B5** (1988), 1391-1395.
16. Sarid, D.: The nonlinear propagation constants of a surface plasmon, *Appl. Phys. Lett.* **39** (1981), 889-891.
17. Stegeman, G.I., and Seaton, C.T.: Nonlinear surface plasmons guided by thin metal films, *Opt.Lett.* **9** (1984), 235-237.
18. Lederer, F., and Mihalache, D.: An additional kind of nonlinear s-polarized surface plasmon polaritons, *Solid State Commun.* **59** (1986), 151-153.
19. Michalache, D., Mazilu, D., and Lederer, F.: Nonlinear TE-polarized surface plasmon polaritons guided by metal films, *Opt.Commun.* **59** (1986), 391-394.
20. Mayer, A.P., Maradudin, A.A., Garcia-Molina, R., and Boardman, A.D.: Nonlinear surface electromagnetic waves on a periodically corrugated surface, *Opt.Commun.* **72** (1989), 244-248.
21. Fedyanin, V.K., and Mihalache, D.: Strongly non-linear surface polaritons, *Prepr. JINR*, Dubna, E17-81-121 (1981).
22. Akhmediev, N.N.: Nonlinear theory of surface polaritons, *Zh.Eksp.Teor.Fiz.* **84** (1983), 1907-1917.
23. Boardman, A.D., Yu.V.Gulyaev, and Nikitiv, S.A.: Nonlinear surface magneto-acoustic waves, *Zh.Eksp.Teor.Fiz.* **95** (1989), 2140-2150.
24. Boardman, A.D., Shabat, M.M., and Wallis R.F,: Nonlinear magneto-dynamic waves on magnetic materials, *Phys.Rev.* **B 41** (1990), 717-730.
25. Khadzhi, P.I., and Kiseleva, E.S.: The dispersion relations of the nonlinear surface waves due to the exciton and biexciton formation, *Phys.Stat.Solidi* **B 147** (1988), 741-745.
26. Wachter, C., Langbein, U., and Lederer, F.: Nonlinear waves guided by graded-index films, *Appl.Phys.* **B 42** (1987), 161-164.
27. Boardman, A.D., Twardowski, and Wright E.M.: The effect of diffusion on surface-guided nonlinear TM waves: A finite element approach, *Opt. Commun.* **74** (1990), 347-352.
28. Mihalache, D., and Ruo Peng, W.: Nonlinear surface-guided waves in semi-infinite superlattice media, *Phys.Lett.* **A 132** (1988), 59-63.
29. Mihalache, D., Ruo Peng, W., and Corciovei, A.: TE-polarized nonlinear surface waves in semi-infinite superlattice media, *Rev.Roum.Phys.* **33** (1988), 1191-1195.
30. Akhmediev, N.N., Korneev, V.I., and Kuz'menko, Yu.V.: On stability of nonlinear surface waves, *Pis'ma Zh.Tech.Fiz.* **10** (1984), 780-784.
31. Miranda, J., Andersen, D.R., and Skinner, S.R.: Stability analysis of stationary nonlinear guided waves in self-focusing and self-defocusing Kerr-like layered media, *Phys.Rev.* **A 46** (1992), 5999-6001.
32. Varatharajah, P., Aceves, A., Moloney, J.V., Heatley, D.R., and Wright, E.M.: Stationary nonlinear surface waves and their stability in diffusive Kerr media, *Opt.Lett.* **13** (1988), 690-692.
33. Stegeman, G.I., Seaton, C.T., Ariyasu, J., Wallis, R.F., and Maradudin, A.A.: Nonlinear electromagnetic waves guided by a single interface, *J.Appl.Phys.* **58** (1985), 2453-2459.
34. Nesterov L.A.: Total internal refraction and trapping of planar soliton by nonlinear interface. *Opt.Spektrosk.* **64** (1988) 1166-1168.
35. Bordman, A.D., Cooper, G.S., Maradudin, A.A., and Shen, T.P.: Surface-polariton solitons, *Phys.Rev.* **A 34** (1986), 8273-8278.
36. Hendow S.T., and Ujihara K.: Excitation of surface polaritons by nondegenerate four-wave mixing of evanescent waves, *Opt. Commun.* **45** (1983), 138-142.
37. Glass, N.E., and Rogovin D.: Surface-polariton and guided-wave excitation in thin-film Kerr media, *Phys.Rev.* **B 40** (1989), 1511-1520.

38. Elphick, Ch., and Meron, E. : Localized structures in surface waves. *Phys.Rev.* **A 40** (1989), 3226-3229.
39. Michalache, D., and Totia, H.: On TE-polarized nonlinear waves guided by dielectric layered structures, *Solid State Commun.* **54** (1985), 175-177.
40. Michalache, D., and Mazilu, D.: Calculations of TM-polarized nonlinear waves guided by thin dielectric films, *Appl.Phys.* **B41** (1986), 119-123.
41. Leine, L., Wachter, C., Langbein, U., and Lederer, F.: Evolution of nonlinear guided optical fields down a dielectric film with a nonlinear cladding, *J.Opt.Soc. Amer.* **B5** (1988), 547-558.
42. Ponath, H.-E., Trutschel, U., Langbein, U., and Lederer, F.: Cross trapping of two counter propagating nonlinear guided waves, *J.Opt.Soc.Amer.* **B5** (1988), 539-546.
43. Jakubszyk, Z., Jerominek, H., Tremblay, R., and Delisle, C.: Beam propagation analysis of modal properties of optical nonlinear waveguides, J.de Phys. **49**, Colloq. C2; Suppl. (1988), 311-314.
44. Jakubszyk, Z., Jerominek, H., Tremblay, R., and Delisle, C.: Power-dependent attenuation of TE waves propagating in optical nonlinear waveguiding structures, *IEEE J.Quant.Electron.* **QE-23** (1987), 1921-1927.
45. Robbins, D.J.: TE modes in a slab waveguide bounded by nonlinear media, *Opt.Commun.* **47** (1983), 309-312.
46. Lederer, F., Langbein, U., and Ponath H.-E.: Nonlinear waves guided by a dielectric slab. I. TE-polarization, *Appl.Phys.* **B 31** (1983), 69-73.
47. Lederer, F., Langbein, U., and Ponath H.-E.: Nonlinear waves guided by a dielectric slab. II. TM-polarization, *Appl.Phys.* **B 31** (1983), 187-190.
48. Langbein, U., Lederer, F., Ponath H.-E., and Trutschel, U.: Dispersion relations for nonlinear guided waves, *J.Mol.Struct.* **115** (1984), 493-496.
49. Langbein, U., Lederer, F., Ponath H.-E., and Trutschel, U.: Analysis of the dispersion relation of nonlinear slab-guided waves. Part.II. Symmetric configuration, *Appl.Phys.* **B 38** (1985), 263-268.
50. Stegeman, G.I., Seaton, C.T., Chilwell, J., and Smith, S.D.: Nonlinear waves guided by thin films, *Appl.Phys.Lett.* **44** (1984), 830-832.
51. Boardman, A.D., Egan, P., and Shivarova, A.: TE modes of a layered nonlinear optical waveguide, *Appl.Sci.Research.* **41** (1984), 345-353.
52. Mihalache, D., Nazmitdinov, R.G., and Fedyanin, V.K.: p-Polarized nonlinear surface waves in symmetric layered structures, *Phys.Scr.* **29** (1984), 269-275.
53 Mihalache, D., and Mazilu, D.: TM-Polarized nonlinear waves guided by asymmetric dielectric layered structures, *Appl.Phys.* **B37** (1985), 107-113.
54. Leine L., Wachter Ch., Langbein, U., and Lederer, F.: Propagation phenomena of nonlinear film-guided waves: a numerical analysis, *Opt. Lett.***11** (1986), 590-592.
55. Akhmediev, N.N., Boltar', K.O., and Eleonskii V.M.: Optical waveguide from the dielectric with nonlinear permeability, *Opt.Spektrosk* **53** (1982), 906-909.
56. Akhmediev, N.N., Boltar', K.O., and Eleonskii V.M.: Optical waveguide from the dielectric with nonlinear permeability. Nonsymmetrical profile of refraction index, *Opt.Spektrosk* **53** (1982), 1097-1103.
57. Langbein, U., Lederer, F., and Ponath H.-E.: A new type of nonlinear slab-guided waves, *Opt.Commun.* **46** (1983), 167-169.
58. Mihalache, D., and Corciovel, A.: Cnoidal waves in asymmetric three-layer dielectric structures, *Rev.Roum.Phys.* **30** (1985), 699-708.
59. Boardman, A.D., and Egan, P.: Optically nonlinear waves in thin films, *IEEE J.Quant.Electron.* **QE-22** (1986), 319-324.
60. Chen Wei, Maradudin, A.A.: S-polarized guided and surface electromagnetic waves supported by a nonlinear dielectric film, *J.Opt.Soc.Amer.* **B5** (1988), 529-538.
61. Al-Bader, S.J.: TM waves in nonlinear saturable thin films: A multilayred approach, *J.Lightwave Technol.* **LT-7** (1989), 717-725.

62. Ogusu, K.: TE waves in a symmetric dielectric slab waveguide with a Kerr-like nonlinear permittivity, *Opt. and Quant.Electr.* **19** (1987), 65-72.
63. Ogusu, K.: Nonlinear TE waves guided by graded-index planar waveguides, *Opt. Commun.* **63** (1987), 380-384.
64. Ogusu, K.: Computer analysis of general nonlinear planar waveguides, *Opt. Commun.* **64** (1987), 425-430.
65. Hayata, K., Misawa, A., and Koshida, M.: Nonstationary simulation of nonlinearly coupled TE-TM waves propagating down dielectric slab structures by the step-by-step finite-element method, *Opt. Lett.***15** (1990), 24-26.
66. Ogusu, K.: Stability of a new type of stationary waves guided by a nonlinear hollow waveguide, *Opt.Commun.* **83** (1991), 260-264.
67. Kumar, Ajit, and Sodha, M.S.: Transverse-magnetic-polarized nonlinear modes guided by a symmetric five-layer dielectric structure, *Opt.Lett.* **12** (1987), 352-354.
68. Trutschel, U., Lederer, F., and Golz, M.: Nonlinear guided waves in multilayer systems, *IEEE J.Quant.Electron.* **QE-25** (1989), 194-200.
69. Ogusu, K.: Analysis of non-linear multilayer waveguides with Kerr-like permittivities, *Opt. and Quant.Electron.* **21** (1989), 109-116.
70. Mihalache, D., Mazilu, D., Bertolotti, M., and Sibilia, C.: Exact solution for nonlinear thin-film guided waves in higher-order nonlinear media, *J.Opt.Soc.Amer.* **B5** (1988), 565-570.
71. Ariyasu, J., Seaton, C.T., Stegeman, G.I., Maradudin, A.A., and Wallis, R.F.: Nonlinear surface polaritons guided by metal films, *J.Appl.Phys.* **58** (1985), 2460-2466.
72. Etrich, C.M., Mitchell, C.S., and Moloney, J.V.: Propagation and stability characteristics of nonlinear guided waves in saturable nonlinear media, *J.de Phys.* **49**, Colloq., (1988), 315-317.
73. Stegeman, G.I., Wright, E.M., Seaton, C.T., Moloney, J.V., Shen, T.-P., Maradudin, A.A., and Wallis, R.F.: Nonlinear slab-guided waves in non-Kerr-like media, *IEEE J.Quant.Electr.* **QE-22** (1986), 977-983.
74. Mihalache, D., and Mazilu, D.: TM-polarized nonlinear slab-guided waves in saturable media, *Solid State Commun.* **60** (1986), 397-399.
75. Al-Bader, S.J., and Jamid, H.A.: Guided waves in nonlinear saturable self-focusing thin films, *IEEE J.Quant.Electron.* **QE-23** (1987), 1947-1955.
76. Al-Bader, S.J., and Jamid, H.A.: Nonlinear waves in saturable self-focusing thin films bounded by linear media, *IEEE J.Quant.Electron.* **QE-24** (1988), 2052-2058.
77. Van Wood, E. , Evans, E.D., and Kenan, R.P.: Soluble saturable refractive-index nonlinearity model, *Opt.Commun.* **69** (1988), 156-160.
78. Akhmediev, N.N., Nabiev, R.F., and Popov, Yu.M.: Stripe nonlinear waves in a symmetrical planar structure, *Opt.Commun.* **72** (1989), 190-194.
79. Langbein, U., Lederer, F., and Ponath H.-E.: Generalized dispersion relations for nonlinear slab-guided wave, *Opt.Commun.* **53** (1985), 417-420.
80. Langbein, U., Lederer, F., and Ponath H.-E.: Amplification of slab-guided modes, *Opt. Acta* **31** (1984), 1141-1149.
81. Ariyasu, J., Seaton, C.T., Stegeman, G.I., and Moloney, J.V.: New theoretical developments in nonlinear guided waves: Stability of TE_1 branches, *IEEE J.Quant.Electr.* **QE-22** (1986), 984-987.
82. Moloney, J.V.: Modulation instability of two-transverse-dimensional surface polariton waves in nonlinear dielectric waveguides, *Phys.Rev.* **A36** (1987), 4563-4566.
83. Al-Bader, S.J., and Jamid, H.A.: Stability waves guided by a nonlinear self-focusing saturable film bounded by linear media, *Opt. Commun.* **66** (1988), 88-92.
84. Hayata, K., and Koshiba, M.: Self-focusing instability and chaotic behavior of nonlinear optical waves guided by dielectric slab structures, *Opt.Lett.* **13** (1988), 1041-1043.
85. Martijn de Sterke, C., and Sipe, J.E.: Polarization instability in a waveguide geometry, *Opt.Lett.* **16** (1991), 202-204.

86. Moloney, J.V., Ariyasu, J., Seaton, C.T., and Stegeman, G.I.: Numerical evidence for nonstationary, slab-guided waves, *Opt.Lett.* **11** (1986), 315-317.
87. Gubbels, M.A., Wright, E.M., Stegeman, G.I., and Seaton, C.T.: Numerical study of soliton emission from a nonlinear waveguide, *J.Opt.Soc.Amer.* **B 4** (1987), 1837-1842.
88. Heatley, D.R., Wright, E.M., and Stegeman, G.I.: Solitary wave emission from a nonlinear slab waveguide in three dimensions, *Appl.Phys.Lett.* **56** (1990), 215-217.
89. Heatley, D.R., Wright, E.M., and Stegeman, G.I.: Numerical calculations of spatially localized wave emission from a nonlinear waveguide: Low-level saturable media, *J.Opt.Soc.Amer.* **B 7** (1990), 990-997.
90. Wright, E.M., Heatley, D.R., and Stegeman, G.I.: Emission of spatial solitons from nonlinear waveguides, *Phys.Rept.* **C 194** (1990), 309-323.
91. Vach, H., Seaton, C.T., Stegeman, G.I., and Khoo, I.C.: Observation of intensity-dependent guided waves, *Opt. Lett.***9** (1984), 238-240.
92. Seaton, C.T., Xu Mai, Stegeman, G.I., and Winful, H.G.: Nonlinear guided wave applications, *Opt.Engin.* **24** (1985), 593-599.
93. Jones, C.K.R.T., and Moloney, J.V.: Instability of standing waves in nonlinear optical waveguides, *Phys.Lett.* **A 117** (1986), 175-180.
94. Mihalache, D., Nazmitdinov, R.G., and Fedyanin, V.K.: Nonlinear optical waves in layered structures, *Fiz. Elem.Chast.and Atom Yadr.* **20** (1989), 198-253 [Russian]
95. Shen, Y.R.: Nonlinear interaction of surface polaritons, *Phys.Rept.* **C 194** (1990), 303-308.
96. Stegeman, G.I., and Seaton, C.T.: Nonlinear Integrated Optics, *J.Appl.Phys.* **58** (1985), R57-R78.
97. Maradudin, A.A.: s-Polarized nonlinear surface polaritons. *Zs.Phys.* **B41** (1981), 341-348.
98. *Guided-Wave Optoelectronics*, ed. T. Tamir., Springer-Verlag,, Heidelberg, 1988.
99. Langbein, U., Lederer, F., Ponath H.-E., and Trutschel, U.: Analysis of the dispersion relations of nonlinear slab-guided waves. I. Asymmetrical configuration. *Appl.Phys.* **B 36** (1985), 187-193.
100. Boardman, A.D., and Egan, P.: S-polarized waves in a thin dielectric film asymmetrically bounded by optically nonlinear media, *IEEE J.Quant.Electron.* **QE-21** (1985), 1701-1713,
101. Boardman, A.D., and Egan, P.: Theory of optical hysteresis for TE guided modes, *Phil.Trans.Royl Soc. London*, **A 313** (1984), 363-369.
102. *Higher Transcendental Functions*, eds. H.Bateman and A.Erdelyi, McGraw-Hill, New York, 1955, vol. 3.
103. Mihalache, D., and Fedyanin, V.K.: P-polarized nonlinear surface and connected waves in layered structures. *Teor.Mat.Fiz.* **54** (1983), 443-455.
104. Tien, P.K.: Integrated optics and new wave phenomena in optical waveguides, *Rev.Mod.Phys.* **49** (1977), 361-420
105 Jensen, S.M.: The nonlinear coherent coupler, *IEEE J.Quant.Electron.* **QE-18** (1982), 1580-1583.
106. Sarid, D., and Sargent, M III,: Tunable nonlinear directional coupler, *J.Opt. Soc. Amer.* **72** (1982), 835-838.
107. Maier A.A., Optical transistors and bistability elements on base of nonlinear light transfer by systems with unidirectional connected waves, *Kvant .elektron.* **9** (1982), 2296-2302 (in Russia)
108. Trillo, S., and Wabnitz, S.: Nonlinear nonreciprocity in a coherent mistached directional coupler, *Appl.Phys.Lett.* **49** (1986), 752-754.
109. Ankiewicz, A.: Interaction coefficient in detuned nonlinear couplers, *Opt. Commun.* **66** (1988), 311-314
110. Mitchell, D.J., and Snyder, A.W.: Modes of directional couplers: building blocks for physical insight, *Opt.Lett.* **14** (1989), 1143-1145.
111. Chang, H., and Tripathi, V.K.: Extended matrix method for the analysis of nonlinear directional couplers with saturable coupling media, *Opt.Engin.* **32** (1993), 735-738.
112. Peng G.-D., and Ankiewicz, A.: Intensity-depending phase shifts in nonlinear coupling devices, *J.Mod.Opt.* **37** (1990), 353-365.

113. Yuan, L.-P.: A unified approach for the coupled-Mode analysis of Nonlinear planar optical couplers, *IEEE J.Quant.Electr.* **QE-30** (1994), 126-133.
114. Bertolotti, M, Fazio, E,. Ferrari, A., Michelotti, F., Righini, G.C., and Sibilia, C.: Nonlinear coupling in a planar waveguide, *Opt. Lett.* **15** (1990), 425-427.
115. Townsend P.D., Baker, G.L., Jackel, J.L., Shelbume, J.A., III, and Elemad, S.: Polydiacetylene-based directional couplers and grating couplers: Linear and nonlinear transmission properties and all-optical switching phenomena, *Proc.Soc. Photo-Opt. Instrum.Eng.* **1147** (1990), 256-264.
116. Aitchison, J.S., Villeneuve A., and Stegeman G.I.: *Opt.Lett.* **18** (1993), 1153-1155.
117. Artigas, D., and Dios, F.: Phase space description of nonlinear directional couplers, *IEEE J.Quant.Electr.* **QE-30** (1994), 1587-1595.
118. Leutheuser, V., Langbein, U., and Lederer, F.: Optical response of a nonlinear bend directional coupler, *Opt. Commun.* **75** (1990), 251-255.
119. Jackel, J.L.: Effects of inhomogeneities on nonlinear integrated optical components, *IEEE J.Quant.Electr.* **QE-26** (1990), 622-626.
120. Stegeman, G.I., Seaton, C.T., Ironside, C.N., Cullen, T., and Walker, A.C.: Effects of saturation and loss on nonlinear directional couplers, *Appl.Phys.Lett.* **50** (1987), 1035-1037.
121. Stegeman, G.I., Caglioti, E., Trillo, S., and Wabnitz, S.: Parameter trade-offs in nonlinear directional couplers: Two level saturable nonlinear media, *Opt. Commun.* **63** (1987), 281-284.
122. Caglioti, E., Trillo, S., Wabnitz, S., and Stegeman, G.I.: Limitations to all-optical switching using nonlinear couplers in the presence of linear and nonlinear absorption and saturation, *J.Opt.Soc.Amer.* **B5** (1988), 472-482.
123. Caglioti, E., Trillo, S., Wabnitz, S., Daino, B., and Stegeman, G.I.: Power-dependent switching in a coherent nonlinear directional coupler in the presence of saturation, *Appl.Phys.Lett.* **51** (1987), 293-295.
124. Heatley, D.R., Wright, E.M., Ehrlich, J., and Stegeman, G.I.: Nonlinear directional coupler with a diffusive Kerr-type nonlinearity, *Opt.Letts.* **13** (1988), 419-421.
125. Wright, E.M., Heatley, D.R., Stegeman, G.I., and Blow, K.J.: Variation of the switching power with diffusion length in a nonlinear directional coupler, *Opt. Commun.* **73** (1989), 385-392.
126. DeLong, K.W., and Stegeman, G.I.: Two-photon absorption as a limitation to all-optical waveguide switching in semiconductors, *Appl.Phys.Lett.* **57** (1990), 2063-2064.
127. Goldburt, E.S., and Russell, P.St.J.: Nonlinear single-mode fiber coupler using liquid crystal, *Appl.Phys.Lett.* **46** (1985), 338-340.
128. Daino, B., Gregori, G., and Wabnitz, S.: Stability analysis of nonlinear coherent coupling, *J.Appl.Phys.* **58** (1985), 4512-4514.
129. Silberberg, Y., and Stegeman, G.I.: Nonlinear coupling of waveguide modes, *Appl.Phys.Lett.* **50** (1987), 801-803.
130. Wabnitz, S., Wright, E.M., Seaton, C.T., and Stegeman, G.I.: Instabilities and all-optical phase-controlled switching in a nonlinear directional coherent coupler, *Appl.Phys.Lett.* **49** (1986), 838-840.
131. Stegeman, G.I., Assanto, G., Zanoni, R., Seaton, C.T., Garmire, E., Maradudin, A.A., Reinisch, R., and Vitrant, G.: Bistability and switching in nonlinear prism coupling, *Appl.Phys.Lett.* **52** (1988), 869-871.
132. Lederer, F., Bertolotti, M., Sibilia, C., and Leutheuser, V.: An external controled nonlinear directional coupler, *Opt. Commun.* **75** (1990), 246-250.
133. Catchmark, J.M., and Christodoulides, D.N.: Switching in the far-field of a nonlinear directional coupler, *Opt. Commun.* **84** (1991), 14-17.
134. Li Kam Wa, P., Miller, A., Park, C.B., Roberts, J.S., and Robson, P.N.: All-optical switching of picosecond pulses in a GaAs quantum well waveguide coupler, *Appl.Phys.Lett.* **57** (1990), 1846-1848.
135. Christodoulides, D.N., and Joseph, R.I.: Discrete self-focusing in nonlinear arrays of coupled waveguides, *Opt.Letts.* **13** (1988), 794-796.

136. Kivshar, Yu.S.: Self-localization in arrays of defocusing waveguides, *Opt.Letts.* **18** (1993), 1147-1149.
137. Aceves, A.B., Luther, G.G., DeAngelis, C., Rubenchik, A.M., and Turitsyn, S.K.: Energy localization in nonlinear fiber arrays: collapse-effect compressor, *Phys.Rev.Lett.* **79** (1995), 73-76.
138. Schmidt-Hattenberger, C., Trutschel, U., and Lederer, F.: Nonlinear switching in multiple-core couplers, *Opt.Letts.* **16** (1991), 294-296.
139. Mann, M., Trutschel, U., Lederer, F., Leine, L., and Wachter, C.: Nonlinear leaky waveguide modulator, *J.Opt.Soc.Amer.* **B8** (1991), 1612-1617.
140. Buryak, A.V., Akhmediev, N.N.: Stationary pulse propagation in N-core nonlinear fiber arrays, *IEEE J.Quant.Electron.* **QE-31** (1995), 682-688.
141. Darmanyan, S., Relke, I., and Lederer, F.: Instability of continuous waves and rotating solitons in waveguide arrays, *Phys.Rev.* **E55** (1997), 7662-7668.
142. Darmanyan, S., Kobyakov, A., and Lederer, F.: Stability of strongly localized excitation in discrete media with cubic nonlinearity, *Zh.Eksp.Teor.Fiz.* **113** (1998), 1253-1261 [in Russia].
143. Mecozzi, A.: Parametric amplification and squeezed-light generation in a nonlinear directional coupler, *Opt.Letts.* **13** (1988), 925-927.
144. Elachi, C., and Yeh, P.: Periodic structures in integrated optics, *J.Appl.Phys.* **44** (1973), 3146-3152.
145. Yeh, P., and Yariv, A.: Bragg reflection waveguides, *Opt. Commun.* **19** (1976), 427.
146. Stoll, H., and Yariv, A.: Coupled mode analysis of periodic dielectric waveguides, *Opt. Commun.* **8** (1973), 5.
147. Rosanov, N.N., and Smirnov, V.A.: On the theory of plane wave propagation in nonlinear layered systems, *Pis'ma Zh.Teh.Fis.* **5** (1979), n.9, 544-548.
148. Okuda, M., Toyota, M., and Onaka, K.: Saturable optical resonators with distributed Bragg-reflectors, *Opt. Commun.* **19** (1976), 138-142.
149. Trillo, S., Wabnitz, S., and Stegeman, G.I.: Nonlinear codirectional guided wave mode conversion in grating structures, *J.Lightwave Technol.* **6** (1988), 971-976.
150. Grabowski, M., and Hawrylak, P.: Wave propagation in a nonlinear periodic medium, *Phys.Rev.* **B41** (1990), 5783-5791.
151. Assanto, G., Marques, M.B., and Stegeman, G.I.: Grating coupling of light pulses into third-order nonlinear waveguides, *J.Opt.Soc.Amer.* **B8** (1991), 553-561.
152. Iizuka, T.: Envelope soliton of the Bloch wave in nonlinear periodic systems, *J.Phys.Soc.Japan* **63** (1994), 4343-4349.
153. Radic, S., George, N., and Agrawal, G.P.: Analysis of nonuniform nonlinear distributed feedback-structures: Generalized transfer matrix method, *IEEE J.Quant.Electron.* **QE-31** (1995), 1326-1336.
154. Winful, H.G., Marburger, J.H., and Garmire, E.: Theory of bistability in nonlinear distributed feedback-structures, *Appl.Phys.Lett.* **35** (1979), 379-381.
155. Winful, H.G., and Cooperman, G.D.: Self-pulsing and chaos in distributed feedback bistable optical devices, *Appl.Phys.Lett.* **40** (1982), 298-300.
156. Ehrlich, J.E., Assanto, G., and Stegeman, G.I.: Butterfly bistability in grating coupler thin film waveguides, *Opt. Commun.* **75** (1990), 441-446.
157. Assanto, G., Ehrlich, J.E., and Stegeman, G.I.: Feedback-enhanced bistability in grating coupled into InSb waveguides, *Opt.Letts.* **15** (1990), 411-413.
158. Shi, Ch.-X.: Optical bistability in reflective fiber gratings, *IEEE J.Quant.Electron.* **QE-31** (1995), 2093-2100.
159. Christodoulides, D.N., and Joseph, R.I.: Slow Bragg solitons in nonlinear periodic structures, *Phys.Rev.Lett.* **62** (1989), 1746-1749.
160. Marnijn de Sterke, C., and Sipe, J.E.: Extensions and generalizations of an envelope-function approach for the electrodynamics of nonlinear periodic structures, *Phys.Rev.* **A39** (1989), 5163-5178.
161. Thirring, W.E.: A soluble relativistic field theory, *Ann.Phys.* **3**, (1958), 91-112.

162. Mikhailov, A.V.: On integrability of the two-dimensional Thirring model, *Pis'ma Zh.Eksp.Teor.Fis.* **23** (1976), 356-358.
163. Kaup, D.J., and Newell, A.C.: On the Coleman correspondence and the solution of the massive Thirring model, *Lett. Nuovo Cimento* **20** (1977), 325-331.
164. Aceves, A.B., and Wabnitz, S.: Self-induced transparency solitons in nonlinear refractive pereodic media, *Phys.Letts.* **A 141** (1989), 37-42.
165. Malomed, B.A., and Tasgal, R.S.: Vibration modes of a gap soliton in a nonlinear optical medium, *Phys.Rev.* **E 49** (1994), 5787-5796.
166. Bjorken, J.D., and Drell, S.D.: *Relativistic Quantum Mechanics;* McGraw-Hill Book Company, New York, 1965.
167. Toyama, F.M., Hosono, Y., Ilyas, B., and Nogami, Y.: Reduction of the nonlinear Dirac equation to a nonlinear Schrodinger equation with a correction terms, *J.Phys.* **A 27** (1994), 3139-3148.
168. So, V.C.Y., Normandin, R., and Stegeman G.I.: Field analysis of harmonic generation in thin-film integrated optics, *J.Opt.Soc.Amer.* **69** (1979), 1166-1171.
169. Christodoulides, D.N., and Joseph, R.I.: Discrete self-focusing in non-linear arrays of coupled waveguides, *Opt.Lett.* **13** (1988), 794-796.
170. Cai, D., Bishop, A.R., and Gronbech-Jensen, N.: Localised states in discrete nonlinear Schrödinger equations, *Phys.Rev.Lett.* **72** (1994), 591-595.
171. Darmanyan, S., Kobyakov, A., and Lederer, F.: Stability of strongly localised excitation in discrete media with cubic nonlinearity. *Zh.Eksp.Teor.Fiz.* **113** (1998), 1253-1261.

THIN FILM OF RESONANT ATOMS: A SIMPLE MODEL OF NONLINEAR OPTICS

A thin film of resonant atoms with a thickness much less than the wavelength of incident light represents a very comprehensive model for studies of nonlinear surface waves [1,2] (see section 3.3) and nonlinear reflection of ultrashort optical pulses [3-5], which can be treated analytically. These phenomena are typical for Hamiltonian systems. Moreover, the model of a thin film of resonant atoms makes it possible to deal also with another class of nonlinear optical phenomena associated with nontrivial dynamics of open dissipative systems. The most familiar phenomena of this kind are optical bistability [6,7] and spontaneous pulsations (self-pulsations) [6,8]. The various types of coherent transients exhibiting new interesting peculiarities appear in the field of interactions of resonant light pulses with thin films [9-19]. On the other hand, thin film is a simple example of low dimensional systems attracting great attention in the last years.

The description of nonlinear regime of resonant interaction of incident light wave with a thin film is characterised by two essential features, which considerably differ the treatment of optical resonance in a thin film from that of extended medium. The first feature consists in accounting for a specific Lorentz field, which describes the distinction of the local field acting on an atom from the macroscopic field in the medium [20,7]. The account for the Lorentz field leads to the dynamical change of the condition of resonant interaction, namely, the detuning from resonance under coincidence of wave and atomic transition frequencies becomes proportional to both the Lorentz parameter and the population difference. This reveals as a specific type of relaxation. In the time interval less than the time of the irreversible relaxation the Rabi oscillations will be suppressed. A stationary bistable regime of interaction of an ultrashort pulse of a rectangular form with the thin film arises. Another manifestation of the Lorentz field occurs in the properties of differential equations of the model under consideration. The Lorentz field destroys the property of exact integrability of these equations. We should remind that in extended medium (except for complete optically dense medium) the effect of Lorentz field is sufficiently small.

We can ignore the Lorentz field and analyse the model equations since they coincide in some cases with the equations describing the transmission of a plane wave through a low-Q Fabry-Perot resonator with resonant atoms within the mean field approximation [6,21]. There are some other situations in which the model equations without the allowance for the Lorentz field are valid. In this model situation the inverse scattering transform method can be applied to study the nonlinear reflection of ultrashort optical pulses [4,5]. In the case of double resonance one can satisfy the condition of simulton formation in the interaction of two-frequency ultrashort pulse with the thin film by selecting the appropriate incidence angles. Attention should be drawn to the applicability of the IST method to the system of ordinary differential equations describing the reflection of pulses from the thin film.

The second feature of nonlinear effects in the thin film reveals both in two-photon and higher order interactions. There is no requirement for the condition of spatially wave synchronism in the thin film. Therefore, with the account of reverse influence of resonant atoms on the incident light wave, it is necessary to include into consideration the effect of third harmonic generation along with the Raman scattering of this harmonics.

In section 9.1 we will derive the main relations between the amplitudes of incident, reflected and refractive waves and film polarisation. Section 9.2 is devoted to the description of optical bistability and self-pulsation in one-photon resonant interactions. The two-photon interaction of incident light wave with thin film is considered in section 9.3. In section 9.4 we apply the method of the inverse scattering transform to examine the passage of an ultrashort two-frequency light pulse through a thin film under condition of double resonance.

9.1. Thin film at the interface of two dielectric media

Let a thin film of atoms resonantly interact with electromagnetic wave locate on the interface of two dielectric media in the x-plane. The dielectric media adjoining the film is characterised by the dielectric constant ε_1 for $x<0$ and ε_2 for $x>0$. Axis z is chosen to be in the interface plane. The resonant atoms are described in the framework of the effective Hamiltonian model.

Since the interface is planar, the system of Maxwell equations breaks up into two independent systems for TE waves $\vec{E}=(0,E_y,0)$, $\vec{H}=(H_x,0,H_z)$, and TM waves $\vec{E}=(E_x,0,E_z)$, $\vec{H}=(0,H_y,0)$.

The reflected wave runs back in region $x<0$ direction, whereas the refracted (transmitted) wave propagates in $x>0$ direction. We present the field strengths $\vec{E}$, $\vec{H}$ and polarisation of resonant atoms inside the film in the form

$$\vec{E}(x,z,t)=\int_{-\infty}^{\infty}\frac{d\omega}{2\pi}\frac{d\beta}{2\pi}\exp(-i\omega t+i\beta z)\tilde{\vec{E}}(x,\beta,\omega),$$

$$\vec{H}(x,z,t)=\int_{-\infty}^{\infty}\frac{d\omega}{2\pi}\frac{d\beta}{2\pi}\exp(-i\omega t+i\beta z)\tilde{\vec{H}}(x,\beta,\omega),$$

$$\vec{P}(z,t)=\int_{-\infty}^{\infty}\frac{d\omega}{2\pi}\frac{d\beta}{2\pi}\exp(-i\omega t+i\beta z)\tilde{\vec{P}}(\beta,\omega).$$

Outside the film the Fourier components of vectors $\vec{\widetilde{E}}(x,\beta,\omega)$ и $\vec{\widetilde{H}}(x,\beta,\omega)$ are defined by the Maxwell equation, and at $x=0$ they follow from the continuity conditions. As a result we obtain the system of equations

for *TE*-mode

$$\frac{d^2\widetilde{E}}{dx^2}+\left(k^2\varepsilon_j-\beta^2\right)\widetilde{E}=0,$$
$$\widetilde{H}_x=-(\beta/k)\widetilde{E},\quad \widetilde{H}_z=-\frac{i}{k}\frac{d\widetilde{E}}{dx}\equiv\widetilde{H},\ \widetilde{E}\equiv\widetilde{E}_y \tag{9.1.1a}$$

with boundary conditions

$$\widetilde{E}(x=0-)=\widetilde{E}(x=0+),\quad \widetilde{H}_z(x=0+)-\widetilde{H}_z(x=0-)=4\pi ik\widetilde{P}_y(\beta,\omega) \tag{9.1.1b}$$

for *TM*-mode

$$\frac{d^2\widetilde{H}}{dx^2}+\left(k^2\varepsilon_j-\beta^2\right)\widetilde{H}=0,$$
$$\widetilde{E}_x=-(\beta/k\varepsilon_j)\widetilde{H},\quad \widetilde{E}_z=\frac{i}{k\varepsilon_j}\frac{d\widetilde{H}}{dx},\ \widetilde{H}=\widetilde{H}_y \tag{9.1.2a}$$

with boundary conditions

$$\widetilde{E}_z(x=0-)=\widetilde{E}_z(x=0+)\equiv\widetilde{E}(\beta,\omega),\quad \widetilde{H}(x=0+)-\widetilde{H}(x=0-)=4\pi ik\widetilde{P}_z(\beta,\omega) \tag{9.1.2b}$$

Here, $j=1,2$ and $k=\omega/c$. Outside the film the solution of eqs. (9.1.1a) and (9.1.2a) with the account of asymptotic behaviour takes the form

for *TE*-mode

$$\widetilde{E}(x,\beta,\omega)=\begin{cases}A(\beta,\omega)\exp(iq_1x)+B(\beta,\omega)\exp(-iq_1x), & x<0\\ C(\beta,\omega)\exp(iq_2x), & x>0\end{cases}$$

$$\widetilde{H}(x,\beta,\omega)=\begin{cases}q_1k^{-1}\{A(\beta,\omega)\exp(iq_1x)-B(\beta,\omega)\exp(-iq_1x)\}, & x<0\\ q_2k^{-1}C(\beta,\omega)\exp(iq_2x), & x>0\end{cases};$$

for *TM*-mode

$$\widetilde{H}(x,\beta,\omega)=\begin{cases}A(\beta,\omega)\exp(iq_1x)+B(\beta,\omega)\exp(-iq_1x), & x<0\\ C(\beta,\omega)\exp(iq_2x), & x>0\end{cases}$$

$$\widetilde{E}(x,\beta,\omega)=\begin{cases}-q_1(k\varepsilon_1)^{-1}\{A(\beta,\omega)\exp(iq_1x)-B(\beta,\omega)\exp(-iq_1x)\}, & x<0\\ -q_2(k\varepsilon_2)^{-1}C(\beta,\omega)\exp(iq_2x), & x>0\end{cases};$$

where $q_j=\sqrt{k^2\varepsilon_j-\beta^2}$, $j=1,2$.

Boundary conditions at $x=0$ (9.1.1b) and (9.1.2b) provide the relationship between the amplitudes of incident *A*, reflected *B*, refracted (transmitted) *C* waves and the film polarisation $P_S=\widetilde{P}_y$ (*TE*-mode) and $P_S=\widetilde{P}_z$ (*TM*-mode)

for *TE*-mode

$$C(\beta,\omega)=\frac{2q_1}{q_1+q_2}A(\beta,\omega)+i\frac{4\pi k^2}{q_1+q_2}P_S(\beta,\omega),$$

$$B(\beta,\omega)=\frac{q_1-q_2}{q_1+q_2}A(\beta,\omega)+i\frac{4\pi k^2}{q_1+q_2}P_S(\beta,\omega). \quad (9.1.3)$$

for *TM*-mode

$$C(\beta,\omega)=\frac{2\varepsilon_2q_1}{\varepsilon_2q_1+\varepsilon_1q_2}A(\beta,\omega)-i\frac{4\pi k\varepsilon_2q_1}{\varepsilon_2q_1+\varepsilon_1q_2}P_S(\beta,\omega)$$

$$B(\beta,\omega)=\frac{\varepsilon_2q_1-\varepsilon_1q_2}{\varepsilon_2q_1+\varepsilon_1q_2}A(\beta,\omega)+i\frac{4\pi k\varepsilon_1q_2}{\varepsilon_2q_1+\varepsilon_1q_2}P_S(\beta,\omega) \quad (9.1.4)$$

Instead of (9.1.4), we need the relations between the electric-field amplitudes $\widetilde{E}$. Defining the electric field amplitude of incident wave $\widetilde{E}^{in}$ as

$$\widetilde{E}^{in}(\beta,\omega)=-q_1A/k\varepsilon_1,$$

instead of (9.1.4) we obtain

$$\widetilde{E}^{tr}(0,\beta,\omega)=\frac{2q_2\varepsilon_1}{\varepsilon_2q_1+\varepsilon_1q_2}\widetilde{E}^{in}(\beta,\omega)+\frac{4\pi iq_1q_2}{\varepsilon_2q_1+\varepsilon_1q_2}P_S(\beta,\omega). \quad (9.1.5)$$

It is convenient to introduce the notation for Frenel coefficient of transmission and for the coupling constant as follows

$$T_{TE}(\beta,\omega)=\frac{2q_1}{q_1+q_2},\quad \kappa_{TE}(\beta,\omega)=\frac{4\pi k^2}{q_1+q_2},$$
$$T_{TM}(\beta,\omega)=\frac{2\varepsilon_1 q_2}{\varepsilon_2 q_1+\varepsilon_1 q_2},\quad \kappa_{TM}(\beta,\omega)=\frac{4\pi q_1 q_2}{\varepsilon_2 q_1+\varepsilon_1 q_2}. \tag{9.1.6}$$

The combination of the results (9.1.3) and (9.1.5) into a single expression yields

$$\widetilde{E}_A^{tr}(\beta,\omega)=T_A(\beta,\omega)\widetilde{E}_A^{in}+i\kappa_A(\beta,\omega)\widetilde{P}_A(\beta,\omega). \tag{9.1.7}$$

Here, for *TE*-mode, $A=TE$ and

$$\widetilde{E}_A^{tr}=\widetilde{E}_y^{tr},\quad \widetilde{E}_A^{in}=\widetilde{E}_y^{in},\quad \widetilde{P}_A=\widetilde{P}_y,$$

and for *TM*-mode, $A=TM$ and

$$\widetilde{E}_A^{tr}=\widetilde{E}_z^{tr},\quad \widetilde{E}_A^{in}=\widetilde{E}_z^{in},\quad \widetilde{P}_A=\widetilde{P}_z.$$

The indices *tr* and *in* correspond to transmitted and incident waves.

Relation (9.1.7) is general. It is independent from the type of resonant atom and from the frequency spectrum of an incident wave, and is valid for both plane and nonplane waves. Though we shall consider here only the case when there is no total internal reflection $\varepsilon_1 \le \varepsilon_2$, expression (9.1.7) can be extended to include the case $\varepsilon_1 > \varepsilon_2$. At an incidence angle exceeding the critical angle of total internal reflection the quantity q_2 in (9.1.7) should be replaced for $i(\beta^2-k^2\varepsilon_2)^{1/2}=iq'_2$.

Equations for slowly varying amplitudes can be derived in the same manner as it was done in section 1.3.1. By neglecting the dispersion of coefficients (9.1.6) and using equations like (1.3.4), one can obtain the same as (9.1.7) relation between slowly varying amplitudes of electric fields and film polarisation. The parameters of the corresponding monochromatic waves will be the arguments in coefficients (9.1.6).

Below we will consider the refracted wave in the case $\varepsilon_1 \le \varepsilon_2$ when the total internal reflection is absent at any incidence angle $\theta^{in}=\arccos(q_1/k\sqrt{\varepsilon_1})$, and the refractive angle θ_{tr} is determined by Snell relation:

$$\sin\theta^{tr}=(\beta/k\sqrt{\varepsilon_2})=\sqrt{\varepsilon_1/\varepsilon_2}\sin\theta^{in}.$$

If the film polarisation is found, the eq.(9.1.7) determines the fields in the whole space. Eq.(9.1.7) is the exact one because there are no assumptions about the variation of the field amplitude. To find film polarisation it is necessary to choose a specific model of resonant medium.

9.2. Optical bistability and self-pulsation in thin film under the condition of one-photon resonance

We assume that a thin film of two-level atoms, evaporated on an insulating substrate of thickness L, is illuminated normally by a plane electromagnetic wave (Fig.9.2.1) and the electric field of this wave is

$$E = \mathcal{E}_I \exp[i(kx - \omega t)] + c.c.,\ x<0.$$

The equations describing the resonant interaction of light with such a film follows from (1.1.12) and (9.1.7) under the assumption of slowly varying field amplitude. If we allow for the wave reflected from $x = L$ surface of the insulating substrate, we can rewrite these equations in the following dimensionless form

$$\left(\frac{d}{d\tau} - i\Delta + \gamma_0\right)p(\tau) = i\varepsilon(\tau)n(\tau), \tag{9.2.1a}$$

$$\left(\frac{d}{d\tau} + \gamma\right)n(\tau) = 2i\{\varepsilon^*(\tau)p(\tau) - \varepsilon(\tau)p^*(\tau)\} + \gamma, \tag{9.2.1b}$$

$$\varepsilon(\tau) = \psi(\tau) + Re^{is}\psi(\tau - \tau_0), \tag{9.2.2a}$$

$$p(\tau) = -i\{\psi(\tau) - a(\tau) - \nu Re^{is}\psi(\tau - \tau_0)\}, \tag{9.2.2b}$$

where

$$n = (\rho_{aa} - \rho_{bb})N_0^{-1},\ p = \rho_{ba}d_{ba}^*\left|d_{ba}^*\right|^{-1}N_0^{-1}\exp(i\omega t),\ a(\tau) = 2E_I\big|_{x=0}\{(1+n_D)\}^{-1},$$

$$\varepsilon(\tau) = (\mathcal{E}_I + \mathcal{E}_R)\big|_{x=0}\varepsilon_0^{-1} = (\mathcal{E}_F + \mathcal{E}_B)\big|_{x=0}\varepsilon_0^{-1},\ \psi(\tau) = \mathcal{E}_F\big|_{x=0}\varepsilon_0^{-1},\ \nu = (n_D - 1)(1+n_D)^{-1},$$

$$\tau_0 = 2Ln_D/ct_0,\ \varepsilon_0 = \hbar/t_0|d_{ba}|,\ t_0 = (1+n_D)\hbar c\{2\pi\omega N_0|d_{ba}|^2\}^{-1},\ n_D = \sqrt{\varepsilon_2},\ \varepsilon_1 = 1,$$

$$N_0 = N_a - N_b,\ \tau = t/t_0.$$

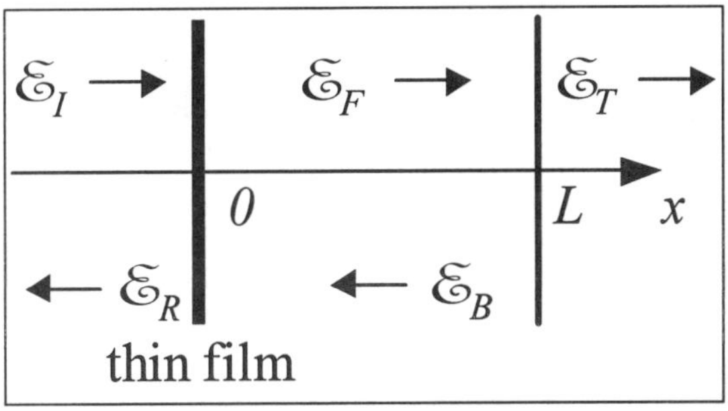

Fig.9.2.1. Geometry of the resonant interaction of light with thin film at *x=0* and reflected surface at *x=L*.

Here, R is the reflection coefficient of the $x=L$ surface allowing for the absorption of the signal in the region $0 < x < L$; Δ represents the dimensionless defect of resonance, γ_0 is the rate of relaxation of the polarisation and γ is the rate of relaxation of the difference between the populations of the upper and lower levels of a resonant atom. The values N_a and N_b are the stationary surface density of atoms at corresponding levels. Function $\varepsilon(\tau)$ describes the dimensionless amplitude of the electric film in the film, which consists of the incident and reflected waves or, equivalently, of the transmitted (forward) and the returned (backward) waves; $a(\tau)$ is proportional to the amplitude of the incident wave. The quantity τ_0 is the dimensionless time taken by the signal to travel from the $x=0$ plane to the $x=L$ plane and back again, whereas s represents the phase shift due to such motion, including a contribution of a possible change due to reflection by $x=L$ surface.

Equations (9.2.1) and (9.2.2) do not take into account the Lorentz field. We can consider them as fruitful model equations, but there are situations where these equations are valid. A thin film of metals under conditions of dimensional quantisation in a transverse quantising magnetic field can be described by eqs. (9.2.1) and (9.2.2). If $R=0$, eqs. (9.2.1) and (9.2.2) become identical with the equations describing a transmission of a plane light wave through a low-Q Fabri-Perot resonator containing a resonant atom in the mean field approximation. The mean field approximation consists of averaging the required values over the length of resonant medium inside the resonator cavity and in usual substitution of the average of production for the production of mean values.

We allow for the Lorenz field in section 8.2.5. Below, in sections 8.2.1-8.2.3 we will discuss optical bistability and self-pulsation on the basis of recurrence relationships derived from eqs. (9.2.1) and (9.2.2). Then, we consider the influence of phase-sensitive thermostat on optical bistability. Finally, we point the attention to modifications in optical bistability due to the allowance for the Lorentz field. The more striking result in this field is the bistable regime of interaction of ultrashort light pulse with a thin film of resonant atoms.

9.2.1. THE MAPPING EQUATION

Equations (9.2.1) and (9.2.2) represent an extremely complex system of equations which cannot be investigated in the general form. However, if the rates γ and γ_0 of the relaxation processes are sufficiently high,

$$\gamma >> 1/\tau_0, \quad \gamma_0 >> 1/\tau_0, \tag{9.2.3}$$

so that during the time τ_0 of the feedback mechanism action all relaxation processes in the thin film finish, i.e., the system becomes the instantaneous-response system, and all the other changes occur on the scale of τ_0. Then the derivatives in eqs. (9.2.1) can be ignored. We can reduce eqs. (9.2.1) and (9.2.2) to their singular limits. Quantities

$$\psi_n = \psi(n\tau_0), \quad a_n = a(n\tau_0), \quad n = 0, 1, 2, ...$$

are described by the recurrence relationships deduced from the expression

$$a_{n+1} = \psi_{n+1} - \nu Re^{is}\psi_n + \frac{(\gamma_0 + i\Delta)(\psi_{n+1} + Re^{is}\psi_n)}{\gamma_0^2 + \Delta^2 + 4\left|\psi_{n+1} + Re^{is}\psi_n\right|^2 \gamma_0/\gamma}. \tag{9.2.4}$$

The implicit form (9.2.4) of point mapping distinguishes the model under consideration from the other models of optical systems that can be reduced to mapping. For simplicity, we shall consider only the case of pure absorption characterised by $\Delta = 0$, a monochromatic incident wave $a_n = a = const$, and two quantities $s = 2\pi m$ and $s = 2\pi(m - 1/2)$, of the phase shift, where *m=1,2,...* Then eq.(9.2.4), expressed in terms of variables $x_n = 2\psi_n/(\gamma\gamma_0)^{1/2}$ and parameter $\alpha = 2a/(\gamma\gamma_0)^{1/2}$, yields

$$\alpha = x_{n+1} - \nu R x_n + \frac{\gamma_0^{-1}(x_{n+1} + Rx_n)}{1 + (x_{n+1} + Rx_n)^2} \tag{9.2.5}$$

where *R>0* corresponds to a phase shift $s = 2\pi m$ and *R<0* - $s = 2\pi(m - 1/2)$.

The different regimes of reflection of a monochromatic wave by a thin film of resonant atom under conditions defined by eq. (9.2.3) are closely related to stable fixed points and cycles of mapping represented by eqs. (9.2.4) and (9.2.5). Many relationships governing the appearance of stable cycles of one dimensional mappings $x_{n+1} = f(x_n)$ are determined by the presence and position of extrema of function $f(x)$. We omit corresponding expressions since they have an implicit form.

9.2.2. OPTICAL BISTABILITY

The fixed points $\bar{x} = f(\bar{x})$ of the mapping of eq.(9.2.5) satisfy the following equation:

$$\alpha = \bar{x}(1-\nu R) + \frac{(1+R)\gamma_0^{-1}\bar{x}}{1+(1+R)^2\bar{x}^2}. \tag{9.2.6}$$

Using the notation

$$Y = \alpha(1+R)/(1-\nu R), \quad X = (1+R)\bar{x},$$

$$C = (1+R)\{2\gamma_0(1-\nu R)\}^{-1}$$

we find that eq.(9.2.6) reduces to the familiar expression $Y = X(1+2C/(1+X^2))$ of the mean field theory of passive optical cavities containing resonant atoms. If *C>4*, i.e., when

$$\frac{1+R}{1-\nu R}\frac{2\pi\omega N_0|d_{ba}|^2}{(1+n_D)\hbar c\gamma_0/\tau_0} > 4, \tag{9.2.7}$$

a given value of the amplitude of the incident wave from the interval

$$Y_+ < Y < Y_-, \quad Y_\pm = \frac{[C-1\pm(C^2-4C)^{1/2}]^{1/2}}{C\pm(C^2-4C)^{1/2}}[3C\pm(C^2-4C)^{1/2}] \tag{9.2.8}$$

corresponds to three fixed points of mapping (9.2.5). We shall denote them by x_L, x_M and x_H in accordance with their value $x_L < x_M < x_H$. One can easily show that

$$\left.\frac{d\alpha}{d\bar{x}}\right|_{\bar{x}=x_L} > 0, \quad \left.\frac{d\alpha}{d\bar{x}}\right|_{\bar{x}=x_M} < 0, \quad \left.\frac{d\alpha}{d\bar{x}}\right|_{\bar{x}=x_H} > 0,$$

so that for the same value of α we have

$$\left.\frac{d\alpha}{d\bar{x}}\right|_{\bar{x}=x_L} > -\left.\frac{d\alpha}{d\bar{x}}\right|_{\bar{x}=x_M}. \tag{9.2.9}$$

Everywhere with the exception of a narrow ($\sim C^{-1}$ when *C>>1*) interval of parameters Y near Y_-, we also have the inequality

$$-\left.\frac{d\alpha}{d\bar{x}}\right|_{\bar{x}=x_M} \geq \left.\frac{d\alpha}{d\bar{x}}\right|_{\bar{x}=x_H}. \tag{9.2.10}$$

The stability of these fixed points can be determined by considering small deviations $\chi_n = x_n - \bar{x}$. Linearisation of eq.(9.2.5) in respect of these deviations gives

$$\chi_{n+1} = R\frac{\nu - \theta}{1+\theta}\chi_n, \tag{9.2.11}$$

where

$$\theta = \gamma_0^{-1}\frac{1-\bar{x}^2(1+R)^2}{[1+\bar{x}^2(1+R)^2]^2}.$$

One can easily show that the instability condition of the fixed points $|R(\nu-\theta)(1+\theta)| > 1$ is equivalent to the inequality $d\alpha / d\bar{x} < 0$. Therefore, in the regions defined by eqs. (9.2.7) and (9.2.8) we have bistable reflection of a monochromatic wave and the dependence of the reflected wave on the incident one is characterised by a hysteresis (Fig.9.2.2a). In other words there is a range of intensities of the incident radiation with the following property. The radiation can pass through the film in the regime of either high or low transmission of the film with two correspondingly different values of the intensity of transmitted radiation. Implementation of an explicit feedback mechanism (involving reflection from the x=L plane) reduces the critical density of resonant atoms, necessary for bistable reflection of a light wave from a thin film, by a factor $(1+R)/(1-\nu R)$ if $R>0$ and increases this density by a factor $(1+\nu|R|)/(1-|R|)$ if $R<0$.

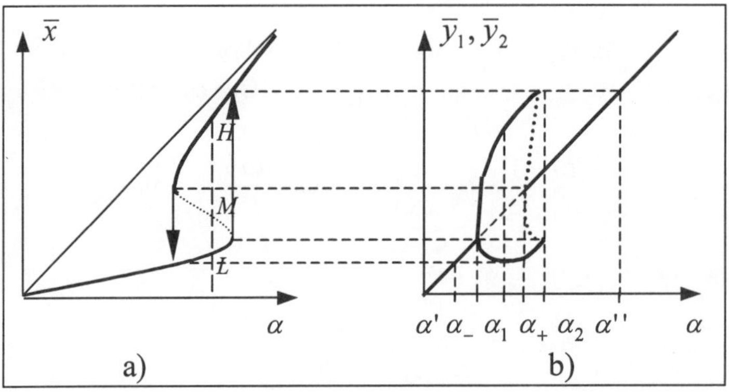

Fig.9.2.2

More traditional and detail description of optical bistability and its applications can be found in the book [22].

9.2.3. SELF-PULSATION

Outside the range defined by eq.(9.2.7), the reflection of a monochromatic wave is a steady-state process if *R>0* and the dependence of the intensity of the reflected wave on the intensity of the incident wave is single-valued. This situation is different from the case when *R<0*. The easiest analysis can be carried out most for the values of *R* close to *-1*. We shall consider only this case characterised by $R \to -1$. Then we have $\bar{x} \approx \alpha(1+\nu)^{-1}$. It follows from eq.(9.2.11) that in the interval

$$\alpha_- < \alpha < \alpha_+, \quad \alpha_\pm = \frac{1+\nu}{1-|R|}[C_1 - 1 \pm (C_1^2 - 4C_1)^{1/2}]^{1/2}$$

with

$$C_1 = \frac{1}{(1-\nu)\gamma_0} = \frac{2}{1-\nu}\frac{2\pi\omega N_0 |d_{ba}|^2}{(1+n_D)\hbar c\gamma_0/\tau_0} > 4 \qquad (9.2.12)$$

the only fixed point of mapping (9.2.5) becomes unstable. We shall consider the fixed points $\bar{x}_1$ and $\bar{x}_2$ of double iteration of mapping (9.2.5) forming the 2-cycle of the mapping (see Refs.[23,24]):

$$\alpha = \bar{x}_2 - \nu R\bar{x}_1 + \frac{\gamma_0^{-1}(\bar{x}_2 + R\bar{x}_1)}{1+(\bar{x}_2 + R\bar{x}_1)^2}, \quad \alpha = \bar{x}_1 - \nu R\bar{x}_2 + \frac{\gamma_0^{-1}(\bar{x}_1 + R\bar{x}_2)}{1+(\bar{x}_1 + R\bar{x}_2)^2}.$$

If we introduce the notation $\bar{y}_1 = (\bar{x}_1 + R\bar{x}_2)/(1+|R|)$ and $\bar{y}_2 = (\bar{x}_2 + R\bar{x}_1)/(1+|R|)$, we find that

$$(1-\nu|R|)\bar{y}_1 + \frac{\gamma_0^{-1}(1+|R|)\bar{y}_1}{1+\bar{y}_1^2(1+|R|)^2} = (1-\nu|R|)\bar{y}_2 + \frac{\gamma_0^{-1}(1+|R|)\bar{y}_2}{1+\bar{y}_2^2(1+|R|)^2}. \qquad (9.2.13)$$

and in the limit $R \to -1$ we have

$$\bar{y}_1 + \bar{y}_2 = \alpha(1-|R|)(1+\nu)^{-1}.$$

The solution of these equations is identical with the fixed points of mapping (9.2.5):

$$\beta = y_{n+1} - \nu|R|y_n + \frac{\gamma_0^{-1}(y_{n+1} + |R|y_n)}{1+(y_{n+1} + |R|y_n)^2},$$

which describes the bistable reflection of a wave at a certain amplitude β from the same film of resonant atoms. But in this case the phase shift of the reflected signal is a multiple of 2π. The value β is governed by the condition (9.2.13). It follows that double iteration of mapping (9.2.5) in the limit $R \to -1$ in the range defined by (9.2.12) is characterised by a nontrivial fixed point forming a 2-cycle $(\bar{y}_1, \bar{y}_2)$ of the original mapping. Three variants of 2-cycle are possible: (y_L, y_M), (y_L, y_H) and (y_M, y_H). We can determine the range of parameter α, where this one or another 2-cycle exist (Fig.9.2.2b), using the inequalities (9.2.9) and (9.2.10). For example, in the interval

$$\min(y_L + y_M) \leq y_L + y_M \leq \max(y_L + y_M)$$

the quantities y_L, y_M, $y_L + y_M$ depend continuously and monotonically on β and eqs. (9.2.9) and (9.2.10) are satisfied. Therefore, each parameter α from the interval

$$\alpha_- = \min\frac{(y_L + y_M)(1+\nu)}{(1-|R|)} \leq \alpha \leq \max\frac{(y_L + y_M)(1+\nu)}{(1-|R|)} \equiv \alpha_1 = \frac{\alpha_+ + \alpha'}{2}$$

corresponds to one and the same one pair (y_1, y_2), namely (y_L, y_M), so that $y_L + y_M = \alpha(1-|R|)(1+\nu)^{-1}$. At the point $\alpha = \alpha_1$ this 2-cycle transforms continuously to the 2-cycle composed of the points (y_L, y_H):

$$y_L + y_H = \alpha(1-|R|)(1+\nu)^{-1},$$

which is defined in the interval

$$\alpha_1 \leq \alpha \leq \alpha_2 \equiv \max\frac{(y_L + y_H)(1+\nu)}{(1-|R|)} \approx \frac{\alpha_- + \alpha''}{2}.$$

At the same time in the range $\alpha_+ \leq \alpha \leq \alpha_2$ there is a 2-cycle of points (y_M, y_H). It should be emphasised that the values of y_H in each pair (y_H, y_L) and (y_M, y_H) are different, since they correspond to different values of parameter β. The interval $\alpha_- \leq \alpha \leq \alpha_2$, where the 2-cycle of the original mapping (9.2.5) exists is much wider than the interval $\alpha_- \leq \alpha \leq \alpha_+$ of instability of the fixed point of the same mapping (Fig.9.2.2b).

The physical conclusions can be drawn from the obtained results provided we know the stability of the 2-cycles. One can easily show that the 2-cycles composed of (y_L, y_M) and (y_H, y_L) are stable and the 2-cycle (y_H, y_M) is unstable.

Thus, in the absence of steady-state regime, $\alpha_- \leq \alpha \leq \alpha_+$ the only stable regime of reflection of a monochromatic wave by a thin film characterised by $R \to -1$ is that of self-pulsations with a period $2\tau_0$ and an amplitude , according to eq.(9.2.2a), proportional to $|y_1 - y_2|$. At the point $\alpha = \alpha_-$ the regime loses stability in a soft manner, whereas at the point $\alpha = \alpha_+$ and $\alpha = \alpha_2$ the change in the reflection regime is abrupt. In the range $\alpha_+ \leq \alpha \leq \alpha_2$ there is a characteristic hysteresis of the steady-state pulsating regimes.

Note, that the above description of self-pulsations is quite analogous to the multi-mode instabilities in the resonator with resonant atoms [25].

9.2.4. OPTICAL BISTABILITY IN PHASE-SENSITIVE THERMOSTAT

Absorption bistability considered above depends strongly on the character of the relaxation of a quantum system. In the field of broadband squeezed light the relaxation of atom significantly changes (see Appendix A) . It will be shown below that the optical bistability effect acquires new features when the film is additionally irradiated with a resonant radiation in a squeezed state. In this case the bistable regime of the transmission (reflection) of the film depends not only on the parameters of the film but also on the phase difference between the coherent and squeezed waves. Moreover, besides hysteresis in the amplitude of transmitted radiation as a function of the amplitude of incident radiation, there appears a bistable dependence of the amplitude and phase of the transmitted coherent wave on the phase of the incident coherent wave. In turn, this changes very substantially the character of the transmission of phase-modulated pulses by the thin film. The other name for squeezed light - phase-sensitive thermostat - is thereby strikingly manifested.

To show clearly the mentioned role of squeezed light we put $\nu = R = 0$ and rewrite the main equations (9.2.1) and (9.2.2) to include the effect of light squeezing

$$\left(\frac{d}{d\tau} - i\Delta + \gamma_0\right)p = i\varepsilon n - \delta p^*,, \tag{9.2.14a}$$

$$\left(\frac{d}{d\tau} + \gamma\right)n = 2i(\varepsilon^* p - \varepsilon p^*) + \gamma\,, \tag{9.2.14b}$$

$$\varepsilon = a + (\xi + i)p\,, \tag{9.2.15}$$

These equations differ from eqs.(9.2.1) and (9.2.2) by the presence of the relaxation term $-\delta p^*$, which is zero in the absence of squeezing. However, this term changes the character of relaxation by increasing the rate of relaxation of the real part of the polarisation amplitude p and decreasing the relaxation of the imaginary part of p. This, in turn, gives rise to new resonant interactions of coherent field, which depend on the phase of this field. We emphasise that the quantity δ in eq. (9.2.14a) is assumed to be real, so that the phase φ of the amplitude of coherent field

$$a = |a|e^{i\varphi}$$

represents the phase difference between the coherent and squeezed fields. Moreover, it should be noted that $\delta \le \sqrt{\gamma_0(\gamma_0+1)}$. The equality corresponds to the ideal squeezing and the absence of the other mechanism of relaxation except for relaxation in a squeezed field.

What else distinguishes eqs.(9.2.14) and (9.2.15) from eqs.(9.2.1) and (9.2.2) is the presence of parameter $\xi = 2\lambda\!\!\!^{-}\xi_0/3\ell$,where $\xi_0 \sim 1$, $\lambda\!\!\!^{-} = 1/k$, and ℓ is the thickness of the film. These parameters are responsible for the Lorentz field that we will discuss in the next section.

The following equation below follows from eqs.(9.2.14) and (9.2.15). It describes the relaxation of the square of the "Bloch vector":

$$\frac{d}{d\tau}\left(|p|^2 + \frac{1}{4}n^2\right) = -2\gamma_0|p|^2 - \delta(p^2 + p^{*2}) + \frac{1}{2}\gamma n(1-n).$$

It is obvious that in the stationary state

$$p'^2\frac{\gamma_0+\delta}{\gamma} + p''^2\frac{\gamma_0-\delta}{\gamma} + \frac{1}{4}n^2 = \frac{1}{4}n,\ p' = \operatorname{Re} p,\quad p'' = \operatorname{Im} p. \qquad (9.2.16)$$

The simplest way to observe the effect produced by squeezing the electromagnetic field of thermostat is to consider the exact resonance $\Delta = 0$. The Lorentz field effect is also neglected by assuming $\xi = 0$. The stationary solutions of the system (9.2.14)-(9.2.15) can be easily found in the two cases: when the phase of a coherent field is equal to the phase of the squeezed thermostat and when the phase of coherent wave is shifted relative by the phase of the squeezed thermostat by $\pi/2$.

When the phase of coherent incident wave is equal to the phase of the squeezed field $\varphi = 0$ (or $\varphi = \pi$), the field inside the film also has the same phase. The polarisation p is then pure imaginary quantity $p^* = -p$. The stationary solution of the system (9.2.14)-(9.2.15) has the form

$$a = \varepsilon[1 + (\gamma_0 - \delta + 4|\varepsilon|^2/\gamma)^{-1}], \qquad (9.2.17)$$

which is identical to the familiar expression $Y = X(1 + 2C/(1+X^2))$ with the cooperative parameter $C = C_0 = \frac{1}{2}(\gamma_0 - \delta)^{-1}$. One can see that the cooperative parameter C itself depends on the degree of squeezing the field: the stronger the squeezing, the more favourable are the conditions for formation of bistable regime.

When the phase of incident coherent wave is shifted relative to the squeezed field by $\pi/2$ (or $3\pi/2$), the polarisation of the film is real. Then it is easy to find that

$$a = \varepsilon[1 + (\gamma_0 + \delta + 4|\varepsilon|^2/\gamma)^{-1}].$$

In the current case the cooperative parameter is $C = C_{\pi/2} = \frac{1}{2}(\gamma_0 + \delta)^{-1}$ and the squeezing of thermostat makes it more difficult to reach a bistable regime.

In the case of arbitrary phase shift the coherent field in the film $\varepsilon = |\varepsilon| e^{i\psi}$ can be expressed in terms of stationary overpopulation n of energy levels as

$$|\varepsilon| = |a| \left[\left(\frac{\cos\varphi}{1 + 2C_0 n} \right)^2 + \left(\frac{\sin\varphi}{1 + 2C_{\pi/2} n} \right)^2 \right]^{1/2},$$

$$\tan\psi = \frac{1 + 2C_0 n}{1 + 2C_{\pi/2} n} \tan\varphi, \qquad (9.2.18)$$

where n is the solution of the equation

$$(n-1)(1 + 2C_0 n)^2 (1 + 2C_{\pi/2} n)^2 +$$
$$+ \frac{8n|a|^2}{\gamma} [C_0 (1 + 2C_{\pi/2} n)^2 \cos^2\varphi + C_{\pi/2} (1 + 2C_{\pi/2} n)^2 \sin^2\varphi] = 0.$$

It is evident from eq.(9.2.18) that the absolute value of the field amplitude in the film varies strongly with the phase φ of incident field, vanishing in the absence of squeezing ($\delta = 0$).

The simple situations studied above make it possible to predict the existence of unique bistable regimes for transmission of a coherent wave through a thin film of resonant atoms. Evidently the first regime is bistable transmission of a coherent wave through the thin film depending on the phase φ of the wave (more accurately on the phase difference between the coherent and squeezed fields), since for the same values of the parameters γ_0 and δ the parameters of the system are that bistability occurs for $\varphi = 0$ (*C>4*) and instability occurs for $\varphi = \pi/2$ (*C<4*). If we choose a value of a corresponding to the transmission of a coherent wave through the film for $\varphi = 0$ under conditions of low transmission, then as φ increases, for some value of $\varphi < \pi/2$ the transition to a regime of high transmission takes place. This is illustrated in Fig. 9.2.3. The figure also illustrates the characteristic dependence of the magnitude of the field amplitude on the phase φ in the regions where there is no bistability. We also note the bistable dependence of the phase of the field in the film on the absolute value of the amplitude of the incident coherent field (Fig. 9.2.4). Obviously, these effects are completely determined by the presence of the phase-sensitive thermostat.

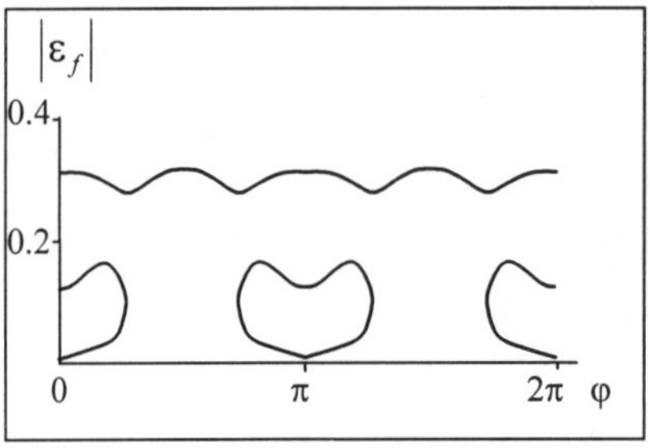

*Fig.9.*2.3. Magnitude of field amplitude in the film as a function of the phase of incident coherent wave. Here $C_0 = 20$, $C_{\pi/2} = 4$, $|a| = 0.42$, $\xi = \Delta = 0$, $\gamma = 2\gamma_0$.

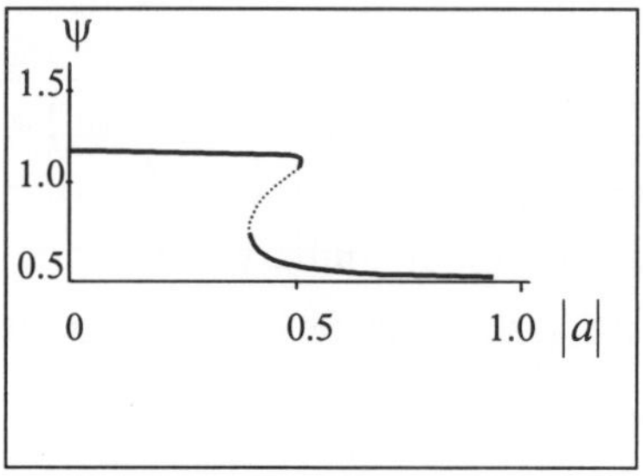

*Fig.9.*2.3. Bistable behaviour of the phase of the field in the film as a function of the absolute value of the amplitude of the incident coherent wave. $C_0 = 20$, $C_{\pi/2} = 4$, $\varphi = 0.5$, $\xi = \Delta = 0$, $\gamma = 2\gamma_0$.

The other regime is associated with the variation of the degree of squeezing of the additional irradiation. Choosing the parameters so that for $\delta = 0$ we have $C<4$ and the cooperative parameter satisfies $C> 4$ as δ increases, we obtain a bistable passage of a coherent wave through the resonant film depending on the degree of squeezing of the irradiation (Fig. 9.2.5). When these factors act together, the bistability picture becomes even more complicated.

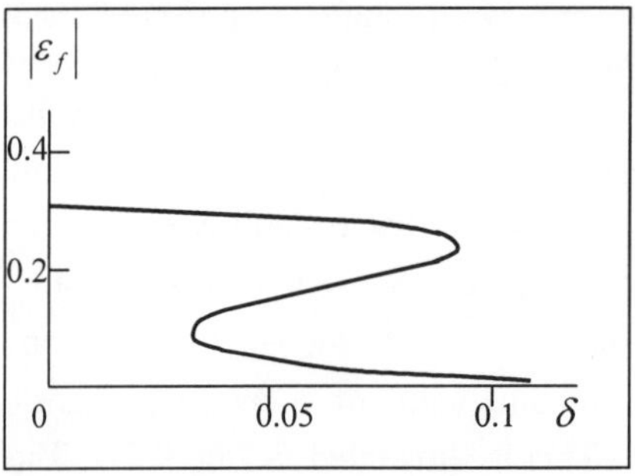

Fig.9.2.5. Amplitude of the coherent field in the film as a function of the degree of squeezing of the thermostat. Here $\gamma_0 = 0.128$, $\varphi = 0$, $|a| = 0.5$, $\xi = \Delta = 0$, $\gamma = 2\gamma_0$.

9.2.5. THE ROLE OF THE LORENTZ FIELD

It is convenient to write the equation (9.2.14a) with reference to (9.2.15) in the form

$$\frac{d}{d\tau}p = (i\Delta + \xi n)p - (\gamma_0 + n)p + ian - \delta p^*, \tag{9.2.19}$$

which demonstrates clearly the role of the Lorentz field, resulting in a shift of the resonant frequency by an amount $-\xi n$, as well as the role of the field reradiated by the atoms in the film. The latter gives rise to collective relaxation of polarisation (the term $-np$). These terms essentially determine the positive feedback mechanism which is necessary for the formation of bistable regimes.

cw-operation

We first discuss the changes in optical bistability. With the allowance for the Lorentz field ($\xi \neq 0$) and a detuning from resonance ($\Delta \neq 0$) the simple formulae like (9.2.17), demonstrating clearly the effect of the squeezing of thermostat under conditions of a bistable regime formation cannot be obtained. However, even here the squeezing of the thermostat gives rise to diametrically opposite effects in the typical cases when the phase of coherent wave is either unshifted ($\varphi = 0$) or shifted by $\pi/2$ ($\varphi = \pi/2$) relative to the squeezed thermostat. It is convenient to investigate the stationary solution (9.2.14) and (9.2.15) by means of eqs. (9.2.16) and

$$p' = -\frac{n|a|[(\gamma_0 + n + \delta)\sin\varphi + (\Delta + \xi n)\cos\varphi]}{(\gamma_0 + n)^2 - \delta^2 + (\Delta + \xi n)^2},$$

$$p'' = \frac{n|a|[(\gamma_0 + n + \delta)\cos\varphi - (\Delta + \xi n)\sin\varphi]}{(\gamma_0 + n)^2 - \delta^2 + (\Delta + \xi n)^2}.$$

When the phase of the coherent wave is equal to the phase of the squeezed field ($\varphi = 0$ or $\varphi = \pi$), we have

$$|a|^2 = \frac{\gamma(1-n)}{4n} \frac{[(\gamma_0 + n + \delta)(\gamma_0 + n - \delta) + (\Delta + \xi n)^2]^2}{(\gamma_0 + \delta)(\Delta + \xi n)^2 + \delta^2 + (\gamma_0 - \delta)(\gamma_0 + n + \delta)^2}. \tag{9.2.20}$$

Making the simultaneous replacements $\gamma_0 + \delta \to \gamma_0 - \delta$ and $\gamma_0 - \delta \to \gamma_0 + \delta$, we find that the result for a $\pi/2$ shift of the phase of the coherent wave relative to the phase of the squeezed wave, i.e., $\varphi = \pi/2$ or $\varphi = 3\pi/2$, follows from eq. (9.2.20). Figure 9.2.6 illustrates the typical changes produced by the squeezing of the thermostat ($\delta \neq 0$), in the plots of $|a|^2$ versus n for $\varphi = 0$ (curve 1) and $\varphi = \pi/2$ (curve 2), compared with the case of no squeezing ($\delta = 0$, curve 3). As with the absence of a Lorentz field, squeezing of the thermostat expands the region of bistability when the phases of the coherent and squeezed fields are equal to each other and it narrows the region of bistability in the case of a $\pi/2$ phase shift. In summary, the characteristic features of the optical bistability of the thin film, which are caused by the additional irradiation of the film with a wave of resonant squeezed light, are practically independent from the Lorentz field and the detuning from resonance.

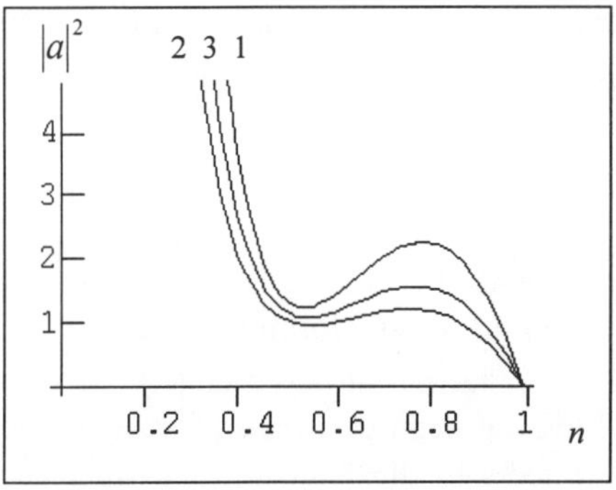

Fig.9.2.6. Shift of the plot of $|a|^2$ versus n caused by squeezing of thermostat $\delta = 0.4$ in the cases $\varphi = 0$ (curve 1) and $\varphi = \pi/2$ (curve 2) relative to the curve 3, characterising $|a|^2$ as a function of n in the absence of squeezing. Here $\gamma = 2\gamma_0 = 2$, $\Delta = -5$, $\xi = 10$.

Pulse dynamics

The Lorentz field is most pronounced in the interaction of an ultrashort pulse with a thin film. Let the form of the USP be rectangle. The relaxation process as well as the effect of squeezing are negligible small during the pulse. In the usual situation of extended medium the nutation oscillations arise near the leading edge of the USP. But in the thin film the Lorentz field provides the collective mechanism of relaxation according to (9.2.19). That, in its turn, leads to the transient stationary regime of the pulse transmission with stationary state of field amplitude ε, level populations, and n. This stationary regime exhibits bistability.

From (9.2.9) and the conservation law for Bloch vector we find that the transmission coefficient $T = |\varepsilon/a|^2$ for stationary state is determined by a stationary value of n:

$$T = \Delta^2[(\Delta + \xi n_{st})^2 + n_{st}^2]^{-1}, \tag{9.2.21}$$

where n_{st} obeys the equation

$$(1 - n_{st}^2)[(\Delta + \xi n_{st})^2 + n_{st}^2] - 4a^2 n_{st}^2 = 0. \tag{9.2.22}$$

We note that this is a fourth-order equation in n_{st}, while in the case of ordinary stationary bistability in section (9.2.2) the corresponding equation is of the third order.

Here we investigate eq.(9.2.22) only in the simplest case characterised by such parameters for which level populations remain close to the initial values. Then it is convenient to introduce the dimensionless population of the upper energy level r so that for stationary regime $r_{st} = (1 - n_{st})/2 << 1$. Eq.(9.2.22) transforms to the following

$$r_{st}\{[(\Delta + \xi) - 2\xi r_{st}]^2 + 1\} = a^2 . \qquad (9.2.23)$$

Eq.(9.2.23) has either a single real root or three real roots. In the first case function $r_{st} = r_{st}(a)$ is single valued. In the second case, however, three different set of populations of atomic energy levels correspond to a certain value of a. This latter case reveals the possibility of a bistable behaviour of the system.

The condition of obtaining three real roots can be found by determining the zeros of the derivative da^2 / dr_{st}. Introducing the notation $\mu = \Delta + \xi - 2\xi r_{st}$ we obtain

$$3\mu^2 - 2(\Delta + \xi)\mu + 1 = 0 .$$

This equation will have three different roots if

$$\Delta + \xi > \sqrt{3} .$$

In this case the plot of the function $r_{st} = r_{st}(a)$ has a wrinkle. The stability of this type of bistable regime should be studied to understand how it can be realised. The further details can be found in ref.[11].

9.3. The features of two-photon resonance

Let us consider in more details the case of degenerate two-photon resonance, when the double carrier frequency of an ultrashort light pulse is close to the frequency of optically forbidden transition (case B of sections 1.1.1-1.1.3). In section 1.1.3 it is shown that in this case the wave with the carrying frequency ω_0 generates a polarisation of resonant medium on the frequencies ω_0 and $3\omega_0$. Therefore, with account for the reverse reaction of the medium, two fields

$$E_{a1} = \mathcal{A}_1 \exp(-i\Phi_0) + k.c., \quad E_{a3} = \mathcal{A}_3 \exp(-3i\Phi_0) + k.c., \qquad (9.3.1)$$
$$\Phi_0 = \omega_0 t - \beta_0 z , \; 2\omega_0 \approx \omega_{ca}$$

affect atoms of a film under conditions of two-photon and Raman resonances (Fig. 9.3.1).

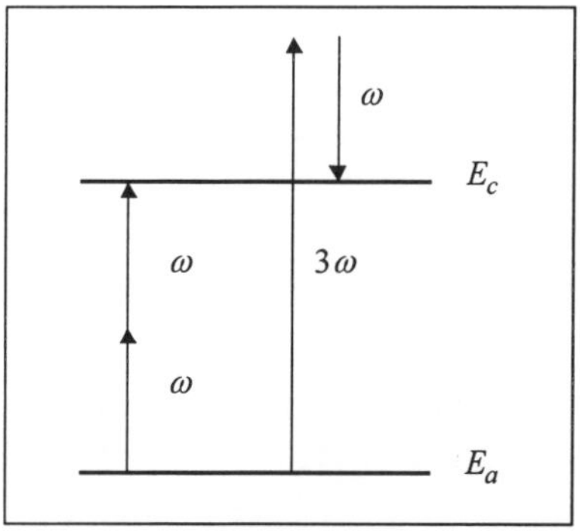

Fig.9.3.1

The indices a and c denote the resonant levels E_a and E_c. The fields (9.3.1) differ from the macroscopic electric fields inside the film by a value of the Lorentz field. We remind that z-axis lies in the plane of a thin film.

Let us derive the expression for polarisation of a film, which arises under the action of the fields (9.3.1). Atomic polarisation is determined by the standard formula (1.1.16):

$$P(t,x) = Sp(\rho d). \tag{9.3.2}$$

Here we normalise the density matrix of the film atoms to make $Sp\rho$ be a surface density of resonant atoms N_0. The density matrix ρ obeys the usual equation (1.0.3)

$$i\hbar\frac{d\rho}{dt} = [H_0 - E_{atom}d, \rho]. \tag{9.3.3}$$

Notation $E_{atom} = E_{a1} + E_{a2}$ is used for the electric field acting on atom.

We apply the method of unitary transformation of Hamiltonian which is slightly modified compared with section 1.1.2. Expansion of both matrix S, which determines the transformed density matrix $\tilde{\rho} = e^{-iS}\rho e^{iS}$, and the transformed Hamiltonian $\tilde{H} = e^{-iS}H_0e^{iS} - e^{-iS}E_{atom}de^{iS} - i\hbar e^{-iS}\frac{\partial}{\partial t}e^{iS}$ yields:

$$S = S^{(1,1)} + S^{(2,0)} + S^{(0,2)}\ldots, \quad \tilde{H} = \tilde{H}^{(0,0)} + \tilde{H}^{(1,0)} + \tilde{H}^{(0,1)} + \tilde{H}^{(2,0)} + \ldots \tag{9.3.4}$$

where $S^{(n,m)}$ and $\tilde{H}^{(n,m)}$ are the terms of n-th order on field E_{a1} and of m-th order of field E_{a3}. The relationships between them are defined by the expressions

$$\tilde{H}^{(0,0)} = H_0,$$

$$\tilde{H}^{(1,0)} = -E_{a1}d - i\left[S^{(1,0)}, H_0\right] + \hbar\frac{\partial}{\partial t}S^{(1,0)},$$

$$\tilde{H}^{(0,1)} = -E_{a2}d - i\left[S^{(0,1)}, H_0\right] + \hbar\frac{\partial}{\partial t}S^{(0,1)},$$

$$\tilde{H}^{(2,0)} = \frac{i}{2}\left[S^{(1,0)}, E_{a1}d\right] - \frac{i}{2}\left[S^{(1,0)}, \tilde{H}^{(1,0)}\right] - i\left[S^{(2,0)}, H_0\right] + \hbar\frac{\partial}{\partial t}S^{(2,0)},$$

$$\widetilde{H}^{(1,1)} = \frac{i}{2}\left[S^{(1,0)}, E_{a2}d\right] + \frac{i}{2}\left[S^{(0,1)}, E_{a1}d\right] - \frac{i}{2}\left[S^{(0,1)}, \widetilde{H}^{(1,0)}\right] - \frac{i}{2}\left[S^{(1,0)}, \widetilde{H}^{(0,1)}\right] - i\left[S^{(1,1)}, H_0\right] + \hbar\frac{\partial}{\partial t}S^{(1,1)},$$

$$\widetilde{H}^{(0,2)} = \frac{i}{2}\left[S^{(0,1)}, E_{a2}d\right] - \frac{i}{2}\left[S^{(0,1)}, \widetilde{H}^{(0,1)}\right] - i\left[S^{(0,2)}, H_0\right] + \hbar\frac{\partial}{\partial t}S^{(0,2)},$$

...

The same routine as in section 1.1.2 gives

$$S^{(1,0)}_{\alpha\alpha'} = -\frac{id_{\alpha\alpha'}}{\hbar}\left(\frac{\mathcal{A}_1 e^{-i\Phi_0}}{\omega_{\alpha\alpha'} - \omega_0} + \frac{\mathcal{A}_1^* e^{i\Phi_0}}{\omega_{\alpha\alpha'} + \omega_0}\right),$$

$$S^{(0,1)}_{\alpha\alpha'} = -\frac{id_{\alpha\alpha'}}{\hbar}\left(\frac{\mathcal{A}_3 e^{-i3\Phi_0}}{\omega_{\alpha\alpha'} - 3\omega_0} + \frac{\mathcal{A}_3^* e^{i3\Phi_0}}{\omega_{\alpha\alpha'} + 3\omega_0}\right).$$

Effective Hamiltonian is chosen in the form

$$H^{eff} = \widetilde{H}^{(0,0)} + \widetilde{H}^{(2,0)} + \widetilde{H}^{(1,1)} + \widetilde{H}^{(0,2)}$$

or

$$H^{eff}_{ca} = -[\frac{1}{2}\mathcal{A}_1^2\Pi_{ca}(\omega_0) + \mathcal{A}_3\mathcal{A}_1^*\Pi_{ca}(-\omega_0)]e^{-i2\Phi_0} = H^{eff*}_{ac},\ H^{eff}_{\alpha\alpha} = E_\alpha + E^{St}_\alpha, \tag{9.3.5}$$
$$E^{St}_\alpha = |\mathcal{A}_1|^2\Pi_\alpha(\omega_0) + |\mathcal{A}_3|^2\Pi_\alpha(3\omega_0),\quad \alpha = a, c, \sigma,$$

The expressions for the parameters of two-photon interaction $\Pi_\alpha(\omega_0)$, $\Pi_{ca}(\omega_0)$,..., are given by the formulae (1.1.11).

The polarisation of a film is expressed by eqs.(1.1.16) and (1.1.17) in terms of the transformed density matrix and effective operator of the dipole moment D:

$$D_{ac} = \mathcal{A}_1\Pi^*_{ca}(-\omega_0)e^{-i\Phi_0} + \mathcal{A}_1^*\Pi^*_{ca}(\omega_0)e^{i\Phi_0} + \mathcal{A}_3\Pi^*_{ca}(-3\omega_0)e^{-i3\Phi_0} + \mathcal{A}_3^*\Pi^*_{ca}(3\omega_0)e^{i3\Phi_0}$$

$$D_{aa} = -\mathcal{A}_1\Pi_a(\omega_0)e^{-i\Phi_0} - \mathcal{A}_3\Pi_a(3\omega_0)e^{-i3\Phi_0} + k.c., \tag{9.3.6}$$

$$D_{cc} = -\mathcal{A}_1\Pi_c(\omega_0)e^{-i\Phi_0} - \mathcal{A}_3\Pi_c(3\omega_0)e^{-i3\Phi_0} + k.c..$$

In the derivation of film polarisation we will neglect both the nonresonant terms and the terms proportional to $\mathcal{A}_3$ in the effective dipole moment operator, assuming that

$$|\mathcal{A}_3| << |\mathcal{A}_1|. \tag{9.3.7}$$

Then, we receive

$$P(t,z) = \mathcal{P}_1(t,z)\exp(-i\Phi_0) + \mathcal{P}_3(t,z)\exp(-3i\Phi_0) + c.c.,$$

$$\mathcal{P}_1(t,z) = -(\Pi_a(\omega_0)\widetilde{\rho}_{aa} + \Pi_c(\omega_0)\widetilde{\rho}_{cc})\mathcal{A}_1 + \Pi^*_{ca}(\omega_0)R\mathcal{A}_1^*,\ \mathcal{P}_3(t,z) = \Pi^*_{ca}(-\omega_0)R\mathcal{A}_1,$$

$$R = \widetilde{\rho}_{ca}\exp(2i\Phi_0).$$

Let us introduce the variable $N = \widetilde{\rho}_{aa} - \widetilde{\rho}_{cc}$. Since $\widetilde{\rho}_{aa} + \widetilde{\rho}_{cc} = const$, the above expressions provide

$$\mathcal{P}_1(t,z) = -(\Pi_+(\omega_0) + \Pi_-(\omega_0)N)\mathcal{A}_1 + \Pi^*_{ca}(\omega_0)R\mathcal{A}_1^*, \tag{9.3.8}$$

where

$$\Pi_+(\omega_0) = \frac{1}{2}(\Pi_a(\omega_0) + \Pi_c(\omega_0)),\ \Pi_-(\omega_0) = \frac{1}{2}(\Pi_a(\omega_0) - \Pi_c(\omega_0)).$$

We have put $\widetilde{\rho}_{aa} + \widetilde{\rho}_{cc} = 1$.

Quantities R and N are determined by the Bloch equation

$$\frac{\partial R}{\partial t} = i\Delta R + i\Lambda N,\quad \frac{\partial N}{\partial t} = 2i(\Lambda^* R - \Lambda R^*), \tag{9.3.9}$$

with the following parameters

$$\Lambda = [\frac{1}{2}\mathcal{A}_1^2\Pi_{ca}(\omega_0) + \mathcal{A}_3\mathcal{A}_1^*\Pi_{ca}(-\omega_0)]\hbar^{-1},$$

$$\Delta = 2\omega_0 - \omega_{ca} - \{|\mathcal{A}_1|^2[\Pi_c(\omega_0) - \Pi_a(\omega_0)] + |\mathcal{A}_3|^2[\Pi_c(3\omega_0) - \Pi_a(3\omega_0)]\}\hbar^{-1}. \tag{9.3.10}$$

The strength of an electric field acting on atom E_{atom} is defined by a field in a film E_f and film polarisation P:

$$E_{atom} = E_f + \xi P$$

where

$$E_f = \mathcal{E}_1\exp(-i\omega_0 t) + \mathcal{E}_3\exp(-i3\omega_0 t) + c.c..$$

For slowly varying amplitudes we have

$$\mathcal{A}_1 = \mathcal{E}_{f1} + \xi\,\mathcal{P}_1\,, \quad \mathcal{A}_3 = \mathcal{E}_{f3} + \xi\,\mathcal{P}_3\,,$$
$$\mathcal{E}_{f1}(t) = T(\beta_0,\omega_0)\mathcal{E}_{in}(t) + i\kappa(\beta_0,\omega_0)\mathcal{P}_1(t)\,,$$
$$\mathcal{E}_{f3}(t) = i\kappa(3\beta_0,3\omega_0)\mathcal{P}_3(t)\,.$$

Here the notation (9.1.6) is used and index *TE* is omitted. Vector $\beta_0\vec{e}_z + p_0\vec{e}_x$ is the wave vector of the incident resonant wave with amplitude $\mathcal{E}_{in}$. With reference to the relation (9.3.8) we can finally obtain the connection equation to complete system (9.3.9):

$$\mathcal{A}_1 = T(\beta_0,\omega_0)U^{-1}\left\{\mathcal{E}_{in} + (\xi - i\kappa)(\Pi_+(\omega_0) + \Pi_-(\omega_0)N)\mathcal{E}_{in} + (\xi + i\kappa)\Pi^*_{ca}(\omega_0)R\mathcal{E}^*_{in}\right\}, \tag{9.3.11}$$
$$U = (1 - \xi(\Pi_+(\omega_0) + \Pi_-(\omega_0)N))^2 + \kappa^2(\Pi_+(\omega_0) + \Pi_-(\omega_0)N)^2 - (\xi^2 + \kappa^2)\left|\Pi_{ca}(\omega_0)R\right|^2.$$

The condition of phase synchronism in an optically thick medium in the case of a thin film looks like a rule governing the angles of emission of third harmonic wave from the film. The relations between the angles of incident (*in*) reflected (*ref*) and transmitted (*tr*) have the form

$$n_1(3\omega_0)\sin\theta^{ref}_{3\omega} = n_1(\omega_0)\sin\theta^{in}_{\omega}\,, \tag{9.3.12a}$$
$$n_2(3\omega_0)\sin\theta^{tr}_{3\omega} = n_1(\omega_0)\sin\theta^{in}_{\omega}\,, \tag{9.3.12b}$$
$$n_2(3\omega_0)\sin\theta^{tr}_{3\omega} = n_2(\omega_0)\sin\theta^{tr}_{\omega}\,, \tag{9.3.12c}$$

First we consider the normal incidence of ultrashort light pulse on a thin film. The pulse duration is about a period of Rabi oscillations. The numerical solution of the equations (9.3.9)-(9.3.11) shows that the form of the transmitted pulse on the main frequency ω_0 repeats the form of incident ultrashort pulse. The transmission coefficient differs from calculated by the Frenel formula by less than 1%. The reason of this is that the Lorentz field destroys the conditions of exact resonance resulting in a weak excitation of resonant atoms. As the consequence, the insignificantly small response appears. The harmonic signal is of the order of size 10^{-4} of the incident amplitude (Fig.9.3.2). The increase of amplitude of the incident signal does not change the Frenel character of pulse reflection. The signal of a harmonic oscillates (Fig.9.3.3 as the result of fast evolution of Bloch vector.

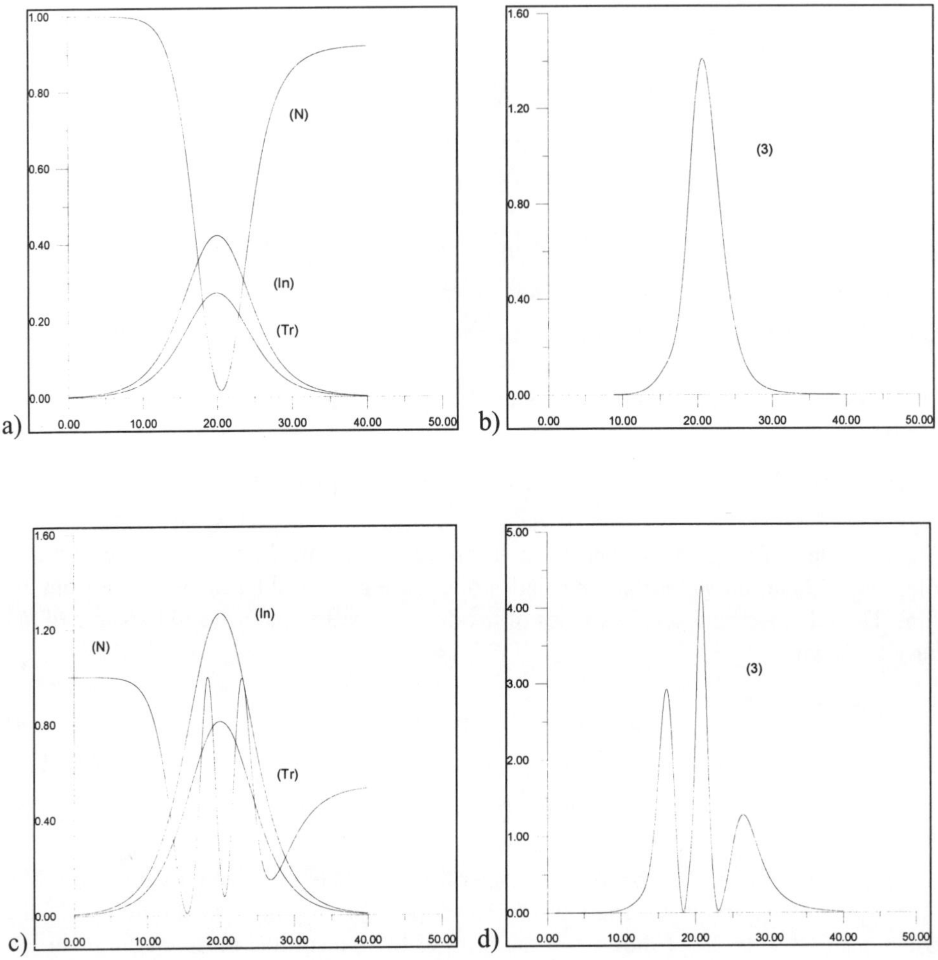

Fig.9.3.2. The envelopes of incident (In), transmitted (Tr), third harmonic (3) waves and population level difference (N). The incident pulse has the form $a_{in}(\tau) = a_0 \operatorname{sech}[(\tau - \tau_m)/\tau_p]$ with $\tau_p = 6$, and $\tau_m = 20$. The pulse amplitude is $a_0 = 0{,}65$ for a) and b) and $a_0 = 1{,}13$ for c) and d).

Now we shall consider the normal incidence of an ultrashort pulse with the duration is much longer than a period of Rabi oscillations. The result of numerical simulation for a characteristic set of parameters is depicted in Fig.9.3.3. As it was mentioned above the form of transmitted signal repeats the form of incident pulse, but it is less in magnitude than it should be by Frenel formula. The essential difference from a previous case is that the oscillations of atom population decay. The level population tends to a stationary value (rather than the equilibrium one). It is important to emphasise that the times of irreversible relaxation in the system are much longer than the pulse duration. The pulse behaviour becomes clearer if to write Bloch equations as

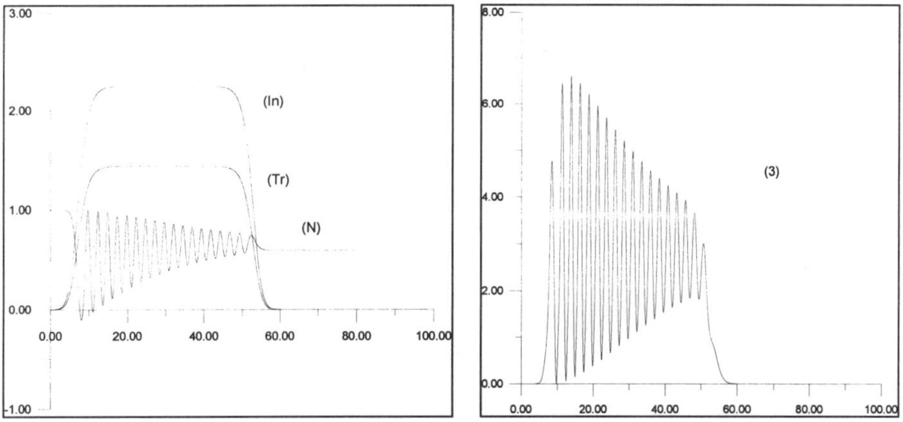

Fig.9.3.3. The dimensionless intensity of incident (In), transmitted (Tr), third harmonic (3) waves (averaged over a period of time of fast field oscillations), and population level difference (*N*). The incident pulse has the form of plateau $a_{in}(\tau) = a_0\{\tanh[2(\tau - \tau_m)/\tau_p] - - \tanh[2(\tau - \tau_m - \tau_w)/\tau_p\}$ with $a_0 = 0{,}75$, $\tau_p = 6$, $\tau_m = 6$, and $\tau_w = 8$. The ordinate axis factor in the plot of third harmonic is 10^{-4}.

$$\frac{\partial R}{\partial t} \sim i(\Delta + \mathrm{Re}(f)N)R + i(g_1 + g_2 N)N - \mathrm{Im}(f)NR \ ,$$

where f, g_1 and g_2 are some functions dependent on an incident field and film parameters according to the general equations (9.3.9)-(9.3.11). One can see that the Lorentz field results in an effective mechanism of relaxation which makes the polarisation and level populations reach a stationary regime. This phenomenon manifests especially strikingly in the form of the third harmonic signal (Fig.9.3.4): the peak modulation near the leading front of the third harmonic is replaced by a stationary mode of harmonic generation. Once again we emphasise, that in a considered case the phase synchronism for the third harmonic generation determines a direction of harmonic radiation from the film. It has no effect on efficiency of harmonic generation.

9.4. Exactly integrable model of double resonance

When ignoring the Lorentz field and the irreversible relaxation, the inverse scattering transform method in some cases can be applied to study the transmission of ultrashort light pulses through the thin film of resonant atoms. For these purposes the double resonance model seems to be the most general model of the resonant interaction. We will show Below how the IST method allows to investigate the reflection (transmission) of two-frequency light pulses by a thin film containing the three-level atoms.

We consider the incidence of the plane front pulses of frequencies ω_1 and ω_2

$$\vec{E}^{(in)} = \mathcal{E}_1^{(in)} \exp[i(\beta_1 z + p_1 x - \omega_1 t)] + \mathcal{E}_2^{(in)} \exp[i(\beta_2 z + p_2 x - \omega_2 t)] + c.c.$$

on a thin film containing three-level atoms under condition of double resonance

$$\omega_1 \approx (E_a - E_b)/\hbar, \quad \omega_2 \approx (E_c - E_b)/\hbar .$$

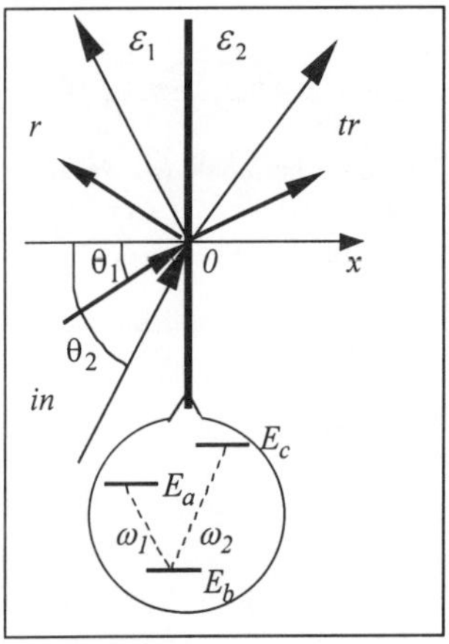

Fig.9.4.1. Geometry of the incidence of two pulses on thin film under condition of double resonance.

Optically allowed atomic transitions forms *V*-configuration (see Fig.9.4.1).

We write the relation between the amplitudes of transmitted pulses using (9.1.7) as

$$\begin{aligned} \mathcal{E}_{A1}^{tr}(t) &= T_A(\beta_1, \omega_1)\mathcal{E}_A^{in}(t) + i\kappa_A(\beta_1, \omega_1)\mathcal{P}_{A1}(t), \\ \mathcal{E}_{B2}^{tr}(t) &= T_B(\beta_2, \omega_2)\mathcal{E}_{B2}^{in}(t) + i\kappa_A(\beta_2, \omega_2)\mathcal{P}_{B2}(t), \end{aligned} , \tag{9.4.1}$$

where $\mathcal{E}_{Aj}^{tr}(t)$ and $\mathcal{P}_{Aj}(t)$ denote as usual the slowly varying amplitudes of the electric field and film polarisation at the carrier frequency ω_j, $j=1,2$. Quantities β_1, and β_2 are given by the incidence angles of the corresponding waves.

The added index *A* (or *B*) indicates the type of electromagnetic mode. Various combinations of different field modes of a two-frequency pulse are possible: $A=B=TE$ or TM and $A=TE$, $B=TM$ or, conversely, $A=TM$ and $B=TE$.

Polarisation of the thin film is determined by

$$\mathcal{P}_{A1} = R_{10} d_{Aba}\,, \quad \mathcal{P}_{B2} = R_{20} d_{Bbc}\,, \tag{9.4.2}$$

where the values R_{10} and R_{20} obey the eqs.(2.6.3):

$$\begin{aligned}
&\frac{\partial R_{10}}{\partial t} = i\Delta R_{10} + i\Lambda_1 (R_{00} - R_{11}) - i\Lambda_2 R_{12}, \quad \frac{\partial R_{20}}{\partial t} = i\Delta R_{20} + i\Lambda_2 (R_{00} - R_{22}) - i\Lambda_1 R_{21},\\
&\frac{\partial R_{21}}{\partial t} = i\Lambda_2 R_{01} - i\Lambda_1^* R_{20}, \quad \frac{\partial R_{00}}{\partial t} = -i(\Lambda_1 R_{01} - \Lambda_1^* R_{10}) - i(\Lambda_2 R_{02} - \Lambda_2^* R_{20}),\\
&\frac{\partial R_{11}}{\partial t} = i(\Lambda_1 R_{01} - \Lambda_1^* R_{10}), \quad \frac{\partial R_{22}}{\partial t} = i(\Lambda_2 R_{02} - \Lambda_2^* R_{20}).
\end{aligned} \tag{9.4.3}$$

We consider the exact resonance with two-photon transition $\Delta_1 = \Delta_2 = \Delta$. The relaxation terms were omitted since the pulses are ultrashort ones. We introduced index A or B in the matrix elements of the dipole moment operator to indicate its appropriate component.

Rupasov and Yudson [4] pointed out that an expression such as (9.4.1) can be replaced by a differential equation with a singular right-hand side. This method can be applied here, too. We introduce auxiliary functions $\mathcal{E}_{Aj}(x,t)$ satisfying the equations

$$\frac{\partial \mathcal{E}_{Aj}}{\partial x} = 2i\kappa_A(\beta_j, \omega_j)\delta(x)\mathcal{P}_{Aj}(t) \tag{9.4.4}$$

For $x = 0$ these functions are defined by the relation

$$\mathcal{E}_{Aj}(0,t) = \tfrac{1}{2}[\mathcal{E}_{Aj}(0+,t) + \mathcal{E}_{Aj}(0-,t)]\,. \tag{9.4.5}$$

Integrating (9.4.4) from $x = -\infty$ to $x=-\zeta$ ($\zeta<<1$) yields

$$\mathcal{E}_{Aj}(x = \zeta, t) = \mathcal{E}_{Aj}(x = -\infty, t) \equiv \mathcal{E}_{Aj}^{(-)}(t)\,.$$

If to integrate eq.(9.4.4) from $x = -\zeta$ to $x = \zeta$, one can obtain

$$\mathcal{E}_{Aj}(x = \zeta, t) = \mathcal{E}_{Aj}(x = -\zeta, t) + 2i\kappa_A(\beta_j, \omega_j)\mathcal{P}_{Aj}(t)\,.$$

In view of (9.4.5) it follows that

$$\mathcal{E}_{Aj}(0,t) = \mathcal{E}_{Aj}(x = -\zeta, t) + i\kappa_A(\beta_j, \omega_j)\mathcal{P}_{Aj}(t)$$

This expression coincides with (9.4.1) if the identities below hold

$$\mathcal{E}_{Aj}(0,t) \equiv \mathcal{E}_{Aj}^{tr}(t), \quad \mathcal{E}_{Aj}(x=-\zeta,t) = T_A(\beta_j,\omega_j)\mathcal{E}_{Aj}^{in}(t).$$

The initial problem of the refraction of ultrashort pulses by the film can be solved on the basis of eqs.(9.4.1)-(9.4.4) by means of the following algorithm: a) one should determine the initial conditions (9.4.4) by a specific value of $\mathcal{E}_{Aj}^{in}(t)$.; b) one should solve (9.4.4) for the indicated initial condition for $\mathcal{E}_{Aj}(0-,t)$ and $\mathcal{P}_{Aj}(t)$ to obtain $\mathcal{E}_{Aj}(x,t)$ for *x>0* and consequently $\mathcal{E}_{Aj}(0+,t)$; c) one should determine the envelopes $\mathcal{E}_{Aj}^{tr}(t)$. of the transmitted ultrashort pulse in accordance with (9.4.5):

$$\mathcal{E}_{Aj}^{tr}(t) = \tfrac{1}{2}[T_A(\beta_j,\omega_j)\mathcal{E}_{Aj}^{in}(t) + \mathcal{E}_{Aj}(0+,t)]\,. \tag{9.4.6}$$

For the equations that specify the evolution of the state of three-level atoms, we should take (9.4.3) with $\Lambda_1 = \mathcal{E}_{A1}^{tr}d_{ab}/\hbar$ and $\Lambda_2 = \mathcal{E}_{B2}^{tr}d_{cb}/\hbar$; the definitions of the remaining variables are the same as before. It can be directly verified that if the condition

$$|d_{Aab}|^2\kappa_A(\beta_1,\omega_1) = |d_{Bcb}|^2\kappa_B(\beta_2,\omega_2) \tag{9.4.7}$$

is valid, the system of equations (9.4.3) and (9.4.4) is a zero-curvature condition for the *U-V* pair

$$U = i\begin{pmatrix} -\lambda & \Lambda_1 & \Lambda_2 \\ \Lambda_1^* & \lambda & 0 \\ \Lambda_2^* & 0 & \lambda \end{pmatrix}, \tag{9.4.8}$$

$$V = \frac{i\tau_s\delta(x)}{\Delta+2\lambda}\begin{pmatrix} -(2R_{00}-R_{11}-R_{22})/3 & R_{10} & R_{20} \\ R_{10}^* & (R_{00}-2R_{11}+R_{22})/3 & -R_{21} \\ R_{20}^* & -R_{21}^* & (R_{00}-2R_{22}+R_{11})/3 \end{pmatrix}, \tag{9.4.9}$$

where

$$\tau_s^{-1} = N_0|d_{12}|^2\kappa_A(\beta_1,\omega_1)\hbar^{-1}.$$

The main difference of *U-V* pair from that in extended medium is the presence of *δ*-function in the expression for *V*.

Condition (9.4.7) is similar to the one of the existence of a simulton regime of propagation of a two-frequency USP in a semi-infinite homogeneous medium (see sections 3.2.1 and 3.2.7) Before we consider the solution of the system (9.4.3) and (9.4.4), it is worth to analyse in more detail the "simulton" condition (9.4.7), which appears in the problem under investigation. It is convenient to introduce two auxiliary functions $F_{TE}(\theta)$ and $F_{TM}(\theta)$ of the incidence angle $\theta_j = \theta^{in}_{\omega_j}$ of each of the carrier waves of the two-frequency pulse:

$$F_{TE}(\theta) = \cos\theta + (\eta^2 + \cos^2\theta)^{1/2},$$

$$F_{TM}(\theta) = \frac{1}{\cos\theta} + \frac{1+\eta^2}{(\eta^2 + \cos^2\theta)^{1/2}},$$

where $\eta^2 = (\varepsilon_2 - \varepsilon_1)/\varepsilon_2 > 0$. Condition (9.4.7) can be expressed in terms of the following functions:

for $A=TE$, $B=TE$,

$$F_{TE}(\theta_1) = \left|d^y_{bc}/d^y_{ba}\right|^2 (\omega_1/\omega_2) F_{TE}(\theta_2);$$

for $A=TM$, $B=TM$

$$F_{TM}(\theta_1) = \left|d^z_{bc}/d^z_{ba}\right|^2 (\omega_1/\omega_2) F_{TM}(\theta_2);$$

for $A=TE$, $B=TM$

$$F_{TE}(\theta_1) = \left|d^y_{bc}/d^z_{ba}\right|^2 (\omega_1/\omega_2) F_{TM}(\theta_2).$$

We denote the factor preceding the function $F_B(\theta_2)$ by ξ. The question arises whether the appropriate incidence angles θ_1 and θ_2 can be found to satisfy the simulton condition (9.4.7), which reduces to finding the solution of the equation

$$F_A(\theta_1) = \xi F_B(\theta_2). \tag{9.4.10}$$

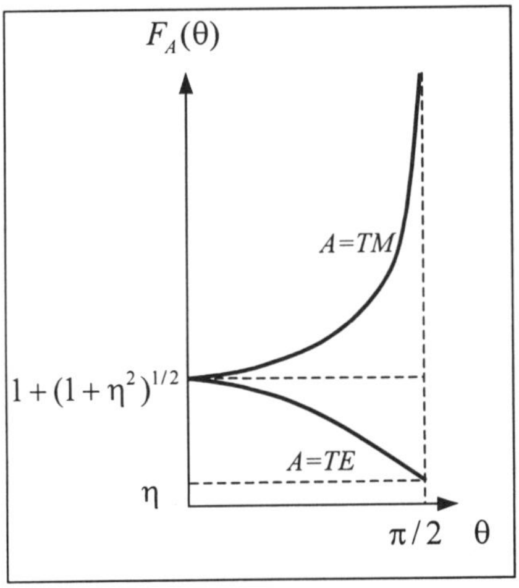

Fig.9.4.1. Plots of auxiliary functions $F_A(\theta)$.

Figure 9.4.2 shows the plots of the functions $F_{TE}(\theta)$ and $F_{TM}(\theta)$. Since the function $F_{TM}(\theta)$ unlimitedly increases as $\theta \to \pi/2$, it can be shown that eq.(9.4.10) has a solution for any parameter ξ for $A=B=TM$. This is the most favourable case for satisfying the simulton condition. For $A=TE$ and $B=TM$, there is a value $\theta_1=\theta_2$ at $\xi<1$, for which (9.4.10) is satisfied, and for given θ_2 one can obtain $\theta_1(\neq\theta_2)$ for which (9.4.10) is also satisfied. Equation (9.4.10) cannot be solved for $\xi>1$. In this case it is necessary to change the wave types, i.e., to choose $A=TM$ and $B=TE$. Then there must be θ_1 and θ_2 pairs when (9.4.10) holds.

The least favourable choice is $A=B=TE$. For the given dielectric constants difference $\varepsilon_2-\varepsilon_1$ there is an interval $\xi_{\min}\le\xi\le\xi_{\max}$ where for given θ_2 one can find the value of θ_1 satisfying simulton condition. Here

$$\xi_{\min}=\eta/[1+(1+\eta^2)^{1/2}],\quad \xi_{\max}=\xi_{\min}^{-1}.$$

In this case it is preferable to use the media with ε_2-ε_1 <<ε_2. The solution of (9.4.10) for $A=B=TE$ can be written explicitly as

$$\cos\theta_1=\eta\,\mathrm{sh}\{\ln[\xi F_{TE}(\theta_2)/\eta]\}.$$

Similarly, for $A=TE$ and $B=TM$ we have

$$\cos\theta_1=\eta\,\mathrm{sh}\{\ln[\xi F_{TM}(\theta_2)/\eta]\}.$$

Let us consider now the solution of eqs. (9.4.3) and (9.4.4) by the IST method with the assumption that the three-level system returns to the initial state after the passing of the USP. That means that at $t\to\pm\infty$ matrix $V(\lambda,t)$ transforms into diagonal matrix $V^{(\pm)}(\lambda)$:
$V^{(+)}=V^{(-)}$

$$V^{(+)}=V^{(-)}=\frac{i\delta(x)}{3\tau_s}\frac{1}{\Delta+2\lambda}diag\left\{-(2R_{00}^0-R_{11}^0-R_{22}^0),R_{00}^0-2R_{11}^0+R_{22}^0,R_{00}^0-2R_{22}^0+R_{11}^0\right\},$$

where quantities R with the upper zero index correspond to $t\to-\infty$.

For the given values of $\mathcal{E}_{Aj}^{in}(t)$ or the normalised values $\Lambda_j^{in}(t)$ the spectral problem for $U(\lambda)$ can be solved to determine the scattering data - the transition matrix $T^{in}(\lambda,x<0)$ and the eigenvalues (discrete spectrum) $\{\lambda_n\}^{in}$. We can use the reasoning of ref. [4] to find $T^{out}(\lambda)=T(\lambda,x>0)$ and $\{\lambda_n\}^{out}$ and to generalise it for covering the case analysed here. If we have $V^{(+)}=V^{(-)}$, then $T(\lambda,x)$ is defined by the equation

$$\partial_x T=[T,V^{(-)}]. \tag{9.4.11}$$

Since $V^{(-)}$ is diagonal, the diagonal elements of matrix T are independent of x; consequently

$$T_{nn}^{out}(\lambda) = T_{nn}^{in}(\lambda). \tag{9.4.12}$$

Then we have $\{\lambda_n\}^{in}=\{\lambda_n\}^{out}$. For the off-diagonal elements T_{nm}

$$\partial_x T_{nm} = (V_m^{(-)} - V_n^{(-)})T_{nm}, \tag{9.4.13}$$

if $V^{(-)}(x,\lambda)=\text{diag}(V_1^{(-)}, V_2^{(-)}, V_3^{(-)})$.What makes the situation unusual is that $V_n^{(-)}$ contains delta function. The definition of $T_{nm}(\lambda,x)$ is completed at the point $x=0$:

$$T_{nm}(\lambda,0) = \tfrac{1}{2}\{T_{nm}(\lambda,0+) + T_{nm}(\lambda,0-)\}. \tag{9.4.14}$$

We put

$$V_m^{(-)} - V_n^{(-)} = 2\delta(x)\Gamma_{nm}. \tag{9.4.15}$$

Integrating of (9.4.13), in view of (9.4.15), with respect to x from $x \to -\infty$ to $x = -\zeta$ $(\zeta > 0, \zeta << 1)$ yields

$$T_{nm}(x=-\zeta) = T_{nm}(x=-\infty) = T_{nm}^{in}(\lambda).$$

Integrating of (9.4.13) from $x=-\zeta$ to $x=\zeta$ yields

$$T_{nm}(\lambda, x=\zeta) = T_{nm}(\lambda, x=-\zeta) + 2\Gamma_{nm}(\lambda)T_{nm}(\lambda,0).$$

The above equality with allowance for (9.4.14) provides

$$T_{nm}(\lambda, x=\zeta) = \frac{1+\Gamma_{nm}(\lambda)}{1-\Gamma_{nm}(\lambda)} T_{nm}^{in}(\lambda). \tag{9.4.16}$$

Since the right-hand side of (9.4.16) does not depend on x, we have

$$T_{nm}^{out}(\lambda) = T_{nm}(\lambda, x=0+) = \frac{1+\Gamma_{nm}(\lambda)}{1-\Gamma_{nm}(\lambda)} T_{nm}^{in}(\lambda). \tag{9.4.17}$$

The solutions of eqs.(9.4.3) and (9.4.4) are the solutions of the inverse problem coupled with the scattering data $\{\lambda_n\}^{out}$, $T_n^{out}(\lambda)$ and eq.(9.4.6). In order to reconstruct the $\Lambda_j(t,x)$ and then to build $U(\lambda,x,t)$, we need the quantity $r_j(\lambda,x)$ to be defined as

$$r_1(\gamma) = T_{12}(\lambda)/T_{11}'(\lambda), \quad r_2(\lambda) = T_{13}(\lambda)/T_{11}'(\lambda),$$

where $T_{11}'(\lambda) = dT_{11}'(\lambda)/d\lambda$. With regard to $V^{(-)}$, the necessary quantities Γ_{12} and Γ_{13} can be found:

$$\Gamma_{12}(\lambda) = \frac{i}{2\tau_s}\frac{R_{00}^0 - R_{11}^0}{\Delta + 2\lambda}, \quad \Gamma_{13}(\lambda) = \frac{i}{2\tau_s}\frac{R_{00}^0 - R_{22}^0}{\Delta + 2\lambda}. \tag{9.4.18}$$

In addition, at the discrete-spectrum points we can represent r_j^{in} in the form

$$r_j^{in}(\lambda_n) = 2\eta_n l_j^{(n)} \exp[2\eta_n t_{0n}], \quad j = 1, 2, \tag{9.4.19}$$

where

$$\lambda_n = \zeta_n + i\eta_n, \quad t_{0n} = (2\eta_n)^{-1} \ln\left|r_0^{in}(\lambda_n)\right|,$$

$$l_j = r_j^{in}(\lambda_n)/r_0^{in}(\lambda_n), \quad r_0^{in}(\lambda_n) = \left[\sum_j r_j^{in}(\lambda_n) r_j^{in*}(\lambda_n)\right]^{1/2}.$$

By introducing new notations

$$\frac{i}{2\tau_s}\frac{R_{00}^0 - R_{jj}^0}{\Delta + 2\lambda} = R_j^{(n)} + iI_j^{(n)},$$

we have

$$r_j^{out}(\lambda_n) = 2\eta_n l_j^{(n)} \exp(2\eta_n t_{0n}) \frac{1 + R_j^{(n)} + iI_j^{(n)}}{1 - R_j^{(n)} - iI_j^{(n)}}. \tag{9.4.20}$$

If the inhomogeneously broadened line is symmetric in shape and if the USP incident on the interface is so that ζ_n=0, then $I_j^{(n)}$=0 and

$$r_j^{out}(\lambda_n) = 2\eta_n l_j^{(n)} \exp[2\eta_n (t_{0n} + t_{jn})], \tag{9.4.21}$$

where

$$t_{jn} = (2\eta_n)^{-1} \ln[(1 + R_j^{(n)})/(1 - R_j^{(n)})]. \tag{9.4.22}$$

It is useful to compare (9.4.19) and (9.4.21) with the analogous expressions from the theory of self-induced transparency. This yields a rule how to obtain the results we need from those already known. Namely, wherever the path Z covered by the pulse occurs in an expression, the substitution $R_j Z \to 2\eta_n t_{jn}$ provides an expression relevant to the problem considered here.

Let for example the incident USP have an envelope $\mathcal{E}_{Aj}^{in}(t)$ so that

$$(d_{Aba}/\hbar) T_A(\beta_j, \omega_j) \mathcal{E}_{Aj}^{in}(t) = 2\eta\, l_j \operatorname{sech}[2\eta(t - t_0)]. \tag{9.4.23}$$

The solution of the spectral problem is then known:

$$\lambda_1 = i\eta, \quad T_{11}^{in} = \frac{\lambda - i\eta}{\lambda + i\eta}, \quad T_{12}^{in} = T_{13}^{in} = T_{21}^{in} = T_{31}^{in} = 0$$

(the remaining matrix elements are unnecessary); $r_j^{in}(\lambda_1)=2hl_j\exp(2\eta t_0)$. The solution (9.4.4) with $x>0$ generated by the IST method, can be written by means of the above formulated rule:

$$\Lambda_j(t,x>0) = \frac{d_{12}}{\hbar}\mathcal{E}_{Aj}(t,x>0) = 2\eta l_j \exp[2\eta(t_0 - t + t_{j1})]D^{-1},$$
$$D = 1 + \exp[4\eta(t_0 - t)]\{l_1^2 \exp(4\eta t_{11}) + l_2^2 \exp(4\eta t_{21})\} \tag{9.4.24}$$

For the V configuration of the energy levels we have $R_1^{(1)}=R_2^{(1)}$, since if all the atoms are in the ground state at $t=-\infty$, then $R_{11}^0 = R_{22}^0$. Expression (9.4.24) simplifies to

$$\Lambda_j(t,x>0) = 2\eta l_j \operatorname{sec}h[2\eta(t_0 - t + t_1)], \tag{9.4.25}$$

where

$$t_1 = t_{11} = t_{21} = (2\eta)^{-1}\ln[(1+R_1^{(1)})/(1-R_1^{(1)}),$$
$$R_1^{(1)} = \frac{\eta R_{00}^0}{2\tau_s}\frac{1}{4\eta^2 + \Delta^2}.$$

With the references to (9.4.6), (9.4.23), and (9.4.25) the envelope of the refracted USP has the form

$$\mathcal{E}_{Aj}^{tr}(t) = \tfrac{1}{2}T_A(\beta_j,\omega_j)[\mathcal{E}_{Aj}^{in}(t) + \mathcal{E}_{Aj}^{in}(t - t_1)]. \tag{9.4.26}$$

Thus, the refracted USP consists of two subpulses - the first one is due to the dielectric constants step (Fresnel subpulse) and the second one is due to the resonant atom. The latter is delayed in time relative to the first. Note the logarithmic dependence of the delay time t_1 on the parameter $R_1^{(1)}$. The value of this parameter depends both on the peak intensity of the electric field of the incident USP and on the incidence angle (via τ_s).

For a Λ configuration of the resonance levels with atoms initially in the ground state we have

$$\Lambda_2(t,x>0) = 2\eta l_2 \frac{\exp[2\eta(t_0 - t)]}{1 + \exp[4\eta(t_0 - t)]\{l_1^2 \exp(4\eta t_{11}) + l_2^2\}},$$
$$\Lambda_1(t,x>0) = 2\eta l_1 \frac{\exp[2\eta(t_0 - t + t_{11})]}{1 + \exp[4\eta(t_0 - t)]\{l_1^2 \exp(4\eta t_{11}) + l_2^2\}}.$$

These envelopes do not have the shape of a hyperbolic secant, and consequently do not duplicate the shape of the incident pulse. In addition, their time delays are different. This constitutes the difference from the case of two-level atoms. In general, if the atoms of a thin film are pre-excited in the initial state, the delays of the frequency components of the USP are not equal.

9.5. Conclusion

The main results concerning the role of the local field acting on atom have been obtained here with the help of phenomenological description of the Lorentz field. This description supposes three contributions to the local field acting on atom. The first is the field formed out of the medium. The second is due to the interaction of the atom with its surroundings within some sphere of radius r_0 . The third is depolarised field from bounded charges at surface of the sphere of radius r_0. It is precisely this third field that is initially named as Lorentz field. Parameter r_0 can be essential if $2r_0 \sim \ell$, where ℓ is the width of the film. One can see from Fig.9.5.1 that only a part of sphere participates in the creation of the Lorentz field in the case $2r_0 \sim \ell$. In addition, the second part of the local field changes with the variation of the atom position within the thin film. Thus, in the case $2r_0 \sim \ell$ atoms located at the different distances from the surface are exposed to different Lorentz fields. The resonant wave incident on the film causes the dynamical shift of the detunings from resonance, which are different for various atoms. This determines the effect of dynamical inhomogeneous broadening of a spectral line. The described inhomogeneous broadening can determine the formation of various transients in a thin film such as nutation echoes, for example. The more accurate approach to the mentioned problem includes the dipole-dipole interactions between atoms [27-29].

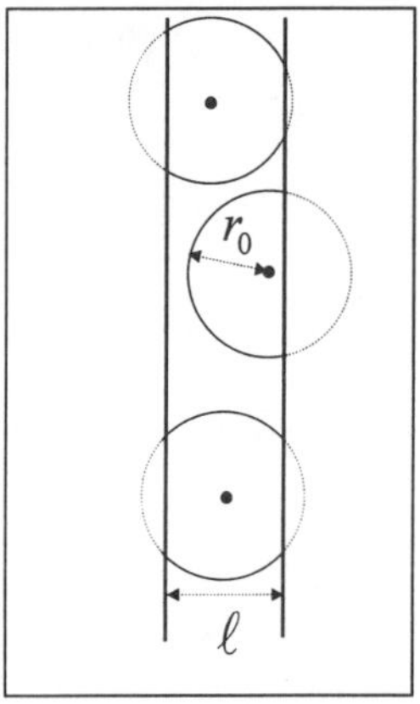

Fig.9.5.1.

A set of spatially separated thin films represents a new important model of nonlinear optics. The propagation of light in such nonlinear structures is accompanied by the distributed optical feedback that leads to diverse effects both in pulse dynamics and in the *cw* operation. In addition to optical bistability and self-pulsations, the band gap solitons [30-32] and symmetry breaking in distributed coupling [33] of counterpropagating light beams were predicted. An interplay of two effects, i.e., the symmetry breaking and transverse modulation instability, may lead to the nonsymmetrical patterns for the opposite directions of propagation. This situation was analysed by Logvin [34] in the simple case of two thin films separated by a large distance (Fig.9.5.2a). When we have many thin films separated with the distance equal to the wavelength of an exciting wave

(Fig.9.5.2b), the Brag diffraction in linear regime determines a forbidden band gap. In nonlinear regime of light interaction with a film of such set, the band gap solitary waves arise. The analysis of these waves is now of importance in connection with the problem

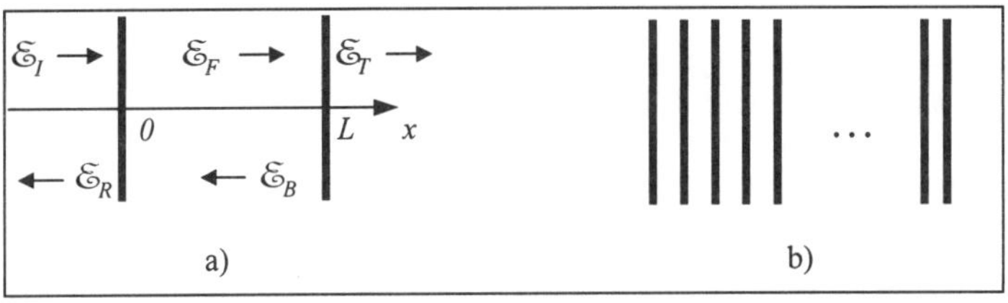

Fig.9.5.2

of localised photons in photonic band-gap materials [35,36]. The consideration of usual transients in the sets of thin films will continue evoking interest [37,38].

References

1. Agranovich, V.M., Rupasov, V.I., and Chernyak, V.Y.: Self-induced transparency of surface polaritons,. *Pis'ma Zh.Eksp.Teor.Fiz.* **33** (1981), 196-199 [*JETP Lett.* **33** (1981), 185].
2. Agranovich, V.M., Chernyak, V.Y., and Rupasov, V.I.: Self-induced transparency in wave guides, *Opt.Commun.* **37** (1981), 363-365.
3. Rupasov, V.I., and Yudson, V.I.: On the boundary problems of nonlinear optics of resonant media, *Kvant.Elektron.* **9** (1982), 2179-2186 [*Sov.J.Quantum Electron.* **12** (1982) 415-419].
4. Rupasov, V.I., and Yudson, V.I.: Nonlinear resonant optics of thin films: the method of inverse scattering transformation, *Zh.Eksp.Teor.Fiz.* **93** (1987), 494-499 [*Sov.Phys. JETP* **66** (1987), 282-285].
5. Basharov, A.M., Maimistov, A.I., and Manykin, E.A.: Exactly integrable models of resonant interaction of light with a thin film of three-level particles, *Zh.Eksp.Teor.Fiz.* **97** (1990), 1530-1543 [*Sov.Phys. JETP* **70** (1990), 864-871].
6. Basharov, A.M.: Thin film of resonant atoms: a simple model of optical bistability and self-pulsation, *Zh.Eksp.Teor.Fiz.* **94** (1988), 12-18 [*Sov.Phys. JETP* **67** (1988), 1741-1744].
7. Benedict, M.G., Zaitsev, A.I., Malysheev, V.A., and Trifonov, E.D.: Resonatorless bistability by ultrashort pulse transmission through thin layer with resonant two-level centers, *Opt. Spektrosk.* **68** (1990), 812-817 [*Opt.Spectrosc.* 68 (1990), 473-476].
8. Logvin, Yu.A., Samson, A.M., and Turovets, S.I.: Instabilities and chaos in bistable thin film of two-level atoms, *Kvant.Elektron.* **17** (1990), 1521-1524 [*Sov.J.Quantum Electron.* **20** (1982) 1425-1427].
9. Samson, A.M., Logvin, Yu.A., and Turovets, S.I.: Interaction of short light pulses with inverted thin film of resonant atoms, *Kvant.Elektron.* **17** (1990), 1223-1226.
10. Lambruschini, C.L.P.: Cooperative effects in a physically adsorbed monolayer of two-level atoms, *J.Mod.Opt.* **37** (1990), 1175-1183.
11. Benedict, M.G., Malysheev, V.A., Trifonov, E.D., and Zaitsev, A.I.: Reflection and transmision of ultrashort light pulses through a thin resonant medium: local-field effects, *Phys. Rev. A* **43** (1991), 3845-3853.
12. Samson, A.M., Logvin, Yu.A., and Turovets, S.I.: Induced superradiance in a thin film of two-level atoms, *Opt.Commun.* **78**, (1990), 208-212.

13. Kaup, D.J., and Malomed, B.A.: The resonant three-wave interaction in the an inhomogeneous medium, *Phys.Lett. A* **169** (1992), 335-340 .
14. Zacharov, S.M., and Manykin, E.A.: Spatially synchronism of photon echo in thin resonant layer at the boundary of two media, *Opt. Spektrosk.* **63** (1987), 1069-1072.
15. Zacharov, S.M., and Manykin, E.A.: Interaction of ultrashort light pulses with thin film of surface atoms at two-photon resonance, *Zh.Eksp.Teor.Fiz.* **95** (1989), 800-806.
16. Zakharov, S.M., and Manykin, E.A.: Nonlinear interaction of light with thin film of surface atoms, *Zh.Eksp.Teor.Fiz.* **105** (1994), 1053-1065.
17. Khadzhi, P.I., and Gaivan, S.L.: Nonlinear interaction of ultrashort light pulses with a thin semiconductor film at two-photon excitation of excitons, *Kvant.Elektron.***22** (1995), 929-935.
18. Khadzhi, P.I., and Gaivan, S.L.: Nonlinear transmission of light by a thin semiconductor film in the exciton resonance region, *Kvant.Elektron.***24** (1997), 546-550 [*Sov.J.Quantum Electron.* **27** (1997) 532-535].
19. Vanagas, E., and Maimistov, A.I.: The reflection of ultrashort light pulses from nonlinear boundary of two dielectric media, *Opt. Spektrosk.* **84** (1998), 301-306.
20. Ben-Aryeh, Y., Bowden, C.M., and Englund, J.C.: Intrinsic optical bistability in collections of spatially distributed two-level atoms. *Phys.Rev.* **A 34** (1986), 3917-3926.
21. Basharov, A.M.: Photonics: *Self-pulsation and chaos in optical systems*, Engineering-Physics Institute, Moscow, 1987.
22. Gibbs, H.M.: *Optical bistability: controlling light with light*, Academic Press, New York, 1985.
23. Collet, P., and Eckmann, J.P.: *Iterated maps on the interval as dynamical systems*, Birkhauser, Boston, 1980.
24. Schuster, H.G.: *Deterministic chaos*, Physik-Verlag, Weinheim, 1984.
25. Carmichael, H.J.: Optical bistability and multimode instabilities, *Phys.Rev.Lett.* **52** (1984) 1292-1295.
26. Basharov, A.M.: Optical bistability of a thin film of resonant atoms ina phase-sensitive thermostat, *Zh.Eksp.Teor.Fiz.* **108** (1995), 842-850 [*Sov.Phys. JETP* **81** (1995), 459-463].
27. Gadomsky, O.N., Krutitsky, K.V.: Near-field effect in surface optics, *JOSA B* **13** (1993) 1679-1691.
28. Gadomsky, O.N., Krutitsky, K.V.: Near-field effect and spatial distribution of spontaneous photons near surface, *Zh.Eksp.Teor.Fiz.* **106** (1994), 936-955 [*Sov.Phys. JETP* **79** (1994), 513-523].
29. Gadomsky, O.N., Krutitsky, K.V., and Gadomskia, I.V.: The effect of near field in surface optics, *Opt. Spektrosk.* **84** (1998), 780-785.
30. Mantsyzov, B.I., Kuz'min, R.N.: Coherent interaction of light with discrete periodic resonant medium, *Zh.Eksp.Teor.Fiz.* **91** (1986), 65-77 [*Sov.Phys. JETP* **64** (1986), 37-46].
31. Mantsyzov, B.I.: Gap 2π-pulse with inhomogeneous broadening line and oscillating solitary wave, *Phys.Rev.A* **51** (1995) 4939-4947.
32. Chen, W., and Mills, D.L.: Gap solitons and nonlinear optical response of superlattices, *Phys.Rev.Lett.* **58** (1987) 160-163.
33. Peschel, T., Peschel, U., and Lederer, F.: Bistability and symmetry breaking in distributed coupling of counterpropagating beams into nonlinear waveguides, *Phys.Rev.A* **50** (1994) 5153-5163.
34. Logvin, Yu.A.: Nonreciprocal optical patterns due to symmetry breaking, *Phys.Rev.A* **57** (1998) 1219-1222.
35. Neset, A, and Sajeev, J.: Optical solitary waves in two- and three- dimensional nonlinear photonic band gap structures, *Phys.Rev. E* **57** (1998), 2287-2320.
36. Conti, C., Trillo, S., and Assanto, G.: Optical gap solitons via second-harmonic generation: exact solitary solutions, *Phys.Rev. E* **57** (1998), R1251-R1255.
37. Zakharov, S.M., and Goriachev, V.A.: Dynamic of ultrashort pulse propagation through thin film's resonator structure, *Kvant.Elektron.***24** (1997), 193-208.
38. Logvin, Yu.A., Samson, A.M.: Light transmission through a system of two bistable thin films, *Zh.Eksp.Teor.Fiz.* **102** (1992), 472-481 [*Sov.Phys. JETP* **75** (1990), 250-257].

THE DENSITY MATRIX EQUATION OF A SYSTEM IN BROADBAND THERMOSTAT

We describe a systematic way for obtaining the density matrix equation and relaxation operator. We would stress at the outset that this approach is valid only for such thermostats which are described by Bose-Einstein statistics and can be simulated by δ-correlated processes. Our examples are the spontaneous atomic relaxation in an external monochromatic nonresonant wave and the relaxation in resonant broadband squeezed light.

It is commonly assumed that the radiative spontaneous relaxation is due to one-photon transitions with the absorption or emission of resonant vacuum photons. Analysis of the spontaneous atomic relaxation in an external coherent field usually involves solving an equation for the density matrix with given spontaneous relaxation constants and then determining the effective relaxation times [1]. But in addition to one-photon processes, there is a complex hierarchy of multiphoton spontaneous processes in the external field, when the transitions in an atom that lead to additional relaxation occur with the absorption of one or several photons from the coherent field simultaneously with the absorption or emission of a vacuum photon [2]. Owing to the intensity of the external field, such processes may be pronounced and can dominate ordinary one-photon spontaneous relaxation. Obviously, when the intensity of the coherent field is high and the role of these processes is important, one must re-examine even the ordinary two-level model of an atom, since for any such field, either resonant or nonresonant to two selected energy levels, other energy levels and an appropriate frequency of the vacuum photon can always be found, with the result that the relaxation transitions from the two selected levels to the other levels involving photons of both coherent and vacuum fields should be essential. The respective relaxation constant will depend here on the intensity of the coherent field in the starting equations for the density matrix of the atom and must be taken into account in the N-level approximation. Needless to say, this constant will affect many optical phenomena.

Another case when the external field changes essentially the relaxation picture is the interaction of atom with a broadband squeezed light [3]. The real and imaginary parts of atomic polarisation decay in the resonant squeezed light with different constant that affects many optical phenomena in coherent and squeezed fields. For example, absorption of a weak signal by a two-level system in squeezed and coherent intense waves is accompanied by new narrow resonances [4-6]; new types of photon echoes arise under the action of separated in time pulses of coherent and squeezed light [7]; the simultaneous effect of coherent and squeezed light on optical resonator and thin film containing resonant atoms produces phase sensitive optical bistability [8-10].

In this Appendix the above relaxation mechanisms in external fields are taken into account systematically. For the simplest case of one- and two-photon relaxations an equation is derived for the atomic density matrix containing, in addition to the usual Einstein relaxation constant, relaxation parameters that depend on the intensity of the external coherent electromagnetic field. The equation is generalised to the case of a squeezed vacuum, whose photons participate only in two-photon relaxation transitions. Relaxation constants that allow for these processes are obtained for the three-level model of an atom. Finally, we discuss generalisation of relaxation operator for squeezed field to allow for atomic level degeneracy and polarisation state of squeezed wave.

In connection with the problems of nonlinear optics in coherent and squeezed light we would like to note that the standard semiclassical approach, which has been used with great success in laser physics, can be applied after a suitable generalisation to a wide class of problems that involve an intense broadband optical field in a squeezed state as well. In this latter case, as Gardiner and his co-authors showed [11-13], the non-classical properties of the pump are "hidden" in the relaxation part of the quantum mechanical equations for the atomic density matrix.

We first obtain the effective Hamiltonian taking into account classical and quantum fields (section A1.1.1). Then, in section A1.1.2 we summarise necessary information on the Ito's quantum stochastic differential equation [11,14]. In section A1.1.3 we derive the equation for density matrix. Expressions for relaxation operator of three-level system are given in section A1.1.4. The master equation for the atomic density matrix in the case of squeezed thermostat fields is obtained in section A1.1.5. The generalisation of relaxation operator accounting level degeneracy and polarisation state of squeezed field is presented in section A1.1.6.

A1.1. The effective Hamiltonian of spontaneous processes in an external coherent field

We describe the system consisting of an atom, the field of a classical electromagnetic wave of frequency ν and electric field strength

$$E = \mathcal{E}e^{-i\nu t} + c.c., \tag{A1.1.1}$$

and the vacuum electromagnetic field, by the Hamiltonian

$$H = H_a + H_b + V_{\text{coh.int}} + V_b, \quad H_a = \sum_{\alpha} E_\alpha a_\alpha^+ a_\alpha, \quad H_b = \int d\omega\, \hbar\omega\, b_\omega^+ b_\omega,$$

$$V_{\text{coh.int}} = -\mathcal{E}e^{i\nu t}\sum d_{\alpha\alpha'} a_\alpha^+ a_{\alpha'} + H.c., \quad V_b = -i\sum \int d\omega\, K(\omega) d_{\alpha\alpha'} a_\alpha^+ a_{\alpha'} b_\omega + H.c., \tag{A1.1.2}$$

where H_a is the Hamiltonian of the isolated atom, H_b is the Hamiltonian of the photon thermostat, $V_{\text{coh.int}}$ is the operator of the interaction of the atom with field (A1.1.1), and V_b is the operator of the interaction of the atom with the thermostat photons. We have introduced the following notation: a_α^+ and a_α are the creation and annihilation operators of the atom in a state with energy E_α satisfying the commutation relations $[a_\alpha, a_{\alpha'}^+] = \delta_{\alpha\alpha'}$, b_ω^+ and b_ω are the creation and annihilation operators of photon of frequency ω, with $[b_\omega, b_{\omega'}^+]=\delta(\omega-\omega')$; $d_{\alpha\alpha'}$. is the matrix element of the operator of the dipole moment between states with energies E_α and $E_{\alpha'}$; and $K(\omega)$ is the coupling constant. We have ignored polarisation effects and recoil and employed the one-dimensional approximation, while the interaction with fields is considered in the electric dipole approximation.

The initial Hamiltonian H describes the evolution of the density matrix ρ of the entire system:

$$i\hbar\partial\rho/\partial t = [H,\rho]. \tag{A1.1.3}$$

To identify the effective terms responsible for the hierarchy of the spontaneous processes under discussion in this Hamiltonian, we perform the following unitary transformation:

$$\tilde{\rho}=e^{-iS}\rho e^{iS}, \quad i\hbar\partial\tilde{\rho}/\partial t = [\tilde{H},\tilde{\rho}], \quad \tilde{H} = e^{-iS}He^{iS} - i\hbar e^{-iS}\partial e^{iS}/\partial t. \tag{A1.1.4}$$

We write S and $\tilde{H}$ in the form of series in powers of the strengths of the coherent and vacuum fields:

$$S = S^{(10)} + S^{(01)} + S^{(11)} + \ldots,$$
$$\tilde{H} = \tilde{H}^{(00)} + \tilde{H}^{(10)} + \tilde{H}^{(01)} + \tilde{H}^{(11)} + \ldots,$$

where the left index in each pair of superscripts refers to the external coherent field and the right index to the vacuum field. Obviously,

$$\begin{aligned}
\tilde{H}^{(00)} &= H_a + H_b,\\
\tilde{H}^{(10)} &= V_{coh.\text{int}} - i[S^{(10)},\tilde{H}^{(00)}] + \hbar\partial S^{(10)}/\partial t,\\
\tilde{H}^{(01)} &= V_b - i[S^{(01)},\tilde{H}^{(00)}] + \hbar\partial S^{(01)}/\partial t,\\
\tilde{H}^{(11)} &= -(i/2)[S^{(01)},V_{\text{coh.int}}] - (i/2)[S^{(10)},V_b] - (i/2)[S^{(01)},\tilde{H}^{(10)}] - (i/2)[S^{(10)},\tilde{H}^{(01)}] -\\
&\quad - i[S^{(11)},H^{(00)}] + \hbar\partial S^{(11)}/\partial t,\ldots
\end{aligned} \tag{A1.1.5}$$

Now we must require that $\tilde{H}^{(10)}$ vanish since the external electromagnetic field (A1.1.1) is assumed to be out of resonance with the atom. The quantity $\tilde{H}^{(01)}$ must contain only terms with the appropriate frequency dependence, corresponding to absorption and emission of resonant vacuum photons:

$$\widetilde{H}^{(01)} = -i\sum \int d\omega_l(\alpha\alpha')K[\omega_l(\alpha\alpha')]d_{\alpha\alpha'}a^+_\alpha a_{\alpha'}b_{\omega_l(\alpha\alpha')}\theta(E_\alpha - E_{\alpha'}) + H.c. \qquad \text{(A1.1.6)}$$

According to ref. [15], the frequency spectrum of the vacuum photons is divided into the independent individual source of noise resonantly related to each atomic transition whose frequencies are denoted by $\omega_l(\alpha\alpha')$. Here $\omega_l(\alpha\alpha')=\omega_l(\alpha'\alpha)$, and the central frequency of this noise source, $\overline{\omega}(\alpha\alpha')$, is equal to $|\omega_{\alpha\alpha'}|$, where $\omega_{\alpha\alpha'} = (E_\alpha - E_{\alpha'})/\hbar$. We must also bear in mind that all the sources of noise mentioned above act as a single source in relation to an atomic transition out of resonance with these sources. We have also used the standard notation for the unit step function, that is, $\theta(x)=1$ for $x>0$ and $\theta(x)=0$ for $x<0$. Then, assuming that the coherent field is switched on adiabatically, we get

$$S^{(10)} = -i\sum \frac{d_{\alpha\alpha'}\mathcal{E}e^{-ivt}a^+_\alpha a_{\alpha'}}{\hbar(\omega_{\alpha\alpha'} - v)} + H.c.,$$

$$S^{(01)} = \sum \int d\omega \frac{K(\omega)d_{\alpha\alpha'}a^+_\alpha a_{\alpha'}b_{\omega_l(\alpha\alpha')}\theta(E_{\alpha'} - E_\alpha)}{\hbar(\omega_{\alpha\alpha'} - \omega)} +$$
$$+\sum \int d\omega_l(\alpha\alpha') \frac{K(\omega_l(\alpha\alpha'))d_{\alpha\alpha'}a^+_\alpha a_{\alpha'}b_{\omega_l(\alpha\alpha')}\theta(E_{\alpha'} - E_\alpha)}{\hbar[\omega_{\alpha\alpha'} - \omega_l(\alpha\alpha')]} + H.c.,$$

where the prime on the integral sign indicates the absence of terms with resonant denominators.

We now substitute these expressions into the formula for $\widetilde{H}^{(11)}$ and retain only the terms with the correct frequency dependence (by defining $\widetilde{S}^{(11)}$ appropriately), which in the final analysis corresponds to the approximation of slowly varying amplitudes:

$$\widetilde{H}^{(11)} =$$
$$= -\frac{i}{2}\sum \int d\omega_\theta(\alpha\gamma)K((\omega_\theta(\alpha\gamma))\mathcal{E}e^{-ivt}a^+_\alpha a_\gamma b_{\omega_\theta(\alpha\gamma)}[\Pi_{\alpha\gamma}(v) + \Pi_{\alpha\gamma}(\omega_\theta(\alpha\gamma))]\theta(E_\alpha - E_\gamma) +$$
$$+\frac{i}{2}\sum \int d\omega_q(\alpha\gamma)K(\omega_q(\alpha\gamma))\mathcal{E}e^{-ivt}a^+_\alpha a_\gamma b^+_{\omega_q(\alpha\gamma)}[\Pi_{\alpha\gamma}(v) + \Pi_{\alpha\gamma}(-\omega_q(\alpha\gamma))]\theta(E_\gamma - E_\alpha) +$$
$$+\frac{1}{2}\sum \int d\omega_q(\alpha\gamma)K(\omega_q(\alpha\gamma))\mathcal{E}e^{-ivt}a^+_\alpha a_\gamma b^+_{\omega_q(\alpha\gamma)}[\Pi_{\alpha\gamma}(v) + \Pi_{\alpha\gamma}(-\omega_p(\alpha\gamma))]\theta(E_\alpha - E_\gamma) +$$
$$+\frac{i}{2}\sum \int d\omega_v K(\omega_v)\mathcal{E}^* e^{ivt}a^+_\alpha a_\alpha b_{\omega_v}[\Pi_\alpha(v) + \Pi_\alpha(\omega_v)] + H.c.$$

(A1.1.7)

The noise source ω_θ>($\alpha\gamma$) with the central frequency $\overline{\omega}_\theta(\alpha\gamma)$ is in two-photon resonance with the two-photon (optically forbidden) transition $E_\alpha \to E_\gamma$, that is, $\overline{\omega}_\theta(\alpha\gamma)+v=(E_\alpha-E_\gamma)/\hbar$ $(E_\alpha>E_\gamma)$; the noise source $\omega_q(\alpha\gamma)$ with central frequency $\overline{\omega}_q(\alpha\gamma)$ and the noise source $\omega_p(\alpha\gamma)$ with central frequency $\overline{\omega}_p(\alpha\gamma)$ are in combination resonance with the two-photon transitions $E_\alpha \to E_\gamma$, that is, $\overline{\omega}_q(\alpha\gamma)-v=(E_\alpha-E_\gamma)/\hbar$ $(E_\alpha>E_\gamma)$ and $v-\overline{\omega}_p(\alpha\gamma)=(E_\alpha-E_\gamma)/\hbar$ $(E_a>E_\gamma)$; and the central frequency $\overline{\omega}_v$ of the noise source ω_v coincides with the frequency v of the nonresonant wave (A1.1.1). Here we have used the following notation:

$$\Pi_{\alpha\gamma}(\omega)=\sum_\beta \frac{d_{\alpha\beta}d_{\beta\gamma}}{\hbar}\left(\frac{1}{\omega_{\beta\alpha}+\omega}+\frac{1}{\omega_{\beta\gamma}+\omega}\right),$$

determines the effective dipole moment of the two-photon transition, and

$$\Pi_\alpha(\omega)=\sum_{\alpha'}\frac{|d_{\alpha\alpha'}|^2}{\hbar}\left(\frac{1}{\omega_{\alpha\alpha'}+\omega}+\frac{1}{\omega_{\alpha\alpha'}-\omega}\right),$$

determines the Stark shift of the level E_α. Since

$$\Pi_{\alpha\gamma}(v)=\Pi_{\alpha\gamma}(\overline{\omega}_\theta(\alpha\gamma)),\quad \Pi_{\alpha\gamma}(v)=\Pi_{\alpha\gamma}(-\overline{\omega}_q(\alpha\gamma)),\quad \Pi_{\alpha\gamma}(v)=\Pi_{\alpha\gamma}(-\overline{\omega}_p(\alpha\gamma)),$$

and ignoring the frequency dependence of the Π parameters, we can write

$$\begin{aligned}\widetilde{H}^{(11)} &= -i\sum\int d\omega_\theta(\alpha\gamma)K((\omega_\theta(\alpha\gamma))\mathcal{E}e^{-ivt}a_\alpha^+a_\gamma b_{\omega_\theta(\alpha\gamma)}\Pi_{\alpha\gamma}(v)\theta(E_\alpha-E_\gamma)+\\ &+i\sum\int d\omega_q(\alpha\gamma)K(\omega_q(\alpha\gamma))\mathcal{E}e^{-ivt}a_\alpha^+a_\gamma b^+_{\omega_q(\alpha\gamma)}\Pi_{\alpha\gamma}(v)\theta(E_\gamma-E_\alpha)+\\ &+i\sum\int d\omega_p(\alpha\gamma)K(\omega_p(\alpha\gamma))\mathcal{E}e^{-ivt}a_\alpha^+a_\gamma b^+_{\omega_p(\alpha\gamma)}[\Pi_{\alpha\gamma}(v)\theta(E_\alpha-E_\gamma)+\\ &\quad+i\sum\int d\omega_v K(\omega_v)\mathcal{E}e^{-ivt}a_\alpha^+a_\gamma b^+_{\omega_v(\alpha\gamma)}+H.c.\end{aligned} \tag{A1.1.7'}$$

The term $\widetilde{H}^{(20)}$ is obtained in the same way as in the classical case and describes the Stark shift of atomic levels:

$$\widetilde{H}^{(20)}=\sum|\mathcal{E}|^2\Pi_\alpha(\omega)a_\alpha^+a_\alpha. \tag{A1.1.8}$$

Thus, the effective Hamiltonian of the atomic system and vacuum photons in the field of a classical coherent electromagnetic wave can be written as the sum of four terms determined by eqs. (A1.1.6)-(A1.1.8):

$$\widetilde{H} = \widetilde{H}^{(00)} + \widetilde{H}^{(01)} + \widetilde{H}^{(11)} + \widetilde{H}^{(20)}. \tag{A1.1.9}$$

The first term $\widetilde{H}^{(00)}$ is the sum of the Hamiltonians of the atomic and photon subsystems isolated from each other and from the other fields. The term $\widetilde{H}^{(20)}$ characterises the Stark shifts of the levels in the field of the nonresonant wave (1). Both $\widetilde{H}^{(01)}$ and $\widetilde{H}^{(11)}$ correspond to spontaneous relaxation processes in the slowly-varying-amplitude approximation, with $\widetilde{H}^{(01)}$ corresponding to one-photon relaxation and $\widetilde{H}^{(11)}$ to two-photon relaxation involving thermostat photon and the coherent wave. It is convenient to think of the effective Hamiltonian (A1.1.9) as the sum of the Hamiltonian of the atomic system in the external coherent nonresonant field H_{sys}, the vacuum-photon Hamiltonian H_b, and the operator of one- and two-photon interaction of the atom with the vacuum photons:

$$\widetilde{H} = H_{sys} + H_b + H_{\text{int}}, \tag{A1.1.9'}$$

where $H_{sys}=H_a+H^{(20)}$ and $H_{\text{int}} = \widetilde{H}^{(01)} + \widetilde{H}^{(11)}$, and for the sake of brevity we write

$$H_{\text{int}} = -i\hbar \sum \int d\omega_j K(\omega_j)[f_j(t)R_j^+ b_{\omega_j} - f_j^*(t)R_j b_{\omega_j}^+]. \tag{A1.1.10}$$

In such expressions the sum index j numbers the atomic transitions and all the above-mentioned noise sources (Fig. A1.1.1) that are in resonance both with optically allowed transitions ($j=l$ and $\omega_l \approx \omega_{\alpha'\alpha} > 0$) and with two-photon transitions ($j=\theta$ for the double resonance $\omega_\theta + \nu \approx \omega_{\gamma\alpha} > 0$ and $j=q$ and $j=p$ for the combination resonances $\omega_q - \nu \approx \omega_{\gamma\alpha} > 0$ and $\nu - \omega_p \approx \omega_{\gamma\alpha} > 0$).The operators R_j and the parameters $f_j(t)$ for the respective transitions are

$$R_l = a_\alpha^+ a_{\alpha'} \theta(E_{\alpha'} - E_\alpha), \quad f_l(t) = d_{\alpha'\alpha} / \hbar,$$

$$R_\theta = a_\alpha^+ a_\gamma \theta(E_\gamma - E_\alpha), \quad f_\theta(t) = \mathcal{E} e^{-i\nu t} \Pi_{\alpha\gamma}(\nu)/\hbar,$$

$$R_q = a_\alpha^+ a_\gamma \theta(E_\gamma - E_\alpha), \quad f_q(t) = \mathcal{E}^* e^{-i\nu t} \Pi_{\gamma\alpha}(-\nu)/\hbar,$$

$$R_p = a_\gamma^+ a_\alpha \theta(E_\gamma - E_\alpha), \quad f_p(t) = \mathcal{E}^* e^{-i\nu t} \Pi_{\alpha\gamma}(-\nu)/\hbar,$$

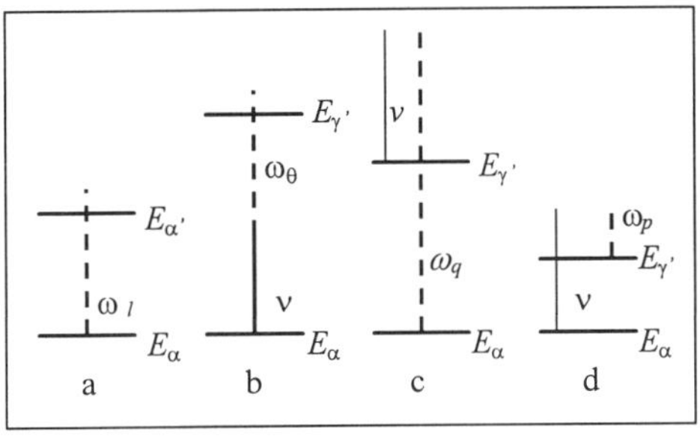

Fig.A1.1.1. The energy levels and noise sources participating in (a) one-photon and (b)-(d) two-photon relaxation processes.

In addition, j "incorporates" the noise source ω_ν at the frequency of the nonresonance wave (1) ($j=\nu$ at $\omega_\nu \approx \nu$). In contrast to the noise sources ω_l , ω_θ , ω_q , and ω_p , the ω_ν photons cause random variations to occur in the Stark frequency shifts rather than to participate in atomic transition. For this source,

$$R_\nu = \sum a_\alpha^+ a_\alpha \Pi_\alpha(\nu), \quad f_\nu(t) = -\mathfrak{E}^* e^{i\nu t} / \hbar.$$

If in addition to the nonresonant field (A1.1.1) there is a resonant classical electromagnetic field acting on the atomic system, the atomic Hamiltonian H_{sys} incorporates the term that allows for this resonant interaction in the slowly varying amplitude approximation. This introduces additional terms into $S^{(10)}$ that contribute to (A1.1.10) and to the Stark shifts of the resonant levels. In relation to nonresonant transitions some of these terms are the same as those considered above but are determined by this resonance frequency. The other fraction is different for resonant transitions and, besides, is the cause of the Bloch-Siegert shift. Often, however, these terms can be ignored.

A1.2. The quantum equations of Langevin and Ito

We go over to the Heisenberg representation and write the equation of motion for an operator of the atomic system, say A:

$$A = -\frac{i}{\hbar}[A, H_{sys}] - \sum \int d\omega_j K(\omega_j) \{ f_j(t)[A, R_j^+] b_{\omega_j} - f_j^*(t)[A, R_j] b_{\omega_j}^+ \}.$$

In this equation we replace b_{ω_j} with the integral representation

$$b_{\omega_j} = \exp[-i\omega_j(t-t_0)] b_{0\omega_j} + K(\omega_j) \int_{t_0}^{t} f_j^*(t') R_j(t') \exp[-i\omega_j(t-t')] dt',$$

where $b_{0\omega_j}$ is the value b_{ω_j} at the initial time t_0, and apply the Markov approximation [16]

$$K(\omega_j) = (\chi_j/2\pi)^{1/2}. \tag{A1.2.1}$$

Assuming that

$$\int_{t_0}^{t} R_j(t')\delta(t-t')dt' = \frac{1}{2}R_j(t),$$

we arrive at the quantum Langevin equation

$$\begin{aligned} A = -\frac{i}{\hbar}[A, H_{sys}] - \sum\Big\{[A, R_j^+]\Big(\chi_j^{1/2} b_{(in)j}(t) f_j(t) + \frac{\chi_j}{2}\left|f_j(t)\right|^2 R_j\Big) \\ -\Big(\chi_j^{1/2} b_{(in)j}^+(t) f_j(t) + \frac{\chi_j}{2}\left|f_j(t)\right|^2 R_j^+\Big)[A, R_j]\Big\}. \end{aligned} \tag{A1.2.2}$$

Here for each noise source we have introduced the in-fields

$$b_{(in)j}(t) = \frac{1}{(2\pi)^{1/2}} \int d\omega_j \exp[-i\omega_j(t-t_0)] b_{0\omega_j}, \tag{A1.2.3}$$

which satisfy the commutation relations

$$[b_{(in)j}(t), b_{(in)j'}^+(t')] = \delta_{jj'}\delta(t-t'), \quad [b_{(in)j}(t), b_{(in)j'}(t')] = 0$$

Note that $\sqrt{\chi_j} b_{(in)j}(t) f_j(t) + \frac{1}{2}\chi_j \left|f_j(t)\right|^2 R_j$ commutes with any atomic operator. We assume that the in-fields are sources of white noise:

$$\begin{aligned} Tr\{\rho_{(in)} b_{(in)j}^+(t) b_{(in)j'}(t')\} \equiv \left\langle b_{(in)j}^+(t) b_{(in)j'}(t')\right\rangle = N_j \delta_{jj'}(t-t'), \\ Tr\{\rho_{(in)} b_{(in)j}(t) b_{(in)j'}^+(t')\} \equiv \left\langle b_{(in)j}(t) b_{(in)j'}^+(t')\right\rangle = (N_j+1)\delta_{jj'}\delta(t-t'), \end{aligned} \tag{A1.2.4}$$

$$\left\langle b_{(in)j}(t) b_{(in)j'}(t')\right\rangle = \left\langle b_{(in)j}^+(t) b_{(in)j'}^+(t')\right\rangle = 0, \tag{A1.2.5}$$

which corresponds to

$$\left\langle b_{0\omega_j}^+ b_{0\omega'_j}\right\rangle = N_j \delta(\omega_j - \omega'_j), \quad \left\langle b_{0\omega_j} b_{0\omega_{j'}}^+\right\rangle = (N_j+1)\delta(\omega_j - \omega'_j),$$

$$\left\langle b_{0\omega_j} b_{0\omega_{j'}}\right\rangle = \left\langle b_{0\omega_j}^+ b_{0\omega_{j'}}^+\right\rangle = 0,$$

with $\rho_{(in)}$ the density matrix of the in-fields. If we also assume that $N_j = \overline{N}(\omega_j)$, where $\overline{N}(\omega) = 1/[\exp(\hbar\omega/kT) - 1]$, eqs. (A1.2.4) will correspond to the case of a photon thermostat of a temperature T.

According to ref. [11], we can also examine out-fields and the respective quantum Langevin equation reversed in time. But this will not be needed here. Barchielli [14] suggested a more elegant approach for obtaining out-fields.

We introduce the quantum Wiener processes $B_j(t,t_0)$ in the following manner:

$$B_j(t,t_0) = \int_{t_0}^{t} b_{(in)j}(t')dt', \tag{A1.2.6}$$

with $[B_j(t,t_0), B^+_{j'}(t,t_0)] = (t-t_0)\delta_{jj'}$. Defining the Ito integral and differential in the usual way [16], we have the following basic formulas of stochastic analysis:

$$\begin{gathered} dB^+_j(t)dB_{j'}(t) = N_j\delta_{jj'}dt, \quad dB_j(t)dB^+_{j'}(t) = (N_j+1)\delta_{jj'}dt, \\ dB_j(t)dB_{j'}(t) = dB^+_j(t)dB^+_{j'}(t)dtddt = dtdB^+_j(t) = dtdB_j(t) = 0. \end{gathered} \tag{A1.2.7}$$

To obtain Ito's quantum stochastic differential equation from the quantum Langevin equation (12), we must replace $b_{(in)j}(t)dt$ with $dB_j(t)$ and then add terms that will guarantee the validity in the obtained equation of Ito's rule of differentiation; namely, that

$$d(A_1A_2) = (dA_1)A_2 + A_1(dA_2) + (dA_1)(dA_2)$$

for each pair of atomic operators A_1 and A_2. We seek the quantum Ito equation for the atomic operator A in the form

$$dA = -\frac{i}{\hbar}[A, H_{sys}]dt - \sum\left\{[A, R^+_j]\chi_j^{1/2} f_j(t)dB_j(t) - \chi_j^{1/2} f^*_j(t)[A, R_j]dB^+_j(t)\right\} + Idt,$$

where the additional term Idt guarantees the satisfaction Ito's rule of differentiation. We seek this term in the form

$$I = \sum\psi_{1j}(R^+_j[A,R_j] + [R^+_j, A]R_j)(1+N_j) + \sum\psi_{2j}(R_j[A,R^+_j] + [R_j, A]R^+_j)N_j.$$

The structure of I is suggested by the form of the expression $(dA_1)(dA_2)$ when $I_1 = I_2 = 0$, where I_1dt and I_2dt are the corresponding terms of the Ito equation for the atomic operator A_1 and A_2. We have used the following expression for the Heisenberg operator $b_{\omega_j} = \exp[-i\omega_j(t-t_0)]b_{0\omega_j}$ and in the quantum Langevin equation (A1.2.2) we have omitted the terms proportional to χ_j describing the reverse influence of atom. This corresponds to the natural interpretation of the thermostat as a reservoir without inverse influence of a system.

It is not difficult to show that

$$\psi_{1j} = \psi_{2j} = \tfrac{1}{2}\chi_j |f_j(t)|^2 .$$

This leads to the quantum Ito equation

$$\begin{aligned} dA = -\frac{i}{\hbar}[A, H_{sys}]dt - \sum \frac{\chi_j}{2}|f_j(t)|^2 (N_j+1)(R_j^+[A,R_j] + [R_j^+, A]R_j)dt + \\ + \sum \frac{\chi_j}{2}|f_j(t)|^2 N_j (R_j[A,R_j^+] + [R_j, A]R_j^+)dt + \\ + \sum \chi_j^{1/2} |f_j^*(t)|^2 [A,R_j]dB_j^+(t) - \sum \chi_j^{1/2} |f_j(t)|^2 [A,R_j^+]dB_j(t). \end{aligned} \quad \text{(A1.2.8)}$$

The structure of this equation coincides with that of the Ito equation of refs. [11,14], but because of the factor $|f_j(t)|^2$ the terms with j=θ, j=q, j=p, and j=ν are proportional to the intensity of the external field (A1.1.1) and describe the spontaneous processes of two-photon relaxation with simultaneous participation of the coherent wave (A1.1.1) and vacuum photons.

A1.3. The equation for the atomic density matrix

In the Ito equation the coefficients of $dB_j(t)$ and $dB_j^+(t)$ are nonanticipating operator functions. Hence, averaging (A1.2.8), we get

$$\begin{aligned} \frac{\langle dA(t)\rangle}{dt} = -\left\langle \frac{i}{\hbar}[A, H_{sys}]\right\rangle - \left\langle \sum \frac{\chi_j}{2}|f_j(t)|^2 ([A,R_j^+]R_j - R_j^+[R_j, A])\right\rangle \\ - \left\langle \sum \frac{\chi_j}{2}|f_j(t)|^2 N_j ([[A,R_j^+],R_j] + [[A,R_j],R_j^+])\right\rangle. \end{aligned}$$

Here the average of an atomic operator in the Heisenberg representation is defined in terms of the density matrix

$$\rho(t) = \mathrm{Tr}_b\{\exp[i\widetilde{H}(t-t_0)/\hbar]\rho_{sys}(t_0)\otimes\rho_b(t_0)\exp[-i\widetilde{H}(t-t_0)/\hbar]\}$$

as

$$\langle A(t)\rangle = \mathrm{Tr}\{A(t_0)\rho(t)\}$$

(initially, at $t=t_0$, the density matrix is equal to the product of the density matrix $\rho_{sys}(t_0)$ of the atomic system and that of the photon thermostat, $\rho_b(t_0)$). Here

$$\frac{\langle dA(t)\rangle}{dt}=\frac{d\langle A(t)\rangle}{dt}=\mathrm{Tr}\left\{A(t_0)\left(\frac{1}{\hbar}[\rho,H_{sys}]+\right.\right.$$

$$+\sum\frac{\chi_j}{2}|f_j(t)|^2(N_j+1)(2R_j\rho R_j^+-\rho R_j^+R_j-R_j^+R_j\rho) \tag{A1.3.1}$$

$$\left.\left.+\sum\frac{\chi_j}{2}|f_j(t)|^2N_j(2R_j^+\rho R_j-\rho R_jR_j^+-R_jR_j^+\rho)\right)\right\}.$$

Since

$$\frac{d\langle A(t)\rangle}{dt}=\mathrm{Tr}\left\{A(t_0)\frac{d\rho(t)}{dt}\right\}, \tag{A1.3.2}$$

and the above equations are valid for any atomic operator A, comparison of (A1.3.1) and (A1.3.2) yields the following equation for the atomic density matrix:

$$\begin{aligned}\frac{d\rho}{dt}=\frac{i}{\hbar}[\rho,H_{sys}]+\sum\frac{\chi_j}{2}|f_j(t)|^2(N_j+1)(2R_j\rho R_j^+-\rho R_j^+R_j-R_j^+R_j\rho)+\\+\sum\frac{\chi_j}{2}|f_j(t)|^2N_j(2R_j^+\rho R_j-\rho R_jR_j^+-R_jR_j^+\rho).\end{aligned} \tag{A1.3.3}$$

A1.4. The relaxation operator for three-level models of an atom

It has proved expedient to write the equation for the transformed atomic density matrix $\widetilde{\rho}$ in terms of the relaxation operator $\widehat{\Gamma}$ as follows:

$$i\hbar\left(\frac{\partial}{\partial t}+\widehat{\Gamma}\right)\widetilde{\rho}=[\widetilde{H}_s,\widetilde{\rho}]. \tag{A1.4.1}$$

Here the Hamiltonian $\widetilde{H}_s$ of the atomic system may include, in addition to $\widetilde{H}_{sys}$ the interaction of the atom with resonant (classical) electromagnetic fields. We assume that the appropriate unitary transformation (see section 1.1.1) has been applied to the initial density matrix and the additional terms in the initial Hamiltonian that describe this resonant interaction with classical electromagnetic fields.

The simplest model of an atom allowing for spontaneous relaxation in an external field contains three levels, say E_a, E_b, and E_c, that form two adjacent optically allowed transitions E_b-E_a and E_b-E_c and an optically forbidden (two-photon) transition E_c-E_a.

Three different configurations are possible here, Λ, *V*, and θ, and in each configuration the frequency ν of the coherent wave (A1.1.1) can be either lower or higher than the frequency ω_{ca} of the optically forbidden transition. Below for these six three-level models of an atom (Fig. A1.4.1) we list the expressions for the matrix elements of the relaxation operator $(\widehat{\Gamma}\rho)_{\alpha\alpha'}$. The matrix elements of the other terms in eq. (A1.4.1) are well-known (see section 1.1).

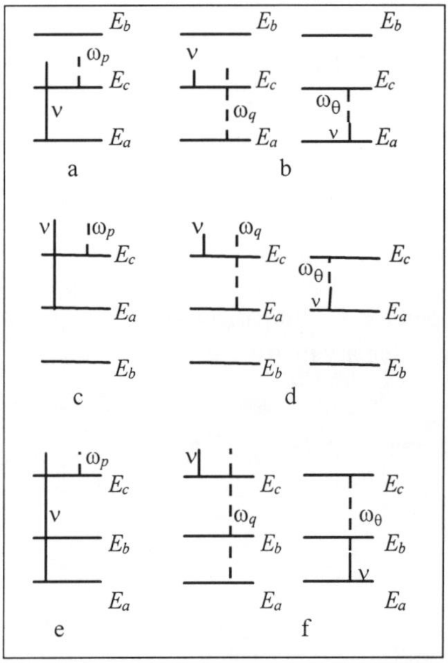

Fig. A1.4.1. Three-level models of an atom and two-photon relaxation transitions in which the photons of the monochromatic and noise fields participate.

We write the matrix elements of the relaxation operator in a form that reflects the "closed" nature of the three-level system (i.e., the fact that the number of three-level atoms does not change):

$$(\widehat{\Gamma}\rho)_{aa} = (\Gamma_a^{(b)} + \Gamma_a^{(c)})\rho_{aa} - \Gamma_b^{(a)}\rho_{bb} - \Gamma_c^{(a)}\rho_{cc},$$

$$(\widehat{\Gamma}\rho)_{bb} = (\Gamma_b^{(a)} + \Gamma_b^{(c)})\rho_{bb} - \Gamma_a^{(b)}\rho_{aa} - \Gamma_c^{(b)}\rho_{cc},$$

$$(\widehat{\Gamma}\rho)_{cc} = (\Gamma_c^{(a)} + \Gamma_c^{(b)})\rho_{cc} - \Gamma_a^{(c)}\rho_{aa} - \Gamma_b^{(c)}\rho_{bb},$$

$$(\widehat{\Gamma}\rho)_{ba} = (\Gamma_{ba}^{(ba)} + \Gamma_{ba}^{(bc)} + \Gamma_{ba}^{(ca)} + \Gamma_{ba})\rho_{ba},$$

$$(\widehat{\Gamma}\rho)_{cb} = (\Gamma_{cb}^{(cb)} + \Gamma_{cb}^{(ca)} + \Gamma_{cb}^{(ba)} + \Gamma_{cb})\rho_{cb},$$

$$(\widehat{\Gamma}\rho)_{ca} = (\Gamma_{ca}^{(ca)} + \Gamma_{ca}^{(cb)} + \Gamma_{ca}^{(ba)} + \Gamma_{ca})\rho_{ca},$$

$$\Gamma_a^{(c)} = 2\Gamma_{ba}^{(ca)},\ \Gamma_c^{(a)} = 2\Gamma_{cb}^{(ca)},\ \Gamma_b^{(c)} = 2\Gamma_{ba}^{(cb)},\ \Gamma_c^{(b)} = 2\Gamma_{ca}^{(cb)},\ \Gamma_a^{(b)} = 2\Gamma_{ca}^{(ba)},\ \Gamma_b^{(a)} = 2\Gamma_{cb}^{(ba)}.$$

In the formulas referring to the off-diagonal elements of the density matrix, the order of the indices in each pair of lower (upper) indices is unimportant. We write the constants $\Gamma_{\alpha\beta}$, $\alpha \neq \beta$ (i.e., Γ_{ca}, Γ_{cb}, and Γ_{ba}) in the unified form

$$\Gamma_{\alpha\beta} = \chi_{\nu}|\mathcal{E}|^2[\Pi_\alpha(\nu) - \Pi_\beta(\nu)]^2(N_\nu + 1/2)/\hbar^2. \qquad \text{(A1.4.2)}$$

and the remaining constants for each possible configuration of the three-level system as follows:

1. Λ-configuration $E_a<E_c<E_b$ with $\nu>\omega_{ca}$ (Fig. A1.4.1a):

$$\Gamma_{ba}^{(ba)} = \chi_{ba}|d_{ba}/\hbar|^2(N_{ba}+1/2),\quad \Gamma_{ba}^{(bc)} = \chi_{bc}|d_{bc}/\hbar|^2(N_{bc}+1)/2,$$
$$\Gamma_{ba}^{(ca)} = \chi_{p}|\mathcal{E}\Pi_{ac}(-\nu)|^2(N_p+1)/2\hbar^2,$$
$$\Gamma_{bc}^{(bc)} = \chi_{bc}|d_{bc}/\hbar|^2(N_{bc}+1/2),\quad \Gamma_{bc}^{(bc)} = \chi_{ba}|d_{ba}/\hbar|^2(N_{ba}+1)/2,$$
$$\Gamma_{bc}^{(ca)} = \chi_{p}|\mathcal{E}\Pi_{ac}(-\nu)|^2N_p/2\hbar^2,$$
$$\Gamma_{ca}^{(ca)} = \chi_{p}|\mathcal{E}\Pi_{ac}(-\nu)|^2(N_p+1/2)/\hbar^2,$$
$$\Gamma_{ca}^{(cb)} = \chi_{bc}|d_{bc}/\hbar|^2N_{bc}/2,\quad \Gamma_{ca}^{(ba)} = \chi_{ba}|d_{ba}/\hbar|^2N_{ba}/2;$$

2. Λ-configuration $E_a<E_c<E_b$ with $\nu<\omega_{ca}$ (Fig. A1.4.1b)

$$\Gamma_{ba}^{(ba)} = \chi_{ba}|d_{ba}/\hbar|^2(N_{ba}+1/2),\quad \Gamma_{ba}^{(bc)} = \chi_{bc}|d_{bc}/\hbar|^2(N_{bc}+1)/2,$$
$$\Gamma_{ba}^{(ca)} = (\chi_{q}|\mathcal{E}\Pi_{ac}(-\nu)|^2N_q + \chi_\theta|\mathcal{E}\Pi_{ac}(\nu)|^2N_\theta)/2\hbar^2,$$
$$\Gamma_{bc}^{(bc)} = \chi_{bc}|d_{bc}/\hbar|^2(N_{bc}+1/2),\quad \Gamma_{bc}^{ba} = \chi_{ba}|d_{ba}/\hbar|^2(N_{ba}+1)/2,$$
$$\Gamma_{bc}^{(ca)} = [\chi_{q}|\mathcal{E}\Pi_{ac}(-\nu)|^2(N_q+1) + \chi_\theta|\mathcal{E}\Pi_{ac}(\nu)|^2(N_\theta+1)]/2\hbar^2,$$
$$\Gamma_{ca}^{(ca)} = [\chi_{q}|\mathcal{E}\Pi_{ca}(-\nu)|^2(N_q+1) + \chi_\theta|\mathcal{E}\Pi_{ca}(\nu)|^2(N_\theta+1/2)]/\hbar^2,$$
$$\Gamma_{ca}^{(bc)} = \chi_{bc}|d_{bc}/\hbar|^2N_{bc}/2,\quad \Gamma_{ca}^{(ba)} = \chi_{ba}|d_{ba}/\hbar|^2N_{ba}/2;$$

3. *V*-configuration $E_b<E_a<E_c$ with $\nu>\omega_{ca}$ (Fig. A1.4.1c):

$$\Gamma_{ab}^{(ab)} = \chi_{ab}|d_{ab}/\hbar|^2(N_{ab}+1/2),\quad \Gamma_{ab}^{(cb)} = \chi_{cb}|d_{cb}/\hbar|^2N_{cb}2,$$
$$\Gamma_{ab}^{(ca)} = \chi_{p}|\mathcal{E}\Pi_{ac}(-\nu)|^2(N_p+1)/2\hbar^2,$$
$$\Gamma_{cb}^{(cb)} = \chi_{cb}|d_{cb}/\hbar|^2(N_{cb}+1/2),\quad \Gamma_{cb}^{(ab)} = \chi_{ab}|d_{ab}/\hbar|^2N_{ab}/2,$$
$$\Gamma_{cb}^{(ca)} = \chi_{p}|\mathcal{E}\Pi_{ac}(-\nu)|^2N_p/2\hbar^2,$$
$$\Gamma_{ca}^{(ca)} = \chi_{p}|\mathcal{E}\Pi_{ac}(-\nu)|^2(N_p+1/2)/\hbar^2,$$
$$\Gamma_{ca}^{(cb)} = \chi_{bc}|d_{bc}/\hbar|^2(N_{cb}+1)/2,\quad \Gamma_{ca}^{(ab)} = \chi_{ab}|d_{ab}/\hbar|^2(N_{ab}+1)/2;$$

4. V-configuration $E_b<E_a<E_c$ with $\nu<\omega_{ca}$ (Fig. A1.4.1d):

$$\Gamma_{ab}^{(ab)} = \chi_{ab}|d_{ab}/\hbar|^2(N_{ab}+1/2), \quad \Gamma_{ab}^{(cb)} = \chi_{cb}|d_{cb}/\hbar|^2 N_{cb}/2,$$
$$\Gamma_{ab}^{(ca)} = (\chi_q|\mathcal{E}\Pi_{ca}(-\nu)|^2 N_q + \chi_\theta|\mathcal{E}\Pi_{ca}(\nu)|^2 N_\theta)/2\hbar^2,$$
$$\Gamma_{cb}^{(cb)} = \chi_{cb}|d_{cb}/\hbar|^2(N_{cb}+1/2), \quad \Gamma_{cb}^{(ab)} = \chi_{ab}|d_{ab}/\hbar|^2 N_{ab}/2,$$
$$\Gamma_{cb}^{(ca)} = [\chi_q|\mathcal{E}\Pi_{ca}(-\nu)|^2(N_q+1) + \chi_\theta|\mathcal{E}\Pi_{ca}(\nu)|^2(N_\theta+1)]/2\hbar^2,$$
$$\Gamma_{ca}^{(ca)} = [\chi_q|\mathcal{E}\Pi_{ca}(-\nu)|^2(N_q+1/2) + \chi_\theta|\mathcal{E}\Pi_{ca}(\nu)|^2(N_\theta+1/2)]/\hbar^2,$$
$$\Gamma_{ca}^{(cb)} = \chi_{cb}|d_{cb}/\hbar|^2(N_{cb}+1)/2, \quad \Gamma_{ca}^{(ab)} = \chi_{ab}|d_{ab}/\hbar|^2(N_{ab}+1)/2;$$

5. θ-configuration $E_a<E_b<E_c$ with $\nu>\omega_{ca}$ (Fig. A1.4.1e):

$$\Gamma_{ba}^{(ba)} = \chi_{ba}|d_{ba}/\hbar|^2(N_{ba}+1/2), \quad \Gamma_{ba}^{(cb)} = \chi_{cb}|d_{cb}/\hbar|^2 N_{cb}/2,$$
$$\Gamma_{ba}^{(ca)} = \chi_p|\mathcal{E}\Pi_{ac}(-\nu)|^2(N_p+1)/2\hbar^2,$$
$$\Gamma_{cb}^{(cb)} = \chi_{cb}|d_{cb}/\hbar|^2(N_{cb}+1/2), \quad \Gamma_{cb}^{(ba)} = \chi_{ba}|d_{ba}/\hbar|^2 N_{ba}/2,$$
$$\Gamma_{cb}^{(ca)} = \chi_p|\mathcal{E}\Pi_{ac}(-\nu)|^2 N_p/2\hbar^2,$$
$$\Gamma_{ca}^{(ca)} = \chi_p|\mathcal{E}\Pi_{ac}(-\nu)|^2(N_p+1/2)/\hbar^2,$$
$$\Gamma_{ca}^{(cb)} = \chi_{bc}|d_{bc}/\hbar|^2(N_{cb}+1)/2, \quad \Gamma_{ca}^{(ba)} = \chi_{ba}|d_{ba}/\hbar|^2 N_{ba}/2;$$

6. θ-configuration $E_a< E_c < E_b$ with $\nu < \omega_{ca}$ (Fig. A1.4.1f):

$$\Gamma_{ba}^{(ba)} = \chi_{ba}|d_{ba}/\hbar|^2(N_{ba}+1/2), \quad \Gamma_{ba}^{(cb)} = \chi_{cb}|d_{cb}/\hbar|^2 N_{cb}/2,$$
$$\Gamma_{ab}^{(ca)} = [(\chi_q|\mathcal{E}\Pi_{ca}(-\nu)|^2 N_q + \chi_\theta|\mathcal{E}\Pi_{ca}(\nu)|^2 N_\theta]/2\hbar^2,$$
$$\Gamma_{cb}^{(cb)} = \chi_{cb}|d_{cb}/\hbar|^2(N_{cb}+1/2), \quad \Gamma_{cb}^{(ba)} = \chi_{ba}|d_{ba}/\hbar|^2(N_{ba}+1)/2,$$
$$\Gamma_{cb}^{(ca)} = [\chi_q|\mathcal{E}\Pi_{ca}(-\nu)|^2(N_q+1) + \chi_\theta|\mathcal{E}\Pi_{ca}(\nu)|^2(N_\theta+1)]/2\hbar^2,$$
$$\Gamma_{ca}^{(ca)} = [\chi_q|\mathcal{E}\Pi_{ca}(-\nu)|^2(N_q+1/2) + \chi_\theta|\mathcal{E}\Pi_{ca}(\nu)|^2(N_\theta+1/2)]/\hbar^2,$$
$$\Gamma_{ca}^{(cb)} = \chi_{cb}|d_{cb}/\hbar|^2(N_{cb}+1)/2, \quad \Gamma_{ca}^{(ba)} = \chi_{ba}|d_{ba}/\hbar|^2 N_{ba}/2,$$

Here, we have used $K(\omega_{\alpha\alpha'}) = (\chi_{\alpha\alpha'}/2\pi)^{1/2}$ for $\chi_{ba}, \chi_{ab}, \chi_{cb}$ and χ_{bc}.

A method for determining the relaxation constant $\Gamma_{ca}^{(ca)}$ by photon echoes is presented in ref.[2]. It is convenient to carry out such investigations using the Raman scattering described in section 2.6.2.

A1.5. Generalisation for squeezed fields

The quantum Ito equation as well as the equation for the atomic density matrix can easily be generalised to the case that lately has been a topic of intensive investigation, where a vacuum field (with $j=l$, θ, q, or p) or several such fields are in a "squeezed" state [16,17]. In refs. [11-14] and other papers the relaxation of an atomic system was considered as being in a squeezed vacuum in resonance with optically allowed transitions. The above reasoning, however, suggests that after the transformed Hamiltonian has been written in the form (A1.1.9') and (A1.1.10), the general analysis of ordinary one-photon relaxation and of the mechanism of two-photon relaxation can be conducted in an analogous way. The values ω_s denote the frequencies of the fields of the squeezed vacuum which are in one-photon resonance with optically allowed transitions ($\omega_s \approx \omega_{\alpha'\alpha} > 0$) and in two-photon resonance with optically forbidden transitions ($\omega_s + \nu \approx \omega_{\gamma\alpha} > 0$ for a double resonance and $\omega_s - \nu \approx \omega_{\gamma\alpha} > 0$ and $\nu - \omega_s \approx \omega_{\gamma\alpha} > 0$ for Raman resonances). The definition of the respective operators R_s and parameters $f_s(t)$ remains unchanged. For ideal squeezing the difference from the case of ideal white noise and thermostat consists in replacing condition (A1.2.5) with

$$\left\langle b_{(ib)s}(t)b_{(ib)s'}(t')\right\rangle = M_s\delta_{ss'}\delta(t-t'),\ \left\langle b^+_{(ib)s}(t)b^+_{(ib)s'}(t')\right\rangle = M^*_s\delta_{ss'}\delta(t-t'),\quad \text{(A1.5.1)}$$

and, in deriving the Ito equation, instead of

$$dB_s(t)dB_{s'}(t) = dB^+_s(t)dB^+_{s'}(t) = 0$$

we must assume that

$$dB_s(t)dB_{s'}(t) = M_s\delta_{ss'}dt,\quad dB^+_s(t)dB^+_{s'}(t) = M^*_s\delta_{ss'}dt,$$

where the parameters M_s and N_s obey the condition $|M_s|^2 \le N_s(N_s+1)$, with the equality attained for squeezed light exiting from an ideal degenerate parametric amplifier [18]. Note that the parameters M_s contain a rapidly oscillating term:

$$M_s = \widetilde{M}_s \exp(-2i\,\overline{\omega}_s t),$$

where $\widetilde{M}_s$ does not depend on time, and $\overline{\omega}_s$ is the central frequency of the squeezed source.

The following result can easily be obtained:

$$\begin{aligned}\frac{d\rho}{dt} = \frac{i}{\hbar}[\rho, H_{sys}] + \sum \frac{\chi_j}{2}|f_j(t)|^2 (N_j+1)(2R_j\rho R_j^+ - \rho R_j^+ R_j - R_j^+ R_j\rho) - \\ - \sum \frac{\chi_j}{2}|f_j(t)|^2 N_j(2R_j^+\rho R_j - \rho R_j R_j^+ - R_j R_j^+\rho) - \\ - \sum \frac{\chi_j}{2}|f_j(t)|^2 M_s(2R_s^+\rho R_s^+ - \rho R_s^+ R_s^+ - R_s^+ R_s^+\rho) - \\ - \sum \frac{\chi_j}{2}|f_j(t)|^2 M_s^*(2R_s\rho R_s - \rho R_s R_s - R_s R_s\rho)\end{aligned} \tag{A1.5.2}$$

where in the sums over j we have included the summation over s.

A1.6. Relaxation operator for degenerate atomic levels

We generalise the result of ref.[11,14] to account for Zeeman level structure and state of squeezed wave polarisation. We follow the paper [19].

A1.6.1 INITIAL EQUATIONS

We describe the system, which consists of an ensemble of noninteracting atoms and a classical electromagnetic field $\vec{E}$, using the ordinary semiclassical approach in the electric-dipole approximation by means of the Hamiltonian

$$H_{sys} = \sum_{\vec{p}\alpha} (E_\alpha + \hbar^2 p^2 / 2M) a^+_{\vec{p}\alpha} a_{\vec{p}\alpha} - \frac{1}{V} \sum_{\vec{p}\vec{q}\alpha\alpha'} \vec{E}(\vec{q},t)\vec{d}_{\alpha\alpha'} a^+_{\vec{p}\alpha} a_{\vec{p}-\vec{q}\alpha'} \tag{A1.6.1}$$

and the Maxwell equation

$$\left(\Delta - \frac{1}{c^2}\frac{\partial^2}{\partial t^2}\right)\vec{E}(\vec{r},t) = \frac{4\pi}{c^2}\frac{\partial^2}{\partial t^2}\vec{P}. \tag{A1.6.2}$$

Here $a_{\vec{p}\alpha}$ is an operator that annihilates an atom with momentum $\hbar\vec{p}$ and mass M in the quantum state $|\alpha\rangle$, V is the system volume. The labels α,α' enumerate eigenstates of the Hamiltonian H_0 of an isolated atom. For an atomic (molecular) gas, these states are distinguished from one another by their energy E_α, and total angular momentum j_α and/or its projection m_α onto the axis of quantisation; the energy levels are degenerate with respect to different orientations of the total angular momentum. We denote by $\vec{d}_{\alpha\alpha'}$ the matrix elements of the dipole moment operator of an atom, and write the matrix elements of a given atomic operator *A*, in the following equivalent forms:

$$\langle\alpha|A|\alpha'\rangle = \langle E_\alpha, j_\alpha, m_\alpha|A|E_{\alpha'}, j_{\alpha'}, m_{\alpha'}\rangle = A_{\alpha\alpha'} = A_{m_\alpha m_{\alpha'}} .$$

The quantity $\vec{E}(\vec{q},t)$ is the spatial Fourier component of the electric field intensity:

$$\vec{E}(\vec{q},t) = \int \vec{E}(\vec{r},t)\exp(-i\vec{q}\vec{r})d\vec{r} .$$

The polarisation $\vec{P}$ of the atomic medium is defined in the usual way:

$$\vec{P} = N\int Sp(\rho\vec{d})d\vec{v} , \qquad \text{(A1.6.3)}$$

where ρ is the density matrix of atoms moving with velocities in the range from **v** to **v**+*d***v**, and *N* is the density of two-level atoms.

We derive equations for the atomic density matrix ρ for the case in which the atoms also interact with a quantized one-dimensional electromagnetic field propagating along the *z* axis. For simplicity we neglect the recoil during the absorption/emission of a photon, Raman effects with the participation of the classical and quantum fields, and Stark shifts of the levels. We first obtain an equation for the density matrix of a single atom at rest, which we then generalise to an ensemble of atoms. In this case we treat the atom as a two-level system with quantum numbers E_a, j_a, m and E_b, j_b, μ, which characterise the upper and lower levels, respectively. The total Hamiltonian of this problem has the form

$$H = H_s + H_b + V_b , \qquad \text{(A1.6.4)}$$

$$H_s = \sum_m E_a a_m^+ a_m + \sum_\mu E_b a_\mu^+ a_\mu - \sum_{m\mu} \vec{E}(\vec{r},t)(\vec{d}_{m\mu} a_m^+ a_\mu + \vec{d}_{\mu m} a_\mu^+ a_m),$$

$$H_b = \sum_\lambda \int d\omega\hbar\omega b_{\lambda\omega}^+ b_{\lambda\omega} , \; V_b = -i\hbar \sum_{\lambda\mu m} \int d\omega K(\omega)(\vec{e}_{\lambda\omega}\vec{d}_{\mu m}) a_\mu^+ a_m b_{\lambda\omega} + H.c.. $$

Here $b_{\lambda\omega}^+$ and $b_{\lambda\omega}$ are creation and annihilation operators of a photon of frequency ω with a polarisation unit vector $\vec{e}_{\lambda\omega}$ and *K*(ω) is the coupling constant. In deriving the operator V_b for the interaction of an atom with a quantum field, we use the rotating-wave approximation.

A1.6.2. QUANTUM ITO EQUATION

We assume that the quantized electromagnetic field represents ideal squeezed white noise, which implies that

$$\langle b^+_\lambda(t)b_{\lambda'}(t')\rangle = N_{ph}\tau_{\lambda\lambda'}\delta(t-t'),\ \langle b_{\lambda'}(t)b^+_\lambda(t')\rangle = (\delta_{\lambda\lambda'} + N_{ph}\tau_{\lambda\lambda'})\delta(t-t'),$$
$$\langle b_\lambda(t)b_{\lambda'}(t')\rangle = M_{ph}\eta_{\lambda\lambda'}\delta(t-t'),\ \langle b^+_{\lambda'}(t')b^+_\lambda(t)\rangle = M^*_{ph}\eta^*_{\lambda\lambda'}\delta(t-t'), \tag{A1.6.5}$$

where

$$b_\lambda(t) = \frac{1}{\sqrt{2\pi}}\int d\omega \exp(-i\omega(t-t_0))b_{\lambda\omega}(t_0), \tag{A1.6.6}$$

the angle brackets imply averaging over the squeezed photon heat bath, the matrices $\tau_{\lambda\lambda'} = \tau^*_{\lambda'\lambda}$ and $\eta_{\lambda\lambda'} = \eta_{\lambda'\lambda}$ characterise the polarisation state of the squeezed field, the parameter $|M_{ph}|$ represents the degree of compression, and N_{ph} represents the intensity of this field. We also assume that $M_{ph} = |M_{ph}|\exp(-2i\overline{\omega}t)$, where the centre frequency of the squeezed heat bath is denoted by $\overline{\omega}$. For a squeezed field prepared in an ideally degenerate parametric amplifier, the quantity $\overline{\omega}$ is the centre frequency of the amplifier. Generally speaking, the centre frequency $\overline{\omega}$ can be separated into right- and left-circularly-polarised photons. All such features are taken into account by the phases of the quantities $\eta_{\lambda\lambda'}$ and the off-diagonal elements $\tau_{\lambda\lambda'}$. The annihilation operator for photons $b_{\lambda\omega}(t_0)$ is referred to a certain initial time t_0.

We write the Heisenberg equation of motion for a given atomic operator A:

$$\dot{A} = -\frac{i}{\hbar}[A, H_s] + \sum_{\lambda q}\int d\omega\, K(\omega)[A, R^q_+]e^q_{\lambda\omega}b_{\lambda\omega} - \sum_{\lambda' q'}\int d\omega\, K(\omega)[A, R^{q'}_-]e^{q'*}_{\lambda'\omega}b^+_{\lambda'\omega}\,. \tag{A1.6.7}$$

Here we have introduced the operators

$$R^q_+ = \sum_{\mu m} a^+_\mu a_m d^{-q}_{\mu m},\ R^q_- = -\sum_{\mu m} a^+_m a_\mu d^q_{m\mu} = (R^q_+)^+ .$$

The labels q, q', etc., specify the spherical components of a vector (see section 1.4).

We assume that there is no feedback of the atoms to the photon thermostat, so that the Heisenberg operators for annihilation of photons in (A1.6.7) are given by the expression for free evolution

$$b_{\lambda\omega} = b_{\lambda\omega}(t_0)\exp(-i\omega(t-t_0)),$$

and that the operator $b_{\lambda\omega}$ commutes with the atomic operators referred to the same instant of time.

Following section A1.2, we assume that the coupling constant $K(\omega)$ does not depend on frequency ω, which corresponds to the Markov approximation:

$$K(\omega) = \sqrt{\chi/2\pi}\,.$$

Moreover, we assume that the polarisation vector $\vec{e}_{\lambda\omega}$ also does not depend on the frequency ω: $\vec{e}_{\lambda\omega} = \vec{e}_{\lambda}$. Then the quantum Langevin equation follows from eq. (A1.6.7):

$$\dot{A} = -\frac{i}{\hbar}[A,H_s] + \sqrt{\chi}\sum_{\lambda q}[A,R_+^q]e_\lambda^q b_\lambda(t) - \sqrt{\chi}\sum_{\lambda' q'}[A,R_-^{q'}]e_{\lambda'}^{q'*}b_{\lambda'}^+(t)\,. \tag{A1.6.8}$$

The quantum Wiener processes $B_\lambda(t,t_0)$ are

$$B_\lambda(t,t_0) = \int_{t_0}^{t} dt'\, b_\lambda(t')\,,$$

where $[B_\lambda(t,t_0), B_{\lambda'}^+(t,t_0)] = (t-t_0)\delta_{\lambda\lambda'}$. Defining the Ito integral and differential in the standard way, we seek the quantum Ito equation for the atomic operator A in the form

$$dA = -\frac{i}{\hbar}[A,H_s]dt + \sqrt{\chi}\sum_{\lambda q}[A,R_+^q]e_\lambda^q dB_\lambda(t) - \sqrt{\chi}\sum_{\lambda' q'}[A,R_-^{q'}]e_{\lambda'}^{q'*}dB_{\lambda'}^+(t) + I dt\,, \tag{A1.6.9}$$

where

$$dB_\lambda^+(t)dB_{\lambda'}(t) = N_{ph}\tau_{\lambda\lambda'}dt, \quad dB_{\lambda'}(t)dB_\lambda^+(t) = (\delta_{\lambda\lambda'} + N_{ph}\tau_{\lambda\lambda'})dt\,,$$

$$dB_\lambda(t)dB_{\lambda'}(t) = M_{ph}\eta_{\lambda\lambda'}dt, \quad dB_{\lambda'}^+(t)dB_\lambda^+(t) = M_{ph}^*\eta_{\lambda\lambda'}^*dt\,,$$

$$dtdt = dtdB_\lambda(t) = dtdB_\lambda^+(t) = dB_\lambda(t)dt = dB_\lambda^+(t)dt = 0\,,$$

$$I = \tfrac{1}{2}\chi\sum_{qq'}\left(R_+^q[A,R_-^{q'}] + [R_+^q,A]R_-^{q'}\right)\left(\delta_{qq'} + T_{q'q}\right) + \tfrac{1}{2}\chi\sum_{qq'}\left(R_-^q[A,R_+^{q'}] + [R_-^q,A]R_+^{q'}\right)T_{qq'} -$$
$$-\tfrac{1}{2}\chi\sum_{qq'}\left(R_+^q[A,R_+^{q'}] + [R_+^q,A]R_+^{q'}\right)M_{q'q} - \tfrac{1}{2}\chi\sum_{qq'}\left(R_-^q[A,R_-^{q'}] + [R_-^q,A]R_-^{q'}\right)M_{q'q}^*,$$

$$T_{q'q} = N_{ph}\sum_{\lambda\lambda'}e_\lambda^q e_{\lambda'}^{q'*}\tau_{\lambda'\lambda} = T_{qq'}^*, \quad M_{qq'} = M_{ph}\sum_{\lambda\lambda'}e_\lambda^q\eta_{\lambda\lambda'}e_{\lambda'}^{q'} = M_{q'q}\,,$$

A1.6.3. RELAXATION OPERATOR

It follows from the eq.(A1.6.9) and section A1.3:

$$\hat{\Gamma}\rho =$$

$$= -\frac{\chi}{2}\sum_{qq'}\left(2R_-^{q'}\rho R_+^{q} - \rho R_+^{q} R_-^{q'} - R_+^{q} R_-^{q'}\rho\right)\left(\delta_{qq'} + T_{q'q}\right) - \frac{\chi}{2}\sum_{qq'}\left(2R_+^{q'}\rho R_-^{q} - \rho R_-^{q} R_+^{q'} - R_-^{q} R_+^{q'}\rho\right)T_{qq'} +$$

$$+ \frac{\chi}{2}\sum_{qq'}\left(2R_+^{q'}\rho R_+^{q} - \rho R_+^{q} R_+^{q'} - R_+^{q} R_+^{q'}\rho\right)M_{q'q} + \frac{\chi}{2}\sum_{qq'}\left(2R_-^{q'}\rho R_-^{q} - \rho R_-^{q} R_-^{q'} - R_-^{q} R_-^{q'}\rho\right)M_{q'q}^{*}.$$

(A1.6.10)

The matrix elements of the relaxation operator have the form

$$\left(\hat{\Gamma}\rho\right)_{mm'} = \chi\sum_{\mu\mu'qq'} d_{m\mu}^{q'}\rho_{\mu\mu'}d_{\mu'm'}^{-q}\left(\delta_{qq'} + T_{q'q}\right) - \frac{\chi}{2}\sum_{m''\mu qq'}\rho_{mm''}d_{m''\mu}^{q}d_{\mu m'}^{-q'}T_{qq'} - \frac{\chi}{2}\sum_{m''\mu qq'} d_{m\mu}^{q}d_{\mu m''}^{-q'}\rho_{m''m'}T_{qq'},$$

$$\left(\hat{\Gamma}\rho\right)_{\mu\mu'} = -\frac{\chi}{2}\sum_{m\mu''qq'}\rho_{\mu\mu''}d_{\mu''m}^{-q}d_{m\mu'}^{q'}\left(\delta_{qq'} + T_{q'q}\right) - \frac{\chi}{2}\sum_{m\mu''qq'} d_{\mu m}^{-q}d_{m\mu''}^{q'}\rho_{\mu''\mu'}\left(\delta_{qq'} + T_{q'q}\right) +$$

$$+ \chi\sum_{mm'qq'} d_{\mu m}^{-q'}\rho_{mm'}d_{m'\mu'}^{q}T_{qq'},$$

(A1.6.11)

$$\left(\hat{\Gamma}\rho\right)_{\mu m} = -\frac{\chi}{2}\sum_{m'\mu'qq'} d_{\mu m'}^{-q}d_{m'\mu'}^{q'}\rho_{\mu'm}\left(\delta_{qq'} + T_{q'q}\right) - \frac{\chi}{2}\sum_{m'\mu'qq'}\rho_{\mu m'}d_{m'\mu'}^{q}d_{\mu'm}^{-q'}T_{qq'} + \chi\sum_{m'\mu'qq'} d_{\mu m'}^{-q'}\rho_{m'\mu'}d_{\mu'm}^{-q}M_{q'q},$$

$$\left(\hat{\Gamma}\rho\right)_{m\mu} = -\frac{\chi}{2}\sum_{m'\mu'qq'}\rho_{m\mu'}d_{\mu'm'}^{-q}d_{m'\mu}^{q'}\left(\delta_{qq'} + T_{q'q}\right) - \frac{\chi}{2}\sum_{m'\mu'qq'} d_{m\mu'}^{q}d_{\mu'm'}^{-q'}\rho_{m'\mu}T_{qq'} + \chi\sum_{m'\mu'qq'} d_{m\mu'}^{q'}\rho_{\mu'm'}d_{m'\mu}^{q}M_{q'q}^{*}.$$

When $T_{qq'}=M_{qq'}=0$ and the summation with respect to q includes the value $q=0$, the relaxation operator (A1.6.11) coincides with the relaxation operator for spontaneous radiative decay under conditions where the energy level E_a is the ground state If $M_{qq'}=0$ while $T_{qq'} \neq 0$ and $q \neq 0$, then (A1.6.11) describes relaxation processes in the field of a noisy light wave that is structured and polarised in a certain way. Interaction with such a wave alters the distribution of atoms with respect to the Zeeman sublevels from the case of spontaneous radiative decay. Although squeezing of the electromagnetic wave directly affects only the optical coherence matrix, which determines transitions between Zeeman sublevels of different energy levels, in the final analysis this also affects the equilibrium distribution of atoms in the field of a coherent wave with respect to Zeeman sublevels of both the upper and lower energy levels. Moreover, the rate of relaxation of the real and imaginary parts of each optical coherence matrix element is explicitly changed, which in turn leads to different attenuation laws for the imaginary and real parts of each projection of the atomic polarisation (see the next section).

It is important to emphasise that by virtue of the geometry of the quantized field chosen by us, the relaxation operator $\hat{\Gamma}\rho$ does not contain terms with q=0 if we do not specifically require it to, and consequently does not take into account the spontaneous decay of an atom with emission of a photon perpendicular to the direction of propagation of the wave. Such processes, and also other channels for relaxation and/or cases in which the lower energy level E_a is not the ground state, can be described with the help of the simplest generalisation of eq. (A1.6.10):

$$\frac{d\rho}{dt} + \hat{\Gamma}\rho + \hat{\Gamma}_0\rho = \frac{i}{\hbar}[\rho, H_s], \qquad \text{(A1.6.10')}$$

where we have introduced a phenomenological relaxation operator $\hat{\Gamma}_0\rho$ with the following matrix elements:

$$\left(\hat{\Gamma}_0\rho\right)_{mm'} = \gamma_1\rho_{mm'} - W_{mm'}, \quad \left(\hat{\Gamma}_0\rho\right)_{\mu\mu'} = \gamma_2\rho_{\mu\mu'} - W_{\mu\mu'}, \quad \left(\hat{\Gamma}_0\rho\right)_{\mu m} = \gamma_{21}\rho_{\mu m} \quad .$$

Here γ_1, γ_2 and γ_{21} are relaxation constants, while the terms $W_{mm'}$ and $W_{\mu\mu'}$ take pumping processes into account. As usual, we assume for simplicity that

$$W_{mm'} = \frac{\gamma_1 N_1 \delta_{mm'}}{2j_a + 1}, \quad W_{\mu\mu'} = \frac{\gamma_2 N_2 \delta_{\mu\mu'}}{2j_b + 1}$$

where N_1 and N_2 are the steady-state densities of atoms that populate the energy levels E_a and E_b. Here and then we assume the density matrix is normalised by the atomic density.

If the matrix ρ describes atoms (molecules) moving with velocities lying in the interval from $\vec{v}$ to $\vec{v} + d\vec{v}$, then it is necessary to replace the derivative $d\rho/dt$ by $\left(\frac{\partial}{\partial t} + \vec{v}\nabla\right)\rho$ in eqs. (A1.6.10) and (A1.6.10'), and to make the following replacement in the matrix elements of the operator $\hat{\Gamma}_0\rho$: $N_1 \to N_1F(v), \quad N_2 \to N_2F(v)$, where $F(v)$ is the distribution function of the atoms with respect to velocity (see the end of section 1.3.4 and Appendix 2).

A1.6.4. RELAXATION OPERATOR FOR SMALL VALUES OF THE TOTAL ATOMIC ANGULAR MOMENTUM

The simplest form of the relaxation operator is obtained for atomic transitions between levels with small values of the angular momentum: j_a=j_b=1/2 and j_a=0=j_b-1

For a resonant atomic transition with $j_a = j_b = 1/2$ we have

$$\hat{\Gamma}\rho^{a}_{1/2\,1/2} = \frac{\chi|d_{ba}|^{2}}{3}\left(\rho^{a}_{1/2\,1/2}T_{11} - \rho^{b}_{-1/2\,-1/2}(1+T_{11})\right),$$
$$\hat{\Gamma}\rho^{a}_{-1/2\,-1/2} = \frac{\chi|d_{ba}|^{2}}{3}\left(\rho^{a}_{-1/2\,-1/2}T_{-1-1} - \rho^{b}_{1/2\,1/2}(1+T_{-1-1})\right), \qquad \text{(A1.6.12)}$$

$$\hat{\Gamma}\rho^{a}_{1/2\,-1/2} = \frac{\chi|d_{ba}|^{2}}{6}\rho^{a}_{1/2\,-1/2}(T_{11}+T_{-1-1}) - \frac{\chi|d_{ba}|^{2}}{3}\rho^{b}_{-1/2\,1/2}T_{1-1},$$
$$\hat{\Gamma}\rho^{b}_{1/2\,1/2} = \frac{\chi|d_{ba}|^{2}}{3}\left(\rho^{b}_{1/2\,1/2}(1+T_{-1-1}) - \rho^{a}_{-1/2\,-1/2}T_{-1-1}\right), \qquad \text{(A1.6.13)}$$
$$\hat{\Gamma}\rho^{b}_{-1/2\,-1/2} = \frac{\chi|d_{ba}|^{2}}{3}\left(\rho^{b}_{-1/2\,-1/2}(1+T_{11}) - \rho^{a}_{1/2\,1/2}T_{11}\right),$$

$$\hat{\Gamma}\rho^{b}_{1/2\,-1/2} = \frac{\chi|d_{ba}|^{2}}{6}\rho^{b}_{1/2\,-1/2}(2+T_{11}+T_{-1-1}) - \frac{\chi|d_{ba}|^{2}}{3}\rho^{a}_{-1/2\,1/2}T_{1-1},$$

$$\hat{\Gamma}\rho^{ba}_{1/2\,-1/2} = \frac{\chi|d_{ba}|^{2}}{6}\rho^{ba}_{1/2\,-1/2}(1+2T_{-1-1}) + \frac{\chi d^{2}_{ba}}{3}\rho^{ba*}_{1/2\,-1/2}M_{-1-1},$$
$$\hat{\Gamma}\rho^{ba}_{-1/2\,1/2} = \frac{\chi|d_{ba}|^{2}}{6}\rho^{ba}_{-1/2\,1/2}(1+2T_{11}) + \frac{\chi d^{2}_{ba}}{3}\rho^{ba*}_{-1/2\,1/2}M_{11}, \qquad \text{(A1.6.14)}$$

$$\hat{\Gamma}\rho^{ba}_{1/2\,1/2} = \frac{\chi|d_{ba}|^{2}}{6}\rho^{ba}_{1/2\,1/2}(1+T_{11}+T_{-1-1}) + \frac{\chi d^{2}_{ba}}{3}\rho^{ba*}_{-1/2\,-1/2}M_{-11},$$
$$\hat{\Gamma}\rho^{ba}_{-1/2\,-1/2} = \frac{\chi|d_{ba}|^{2}}{6}\rho^{ba}_{-1/2\,-1/2}(1+T_{11}+T_{-1-1}) + \frac{\chi d^{2}_{ba}}{3}\rho^{ba*}_{1/2\,1/2}M_{1-1}.$$

Here, the density matrices are furnished with appropriate labels that describe atomic levels b,a and transitions between them for the sake of convenience.

When linearly polarised coherent light propagates in such a medium, interacting with the system of Zeeman sublevels of the resonant atomic levels, two independent subsystems separate out, one of which interacts with left-handed circularly-polarised photons of the coherent field only, the other with right-handed circularly-polarised photons only. If the atomic medium is initially in thermodynamic equilibrium, then only relaxation operators (A1.6.12)-(A1.6.14) appear to be important.

We now discuss the relaxation of the amplitude

$$p^q = \sum_{\mu m} \rho_{\mu m} d^q_{m\mu} \exp(i\omega_\Gamma t) \ , \qquad \text{(A1.6.15)}$$

of the atomic polarisation

$$\vec{P} = \sum_{\mu m} \rho_{\mu m} \vec{d}_{m\mu} + k.c. .$$

It is not difficult to show that the corresponding relaxation operators for the real (p^q_{real}) and imaginary (p^q_{im}) parts of the polarisation amplitude have the form

$$\hat{\Gamma} p^q_{real} = \frac{\chi |d_{ba}|^2}{3} p^q_{real}\left(T_{qq} + |M_{qq}| + \frac{1}{2}\right), \ \hat{\Gamma} p^q_{im} = \frac{\chi |d_{ba}|^2}{3} p^q_{im}\left(T_{qq} - |M_{qq}| + \frac{1}{2}\right).$$

(repeated indices here do not imply summation!) Since $|M_{qq}| \le \sqrt{T_{qq}(T_{qq}+1)}$, when $T_{qq} >> 1$, and under certain conditions where $|M_{qq}| \approx T_{qq}$, it is clear that there is appreciable suppression of the relaxation of p^q_{im} and enhancement of the attenuation rate of p^q_{real}:

$$\hat{\Gamma} p^q_{im} \approx \frac{\chi |d_{ba}|^2}{6} \cdot \frac{1}{T_{qq}} \cdot p^q_{im} \approx 0, \ \hat{\Gamma} p^q_{real} \approx 2\frac{\chi |d_{ba}|^2}{3} \cdot T_{qq} \cdot p^q_{real} .$$

The picture we obtain of the density relaxation agrees with the solution of the problem without taking into account degeneracy of the resonant levels and polarisation states of the coherent and squeezed waves.

Entirely different relaxation effects occur for the atomic transition $j_a=0=j_b-1$. We first write the following expressions for the relaxation operator:

$$\hat{\Gamma} \rho^a = \frac{\chi |d_{ba}|^2}{3} \rho^a \sum_q T_{qq} - \frac{\chi |d_{ba}|^2}{3} \left(\sum_q \rho^b_{qq} + \sum_{qq'} \rho^b_{qq'} T_{-q-q'} \right), \qquad \text{(A1.6.16)}$$

$$\left(\hat{\Gamma} \rho^{ba}\right)_{\mu 0} = \frac{\chi |d_{ba}|^2}{6} \rho^{ba}_{\mu 0}\left(1 + \sum_q T_{qq}\right) + \frac{\chi |d_{ba}|^2}{6} \sum_q \rho^{ba}_{q0} T_{-q-\mu} + \frac{\chi d^2_{ba}}{3} \sum_q \rho^{ab}_{0q} M_{-\mu-q} .$$

For a squeezed field with the simplest polarisation structure

$$T_{qq'} = \delta_{qq'} N_q, \quad M_{qq'} = \delta_{qq'} M_q, \quad |M_q| \le \left(N_q (N_q + 1)\right)^{1/2} \qquad \text{(A1.6.17)}$$

the relaxation operator for the transition $j_a=0=j_b-1$ takes the form

$$\begin{aligned}\hat{\Gamma}\rho^a &= \frac{\chi|d_{ba}|^2}{3}\rho^a(N_{-1}+N_1)-\frac{\chi|d_{ba}|^2}{3}\left(\rho^b_{-1-1}(1+N_1)+\rho^b_{11}(1+N_{-1})\right),\\ \left(\hat{\Gamma}\rho^b\right)_{\mu\mu'} &= \frac{\chi|d_{ba}|^2}{3}\rho^b_{\mu\mu'}+\frac{\chi|d_{ba}|^2}{6}\left(\rho^b_{\mu\mu'}N_{-\mu'}+\rho^b_{\mu\mu'}N_{-\mu}\right)-\frac{\chi|d_{ba}|^2}{3}\rho^a\delta_{\mu\mu'}N_{-\mu},\\ \left(\hat{\Gamma}\rho^{ba}\right)_{\mu 0} &= \frac{\chi|d_{ba}|^2}{6}\rho^{ba}_{\mu 0}(1+N_{-1}+N_1)+\frac{\chi|d_{ba}|^2}{6}\rho^{ba}_{\mu 0}N_{-\mu}+\frac{\chi d^2_{ba}}{3}\rho^{ab}_{0\mu}M_{-\mu}.\end{aligned} \quad (A1.6.18)$$

From the last of eqs. (A1.6.18) we obtain the next term, which determines the attenuation of the real and imaginary parts of the spherical components of the atomic polarisation amplitude vector for the transition $j_a=0=j_b-1$ (when (21) is satisfied and the phases of the quantities M_{-1} and M_1 are equal):

$$\begin{aligned}\hat{\Gamma}p^q_{real} &= \frac{\chi|d_{ba}|^2}{6}\left(1+N_1+N_{-1}+N_q+2|M_q|\right)p^q_{real},\\ \hat{\Gamma}p^q_{im} &= \frac{\chi|d_{ba}|^2}{6}\left(1+N_1+N_{-1}+N_q-2|M_q|\right)p^q_{im}.\end{aligned} \quad (A1.6.19)$$

It is clear that the suppression of relaxation of the imaginary part of the polarisation is maximised only when a squeezed field is used, consisting of photons with the same circular polarisation. For example when $N_{-1}=M_{-1}=0$, $|M_1|=\sqrt{N_1(N_1+1)}$, $N_{-1}>>1$, we can say that $\hat{\Gamma}p^q_{im}\approx\frac{\chi|d_{ba}|^2}{6}\cdot\frac{1}{N_1}\cdot p^q_{im}\approx 0$ compared to $\hat{\Gamma}p^q_{real}\approx 2\frac{\chi|d_{ba}|^2}{3}\cdot N_1\cdot p^q_{real}$.

Because of the two available possible transitions from level E_a to sublevels of level E_b, with μ=1 and μ=-1, the smallest possible relaxation is determined by the expression

$$\hat{\Gamma}p^q_{im}=\frac{\chi|d_{ba}|^2}{6}(N_{\min}+\frac{1}{N_{\max}})p^q_{im},$$

for the case of both left- and right-hand polarised photons are in the squeezed field. Here $N_{min}=\min\{N_1, N_{-1}\}$, $N_{max}=\max\{N_1, N_{-1}\}$. Thus, we may conclude that for an arbitrary polarisation state of the squeezed field, the Zeeman structure of the resonance levels increases the smallest possible value for the relaxation (compared to the nondegenerate case and the specific case $j_a=j_b=1/2$) of any quadrature component of the atomic polarisation amplitude.

A1.6.5. RELAXATION OPERATOR FOR THE ATOMIC ANGULAR MOMENTUM: QUASICLASSICAL DESCRIPTION

Let us assume that $j_a >> 1, j_b >> 1$. Then we can use the semiclassical representation of the $3j$ symbol [17] and assume that the matrix elements of the spherical components of the dipole moment operator are given by

$$d^q_{\mu m} \approx \frac{d_{ba}}{\sqrt{2j}} \mathcal{D}^1_{q j_b - j_a}(0,\vartheta,0)\delta_{\mu - mq}, \quad d^q_{m\mu} \approx \frac{d^*_{ba}}{\sqrt{2j}}(-1)^q \mathcal{D}^1_{-q j_b - j_a}(0,\vartheta,0)\delta_{m q+\mu},$$

where $j=(j_a+j_b)/2$, $\cos\vartheta \approx K/j$, $K=(\mu+m)/2$, while $\mathcal{D}^\kappa_{qq'}(\alpha,\beta,\gamma)$ is the Wigner D-function.

Following Nasyrov and Shalagin [20], let us go to a new representation for the matrix elements of the atomic density matrix:

$$\begin{aligned}
\rho_a(\vartheta,\varphi) &= \sum_\xi \rho^a_{K+\xi/2\, K-\xi/2} \exp(-i\xi\varphi), \quad K=(m+m')/2, \quad \xi=(m-m'); \\
\rho_b(\vartheta,\varphi) &= \sum_\xi \rho^b_{K+\xi/2\, K-\xi/2} \exp(-i\xi\varphi), \quad K=(\mu+\mu')/2, \quad \xi=(\mu-\mu'); \\
\rho_{ba}(\vartheta,\varphi) &= \sum_\xi \rho^{ba}_{K+\xi/2\, K-\xi/2} \exp(-i\xi\varphi), \quad K=(\mu+m)/2, \quad \xi=(\mu-m);
\end{aligned} \tag{A1.6.20}$$

with $\cos\vartheta \approx K/j, j=(j_a+j_b)/2$ everywhere. Let us assume that the effective interval where the density matrix is nonzero with respect to the variable ξ is much smaller than the characteristic scale of variation of the density matrix with respect to the variable K. Then the summation over ξ in eq. (A1.6.20) can be extended to infinity and

$$\rho_{K+\xi/2\, K-\xi/2} = \frac{1}{2\pi}\int_0^{2\pi} \exp(i\xi\varphi)\rho(\vartheta,\varphi)d\varphi .$$

Thus, the transformation (A1.6.20) can be considered as analogous to the Wigner transformation for the incoming degrees of freedom. Assuming that quantities of type K are the same when they are in the same expression, we obtain semiclassical representations of the relaxation operators:

$$\begin{aligned}
\hat{\Gamma}\rho_a(\vartheta,\varphi) &= \frac{\chi|d_{ba}|^2}{2j}\rho_a(\vartheta,\varphi)\mathcal{T}_\sigma(\vartheta,\varphi) - \frac{\chi|d_{ba}|^2}{2j}\rho_b(\vartheta,\varphi)\big(\tau_\sigma(\vartheta)+\mathcal{T}_\sigma(\vartheta,\varphi)\big), \\
\hat{\Gamma}\rho_b(\vartheta,\varphi) &= \frac{\chi|d_{ba}|^2}{2j}\rho_b(\vartheta,\varphi)\big(\tau_\sigma(\vartheta)+\mathcal{T}_\sigma(\vartheta,\varphi)\big) - \frac{\chi|d_{ba}|^2}{2j}\rho_a(\vartheta,\varphi)\mathcal{T}_\sigma(\vartheta,\varphi), \\
\hat{\Gamma}\rho_{ba}(\vartheta,\varphi) &= \frac{\chi d^2_{ba}}{4j}\rho_{ba}(\vartheta,\varphi)\big(\tau_\sigma(\vartheta)+2\mathcal{T}_\sigma(\vartheta,\varphi)\big) + \frac{\chi|d_{ba}|^2}{2j}\rho^*_{ba}(\vartheta,\varphi)\mathcal{T}_\sigma(\vartheta,\varphi).
\end{aligned} \tag{A1.6.21}$$

The angles ν and φ characterise the direction of the total angular momentum, while the quantity ($\sigma=j_a-j_b$) identifies the type of resonant transition. The following functions are introduced:

$$\mathcal{T}_0(\vartheta,\varphi)=\tfrac{1}{2}\sin^2\vartheta\left(T_{11}+T_{-1-1}\right)-\tfrac{1}{2}\sin^2\vartheta\left(e^{-2i\varphi}T_{1-1}+e^{2i\varphi}T_{-11}\right),$$

$$\mathcal{T}_{\pm1}(\vartheta,\varphi)=\tfrac{1}{4}\left(1+\cos^2\vartheta\right)\left(T_{11}+T_{-1-1}\right)\mp\tfrac{1}{2}\cos\vartheta\left(T_{11}-T_{-1-1}\right)+\tfrac{1}{4}\sin^2\vartheta\left(e^{-2i\varphi}T_{1-1}+e^{2i\varphi}T_{-11}\right),$$

$$\mathfrak{T}_0(\vartheta,\varphi)=\tfrac{1}{2}\sin^2\vartheta(M_{11}e^{2i\varphi}+M_{-1-1}e^{-2i\varphi})-\tfrac{1}{2}\sin^2\vartheta(M_{1-1}+M_{-11}),$$

$$\mathfrak{T}_{\pm1}(\vartheta,\varphi)=$$
$$=\tfrac{1}{4}(1+\cos^2\vartheta)(M_{11}e^{2i\varphi}+M_{-1-1}e^{-2i\varphi})\mp\tfrac{1}{2}\cos\vartheta(M_{11}e^{2i\varphi}-M_{-1-1}e^{-2i\varphi})+\tfrac{1}{4}\sin^2\vartheta(M_{1-1}+M_{-11}),$$

$$\tau_0(\vartheta)=\sin^2\vartheta,\ \tau_{\pm1}(\vartheta)=\tfrac{1}{2}(1+\cos^2\vartheta).$$

A1.6.6. THE MAXWELL-BLOCH EQUATIONS

Assume that an atom that interacts resonantly both with the coherent field

$$\vec{E}=\vec{\mathcal{E}}\exp[i(kz-\omega t)]+k.c., \tag{A1.6.22}$$

and with the squeezed field (A1.6.5) is characterised by many values of the angular momentum, so that in the resonance approximation Eq. (A1.6.10') can be written in the form

$$\begin{aligned}
&\left(\frac{d}{dt}+\gamma_1\right)R_1(\vartheta,\varphi)=\gamma_1\frac{N_1}{2j}+\frac{i|\mathcal{E}|}{\hbar}\left(R_{21}^*(\vartheta,\varphi)\mathcal{D}_\sigma(\vartheta,\varphi)-R_{21}(\vartheta,\varphi)\mathcal{D}_\sigma^*(\vartheta,\varphi)\right)-\\
&\qquad-\gamma R_1(\vartheta,\varphi)\mathcal{T}_\sigma(\vartheta,\varphi)+\gamma R_2(\vartheta,\varphi)\left(\tau_\sigma(\vartheta)+\mathcal{T}_\sigma(\vartheta,\varphi)\right),\\
&\left(\frac{d}{dt}+\gamma_2\right)R_2(\vartheta,\varphi)=\gamma_2\frac{N_2}{2j}-\frac{i|\mathcal{E}|}{\hbar}\left(R_{21}^*(\vartheta,\varphi)\mathcal{D}_\sigma(\vartheta,\varphi)-R_{21}(\vartheta,\varphi)\mathcal{D}_\sigma^*(\vartheta,\varphi)\right)-\\
&\qquad-\gamma R_2(\vartheta,\varphi)\left(\tau_\sigma(\vartheta)+\mathcal{T}_\sigma(\vartheta,\varphi)\right)+\gamma R_1(\vartheta,\varphi)\mathcal{T}_\sigma(\vartheta,\varphi),\\
&\left(\frac{d}{dt}-i\Delta+\gamma_{21}\right)R_{21}(\vartheta,\varphi)=\frac{i}{\hbar}\frac{|d_{ba}|^2|\mathcal{E}|}{2j}(R_2(\vartheta,\varphi)-R_1(\vartheta,\varphi))\mathcal{D}_\sigma(\vartheta,\varphi)-\\
&\qquad-\frac{\gamma}{2}R_{21}(\vartheta,\varphi)\left(\tau_\sigma(\vartheta)+2\mathcal{T}_\sigma(\vartheta,\varphi)\right)-\gamma R_{21}^*(\vartheta,\varphi)\mathfrak{f}_\sigma(\vartheta,\varphi).
\end{aligned}\tag{A1.6.23}$$

Here the following notation has been used:

$$R_1(\vartheta,\varphi)=\rho_a(\vartheta,\varphi),\ R_2(\vartheta,\varphi)=\rho_b(\vartheta,\varphi),\ R_{21}(\vartheta,\varphi)=\rho_{ba}(\vartheta,\varphi)\frac{d^*_{ba}}{\sqrt{2j}}\exp[i(\omega t-kz)],\ \Delta=\omega-\omega_0,$$

$$\gamma=\frac{\chi|d_{ba}|^2}{2j},\ \mathcal{D}_\sigma(\vartheta,\varphi)=\sum_q \mathcal{D}^1_{-q\sigma}(0,\vartheta,0)e^{iq\varphi}l^q,\ l^q=\mathcal{E}^q/|\mathcal{E}|,$$

the function $f_\sigma(\vartheta,\varphi)$ differs from the function $\mathcal{F}_\sigma(\vartheta,\varphi)$ by the substitution of $M_{qq'}=m_{qq'}\exp[2i(kz-\omega t)]$ for $m_{qq'}$. Depending on the relation between frequencies ω and ω_Γ, the quantity $m_{qq'}$ can be an oscillatory function.

In a semiclassical description of the angular momentum, the spherical component of the atomic polarization p^q (A1.6.15) at $\omega_\Gamma=\omega$ is given by the expression

$$p^q=-\frac{j}{2\pi}\int_0^{2\pi}d\varphi\int_0^{\pi}\sin(\vartheta)d\vartheta\mathcal{D}^1_{-q\sigma}(0,\vartheta,0)e^{-iq\varphi}R_{21}(\vartheta,\varphi). \qquad \text{(A1.6.24)}$$

Applications of eqs.(A1.6.23) are presented in ref.[19].

References

1. Butylkin, V.S., Kaplan, A.E., Khromopulo, Yu.G., and Yakubovich, E.I.: *Resonant nonlinear interaction of light with matter*, Springer, Berlin, 1989.
2. Basharov, A.M.: Two-photon atomic relaxation in the field of nonresonant electromagnetic wave, *Zh.Eksp.Teor.Fiz.* **102** (1992), 1126-1139 [*Sov.Phys. JETP* **75** (1992), 611-618].
3. Gardiner, C.W.: Inhibition of atomic phase decays by squeezed light: A direct effect of squeezing, *Phys.Rev.Lett.* **56** (1986), 1917-1920.
4. Carmichael, H.J., Lane, A.S., and Walls, D.F.: Resonance fluorescence from an atom in a squeezed vacuum. *Phys.Rev.Lett.* **58** (1987), 2539-2542.
5. Ritsch, H., and Zoller, P.: Absorption spectrum of a two-level system in a squeezed vacuum, *Optics Communications* **64** (1987), 523-528.
6. An, S., Sargent III, M., and Walls, D.F.: Effects of a squeezed vacuum on probe absorption spectra. *Optics Communications* **67** (1988), 373-377.
7. Parkins, A.S., and Gardiner, C.W.: Photon echoes with coherent and squeezed pulses. *Phys.Rev.A* **40** (1989), 2534-2538.
8. Galatola, P., Lugiato, L.A., Porreca, M.G., and Tombesi, P.: Optical switching by variation of the squeezing phase, *Optics Communications* **81** (1991), 175-178.
9. Bergou, J., and Zhao, D.: Effect of a squeezed vacuum input on optical bistability. *Phys.Rev.A* **52** (1995), 1550-1560.
10. Basharov, A.M.: Optical bistability of a thin film of resonant atoms in a phase-sensitive thermostat, *Zh.Eksp.Teor.Fiz.* **108** (1995), 842-850 [*Sov.Phys. JETP* **81** (1995), 459-463].
11. Gardiner, C.W., and Collett, M.J.: Input and output in damped quantum systems: Quantum stochastic differential equations and the master equation, *Phys.Rev.A* **31** (1985), 3761-3774.

12. Gardiner, C.W., Parkins, A.S., and Zoller, P.: Wave-function quantum stochastic differential equations and quantum-jump simulation methods, *Phys.Rev.A*, **46** (1992), 4363-4381.
13. Gardiner, C.W., and Parkins, A.S.: Driving atom with light of arbitrary statistics. *Phys.Rev.A* **50** (1994), 1792-1806.
14. Barchielli, A.: Measurement theory and stochastic differential equations in quantum mechanics, *Phys.Rev.A* **34** (1986), 1642-1649.
15. Lax, M.: Quantum noise IV. Quantum theory of noise sources, *Phys.Rev.* **145** (1966), 110-129.
16. Gardiner, C.W.: *Quantum noise.* Springer, Berlin, 1991.
17. Caves, C.M., Schumaker, B.L.: New formalism for two-photon quantum optics. I. Quadrature phases and squeezed states, *Phys.Rev.A* **31** (1985), 3068-3092; New formalism for two-photon quantum optics. II. Mathematical foundation and compact notation, *ibid.*, 3093-3111.
18. Collett, M.J., and Gardiner, C.W.: *Phys.Rev.A* **30** (1984), 1386.
19. Basharov, A.M.: Semiclassical approach to resonance-optics polarization effects in coherent and squeezed fields, *Zh.Eksp.Teor.Fiz.* **84** (1997), 13-23 [*Sov.Phys. JETP* **111** (1997), 25-43].
20. Nasyrov, K.A., and Shalagin, A.M.: Interaction between intense radiation and atoms or molecules experiencing classical rotary motion, *Zh.Eksp.Teor.Fiz.* **81** (1981), 1649-1663 [*Sov.Phys. JETP* **54** (1981), 877].

THE DENSITY MATRIX EQUATION FOR A GAS MEDIUM

Let us describe atoms of a gas by ψ-operator

$$\psi(\vec{r}) = \frac{1}{\sqrt{V}} \sum_{\vec{p}\alpha} e^{i\vec{p}\vec{r}} |\alpha\rangle a_{\vec{p}\alpha},$$

where $a_{\vec{p}\alpha}$ is the annihilation operator of atom in quantum state $|\alpha\rangle$ with the moment $\hbar\vec{p}$, V is the volume of a system. The annihilation and creation operators obey Bose commutation relations for definiteness $[a_{\vec{p}\alpha}, a^{+}_{\vec{p}'\alpha'}] = \delta(\vec{p}-\vec{p}')\delta_{\alpha\alpha'}$, $[a_{\vec{p}\alpha}, a_{\vec{p}'\alpha'}] = 0$. The Hamiltonian $\mathcal{H}$ of the system interacting with electromagnetic field $\vec{E}$ in electro-dipole approximation is defined as

$$\mathcal{H} = \int d\vec{r}\, \psi^{+}(\vec{r}) \left\{ -\frac{\hbar^2 \Delta}{2M} + H_0 - \vec{E}\vec{d} \right\} \psi(\vec{r})$$

or

$$\mathcal{H} = \mathcal{H}_0 + \mathcal{V}, \tag{A2.1}$$

$$\mathcal{H}_0 = \sum_{\vec{p}\alpha} \varepsilon_{p\alpha} a^{+}_{\vec{p}\alpha} a_{\vec{p}\alpha}, \; \varepsilon_{p\alpha} = \varepsilon_p + E_\alpha, \; \varepsilon_p = \hbar^2 p^2 / 2M, \tag{A2.2}$$

$$\mathcal{V} = -\frac{1}{V} \sum_{\vec{p}\alpha\vec{p}'\alpha} \left\{ \int \vec{E}(\vec{r},t) e^{i(\vec{p}-\vec{p}')\vec{r}} d\vec{r} \right\} \vec{d}_{\alpha'\alpha} a^{+}_{\vec{p}'\alpha'} a_{\vec{p}\alpha} = -\frac{1}{V} \sum_{\vec{p}\vec{q}\alpha\alpha'} \vec{E}(\vec{q},t) \vec{d}_{\alpha\alpha'} a^{+}_{\vec{p}\alpha} a_{\vec{p}-\vec{q}\,\alpha'}. \tag{A2.3}$$

Here ε_p is the kinetic energy of atom with mass M, $\vec{E}(\vec{q},t)$ is the spatial Fourier component of the electric field strength,

$$\vec{E}(\vec{q},t) = \int d\vec{r}\, \vec{E}(\vec{r},t) e^{-i\vec{q}\vec{r}}.$$

The dipole-dipole interaction between atoms is neglected that is valid for the sufficiently rarefied gas $n\lambda\!\!\!^{-3} \ll 1$.

The atomic density matrix $\wp$ obeys the usual equation

$$i\hbar \frac{\partial \wp}{\partial t} = [\mathcal{H}, \wp], \tag{A2.4}$$

whereas the electric field strength $\vec{E}$ obeys the Maxwell equation

$$\left(\Delta - \frac{1}{c^2}\frac{\partial^2}{\partial t^2}\right)\vec{E} = \frac{4\pi}{c^2}\frac{\partial^2 \vec{P}}{\partial t^2}, \tag{A2.5}$$

where the polarisation is

$$\vec{P} = Sp(\wp\vec{\mathcal{P}}). \tag{A2.6}$$

The operator of polarisation in secondary quantisation is

$$\vec{\mathcal{P}}(\vec{r}) = \int d\vec{r}'\psi^+(\vec{r}')\{\vec{d}\delta(\vec{r}-\vec{r}')\}\psi(\vec{r}') = \psi^+(\vec{r})\vec{d}\psi^+(\vec{r}). \tag{A2.7}$$

In order to establish the relationship between atomic density matrix $\wp$ (which is N-particle) and one-particle density matrix $\rho = \rho(\vec{r},\vec{v},t)$ describing atoms moving with velocity $\vec{v}$, we consider the polarisation of the medium. Atomic polarisation of the medium in terms of a one-particle density matrix is

$$\vec{P} = \int d\vec{v} Sp(\rho\vec{d}) = \sum_{\alpha\alpha'}\int d\vec{v}\rho_{\alpha\alpha'}\vec{d}_{\alpha'\alpha}.$$

The same expression in terms of (A2.6) and (A2.7) has the form

$$\vec{P} = Sp(\wp\sum_{\substack{\vec{p}\vec{p}'\\ \alpha\alpha'}}\frac{1}{V}e^{-i\vec{p}'\vec{r}}\vec{d}_{\alpha'\alpha}e^{i\vec{p}\vec{r}}a^+_{\vec{p}'\alpha'}a_{\vec{p}\alpha}).$$

Introducing new variables $M\vec{v} = \hbar(\vec{p}+\vec{p}')/2$ and $\vec{\kappa} = \vec{p}-\vec{p}'$, we have

$$\vec{P} = Sp(\wp\sum_{\substack{\vec{v}\vec{\kappa}\\ \alpha\alpha'}}\frac{1}{V}e^{i\vec{\kappa}\vec{r}}\vec{d}_{\alpha'\alpha}a^+_{\frac{M\vec{v}}{\hbar}-\frac{\vec{\kappa}}{2}\,\alpha'}a_{\frac{M\vec{v}}{\hbar}+\frac{\vec{\kappa}}{2}\,\alpha}) = Sp(\wp\sum_{\substack{\vec{v}\\ \alpha\alpha'}}\int\frac{d\vec{\kappa}}{(2\pi)^3}e^{i\vec{\kappa}\vec{r}}a^+_{\frac{M\vec{v}}{\hbar}-\frac{\vec{\kappa}}{2}\,\alpha'}a_{\frac{M\vec{v}}{\hbar}+\frac{\vec{\kappa}}{2}\,\alpha}\vec{d}_{\alpha'\alpha}).$$

We obtain the one-particle density matrix by comparing two representations of atomic polarisation:

$$\rho_{\alpha\alpha'}(\vec{r},\vec{v},t) = \int\frac{d\vec{\kappa}}{(2\pi)^3}e^{i\vec{\kappa}\vec{r}}Sp(\wp a^+_{\frac{M\vec{v}}{\hbar}-\frac{\vec{\kappa}}{2}\,\alpha'}a_{\frac{M\vec{v}}{\hbar}+\frac{\vec{\kappa}}{2}\,\alpha}) \tag{A2.8}$$

Now we exclude the kinetic operator from the Hamiltonian by the following transformation

$$\overline{\wp} = e^{i\mathcal{K}t/\hbar}\wp e^{-i\mathcal{K}t/\hbar}, \quad \mathcal{K} = \sum_{\vec{p}\alpha}\varepsilon_p a^+_{\vec{p}\alpha}a_{\vec{p}\alpha}.$$

Then,

$$i\hbar \frac{\partial \overline{\wp}}{\partial t} = [\overline{\mathfrak{H}}, \overline{\wp}], \tag{A2.9}$$

where

$$\overline{\mathfrak{H}} = e^{i\mathfrak{K}t/\hbar}\mathfrak{H}e^{-i\mathfrak{K}t/\hbar} - i\hbar e^{i\mathfrak{K}t/\hbar}\frac{\partial}{\partial t}e^{-i\mathfrak{K}t/\hbar} = \overline{\mathfrak{H}_0} + \overline{\mathfrak{V}},$$

(A2.10)

$$\overline{\mathfrak{H}_0} = \sum_{\vec{p}\alpha} E_\alpha a^+_{\vec{p}\alpha} a_{\vec{p}\alpha}, \quad \overline{\mathfrak{V}} = -\frac{1}{V}\sum_{\vec{p}\vec{q}\alpha\alpha'} \vec{E}(\vec{q},t)\vec{d}_{\alpha\alpha'}\overline{a}^+_{\vec{p}\alpha}\overline{a}_{\vec{p}-\vec{q}\,\alpha'},$$

$$\overline{a}_{\vec{p}\alpha} = e^{i\mathfrak{K}t/\hbar}a_{\vec{p}\alpha}e^{-i\mathfrak{K}t/\hbar} = a_{\vec{p}\alpha} + i[\mathfrak{K}, a_{\vec{p}\alpha}] - \frac{1}{2}[\mathfrak{K},[\mathfrak{K},a_{\vec{p}\alpha}]] + \ldots = a_{\vec{p}\alpha}\exp(-i\varepsilon_{\vec{p}}t/\hbar).$$

The expression for the one-particle density matrix becomes as

$$\rho_{\alpha\alpha'}(\vec{r},\vec{v},t) = \int \frac{d\vec{\kappa}}{(2\pi)^3} e^{i\vec{\kappa}\vec{r}} Sp(\overline{\wp}\,\overline{a}^+_{\frac{M\vec{v}}{\hbar}-\frac{\vec{\kappa}}{2}\,\alpha'}\overline{a}_{\frac{M\vec{v}}{\hbar}+\frac{\vec{\kappa}}{2}\,\alpha}). \tag{A2.8$'$}$$

Let us consider the derivative $\dfrac{\partial \rho_{\alpha\alpha'}}{\partial t}$:

$$i\hbar\frac{\partial\rho_{\alpha\alpha'}}{\partial t} = \int\frac{d\vec{\kappa}}{(2\pi)^3}e^{i\vec{\kappa}\vec{r}}Sp\left\{\frac{\partial\overline{\wp}}{\partial t}\overline{a}^+_{\frac{M\vec{v}}{\hbar}-\frac{\vec{\kappa}}{2}\,\alpha'}\overline{a}_{\frac{M\vec{v}}{\hbar}+\frac{\vec{\kappa}}{2}\,\alpha} + \overline{\wp}(\varepsilon_{\frac{M\vec{v}}{\hbar}+\frac{\vec{\kappa}}{2}} - \varepsilon_{\frac{M\vec{v}}{\hbar}-\frac{\vec{\kappa}}{2}})\overline{a}^+_{\frac{M\vec{v}}{\hbar}-\frac{\vec{\kappa}}{2}\,\alpha'}\overline{a}_{\frac{M\vec{v}}{\hbar}+\frac{\vec{\kappa}}{2}\,\alpha}\right\}.$$

We assume that the density matrix ρ is the slowly varying function along coordinate in comparison with $\exp(iM\vec{v}\vec{r}/\hbar)$. Then,

$$\varepsilon_{\frac{M\vec{v}}{\hbar}+\frac{\vec{\kappa}}{2}} - \varepsilon_{\frac{M\vec{v}}{\hbar}-\frac{\vec{\kappa}}{2}} \approx \hbar\vec{v}\vec{\kappa}. \tag{A2.11}$$

Simple calculation gives

$$i\hbar(\frac{\partial}{\partial t} + \vec{v}\nabla)\rho_{\alpha\alpha'}(\vec{r},\vec{v},t) =$$

$$= \int\frac{d\vec{\kappa}}{(2\pi)^3}e^{i\vec{\kappa}\vec{r}}Sp\left\{\overline{\wp}\left(\overline{a}^+_{\frac{M\vec{v}}{\hbar}-\frac{\vec{\kappa}}{2}\,\alpha'}[\overline{a}_{\frac{M\vec{v}}{\hbar}+\frac{\vec{\kappa}}{2}\,\alpha},\overline{\mathfrak{H}}] + [\overline{a}^+_{\frac{M\vec{v}}{\hbar}-\frac{\vec{\kappa}}{2}\,\alpha'},\overline{\mathfrak{H}}]\overline{a}_{\frac{M\vec{v}}{\hbar}+\frac{\vec{\kappa}}{2}\,\alpha}\right)\right\}. \tag{A2.12}$$

In eq.(A2.12) we do not specify the form of the Hamiltonian $\mathcal{H}$. Unique requirement for $\mathcal{H}$ is the determination of atomic dynamics according to eq.(A2.9). The Hamiltonian (A2.10) defines the following equation for the density matrix

$$i\hbar\left(\frac{\partial}{\partial t}+\vec{v}\nabla\right)\rho(\vec{r},\vec{v},t)=[H_0,\rho(\vec{r},\vec{v},t)]+$$
$$+\int\frac{d\vec{\kappa}}{(2\pi)^3}e^{i\vec{\kappa}\vec{r}}\left\{-\vec{E}(\vec{k},t)\vec{d}\rho(\vec{r},\vec{v}-\frac{\hbar\vec{\kappa}}{2M},t)+\rho(\vec{r},\vec{v}+\frac{\hbar\vec{\kappa}}{2M},t)\vec{E}(\vec{k},t)\vec{d}\right\}. \tag{A2.13}$$

Thus the electromagnetic wave entangles atomic states with the moments distinguished by the moment of photon $\hbar\vec{\kappa}$ due to recoil in radiative processes. If we neglect the recoil, we obtain

$$i\hbar\left(\frac{\partial}{\partial t}+\vec{v}\nabla\right)\rho(\vec{r},\vec{v},t)=[H_0-\vec{E}\vec{d},\rho(\vec{r},\vec{v},t)]. \tag{A2.14}$$

The equation (A2.14) can be obtained directly in the described way by taking the Hamiltonian (A2.1) with interaction operator in the form

$$\mathcal{V}=-\sum_{\vec{p}\,\alpha\alpha'}\vec{E}(\vec{r},t)\vec{d}_{\alpha\alpha'}a^{+}_{\vec{p}\alpha}a_{\vec{p}\,\alpha'}\,.$$

ADIABATIC FOLLOWING APPROXIMATION

The dynamics of a state of the two-level atom is described by the system of Bloch equations for slowly varying envelope of off-diagonal elements of the density matrix $p(\tau,\zeta)$ and for the population difference $R(\tau,\zeta)$ (1.1.12). We shall write this system in a normalised form

$$\frac{\partial p}{\partial \tau} = -i\delta p + iqR\,, \qquad \text{(A3.1a)}$$

$$\frac{\partial R}{\partial \tau} = \frac{i}{2}\left(q^* p - qp^*\right). \qquad \text{(A3.1b)}$$

Here $\delta = \Delta\omega t_{p0}$ is the normalised frequency detuning from the resonance line centre, q is the normalised USP envelope, $q = d\mathcal{E}t_{p0}/2\hbar$.

When $\Delta\omega \neq 0$ this system of equations can be solved by the adiabatic following approximation (AFA) method. It was offered by Grischkowsky [1] in his studies of the coherent transient processes. Originally the main results of AFA were obtained from a certain geometric picture, which presents the motion of the Bloch vector around the 'pseudofield' direction in the framework of the Feynman-Vernon-Hellwarth's vector model [2] or the Bloch's 'pseudospin' model [3]. Under condition of sufficiently large frequency detuning $\Delta\omega$, only this detuning will define an angular velocity of the Bloch vector rotating. The term in the equation (A3.1), which is proportional to electrical field of USP, will provide the corrections to the angular velocity. Geometrically the Bloch vector is nearly collinear to the 'pseudofield' vector and it follows it, when the envelope of the USP varies in time and in space. A more formal way to solve Bloch equations was offered by Crisp in [4], where Grischkowsky's results were obtained as a specific case. The essence of the approach, used by Crisp, is the solution of equations (A3.1) by perturbation theory with the Rabi frequency to the frequency detuning $|\Delta\omega|$ ratio serving as a small parameter. Below we shall consider this Crisp's approach as far as it is completely analytical and it does not lean upon the model of "pseudospin ".

Formal integration of equation (A3.1a) results in:

$$p(\tau) = i\int_{-\infty}^{\tau} q(\tau')R(\tau')\exp\left[-(\gamma + i\delta)(\tau-\tau')\right]d\tau' =$$

$$= i\int_{0}^{\infty} q(\tau - s)R(\tau - s)\exp\left[-(\gamma + i\delta)s\right]ds\,. \qquad \text{(A3.2)}$$

Here the attenuation coefficient γ is introduced for generality. The product $q(\tau - s)R(\tau - s)$ can be represented by a Taylor series:

$$q(\tau - s)R(\tau - s) = \sum_{m=0}^{\infty} \frac{(-1)^m}{m!} s^m \frac{\partial^m (qR)}{\partial \tau^m}.$$

Substitution of this expression into equation (A3.2) and integration with respect to s gives the following series representation of $p(\tau)$:

$$p(\tau) = \frac{i}{\gamma + i\delta} \sum_{m=0}^{\infty} \left(\frac{-1}{\gamma + i\delta} \right)^m \frac{\partial^m (qR)}{\partial \tau^m}. \qquad \text{(A3.3)}$$

In the case of USP we can now assume that $\gamma = 0$.

If we retain only the first term of the series (A3.3) and assume that $R = R_0 = R(\tau \to -\infty)$ (in a non-inverted medium we have $R_0 = 1$), the result is similar to one obtained from the linear theory of dispersion:

$$p^{(0)}(\tau) = R_0 q(\tau)/\delta.$$

A non-linear generalisation of this theory can be done if we take into account the exact relationship that follows from the Bloch equations:

$$|p|^2 + R^2 = R_0^2. \qquad \text{(A3.4)}$$

If in this expression we substitute $p^{(1)} = qR/\delta$, then one can find the expression for $R = R^{(1)}$:

$$R^{(1)}(\tau) = R_0 \left(1 + |q(\tau)|^2 / \delta^2\right)^{-1/2} \qquad \text{(A3.5a)}$$

and it follows

$$p^{(1)}(\tau) = [R_0 q(\tau)/\delta]\left(1 + |q(\tau)|^2 / \delta^2\right)^{-1/2}. \qquad \text{(A3.5b)}$$

The variation of $p^{(1)}(\tau)$ with time (and in space) is determined only by the USP envelope $q(\tau,\zeta)$. That means that the function $p(\tau,\zeta)$ *follows* the function $q(\tau,\zeta)$.

An even more rigorous expression for $p(\tau,\zeta)$ can be derived by taking into account the following summand in the series (A3.3).

$$p(\tau) = \frac{1}{\delta} \left\{ Rq + \frac{i}{\delta} \frac{\partial (Rq)}{\partial \tau} \right\}.$$

Derivative $\partial R/\partial\tau$ can be expressed from the equation (A3.1b) so, that

$$p(\tau)=\frac{1}{\delta}\left\{Rq+\frac{iR}{\delta}\frac{\partial q}{\partial\tau}+\frac{q}{2\delta}(p^*q-pq^*)\right\},$$

$$p^*(\tau)=\frac{1}{\delta}\left\{Rq^*-\frac{iR}{\delta}\frac{\partial q^*}{\partial\tau}-\frac{q^*}{2\delta}(p^*u-pu^*)\right\}.$$

If to replace R in these expressions with $R^{(1)}$ from expression (A3.5a), then the above system is converted to a linear system of equations for p and p^*, the solution of which can obtained by usual methods. Thereby, we can find an expression for $p^{(2)}$:

$$p^{(2)}(\tau)=\frac{R_0}{\delta}\left(1+\frac{|q(\tau)|^2}{\delta^2}\right)^{-3/2}\left\{\left[q+\frac{i}{\delta}\frac{\partial q}{\partial\tau}\right]\left(1+\frac{|q(\tau)|^2}{2\delta^2}\right)+\right.$$
$$\left.+\frac{u^2}{2\delta^2}\left[q^*+\frac{i}{\delta}\frac{\partial q^*}{\partial\tau}\right]\right\}. \qquad \text{(A3.6)}$$

This result represents a certain generalisation of the known AFA formulas from [1,4,5], which can by derived from (A3.6) by omitting the second term in the square brackets and expanding $p^{(2)}$ into the real and imaginary parts.

The validity of the approximate expression (A3.6) is attributed by the condition of convergence of the series (A3.3). If we additionally postulate that the amplitude of an USP is so small that the Rabi frequency is much less than the frequency detuning $|\delta|$, then expression (A3.6) can be simplified by expanding it into powers of $\theta=|q|/\delta$ and by retaining only terms of order θ^4. In this case we have

$$p^{(2)}(\tau)=\frac{R_0}{\delta}\left\{q+\frac{i}{\delta}\frac{\partial q}{\partial\tau}-\frac{|q|^2q}{2\delta^2}-\frac{i}{\delta^3}\frac{\partial(|q|^2q)}{\partial\tau}-\frac{3}{2\delta^4}|q|^4q\right\}. \qquad \text{(A3.7)}$$

Let us consider the reduced wave equation

$$\frac{\partial q}{\partial\zeta}=i\langle p\rangle .$$

Definitions $\kappa_n=\langle R_0\delta^{-n}\rangle$ and expression (A3.7) allow to write this equation as

$$i\frac{\partial q}{\partial\zeta}=-\kappa_1q-i\kappa_2\frac{\partial q}{\partial\tau}+\frac{1}{2}\kappa_3|q|^2q+\frac{i}{2}\kappa_4\frac{\partial}{\partial\tau}(|q|^2q)+\frac{3}{2}\kappa_5|q|^4q. \qquad \text{(A3.8)}$$

If we re-define the variables as

$$q(\tau,\zeta)=u(\eta,\xi)\exp[i\kappa_1\zeta], \qquad \eta=\tau-\kappa_2\zeta, \quad \xi=\zeta,$$

then the equation (A3.8) becomes

$$i\frac{\partial u}{\partial \xi}=\frac{1}{2}\kappa_3\,|u|^2\,u+\frac{i}{2}\kappa_4\frac{\partial}{\partial \tau}(|u|^2\,u)+\frac{3}{2}\kappa_5\,|u|^4\,u. \tag{A3.9}$$

Thus we obtain the dispersionless generated non-linear Schrödinger equation that was discussed in section 5.2.1. Introduction the real functions a and φ so as $u=a\exp(i\varphi)$ results in

$$\frac{\partial a}{\partial \xi}-\frac{3}{2}\kappa_4 a^2\frac{\partial a}{\partial \eta}=0, \tag{A3.10a}$$

$$\frac{\partial \varphi}{\partial \xi}-\frac{1}{2}\kappa_4 a^2\frac{\partial \varphi}{\partial \eta}=-\frac{1}{2}\kappa_3 a^2-\frac{1}{2}\kappa_5 a^4. \tag{A3.10b}$$

Equation (A3.10a) as known, describes the shock-wave formation or self-steeping phenomenon [6]. Solution of this equation can be written as

$$a(\eta,\xi)=f_0\left(\eta+\frac{3}{2}\kappa_4\xi a^2(\eta,\xi)\right), \tag{A3.11}$$

with

$$a(\eta,\xi=0)=f_0(\eta).$$

If the transition frequency of the two-level atoms is approximately equal or sufficiently close to double frequency of the carrier wave of the USP, then they say about the two-photon resonance transition. In such situation many coherent effects are possible, which are analogous to those under one-photon transition [6]. The investigation of these effects can also be carried out by using the adiabatic following approximation [7-11]. Shock wave formation under two-photon resonance conditions investigated in [12] without AFA.

Bloch equations for two-photon resonance transition (4.4.12b,c) can be written in the form

$$\frac{\partial p}{\partial \tau}=-i(\delta+\delta_{St})p+iRq^2, \tag{A3.12a}$$

$$\frac{\partial n}{\partial \tau}=\frac{i}{2}(pq^{*2}-p^{*}q^2), \tag{A3.12b}$$

where $\delta = (\omega_{21} - \omega_0)t_{p0}$ is a normalised frequency detuning. Normalised Stark shift $\delta_{st} = 2b|q|^2$ was defined in section 4.4.1.

The system of equation (A3.12) is completely identical to (A3.1) provided we assign $q(\tau,\zeta)$ as $q^2(\tau,\zeta)$ and ignore the Stark shift, i.e., $|\delta_{st}| << |\delta|$. (This condition is in agreement with the assumption of smallness of the Rabi frequencies compared with the frequency detuning from the resonance $|q^2| << |\delta|$.) For this reason we can use the expressions (A3.6) in order to find the formula for the polarisation of resonance atoms. However, if we retain only the terms whose order of smallness is θ^3, where $\theta = |q^2|/\delta$, then the following formulae should be used

$$p^{(2)}(\tau) = \frac{R_0}{\delta}\left\{q + \frac{i}{\delta}\frac{\partial q^2}{\partial \tau} - \frac{|q|^4 q^2}{2\delta^2}\right\},$$
$$R^{(2)}(\tau) = R_0 -.\frac{R_0|q|^4}{2\delta^2} - \frac{iR_0}{2\delta^3}\left(q^{*2}\frac{\partial q^2}{\partial \tau} - q^2\frac{\partial q^{*2}}{\partial \tau}\right). \quad \text{(A3.13)}$$

The equation, describing the evolution of the envelope of an ultrashort pulse under adopted approximation, can be obtained from (4.4.12a) and (A3.13):

$$i\frac{\partial q'}{\partial \zeta} + \frac{1}{2}\kappa_1|q'|^2 q' + \frac{1}{2}\kappa_2 b|q'|^4 q' - \frac{1}{4}\kappa_3|q'|^6 q' +$$
$$+ i\kappa_2|q'|^2\frac{\partial q'}{\partial \tau} + i\kappa_3 b\left(2|q'|^4\frac{\partial q'}{\partial \tau} - |q'|^2 q'\frac{\partial |q'|^2}{\partial \tau}\right) = 0. \quad \text{(A3.14)}$$

In the equation (A3.14) coefficients $\kappa_n, n = 1,2,3$ is defined as above. The functions q' is $q' = q(\tau,\zeta)\exp(-ibR_0\zeta)$.

If the absorption line is described by the symmetric form-factor, then $\kappa_1 = \kappa_3 = 0$ and the equation (A3.14) can be rewritten as

$$i\frac{\partial q'}{\partial \zeta} + \frac{1}{2}\kappa_2 b|q'|^4 q' + i\kappa_2|q'|^2\frac{\partial q'}{\partial \tau} = 0. \quad \text{(A3.15)}$$

As above, we can consider the real system of equations that leads from (A3.15)

$$\frac{\partial a}{\partial \zeta} + \kappa_2 a^2\frac{\partial a}{\partial \tau} = 0, \quad \text{(A3.16a)}$$

$$\frac{\partial \varphi}{\partial \zeta} + \kappa_2 a^2\frac{\partial \varphi}{\partial \tau} = \frac{1}{2}b\kappa_2 a^4. \quad \text{(A3.16b)}$$

Once again the equation describing the shock wave formation (A3.16a) is found. Thus, we are concluded that the principal effect following from AFA condition is the optical shock wave formation.

References

1. Grischkowsky, D.: Adiabatic following and slow optical pulse propagation in rubidium vapor, *Phys.Rev.* **A 7** (1973), 2096-2102.
2. Feynman, R.P., Vernon, F.L.,Jr., and Hellwarth, R.W.: Geometrical representation of the Schrödinger equation for solving maser problems. *J.Appl.Phys.* **28** (1957), 49-52.
3. Macomber J.D., *The dynamics of spectroscopic transitions*, John Wiley and Sons, New York, London, Sydney, Toronto, 1976.
4. Crisp, M.D.: Adiabatic-following approximation, *Phys.Rev.* **A 8** (1973), 2128-2135.
5. Allen, L., and Eberly, J.H.: *Optical Resonance and Two-Level Atoms*, Wiley, New York, 1975.
6. Takatsuji, M.: Theory of coherent two-photon resonance, *Phys. Rev.* **A11** (1975), 619-624.
7. Grischkowsky, D. Courtens, E., and Armstrong J.A.: Observation of self-steeping of optical pulses with possible shock formation, *Phys.Rev.Letts.***31** (1973), 422-425.
8. Grischkowsky, D.: Optical pulse compression, *Appl.Phys.Letts.* **25** (1974), 566-568.
9. Grischkowsky, D., Loy, M.M.T., and Liao, P.F.: Adiabatic following model for two-photon transition: nonlinear mixing and pulse propagation, *Phys.Rev.* **A 12** (1975), 2514-2533.
10. Loy, M.M.T Observation of population inversion by optical adiabatic rapid passage, *Phys.Rev.Lett.* **32** (1974), 814-817.
11. Lehmberg, R.H., and Reintjes, J.: Generalised adiabatic following approximation. *Phys.Rev.* **A 12** (1975), 2574-2583
12. DeMartini, F., and Giuliani, C.: Optical shocks in nonlinear two-photon interactions, *Opt.Commun.* **11** (1974), 42-45.

APPENDIX 4

RELATION BETWEEN EXACTLY INTEGRABLE MODELS IN RESONANCE OPTICS

The exactly integrable models of resonance optics corresponding to different resonance conditions are studied as a rule independent of one another. This is perfectly natural from the mathematical viewpoint, since both the evolution equations and their Lax representations needed to apply the IST method, differ substantially for different models. Yet, various optical-resonance theories can be developed in a single manner by using a unitary transformation of the total Hamiltonian of the system (see chap. 1). It is interesting to consider the relation between the exactly integrable models of optical resonance. It should be noted that the gauge equivalence of certain exactly integrable theories had been discussed earlier (see sec. 4.5). As to optical-resonance theories, it must be emphasised that they cannot be mutually gauge equivalent, since each of them is equivalent to a principal chiral field on different groups, for example $SL(3)$ and $SU(2)$ for models of propagation of two-frequency USP in double (Fig. A4.1.a) and Raman (Fig. A4.l.b) resonance, respectively.

It is shown in this Appendix that the unitary transformation, just as it connects the Hamiltonians of different optical-resonance models, makes it possible to obtain from the Lax representation of one exactly integrable theory a Lax representation of another theory. The operators of the U-V pair of the IST method serve as the Hamiltonian under transformation in the unitary transformation method of Chapter 1. It becomes possible to construct new exactly integrable models of nonlinear optics simultaneously with their Lax representation. As an example, we consider exactly integrable models of double resonance with resonance energy levels that are nondegenerate and degenerate in different orientations of the total angular momentum. This choice is not accidental. One can see here quite clearly the role of the unitary transformation that excludes adiabatically, when applied to the Hamiltonian of a three-level system, a common level of adjacent optically allowed transitions (level E_b Fig. A4.1.a), transforming the system to a Raman resonance with an optically forbidden transition. The integrability of polarisation models of Raman resonance by the IST has been proved. Our treatment is based on the paper [1].

Section A4.1 describes in detail the use of the proposed method for double resonance with nondegenerate levels, and repeats the results previously obtained for Raman resonance by differential-geometry methods [2,3]. Instead of using the results of Chapter 1, we develop the unitary transformation method as applied to three-level system from the first principles. In the next section the Lax representations of polarisation models of Raman resonance with nondegenerate and threefold degenerate energy level are obtained.

A4.1. The unitary-transformation method for three-level system

We obtain first the equations that describe the propagation of two-frequency USP having an electric field of the form

$$\vec{E} = \vec{\mathcal{E}}_1 c^{-i\Phi_1} + \vec{\mathcal{E}}_2 c^{-i\Phi_2} + c.c., \quad \Phi_j = \omega_j t - k_j z, \quad j = 1,2 \tag{A4.1.1}$$

in a half-space $z > 0$ filled with three-level particles. We start with the classical Maxwell equations

$$\left(\frac{\partial^2}{\partial z^2} - \frac{1}{c^2}\frac{\partial^2}{\partial t^2}\right)\vec{E} = \frac{4\pi}{c^2}\frac{\partial^2}{\partial t^2}\vec{P} \tag{A4.1.2}$$

and the quantum-mechanical equation for the density matrix ρ:

$$i\hbar\frac{\partial}{\partial t}\rho = [H_0 - \vec{E}\vec{d}, \rho] \tag{A4.1.3}$$

where H_0 and $\vec{d}$ are the Hamiltonian and dipole-moment operator of the three-level particle, and $\vec{P}$ is the polarisation of the medium

$$\vec{P} = Sp\rho\vec{d}\ . \tag{A4.1.4}$$

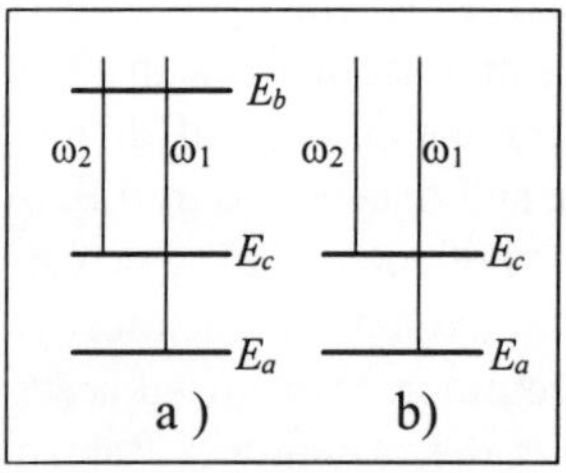

Fig.A4.1.1. Energy-level structure for double (a) and Raman (b) resonances.

The density matrix is normalised to the density N of the three-level particles, Spρ=N.

We have left out of (A4.1.3) the relaxation terms by virtue of the short duration of the USP. Furthermore, we neglect the inhomogeneous broadening of the spectral lines. We assume also that the energy levels E_a, E_b, and E_c form a Λ configuration: the transitions $E_b \to E_a$ and $E_b \to E_c$ are optically allowed, $E_c \to E_a$ is optically forbidden, and $E_a < E_c < E_b$. In the present section we neglect the energy-level degeneracy and the USP polarisation.

We transform the density matrix using the unitary operator e^{iS} (see sec. 1.1.1):

$$\widetilde{\rho} = e^{-iS} \rho e^{iS}.$$

The equation for the transformed density matrix $\widetilde{\rho}$,

$$i\hbar \frac{\partial}{\partial t} \widetilde{\rho} = [\widetilde{H}, \widetilde{\rho}], \tag{A4.1.3'}$$

is determined by the Hamiltonian

$$\widetilde{H} = e^{-iS} H_0 e^{iS} - e^{-iS} E d e^{iS} - i\hbar e^{-iS} \frac{\partial}{\partial t} e^{iS},$$

which we expand in a usual manner:

$$\widetilde{H} = H_0 - i[S, H_0] - \tfrac{1}{2}[S,[S,H_0]] -$$
$$\ldots - Ed + i[S, Ed] + \tfrac{1}{2}[S,[S,Ed]] + \ldots - i\hbar e^{-iS} \frac{\partial}{\partial t} e^{iS}.$$

The polarisation (A4.1.4) of the medium also takes a similar form

$$P = Sp\{\widetilde{\rho}(d - i[S,d] - \tfrac{1}{2}[S,[S,d]] - \ldots)\} \tag{A4.1.4'}$$

We represent S and $\widetilde{H}$ by series in powers of the electric field intensity:

$$S = S^{(1)} + S^{(2)} + \ldots, \quad \widetilde{H} = \widetilde{H}^{(0)} + \widetilde{H}^{(1)} + \widetilde{H}^{(2)} + \ldots. \tag{A4.1.5}$$

Here

$$\begin{aligned} \widetilde{H}^{(0)} &= H_0, \\ \widetilde{H}^{(1)} &= -Ed - i[S^{(1)}, H_0] + \hbar \frac{\partial}{\partial t} S^{(1)}, \\ \widetilde{H}^{(2)} &= \frac{i}{2}[S^{(1)}, Ed] - \frac{i}{2}[S^{(1)}, \widetilde{H}^{(1)}] - i[S^{(2)}, H_0] + \hbar \frac{\partial}{\partial t} S^{(2)}. \end{aligned} \tag{A4.1.6}$$

The succeeding simplifications differ here and depend on the USP propagation conditions.

For the double resonance

$$\omega_1 \approx \omega_{ba}, \quad \omega_2 \approx \omega_{bc}, \quad \omega_{ba} = (E_b - E_a)\hbar^{-1}, \quad \omega_{bc} = (E_b - E_c)\hbar^{-1},$$

we stipulate that the only nonzero matrix element of the operator $\widetilde{H}^{(1)}$ should be the following:

$$\widetilde{H}^{(1)}_{ba} = -\mathcal{E}_1 d_{ba} e^{-i\Phi_1} = \widetilde{H}^{(1)*}_{ab}, \quad \widetilde{H}^{(1)}_{bc} = -\mathcal{E}_2 d_{bc} e^{-i\Phi_2} = \widetilde{H}^{(1)*}_{cb},$$

and that the operator $\widetilde{H}^{(1)}$ should be diagonal and contain no oscillating exponentials $\exp(\pm i\Phi_j)$. We obtain then from (A4.1.6)

$$S^{(1)}_{ba} = -\frac{id_{ba}}{\hbar}\left(\frac{\mathcal{E}_2 e^{-i\Phi_2}}{\omega_{ba} - \omega_2} + \frac{\mathcal{E}_2^{*} e^{i\Phi_2}}{\omega_{ba} + \omega_2} + \frac{\mathcal{E}_1^{*} e^{-i\Phi_1}}{\omega_{ba} + \omega_1}\right),$$

$$S^{(1)}_{bc} = -\frac{id_{bc}}{\hbar}\left(\frac{\mathcal{E}_1 e^{-i\Phi_1}}{\omega_{bc} - \omega_1} + \frac{\mathcal{E}_1^{*} e^{i\Phi_1}}{\omega_{bc} + \omega_1} + \frac{\mathcal{E}_2^{*} e^{-i\Phi_2}}{\omega_{bc} + \omega_2}\right).$$

We do not present here the expressions for $S^{(2)}$, which will not be needed. It is important that the $S^{(1)}$ and $S^{(2)}$ contain no resonant denominators and confirm that the assumptions made concerning $\widetilde{H}$ are not contradictory.

The effective double-resonance operator defined in this manner

$$H^D = \widetilde{H}^{(0)} + \widetilde{H}^{(1)} + \widetilde{H}^{(2)}$$

has the matrix elements

$$H^D_{ba} = -\mathcal{E}_1 d_{ba} e^{-i\Phi_1} = H^{D*}_{ab}, \quad H^D_{bc} = -\mathcal{E}_2 d_{bc} e^{-i\Phi_2} = H^{D*}_{cb}, \quad H^D_{\alpha\alpha} = E_\alpha + E^{St}_\alpha,$$

$$E^{St}_a = |\mathcal{E}_2|^2 \Pi_a(\omega_2) - |d_{ba}\mathcal{E}_1|^2 / \hbar(\omega_{ba} + \omega_1), \quad E^{St}_c = |\mathcal{E}_1|^2 \Pi_c(\omega_1) - |d_{bc}\mathcal{E}_2|^2 / \hbar(\omega_{bc} + \omega_2),$$

$$E^{St}_b = -|\mathcal{E}_1|^2 \Pi_c(\omega_1) - |\mathcal{E}_2|^2 \Pi_a(\omega_2) + |d_{ba}\mathcal{E}_1|^2 / \hbar(\omega_{ba} + \omega_1) + |d_{bc}\mathcal{E}_2|^2 (\omega_{bc} + \omega_2),$$

where

$$\Pi_\alpha(\omega) = \sum_{\alpha'} \frac{|d_{\alpha\alpha'}|^2}{\hbar}\left(\frac{1}{\omega_{\alpha\alpha'} + \omega} + \frac{1}{\omega_{\alpha\alpha'} - \omega}\right), \quad \omega_{\alpha\alpha'} = (E_\alpha - E_{\alpha'})\hbar^{-1},$$

$\alpha, \alpha' = a, b, c,$ the quantities E^{St}_α are the Stark shifts of the levels, and their terms that do not contain the factor Π_α are the Bloch-Siegert shifts.

The main contribution to the polarisation (A4.1.4') of the medium is made by the first term, namely $P = Sp\tilde{\rho} d$. Transforming (A4.1.2) and (A4.1.3') to slowly changing variables and reducing them to dimensionless form, we obtain evolution equations that can be written here for convenience in the form

$$i\frac{\partial \hat{E}}{\partial \zeta} = -\frac{1}{2}[\hat{R}, \hat{J}], \quad i\frac{\partial \hat{R}}{\partial \tau} = -\frac{\delta}{2}[\hat{R}, \hat{J}] + [\hat{R}, \hat{E}]. \qquad \text{(A4.1.7)}$$

This equation contains new independent variables $\zeta = z/ct_0$ and $\tau = (t - z/c)/t_0$ and also

$$\hat{J} = \begin{pmatrix} -1 & 0 & 0 \\ 0 & 1 & 0 \\ 0 & 0 & 1 \end{pmatrix}, \quad \hat{R} = \begin{pmatrix} r^b & r^{bc} & r^{ba} \\ r^{bc*} & r^c & r^{ca} \\ r^{ba*} & r^{ca*} & r^a \end{pmatrix}, \quad \hat{E} = \begin{pmatrix} 0 & \varepsilon_2 & \varepsilon_1 \\ \varepsilon_2^* & 0 & 0 \\ \varepsilon_1^* & 0 & 0 \end{pmatrix},$$

$$r^\alpha = \tilde{\rho}_{\alpha\alpha} N^{-1}, \quad r^{ba} = \tilde{\rho}_{ba} N^{-1} e^{i\Phi_1}, \quad r^{bc} = \tilde{\rho}_{bc} N^{-1} e^{i\Phi_2}, \quad r^{ca} = \tilde{\rho}_{ca} N^{-1} e^{i(\Phi_1 - \Phi_2)},$$

$$\varepsilon_j = t_0 d_{ab} \mathcal{E}_j \hbar^{-1}, \quad t_0 = (\hbar / 2\pi\omega_1 |d_{ba}|^2 N)^{1/2}, \quad \delta = (\omega_1 - \omega_{ba}) t_0 = (\omega_2 - \omega_{bc}) t_0 .$$

With the aim of further analysis of the exactly integrable case, we neglect the Stark level shifts in (A4.1.7) and assume that

$$\omega_1 |d_{ba}|^2 = \omega_2 |d_{bc}|^2 . \qquad \text{(A4.1.8)}$$

In the case of Raman resonance

$$\omega_1 - \omega_2 \approx \omega_{ca}, \quad \omega_{ca} = (E_c - E_a)\hbar^{-1}$$

we must put $H^{(1)} = 0$ and $\tilde{H}^{(2)}_{ba} = \tilde{H}^{(2)}_{bc} = 0$, retain in the expressions for $\tilde{H}^{(2)}_{\alpha\alpha}$ only the terms that do not contain rapidly varying exponentials, and retain in $\tilde{H}^{(2)}_{ca}$ only the term proportional to $\exp[-i(\Phi_1 - \Phi_2)]$. The unitary transformation is then determined by the matrix

$$S^{(1)}_{\alpha\alpha'} = -\frac{i d_{\alpha\alpha'}}{\hbar} \sum_{j=1,2} \left(\frac{\mathcal{E}_j e^{-i\Phi_j}}{\omega_{\alpha\alpha'} - \omega_j} + \frac{\mathcal{E}_j^* e^{i\Phi_j}}{\omega_{\alpha\alpha'} + \omega_j} \right),$$

and the matrix elements of the effective Raman-resonance Hamiltonian

$$H^R = \tilde{H}^{(0)} + \tilde{H}^{(2)},$$

take the form

$$H^R_{ca} = -\mathcal{E}_1 \mathcal{E}_2^* \Pi_{ca}(\omega_1) e^{-i(\Phi_1 - \Phi_2)} = H^{R*}_{ac}, \quad H^R_{\alpha\alpha} = E_\alpha + E^{St}_\alpha,$$
$$E^{St}_\alpha = |\mathcal{E}_1|^2 \Pi_\alpha(\omega_1) + |\mathcal{E}_2|^2 \Pi_\alpha(\omega_2),$$

where

$$\Pi_{ca}(\omega) = \sum_\alpha \frac{d_{c\alpha} d_{\alpha a}}{\hbar} \left(\frac{1}{\omega_{\alpha c} + \omega} + \frac{1}{\omega_{\alpha a} - \omega} \right)$$

with $\Pi_{ca}(\omega_1) = \Pi_{ca}(-\omega_2)$, and $\Pi_\alpha(\omega)$ the same as in the case of double resonance.

The polarisation of the medium in Raman resonance is determined by the second term of (A4.1.4'):

$$P = \widetilde{\rho}_{ca} D_{ac} + c.c. + \widetilde{\rho}_{aa} D_{aa} + \widetilde{\rho}_{cc} D_{cc},$$

where $D = i[d, S^{(1)}]$ denotes the effective dipole moment,

$$D_{ac} = \sum_{j=1,2} [\mathcal{E}_j \Pi_{ca}^*(-\omega_j) e^{-i\Phi_j} + \mathcal{E}_j^* \Pi_{ca}^*(\omega_j) e^{-i\Phi_j}] = D_{ca}^*$$

$$D_{\alpha\alpha} = -\sum_{j=1,2} \mathcal{E}_j \Pi_\alpha(\omega_j) e^{-i\Phi_j} + c.c..$$

The desirable evolution equations are obtained from (A4.1.2) and (A4.1.3') by going to slowly varying amplitudes and neglecting the reaction of the waves generated at the combined $2\omega_1$-ω_2 and $|2\omega_1$-$\omega_2|$ frequencies. It is important to emphasise that the matrix elements H^R have turned out to be interrelated by virtue of the restriction of the initial particle model to three levels.

The Hamiltonian H^R can also be obtained by a unitary transformation and from H^D:

$$H^R = e^{-i\widetilde{S}} H^D e^{i\widetilde{S}} - i\hbar e^{-i\widetilde{S}} \frac{\partial}{\partial t} e^{i\widetilde{S}}. \tag{A4.1.9}$$

As a rule, however, Stark levels are not taken into account when double resonance is considered. This leads to an effective Hamiltonian $\widetilde{H}^R$ and an effective dipole moment $\widetilde{D}$ with somewhat different parameters:

$$\widetilde{H}^R_{ca} = -\mathcal{E}_1 \mathcal{E}_2^* d_{cb} d_{ba} [2\hbar(\omega_{bc} - \omega_2)]^{-1} \exp[-i(\Phi_1 - \Phi_2)] = \widetilde{H}^{R*}_{ac},$$

$$\widetilde{H}^R_{aa} = E_a - |\mathcal{E}_1 d_{ba}|^2 / \hbar(\omega_{bc} - \omega_2),$$

$$\widetilde{H}^R_{cc} = E_c |\mathcal{E}_2 d_{bc}|^2 / \hbar(\omega_{bc} - \omega_2),$$

$$\widetilde{D}_{ac} = d_{ab} d_{bc} (\mathcal{E}_2 e^{-i\Phi_2} + \mathcal{E}_1^* e^{i\Phi_1}) / \hbar(\omega_{bc} - \omega_2),$$

$$\widetilde{D}_{aa} = |d_{ab}|^2 \mathcal{E}_1 e^{-i\Phi_1} / \hbar(\omega_{bc} - \omega_2) + c.c.,$$

$$\widetilde{D}_{cc} = |d_{cb}|^2 \mathcal{E}_2 e^{-i\Phi_2} / \hbar(\omega_{bc} - \omega_2) + c.c..$$

It is recognised here that by virtue of the Raman-resonance condition we have $\omega_{ba} - \omega_1 \approx \omega_{bc} - \omega_2$.

We do not need the evolution equations for Raman resonance, since we obtain their Lax representation from the Lax representation of the double-resonance equations (A4.1.7). The ensuing restrictions are determined entirely by the relation between the parameters of the Hamiltonian $\widetilde{H}^R$ and by condition (A4.1.8).

We proceed now to analyse the Lax representation of the double-resonance equations. Eqs. (A4.1.7) are the condition for the compatibility of the solutions of a system of third-order linear equations

$$\frac{\partial}{\partial\tau}q = \hat{U}q, \quad \frac{\partial}{\partial\zeta}q = \hat{V}q. \qquad \text{(A4.1.10)}$$

Here, we write the Lax operators $\hat{U}$ and $\hat{V}$ in the form

$$\hat{U} = i(\lambda - \delta/2)\hat{J} + i\hat{E}, \quad \hat{V} = i(2\lambda)^{-1}\hat{R}. \qquad \text{(A4.1.11)}$$

Recall that the evolution equations (A4.1.7) are obtained from the zero-curvature representation

$$\frac{\partial\hat{U}}{\partial\zeta} = \frac{\partial\hat{V}}{\partial\tau} + [\hat{V},\hat{U}]$$

(the conditions for the compatibility of the solutions of (A4.1.10)) after substituting in them (A4.1.11) and equating the expressions for equal powers of the spectral parameter λ.

We transform the auxiliary equations (A4.1.10) in a manner similar to the one used above for eq. (A4.1.3):

$$\tilde{q} = e^{-iQ}q, \quad \frac{\partial}{\partial\tau}\tilde{q} = \tilde{\hat{U}}\tilde{q}, \quad \frac{\partial}{\partial\zeta}\tilde{q} = \tilde{\hat{V}}\tilde{q},$$

$$\tilde{\hat{U}} = e^{-iQ}\hat{U}e^{iQ} - e^{-iQ}\frac{\partial}{\partial\tau}e^{iQ}, \quad \tilde{\hat{V}} = e^{-iQ}\hat{V}e^{iQ} - e^{-iQ}\frac{\partial}{\partial\zeta}e^{iQ}. \qquad \text{(A4.1.12)}$$

We represent the new Lax operators and the matrix Q by series in powers of the field $\hat{E}$ (just as in Eq. (A4.1.5)):

$$\tilde{\hat{U}} = \tilde{\hat{U}}^{(0)} + \tilde{\hat{U}}^{(1)} + \tilde{\hat{U}}^{(2)} + \ldots, \quad \tilde{\hat{V}} = \tilde{\hat{V}}^{(0)} + \tilde{\hat{V}}^{(1)} + \ldots,$$

$$Q = Q^{(1)} + Q^{(2)} \ldots$$

$$\tilde{\hat{U}}^{(0)} = i(\lambda - \delta/2)\hat{J}, \qquad \text{(A4.1.13)}$$

$$\tilde{\hat{U}}^{(1)} = -i[Q^{(1)},\tilde{\hat{U}}^{(0)}] + i\hat{E} - i\frac{\partial}{\partial\tau}Q^{(1)},$$

$$\tilde{\hat{U}}^{(2)} = \frac{1}{2}[Q^{(1)},\hat{E}] - \frac{i}{2}[Q^{(1)},\tilde{\hat{U}}^{(1)}] - i[Q^{(2)},\tilde{\hat{U}}^{(0)}] - i\frac{\partial}{\partial\tau}Q^{(2)}.$$

Assume that the common level E_b of the adjacent optically allowed transitions $E_b \to E_a$ and $E_b \to E_c$ is not at resonance with the two-frequency USP, or more accurately, that the detuning δ from the resonance is much larger than the spectral width of the USP. This allows us to require that the effective operator

$$\widehat{U}^e = \widehat{\widetilde{U}}^{(0)} + \widehat{\widetilde{U}}^{(1)} + \widehat{\widetilde{U}}^{(2)}$$

have the block form

$$\widehat{U}^e = \begin{pmatrix} \cdot & 0 & 0 \\ 0 & \cdot & \cdot \\ 0 & \cdot & \cdot \end{pmatrix}. \tag{A4.1.14}$$

To this end we must have

$$Q_{11}^{(1)} = Q_{22}^{(1)} = Q_{33}^{(1)} = Q_{23}^{(1)} = Q_{32}^{(1)} = 0$$

$$Q_{12}^{(1)} = \int_{-\infty}^{\tau} \varepsilon_2 e^{i(2\lambda-\delta)(\tau'-\tau)} d\tau', \quad Q_{21}^{(1)} = \int_{-\infty}^{\tau} \varepsilon_2^* e^{-i(2\lambda-\delta)(\tau'-\tau)} d\tau',$$

$$Q_{13}^{(1)} = \int_{-\infty}^{\tau} \varepsilon_1 e^{i(2\lambda-\delta)(\tau'-\tau)} d\tau', \quad Q_{31}^{(1)} = \int_{-\infty}^{\tau} \varepsilon_1^* e^{-i(2\lambda-\delta)(\tau'-\tau)} d\tau',$$

and the principal role is played here, by virtue of the assumption made, by terms resulting from integration by parts:

$$Q_{12}^{(1)} = \frac{\varepsilon_2}{i(2\lambda-\delta)}, \quad Q_{21}^{(1)} = -\frac{\varepsilon_2^*}{i(2\lambda-\delta)}, \quad Q_{13}^{(1)} = \frac{\varepsilon_1}{i(2\lambda-\delta)}, \quad Q_{31}^{(1)} = -\frac{\varepsilon_1^*}{i(2\lambda-\delta)}.$$

Defining $Q^{(2)}$ as the solution of the equation

$$\frac{1}{2}[Q^{(1)} - \widetilde{Q}^{(1)}, \widehat{E}] - i[Q^{(2)}, \widehat{L}^{(0)}] - i\frac{\partial}{\partial\tau}Q^{(2)} = 0,$$

we obtain

$$\widehat{U}^e = i(\lambda - \delta/2)\widehat{J} + \tfrac{1}{2}[\widetilde{Q}^{(1)}, \widehat{E}], \tag{A4.1.15}$$

$$\widehat{U}^e = i\begin{pmatrix} -(\lambda-\delta/2) - \dfrac{(|\varepsilon_1|^2 + |\varepsilon_2|^2)}{(2\lambda-\delta)} & 0 & 0 \\ 0 & \lambda - \delta/2 + \dfrac{|\varepsilon_2|^2}{(2\lambda-\delta)} & \varepsilon_1\varepsilon_2^*/(2\lambda-\delta) \\ 0 & \varepsilon_1^*\varepsilon_2/(2\lambda-\delta) & \lambda-\delta/2+\dfrac{|\varepsilon_1|^2}{(2\lambda-\delta)} \end{pmatrix}.$$

As to the operator $\widehat{V}$ we find that, with the accuracy to the first order in the field inclusive and with allowance for the evolution equations (A4.1.7), the effective terms

$$\widehat{V}^e = \widehat{V} - i[\widetilde{Q}^{(1)}, \widehat{V}] - i\frac{\partial}{\partial \zeta}\widetilde{Q}^{(1)} \tag{A4.1.16}$$

also take the block form (A4.1.14).

We now reduce $\widehat{U}^e$ and $\widehat{V}^e$ to 2×2 matrices consisting of $\widehat{U}^e$ and $\widehat{V}^e$ elements with indices 2 and 3. We remove the inessential identical diagonal elements $i(\lambda-\delta/2)$ from $\widehat{U}^e$, and confine the terms of zeroth and first (which equals zero) orders in the field in $\widehat{V}^e$. This results in the following Lax operators:

$$\widehat{U}^R = \frac{i}{2\lambda - \delta}\begin{pmatrix} |\varepsilon_2|^2 & \varepsilon_1 \varepsilon_2^* \\ \varepsilon_1^* \varepsilon_2 & |\varepsilon_1|^2 \end{pmatrix}, \quad \widehat{V}^R = \frac{i}{2\lambda}\begin{pmatrix} r^c & r^{ca} \\ r^{ca*} & r^a \end{pmatrix}, \tag{A4.1.17}$$

which are the Lax representations of the Raman-resonance equations with Hamiltonian $\widetilde{H}^R$, with account taken of the condition (A4.1.8). This can be easily verified directly. In addition, the Lax representation (A4.1.17) coincides with Steudel's result in which we put $f = 0$ and redesignate the spectral parameter. It must also be noted that after determining (A4.1.17) it is easy to dispense with the restrictions of the model and obtain the results of ref. [2] with $f \neq 0$.

The above derivation of the Lax representation of the Raman-resonance equations seems quite natural if it is noted that the role of the operator $\widehat{U}$ in the double-resonance problem is played by H^D (after neglecting the Stark shifts of the levels and separating the slowly changing variables), while the first equation of (10) is an abbreviated Schrodinger equation [4]. The requirement (A4.1.14) has therefore turned out to be perfectly analogous to the assumptions that reduce H^D to $\widetilde{H}^R$ under the transformation (A4.1.9). What remains surprising is only the corresponding transformation of the operator $\widehat{V}$. It must also be emphasised that direct substitution of the abbreviated combination-resonance Hamiltonian $\widetilde{H}^R$ for the operator $\widehat{U}$ does not make it possible to find a Lax representation in analogy with refs. [6,7] according to calculations [5].

A4.2. Lax representation of polarisation models of Raman resonance

Following the proposed method, we discuss now the theory of propagation of arbitrarily polarised USP (eq. (A4.1.1)) under Raman-resonance conditions on the basis of the exactly integrable double-resonance polarisation models. It is shown in ref. 7 that in double resonance of two-frequency USP with energy levels E_a, E_b, and E_c. characterised by angular momenta $J_a = J_c = J_b - 1 = 0$ and $J_a = J_c = J_b + 1 = 1$, the evolution equations can be integrated by the ISP method and are the conditions for the compatibility of the solutions of systems of auxiliary linear equations (A4.1.10) of the fourth and fifth order, respectively, with Lax operators of form (A4.1.11) but with matrices $\hat{J}$, $\hat{E}$, and $\hat{R}$ of their own.

In the case $J_a = J_c = J_b - 1 = 0$ we have

$$\hat{J} = diag(-1,-1,1,1),\ \hat{E} = -\begin{pmatrix} 0 & 0 & \varepsilon_1^{-1} & \varepsilon_2^{-1} \\ 0 & 0 & \varepsilon_1^{1} & \varepsilon_2^{1} \\ \varepsilon_1^{-1*} & \varepsilon_1^{1*} & 0 & 0 \\ \varepsilon_2^{-1*} & \varepsilon_2^{1*} & 0 & 0 \end{pmatrix},\ \hat{R} = \begin{pmatrix} \mu_{-1-1} & \mu_{1-1} & p_1^{-1} & p_2^{-1} \\ \mu_{-11} & \mu_{11} & p_1^{1} & p_2^{1} \\ p_1^{-1*} & p_1^{1*} & m & r^* \\ p_2^{-1*} & p_2^{1*} & r & v \end{pmatrix}, \tag{A4.2.1}$$

$$\mu_{qq'} = \tilde{\rho}^{bb}_{-q'-q} N^{-1},\quad m = \tilde{\rho}^{aa}_{00} N^{-1},\quad v = \tilde{\rho}^{cc}_{00} N^{-1},$$

$$r = \tilde{\rho}^{ca}_{00} N^{-1} e^{i(\Phi_1 - \Phi_2)},\quad p_1^q = \tilde{\rho}^{ba}_{-q0} N^{-1} e^{i\Phi_1},\quad p_2^q = \tilde{\rho}^{bc}_{-q0} N^{-1} e^{i\Phi_2}.$$

In the case $J_a = J_c = J_b - 1 = 0$ we have

$$\hat{J} = diag(-1, 1, 1, 1, 1),$$

$$\hat{E} = -\begin{pmatrix} 0 & \varepsilon_1^{-1} & \varepsilon_1^{1} & \varepsilon_2^{-1} & \varepsilon_2^{1} \\ \varepsilon_1^{-1*} & 0 & 0 & 0 & 0 \\ \varepsilon_1^{1*} & 0 & 0 & 0 & 0 \\ \varepsilon_2^{-1*} & 0 & 0 & 0 & 0 \\ \varepsilon_2^{1*} & 0 & 0 & 0 & 0 \end{pmatrix},\ \hat{R} = \begin{pmatrix} \mu & p_1^{-1} & p_1^{1} & p_2^{-1} & p_2^{1} \\ p_1^{-1*} & m_{-1-1} & m_{-11} & r_{-1-1} & r_{-11} \\ p_1^{1*} & m_{1-1} & m_{11} & r_{1-1} & r_{11} \\ p_2^{-1*} & r^*_{-1-1} & r^*_{1-1} & \nu_{-1-1} & \nu_{-11} \\ p_2^{1*} & r^*_{-11} & r^*_{11} & \nu_{1-1} & \nu_{11} \end{pmatrix}, \tag{A4.2.2}$$

$$\mu = \tilde{\rho}^{bb}_{00} N^{-1},\quad m_{qq'} = \tilde{\rho}^{aa}_{qq'} N^{-1},\quad v_{qq'} = \tilde{\rho}^{cc}_{qq'} N^{-1},$$

$$r_{qq'} = \tilde{\rho}^{ca*}_{q'q} N^{-1} e^{i(\Phi_2 - \Phi_1)},\quad p_1^q = \tilde{\rho}^{ba}_{0q} N^{-1} e^{i\Phi_1},\quad p_2^q = \tilde{\rho}^{bc}_{0q} N^{-1} e^{i\Phi_2}.$$

Here $\varepsilon_j^{\pm 1}$ are the dimensionless spherical components of the vector $\vec{\mathcal{E}}_j$, the superscripts in the density matrix $\tilde{\rho}$ label the matrix elements for transitions between energy levels whose lower indices indicate Zeeman sublevels with different q, $q' = 0, \pm 1$ components along the quantisation axis of the corresponding total angular momentum. For a unified description of various cases $J_a = J_c = J_b - 1 = 0$ and $J_a = J_c = J_b + 1 = 1$, the notation is somewhat different from that in ref. [8].

We transform eqs. (A4.1.10), (A4.1.11) and (A4.2.1), and (A4.2.2) in accordance with (A4.1.12) and (A4.1.13) assuming, just as in the preceding section, that the detuning from the resonance is large enough. We stipulate here that the block form of the effective operator $\widehat{U}^e$ should be the following:

$$\widehat{U}^e = \begin{pmatrix} \cdot & \cdot & 0 & 0 \\ \cdot & \cdot & 0 & 0 \\ 0 & 0 & \cdot & \cdot \\ 0 & 0 & \cdot & \cdot \end{pmatrix}, \quad j_a = j_c = j_b - 1 = 0;$$

$$\widehat{U}^e = \begin{pmatrix} \cdot & 0 & 0 & 0 & 0 \\ 0 & \cdot & \cdot & \cdot & \cdot \\ 0 & \cdot & \cdot & \cdot & \cdot \\ 0 & \cdot & \cdot & \cdot & \cdot \\ 0 & \cdot & \cdot & \cdot & \cdot \end{pmatrix}, \quad j_a = j_c = j_b + 1 = 1.$$

Such a transformation is effected by the matrices

for $J_a = J_c = J_b - 1 = 0$

$$\widetilde{Q}_{13}^{(1)} = -\frac{\varepsilon_1^{-1}}{i(2\lambda-\delta)}, \quad \widetilde{Q}_{31}^{(1)} = \frac{\varepsilon_1^{-1*}}{i(2\lambda-\delta)}, \quad \widetilde{Q}_{23}^{(1)} = -\frac{\varepsilon_1^{1}}{i(2\lambda-\delta)}, \quad \widetilde{Q}_{32}^{(1)} = \frac{\varepsilon_1^{1*}}{i(2\lambda-\delta)},$$

$$\widetilde{Q}_{14}^{(1)} = -\frac{\varepsilon_2^{-1}}{i(2\lambda-\delta)}, \quad \widetilde{Q}_{41}^{(1)} = \frac{\varepsilon_2^{-1*}}{i(2\lambda-\delta)}, \quad \widetilde{Q}_{24}^{(1)} = -\frac{\varepsilon_2^{1}}{i(2\lambda-\delta)}, \quad \widetilde{Q}_{42}^{(1)} = \frac{\varepsilon_2^{1*}}{i(2\lambda-\delta)};$$

for $J_a = J_c = J_b + 1 = 1$ we have

$$\widetilde{Q}_{12}^{(1)} = -\frac{\varepsilon_1^{-1}}{i(2\lambda-\delta)}, \quad \widetilde{Q}_{21}^{(1)} = \frac{\varepsilon_1^{-1*}}{i(2\lambda-\delta)}, \quad \widetilde{Q}_{13}^{(1)} = -\frac{\varepsilon_1^{1}}{i(2\lambda-\delta)}, \quad \widetilde{Q}_{31}^{(1)} = \frac{\varepsilon_1^{1*}}{i(2\lambda-\delta)},$$

$$\widetilde{Q}_{14}^{(1)} = -\frac{\varepsilon_2^{-1}}{i(2\lambda-\delta)}, \quad \widetilde{Q}_{41}^{(1)} = \frac{\varepsilon_2^{-1*}}{i(2\lambda-\delta)}, \quad \widetilde{Q}_{15}^{(1)} = -\frac{\varepsilon_2^{-1}}{i(2\lambda-\delta)}, \quad \widetilde{Q}_{51}^{(1)} = \frac{\varepsilon_2^{1*}}{i(2\lambda-\delta)};$$

the remaining matrix elements are zero.

As a result of the calculations (A4.1.15) and (A4.1.16) of the effective operators $\widehat{U}^e$ and $\widehat{V}^e$ and their reduction we obtain the following Lax operators that realise a zero-curvature representation of the polarisation models of Raman resonance:

for $J_a = J_c = 0$

$$\widehat{U}^R = \frac{i}{2\lambda-\delta}\begin{pmatrix} |\varepsilon_1|^2 & \varepsilon_1^*\varepsilon_2 \\ \varepsilon_1\varepsilon_2^* & |\varepsilon_2|^2 \end{pmatrix}, \quad \widehat{V}^R = \frac{i}{2\lambda}\begin{pmatrix} m & r^* \\ r & \nu \end{pmatrix}, \tag{A4.2.3}$$

for $j_a = j_c = 1$

$$\hat{U}^R = \frac{i}{2\lambda - \delta}\begin{pmatrix} \left|\varepsilon_1^{-1}\right|^2 & \varepsilon_1^{-1*}\varepsilon_1^1 & \varepsilon_1^{-1*}\varepsilon_2^{-1} & \varepsilon_1^{-1*}\varepsilon_2^1 \\ \varepsilon_1^{1*}\varepsilon_1^{-1} & \left|\varepsilon_1^1\right|^2 & \varepsilon_1^{1*}\varepsilon_2^{-1} & \varepsilon_1^{1*}\varepsilon_2^1 \\ \varepsilon_2^{-1*}\varepsilon_1^{-1} & \varepsilon_2^{-1*}\varepsilon_1^1 & \left|\varepsilon_2^{-1}\right|^2 & \varepsilon_2^{-1*}\varepsilon_2^1 \\ \varepsilon_2^{1*}\varepsilon_1^{-1} & \varepsilon_2^{1*}\varepsilon_1^1 & \varepsilon_2^{1*}\varepsilon_2^{-1} & \left|\varepsilon_2^1\right|^2 \end{pmatrix},$$

$$\hat{V}^R = \frac{i}{2\lambda}\begin{pmatrix} m_{-1-1} & m_{-11} & r_{-1-1} & r_{-11} \\ m_{1-1} & m_{11} & r_{1-1} & r_{11} \\ r^*_{-1-1} & r^*_{1-1} & \nu_{-1-1} & \nu_{-11} \\ r^*_{-11} & r^*_{11} & \nu_{1-1} & \nu_{11} \end{pmatrix}. \qquad \text{(A4.2.4)}$$

The Lax representations (A4.2.3) and (A4.2.4) serve as the basis for the investigation, by the ISP method, of the polarisation singularities of the propagation of USP in Raman resonance (the corresponding evolution equations are given in the general form in sec.1.4.1). Following refs. [9-11], it is easy to find the Darbou transformation and the N-soliton equations.

References

1. Basharov, A.M.: Relation between exactly integrable models in resonance optics, *Zh.Eksp.Teor.Fiz.* **97** (1990) 169-178 [*Sov.Phys. JETP* **70** (1990) 94-99].
2. Steudel, H.: Soliton in stimulated Raman scattering and resonant two-photon propagation, *Physica* **D6** (1983) 155.
3. Maimistov, A.I., Manykin, E.A.: Prolongation structure for the reduced Maxwell-Bloch equations describing the two-photon self-induced transparency, *Phys.Lett.A* **95** (1983) 216.
4. Bol'shov, L.A., Likhanskii, V.V., Persiantsev, M.I.: Theory of coherent interaction of light pulses with resonant multilevel media, *Zh.Eksp.Teor.Fiz.* **84** (1983) 903 [*Sov.Phys. JETP* **57** (1983) 524].
5. Maimistov, A.I.: Rigorous theory of self-induced transparency in three-level medium at double resonance, *Kvantovaya Elektron.(Moscow)* **11** (1984) 567 [*Sov.J.Quant.Electron.* **14** (1984) 385].
6. McLaughlin, D.W., Corones, J.: Semiclassical radiation theory and the inverse method, *Phys.Rev.A* **10** (1974) 2051.
7. Haus, H.A.: Physical interpretation of inverse scattering formalism applied to self-induced transparency, *Rev.Mod.Phys.* **51** (1979) 331.
8. Basharov, A.M., Maimistov, A.I.: Polarised solitons in three-level media, *Zh.Eksp.Teor.Fiz.* **94** **n.12** (1988) 61 [*Sov.Phys. JETP* **67** (1988) 2426].
9. Meinel, R.: Backlund transformation and N-soliton solutions for stimulated Raman scattering and resonant two-photon propagation, *Opt.Comm.* **49** (1984) 224.
10. Steudel, H.: N-soliton solutions to degenerate self-induced transparency, *J.Mod.Opt.* **35** (1988) 693.
11. Basharov, A.M., Maimistov, A.I.: Rigorous theory of propagation of polarised ultrashort light pulses under double resonance condition, *J.Quant.Nonlin.Phenom.* **1** (1992), 76-91.

INDEX

Fundamental Theories of Physics

Series Editor: Alwyn van der Merwe, *University of Denver, USA*

1. M. Sachs: *General Relativity and Matter.* A Spinor Field Theory from Fermis to Light-Years. With a Foreword by C. Kilmister. 1982 ISBN 90-277-1381-2
2. G.H. Duffey: *A Development of Quantum Mechanics.* Based on Symmetry Considerations. 1985 ISBN 90-277-1587-4
3. S. Diner, D. Fargue, G. Lochak and F. Selleri (eds.): *The Wave-Particle Dualism.* A Tribute to Louis de Broglie on his 90th Birthday. 1984 ISBN 90-277-1664-1
4. E. Prugovečki: *Stochastic Quantum Mechanics and Quantum Spacetime.* A Consistent Unification of Relativity and Quantum Theory based on Stochastic Spaces. 1984; 2nd printing 1986 ISBN 90-277-1617-X
5. D. Hestenes and G. Sobczyk: *Clifford Algebra to Geometric Calculus.* A Unified Language for Mathematics and Physics. 1984 ISBN 90-277-1673-0; Pb (1987) 90-277-2561-6
6. P. Exner: *Open Quantum Systems and Feynman Integrals.* 1985 ISBN 90-277-1678-1
7. L. Mayants: *The Enigma of Probability and Physics.* 1984 ISBN 90-277-1674-9
8. E. Tocaci: *Relativistic Mechanics, Time and Inertia.* Translated from Romanian. Edited and with a Foreword by C.W. Kilmister. 1985 ISBN 90-277-1769-9
9. B. Bertotti, F. de Felice and A. Pascolini (eds.): *General Relativity and Gravitation.* Proceedings of the 10th International Conference (Padova, Italy, 1983). 1984 ISBN 90-277-1819-9
10. G. Tarozzi and A. van der Merwe (eds.): *Open Questions in Quantum Physics.* 1985 ISBN 90-277-1853-9
11. J.V. Narlikar and T. Padmanabhan: *Gravity, Gauge Theories and Quantum Cosmology.* 1986 ISBN 90-277-1948-9
12. G.S. Asanov: *Finsler Geometry, Relativity and Gauge Theories.* 1985 ISBN 90-277-1960-8
13. K. Namsrai: *Nonlocal Quantum Field Theory and Stochastic Quantum Mechanics.* 1986 ISBN 90-277-2001-0
14. C. Ray Smith and W.T. Grandy, Jr. (eds.): *Maximum-Entropy and Bayesian Methods in Inverse Problems.* Proceedings of the 1st and 2nd International Workshop (Laramie, Wyoming, USA). 1985 ISBN 90-277-2074-6
15. D. Hestenes: *New Foundations for Classical Mechanics.* 1986 ISBN 90-277-2090-8; Pb (1987) 90-277-2526-8
16. S.J. Prokhovnik: *Light in Einstein's Universe.* The Role of Energy in Cosmology and Relativity. 1985 ISBN 90-277-2093-2
17. Y.S. Kim and M.E. Noz: *Theory and Applications of the Poincaré Group.* 1986 ISBN 90-277-2141-6
18. M. Sachs: *Quantum Mechanics from General Relativity.* An Approximation for a Theory of Inertia. 1986 ISBN 90-277-2247-1
19. W.T. Grandy, Jr.: *Foundations of Statistical Mechanics.* Vol. I: *Equilibrium Theory.* 1987 ISBN 90-277-2489-X
20. H.-H von Borzeszkowski and H.-J. Treder: *The Meaning of Quantum Gravity.* 1988 ISBN 90-277-2518-7
21. C. Ray Smith and G.J. Erickson (eds.): *Maximum-Entropy and Bayesian Spectral Analysis and Estimation Problems.* Proceedings of the 3rd International Workshop (Laramie, Wyoming, USA, 1983). 1987 ISBN 90-277-2579-9
22. A.O. Barut and A. van der Merwe (eds.): *Selected Scientific Papers of Alfred Landé. [1888-1975].* 1988 ISBN 90-277-2594-2

23. W.T. Grandy, Jr.: *Foundations of Statistical Mechanics.*
Vol. II: *Nonequilibrium Phenomena.* 1988 ISBN 90-277-2649-3
24. E.I. Bitsakis and C.A. Nicolaides (eds.): *The Concept of Probability.* Proceedings of the Delphi Conference (Delphi, Greece, 1987). 1989 ISBN 90-277-2679-5
25. A. van der Merwe, F. Selleri and G. Tarozzi (eds.): *Microphysical Reality and Quantum Formalism, Vol. 1.* Proceedings of the International Conference (Urbino, Italy, 1985). 1988 ISBN 90-277-2683-3
26. A. van der Merwe, F. Selleri and G. Tarozzi (eds.): *Microphysical Reality and Quantum Formalism, Vol. 2.* Proceedings of the International Conference (Urbino, Italy, 1985). 1988 ISBN 90-277-2684-1
27. I.D. Novikov and V.P. Frolov: *Physics of Black Holes.* 1989 ISBN 90-277-2685-X
28. G. Tarozzi and A. van der Merwe (eds.): *The Nature of Quantum Paradoxes.* Italian Studies in the Foundations and Philosophy of Modern Physics. 1988 ISBN 90-277-2703-1
29. B.R. Iyer, N. Mukunda and C.V. Vishveshwara (eds.): *Gravitation, Gauge Theories and the Early Universe.* 1989 ISBN 90-277-2710-4
30. H. Mark and L. Wood (eds.): *Energy in Physics, War and Peace.* A Festschrift celebrating Edward Teller's 80th Birthday. 1988 ISBN 90-277-2775-9
31. G.J. Erickson and C.R. Smith (eds.): *Maximum-Entropy and Bayesian Methods in Science and Engineering.*
Vol. I: *Foundations.* 1988 ISBN 90-277-2793-7
32. G.J. Erickson and C.R. Smith (eds.): *Maximum-Entropy and Bayesian Methods in Science and Engineering.*
Vol. II: *Applications.* 1988 ISBN 90-277-2794-5
33. M.E. Noz and Y.S. Kim (eds.): *Special Relativity and Quantum Theory.* A Collection of Papers on the Poincaré Group. 1988 ISBN 90-277-2799-6
34. I.Yu. Kobzarev and Yu.I. Manin: *Elementary Particles. Mathematics, Physics and Philosophy.* 1989 ISBN 0-7923-0098-X
35. F. Selleri: *Quantum Paradoxes and Physical Reality.* 1990 ISBN 0-7923-0253-2
36. J. Skilling (ed.): *Maximum-Entropy and Bayesian Methods.* Proceedings of the 8th International Workshop (Cambridge, UK, 1988). 1989 ISBN 0-7923-0224-9
37. M. Kafatos (ed.): *Bell's Theorem, Quantum Theory and Conceptions of the Universe.* 1989 ISBN 0-7923-0496-9
38. Yu.A. Izyumov and V.N. Syromyatnikov: *Phase Transitions and Crystal Symmetry.* 1990 ISBN 0-7923-0542-6
39. P.F. Fougère (ed.): *Maximum-Entropy and Bayesian Methods.* Proceedings of the 9th International Workshop (Dartmouth, Massachusetts, USA, 1989). 1990 ISBN 0-7923-0928-6
40. L. de Broglie: *Heisenberg's Uncertainties and the Probabilistic Interpretation of Wave Mechanics.* With Critical Notes of the Author. 1990 ISBN 0-7923-0929-4
41. W.T. Grandy, Jr.: *Relativistic Quantum Mechanics of Leptons and Fields.* 1991 ISBN 0-7923-1049-7
42. Yu.L. Klimontovich: *Turbulent Motion and the Structure of Chaos.* A New Approach to the Statistical Theory of Open Systems. 1991 ISBN 0-7923-1114-0
43. W.T. Grandy, Jr. and L.H. Schick (eds.): *Maximum-Entropy and Bayesian Methods.* Proceedings of the 10th International Workshop (Laramie, Wyoming, USA, 1990). 1991 ISBN 0-7923-1140-X

44. P.Pták and S. Pulmannová: *Orthomodular Structures as Quantum Logics*. Intrinsic Properties, State Space and Probabilistic Topics. 1991 ISBN 0-7923-1207-4
45. D. Hestenes and A. Weingartshofer (eds.): *The Electron*. New Theory and Experiment. 1991 ISBN 0-7923-1356-9
46. P.P.J.M. Schram: *Kinetic Theory of Gases and Plasmas*. 1991 ISBN 0-7923-1392-5
47. A. Micali, R. Boudet and J. Helmstetter (eds.): *Clifford Algebras and their Applications in Mathematical Physics*. 1992 ISBN 0-7923-1623-1
48. E. Prugovečki: *Quantum Geometry*. A Framework for Quantum General Relativity. 1992 ISBN 0-7923-1640-1
49. M.H. Mac Gregor: *The Enigmatic Electron*. 1992 ISBN 0-7923-1982-6
50. C.R. Smith, G.J. Erickson and P.O. Neudorfer (eds.): *Maximum Entropy and Bayesian Methods*. Proceedings of the 11th International Workshop (Seattle, 1991). 1993 ISBN 0-7923-2031-X
51. D.J. Hoekzema: *The Quantum Labyrinth*. 1993 ISBN 0-7923-2066-2
52. Z. Oziewicz, B. Jancewicz and A. Borowiec (eds.): *Spinors, Twistors, Clifford Algebras and Quantum Deformations*. Proceedings of the Second Max Born Symposium (Wrocław, Poland, 1992). 1993 ISBN 0-7923-2251-7
53. A. Mohammad-Djafari and G. Demoment (eds.): *Maximum Entropy and Bayesian Methods*. Proceedings of the 12th International Workshop (Paris, France, 1992). 1993 ISBN 0-7923-2280-0
54. M. Riesz: *Clifford Numbers and Spinors* with Riesz' Private Lectures to E. Folke Bolinder and a Historical Review by Pertti Lounesto. E.F. Bolinder and P. Lounesto (eds.). 1993 ISBN 0-7923-2299-1
55. F. Brackx, R. Delanghe and H. Serras (eds.): *Clifford Algebras and their Applications in Mathematical Physics*. Proceedings of the Third Conference (Deinze, 1993) 1993 ISBN 0-7923-2347-5
56. J.R. Fanchi: *Parametrized Relativistic Quantum Theory*. 1993 ISBN 0-7923-2376-9
57. A. Peres: *Quantum Theory: Concepts and Methods*. 1993 ISBN 0-7923-2549-4
58. P.L. Antonelli, R.S. Ingarden and M. Matsumoto: *The Theory of Sprays and Finsler Spaces with Applications in Physics and Biology*. 1993 ISBN 0-7923-2577-X
59. R. Miron and M. Anastasiei: *The Geometry of Lagrange Spaces: Theory and Applications*. 1994 ISBN 0-7923-2591-5
60. G. Adomian: *Solving Frontier Problems of Physics: The Decomposition Method*. 1994 ISBN 0-7923-2644-X
61 B.S. Kerner and V.V. Osipov: *Autosolitons*. A New Approach to Problems of Self-Organization and Turbulence. 1994 ISBN 0-7923-2816-7
62. G.R. Heidbreder (ed.): *Maximum Entropy and Bayesian Methods*. Proceedings of the 13th International Workshop (Santa Barbara, USA, 1993) 1996 ISBN 0-7923-2851-5
63. J. Peřina, Z. Hradil and B. Jurčo: *Quantum Optics and Fundamentals of Physics*. 1994 ISBN 0-7923-3000-5
64. M. Evans and J.-P. Vigier: *The Enigmatic Photon*. Volume 1: The Field $\boldsymbol{B}^{(3)}$. 1994 ISBN 0-7923-3049-8
65. C.K. Raju: *Time: Towards a Constistent Theory*. 1994 ISBN 0-7923-3103-6
66. A.K.T. Assis: *Weber's Electrodynamics*. 1994 ISBN 0-7923-3137-0
67. Yu. L. Klimontovich: *Statistical Theory of Open Systems*. Volume 1: A Unified Approach to Kinetic Description of Processes in Active Systems. 1995 ISBN 0-7923-3199-0; Pb: ISBN 0-7923-3242-3

Fundamental Theories of Physics

68. M. Evans and J.-P. Vigier: *The Enigmatic Photon.* Volume 2: Non-Abelian Electrodynamics. 1995 ISBN 0-7923-3288-1
69. G. Esposito: *Complex General Relativity.* 1995 ISBN 0-7923-3340-3
70. J. Skilling and S. Sibisi (eds.): *Maximum Entropy and Bayesian Methods.* Proceedings of the Fourteenth International Workshop on Maximum Entropy and Bayesian Methods. 1996 ISBN 0-7923-3452-3
71. C. Garola and A. Rossi (eds.): *The Foundations of Quantum Mechanics – Historical Analysis and Open Questions.* 1995 ISBN 0-7923-3480-9
72. A. Peres: *Quantum Theory: Concepts and Methods.* 1995 (see for hardback edition, Vol. 57) ISBN Pb 0-7923-3632-1
73. M. Ferrero and A. van der Merwe (eds.): *Fundamental Problems in Quantum Physics.* 1995 ISBN 0-7923-3670-4
74. F.E. Schroeck, Jr.: *Quantum Mechanics on Phase Space.* 1996 ISBN 0-7923-3794-8
75. L. de la Peña and A.M. Cetto: *The Quantum Dice.* An Introduction to Stochastic Electrodynamics. 1996 ISBN 0-7923-3818-9
76. P.L. Antonelli and R. Miron (eds.): *Lagrange and Finsler Geometry.* Applications to Physics and Biology. 1996 ISBN 0-7923-3873-1
77. M.W. Evans, J.-P. Vigier, S. Roy and S. Jeffers: *The Enigmatic Photon.* Volume 3: Theory and Practice of the $\boldsymbol{B}^{(3)}$ Field. 1996 ISBN 0-7923-4044-2
78. W.G.V. Rosser: *Interpretation of Classical Electromagnetism.* 1996 ISBN 0-7923-4187-2
79. K.M. Hanson and R.N. Silver (eds.): *Maximum Entropy and Bayesian Methods.* 1996 ISBN 0-7923-4311-5
80. S. Jeffers, S. Roy, J.-P. Vigier and G. Hunter (eds.): *The Present Status of the Quantum Theory of Light.* Proceedings of a Symposium in Honour of Jean-Pierre Vigier. 1997 ISBN 0-7923-4337-9
81. M. Ferrero and A. van der Merwe (eds.): *New Developments on Fundamental Problems in Quantum Physics.* 1997 ISBN 0-7923-4374-3
82. R. Miron: *The Geometry of Higher-Order Lagrange Spaces.* Applications to Mechanics and Physics. 1997 ISBN 0-7923-4393-X
83. T. Hakioğlu and A.S. Shumovsky (eds.): *Quantum Optics and the Spectroscopy of Solids.* Concepts and Advances. 1997 ISBN 0-7923-4414-6
84. A. Sitenko and V. Tartakovskii: *Theory of Nucleus.* Nuclear Structure and Nuclear Interaction. 1997 ISBN 0-7923-4423-5
85. G. Esposito, A.Yu. Kamenshchik and G. Pollifrone: *Euclidean Quantum Gravity on Manifolds with Boundary.* 1997 ISBN 0-7923-4472-3
86. R.S. Ingarden, A. Kossakowski and M. Ohya: *Information Dynamics and Open Systems.* Classical and Quantum Approach. 1997 ISBN 0-7923-4473-1
87. K. Nakamura: *Quantum versus Chaos.* Questions Emerging from Mesoscopic Cosmos. 1997 ISBN 0-7923-4557-6
88. B.R. Iyer and C.V. Vishveshwara (eds.): *Geometry, Fields and Cosmology.* Techniques and Applications. 1997 ISBN 0-7923-4725-0
89. G.A. Martynov: *Classical Statistical Mechanics.* 1997 ISBN 0-7923-4774-9
90. M.W. Evans, J.-P. Vigier, S. Roy and G. Hunter (eds.): *The Enigmatic Photon.* Volume 4: New Directions. 1998 ISBN 0-7923-4826-5
91. M. Rédei: *Quantum Logic in Algebraic Approach.* 1998 ISBN 0-7923-4903-2
92. S. Roy: *Statistical Geometry and Applications to Microphysics and Cosmology.* 1998 ISBN 0-7923-4907-5

Fundamental Theories of Physics

93. B.C. Eu: *Nonequilibrium Statistical Mechanics.* Ensembled Method. 1998 ISBN 0-7923-4980-6
94. V. Dietrich, K. Habetha and G. Jank (eds.): *Clifford Algebras and Their Application in Mathematical Physics.* Aachen 1996. 1998 ISBN 0-7923-5037-5
95. *Not yet known*
96. V.P. Frolov and I.D. Novikov: *Black Hole Physics.* Basic Concepts and New Developments. 1998 ISBN 0-7923-5145-2; PB 0-7923-5146
97. G. Hunter, S. Jeffers and J-P. Vigier (eds.): *Causality and Locality in Modern Physics.* 1998 ISBN 0-7923-5227-0
98. G.J. Erickson, J.T. Rychert and C.R. Smith (eds.): *Maximum Entropy and Bayesian Methods.* 1998 ISBN 0-7923-5047-2
99. D. Hestenes: *New Foundations for Classical Mechanics (Second Edition).* 1999 ISBN 0-7923-5302-1; PB ISBN 0-7923-5514-8
100. B.R. Iyer and B. Bhawal: *Black Holes, Gravitational Radiation and the Universe.* Essays in Honor of C. V. Vishveshwara. 1999 ISBN 0-7923-5308-0
101. P.L. Antonelli and T.J. Zastawniak: *Fundamentals of Finslerian Diffusion with Applications.* 1999 ISBN 0-7923-5511-3
102. Reserved
103. M.A. Trump and W.C. Schieve: *Classical Relativistic Many-Body Dynamics.* 1999 ISBN 0-7923-5737-X
104. A.I. Maimistov and A.M. Basharov: *Nonlinear Optical Waves.* 1999 ISBN 0-7923-5752-3
105. W. von der Linden, V. Dose, R. Fisher and R. Preuss: *Maximum Entropy and Bayesian Methods.* Proceedings of the 18th International Workshop on Maximum Entropy and Bayesian Methods of Statistical Analysis. 1999 ISBN 0-7923-5766-3

KLUWER ACADEMIC PUBLISHERS – DORDRECHT / BOSTON / LONDON